Turkey Management

By

STANLEY J. MARSDEN, B.S., M.S.
Poultry Husbandman,
The U. S. Department of Agriculture,
Research Center
Beltsville, Maryland

AND

J. HOLMES MARTIN, Ph.D.
Head of Department of Poultry Science
Purdue University, Lafayette, Indiana

Sixth Edition

A Norton Creek Classic

Norton Creek Press
http://www.nortoncreekpress.com

Poultry Management
Sixth Edition

Cover by Robert Plamondon

ISBN 978-1-938099-05-2

Originally published in 1955 by Interstate Printers and Publishers, Inc.

Prior editions published in 1939, 1944, 1945, 1946, and 1949.

To Lottie Marsden through whose aid and comfort to the senior author, this book was made possible.

Table of Contents

Preface to Sixth Edition

In the preface to the first edition (1939) and in subsequent editions (1944, 1945, 1946 and 1949) the authors have enumerated the many persons to whom they are indebted for photographs and tabular material as well as constructive criticisms and careful reading of the proof. To all of them and especially the *Turkey World,* appreciation is expressed again. Further appreciation is expressed to Dr. Harry W. Titus for his permission to use the tables on feedstuffs composition in "The Scientific Feeding of Chickens" and to Dr. H. E. Moses who made many pertinent and valuable suggestions regarding the chapters dealing with diseases.

The first few editions contain rather lengthy lists of references at the end of each chapter. In the interests of conservation of space those are omitted from this edition. Only a few of the salient references are listed and only those discoveries since 1945 referred to by date. For ease of reference the figures and tables are numbered within the chapter; thus Fig. 2.11 is the eleventh figure in Chapter 2.

Every attempt has been made to focus this rather complete revision on the producer and his problems. It is hoped this edition may prove its worth on the turkey grower's shelf. It should be noted that the chapters have been arranged in the order in which the turkey producer would be confronted with the problems involved in turkey management; assuming that he starts with purchased day old poults.

May, 1955. S.J.M.—J.H.M.

CHAPTER 1

STARTING WITH TURKEYS

OUTLINE

Many so-called "secrets" of successful turkey raising have been proposed. However, they are now open secrets, and successful methods of management may be put into practice by anyone who will study turkey production, sanitation, and disease-control problems. Common sense and attention to details in carrying out simple practices to avoid turkey diseases are of much more value than all of the medicines and condiments on the market. To succeed with turkeys one must have a willingness to work, a desire to learn the facts, and the courage to follow sound practices.

Making the Start

There are several successful ways of making a start in the turkey business, the best of which is the purchase of day-old turkey poults. Many hatcheries and turkey raisers are making a specialty of the production and sale of day-old poults and many turkey raisers obtain each year's supply of poults by buying them from these specialists. Of course, the poults cost more than would hatching eggs but the hazards and costs of either natural or artificial incubation are eliminated and there is little danger of introducing diseases or parasites. The later one waits after the first of December to purchase

Fig. 1.1. Small farm breeding unit of Bronze turkeys. (Photo from U. S. Dept. of Agr.)

breeding stock, the less selection he will have, as the breeder usually has culled his flock down pretty well during the Thanksgiving and Christmas marketing seasons. If one desires to enter the business of producing hatching eggs or poults, then the most logical procedure is to purchase eggs or poults and raise the young stock to market age when the breeding birds for the following season should be selected. The chief disadvantage in purchasing breeding stock instead of eggs or poults is that one could bring to his farm certain parasites and diseases that the birds might harbor. If on the other hand, hatching eggs or day-old poults were purchased one would avoid the risks of bringing in such diseases as blackhead and such parasites as intestinal worms and lice. There would, however, still remain the risk of introducing the pullorum, typhoid, and paratyphoid group of diseases unless the parent stock is blood-tested for these diseases and the eggs hatched in incubators free of infection.

Land and Labor Requirements

One question that is frequently asked is, how many birds can be raised per acre of range available? There is no definite answer, since the value of range varies so, depending upon the fertility of the soil, the climate, and other factors. As soon as the turkeys are old enough to "rustle for themselves" in the open (8 to 12 weeks old) they will eat weeds and grass closely from wild areas if pastured in sufficiently large numbers in proportion to the area. A simple procedure consists in placing within a fenced area more turkeys than the green growth will support. Obviously the range should be supplemented with a well-balanced mash mixture and grain. As soon as the area has been pastured bare, within a week or so, the fence should be moved to a new site; not returning to the first area until the second or third year.

There are records of successful turkey raisers who produced as much as 40 tons of turkeys on a half acre of land, using strict confinement. There are also many other cases where turkey raisers have failed to produce a half dozen good birds on a 200-acre farm from 100 or more poults started. Why this great difference? In the first case every known precaution was taken to minimize the hazards from disease, whereas in the other case the turkeys were allowed to roam at will, express their so-called wild nature, and as a consequence pick up disease germs and parasites about the farm, especially those in the chicken and turkey droppings. If one will trace turkey production through the years on any individual farm he will find many cases in which turkeys were raised successfully with chickens. However, if turkey production continued year after year, diseases and parasites made their inroads and finally the turkey raiser either arranged to raise his turkeys away from his chickens or abandoned turkey production.

Labor costs vary with the equipment, management, and other factors difficult to evaluate. It is generally considered, however, that one experienced operator can raise

4,000 "roaster" or 5,000 "broiler" turkeys; or service 1500 to 2000 non-trapnested breeding turkeys or 400 trapnested hens, including the care of the eggs.

Distribution of the Turkey Industry in U.S.

In 1890 the Middle West was the main center of production with Illinois the top State in the Union and Iowa the second. There were smaller centers located in New York, Pennsylvania, Virginia, and Texas. At that time California produced only a few more turkeys than the small State of Maryland. In 1900 and 1910 the picture was still much the same but Texas and Missouri had taken the lead. In 1920, turkey raising was near its lowest ebb; the eastern States, except for Virginia, had almost dropped out of the picture, California began to be noticed, and the Texas-Oklahoma-Missouri area held a commanding lead. By 1930 the main center of production had split two ways: Texas-Oklahoma-Missouri and Minnesota-Dakotas-Montana, while California and Virginia remained smaller centers of turkey growing. By 1940 the long-distance shipping of hatching eggs and poults had become so widespread that several States which showed relatively few breeding birds on hand April 1 were among the top States in the production of market turkeys. Utah, Colorado, and Iowa were in this category. In the meantime, the centers of production established in 1930 remained about the same in 1940 but extended their boundaries, the southwestern area taking in Colorado and Kansas; the west north central area taking in Iowa and Nebraska; California being joined by Oregon, Washington, and Utah; and New York joining Pennsylvania. In addition, a new center of production began to show up in the East North Central States of Ohio, Indiana, Illinois, Wisconsin, and Michigan.

By 1950, other trends were in evidence, namely increased production in the East Coast States from New England to Georgia.

In 1929 the U. S. Department of Agriculture (BAE), to aid in the orderly marketing of the growing turkey industry, first undertook to estimate at the beginning of and during each growing season the probable size of the turkey crop, and to make accurate annual estimates after the crop was marketed. This information along with poult hatchings, and breeding birds blood-tested at monthly intervals (starting in 1953) and other turkey production data is published regularly. These reports may be obtained free by writing to the U. S. Agricultural Marketing Service, Washington 25, D. C. The final estimates of turkeys raised, produced, and breeder hens kept differ somewhat from the U. S. census reports but are believed by the authors to be more accurate. A summary of turkeys raised, by states, at three year intervals, 1929 to 1950, inclusive, will be found in the tables in the Appendix.

Reasons for the growth of the turkey industry in certain areas of the country may vary from what may best be described as "follow-the-leader" to a logical natural development based on favorable climatic conditions, available feed and labor. In the latter category, we may place the southwestern States of Texas and Oklahoma where the soil and climate appear to be well adapted to turkey raising and grains are in abundance. Turkeys can be grown in this area with relatively less attention paid to the maintenance of soil sanitation which is so essential to range turkey production efficiency in most other areas.

In Minnesota the great turkey raising industry developed under the tutelage of Dr. W. A. Billings, extension veterinarian of the University of Minnesota staff. The climate and soil of Minnesota appear to be no better suited for turkey raising than many other sections, but the industry has developed to a remarkable extent in that State through use of the "Billings Plan" of land rotation to prevent soil-borne infections. In the Shenandoah Valley of Virginia, climatic and soil conditions are very well adapted to turkey production, markets are close, and feed is relatively abundant and reasonably priced. In this case also, the pioneer work of Charles W. Wampler of Harrisonburg, who gave up county agricultural agent work to raise turkeys on a commercial scale, probably had considerable to do with the growth of the industry in that area. In the Delmarva peninsula (Delaware and the eastern shore counties of Maryland and Virginia) the sandy soil and nearness to market have favored the early development of a very concentrated turkey industry. In recent years heavy concentrations of turkeys have developed in Iowa, and in the far Western States. In Iowa, the recent sensational development was spearheaded by the operations of the Maplecrest turkey organization headed by A. C. Gingerich at Wellman. In the Mountain and Pacific Coast States, favorable climatic conditions aided by the activities of the late L. E. Cline of the Nevada Extension Service, fostered the extensive development there.

In the Intermountain States of Utah, Idaho and Colorado, the climate, with respect to rainfall and temperature, perhaps approaches the ideal for turkey raising. These areas have relatively dry climates, especially during the turkey growing season, and have cool weather in the fall. In these areas too, the farmers have had to find a crop in which they could compete successfully with more favorably located areas so they took up turkey raising which has been a relatively stable crop from the profit angle for many years.

By utilizing the easily obtainable scientific knowledge of turkey raising it is possible to raise turkeys successfully anywhere in the United States, and this has actually taken place even in the Cotton Belt where climatic conditions are less favorable. Rearing the birds in confinement has made turkey production possible in areas possessing characteristics which heretofore were so unfavorable as to prohibit range production of turkeys. The Northeast, embracing New England, New York, Pennsylvania, and New Jersey, is included in this category. The

East North Central States of Illinois, Wisconsin, Indiana, Ohio and Michigan on the one hand, and the southern states of Georgia and the Carolinas on the other have shown the greatest recent increases in turkey growing with the combination of range and confinement projects most of which are of a commercial size. In the opposite direction is the decline of the Dakotas and Montana as turkey raising States, due apparently to the relatively high profitableness of grain farming there since 1940.

The Turkey Breeding Industry

Since the number of breeder hens kept is a strong indication of the size of the coming year's turkey crop, the Agricultural Marketing Service each year makes estimates of the number of breeding hens on hand January first. In the Appendix are comparisons using U. S. Department of Agriculture figures of the number of breeder hens, along with the size of the turkey crop this year in the United States. Not all of the breeding hens indicated as being on hand were actually kept, but it is believed that in the average year nearly all were. In a few instances, due to unfavorable outlook, larger-than-normal numbers were marketed after January 1 before or during the normal breeding season; as was the case in 1946.

The growth of the United States turkey industry in numbers, tonnage, and dollar value, along with average weights of turkeys marketed and estimates of mortality is shown in tables in the Appendix.

Financing Turkey Raising

Turkey raising in commercial numbers involves considerable financing which may be carried by the grower himself, the Federal Government through Production Credit and Farm Security, feed companies, processing plants, cooperative organizations, banks, finance companies, hatcheries, and landowners or other private individuals. To obtain financing a grower must furnish a detailed statement of assets and liabilities, along with an estimate of costs of the project, which should include cost of poults, feed, transportation, fuel, litter, vaccination, disinfectants, medicines, insurance, and miscellaneous items, the total of which may vary from $2.50 to $7.00 per bird raised, depending on locality, management, and type of turkey raised. In addition to these costs, there are those relating to land, buildings, equipment, and labor.

A common type of partnership agreement is that in which the farmer furnishes land, labor, houses, equipment, and transportation, and the financing agency furnishes poults—all at no charge. The cost of raising the turkeys (feed, litter, fuel, vaccine, disinfectants, insurance, and miscellaneous) is then advanced by the financing agency and deducted from the proceeds of the sale of the turkeys. The profit remaining after this deduction is then divided equally between the two parties.

Another popular type of partnership is based on the farmer's furnishing land, labor, housing, equipment, and transportation, and the financing agency's furnishing poults, feed, insurance, and other rearing necessities. When the turkeys are sold, the grower receives five percent of the total proceeds of the sale and the costs of rearing are deducted from the remaining 95 percent of the proceeds. The profits resulting from this procedure are then divided 60 percent to the grower and 40 percent to the financing agency. If a loss occurs, the financing agency absorbs it but the grower still receives his 5 percent of gross sales.

In any case, a detailed contract in writing should be drawn up and signed. A chattel mortgage note is usually signed by the grower in favor of the financing agency. Good equipment is commonly demanded of the grower and insurance against losses due to bad weather is usually required. The terms of any contract should be thoroughly studied and understood by both parties for their mutual protection.

Landowners sometimes give their hired men a place to live with garden, milk, eggs, meat, and a monthly salary plus 10 percent of the net profits. Contract growing is another form of financing in which the financier furnishes everything and pays the grower a stated amount for each turkey raised, the amount increasing if more than 75 percent of the turkeys are raised.

Insurance

Turkey insurance against the hazards of weather and smothering, drowning, and freezing as a direct consequence of such hazards is now generally available from various insurance organizations. The National Turkey Federation, with headquarters at Mt. Morris, Illinois, a nationwide association of persons and organizations interested in turkeys, sponsors a mutual insurance plan which has gained considerable support. Many growers consider insurance indispensable.

Value of Range

When limited range along with the self feeding of a balanced diet is practiced, the range appears to reduce significantly the quantity of feed eaten by the growing

Fig. 1.2. A portion of a large flock of 20,000 in one Indiana flock. Note water storage tanks. (Photo from J. C. Allen and Son.)

turkeys. However, if the range is counted on to supply vitamins, sunshine, and additional protein, and the diet is cheapened as far as possible by leaving out the high-priced, vitamin-bearing supplements and reducing the protein supplement to one or two of the cheaper kinds, a substantial saving can be made. Where growing turkeys

Fig. 1.3. Wild turkey toms—the lower one was raised in confinement. (Photo from U. S. Dept. of Agr.)

are herded or otherwise managed in such a way as to make the best possible use of range and fed only part-time from self-feeders, the range makes a maximum contribution to the quantity of feed consumed and will supply most of the necessary vitamins and a liberal share of the proteins and carbohydrates at low cost.

Workers at the Oregon Station (1948) in an experiment involving 250 standardbred Bronze turkeys have reported on the value of range. The poults were brooded artificially and fed a complete starting diet for the first 8 weeks when they were taken to the rearing quarters. Lot 1 had a sand yard and was fed a mash-grain basal diet complete except that it contained no green feed of any kind either fresh or dried. Lot 2 had a sand yard and was fed the same diet as that fed to lot 1, with 10 per cent of alfalfa leaves added to the mash. Lot 3 had a dry-land yard and was fed all the chopped, fresh alfalfa the birds would eat twice daily, from 8 weeks of age to October 1st. After that no green feed was supplied except in cured form from the range lot. The basal diet was fed. Lot 4 had access to a yard containing growing alfalfa and was fed the basal diet. The birds were marketed at the age of 8 to 9 months (being older than average). The following table shows the results obtained:

Table 1.1—Effect of pasture on feed consumption (Oregon Station Data)

Lot	Feed consumed per lb. of live turkey produced	Feed cost per lb. of live turkey produced	Average dressed weights	
			Toms	Hens
	Lbs.	Cents	Lbs.	Lbs.
1 (no greens)	6.64	11.27	20.4	13.00
2 (dried greens)	6.50	10.80	22.9	13.30
3 (freshly cut alfalfa)	5.45	9.00	23.1	12.85
4 (alfalfa range)	5.37	8.80	24.3	12.80

The figures indicate that range or cut, succulent green feed reduced feed consumption by about 18% and feed cost by the same percentage; not counting the labor of feeding the green feed. Hens not fed green feed were as large as those receiving it, but the toms were 2½ to 4 pounds lighter.

Rather extensive trials by many practical turkey raisers indicate a saving of from 15 to 20% on feed consumption when luxuriant range is available and not too heavily pastured.

At the Pennsylvania Station (1945), confinement-reared White Holland turkeys given either fresh or dried and soaked lawn clippings grew larger (1 lb. for males, ½ lb. for females), matured faster, graded out better (about twice as many in U. S. "A" grade), and had more yellow in the skin at 28 weeks of age than did similar groups not given green feed. This green feed was high in protein, vitamin A and riboflavin, which apparently improved the already well-balanced diets that were fed. The consumption per pound of gain of the grain and mashes fed was not significantly reduced by the feeding of the green fed and when the dry matter in the green feed was added, the total feed consumption per pound of gain was slightly greater. Whether or not the feeding of lawn clippings would be economically profitable is not shown in the results. The greens-fed turkeys were decidedly larger and in better market condition at 28 weeks. Profitability would probably depend on the cost of the green feed, the labor required to feed it, and whether or not a complete, well-balanced mash and grain diet can be obtained at a competitive price. Grass clippings undoubtedly would be an excellent supplement to an intentionally cheapened and simplified ration or to wartime feeds that might be deficient in vitamins, proteins, or minerals.

Manure

Breeding birds voided a total of 2.4 pounds of fresh manure per bird per week from October 9 to November 6 in tests run by Marsden (1938). On a machine-dried basis this amounts to approximately 10 ounces of dry matter, since fresh turkey manure contains about 75 percent moisture. The same non-producing breeding birds were fed *ad libitum* 2.2 pounds of feed per bird per week. Manure production may be estimated by assuming that 1.1 pounds of fresh manure will be voided by breeding turkeys for each pound of feed eaten. About 28 percent of the total manure produced will be deposited while the birds are on the roosts. The percentage probably would be somewhat more during the long nights and somewhat less during the shorter nights.

Fig. 1.4. Confinement rearing pens 360 feet by 70 feet over all (26 pens 28' by 30'). Maximum capacity 6,500 birds to market age. (Photo from Richard Warren, Tippecanoe County, Indiana.)

Manure production of young White Holland turkeys grown on bluegrass range was determined by the Pennsylvania Station to be about 1.57 lbs. of fresh manure per lb. of feed eaten. In the 16 weeks before marketing, 380 turkeys ate 12.7 tons of grain and mash and produced 19.7 tons of manure, 35% of it in the range shelters, 65% on the range. On a moisture-free basis turkey manure averaged 5.0% nitrogen (N), 2.74% phosphoric acid (P_2O_5), and 1.91% potash (K_2O) and had a value of $3.72 per ton. Old manure from pits and under perches, 10 to 16 weeks old, had a value of $7.13 per ton and averaged to contain 54% as much nitrogen, 138% as much phosphoric acid, and 139% as much potash. Turkey manure has great value in compost for gardens, as top dressing for grass, and as fertilizer for corn, potatoes, tomatoes, and leafy vegetable groups, worked into the soil several weeks before seeding, 3 to 4 tons per acre, preferably mixed equal parts with stable manure. Quick air drying without excessive leaching helps to preserve the fertilizing value. Top dressing the fresh manure with 200 lbs. of superphosphate or 100 lbs. land plaster (gypsum) per ton preserves the nitrogen well but is a profitable practice only if the treated manure is used to best advantage or sold at a comparatively high price.

When to Start Poults

Since it requires 26 to 28 weeks to mature medium- and large-sized turkeys to U. S. Grade A condition when slaughtered, the poults should be hatched 26 to 28 weeks before the turkeys are to be killed for market. Small-type turkeys require 22 to 24 weeks to reach Grade A condition. If early-fall or late-summer marketing is planned, early poults should be obtained, but if marketing is to be done in November or December or later, poults hatched during what is considered the normal hatching season should be obtained. To hold and feed turkeys beyond the U. S. Grade A stage is not economical unless prices are at a highly profitable level or a premium for extra finish can be obtained. Birds held longer carry an extra thick covering of fat (sufficient for the U. S. AA grade). If poults are hatched over a period of several weeks, they should be marketed over a similar period so as to slaughter them when "ripe." It often is practicable to hold early hatched birds in storage until a favorable market develops.

Turkeys can be brooded successfully at any time of year. Hot weather does not seriously retard their growth if proper sanitation is provided. In the central and northern states at least, poults hatched any time from midwinter up to the first of July grow equally well. After July 1, there may be a slight retardation of growth, but the poults can be raised satisfactorily nevertheless.

A new enterprise consists of the production of turkey *broilers* or *fryers*. These are in reality small *roasters* and first came into popular consumption in 1949 after a very small start in 1948. In 1952 still greater numbers of turkey broilers were produced resulting in prices to the grower which barely paid the cost of production. In 1953 prices received were slightly improved in spite of a continued high production; the market being more aware of the virtues of the small turkey roaster. Most turkey broilers so far produced have been Beltsville Small Whites with a small production of Jersey Buffs and White Hollands. In 1953 and 1954 some of the Broad Breasted Bronze turkeys were sold to the broiler market. This practice did not meet with general favor. Standard brooding, rearing, and feeding methods are used to produce turkey broilers. The use of antibiotics in the feed, in appropriate amounts, is recommended in broiler production.

Early Growth and Feathering

The poults when first hatched are covered with soft down and usually have quills showing where the larger wing feathers are to appear. These quills start to grow at once and within a week the poults' primaries and secondaries appear prominently. At about 10 days of age feathers appear where the wings join the body. In about 3 weeks main tail feathers appear. From then on, feather growth is rapid and when the poults are two months old they are well feathered.

About the fifth week, fleshy protuberances called caruncles begin to appear on the neck. These caruncles are bright red, changeable to pale flesh color. The face of the male later becomes changeable to blue. The appearance of the caruncles is termed "shooting the red." It is a stage of no particular significance in the development of the poult.

At about three months the beards (of males) appear though the formative tissue is clearly visible at about 11 weeks.

In artificially brooded poults, except those raised in feather brooders, the primary wing feathers tend to protrude in an unsightly manner for a few weeks, starting at about 1 week of age. This condition is normal. It is not a symptom of poor feeding or management. In hen-brooded, and in feather-brooder-raised poults these feathers do not protrude due to continual rubbing against the feathers of the mother hen or brooder.

Segregation from Chickens

Turkeys and chickens cannot be raised together successfully under farm conditions. Poults must not be allowed to mingle with chickens and other poultry, nor to occupy the same ground unless it has been properly "rested." The reason is that the other poultry, especially chickens, commonly harbor parasites and the causative organisms of blackhead, and some other diseases which affect turkeys. Under modern conditions of management, particularly confinement rearing, where good sanitation is practiced, turkeys and chickens or other poultry sometimes are raised side by side though this is not recommended. In any case however, they should not use the same land or equipment, unless both have been rendered free from contamination. Buildings and equipment such as sunporches can be cleaned and disinfected satisfactorily but land cannot be. Land must be rested for at least 3 years, and in cases where serious outbreaks of diseases or parasites have occurred, four full crop seasons should elapse before returning with turkeys.

PLANS FOR BROODING AND REARING

The common method is to brood the poults on the floor in either stationary or portable brooder houses for 6 to 12 weeks, usually about 8 weeks, and then transfer to limited range, where the poults remain until marketed. Other plans used are: (1) Battery brooding from 10 to 21 days, usually not over 14 days, then floor brooding for about 6 weeks longer, followed by range rearing; (2) battery brooding up to 6 or 8 weeks, followed by confinement rearing; (3) floor brooding up to 7 or 8 weeks followed by confinement rearing, and (4) brooding as in (1) above, followed by rearing in strict confinement. Whatever the method of management, there are a number of cardinal principles which are followed by successful growers. These principles are:

(1) To furnish a favorable environment by means of adequate heat, ample space, sufficient ventilation, and a well-balanced ration.

(2) To protect the poults from weather and natural enemies.

(3) To control diseases and parasites by sanitation.

The entire growing period is divided into, first, *the brooding period* of about 8 weeks, during which considerable equipment and painstaking care are necessary for success, and second, *the rearing or growing period* (9th week to market age), during which proper feed, a sanitary environment, and some protection are the main points to be considered. *The finishing period* is a convenient term to distinguish the last six weeks of the growing period just before marketing. This period is often referred to as the *fattening period.*

Brooder Houses

Elaborate housing is not required anywhere in the United States for successful turkey management. During the brooding period when birds are raised artificially

Fig. 1.5. Moving colony brooder house to clean range. (Photo courtesy James S. Kellerman, Romney, Ind.)

all that is needed is a brooder house that can be heated to a floor temperature of about 70° F. or a little more. Brooder houses similar to those used in chicken brooding are satisfactory. In warm climates the construction can be cheaper than in cold climates. Turkeys brooded in advance of the normal season may require better housing than those brooded in the normal spring season. The general principles of brooder-house construction may be applied directly to turkey brooder houses. A warm dry floor, ample light, sufficient ventilation, protection from vermin, tight walls, and a roof that will retain the heat and protect against the weather, are the requirements. Interior fixtures required are: Brooder, roosts, water fountains, feeders, wire-covered platforms for feeders and waterers, and litter or wire floors. If wire floors are used, the feeders and waterers are placed on them, no special wire platform being needed. A good type of roost for the brooder house is illustrated in the diagram in Fig. 1.12. Three to four inches of roost space should be provided for each poult. Excessive window space is neither necessary nor desirable. One square foot of window space for each ten square feet of floor space is about the correct proportion.

Portable brooder houses are constructed of wood or composition board and placed on skids. If use in subfreezing temperature is contemplated, the floors should be double-boarded with building or tar paper between the layers. The walls are usually single-boarded with tight-fitting material. The roof is commonly made of boards and composition roofing, or less frequently, of metal or plywood. A house 10′ by 12′ or 9′ by 14′ is satisfactory in size. Brooder houses much larger cannot be moved readily and houses much smaller are not economical. These sizes will carry about 250 poults to 8 weeks, with an outside yard or sunporch of about the same size added; without outside quarters, or in weather not permitting the use of outside quarters, the capacity is 150 poults.

Permanent brooder houses in most cases can be constructed on the general principles suggested for portable brooder houses but often have damp-proof concrete floors and sometimes are insulated inside with wallboard or with tongue-and-groove sheathing. In multiple-unit houses the brooder rooms may be 9′ x 14′, 10′ x 12′, 12′ x 14′, or 10′ x 20′ with or without sunporches. Rooms 20′ x 30′, of 3 brooding units each are practical. There is often a service alleyway about 4 ft. wide in the rear of the house. A new and satisfactory type of multiple-unit house stands with its long axis north and south. The service alleyway is in the middle, with pens on either side, one row facing east and the other west. A saving in construction and labor costs of operation is thus accomplished. Brooding equipment offering minimum fire risk should be chosen for use in multiple-unit houses.

Several-purpose houses which can be used for brooding and rearing and even for housing breeding stock are often used. Sometimes a large sunporch is added for use in confinement rearing or the house is moved to a rear-

Fig. 1.6. Roosts on the roof of the shelter are desirable except in late fall when the birds might be subject to severe storms.

ing field to serve as a range shelter with or without the addition of a sun-porch, lean-to, or outdoor roosts. If brooder-house wire floors of permanent construction are desired they should be built in easily removable sections or placed solidly 2 to 3 feet above the ground or floor to allow for cleaning underneath. Brooder-house plans may be secured from commercial organizations, poultry magazines, and from some State Colleges of Agriculture.

Essentials of Brooding

A satisfactory brooding device should provide the following:

(1) *Ample heat*—Heat sufficient to provide a temperature near the center of the hover 1 inch above the floor of 95° F. for dark-plumaged turkeys and 100° F. for white-feathered turkeys; and a floor temperature of the room of about 70° F., or a little higher.

(2) *Uniform controllable heat*—That can be adjusted to the demands of the birds at various stages of growth and under varying weather and management conditions.

(3) *Safety* from fire.

(4) *Ventilation* under the hover and in the brooder room, sufficient to avoid strong odors and meet the requirements of the poults for fresh air.

(5) *Reasonable* first cost and reasonable operating cost.

Heat may be supplied by coal, wood, oil, gas, or electricity, the choice depending partly upon the relative cost and availability of the various fuels. Table 1.2 gives comparative operating costs of the more common types of brooders.

Types of Brooders

(1) **Coal-burning colony brooders.** These brooders supply an abundance of heat and if properly constructed and operated, are dependable, safe, and efficient. Some disadvantages are the labor required to operate them,

lack of flexibility in the heat supply during warm weather, and danger of fire when carelessly operated. Using wire floors or non-inflammable litter such as sand or peat moss with coal brooders is almost necessary in multiple-unit houses. Large coal capacity and a sensitive and durable regulator are points to consider. Some coal-burning stoves do not hold enough coal and some have poor regulators that will not permit the fire to be "throttled down" for warm weather use without danger of going out. Anthracite coal of chestnut size is probably more satisfactory than other kinds, although semi-anthracite, bituminous, and even lignite coal can be used in some stoves.

(2) **Oil-burning colony brooder stoves of air-blast type.** These stoves are meeting with much favor among commercial turkey growers due to the abundance, flexibility, and reliability of the heat supply. They are usually built to burn kerosene, although some will utilize distillate. Many stoves may be connected to the same oil supply. Some disadvantages are higher fuel costs, danger of fire, and danger of gas formation when carelessly operated or poorly constructed. The selection of a well-constructed stove with ample provisions for disposing of gases formed during combustion is most important. Fire risk is reduced by using wire floors or non-inflammable litter. In multiple-unit houses this precaution plus careful, 24-hour attention is necessary if oil brooders are used.

(3) **Electric colony brooders.** Electricity is becoming popular in many sections as a source of heat for brooding turkeys. Where the source of supply is reliable and reasonable in cost (less than 3c per kilowatt-hour) and the weather not severe, electric brooders as the source of heat have been very successful. When supplemented by heat from another source, electric brooders are satisfactory under all conditions provided the source of electric power is reliable. Some disadvantages when used as the sole source of heat are: Lack of sufficient heat to keep the brooding quarters warm and dry enough in damp or cold climates, and susceptibility to power interrup-

tions in some locations. Electric brooders are satisfactory for brooding turkeys only in *warm dry weather* or with *auxiliary heat* unless changes are made to remove the inadequacies that now exist. In the purchase of electric brooders advantage should be taken of equipment especially designed for brooding turkeys which, in some types, includes curtains 6″ to 8″ high around the edge of the canopy. There always should be a small ventilation space between the floor and the lower edge of the curtain.

When electric brooders are used along with auxiliary heat from some other source, the room temperature at the floor level should be maintained near 70° F., the electric brooder supplying additional heat. Some sources of auxiliary heat that have been used successfully are: hot water radiators, hot water pipes on rear wall of the brooder house, hot water under floor, heat in rear half of the floor, heating stoves fired with coal or natural gas, and space heaters.

(4) **Hot-water brooders.** Hot water brooding systems that are properly designed and installed are well adapted to cold weather brooding and for use in multiple-unit houses. They are safe, dependable, durable, and provide a margin of safety in case of heat interruptions. They

Fig. 1.7. These poults were transferred from batteries to these range shelters at 4 weeks of age. Note protected sides. (Photo by R. L. Hogue, Mulberry, Ind.)

may be heated by gas, oil, wood, or coal. Great care should be exercised to select equipment that has proved satisfactory and to install it according to correct engineering principles. A common design for a 24-foot wide house consists of three 1½-inch (inside diameter) pipes carrying water from the heater and three returning it, alternately, about 6″ apart, all on the same level 10″ to 12″ above the floor and about 5 ft. out from the rear wall; roofing or similar material being placed on top of the pipes to form a hover, its lower edge coming within 2 or 3 inches of the floor. Another plan is to place the pipes 8 inches to 9 inches above the floor and use a board platform as a covering upon which litter is placed. The pipes must be properly painted to provide efficient radiation. Circulation is by gravity or pumps.

(5) **Gas brooders.** In the experience of many turkey growers in milder climates, gas brooders are the most desirable type now available. They have many of the same characteristics of the electric brooders, but supply somewhat more heat, usually enough for 200 or more poults in moderate weather. For brooding from 8 to 12 weeks, 25 to 75 cubic feet of natural gas would be required per poult. Either piped or bottled artificial or natural gas may be used. Gas brooders of the conventional design are not the best type to use in cold weather operations unless auxiliary heat can be supplied.

(6) **Wood-burning colony brooders.** Wood is the cheapest fuel available in many localities. Automatically regulated brooder stoves have been designed that utilize wood (preferably green hardwood) to advantage as a satisfactory source of heat. If properly operated, the fire will last from 6 to 8 hours without refueling, and an abundance of heat is produced. One to 2 cubic feet of wood will be required per poult for brooding up to from 8 to 12 weeks of age. The fire hazard with wood-burning brooders is equal to that of coal-burning brooders and all of the necessary precautions must be taken.

Table 1.2—The costs of operating brooders for 10 weeks in the first series of tests and 8 weeks each in the second and third series of tests. (Va. Agr. Exp. Sta. Bull. 306)

Brooder	Fuel cost			Labor cost at 25c per hour			Total operating cost		
	First test	Second test	Third test	First test	Second test	Third test	First test	Second test	Third test
Commercial electric with fan	$16.29	$10.77	....	$2.63	$1.63	....	$18.92	$12.40	....
Commercial electric with natural ventilation	8.49	7.53	$3.96	2.00	1.63	$1.25	10.49	9.16	$5.21
Converted electric	15.99	....	....	2.25	....	....	18.24	....	....
Home-made electric	12.15	11.61	....	2.63	1.63	....	14.78	13.24	....
Coal	4.73	3.31	1.02	13.38	8.25	6.00	18.11	11.56	7.02
Kerosene	30.12	15.84	7.92	8.13	6.00	3.75	38.25	21.84	11.67
Wood	....	5.83	2.07	....	5.75	3.63	....	11.58	5.70

Note: These chicks were brooded in 9′ x 12′ colony brooder houses. First test: Nov. 21, 1935 to Jan. 30, 1936, 200 Barred Rock chicks per brood. Second test: Feb. 11, 1936 to Apr. 7, 1936, 200 Rocks and Reds per brood. Third test: April 28, 1936 to June 23, 1936, 150 Rocks, Reds and Giants per brood.

(7) **Radiant or under-floor heat.** This type of heat is circulated by convection from iron pipes (1″ to 1¼″) set in 3½″ to 4″ of concrete on top of two thicknesses of slater's felt crisscrossed over stone or gravel fill; cinders never being used. Pipes are placed 12″ apart at the rear and 18″ apart at the front of the house and preferably are welded. They are arranged in separately controlled sections, all over the floor area. The concrete should extend at least ⅝ to 1″ above the pipes. Copper tubing of ⅜ to ½″ diameter is preferred by some operators. Its coils are spaced from 8″ apart at the rear of the house to 12″ apart at the front. Only a small amount of sand or other suitable litter should be used on the floor since the litter serves as a heat retardant. Solid partitions should be installed every 40 to 60 feet to break up drafts if the house is long; and in addition, low wire partitions may be used to separate the poults into groups of from 300 to 600 each. Circulating pumps force the 130° F. heated water through the system and back into the boiler. Sufficient floor temperature to keep the poults comfortable (80 to 85° F.) is maintained. To prevent dehydration in young poults, humidifiers or sprinklers may be installed. Sun porches increase the housing capacity. Roosts may or may not be provided. Ventilation, front and rear, is important. In warm weather, radiant heat is not too satisfactory. Its high cost of installation, difficulty in repairing, slow response to temperature changes, and tendency to dry out the poults, are disadvantages which lower its value. However, under-floor heat in the rear half of the brooder house provides good auxiliary heat for gas or electric brooding.

(8) **Brick colony brooders.** Home-made brick brooders, burning green wood, are in common use for brooding poults as well as chicks in some sections of the South. They proved to be quite popular during the 1930's when every effort was made to reduce the cash costs involved in the production of agricultural products. Detailed in-

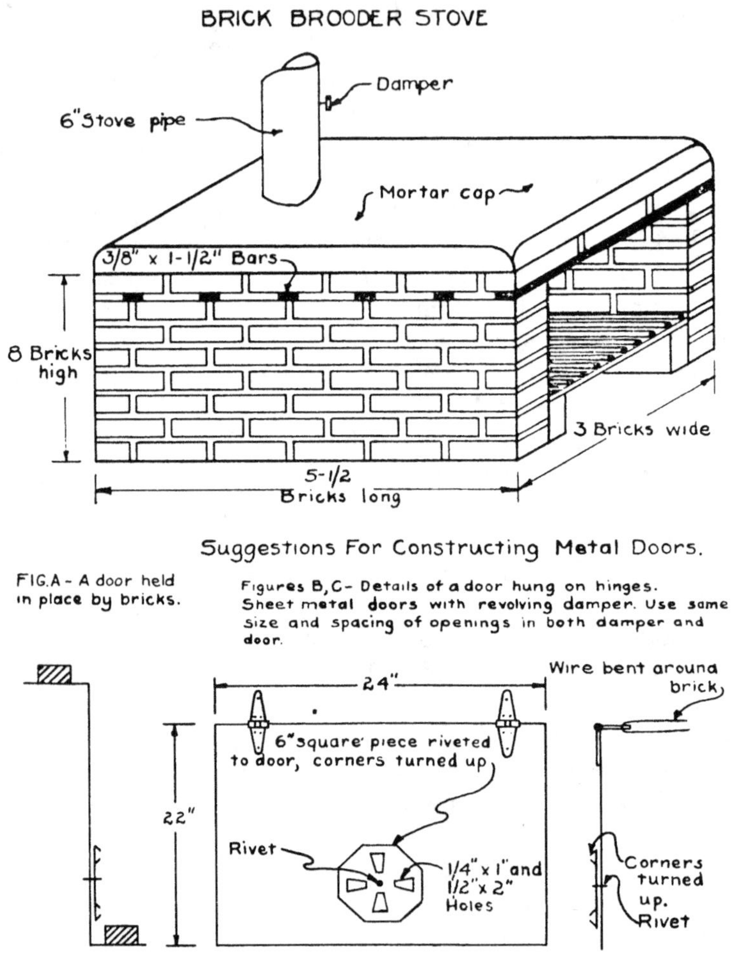

Fig. 1.8. Brick brooder stove for burning wood or coal. The grate and one course of brick may be omitted if coal is not to be used as fuel. (Plans from U. of Kentucky.)

structions for the construction of these brooders may be secured from the University of Kentucky (Lexington).

(9) **Battery brooders.** Many operators brood the poults for only 10 to 14 days in batteries and then trans-

fer them to their regular brooding quarters. Sometimes the poults are raised in batteries with sun porches attached, throughout the brooding period of 8 weeks, and are then transferred to range. Another method used when the weather is warm and dry, is to brood in batteries for about 4 weeks and then transfer direct to range shelters that have three sides protected. Such poults should be "hardened off" while in the batteries. In 1949 and 1950, extensive trials were run in the transferring of poults directly from battery brooders to range shelters at 4 weeks of age. In this case, the brooder room is gradually cooled from about 70° at the beginning to about 40° F. by the fourth week. The thermostats in the battery brooders should be gradually lowered, or the heating unit, if moveable, should be raised. Poults have withstood freezing weather (28 to 30° F.) in unheated shelters at 4 weeks of age if properly "hardened" before transferring. (See Fig. 1.7.)

Battery operation generally requires a room heated to about 70° F. with about 60 percent humidity and a properly designed battery. The battery (especially the feed troughs) and the room, must be well lighted. Feed and water should first be placed inside the battery as well as in the feed and water troughs. Ventilation without drafts should be provided in the battery room, preferably through fans or flues. Ample head room, preferably 12 inches (except under the hover), and ample battery floor space (about 1/3 square foot per bird) should be provided for poults up to 4 weeks of age. The vertical wires of the sides of the battery should be 1 3/8 inches apart. For the floor, one of the best materials is .4″ x .4″ mesh hardware cloth; 1/3″ mesh is satisfactory at first but clogs with droppings after the third week. Half-inch mesh is satisfactory if made of 16-guage wire; if made of light weight wire it may cause leg injuries and loss of wing bands which become caught in it and may be pulled out. If 1/2-inch light-guage wire is used, it is desirable to cover the wire with burlap cloth or sand covered newspapers for the first week. When poults are to be battery brooded to 8 weeks of age, it is well to provide more head room; and a sun porch, preferably the same size as the battery floor space area. Chick batteries work well but only with poults up to 2 weeks of age. Battery operations are highly specialized and the requirements are most exacting. Experience with battery raised poults indicates that where battery operation is efficient, growth is rapid and mortality tends to be low. Losses from disease may be reduced due to the favorable environmental conditions, as to sanitation especially, supplied by the well-built and operated equipment. Battery manufacturer's instructions should be followed carefully. Batteries are widely used by hatcherymen and poult dealers for the production of started poults and for holding poults for a few days prior to delivery.

(10) **Feather brooders.** This type of equipment is intended for use along with supplementary heat. The feather boards are made of tightly bound and glued clusters of turkey feathers inserted into holes in boards. Feather brooders preserve the heat produced by the poults, provide a surface against which the poults may hover, and tend to prevent the poults piling up. However, a room temperature of about 70° F. at poult level is necessary for successful operation. The plumage of feather-brooded poults is smoother than that of poults brooded by other artificial means.

(11) **Lamp brooders.** This type of brooder has been supplanted to a large extent by other types listed. Lamp brooders have small capacity and the danger from fire and the necessity for continual detailed attention make them of little value to the turkey raiser. In addition, the fumes from the lamp, unless carried away by a pipe, are dangerous to the health of the poults. In the opinion of the authors this type of equipment has little place in modern turkey raising. If poults are to be brooded in

small groups, feather brooders, batteries, or small-capacity electric brooders are preferable.

(12) **Hot-Air brooders.** Direct utilization of heated air in turkey brooding is in limited use. F. C. Brown, Poultry Agent at Pittsboro, North Carolina, describes (1951) two such systems as follows: One type heats from an automatic coal stoker and the heat goes into a tobacco-type flue which is 12″ in diameter, the bottom of which is approximately 18″ from the floor. "The heat enters the center of the flue near the middle of the brooder house and then flows toward both ends of the house. About 10 feet from the end of the house, the heat and fumes are exhausted through small stacks. Each stack has a damper in it. This prevents the heat from going to one end of the house. It also has a divider in the "T" joint, which helps to keep the heat going in two directions. The volume of heat is controlled by a thermostat located within a few feet of the heat flue. Thermostats are located in various positions but the operations all attempt to achieve a heat of 95° at the floor for a distance of at least two feet to each side of the pipe. Two sheets of aluminum may be bolted together and hung over the flue to form a hover."

Another type of hot air brooding system is one which uses ducts and hovers and the air is heated by an automatic oil furnace. "This system re-circulates 75% of the air in the room and brings in 25% fresh air from the outside. The re-circulated air is passed through filters, which should be cleaned daily. . . . The air is forced through this system by a Siroca-type blower. There are dampers in each duct, so that the heat going to each hover can be regulated. These ducts are all connected to a central duct which decreases in size as the distance from the heat source increases. The temperature variation between the first and last hover is usually no greater than 2°. These hovers are adjustable. Under each hover is a baffle-plate, which helps to disburse the hot air in all directions in a more uniform manner. This baffle-plate now is constructed of aluminum screening which is more effective than the solid metal plate."

(13) Space heaters. In 1948 in Indiana, space heaters were first used for the brooding of chicks. Since then they have been commonly used throughout Indiana and elsewhere for the brooding of turkey poults as well as

Fig. 1.9. Space heater furnishes ample heat for brooding poults in large groups. The fan is set to run continuously, even though the thermostat shuts off the heat supply. (Photo by J. C. Allen and Son.)

chicks. Space heaters of the type commonly used in garages and stores are adaptable for brooding purposes if placed closer to the floor than would be convenient otherwise. (see Fig. 1.9). A space heater is nothing more than a heating coil or unit with a fan behind it to blow the heat and deliver it to the point desired. The fuel supply may either be bottled gas, which is one of the most convenient to use for poultry, hot water, steam, or electricity. In the case of the hot water or steam units, it is necessary to have a boiler (usually in an adjoining shed) in which either a coal fire or oil fired burner supplies the heat to warm the water which circulates through the pipes to the various units within the poultry house. Space heaters are of two types as regards heat delivery, namely, vertical delivery and horizontal delivery. The vertical delivery units must be hung toward the ceiling of the poultry house since the fan blows the warm air directly to the floor from which it spreads out over the building. The horizontal delivery delivers the current of air in a horizontal position parallel to the floor and it gradually strikes the floor at distances from a few feet up to 15 to 20 feet from the heating unit. It is usually wise to confine the poults to a wire guard (never use cardboard or boards) or poult circle the first few nights to make sure they are kept within the area that is warmest, either underneath the vertical delivery or just a few feet away from the horizontal delivery unit. One Hoosier turkey raiser installed two horizontal delivery units with an input of 135,000 b.t.u. and an output of 108,000 b.t.u.; in a brooder house 49 by 125 feet. In this, he successfully brooded 5,300 poults with a cost of the bottled gas (propane) fuel for the first three weeks of $91.00. This was a considerable saving over the cost of oil for eight nearby colony brooder house units in which he was carrying 1900 poults, at the same season of the year. Much is to be learned about the use of space heaters for brooding poults as well as chicks, but suffice it to say that they have a definite place for those who wish to brood in large numbers. The Purdue (Indiana) Station ran an experiment in a brooder house 54 x 112 feet in which were installed three vertical delivery and two horizontal delivery space heater units. Their experience indicates a number of factors of management peculiar to this type of brooding, the horizontal delivery heaters proving more satisfactory than the vertical.

(14) **Infra-red brooders.** Infra-red heat lamps have come into rather common use in the brooding of both chicks and poults. Bulbs of 250 to 375 watt capacity should be set in a baffle-board or small hover in porcelain sockets and the board placed so that the bottoms

of the lamps are 15 or 16 inches above the litter at the start; raising them 2 inches per week to a maximum of 2 feet. Care should be taken that the watering pans are a good distance from the bulbs so that the birds cannot shake cold water upon the bulbs and thereby crack them. There are numerous types of holders or frames on the market to hold the bulbs, but care should be taken not to use a type in which too much heat is concentrated in a very small area of the brooder house. One bulb never should be used alone for if it should burn out during the night the poults will chill. Batteries of two or more bulbs are advisable. A wafer thermostat is desirable when 3 or more lamps are used. Such thermostats are usually set to cut off all but 1 or 2 of the bulbs in the first response, and the remainder of the bulbs if the heat gets too high. In case of a current interruption care must be taken to supply some other source of heat immediately to the poults. As with other types of electric brooders, infra-red brooders alone do not supply sufficient heat for real cold weather brooding of turkey poults. They have the advantage of low initial cost, but consume larger amounts of electricity than do conventional electric brooders (on a per poult basis) over which they have but few special advantages. Infra-red bulbs can be used in houses which are especially well insulated and have thermo-pane windows in the south exposure of sufficient capacity to absorb a large share of the heat of the sun's rays during the daytime and thereby avoid the necessity of supplying auxiliary heat from some other source.

Litter, Wire Floors and Porches

During the first week or two, sand, gravel, vegetable fossil litter (silica), peat moss and chopped corn cobs are the chief types of litter that are satisfactory. Other types are likely to cause impaction of the digestive tract. Even when peat moss is used it should be covered with sacking material the first few days. To prepare dry, non-moldy corncobs for litter, put them through a hammermill with 1″ roughage screen then through a clipper fanning mill with top screen removed and a cracked-bean screen used on the bottom. This removes husks, cob dust, and cob pulp (or heart) and drops the new cob litter into the cleaned-seed box at the bottom of the mill.

After the poults are 1 or 2 weeks old, clean wheat straw, planer shavings, crushed cobs or peanut hulls, shredded cane, corn stover, or peat moss may be used. These materials may be used throughout if covered 9 weeks with sacking. Sand or fine gravel, if free from silt or clay, so it does not "ball-up" on the poults' toes, makes good litter at all ages. A good way to use it is to put in about 4 inches, dry, at the start of brooding, then rake it daily, removing droppings crusts. The one layer thus suffices for the entire brooding period. Sand also may be used in a thin (½″) layer, replacing or adding to it weekly. Other kinds of litter, such as hay, oats and barley straw, dried beet pulp, cottonseed hulls, oat hulls, splinty shavings, or sawdust, are sometimes used, but are hazardous and are not recommended. Dried beet pulp has been used successfully, but with it, as well as sawdust, care must be taken to avoid all moldy material. When any litter except sand, silica, shredded cane or peat moss is used, it must be added to or fully replaced every 5 to 7 days, as a necessary sanitary precaution. The depth at first should be 1½ inches, adding more so as to use the same material throughout as "built-up" litter, the use of which saves much labor but may increase the disease hazard when dampness is present or if used for more than one brood.

When poults are brooded on concrete, earth, or cold wooden floors with litter, a wooden platform or a piece of some other insulating material should be placed under the hover for the first three weeks, to prevent chilling and floor dampness. A wire platform made of one of the wire flooring materials described also is satisfactory. Such an arrangement is especially important if electric hovers are used. This special equipment usually can be removed if necessary at the end of the third week of brooding, although it often is desirable to retain it throughout.

Wire floors are more sanitary than litter but do not guarantee freedom from disease, although they are a help in disease control. For starting poults, wire of 21-gauge, .4″ x .4″ mesh is very suitable, especially where

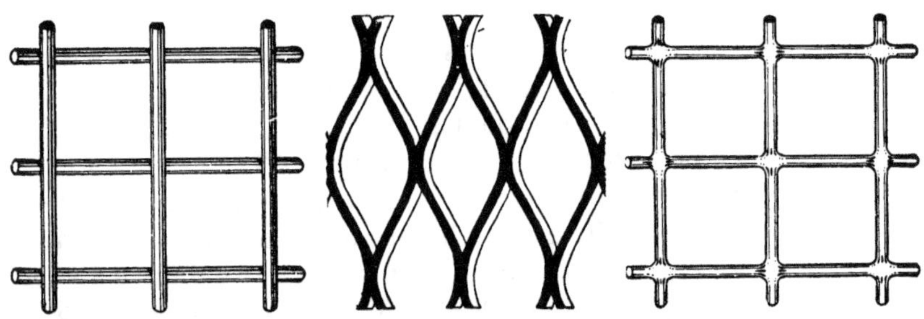

Fig. 1.10. Materials for wire floors. Hardware cloth, (left), electrically welded wire fabric (right) and expanded metal fabric (middle).

it lies on top of heavier wire. When used alone, the ½-inch mesh, 16-gauge hardware cloth, galvanized after weaving, is very good although the ½″ x 2″ and ½″ x 4″ welded 16-gauge wire are more durable. The latter size can be continued in use for older turkeys, even up to market age. A practical method of utilizing wire floors is to make the floor of 1-inch mesh 16-gauge hardware cloth or 1″ x 2″ or 1½″ x 1½″ mesh, 12 gauge galvanized welded wire. At the start, cover about two-thirds of the brooder room, including the area under the hover, with 4/10″ x 4/10″ mesh wire which is rolled up and removed after three or four weeks. After the poults are about four weeks old, square-mesh wire of smaller mesh than ¾″ is difficult to keep clean. Where there is any trouble from floor drafts or from inadequate heat, it is a good practice to cover the wire under and close to the hover, for the first two weeks or so, with newspapers, sacking, or a board platform on which there is a layer of sand or gravel. Wire floors in the brooder room are sometimes preferred because of the advantage gained from requiring no litter and preventing outbreaks of disease, especially blackhead and coccidiosis. However, special precautions must be taken to supply ample heat and the correct type of floor wire with the right amount of ventilation under it to keep the droppings dry. Floor drafts should be guarded against, however, by proper floor construction and the use of floor partitions at reasonable intervals. If sufficient ventilation and reasonably dry air are provided, the droppings under wire floors and other types of platforms need not be cleaned out except at the end of the brooding season.

Wire-floor units about 3′ square and 4″, preferably 6″, high, altho fairly costly, are quite widely used. There are several different kinds of wire flooring normally available but no matter what kind is used, there must be a good but not excessive circulation of air under the wire floors or platforms, otherwise too-frequent cleaning will be necessary. If wire floors are laid on wooden framework the frames must be above the floor a distance sufficient to provide liberal air circulation underneath. Wire sunporches and even brooder house floors are frequently constructed at a height of 2 to 4 feet off the floor or ground. The floors of sunporches used only during the brooding period, may be of 1″ x 1″ or ¾″ x ¾″ mesh, 16-gauge, or special fabrics, or 1″ x 1″, 1¼″ x 1¼″, or 1″ x 2″ slats placed flat-wise about 1 inch apart. For sides and tops of sunporches and also pens used for confinement rearing, the hexagonal mesh is better than the ordinary fencing in which the line wires are of varying distances apart. Where small poults are involved, a 1½ foot strip of 1 inch mesh, 18-gauge wire should be used next to the floor. Above this strip and on top of the porch, the 2″ mesh, 16 to 18-gauge, may be used. The 19-gauge is cheaper but is durable only when used inside the house. *Plaster laths* may be used for floors and sides of the sunporch. The laths are nailed ¾″ apart on three 1″ x 4″ or 2″ x 2″ boards, forming sections, and then painted with two coats of crankcase drainings (used engine oil). For the sides 2′ to 4′ high, lath spacings should be 1¼″.

Sunporches for brooding units are usually equal in area to the brooder room, although some operators cut the porch area down to half or two-thirds.

Wire platforms for use under feeders and waterers are desirable equipment to use in connection with brooding on litter. They contribute greatly to sanitation. Three-fourth-inch mesh on a frame-work made of 1″ x 6″ material, is satisfactory for them, since they usually are not used until poults are 2 weeks of age. Heavy use by the poults will keep them clean, even with well-grown birds.

The use of air-slaked lime, crankcase drainings, or a 3% cresol spray under the wire floors and sunporches is advised if fly-breeding in the droppings becomes a problem. However, a D.D.T. spray or malathion flakes now available, afford the most practical control. Wire floors and platforms can be cleaned by first soaking them in or spraying them with water for an hour or two; then using water under pressure to remove the dirt. It may be necessary to use a brush with wire bristles.

Management Problems

Compared to chicks, poults are more difficult to brood. They do not learn to eat so rapidly; they are more subject to fright, crowding, and piling; they do not take kindly to changes in feed or equipment; and they are more easily chilled. The essentials of success in poult brooding are good equipment and eternal vigilance. Overcrowding is very dangerous, and large units, over 200 poults, are more difficult to handle than are small units. Ample feeding and watering space and sufficient heat must be provided. Getting the poults out into the sunporches or yards is good practice but it must be done carefully and the poults watched continually. If quick and easy access to the brooding room and to the brooder itself is not provided, the poults may crowd together outside in the corners of the porch next to the house. Boards or wire may be used to guide the poults back into the house openings.

Heat. There is no substitute for adequate heat in the brooding quarters. Without it young poults will chill and soon die. The exact temperature under the hover and generally over the floor of the brooder house is of minor importance provided the poults are comfortable and good ventilation is maintained. However, an accurate brooder thermometer should be a part of every brooding unit for a quick check-up of temperature. A wall thermometer placed near the floor away from the brooder is very helpful. In some cases it may be desirable to lower the room temperature at dusk to encourage the poults to settle down next to or under the brooder. If the temperature is too high at dusk the poults sometimes huddle in groups away from the hover. The poults, if comfortable, will be active and contented during the daytime. This is the best test of temperature. Overheating is dangerous to the health of poults. If overheated, the poults will appear listless and quiet but will be spread out well, not grouped together. Panting will be noticed if heat is extreme. If underheated, poults will huddle, peep, and refuse to eat. As a general guide, 70° to 75° F. at the floor of the brooder house is desirable. The yard, sunporch, or the outer edges of a large room provide a stimulating variation in air temperature. The temperature 2 or 3 inches above the floor under the center of the hover should range from 95 to 100° for dark colored poults, and 100 to 105° for white poults. White poults seem to need slightly more heat than dark colored ones; especially during the first week. Hover temperature may be lowered gradually as the poults grow; or it may be left the same and the poults will move away.

Except in electric brooding, enough heat should be provided at night so the poults will form a ring under the outer edge of the hover, leaving the center vacant.

In electric brooding sufficient heat should be provided to keep the poults well spread out under the hover, avoiding all semblance of crowding.

Electric hovers are usually equipped with dim (5 or 7½-watt) attraction lights under the hover. The poults appear to hover better, are less subject to fright and crowding, and learn to eat more readily when attraction lights are used. Such a light may be installed in other types of brooders but care should be taken to use a low-wattage bulb.

Under favorable conditions, well-grown poults will start to roost by 3 weeks. A good share of them will be on the roosts at 4 weeks of age and nearly all of them at 5 weeks. Early roosting aids in solving brooding problems and should be encouraged. Brooder-room temperature must be kept up fairly well at night in order to encourage early roosting, otherwise the poults will not roost or will leave the roosts in favor of the floor. In cool weather heat is needed until the poults are 6 to 8 weeks old, while in warm weather 5 weeks usually is sufficient.

Ventilation. Ventilation is a simple matter in a well-constructed brooder house provided with a brooder capable of producing a steady and ample supply of heat. Electric brooders work best if equipped with an elevated wire floor under the hover, or a ventilating fan that forces a circulation of air under the hover. Window ventilation usually is used, sometimes supplemented by rafter or cupola ventilation in the brooder house. A double door with building paper between the two layers of flooring is desirable to prevent "sweating" of the litter in cold weather in houses with raised floors. A portable floor six feet square may be placed under the hover if the entire floor of the brooder house cannot be covered with double flooring.

If a brooder house is insulated and tightly built, definite provision should be made for ventilation with adjustable buffered outlets near the front or in the roof, and intakes in the front walls. The ordinary brooder house can be ventilated as needed through windows. The curtain of the brooder should not touch the floor (unless it is a wire floor), especially at night, or ventilation under the brooder will be insufficient.

Humidity. Insufficient moisture does not appear to be an important factor in brooding except in very dry climates, or with under-floor heat. Some operators keep open water receptacles on the brooder stoves or heat pipes to raise the humidity to a more desirable level. There is no direct evidence at hand to show that poults suffer from low relative humidity, except with under floor heat during the early brooding stages; when excess dust and dehydration of the poults may occur. Excess humidity is more likely to be a practical problem since it may contribute to damp floors, damp litter, and bad odors. If the operator of a brooder house is able to provide warm, dry floors, reasonably dry litter (when litter is used), fresh air and prevent insect pests from flourishing as a rule he need not concern himself with the relative humidity of the air. An optimum of 60% relative humidity has been suggested by some operators. In any case, warm dry floors and dry litter are essential to success in the floor brooding of turkeys.

Sanitation. If disease can be prevented during the brooding period, an excellent start has been made in turkey raising. Only by sanitation can this be accomplished. If ordinary litter such as corn stover, shavings, or straw is used, it should be added to or fully replaced every 5 to 7 days without fail, preferably twice a week. Users of peat moss and other absorbent kinds of litter are apparently able to obtain satisfactory results even though the same litter is left in the brooder house for several weeks. Wire floors are a great aid to sanitation, if they are of a practical type and are properly handled.

Floor Space and Hover Space

A practical brooding unit is about 200 poults. However commercial operators often brood from 250 to 300 or

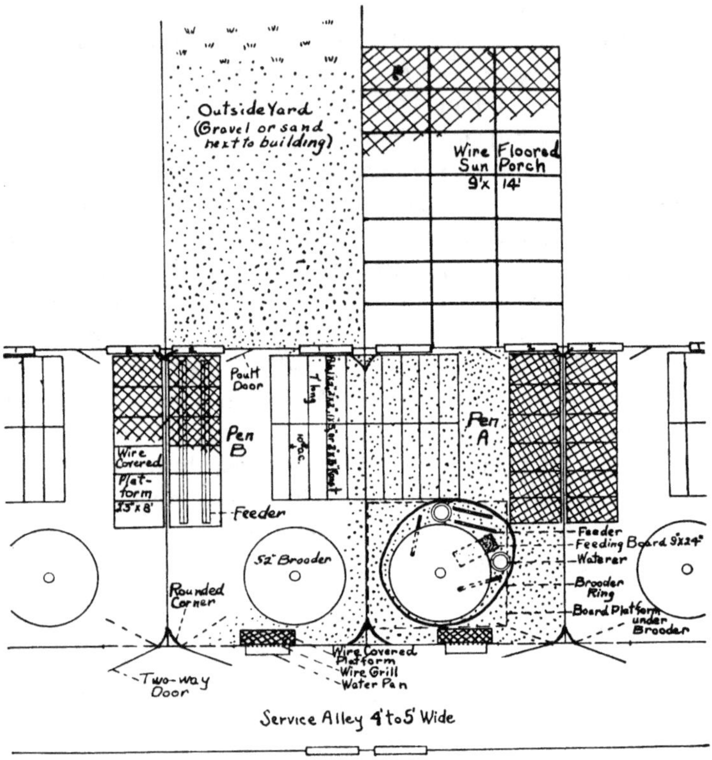

Fig. 1.11. Diagram of floor plan of section of cement-floored turkey brooder house showing arrangement of equipment and two means of providing outdoor environment for the poults. Pen A shows sunporch and the arrangement of equipment for starting poults. Pen B shows an outside runway and the arangement of equipment for poults over two weeks old. Windows each consist of one 30" six pane sash. Those marked 1 slide sidewise; those marked 2 slide up and down. The runway may be 100 to 150 feet long. Instead of litter the floor of the brooding room may be covered with wire. See text for details of house and equipment.

even up to 400 per unit. It is better practice and less risky to limit the brood to 200. About 1 square foot of brooder-house floor or floor and porch space combined should be allowed per poult if the birds are to be brooded for about 8 weeks. If brooded 10 to 12 weeks, floor space of about 1½ sq. ft. per bird, including sunporch or yard space if used, are required. However, if weather is such that the yards or porches cannot be used, 1 sq. ft. to 8 weeks or 1½ to 12 weeks should be provided in the house.

If grown to fryer age, 14 to 16 weeks, 2 sq. ft. per bird is desirable. The usual recommendation is a minimum of 0.5 to 0.7 sq. ft. per bird inside the brooder house with additional required area furnished by yard or porch.

A colony brooder house or a unit in a larger house, either 10′ x 12′ or 9′ x 14′ with a sunporch of the same size will hold up to 250 poults to eight weeks, 165 to 12 weeks, or 125 to 16 weeks. In brooders with canopies, about 10 to 14 square inches of canopy space per poult is needed at the start. A 52-inch canopy will accommodate 150 poults.

Large units of 600 to 700 poults in one large room containing three brooders with brooder rings at the start, are practical. Plenty of heat, large brooders, one square foot of floor space per poult up to eight weeks, and expert care are required.

Arrangement of Equipment

Figure 1.11 shows a workable arrangement of equipment. Roosts should be near the front, or near a source of light, as poults will not go to roost in a dark cor-

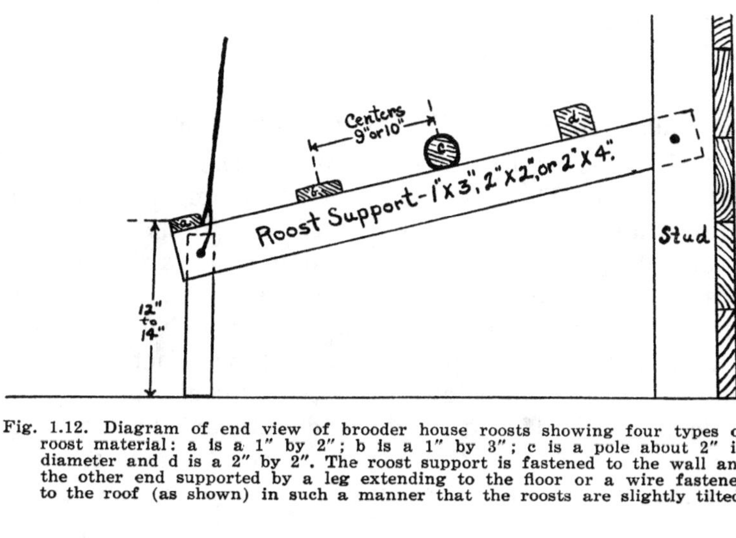

Fig. 1.12. Diagram of end view of brooder house roosts showing four types of roost material: a is a 1" by 2"; b is a 1" by 3"; c is a pole about 2" in diameter and d is a 2" by 2". The roost support is fastened to the wall and the other end supported by a leg extending to the floor or a wire fastened to the roof (as shown) in such a manner that the roosts are slightly tilted.

ner. Frequently rear windows are provided and the roosts placed against the rear wall of the house. The hover may be placed in any part of the room if all edges of it are a foot or more from the wall. No unnecessary equipment should be permitted in a brooder room as it provides encouragement for the poults to crowd and pile. Small sunny areas, heat pipes, and the like are also danger spots. All warm points and surfaces except at the brooder itself should be eliminated. Free access to the brooder from all parts of the brooder room should be provided. The corners, especially those back of the hover, should be rounded, using ½- to 1-inch mesh wire for this purpose. A "brooder ring" made of the same material or other small mesh wire should encircle the hover the first few days. Feeders and waterers are set inside this fence which may be enlarged after the first day or two, then after 4 or 5 days used only back of the hover to keep poults from getting in the corners. Within 8 to 10 days it usually can be removed. Early roosting is an advantage particularly from the standpoint of the prevention of crowding and piling at night. If the ration is adequate, deformed breastbones will not be caused by early roosting. Poles about 2 inches in diameter make ideal brooder-house roosts. In place of poles, slightly tilted 2″ by 2″, 2″ by 3″, 1″ by 2″ or 1″ by 3″ boards may be used.

Range Utilization During the Brooding Period

During the early brooding period, the range is but little utilized. The present tendency among commercial turkey raisers is to use wire or clatted porches in place of outside yards. Many use no porch or yard, and instead supply the required floor space inside the house, which, in these cases, usually is of economical construction. Small gravel-floored yards are satisfactory, provided they are not allowed to become contaminated by prolonged use. If the indoor and outdoor quarters are kept sanitary, such diseases as blackhead and coccidiosis may be easily controlled. The equipment described under confinement rearing can be used for brooding. Poults should be allowed in their outside yards or porches as soon as weather permits. If 1" by 1" or the 1" by 2" mesh wire or slats are used for the porch floor, the poults may be allowed on it at the age of two to three weeks.

A permanent brooder house so arranged that each brooder room is equipped with one good-sized yard containing growing green feed, is very satisfactory pro-

Fig. 1.13. Range rearing is popular in this Shenandoah Valley; 2,500 Beltsville Small White Turkeys on range in Virginia. Note range shelters and hopper-type feeders. (Photo from East Point Turkeys, Elkton, Va.)

vided the runs are clean at the start, are well drained, and are used only for one brooding period of 6 to 12 weeks, or at the most, two such periods, during each calendar year. During the remainder of the year the yards should be "rested" and kept free of all poultry. For each brooding unit of about 150 poults, a yard containing from 500 to 700 square feet is needed under average conditions if the stand of green feed is good to fair. However, the use of green range during the brooding period is becoming rare. Portable brooder houses are quite satisfactory for brooding poults. They should be moved as often as necessary to prevent the occurrence and recurrence of disease and to provide fresh green feed.

If clean, well-drained land is available, only one move at the end of the first four weeks need be made for each brood of poults. Some growers use a small temporary corral at first and move it to a different side of the house once each week for four weeks, then move the house to a clean spot and repeat the operation. This precaution usually is unnecessary under favorable soil and management conditions, but may be a life-saver under unfavorable conditions. The likelihood of diseases and worm infestation may govern the frequency and extent of the moves required. It is this risk that has cut down the use of range during the brooding period (see Fig. 1.13).

There are several advantages in providing clean yards containing growing green feeds during the brooding period: (1) The green feed problem is solved and the resultant saving of feed, (2) if necessary, cheap simpli-

Fig. 1.14. Abundance of room is available in pole sheds during early age of poults. Clean, dry wood shavings are satisfactory for littering in these sheds. (Photo from Browning Turkey Farm, Winchester, Ky.)

fied starting rations may be utilized although not generally recommended, (3) the natural conditions outside tend to offset unfavorable conditions inside, (4) bad habits are less likely to develop, and (5) where a complete feed is difficult or impossible to obtain, the green feed consumed will make it possible to raise good turkeys

Fig. 1.15. This 60′ x 1500′ pole shed covers about 2 acres of earth floor. From 40,000 to 44,000 poults are brooded under gas hovers supplemented with heat lamps. This shed provides slightly over 4 sq. ft. per BBB turkey for 20,000 birds from 10 weeks of age to maturity. (Photo from Browning Turkey Farm, Winchester, Ky.)

on a simplified diet. There are certain disadvantages such as: (1) The difficulty of providing suitable land, (2) The considerable attention that must be given in the management of the yards to prevent diseases and parasites gaining a foot-hold, (3) The short period of use of the brooder range yards, and (4) The extra management problems involved in the utilization of range lots by small poults. It is desirable but not necessary to provide some sort of open air quarters such as yards or porches for turkeys during the brooding period; and it is equally important to carry out a definite plan of sanitation for these quarters. However, as mentioned, a brooder house that is well ventilated and has ample floor space is sufficient.

Pole barns or sheds have increased in popularity in recent years for all types of livestock. Plans of gable or half-moniter type roof pole sheds are available through the Dept. of Agricultural Engineering of the State College or from Farm Management Services. They are readily adapted to confinement rearing of turkeys, and may be used to hold breeding stock. The feeding and management problems are much the same as with other types of confinement equipment (see page 75 and Fig. 1.14). However the advantages of pole barns or sheds are such that they seem likely to supersede porch rearing.

CHAPTER 2

REARING PROBLEMS

OUTLINE

TROUBLE SHOOTING

If turkey poults do not grow rapidly and feather evenly, or if a heavy mortality occurs, the following possible sources of trouble should be considered.

1. **The feed: (seldom at fault)**
   - (a) Poor formula for starting mash:
     1. Too-low protein content—below 22%.
     2. Lack of vitamins or other essential factors.
     3. Too finely ground or sticky when wet.
     4. Too coarsely ground, containing large hulls or particles.
     5. Possession of undesirable ingredients.
     6. Ingredients out of proportion.

Fig. 2.1. Long brooder house with sunporches. Note tip-out feeders. (Courtesy Spar Hill Turkey Farm.)

(b) Staleness, mold, or decomposition of the mixture or its ingredients, especially the animal products and the dried alfalfa products.

(c) Possession of poisonous substances (rare).

2. **Failure of poults to learn to eat.** Check the amount of feeding space and type of feeders; see that some attractive feed such as rolled oats or finely cut, tender green feed is given at the start. Feed poults within 24 hours of hatching. Use bright glass marbles, particularly green ones, on top of feed. At start do not require poults to put their heads through wires or holes—make feed easily accessible.

3. **Failure of poults to learn to drink** or to obtain a sufficient quantity of water. Check number and type of waterers, making sure the poults can actually reach the water. Put founts under hovers at first. Put glass marbles in water. As with the feed, make the water easily accessible to the poults

and get them to drinking within 24 hours after hatching.

4. **Insufficient feeding space.** Provide 2″ per poult.
5. **Overcrowding** under hover or in brooding quarters. Check floor and hover space per poult.
6. **Dampness** under hovers. Check floor construction, ventilation, and heat produced by the brooding device.
7. **Insufficient heat** under hover or in brooding quarters. At the start of brooding, poults require an average brooder-room temperature near the floor of about 70 to 75° F. along with a hovering device providing still more heat, 95 to 105° F.
8. **Unsuitable litter;** clean silt-free sand and gravel are among the few satisfactory types of litter for the first week or two.
9. **Poor ventilation** in broooder house or under hover.
10. **Disease** (have sick or freshly dead birds examined by competent authority).

WATER SUPPLY

During the first 2 or 3 weeks of the brooding period water is conveniently supplied by the ordinary two-piece tip-up "pail and pan" waterer. Small water fountains of the Mason-jar type are especially useful at the start of brooding. The glass-bottomed type is best. Liquid milk is outmoded as a turkey feed. When it must be fed to round out the diet, two-piece tip-up crockery fountains may be used for dispensing it for the first 2 to 4 weeks. If milk is to be given in a galvanized vessel it is wise to first pour sour milk in the pan or bucket and allow it to remain for 24 hours. This milk should then be thrown away and the vessel cleaned well. After 2 to 4 weeks milk or water may be fed in troughs, pans, or bowls provided with wire guards or grills to prevent turkeys getting into or fouling the contents. At least one 2- to

Fig. 2.2. A "tee-pee" over a washtub keeps the water clean. This same equipment may be used for dry mash, filling the tub about half full.

3-gallon waterer or two fruit-jar waterers should be provided for each 60 to 75 poults. This amounts to about ⅔" of watering space per turkey (circumference of a circle is 3 1/7 times diameter). After they have learned to drink (in 3 or 4 days) the number may be reduced to one for every 90 poults or ½" of watering space per bird. At 2 or 3 weeks the number may be reduced still further or the fountains replaced by pans, troughs, or automatic waterers provided with a guard to prevent drowning of poults and fouling of the water. From this time until the end of the brooding period and beyond, the watering facilities should provide about ¼ lineal inch of watering space per bird, have sufficient capacity so frequent refilling is unnecessary, and be easily accessible. Daily washing and rinsing of the waterers is necessary.

During the rearing period, water or milk may be supplied by a variety of devices, the choice depending largely upon the facilities available. Where water can be piped or hauled to the place where it is to be used, the birds can drink from ordinary water pails, wash tubs, or metal or wooden troughs, provided precautions are taken to prevent them from drowning or fouling the containers. Placing them on a screened platform over a "dry well" is a satisfactory plan. Any time after the first few days in the brooder, turkeys will quickly learn to drink through a grill, a wire or wooden guard, or through a fence, provided the openings are large enough for their heads and necks. Rubber or plastic hose is convenient for connecting waterers and placing them where needed. Hand-controlled and float valves are useful where the water supply is ample and continuous. Drainage for overflow and to take care of valve failures should be provided.

Barrel hog waterers are satisfactory. Barrels with trough attached and tightly closed bung to retain water level in the trough are satisfactory. Other satisfactory watering devices are barrels cut in half with alternate staves 6 in. lower to provide openings through which the young turkeys may drink; and metal eave trough or wooden troughs, any length, with a continuous flow of water. One of the best is a bubbler attached to a 6″ or 8″ drain tile, to carry off the overflow. Continuous flow waterers properly installed are freeze-proof even in subzero weather.

Probably no single watering unit, unless it be a long trough, should be counted on to serve more than 250 birds. For example, a flock of 300 turkeys should have 2 units to provide about ¼ lineal inch of watering space per bird. Continued crowding around the waterers indicates insufficient watering space.

The water should always be placed near the mash and grain feeders. Waterers also should be deep enough for the turkeys to immerse their heads almost to the eyes, otherwise feed will be wasted and the birds cannot drink in a satisfactory manner.

Fig. 2.3. A simple way to keep the water clean. This trough is of small capacity and requires frequent filling unless filled automatically from supply.

Natural sources of water generally are to be avoided because certain parasites and diseases may be transmitted thereby. Unless experience or examination has demonstrated the safeness of the natural supply, turkeys should be kept away from lakes, streams, or springs. Marshy areas in particular are to be avoided. Streams have been known to serve as sources of infection, bringing bacteria and protozoan parasites down-stream to uninfected flocks. Almost all water found in nature contains life in various forms and some of these plants and animals are themselves harmful to turkeys or may harbor parasites or disease organisms. In general, only water from a source suitable for human consumption should be supplied to turkeys. Spring water, artesian water, and water from surface wells and from clean lakes and streams may be used but in general it should be carried or piped to the birds and dispensed in a sanitary manner. Cistern water should not be used if pigeons roost on the roof of the building that drains into the cistern. Water containers should be kept clean by daily washing with a brush or clean cloth, followed by rinsing in clean water. Occasional disinfecting with boiling water or germicidal solution, such as 1% sodium orthophenylphenate, is desirable. Temperature of water when consumed should be moderate, neither hot nor very cold.

Snow cannot be depended upon to supply water in winter to growing or laying turkeys, although it may suffice for maintenance at times when it is impracticable or impossible to keep the water supply constantly open. In these cases, a supply of warm water should be placed before the birds twice a day. They will drink it greedily even though plenty of snow is available.

Sanitation around waterers is most important. Wire floors, wire or slatted platforms, gravel, cinders, or other porous material should be placed underneath stationary waterers and the drainage around them should be good. A wire or slatted platform placed over a dry well is an excellent place for a permanent waterer. Occasional cleaning around a permanent source of water is advisable. When portable, the waterers should be moved weekly and placed only in well-drained locations.

FEEDING EQUIPMENT

Satisfactory feeders for turkeys of various ages may be built or purchased. Care should be taken to provide adequate feeder space.

Small feeders for use during the first week may be made of lath or of resawed material 1½ to 2 inches wide. From the second through the fourth week, feeders with 2½-inch sides may be used. From the fifth to the eighth or on through to the twelfth week, feeders, with 5-inch sides are satisfactory. After the age of 8 to 12 weeks, one size of feeder with about 8-inch sides will do. Various types are in use. The trough type is widely used because of the comparatively greater feeding space avail-

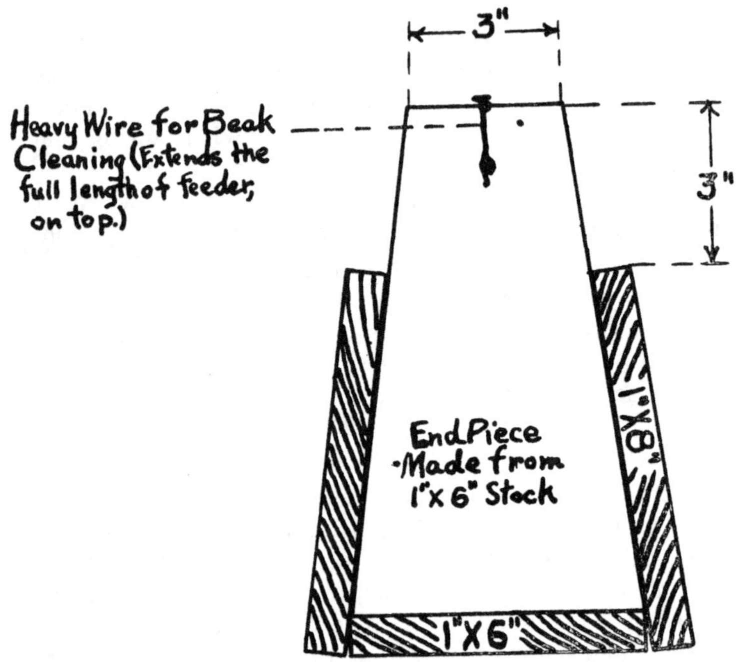

Fig. 2.4. Diagram of an inexpensive, easily constructed, outdoor or indoor feeder for young or adult turkeys. A convenient length is 10′ with two equally spaced center pieces which are patterned after the end pieces, and provide partitions for corn, oats and mash. (Courtesy W. W. Neel.)

able to the birds. Up to 8 weeks provide about 2″, and after 8 weeks, about 2½″ of feeder space per bird. Baby poults do not take readily to feeders that require the birds to put their heads through holes. On or in all of the feeders in a position convenient for beak-cleaning, No. 10 to 12 wire should be stretched tightly to aid in the prevention of feather picking. The wire may be attached with a screen door spring to keep it taut.

After the second week of the brooding period the feeders should be set on wire-covered platforms unless the birds are already on wire. During the rearing period the feeders should be placed on wire. However, if kept on the ground, they should be moved at intervals of a few days to two weeks, depending upon the condition

Fig. 2.5. Range self-feeder of large capacity; plans available from University of Kentucky. (Photo courtesy Browning Turkey Farm.)

of the soil near them. Frequent moving is desirable under all conditions but especially when the birds are ranging on heavy soils and flat land in humid climates. Placing the feeders and waterers on a wire-floored alley in the roosting shelter is probably the most satisfactory

List of Materials Needed for Kentucky Self-Feeder

Quantity	Description	Use
2 pc.	3"x6"x10'—0"	Runners
4 pc.	2"x4"x 4'—0"	Under floor-trough
2 pc.	1"x8"x 8'—0"	For "A" in trough
2 pc.	1"x6"x 8'—0"	Sides of trough
2 pc.	1"x2"x 8'—0"	Lips to trough
6 pc.	2"x4"x14"	Braces—side of trough
2 pc.	2"x4"x 8'—0"	Brace across trough
4 pc.	2"x4"x 3'—0"	Upright studs—ends
2 pc.	2"x4"x 4'—0"	End plates
2 pc.	2"x4"x 8'—0"	Side plates
10 pc.	2"x4"x 5'—0"	Rafters
1 pc.	2"x8"x 3'—0"	Crosspiece for trough
90 Sq. Ft.	Metal or Composition	Roofing
90 Sq. Ft.	Sheeting	Roof
56 Sq. Ft.	Shiplap or Flooring	Ends—trough—hopper

Nails: 2 lbs.—20d, 1 lb.—10d, 2 lbs.—8d, 1 lb.—6d; 2 pr. hinges and screws.

method of dispensing the feed and water. Feeders should either be short, from 4 to 6 feet long, or else have partitions every 2 to 4 feet so as to provide separate compartments for mash and grain feeding. Adjustable partitions are desirable. A very desirable type of range feeder may be made of 1″ x 8″ sides nailed to the edges of a 1″ x 12″ bottom board. The roof consists of a horizontal 12″ board which may be hinged. The opening between the top board and the sides is 7 inches. Hopper-type feeders needing filling only about once a week are becoming popular. The Kentucky Station publishes plans for such a feeder. (See Fig. 2.6.)

V-shaped open trough feeders made of one 10″ and one 11″ board (or one 11″ and one 12″) are cheap to construct and quite useful.

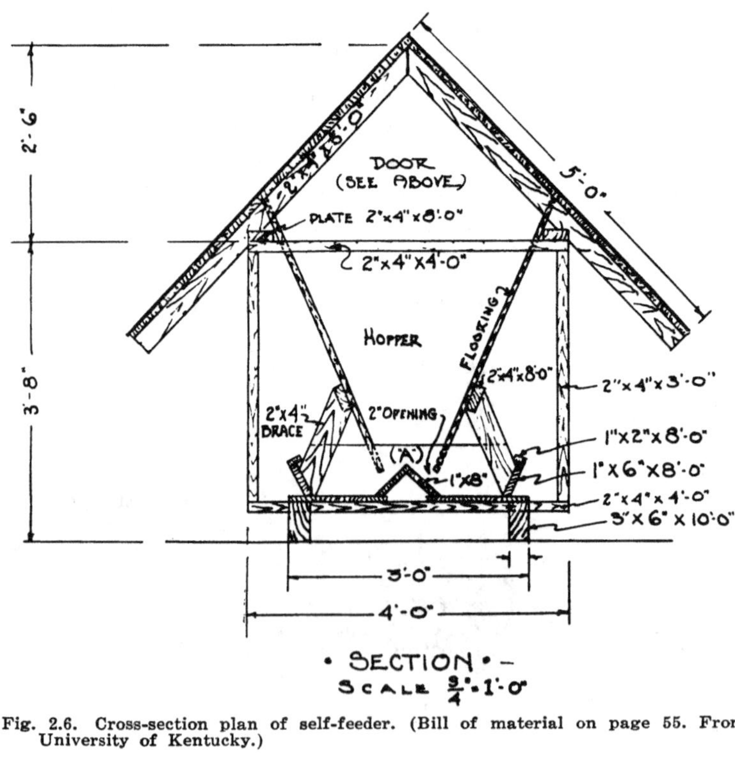

Fig. 2.6. Cross-section plan of self-feeder. (Bill of material on page 55. From University of Kentucky.)

FENCES AND GATES

Rectangular-mesh poultry fencing five feet high is generally used for confining growing turkeys. Regulation 58″ P.&R. (poultry and rabbit) fence is quite satisfactory. For special breeding pens, the six-foot fence is desired as a precaution against mismating. Six-foot wire is also advisable for yards that are small or nar-

Fig. 2.7. A labor-saving method of feeding on range.

row such as brooder house yards, and for yards located on hillsides. In long narrow yards, birds have a tendency to fly length-wise of them and then over the fence. On hilly terrain the birds will fly down hill especially on windy days when they utilize the force of the wind to take off.

On rigid-top gates and on buildings that come within about 8 feet of the ground, antiflies must be placed to prevent birds from alighting on them and getting out of their yards. Even when the wings are clipped, antiflies are desirable to discourage attempts to fly. Antiflies (see Fig. 2.9) are made of lightweight rectangular-mesh poultry wire supported by 2″ x 2″ (or heavier) boards or, where a neat appearance is important, by sections of lightweight angle iron nailed or wired to the buildings, gates, or fences. Often the antiflies must be extended out from the gates or buildings on top of the fences for two or more rods. On gates and fences the antiflies are placed vertically. On buildings it is convenient to fasten them to the sides and eaves which causes them to slant in towards the yards.

Poultry fencing of medium to heavy weight and of good quality is more economical in the long run (except

Fig. 2.8. A waste-proof, vermin-proof, contamination-proof reel-top feeder for inside use. (Photo by U. S. Dept. of Agr.)

for antiflies) than cheap material of light weight. Fencing with top and bottom wires 11- to 12½-gauge and center wires 14½ to 15½-gauge, is advisable. Stays should be 4 to 6 inches apart. In damp climates the best available rust-proofing is economical.

Steel posts are in common usage except in cases where wooden posts are very cheap. Steel posts should have an anchor plate and weigh not under 1½ lbs. per lineal foot. If posts with baked enamel finish are used, the parts under ground should be protected with an asphalt coating.

Painted metal posts are satisfactory only in dry climates. Galvanized posts are preferred in climates of high humidity. Five-foot fencing requires 7-foot line posts and 8-foot gate and corner posts, while 6-foot fencing requires 8- and 9-foot posts, respectively. The line posts may be set a rod apart under ordinary conditions. In very loose soil or where conditions make especially heavy demands on the fencing, the spacing is best reduced to 10 or 12 feet. Special conditions may also necessitate setting the line posts about 2½ feet in the ground instead of 1½ to 2 feet. In ordinary firm soil and for the comparatively short stretches usually involved in poultry fencing, dirt-set corner and end posts are satisfactory and are useful under conditions where setting in concrete is impracticable. Angle-type corner posts set in dirt should have the corner rounded, possess an expanding anchor plate, and have 2 angle braces. Total weight of the 8-foot length should not be under 90 lbs., the 9-foot length 100 lbs. The corresponding end or gate post should have one brace and in the 8-foot length should weigh 60 lbs. or more; 67 lbs. for the 9-foot length. Cement-set posts are required in soft or wet earth and often are preferred to the other kind in situations where it is easy to mix concrete or to obtain ready-mixed concrete.

Fencing used for poults during the brooding period should consist of a strip of 1-inch mesh 19-gauge wire about 1½ or 2 feet wide placed next to the ground or floor. Ordinary square-mesh fencing or 2-inch hexagonal-mesh 18-gauge netting may then be placed above the 1-inch mesh. Fencing for permanent brooder-house yards should be provided with baseboards close to the building and these should be placed close to the ground at all points. Tight stretching with a stretching device is very desirable, however, in all cases.

Gates should in general be the same height as the fences. In locations where they must be opened and closed frequently, a strong, rigid gate, preferably one with a steel frame, is highly desirable. If antiflies are to be placed on the gates, stock gates 4 or 5 feet high are

Fig. 2.9. Portable range shelter 12′ by 19′ to hold 125 turkeys of market age. This shelter provides excellent protection and room for inside feeding. The feeders are filled from the outside. The floor is of 1½″ hexagonal mesh 16 gauge wire. Note antiflies at edge of roof. (Photo by U. S. Dept. of Agr.)

satisfactory. Narrow metal gates 3 feet wide may be used in some places but generally the longer gates are better. Where trucks or wagons must enter, a 12-foot gate is necessary. If they are not to be opened frequently, gates made of the regular poultry fencing with 2″ x 2″ or 2″ x 3″ strips or poles at the ends, serve the purpose well. No antifly is needed on this type of gate and it is much cheaper than the rigid type. It is often advisable to use the cheap gates made of fencing in such locations as to permit entering or turning with a team or tractor for cultivation of the yards. Extra end posts are not needed as these light gates may be wired to the end post or corner post at one end and the line post, the building, or even the fence wire, at the other end. A wire loop may be attached to the post or a tile may be set in the ground to receive and hold the post at the bottom while a flexible wire hook that can easily be unhooked will secure the top. Electric fences are not suitable for confining turkeys, but they are reported as satisfactory for excluding predatory animals. They do not restrain turkeys satisfactorily and tend to cause burns.

SHADE

Shade of some sort is necessary for turkeys of all ages during hot weather. As a rule, special efforts are not made to supply brooder house yards and sunporches with shade. The poults use the house for shade during brooding. The sunshine is important and when it is too bright and hot for the birds they will go to the house which is desirable because feed, water, and protection await them there.

During the rearing period shade presents more of a problem. For birds raised in confinement the problem may be solved by shading from one-fourth to one-third of the sunporch with sacking, metal roofing or boards. However, the shelter itself if large enough and built so it can be opened up to allow a liberal circulation of air, will supply sufficient shade without additional shade on the porch. For birds raised on limited or free range, the range shelter will, if large enough and well-ventilated, supply sufficient shade, but if such shade is not available other means of supplying it must be used.

When outdoor shade is provided, there should be sufficient shade for all the flock without causing the birds to crowd into a small area. Crowding in a shaded area leads to the development of diseases and bad habits. It is better to have the birds congregate in a wire-floored range shelter than in a small shady spot that is likely to become heavily contaminated. On the other hand, very thick shade covering a comparatively large area is also undesirable because the sun has no chance to dry out and disinfect the soil, and no green feed will grow there. Probably the ideal shade for growing and breeding turkeys is that furnished by scattered trees and bushes. A row of trees or shrubs or a strip of corn or sunflowers will afford satisfactorily shade. Alternate strips of corn or sunflowers and a green crop are frequently used to afford shade on turkey ranges. Sudan grass in rows or strips and some types of tall weeds supply fairly good shade.

If the yards are alternated, well drained, and not overcrowded, little concern need be had over contamination of shaded areas. If conditions are not entirely favorable for soil sanitation, the problem of providing sanitary shade should be seriously considered, bearing in mind that slatted or wire-floored range shelter or roofed-over roosts is a good way to provide safe shade. When natural shade cannot be grown without considerable trouble and expense, it is economical and satisfactory to put a roof over the roosts. If these shelters are moved frequently, dropping pits are not needed but if the shelters are not moved frequently, the space under the roosts should be wired in so as to provide protected, sanitary roosting and shading facilities in one operation. Lightweight, low, portable, shed-type frames covered with boards, sheet metal, straw, weeds, cornstalks, or other suitable roughage are satisfactory.

STORM PROTECTION

In the so-called storm belt of middle-western States, turkeys should be marketed prior to November first or protection should be provided by shelters arranged in a semi-circle (Fig. 2.10) and guarded by a snow fence placed 130 to 150 feet away in the direction of the prevailing wind which usually is from the north-northwest. Snow fences may be improvised, if necessary, of woven wire and corn fodder. Turkeys must be trained to use the shelters, which should have solidly closed backs and provide one square foot of space per bird. Where the wind has a long sweep over flat country, a double row of snow fences may be needed. Shelters should not be placed near ditches, ravines, or creeks because turkeys will drift into them and be smothered in the snow. Natural

Fig. 2.10. Shelters aranged in semi-circle for storm protection.

windbreaks such as 15 or more rows of standing corn or a grove of trees are effective. The shelters should be placed close to these natural windbreaks.

PROTECTION FROM PREDATORS

Turkeys of all ages must be protected against thieves and predatory animals at all times. Brooder houses must be built so as to exclude rats and other animal enemies. Marauding dogs are a constant menace in almost all situations. Foxes and coyotes take a heavy and continuous toll of the turkey flocks wherever permitted to do so. Racoons and weasels also contribute to turkey mortality. Tight, well-set poultry fencing affords a fair degree of protection but alone may not be sufficient, because fences cannot be made predator-proof without great expense. For protection during the night an enclosed roosting shelter, a staggered ring of cannon-ball road flares set 30 ft. apart, or the presence of a guard is necessary. The flickering flames of the flares are very effective protection against predatory animals during darkness. However, during the early morning hours the danger is greatest and an attendant should be in the vicinity or else the birds should be confined to fully protected quarters. The latter alternative is somewhat undesirable because the birds are deprived of access to the range at a time when they obtain the most good from it. In addition, unless the protected quarters are large, the birds are likely to start picking feathers and to develop other bad habits while confined in the morning. If they are confined at that or any other time, feed and water should be kept within the enclosure.

Crows, hawks, and owls sometimes cause losses. Protection against owls is afforded by housing, or by lighted flares if the turkeys are not housed. Hawks and crows do not often cause losses in turkeys raised by modern methods but if they are troublesome, traps may be set on top of posts. Strings or wires placed across the top of the yards are said to be quite effective in protecting from hawks and owls.

High roosts alone are of little value because the turkeys leave them at daybreak when danger is greatest and while on the roosts they are not protected against owls and climbing animals. Furthermore, turkeys often will stampede from the roosts if frightened and will scatter over the countryside.

Watchdogs are relied upon for guarding the turkeys in many situations. If the dogs are fully dependable they afford splendid protection especially if human assistance is available when the alarm is sounded. Collies and other sheep herding dogs can be taught to "herd" turkeys.

Continuous guarding is necessary in many localities to protect turkeys from thieves. Watchdogs, electric lighting, and electric alarms are helpful aids in protecting the turkey flock. A tent or other shelter near the roosting quarters of a large turkey flock will provide facilities for an older boy or man who should have a shotgun within easy reach of the cot. Such an arrangement provides over-all protection and for large flocks on range is almost a necessity. The night guard may be the daytime caretaker or attendant. He may be aided by a good dog and by road-flares or other lighting. A cheap radio operated by storage battery, where electricity is not available, may be left tuned in on an all-night station. The frequent sound of the human voice serves to frighten dogs, foxes and coyotes.

Confinement rearing equipment gives almost perfect protection and this is one of its greatest assets. Tattooing the turkeys in the web of the wing makes positive identification possible in event birds are stolen.

HOUSING DURING THE REARING PERIOD

The majority of the turkeys raised in the United States are without overhead shelter after the age of 8 to 12 weeks. The roosts are set in the open. Protection is afforded by watchdogs, guards, fences, lights; and shade is provided by natural or other means. However, the use of range houses is increasing. Prefabricated range shelters are now on the market.

The features desired in a range shelter are: (1) A reasonably tight roof that will shed most of the rainfall; (2) protection from prevailing winds that bring cold, snow, or rain; (3) protection from all predators; (4) properly constructed roosts; (5) sanitation; (6) open-air conditions.

These advantages are provided by a simple shelter constructed of wood, wire, and roofing. Several types of shelters are illustrated in the accompanying photographs. Plans for range shelters may be obtained from the U.S.D.A., at Beltsville, Md., from several of the State Experiment Stations, and from the "Turkey World" magazine published at Mt. Morris, Ill.

There are some advantages in providing feeding and watering facilities inside the shelter. These are: (1) Feed and water are constantly available in all kinds of weather; (2) there is a saving of labor in feeding and watering; (3) there is no wastage due to wind or rain; (4) rotation of yards is much easier; and (5) sanitation around feeders and waterers is easily attained.

Fig. 2.11. This permanent, shed-roof shelter will hold about 150 turkeys to market age. It may be built on a larger scale. Note wire under roosts which should be nailed to their undersides. (Photo by U. S. Dept. of Agr.)

The disadvantages are: (1) There is greater initial housing expense. (2) In some situations it is difficult to prevent flies from breeding in the droppings under the wire floors or platforms or in the litter. Frequent cleaning or treatment with D.D.T., malathion, or disinfectant is then necessary. However, plenty of space and freedom of air movement under wire-floored feeding alleys will remove this difficulty in most cases. (3) When portable shelters are used, inside feeding equipment increases the weight of the shelter considerably.

In general, the advantages outweigh the disadvantages, therefore, inside feeding should be considered in planning the rearing equipment.

Some suggestions to be considered in the building of range and confinement shelters are given herewith:

(1) **Roof:** Aluminum makes an excellent roof, is satisfactory and economical in almost any climate. Overlapping boards make a fairly good roof if lumber is not too knotty or too likely to split. Roofing paper, if the sheathing is tightly boarded underneath to prevent wind damage, is a good type of roof. Used paper-mill belting covered with

asphalt makes a cheap, durable, and waterproof roof. Aluminum roofing reflects the most heat and black roofing paper absorbs the most.

(2) **Walls:** Woven wire, studding, and a few boards usually make up the walls. Studding may be of 2- by 2-inch material in low-eaved, portable buildings; two-by-fours or poles are used in buildings with high eaves. Wire for side walls should be as follows: where wire comes within 3 feet of the ground it should be strong, 2″ x 2″ mesh 14 gauge welded wire, or 1½-inch hexagonal mesh 16-gauge, or 1-inch hexagonal mesh 18-gauge, tightly fastened to exclude dogs; where not within reach of dogs, 1½″ hexagonal mesh 18-gauge, 2-inch mesh 16-gauge, or wire of different weave but equal in strength may be used.

(3) **Floors:** Range-house floors must be strong, durable, and sanitary. The following materials are satisfactory, practical, and economical: 1½″ wooden slats placed 1¼″ apart, 2″ slats placed 1¼″ apart, or 1″ slats placed 1″ apart; 1½-inch mesh 16-gauge hex-weave wire, 1½″ x 1½″, 1″ x 2″, or 1″ x 4″, 11 to 12½-gauge electric-weld wire; 1-inch 12 or 14-gauge hardware cloth; and rubber-covered welded wire or other suitable materials. The wire or slats may be nailed to removable wooden frames of 2 by 4-inch material on the outside and 2 by 4-inch pieces on the inside; or on solid framework if there is cleaning space underneath. The top of each 2-by-4 may be beveled to present an upper surface of not less than ¾-inch, or it may be unbeveled. The wire is nailed only to the 2-by-4 framework, not to the 2-by-4 center supports. A 1 or 1¼″ poultry-wire staple or an 8-penny nail is used to secure each strand of wire (¾-inch poultry wire staples are not large enough). The sections are placed on supporting

framework and held 1½″ apart by nails so the droppings will fall through. They should fit the feeding alley after making allowance for expansion when damp. Five feet by 4′ or 5′ is a good size to use. The center supports should be 1′ apart and parallel to the long axis of the feeding alley. Slats or the heavier self-supporting types of wire need less support than the lighter weight materials. In all cases the top of the floor should be at least 18-inches above the ground or other surface which receives droppings that fall through the floor. Free circulation of air underneath the floor will aid in the drying out of the droppings so that cleaning under the floor need not be frequent in most situations. In most climates, the droppings will need to be cleaned out only at the end of the rearing season.

(4) **Roosts:** Poles 2 to 4″ thick or beveled 2 x 3's flat make the best roosts. Unbeveled 2 x 3's or 2 x 4's flatwise will do. For heavy turkeys they should be set level but for other types they should be tilted at a 15 to 20° angle (12 to 16″ vertical rise to 48″ horizontal measurement). In general the high side of the roosts should be placed toward the light or prevailing wind. Sloping roosts are better than roosts all placed on the same level because the birds go to roost in a more orderly manner. In construction, it is convenient to place the roost supports at the proper angle and nail the roosts directly to them. However if roosts are of the stepladder type with their surfaces horizontal for heavy turkeys, the roost support must be notched or the lower side blocked up. Wire of 1½″ hexagonal mesh 16-gauge should be nailed tightly to their undersides and on top of the roost supports. A more permanent arrangement would be 2″ x 2″ welded wire, 12 gauge or 1½ x 1½″ welded, 12½

or 14 gauge nailed to the undersides of the roosts and on top of the roost supports. Roost supports of 2 x 6 or other suitable material are placed at a distance apart adequate to properly support the roosts and roost wire. As a rule, 6 feet apart gives sufficient strength. Roosts should be placed 20 to 24 inches apart on center (center to center); for large-type turkeys 24″ are preferred. The lowest roost should be placed twelve (12) inches above the ground or floor. (See Fig. 2.12.) Low roosts built in the open all on the same level about 8 to 14″ from the ground are becoming popular. Birds do not walk in the droppings to any great extent when the roosts are built close to the ground. There is no advantage in high roosts. Allowing no roosts at all beyond the brooding stage is considered best for the large broad-breasted types by many growers.

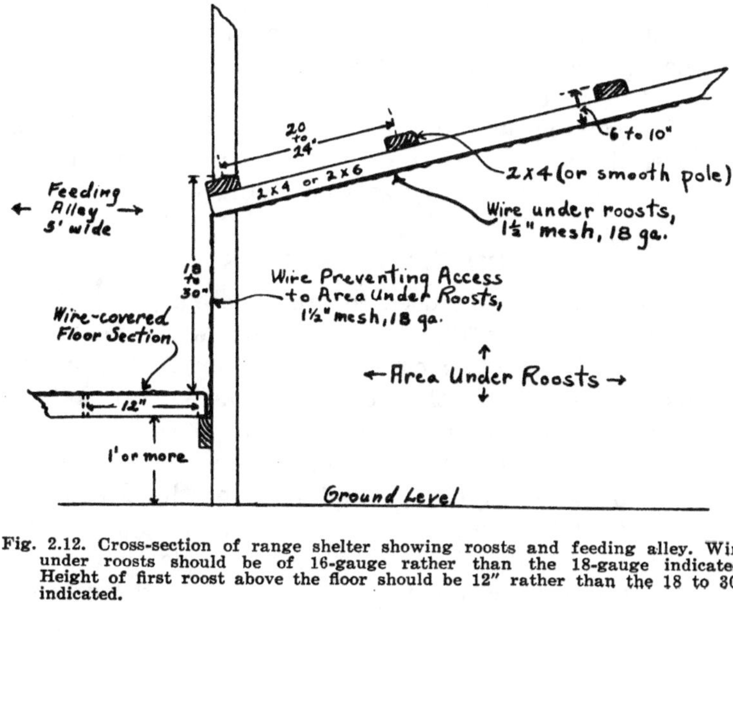

Fig. 2.12. Cross-section of range shelter showing roosts and feeding alley. Wire under roosts should be of 16-gauge rather than the 18-gauge indicated. Height of first roost above the floor should be 12″ rather than the 18 to 30″ indicated.

(5) **Feeding Alley and Walkway:** Five feet is a good width for the feeding alley, allowing one foot for the feeder and two feet on either side. The sides of the feeding alley next to the roosting compartments usually are made of $1\frac{1}{2}$-inch hexagonal mesh 18-gauge wire. The exits of the shelter open into the feeding alley or the walkway. The latter when used, is usually made 24 to 30 inches wide and floored with the same type of wire as the feeding alley. Walkways are used for the convenience of the attendant or to provide proper exits in proper places when multiple yards are used.

GROUND ROOSTING AND BEDDING

For flocks that are roosted on the ground and moved frequently, so-called anchor roosts are provided. These usually are low (1 ft. high) roosts built in easily removable sections and accommodating only a small fraction of the flock. The anchor roosts are moved to the desired bedding grounds; at dusk, they will immediately be filled with turkeys and the remainder of the flock will bed down around them. Bedded turkeys have fewer deformities of the skin and bone of the breast and also have fewer wing and body bruises than do roosted birds. However, they must not be bedded on dirty ground; the bedding area must be dry and be changed every day or two.

ROOSTING CORRALS

Roosting corrals are used to protect turkeys at night. A corral containing 10,000 square feet will hold roosting, feeding, and watering equipment for 1500 turkeys. Roosts are placed in the center of the corral. Where the corral is permanent, a fence about 8 ft. high, well staked to the ground or predator-proofed by laying a 2 ft. strip of wire on the ground next to the fence and attached to it to exclude animals such as dogs and foxes that might kill turkeys and smaller animals that might start a stampede, is desirable. A 40 watt light at each corner affords good protection. The turkeys are shut in the corral just before dusk with feed and water available if they are not to be let out at sunrise; without it if they are, excluding them from the corral during the daytime. For permanent corrals, level roosts 30″ high, or the stepladder type, preferably with wire under them, are best. The fence should be placed 20 feet or more away from the roosts in all directions. Allow 1¼ linear feet of roost space per bird.

CONFINEMENT REARING

Advantages of this system are lower labor cost, better control over the turkeys and better protection from weather, thieves, predators, soil-borne diseases, and parasites. Disadvantages are (1) higher equipment costs; (2) greater tendency toward leg and foot trouble in male turkeys, and toward feather picking, breast blisters and leg calluses in both sexes; and (3) higher feed costs since the environment contributes nothing (except for direct sunshine) to the nutrition of the birds. With confinement equipment it is possible to raise both chickens and turkeys on the same premises. However this combination is not recommended. Three general types of confinement rearing are described.

Platform rearing. Equipment needed consists of a roosting shelter with a sunporch (rearing platform) attached. For mixed sexes of standardbred or large-type stock, 8 to 9 sq. ft. of floor space per bird; for small-type stock, about 7 are needed for best results. With this amount of space allowed, feather-picking usually is not a problem. However most operators prefer to use about 4 square feet per bird and debeak or install bits at 14 weeks or earlier if necessary to prevent losses from feather picking and fighting. Partitioned units to hold 150 to 300 birds each are desirable.

Platform floors should be 6 to 8 feet above ground to allow for cleaning. Sloping ground, for natural air and water drainage, and easy cleaning should be selected where possible. The floor structure, on treated or cement piling, may be 2 x 6's set 24″ apart and braced by 1 or 2″ material only where necessary for strength. The floor wire, preferably 1½ x 1½″ or 1 x 4″ mesh, 11-gauge welded, is laid over the 2 x 6 joists, laced at the selvages with wire or hog rings, and nailed only at the ends. A new rubber-covered welded wire appears to be well adapted for use as flooring for continuous quarters. Joists may be unbeveled though some prefer them beveled to not less than ¾″ surface. Slats made of 1 x 1″, 1 x 2″, or 1 x 3″ material laid flat and spaced about 1¼ inches apart are generally satisfactory for all types of turkeys and for all ages beyond 3 weeks. Where only poults 8 weeks or older are involved, slats may be spaced 2 inches apart, according to some operators. Slats ¾″ wide spaced ¾″ apart have been reported as suitable for all ages of turkeys from day-old to breeding age. Slats for the uncovered portion and wire for the roofed-over section make a good combination. Slats last longer and are more nearly self cleaning if they first are treated with No. 2 oil, then sprayed or rinsed with water before using. Cypress lumber possibly is the best material for slats as it is strong, workable, almost warp-proof, will outlast almost any other material and need not be treated. Fir is good; in fact any available wood can be used. Oak is durable but warps easily and is hard to cut and nail. Rough-sawed material can be used but finished stock is preferable.

The side walls of the platform should be 5 to 7′ high and can be made of 2″ hexagonal mesh 16-gauge galvanized wire or of heavy-weight poultry fencing. For rearing in cold weather, the sidewalls of the roosting section should be solid on 3 sides and the floor should be solid and covered with litter. Feeders (and sometimes waterers) may be hung to the side walls and a 3 x 12″ feeding grill of welded wire placed in front of them. With this arrangement, the birds are serviced from an outside walkway. Either troughs or hoppers set in the pens, which serve as feeders from both sides, are to be preferred; however, the single side hopper-type feeder may be serviced from the outside of the pens. The best type of waterer is a bubbler with drain attached, set inside the enclosure, one per 150 to 200 turkeys. Troughs with automatic control are satisfactory but are not to be used when the temperature drops below freezing at night;

Fig. 2.13. Confinement rearing platform 30 by 40 divided into two 20 by 30 units, each for 100 toms or 125 hens, large type. The ground slopes away rapidly on one side to allow for cleaning. Note low roosts and feeders and waterers hung to the side wall. (Courtesy Spar Hill Turkey Farm.)

unless they are drained each evening. The open section is uncovered or preferably covered with wire; ordinary poultry fencing or 2-inch mesh, 16-gauge hexagonal weave.

Stone yards may be used instead of porches or platforms. They are floored with small washed cobble stones 9 to 12″ deep. The stones may vary in size from that of a chicken egg to cocoanuts. It will take approximately 100 tons to cover an area of 2,000 sq. ft. to the desired depth. About 12 to 15 square feet of yard and shelter space per bird should be provided. Drainage should be good. At the beginning of each season the stones should be inspected and replenished if and where needed. Any grass or weeds within the stoneyard should be removed in order to prevent a build-up of droppings. Feeders and waterers should be placed on wire or slat platforms on the stones, or preferably inside of a wire-floored roosting shelter at one end or the side of the yard. Without these arrangements frequent cleaning must be done.

The roosting section of the platform, or the one attached to the stone yard, should be roofed, have a wire or slat floor, and if feeding and watering are not done from the outside, should provide room for feeders and waterers in addition to the roosts. The best roosting arrangements probably consist of (1) low roosts, about 6″ to 1′ above the wire floor and all on the same level. (2) Stepladder-type roosts, the lowest about 12″ above the floor, the angle of slope about 15-20°; with wire, nailed directly to the undersides and on top of the roost supports so that the space underneath is wired in; the roost wire thus forming the floor of the shelter and the regular floor wire or slats being eliminated under roosts. (3) No roosts at all, the birds roosting on wire or slats and the supporting joists. Where heavy type turkeys are forced to roost on the wire or slat floors, calluses sometimes develop on the back of shanks and hocks. This can be avoided by using litter in the roosting section where no roosts are used. Where stepladder-type roosts are used, they should be enclosed in such a way that the birds are not likely to fly or jump to a wire or slat floor from a height of more than 1 foot, as this is believed to be productive of leg and foot trouble, also breast blisters and calluses.

Rearing on litter. This relatively new development in confinement rearing practice appears to be gaining favor rapidly. It offers the usual advantages of confinement rearing plus low equipment cost and freedom from breast blisters and foot and leg troubles caused by hard, rough surfaces. Growers who have used this method have frequently reported less trouble from disease in spite of the fact that soil borne diseases and parasites might be expected to reproduce rapidly in this type of environment. Frequent re-littering, maintenance of dryness, and rigid precautions against introduction of diseases and parasites are essential features if this type of rearing is to be used successfully. Turkeys which have picked up soil-borne types of infection in the brooder house probably should not be reared on litter until such infection has been eliminated.

Pole sheds are a popular method of construction with home grown timber. The departments of agricultural engineering in the various experiment stations have issued instructions regarding pole type construction for sheds and barns which apply equally well to turkey shelters. Such a shed may be about 17′ high at the peak of the gable roof, about 40′ wide and 5′ high at the eaves, with overhanging roof as a protection from driving rains. Corrugated metal or aluminum is to be preferred for the roof if the pitch is fairly steep. A good roof is essential. The sides may be made of strong hog or sheep fencing, against which feeders should be placed so that they may be filled from the outside. Wire partitions placed every 25′ to break up the flock into groups of 200 or 300 each afford protection from possible losses by stampeding or crowding. From 4 to 5 sq. ft. of floor space should be allowed per mature bird. Even in such cases debeaking or "bitting" is necessary to prevent feather picking. Good waterers are the continuous trough or bubbler with drain attached. Properly installed, such waterers are frost-proof down to zero, since the flowing water keeps the pipes continuously above freezing temperature. A well-drained site for a pole shelter should be selected. If it is fairly level, it should be bulldozed to a level of about a foot above the surrounding terrain. If a side hill is selected, leveling off of the site allowing for drainage ditches to carry off surface water should precede construction. The frame work of poled houses is usually peeled, seasoned poles with a minimum diameter of 6″ at the top. Sometimes square timbers are used, in which they are sawed to 6 x 6″ or 8 x 8″. They are set 30″ to 40″ in the ground preferably on concrete footings; after treatment to avoid dry rot and possible termite infestation. In pole-type construction of such width as desired for turkey shelters, four rows of poles would be desired. Two rows in the interior would support the roof and the two rows, one on each side, for the framing, placed approximately 14′ on center. The poles should be of catalpa, cedar, locust, or some other good rot-resistant wood properly treated for both rot-resistance and termite control. The rafters and plates are usually made of sawed material. Dirt floors are most common in pole type construction; consequently drainage to avoid wet floors is of vital concern (see Fig. 1.14 and 12.5).

These shelters contain no roosts, only feeders, waterers and litter. The feeders may be placed on the side walls. The poults are transferred to the pole shelter at the end of the usual brooding period. These shelters with some modifications may also be used for brooding as well as rearing. In addition, they are adaptable to house breeding stock which may be held over after the marketing season. Because of the lack of foundations and concrete floors, built-up litter is the rule in pole type poultry buildings. For young turkeys, starting with two or three inches of fresh planer shavings on the ground is recommended; then adding about four more applications once each week or 10 days. Straw or hay may be used as litter, as well as other suitable types. Sawdust, because of the danger from molds, must be fresh when it is used and watched more closely. The litter need only be removed once at the end of the season.

Fig. 2.14. Hog ring inserted to reduce feather picking.

Confinement rearing brings special problems foremost among which is feather picking. If about 7 sq. ft. of floor space per small-type bird or 8 or 9 per medium or large type is allowed, feather picking can be controlled fairly well by the practices described elsewhere, but if less is supplied as it usually is, debeaking must be done or an anti-pick device must be placed on each turkey at the age of about 14 weeks. A specially designed bit resembling somewhat a 1¼″ hog ring may be inserted into but not through the nostrils so it hangs loosely between the jaws and holds them slightly open (Fig. 2.14). By the time the bird reaches market age, the beaks tend to close so that feather-picking is possible to some extent. Some anti-pick devices tend to get caught in the wire and to injure the birds. Electric debeaking is practical and effective and is preferred to the bits by many growers.

Deformed keels, legs, feet, and skin on the breast may be prevented largely by the use of well-balanced diets, proper roosts, and proper floors. High roosts and wire flooring that is rough and irregular may cause trouble.

The large broad breasted turkeys are not well adapted to confinement rearing on wire or slats but may be raised successfully that way if equipment and management are the best. A relatively new practice is to equip the sheltered portions of the platform with a solid floor upon which planer shavings, straw, hay, shredded cane, or other suitable litter may be placed, using built up litter.

The best possible diet is needed for confinement-reared turkeys since, as a rule, fresh green feed is not supplied. If succulent green feed is available near-by and can be supplied liberally, it will supplement the mash and grain in the same manner as it does on the range, but of course adds to the labor involved. However the increased labor cost may be slightly offset by feeding some of the less expensive mashes which are recommended for range feeding.

RANGE REARING

Two types of range rearing are in use: (1) limited range or semi-confinement and (2) free range.

Limited range. This type involves the use of fully or partially fenced areas large enough to provide growing green feed. The so-called Minnesota (or Billings) plan which is now widely used consists of moving the birds and rearing equipment once each week or so, to an adjoining clean area of a large field containing growing green feed. Roosts, feeders, and waterers are the minimum requirements for equipment plus 5-foot poultry fence or 4-foot hog fence and steel posts, where land is limited. The posts are placed a rod apart, 8″ deep and on the side of the pen with the fence hung only by a top wire, just so the bottom wire clears the ground. This arrangement is satisfactory to prevent stampedes and also may be set up and taken down more quickly and easily. When ample range is available, thus allowing greater latitude of movement, fences are not needed around the equipment, as it can simply be moved to an adjoining clean area. The birds will follow the feed, water and roosts.

Range shelters built on skids, with roosts under the roof and sometimes on top, are desirable in most situations. Feeders may be placed at right angles to the roosts and on the side nearest the new green feed. The equipment should be placed on skids which have strong tractor hitches, so that they may be moved quickly and easily.

With limited land, only enough should be fenced to give green feed to the flock for a week at a time, under average range conditions. This is approximately 80 sq. rods or one-half acre for each 2,500 turkeys; approximately .033 sq. rods per bird. The birds may be returned to the same section of the field for another weekly period later on during the season after at least a month has elapsed. A fairly large area is desired for this type of rotation and the field should not be used oftener than once every four years. During the intervening three years it should be pastured with some other livestock or plowed up immediately following its use for turkeys and used for cropping one or two years. The latter procedure has the advantage in that the turkey droppings have been well distributed over the field and should increase crop yields the following year. A 4-year rotation of a hay or pasture crop followed by turkeys, then corn, and finally wheat or oats is often followed with excellent success. This plan, if followed rigidly, is most effective in the control of soil-borne diseases. It is practical for small flocks as well as fairly large flocks up to 3,000 birds. For best results all of the turkeys in any one unit should be of the same age.

Adaptations of the Minnesota plan are used quite extensively. For example, some growers will leave the poults on a given area for from two to six weeks, instead of one; or they may be roosted in a permanent corral but the feeders and waterers moved weekly, circling the corral. As in the original plan, growers using adaptations move to a clean area each year, not coming back to an area until it has had at least 3 years' rest These adaptations appear to be satisfactory when environmental conditions are favorable. Among the conditions considered favorable are light soil, good drainage,

Fig. 2.15. Confinement rearing in Massachusetts. (Courtesy of Nelson O. Holland.)

much sunshine, comparatively light precipitation, ample acreage of land, sanitation in and around buildings and equipment, and the exclusion of carriers of infection. In any case turkeys never should range on poorly drained land.

When the birds stay on the same area for extended periods, an acre will provide green feed for 50 to 300 birds, depending on the nature of the vegetation. Average good grass pasture possessing sufficient fertility and receiving enough moisture for good growth of green feed will support about 200 turkeys per acre throughout the normal growing season. If turkeys are permanently yarded one acre would therefore be needed for each 200 turkeys to be raised. If, however, the Minnesota plan of range rotation combined with other farm operations were to be used, an acre would be required for each 50 turkeys using a 4-year crop rotation or an acre for 67 with a 3-year rotation. The Kansas Station reports that in a season of normal rainfall as many as 75 birds may range continuously over an acre of alfalfa without destroying the stand. If the birds are moved frequently, many more per acre could be accommodated.

Range management studies with growing turkeys on ladino clover and grass pasture were made at the Pennsylvania Station. Ladino clover was found to provide the best type of green feed and could be pastured heavily. Fig. 2.16 shows the eagerness with which turkeys range over luxuriant ladino pasture.

Fig. 2.16. Turkeys on ladino clover. (Photo by S. C. Allen and Son.)

Permanently fenced yards along with a permanent wire-floored range shelter are not so widely used as temporary yards but may be satisfactory to use where drainage is excellent and the soil sandy. If this type of rearing is attempted, the birds should be kept out of their own droppings, the feeders moved weekly, and the utmost in sanitation provided around waterers and elsewhere. It is advisable only in arid or semi-arid regions.

Permanent pasture will supply satisfactory conditions with the least trouble in most situations where suitable rotations are practiced and the land not overcrowded. Small fenced areas are hard to cultivate so permanent

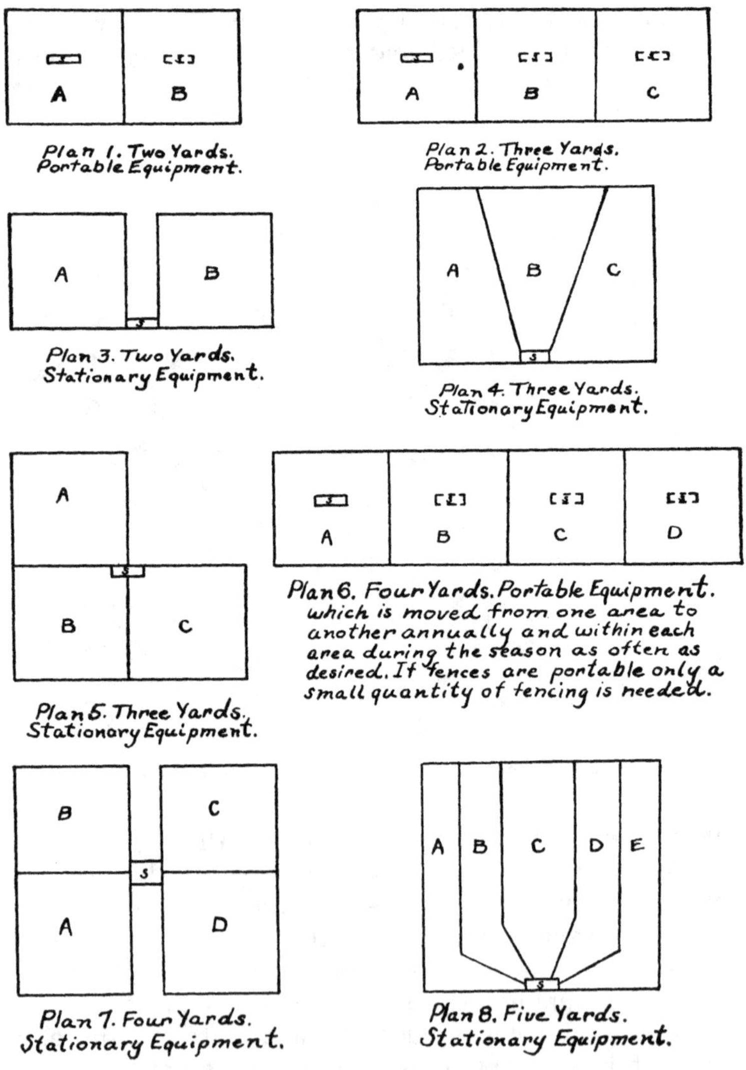

Fig. 2.17. Yard rotation systems.

pasture seems to work best in small units. However, where cultivation can be practiced there is no objection to it, in fact, it may be advisable under some conditions. Rainfall, soil, terrain and other environmental factors will influence the crops grown and the type of rotation and yarding used. Cattle and turkeys make a good combination, since they appear to have no common parasites or diseases.

Free Range. In the true sense of the word, free range turkey growing is not practiced commercially. In order to protect the birds and keep them from straying, restriction of some kind is always provided. Several methods of so-called free-range rearing are in use.

Herding consists of keeping the birds in the flocks under control during the daylight hours and seeing that they assemble and roost together in a group at night. This is usually accomplished by a herder on horseback or on foot, and dogs that have been trained to assist without injury to the birds. Camping facilities such as a wagon, automobile trailer, or a tent that can be moved readily from place to place are usually provided for the turkey herder and lights, particularly road flares, are often set up at night to keep off coyotes and other predatory animals. Portable roosts, feeders, and watering devices are commonly used although permanent, well-protected roosting corrals and shelters also are coming into use. In many cases, only anchor roosts or no roosts at all are provided and sometimes poles that can be moved every few days are laid on the ground for roosts. Roosts all on the same level, 10″ to 14″ off the ground, are commonly used, with a preference for the low type. While the unit may be as large as 5,000 birds, preference is for flocks of 2,000 to 3,000 birds. Sometimes limited feeding can be practiced but full feeding is usually the rule. Through this type of management, careful protection of the flock, desirable soil sanitation, and better utilization of the natural feed may result.

A small-unit range plan can be used in which poults are

reared in groups of up to 500, scattered over a large range area. The birds within a group are of the same age. A simple shelter is used, sufficient in size to protect the poults from the time they are put on range until they are about 3½ months old. Roosts are then set up near the shelter and a portable corral of ordinary heavy weight poultry fencing set up around the shelter, roosts, and feeders at night to give protection from predators. Additional protection afforded by dependable watch dogs, or a night guard, is almost always necessary. Free choice feeding is practiced under this plan. If the poults are not full-fed, the various flocks will intermingle unless they are placed long distances apart. With full-feeding, the units may be placed as close together as 500 yards, the distance depending upon the terrain and the nature of the vegetation. Each unit usually remains in the same general area for the season, but may be moved if disease or range conditions warrant it. The feeders and waterers usually are moved every week or two, circling the roosting corral to keep the birds on relatively clean ground. If properly carried out, this plan provides excellent growing conditions, but will work well only if the land is well drained, cleaned and rotated to prevent any build-up of diseases or parasites.

Another method of free range rearing consists in the use of islands and peninsulas to provide natural boundaries and a bountiful supply of water; unless it is salt water. However, as noted elsewhere, trouble with parasites is frequently encountered when turkeys have access to natural supplies of water such as lakes or streams.

AGE DETERMINATION IN LIVE TURKEYS

Petrides (1942) suggests that in wild turkeys, age may be determined by the two outer primary feathers which are carried through the first year and not shed until the birds molt as yearlings. These chick feathers are pointed at the end and are easily distinguished from the mature feathers which are rounded. In yearling and older birds the white bars go nearly to the ends of the feathers while in young birds they do not go within about 2 inches of the ends. The bursa also is of value in age determination. Another way to tell age is by the wing bar which is about as wide as three fingers in the mature females and has a regular lower edge line while in pullets the bar is not more than two fingers wide and the lower edge line is irregular. In toms the bar is about a finger wider than in the females (see page 500).

In domesticated turkeys, age determination by the above described system is difficult to demonstrate. In fact no method except permanent identification by banding, tatooing or physical altering such as toe punching is completely effective. The rear tip of the keel bone of turkeys up until about 7 months of age is reasonably pliable. It becomes rigid and completely ossified in older birds. The headgear of toms over 10 months of age tends to be heavier and coarser than in the younger males, but such differences are not discernible in the hens. In toms over 8 months of age there is a pronounced flabbiness over the crop (the breast sponge) but again, the difference does not appear in hens although hens over 10 months of age tend to have scaly skin. White turkeys of both sexes over one year of age tend to develop a deep pink color in the shanks compared to the paler and almost white shanks of younger birds. However, these differences in shank color do not occur with enough regularity to be of much value in determining age.

PREVENTION OF FLIGHT

In situations where it is important to keep turkeys within the areas or yards allotted to them, some wing-feather clipping or other treatment is necessary if 5′ or 6′ fences and antiflies are used. The outer ½ to ⅔ of the primary wing feathers may be removed with light weight tin-snips, heavy shears, hedge clippers, or a hatchet and block. With large-type turkeys it is a good practice to clip the primaries of one wing at the age of about 8 weeks or at the time of transfer to the rearing quarters. These feathers should be clipped again at 16 weeks, especially the female. Small-type turkeys of both sexes should be clipped at the time of transfer and again at 16 weeks of age.

Breeding toms should not be clipped at all, or if it is found necessary to do so, only a light clipping should be done. Breeding hens should be clipped except in cases where it is of no serious consequence if they get out of their yards. Only the primaries should be clipped, except in the most persistent cases of flying or fence-climbing, when both primaries and secondaries may be clipped short. Clipping the inner 7 primaries of each wing is preferred by some especially for breeding birds.

Cutting off the last segment of one or both wings with sharp heavy shears or bone clippers (see Fig. 2.18) at 3 to 7 days of age may prove to be a desirable practice as flying is prevented permanently and plucking the feathers for market is rendered easier due to absence of primaries, primary coverts, pinion feathers, and the small contour feathers that are found in the outer wing segment or forearm. The cut is made a very short distance out from the last wing joint, so not to harm the joint itself. Properly made, the cuts heal perfectly and the birds live and grow in a normal manner. However, these stubs are tender and if care is not taken in catching, a disfigured market carcass will result. If the small "thumb" of the wing is cut off (as illustrated) a smooth stub results. If the thumb is not cut, the pinion feathers remain and act to protect the stub to a considerable extent. Only one wing, preferably the right, need be cut to prevent flight. Growers of both large and small-type turkeys report that wing-cut toms are not handicapped as breeders. However at the Beltsville Station, removing both wing tips reduced slightly the weight and finish of BSW turkeys at 24 weeks of age.

Tenotomy (wing notching) consists of serving the main wing tendon. It has been recommended (Mich. Sta.)

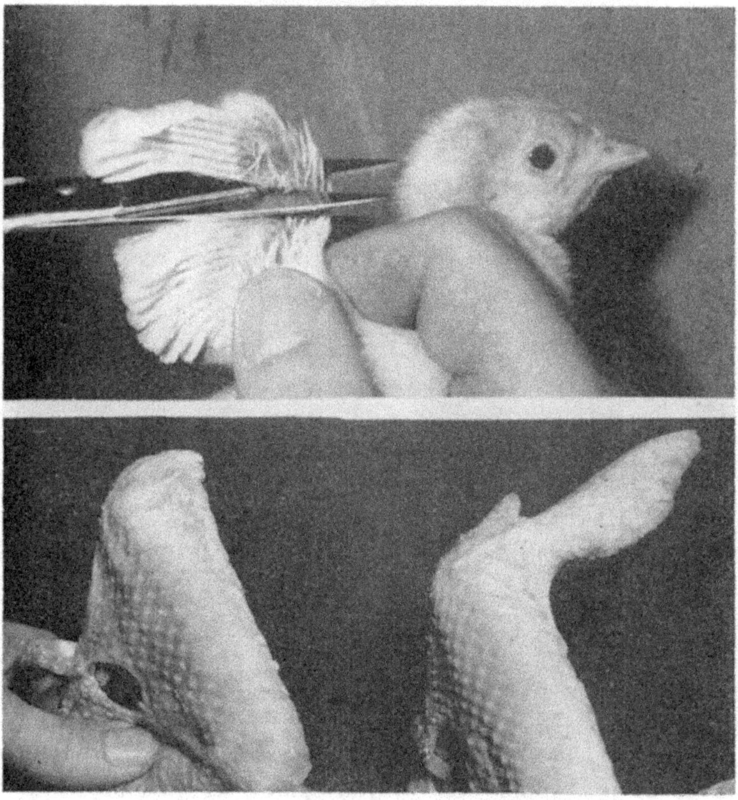

Fig. 2.18. Snipping the end wing segment prevents flying over fences. In lower photo, normal wing tip at right and snipped wing at left.

as a means of preventing flight. To sever the tendon which passes over the outer edge of the last joint of the wing, hold the wing tightly folded within the hand and burn through the skin and the tendon with an electric wood-burning needle or special vertical debeaker blade heated only to a dull red color. Contact should be made at a point a little to the outside of the folded wing where there is a small flat place. When the tendon is cut the operator can feel or hear it snap. Only one wing of the turkey need be treated to prevent flight. Treating both wings leaves the bird balanced even though it cannot fly very well. Wing notching should be done when turkeys are 2 to 3 weeks old and never beyond that age. If done properly it does not cause wings to droop or leave scars that are noticeable at marketing time.

DESNOODING

Cutting off the snood of the young poult (day old to three or four days) has been suggested as an aid in preventing injuries from fighting and in the prevention of erysipelas. Use a strong, sharp-pointed small shears such as surgical or cuticle scissors and clip the snood off close to the head. The result will be no snood at all or a small rounded bump when the turkey grows up. The snoods of day-old turkeys vary considerably in size but size of snood appears to bear no relationship to sex. Day-old poults can be de-snooded by the use of strong thumbnail and finger much as one would remove strawberry stems. Snoods may be clipped up to 3 weeks of age.

DEBEAKING

Removal of part of the upper beak effectively prevents cannibalism. It can be done manually or electrically with a debeaker. To debeak manually, cut a nick in both under sides of the upper beak a quarter of an inch or so back from the tip. The cut should be 1/16 to ⅛ inch deep (even with the roof of the mouth). By pressing down on the tip, the cut end can be broken off. This leaves a little of the "quick" exposed which discourages feather picking for a few weeks, after which the horny beak grows back. If bleeding is extensive, sear with a hot iron.

Electric debeaking is practical, quick, and can be done at any age. Usually there is no need for debeaking until the birds are about 12 to 14 weeks old. At this age if the upper beak only is cut off at a point about half way between the nostril and the tip, the debeaking will be effective until market age. For breeders, debeaking at this point also is effective. Where it is desirable to debeak prior to about 12 weeks of age, the best place to make the cut, according to some writers, is "slightly better than ⅛ inch in front of the nostrils." Some suggest debeaking at 5 weeks, finding the soft spot on the beak (just in front of the nostril) and making the cut there. Still others suggest debeaking at 3 to 14 days, cutting the upper beak only "nearly to the nostril" (never cut or sear the lower beak). This debeaking is effective until market age. The S. D. State College suggests that, to prevent brooder-house cannibalism, debeak at day-old using finger-nail clippers and searing the stump by rubbing against a hot bar. Poults a little older can be debeaked with a knife, as suggested or with side-cutter pliers, followed by searing with a hot soldering iron. In debeaking small poults it is necesary to guard against injury to the tongue. At the Beacon Mills Experimental Farm (1953) severe debeaking was observed to curtail feed consumption, due apparently to continued sensi-

Fig. 2.19. Debeaking should remove the upper beak close to the nostril. This bird was not debeaked closely enough.

tivity of the stub of the beak (especially in cold weather) plus mechanical difficulty in eating grain and pellets. A number of field cases also suggested that severe debeaking increased the incidence of some types of hock trouble. The use of hog-ring type bits was preferred (Fig. 2.14).

DRIVING

Turkeys of all ages may easily be driven from place to place with the aid of light poles or sticks several feet long. With a pole in each hand, one person is able to handle a good sized flock. However, if turkeys are able to fly well, driving is less satisfactory as they may take to the air if frightened. Driving entails much less labor than crating and hauling and if distances are not too great, is easier on the birds. A cloth tied to the end of each driving stick facilitates the control of the flock. Dogs are satisfactory for herding turkeys but they must be well trained and gentle and the turkeys must be accustomed to their presence.

MANAGEMENT OF WILD TURKEYS

Wild turkeys do not differ materially in their nutritive requirements from domestic turkeys. They can be fed the same diets at all ages and for all purposes. If they are to be liberated on open range they should be reared in very large or completely enclosed yards (with wire over top as well as sides). Otherwise they must have the primaries and sometimes part of the secondaries of one wing closely clipped in order to confine them by ordinary fencing. Clipping several times during the growing season is necessary. Unless they are brooded in quarters that are entirely enclosed (which is desirable) the first clipping must be done at the age of about four weeks. Toms as well as hens must be clipped. Toms selected as breeders should have the inner 7 primaries (only) of both wings clipped or allowed to grow their wing feathers part way before mating season if clipped closely.

When wild turkeys are raised for liberation in open unprotected range, game protectors have found that ordinary methods of brooding and rearing tend to produce birds that will not take readily to the wild environment but will mix with domesticated turkeys and will come to farms for feed and protection. As a result of this, the Game Departments of some states are using special methods of rearing wild turkeys for stocking purposes. For information on this subject the reader is referred to the Pennsylvania State Game Commission at Harrisburg, and the U. S. Fish and Wildlife Service, Chicago, Ill.

True wild stock unmixed with the blood of domesticated turkeys is hard to obtain and difficult to identify. "Wildness" alone does not necessarily characterize the pure wild strain. Although genuine wild turkeys and those possessing 50% or more wild blood are noticeably "wild," gentle handling will do much to "tame" the wildest stock in a comparatively short time. On the other hand, domesticated stock if not gently handled and if allowed to range freely, readily becomes decidedly "wild" in behavior.

Wild turkey stock is as susceptible to blackhead and other diseases as the domesticated birds. In fact, wild turkeys in native habitat have suffered, and do suffer, from blackhead. Consequently, in raising wild turkeys for restocking purposes, diseases must be guarded against by soil sanitation and separation from older tame turkeys and from chickens.

CHAPTER 3

FEEDING PRACTICE

OUTLINE

DEFINITIONS OF TERMS

Ad libitum is a Latin phrase meaning at will; in feeding parlance it means available at all times or free choice. It is sometimes abbreviated to ad lib.

Starting mashes or "starters" are fed during the brooding period when poults require the protection and warmth furnished by the brooding hen or by brooding equipment. The first eight weeks usually constitutes the brooding period. Sometimes it is considered through only the first six weeks, whereas others may consider that the poults are actually being brooded up till twelve weeks of age. Starting mashes are characterized by a relatively high protein and vitamin content and the presence of an antibiotic. They are usually fed on an all-mash basis which assures a well-rounded diet for each poult's consumption each day. Starting mashes sometimes are continued as growing feeds with the addition of grains fed free-choice. Starters may be in the form of mash or granules (crushed pellets).

Growing mashes or "growers" are fed during the growing (developing or rearing) period which extends from the end of the brooding period up to market age. They are sometimes referred to as developing feeds and usually are less well fortified with vitamins than the starters, have less protein, and usually do not contain any of the antibiotics. However, the feeding of antibiotics in growing diets may possibly become standard practice as the increased production of the antibiotics makes it possible for them to be secured at a constantly lower price, relative to the other ingredients of the feed. Growing mashes almost always are fed along with grain, both free-choice, on account of the turkey's needs for protein becoming gradually less as the birds grow older. The gradual increase in grain consumption results in a lower total intake of protein. Growing mashes may be fed as dry mash or as pellets.

Finishing or fattening feeds are those fed during the last few weeks of the fattening period, just before marketing. Feed used to fatten adult stock is also included. In the case of young stock the term "developing" or "growing" feed usually includes the finishing (or fattening) feed although the latter term is useful when referring specially to the feeds fed during the latter part of the rearing period from the age of 20 weeks to marketing.

Free choice feeding is a term used when the feed (usually both grain and mash) is kept before the birds at all times. (See ad libitum)

Wintering or breeder maintenance feeds are those fed to prospective breeding stock from market or selection time until laying feeds are fed. They are growing feeds with a relatively higher content of fiber and a low content of corn.

Breeder feeds are those fed from about a month prior to the start of production until production is no longer desired. Breeder mashes are so named because turkeys are always fed for hatching egg production, never for the commercial production of eggs for the table.

Breeder mashes are high in vitamins and usually contain a higher ratio of animal to plant proteins than do the starters or growing feeds. Most breeder mashes now contain from 20 to 22 percent protein and are intended to be fed on the basis of equal parts of grain and mash. However, there is a growing tendency toward the use of "all-mash" breeder diets which may consist entirely of ground material or may contain some cracked or even whole grains. A relatively new idea in breeder feeding is to use a high protein mash (approximately 28 percent) and feed it free-choice with grain. With such high protein mashes the birds have less tendency to eat too much grain. With a 28 percent protein mash, two-thirds of the diet can be grain and still result in proper protein intake of 15 to 17 percent of the total diet. The vitamin B_{12} content of the breeder diet should be relatively high. The addition of antibiotics is frequently of value. Further research is necessary to clarify the exact role of the many antibiotics available.

Summer maintenance feeds are those fed to breeding stock from the time the egg production diet is discontinued until the fall fattening period begins.

Confinement rearing involves confinement of the birds to a very small enclosure. In confinement rearing the environment itself does not supply any feed such as the grass and insects that are secured on range. Fresh air, sunshine, and a limited amount of exercise are all that the environment affords, so a complete diet must be supplied.

Limited range rearing, sometimes referred to as **semi-confinement,** involves confinement of the birds to a moderate sized fenced range lot, or its equivalent in protection from infection and predators if the area is not completely fenced. Turkeys given access to a very large area will not range widely when they are fed ad libitum, hence this type of management constitutes limited range rearing although there may be an unlimited land area available. In this type of rearing, the environment is expected to furnish the vitamin D factor through direct sunshine and to make a significant, although not necessarily extensive, contribution to the diet in the form of growing green feed. The prevailing practice is to feed dry mash or pellets and grain, both ad libitum.

Free range rearing involves the use of a very large range area which is expected to make important contributions to the birds' diet (about 25% or more) in the form of green feed, insects, and other feeds such as waste grain gleaned from stubble fields. The free range plan implies dependence upon the environment for a good share of the birds' feed. There are two general types of free range rearing that are compatible with good business principles, namely (1) the fencing of large areas of land suitable for ranging turkeys, and (2) the herding of turkeys so as to provide continuous protection and to prevent straying. Allowing the birds to roam at will without liberal feeding and protection from enemies is not a satisfactory method under any except the most unusual conditions such as, for instance, the raising of turkeys on an island. Even then the presence of an attendant or proximity to the owner's residence would be necessary to prevent losses by predators and thieves. The necessity for soil sanitation must always be borne in mind, remembering that birds cannot be raised on the same ground year after year with any assurance of success. Separation from adult turkeys and other poultry of all kinds is most important.

THE FREE CHOICE MASH AND GRAIN PLAN FOR FEEDING GROWING TURKEYS

That turkey poults should be fed and watered with the least practicable delay is now well recognized and almost universally practiced. Chilson and Patrick (1946) reported that Bronze poults in groups of 103 each, supplied with feed and water at 24, 48, and 72 hours after hatching, respectively, gave the following results: weight when fed (grams) 57.2, 51.3, and 45.2; weight at 6 weeks in grams, 704.4, 646.6, and 632.2; livability at 6 weeks, percent, 97.2, 90.4, and 72.6; livability at 12 weeks, percent, 95.2, 88.5, and 70.7. The poults were held in poult boxes until fed and watered. This experiment coroborates previous work and indicates that, in accordance with modern practice, it is better to feed and water poults as soon as possible after hatching.

Experiments at the Beltsville Station (1944) showed that lack of water consumption by the young poults during the first week of life caused loose, slimy gizzard linings and high mortality; that consuming feed but no water was more detrimental than consuming neither; and that feed and water or water but no feed resulted in normal gizzard linings. Feeding 3 to 6% of A and D oils in the starting diet caused gizzard erosions and high mortality. It also may be expected to cause fishy flavor in the flesh even up to market age several months later. Poults kept from feed but given water for 1, 2, and 3 days then force-fed and taught to eat and drink showed no ill effects but those so held for 4 days then force-fed suffered high mortality. Without force-feeding the poults held 2 and 3 days doubtless would have shown considerable mortality.

The Washington Station (1945) observed that turkey poults when hatched are almost blind; it requires several days before full vision is attained. They also cannot see objects on or near the floor nearly so well as objects on a level with their heads; green color is the most attractive and light of any kind is likewise attractive. These findings suggest methods of teaching poults to eat and drink, some of which are; raised feeders heaped with feed; all-glass waterers; cut green feed, oatmeal, and colored glass marbles in feed and water; plenty of light; and stringing bright metal objects such as paper clips over watering troughs on starting batteries.

Poults should be placed in the brooding quarters as soon as hatching is complete or within 24 hours thereafter. As soon as possible after hatching is preferred. The poults should be fed and watered immediately, special efforts being desirable during the first two days in the brooder to teach them to eat and drink. When first put under the hover the beak of each poult may be dipped in water, then in mash. Dry starting mash, or possibly crumbles, is placed in numerous little feed troughs made of metal or narrow wood strips such as lath or 2″ strips. These feeders are placed close to and under the hover like spokes in a wheel. In addition, mash may be placed on pie plates, egg case flats, or cup flats which may be reversed and used a second day. These, too, are placed under the outer edge of, and near, the canopy of the brooder. Every two or three hours a little oatmeal (rolled oats), or finely chopped tender green feed, or both, is scattered on top of the mash and on the surface of the drinking water to encourage the poults to eat. Finely chopped tender green feed is undoubtedly the most attractive of all feeds to baby poults. The Maryland Station (1952) found that 5% of corn oil mixed in the starting mash for the first two days stimulated early feed consumption of poults apparently because of its odor. The growth depressing effects of that much corn oil could be overcome by the addition of 15 p.p.m. of procaine penicillin in the starter according to the Ontario Station (1952).

Placing bright, glass marbles on top of the mash in the feeders and in the water helps to teach poults to eat and drink. Other devices for teaching poults to eat have been used successfully. One of these is to place a multiple-perforated shield above the feeders and above the shield a strong light which causes dots of light to illuminate the feed. The commercial growers, however, usually depend on plenty of feeders and waterers well distributed and well lighted combined with ample heat and poults fresh from the hatchery. Putting a few older poults in with day-old birds for 2 or 3 days is helpful. The older poults quickly show the beginners how to eat and drink. However, if disease is present in the older brood this practice is dangerous. Putting chicks in with the poults likewise may be dangerous.

Force-feeding of backward poults by means of a standard 25 cc. pipette and a sloppy mixture of milk or water and a finely ground starting mash is practicable. Two men can force-feed 200 poults per hour. This practice is useful in starting poults over 48 hours old, or those which for any reason will not eat readily. The large end of the pipette is inserted in the poult's throat well below the opening to the lungs, and the feed blown into the crop until it is noticeably distended. Sweet milk alone is satisfactory.

The small lath feeders used at the start should be numerous enough to provide about 2 inches of lineal feeding space per poult. If the feeder is available to the poults on both sides, this amounts to one inch of feeder per poult. After a few days' time or from the start, feeders of a little larger size, with 2-inch sides, can be used, changing at about 2 weeks of age to still larger feeders with 2½ to 3 inch sides which should be placed on wire platforms (unless wire floors are being used). From about 5, to 10 or 12 weeks, feeders with 5 to 5½-inch sides are utilized, followed by the regular full-sized feeders, some of which are described in Chapter 1.

Suitable and adequate watering equipment is highly essential especially at the start. Water fountains should be so designed that the poults cannot get into them and so that the water is within easy reach. About 3 fountains (2 to 3 gal. size) should be provided for each 150 poults at the start, about one fountain for each 60 to 75 poults, or ¾ lineal inch of watering space per bird. If the fountains are smaller, more of them should be provided and some of them should be placed under the hover when practicable to do so. Twice this number of quart or pint size founts may be used at the start but they should be replaced by the 2 or 3 gallon size or by auto-

Fig. 3.1. Automatic water founts or other drinking utensils should be placed close to the mash hoppers. Note wire platform to keep area around founts dry.

matic waterers after 3 or 4 days, when the number may be reduced to a ratio of one 3-gallon waterer to every 85 to 90 poults or ½ lineal inch of watering space per bird. After about four weeks ¼ inch per bird is sufficient. Changes in watering equipment should be made with care because often poults will go without water before they find or learn to use the new and different equipment (see Figs. 3.1 and 3.2).

Fig. 3.2. A watering device that permits ready and easy removing to new range. The final connection from the pipe line is by garden hose. (Photo from Walter E. Grant, Hamilton, Va.)

Fig. 3.3. A watering device that keeps conditions clean and dry. (Photo from Pacific Rural Press.)

Up to the 6th to 8th week the starting mash is kept before the poults along with water. Gravel, sand, or other insoluble grit such as granite or quartz usually is provided as 1% of the mash; or in separate feeders, unless the poults already have it on the floor or in the yards. However, grit is not essential until cracked or whole grains are fed. Young tender finely cut green feed is excellent, and if the adequacy of the diet is in doubt, it should be fed. The greens may be scattered on top of the mash once or twice daily, all the poults will clean up in half an hour. Of course if the poults have access to growing green feed it need not be hand fed, or if a complete diet is fed, the green feed may be omitted entirely. Attempts to utilize eggs, meat, clabber, liquid milk, corn bread, and such products in the starting diet are likely to result in more harm than good and are not recommended except where it is impossible to feed a well-balanced starting mash. They tend to complicate the feeding program, increase the labor, interfere with sanitation, and are generally unsatisfactory. Reliable commercial starting feeds or the ingredients necessary to mix satisfactory feeds at home are available or can be made so anywhere in the United States or Canada. It is especially important that exactly the right combination of good quality ingredients be used in the preparation of starting mashes.

From 7 to 9 weeks to marketing, both the growing mash and hen-size scratch grain (no whole corn until the 16th week) are fed ad libitum. Water and insoluble grit or gravel are provided. If a *complete* growing diet is fed, it is not necessary to add green feed of any kind. If a *semi-complete* diet is fed, green feed of some kind such as hay, cut green feed, or that obtained by the birds from poor or dried-up range, should be provided. If a *simplified* diet is fed, continuous access to growing green feed must be provided. In all feeding, the accumulation of stale mash should be prevented, by forcing the birds to empty the feeders once or twice each week.

Green feed. In general, the feeding of fresh green feed is desirable during the brooding period because it is stimulating to the appetite and is good insurance against certain dietary deficiencies, principally of vitamins A, B, riboflavin, and other factors stimulating to growth and feather development. However, only young, tender, fresh, finely cut alfalfa tops, clover or other tender succulent greens should be fed, or no advantage will come from its use. High labor cost is an important drawback and in addition the right kind of green feed is often difficult to obtain. For these and other reasons, investigators and poultrymen alike have been seeking starting diets that are complete so that green feed need not be given. Most prepared starters are intended for use as complete feeds. However, because vitamin A and some other nutritive factors tend to be unstable, staleness and exposure to air sometimes cause deficiencies to appear in spite of precautions taken in preparation of the feeds. Consequently the feeding of fresh young tender finely cut green feed has a special value as a preventive of vitamin deficiencies. From their experience the authors conclude that the feeding of the right kind of green feed from hatching to marketing is desirable whenever facilities for providing it are available. It is recognized however that such facilities are not always available and in those cases it is suggested that a dependable, well-balanced, and complete starting feed, or reliable ingredients necessary to mix it, be obtained, omitting fresh greens from the diet. In *selecting* a prepared feed, care should be used and the manufacturer's directions followed. In selecting a formula for the mixing of a starter, only high quality fresh ingredients should be used, and the suggestions of the formulator should be followed.

Grain Feeding. At the time grain is first added, it is placed in a small portion of the feeding space. As the birds grow older they will eat more of the grain, until near the end of the growing period the grain should occupy about half of the feeder space. This plan permits the birds to adjust their protein intake according to their needs which they will do if conditions are right, from week-old to marketing. As they approach maturity, less protein is needed, consequently, they eat less of the high-protein mash and more of the low-protein grain.

A combination of whole corn and oats fed free choice along with mash free choice is widely used. At the Michigan Station, turkeys fed whole corn and oats free choice with a 27% protein growing mash, ate (from 9 to 24 weeks) corn 30% and oats 70%. The whole corn was eaten but little, until the poults were 19 weeks old. If the corn is cracked and cleaned, turkeys eat it fairly well at 8 weeks or earlier if it is finely cracked. Mixtures of grains are generally eaten relatively well at all ages. Barley is an excellent grain to be fed by itself or in a mixture, but if offered free choice will be neglected in favor of any of the other common grains except whole corn prior to about 16 weeks. In palatability to turkeys the common grains rank about as follows:—

(1) Oats
(2) Wheat
(3) Grain sorghum
(4) Rye and proso
(5) Cleaned cracked corn
(6) Barley
(7) Whole corn up to about 18 weeks—after that age it ranks next below wheat.

At the U. S. Department of Agriculture Station at Miles City, Montana (1932) standard bred Bronze turkeys were fed four grains free-choice along with liquid milk to drink. They ate the following percentages during the period from 9 to 28 weeks of age: oats 40%, wheat 27%, whole yellow corn 26%, and barley 7%. Oats were preferred throughout except from 25 to 28 weeks inclusive, when corn was favored. At the Michigan Station (1938) Bronze turkeys fed grains free-choice from 9 to 24 weeks of age ate at the following rate: oats 44%, wheat 33%, whole yellow corn 20%, and barley 3%. Turkeys do not always show preference for the different grains in accordance with the foregoing figures. Field observations indicate that turkeys have the ability to discriminate between grains grown on highly fertile soils and unfertilized fields, giving preference to the former.

Another factor is quality of the grains, turkeys preferring clean heavy grain to light-weight grains. They also show preference to clean grain mixtures as against those containing meal, dirt, wheat seeds, or the less palatable grains. For example, turkeys will eat heavy oats better than light oats. On the other hand, they will eat barley at a better-than-average rate, if it contains a considerable quantity of oats or other palatable grains mixed with it.

With a palatable growing mash containing up to about 27% protein any single grain of the seven listed above, except whole corn prior to 18 weeks makes a good grain supplement to offer free choice with the mash. Likewise any combination of grains is satisfactory. However, if the mash fed is high in any one grain it is good practice to use a grain or grains as a supplement that will result in a total feed consumption divided fairly well between two or three grains. This is an advantage according to the West Virginia Station. There is little difference in the ability of the various grains to fatten turkeys—any of them will do a good job, according to Wyoming Station results.

In feeding trials at the Ontario Station (1952) it was observed that a mixture of equal parts corn and oats fed with the standard 20% protein growing mash was eaten well and produced a fine finish with few leg abnormalities in porch-reared turkey toms. The high feeding value, good palatability and worth-while economy of milo, when fed as the only grain, as well as in the mash, (as its chief grain constituent) was demonstrated by the Kansas Station (1950).

When mashes with more than about 27% protein are

fed, the selection of the accompanying grain is more difficult. If oats alone are offered, the turkeys may eat too heavily of the oats. If whole corn or barley alone is offered, they may eat too little. One of the best grain supplements to use with these mashes is a mixture of barley or oats ⅓, wheat ⅓, and either cleaned cracked corn or grain sorghum ⅓ (by weight). Clean cracked corn or grain sorghum and oats equal parts, are good; likewise equal parts by weight of barley and wheat; oats and wheat; or oats and corn. If mixing of the grains is impracticable, a free-choice of oats and wheat works well; likewise wheat and grain sorghums, and oats and cleaned cracked corn to 18 weeks after which, if desired, whole corn may be fed. Proso (hog millet) is a good grain for turkeys and can be fed in place of oats or barley. The same can be said for emmer and spelt. These grains have all been found to be reasonably palatable to turkeys, ranking probably along with the grain sorghums.

Rye can constitute up to half of the grain portion of growing diets, either free-choice or mixed in place of wheat, corn, or grain sorghums. Rye seems to be about as palatable as the grain sorghums.

Toms eat relatively less grain than the hens, in most flocks. At the Washington Station turkeys fed a 40% protein concentrate and grain free choice ate, between 9 and 28 weeks as follows:—toms 26% concentrate and 74% grain; hens 17% concentrate and 83% grain. The higher percentage grain consumption of the hens tends to even up costs of production of the two sexes.

The relative mash-grain consumption of turkeys fed mashes containing different percentages of protein and a grain or a grain mixture, while it follows a general pattern, does show considerable variation, the variation depending on the nature of the mash, the nature of the grain, certain environmental conditions, and apparently in some cases merely the habits of the birds (see table 5.24). Generally speaking, it is true that the higher the percentage of protein in the mash the less mash is eaten during the growing period from eight weeks to market. However, if a palatable mash and an unpalatable grain are fed, the mash consumption will be higher than normal while if a palatable grain and an unpalatable mash are fed the mash consumption will be lower than normal. A palatable grain and a palatable mash or an unpalatable grain and unpalatable mash usually results in what may be considered "normal" mash-grain consumption.

The palatability of the mash seems to depend, for one thing, on how much vegetable protein supplement is used, the more of it being used the more unpalatable the mash becomes. Meat scrap and fish meal seem to be very palatable and mashes with fairly large percentages of one or both are likely to be highly palatable. Other mash ingredients show variations in palatability too, but information concerning them is lacking. Generally speaking, mashes consisting of the ordinary percentages of the common grains, grain by-products, and protein supplements, including fish meal, meat scrap, milk, and with a not-too-high percentage of vegetable protein supplements are palatable. When whole corn alone is fed as the grain portion, very little of it is eaten until the birds are 18 to 20 weeks old, even though the mash fed may be quite unpalatable. Barley, while it is relatively unpalatable and will not be eaten well if other grains are offered free choice, is readily eaten if fed alone. Oats, on the other hand, are extremely palatable and it is difficult to obtain normal mash consumption when they are fed as the only grain. This is especially true if the mash is rather unpalatable. Wheat is quite palatable and when fed as the only grain is eaten in about the right proportion. The same is true of rye, grain sorghum and cracked corn that is free of dust and meal although, as a rule, the birds will eat wheat alone better than they will cracked corn, rye or grain sorghum.

The average 32-percent (approximately) protein-concentrate with animal and vegetable proteins is a reasonably palatable feed. If it is nearly all vegetable protein, however, it is likely to be relatively unpalatable and will not be eaten in adequate quantities if oats are fed as the only grain. With a grain mixture or with cracked corn, or with barley, this mash is consumed in adequate quantity.

As mentioned elsewhere, turkeys fed mashes with 26 percent or more of crude protein and any reasonably palatable grain or grain mixture, both free choice, eat a higher percentage of protein than is necessary for maximum results. It is therefore not a serious matter if grain consumption is somewhat heavier than that shown in Table 5.24.

Pelleting apparently increases the palatability of some mashes, but this fact does not appreciably influence the principles of mash-grain feeding as herein outlined. Pellet feeding is discussed more fully on page 108.

Whether or not concentrates should be fed depends largely upon the relative cost of the concentrate and the grain. This can be secured in a simple manner by taking the estimated pounds per hundredweight (percent) of grain and multiplying it by the cost per hundredweight, doing the same for the mash or concentrate and then adding the two. The sum is the cost per hundredweight of the total feed consumed.

When the protein is the most expensive portion of the diet, as it usually is, the fact that turkeys fed high-protein mashes eat more total protein than necessary indicates a disadvantage for the concentrate and consequently it is by no means to be taken for granted that concentrate feeding is economical even though the gains are relatively cheap and easy to obtain. With a knowledge of price relationships of the various feeds, the simple calculations suggested above make it a relatively easy matter to determine which is the more economical.

Minerals. All balanced turkey mashes contain adequate quantities of the essential minerals or provide for them by calling for supplemental feeds such as calcium carbonate in some form. The addition of mineral supplements not called for is costly and unnecessary, and may even cause trouble. This suggestion applies especially to bonemeal and other feeds high in phosphorus and to prepared mineral mixtures.

WET MASH FEEDING

Although the feeding of all or part of the mash in damp form is practiced to some extent, the authors believe that the extra labor involved is unwarranted. Many experiments as well as trials by practical turkey growers have substantiated this belief.

CRUMBLED OR GRANULATED FEEDS

Crumbles or granules are crushed pellets and are useful for starting poults. The Minnesota Station (1951) reported substantially better early growth of poults started on crumbles than that of poults started on the same feed given as loose mash. The Dominion Experimental Station (Saskatchewan, 1952) found that crumbles did not improve growth of poults when there was no antibiotic in the feed. However, with procaine penicillin (4.41 p.p.m.) the crumbles improved the growth in male Broad Breasted Bronze turkeys 15% at four weeks of age and 10% at eight weeks of age, over the pen fed loose mash; but did not improve the growth of the females. However, crumbles are more expensive than the loose mash and they tend to cause feather picking. There is some evidence which indicates the doubtful value of crumbles. The usual procedure is to offer the granules during the first four weeks or so and then gradually replace them by pellets or loose mash.

PELLET FEEDING

Mash prepared in pellet form has had a favorable reception among turkey raisers. The Washington Station reported that at 24 weeks of age, the cost of feed per pound of gain was slightly lower for the turkeys fed pellets and grain than it was for those fed dry mash and grain. Reduction in feather picking and in the loss of feed through wind blowing was observed. The Oregon Station reported that in 1941 and 1942 in northeastern Oregon, Broad Breasted Bronze turkeys at 29 weeks of age produced equal results on five types of feeding; 20%-protein growing mash and grain free choice; 20%-protein growing pellets and grain free choice; 30%-protein growing concentrate and grain free choice; growing mash and grain free choice plus pellets fed liberally at noon daily, and growing mash and grain free choice plus rolled barley soaked in milk to which alfalfa meal was added, fed daily. The Ontario Station (1949) reported that pelleted growing mashes produced significantly larger Broad Breasted Bronze turkeys at 24 weeks than did the same feeds not pelleted, with greater differences in lots fed 15 and 20% of dehydrated cereal grasses and alfalfa meal combined than in the lot fed 10%. The pellets did not produce a significantly higher feed efficiency than the loose mash. Neither did the lot with 10% dried green feed as compared to those on 15 and 20%. Pelleted feed may be given to turkeys of all ages and classes except poults under four weeks of age for which loose mash or possibly crumbles is needed. The N. D. Station obtained excellent results with pellets. Chick-size pellets were 1/8″ in diameter and 3/8″ long; adult size were 1/4″ in diameter and 1/4″ long. Feed consumption was less than with dry mash by 1/3 lb. per pound of gain to 26 weeks of age. Pellets were readily eaten except during the first week or two. The pelleting of mashes enables turkeys to consume greater quantities and permits more efficient use of high-fiber feeds. Pellets are not so likely to be blown away by the wind as the loose mash; which may account for the more efficient feed conversion indicated in some experiments.

LIQUID MILK FEEDING

The feeding of liquid skimmilk or buttermilk to turkeys is not recommended except in cases where the necessary protein feeds cannot be obtained in dry form. It is competitive only when available at a cost of not more than 5% of the cost of a hundredweight of dried milk, meat scraps, or soybean meal. Two practices are described:

1. **Feeding liquid milk during the brooding period along with a mash.** In this type of feeding it appears best to provide water also, allowing one dish of water to several of milk and keeping both before the poults at all times. The milk may be fed sour or sweet; it appears to make no difference. On general principles, however, it is best to let it sour slightly before it is fed, mixing or stirring to assure feeding both whey and curd. The mash should contain 18 to 20% crude protein if milk and water are both offered. If no water is given, the mash may contain as little as 15% protein.

2. **During the rearing period, feeding liquid milk as the only drink and the only source of supplementary protein by keeping it and whole grains before the birds at all times.** Limited range with access to growing green feed should be provided. The birds may be given free choice of several common grains or one of the following mixtures:

Corn or grain sorghum	100	Corn* or barley	200
Wheat	100	Wheat	100
Oats	100	Oats	100
Barley	100		
Total	400	Total	400

If the poults have been fed liquid milk previously, during the brooding period, they can be "eased over" in about one week's time to the grain and milk diet. If they have not been fed milk previously it is well to allow a somewhat longer period for the transition. Turkeys fed dry mash and milk for the first 8 weeks and grain and milk with access to green range from 9 to 28 weeks inclusive, will consume about 12½ lbs. of liquid milk, ⅓ lb. of

*Cleaned cracked corn to 18 weeks, then whole corn may be fed.

mash, and 3½ lbs. of grain per pound of grain from hatching to 28 weeks of age. About 206 lbs. (24 gal.) of milk will be consumed per turkey during the 20-week rearing period (9th to 28th weeks). Skin color tends to become quite yellow in turkeys fed in this manner.

It is probable that the foregoing plan of feeding liquid milk as the sole source of supplementary protein should be used only in semi-confinement rearing because birds forced to obtain a large share of their living from range will not have the opportunity to drink enough milk; and, also, if on free range they usually will have access to water from some source which would result in inadequate consumption of milk. When turkeys are raised in strict confinement, the liquid milk-grain feeding plan probably would not work well unless tender fresh green feed in liberal quantities is carried to the birds.

The Oregon Station (1948) reported that turkeys fed liquid-milk and grain were produced at a feed cost of ¾ to 1½¢ (per pound of meat produced) less than mash-and-grain fed birds, but that high labor cost and other disadvantages made the plan impracticable. The milk was valued at 25¢ per 100 pounds.

Some disadvantages of liquid milk feeding are: (1) the frequent occurrence of pendulous crops when the milk is fed liberally; (2) the difficulty of providing proper sanitation; and (3) the extra labor involved. Whenever liquid milk is fed, special precautions are necessary to control flies and to keep the environment sanitary.

CONDENSED MILK AND WHEY SUPPLEMENTS

These products, when fed to growing turkeys, are intended to speed up market maturity and produce a fine finish. Very little experimental evidence can be found concerning the economy or need of their use as supplemental feeds. As with breeding stock, growing turkeys fed a fully adequate diet particularly those on good range, should have no need for supplementary feeds. However, when such supplements can be fed with very little extra labor and are not more expensive per pound than mash, their use might be indicated particularly for birds in confinement or on diets the adequacy of which might be doubted.

FEEDING PLANS FOR FREE RANGE REARING

Turkeys intended for free range rearing usually are brooded and fed under favorable conditions for 10 to 12 weeks and then placed on range. The plan is to promote maximum growth so that all the turkeys will be well-grown and able to handle themselves when put on range. Poults brooded with access to growing green feed are in better condition to go on range than are poults brooded in confinement because they have become used to ranging and eating coarse feeds.

Where ordinary green range is available a satisfactory method of feeding is to give, each evening, all the dry growing mash and grains the birds will eat free choice. Feed is put in troughs an hour or two before sunset and the birds allowed to eat their fill. In the morning only a little or none at all is given, the birds being encouraged thus to forage. If the birds are on grain stubble, only the mash, no grain, is fed. The University of California suggests the following growing mash to be used as a supplement to ranges.

	Per cent
Ground grains	57
Good quality sardine fish meal, or meat scrap, or both	15
Wheat bran	10
Alfalfa meal, good quality	13
Steamed bonemeal	2
Ground oyster shell or limestone	2
Salt	1

BULK-DELIVERED FEEDS

The delivering of turkey mash feeds in bulk instead of in sacks leads to a lower cost per ton of feed to the grower. It also helps in reducing labor costs, but in turn, requires certain investments in bulk storage bins and extra large feed hoppers. At present it seems better adapted for turkeys reared under confinement than those reared on the range because of this storage factor. Each grower must determine the feasibility of its use under his particular conditions. In some cases, over long rough hauls, the feeds have been known to "settle out" so that the ingredients fed to the turkeys are not all well blended and mixed as they would be as they came from the mixer into the feed sacks. There is also the factor of the inability of the trucks to make delivery to the feeders or storage bins on range in unfavorable weather. However, the bulk delivery fits in well where automatic feeders are used for the growing poults or the turkey breeders.

AUTOMATIC FEEDERS

These new mechanically operated production line feeding devices may reduce the cost of labor of feeding in brooder houses or in confinement rearing operations. They have now been as widely used and adopted by turkey growers as by broiler raisers, and poultrymen who feed laying hens. Mechanical perfection is difficult to obtain and the maintenance of the special equipment is a factor that must be considered. Several new mechanical feeding devices, on the market since 1950, are being adapted into wider use with turkey feeding. They may find their place where a minimum of labor is available.

FEED CONSUMPTION

Table 3.1 gives detailed data on feed consumption and growth rate for range reared turkeys.

Table 3.1—Growth and Feed Consumption of Broad Breasted Bronze Turkeys

These figures do not include green feed, and were calculated from farms where the turkeys were running on green range and in large flocks. (Data are taken from smoothed curves.)

Age in weeks	Average weight of toms & hens	Gain in weight	FEED CONSUMPTION PER BIRD 26% protein mash from 1 to 28 weeks plus grain after 6 weeks—all self-fed. Mash	Grain		Weekly feed	Cumulative feed	Per cent protein of total ration	Feed required per pound of gain by weeks
	lbs.	lbs.	lbs.	lbs.	Per cent	lbs.	lbs.		lbs.
1	.2	.1	.1	......	......	.1	.1	26	1.0
2	.5	.3	.3	......	......	.3	.4	26	1.5
3	.8	.3	.5	......	......	.5	.9	26	1.7
4	1.2	.4	.7	......	......	.7	1.6	26	1.9
5	1.6	.4	.9	......	......	.9	2.5	26	2.1
6	2.1	.5	1.2	.01	1	1.2	3.7	26	2.2
7	2.7	.6	1.4	.03	2	1.4	5.1	26	2.3
8	3.4	.7	1.5	.08	3	1.6	6.7	26	2.3
9	4.2	.8	1.7	.1	5	1.8	8.5	25	2.4
10	5.0	.8	1.9	.2	8	2.1	10.6	25	2.6
11	5.8	.8	2.0	.3	11	2.3	12.9	24	2.8
12	6.7	.9	2.1	.4	15	2.5	15.4	24	2.9
13	7.6	.9	2.3	.5	19	2.8	18.2	23	3.1
14	8.5	.9	2.3	.7	25	3.0	21.1	23	3.3
15	9.4	.9	2.2	1.1	32	3.3	24.5	22	3.6
16	10.3	.9	2.1	1.4	39	3.5	28.0	21	3.8
17	11.3	1.0	2.0	1.7	47	3.7	31.7	19	4.0
18	12.3	1.0	1.8	2.1	55	3.9	35.6	18	4.2
19	13.2	.9	1.5	2.5	62	4.0	39.6	17	4.4
20	14.1	.9	1.3	2.9	68	4.2	43.8	16	4.7
21	14.9	.8	1.2	3.2	73	4.4	48.2	15	5.1
22	15.7	.8	1.0	3.6	78	4.6	52.8	14	5.6
23	16.5	.8	.9	4.0	80	4.9	57.7	14	6.1
24	17.3	.8	.9	4.3	82	5.2	62.9	13	6.7
25	18.0	.7	.9	4.6	84	5.5	68.4	13	7.5
26	18.6	.7	.9	4.9	85	5.8	74.2	13	8.1
27	19.4	.7	.8	5.2	87	6.0	80.2	12	8.6
28	20.1	.7	.7	5.5	89	6.2	86.4	12	9.0
Total or Average									
0-24	17.2	17.3	33.8	29.1	46	62.9	62.9	18.6	3.6
0-26	18.7	18.7	35.6	38.6	52	74.2	74.2	17.7	4.0
0-28	20.1	20.1	37.1	49.3	57	86.4	86.4	16.9	4.3
9-28	20.1	16.7	30.5	49.2	62	79.7	79.7	16.1	4.8

HOME-MIXED AND PREPARED FEEDS

High quality commercial feeds, if fed while fresh, have several advantages over home-mixed feed: (1) the probability of more uniform and more thorough mixing; (2) saving of labor in mixing; (3) the opportunity of obtaining pelleted mashes which seldom can be produced by the home mixer or small mill; (4) the opportunity to obtain help in financing; and (5) a well assured supply in periods of shortages or stress. Home mixed feeds usually can be produced at less cost per pound and it often is possible to cut the cost further by taking advantage of feed-ingredient price fluctuation. Special diets prepared so as to best utilize fresh greens and other home grown feeds also may result in a rather considerable saving. Most of these advantages also occur when the services of local mills are utilized and in addition, the ingredient supply is likely to be more reliable from the local elevator.

In considering the use of commercial feeds the reputation of the organization and the actual performance of the feed in the field should be taken into account. In the case of home mixing, the final cost, the availability of essential ingredients, and the facilities for mixing are factors to be considered. There is one point that the authors believe should be clearly understood, and this is that properly compounded home-mixed feeds can be made the equal nutritionally of good commercial feeds. Home-mixed feeds when properly blended from high grade ingredients, are not handicapped because of any inferiority or inability to produce rapid growth efficiently. Poorly formulated feeds or those containing inferior ingredients whether home-mixed or commercial are not so efficient as feeds properly mixed on the basis of well-tested formulas. A point that should be stressed is that the selection of a commercial feed is, in every way, as important as the selection of a formula for home-mixing and the ingredients that go into the feed. In each case the feed should meet the specific need for which it is bought or mixed. A starting feed should be fed until the poults are 8 weeks old and a "grower" from then until the birds are marketed or are fed breeder mashes. Freshness likewise is an important qualification for both commercial and home-mixed diets.

If home mixing is resorted to, emphasis should be placed on the following points; (1) selecting a well-tested formula adapted to the purpose intended; (2) obtaining fresh, high quality ingredients; (3) mixing the ingredients thoroughly in the correct proportion; (4) if shovelled over and over as in the mixing of concrete, care should be taken to see that the floor, shovel and shoes are clean and free from poultry manure, as floor mixing presents a possible source of entrance of parasite eggs or disease germs; (5) making no substitutions except on the basis of a full knowledge of all the pertinent facts; (6) obtaining a knowledge of the characteristics, composition, and digestibility of the various feed ingredients; (7) adopting the practice of estimating protein, fiber, mineral, and vitamin content; (8) obtaining a knowledge of the requirements of the birds at different ages and under different conditions, for protein, mineral, fiber and vitamins and of the ability of turkeys to utilize the different feedstuffs. If the last three suggestions are carried out it becomes possible for the grower to make changes in the feeding program that will permit the most economical use of available ingredients and best advantage to be taken of price fluctuations, without sacrificing results. The use of stale feeds, especially animal products and feeds fed for their vitamin A content, should be avoided. All stale products should be placed under suspicion and freshness should be prerequisite to purchase.

HIGH ENERGY DIETS

The value of high energy diets was discovered in 1947. High in nitrogen-free extract, vitamins, and proteins, and low in fiber, these diets or modifications of them have been widely adopted. The value of antibiotics was discovered in 1950. As a result of these findings, most prepared starters and recommended starting formulas are modified high-energy feeds and contain an antibiotic or an antibiotic B_{12} supplement. With these diets improved feed conversion and fast growth are the rule.

Prior to the advent of the turkey fryer (broiler, springer, or light roaster) the content of fish products in starting mashes was determined only by the need for them nutritionally. However, when turkeys are slaughtered at an early age—10 to 18 weeks—the quantities of fish meals, fish solubles and fish oils in the diet must be limited to avoid off-flavor in the flesh. From available experimental data it appears that fish meals and/or solubles should be limited to 5% and fish oils to 0.15% in diets fed to turkey fryers. The use of dry vitamins A and D would eliminate the need for the oils and this seems advisable. Where oils must be fed, those of high vitamin potency should be selected so as to keep the quantity down to about 0.15%.

PURIFIED DIETS

Purified diets are useful in nutritional research but until recently poults could not be grown that way. However, in 1948 the Cornell Station reported a highly satisfactory purified diet containing the following ingredients in percentages:

Ingredients	Percent
Crude casein	35.0
Gelatin	6.5
Cornstarch	40.7
Soybean oil	6.0
Cellophane (100% fiber)	3.5
Salt mixture containing all known required mineral elements	7.0
1500A - 300 D (new units) feeding oil	1.0
Synthetic vitamins	
All those known to be required by poults	0.3
Total	100.0

Mineral elements supplied were calcium (Ca), phosphorus (P), potassium (K), sodium (Na), magnesium

(Mg), iron (Fe), manganese (Mn), iodine (I), chlorine (Cl), copper (Cu), zinc (Zn), cobalt (Co), and sulphur (S). The vitamins supplied were:—choline; inositol; niacin (nicotinic acid); pantothenic acid; riboflavin; thiamin; pyridoxine; 2-methyl-1, 4-napthoquinone (for vitamin K activity); alpha tocopherol (vitamin E); biotin; and folic acid.

FEEDS FOR TURKEY BROILERS

Special diets and feeding methods for the production of turkey broilers are not necessary. The standard starting and growing diets may be used equally as well in broiler production. However, the percentage of fish products should be watched carefully since there is danger of producing off-flavor in the flesh of these immature turkeys which are marketed as fryer-roasters. Some of the starting diets given herein contain more fish products than are advisable if the turkeys are to be marketed as fryer-roasters. Not over 0.15% of fish oil and not over 5% of fish meal, including fish solubles, should be included in any broiler diet.

RECOMMENDED FEED FORMULAS

Starting Diets

It is very important that starting diets for turkeys be most carefully compounded of fresh, high-quality feedstuffs. Attempts to make any substitutions in recommended formulas or changes in recommended methods except where absolutely necessary, should be avoided. A number of feed formulas suggested by various agencies are presented herewith for the reader's consideration.

All the diets recommended are presented in percentage form, and/or in pounds per ton or in pounds per half ton of mixed feed, and the starting diets are all intended to be fed for about the first 8 weeks. Insoluble grit such as coarse sand, gravel, quartz, or granite should be available to the poults. The salt used in all rations should be finely granulated and sifted to remove lumps. Water is the only drink to be given except where liquid milk is specified. All of these starting diets are complete feeds when their supplements, if any, are included and may be fed according to the free-choice feeding program heretofore outlined, with such modifications as may be noted.

Following are a number of recommended starting diets developed in various sections of the United States and Canada.

STARTERS NO. 1, 2, 3, and 4.
THE UNIVERSITY OF CALIFORNIA,
Davis, California

Ingredients	No. 1	No. 2	No. 3	No. 4
	Lbs.	Lbs.	Lbs.	Lbs.
(1) Fish meal, sardine preferred	200	140	150	100
(2) Meat scrap		140	50	100
(3) Soybean oil meal	500	450	500	525
(4) Alfalfa meal[1]	100	100	100	100
(5) Wheat bran	200	200	200	200
(6) Ground wheat	200	200	200	200
(7) Finely ground barley	259	264	264	249
(8) Ground corn	400	400	400	400
((9) Bone meal, special steamed[2]	25	0	30	20
(10) Ground limestone	50	40	40	40
(11) Salt, iodized	10	10	10	10
(12) Manganese sulphate, anhydrous	½	½	½	½
(13) Dried whey	40	40	40	40
(14) Fish oil (2250 A - 300 D)[3]	3	3	3	3
(15) Vitamin D carrier (1500 D per gm.)	1½	1½	1½	1½
(16) Choline chloride	1	1	1	1
(17) Riboflavin supplement[4]	5	5	5	5
(18) Antibiotic B_{12} supplement[5]	5	5	5	5
	2000	2000	2000	2000
Estimated protein content, %	25±	25±	25±	25±

[1]Should contain carotene equivalent to at least 67,000 I.U. of vitamin A per pound.
[2]Other products of equivalent calcium and phosphorus content may be used.
[3]Proportionate amounts of feeding oils of other potencies may be used.
[4]Containing 225 milligrams of riboflavin per pound. A proportionate amount of a product of different potency may be used.
[5]As directed by the manufacturer which may be different from amounts given.

Remarks: To convert to approximate percentages, divide each figure by 2 and point off one place. For example, 10 lbs. of salt ÷ 2 = 5, pointed off one place equals .5% These are all-mash diets to be fed the first 8 weeks only. No. 4 would be entirely satisfactory for feeding turkey fryers, the others might produce some off-flavor in fryers if certain types of fish meal* are used and would be satisfactory as starters for turkeys fed non-fishy growing mashes from 8 weeks to market age of 24 to 28 weeks.

*Sardine or tuna meal would be least likely to cause off-flavor.

STARTER NO. 5. CLEMSON AGRICULTURAL COLLEGE,
Clemson, South Carolina

Ingredients	Pounds
Ground yellow corn	32.00
Wheat middlings	10.00
Alfalfa leaf meal or high-quality meal	5.00
Soybean oil meal	35.00
Fish meal	10.00
Dried whey	3.00
Calcium carbonate	2.00
Bone meal (or equivalent)	2.00
Salt	0.50
Manganese sulphate	0.0125
Vitamin supplement (to supply 2 mgs, riboflavin. 4 mgs. pantothenic acid, 9 mgs. niacin and 90 mgs. choline per pound of mash)	0.10
Vitamin A and D concentrate—(3000 A—600 D_3 per gram)	0.10
D-activated animal sterol—(2,000 D_3 per gram)	0.05
Vitamin B_{12} and antibiotic feed supplement—(to supply 0.0045 mg. B_{12} per pound of mash)	0.25
Total	100.0125

Estimated protein content 26.4%; fiber, 4.4%; Ca, 2.2%; P. 1.26%; vitamin A per lb., 9000 I.U.; riboflavin, 2.76 mg. (2760 micrograms) per lb.; vitamin D, 686 I.C.U. per lb.; and manganese, 41 mg. per lb. (90 parts per million).

Remarks: Feed to poults from day old to 8 weeks of age. Grain may be fed in separate hoppers free choice after 6 weeks.

If fed to turkeys to be marketed as broilers, use sardine fish meal; or reduce the fish meal to 5%, adding 5% of meat scrap, or both.

STARTER NO. 6. CORNELL UNIVERSITY,
Ithaca, New York

(A high efficiency all-mash starting ration to be fed until 4 to 8 weeks of age)

Ingredients	Pounds Per Ton	Approximate Per Cent
Corn, ground yellow	588	29.40
Wheat flour middlings	300	15.00
Oats, Pulverized	100	5.00
Soybean meal, 44% protein	580	29.00
Fish meal, 60% protein	100	5.00
Meat scrap, 50% protein	150	7.50
Skim milk, dried; distillers dried solubles; or dried whey	60	3.00
Yeast, dried brewers'	40	2.00
Alfalfa meal	40	2.00
Ground limestone	30	1.50
Salt, iodized	10	.50
Manganese sulphate (feeding grade, 67% $MnSO_4$)	0.5	.025
Choline chloride	1.5	.075
Niacin	40 gms. per ton	
Alpha tocopheryl acetate supplement to supply	5 gms. per ton	
High-potency vitamin A source to supply	3000 I. U. per lb.	
D-activated sterol to supply	600 I. C. U. per lb.	
Antibiotic feed supplement to supply	10 gms. per ton	
Totals	2000.0	100.000
Approximate Analyses:		
Protein, %		28
Fat, %		3.6**
Fiber, %		4.3
Calcium, %		1.9
Phosphorus, %		0.90
Methionine, %		0.50
Cystine, %		0.40
Riboflavin, mg/lb.		2.3
Niacin, mg/lb.		42.0
Choline, mg/lb.		1028.0
Pantothenic acid, mg/lb.		5.7
Vitamin A, units/lb.		5690.0
Vitamin D, units/lb.		600.0

**If fat contributed by ingredients falls below 3.5%, then stabilized grease or tallow to bring the level to 3.5 to 4.5% should be added.

This mash is high in both available energy content and protein content. Most rapid growth and greatest uniformity are promoted by a ration of high available energy content. With such a ration the turkey poults need a relatively high protein intake. The two characteristics go hand in hand; that is, a high protein level has no advantage except when the ration is high in available energy and, conversely, the high energy content is most advantageous in the presence of a high protein content. Therefore, the ration as presented cannot be greatly altered without decreasing its value. Care should be taken to avoid the use of ingredients very high in fiber, such as light oats, wheat bran and fibrous alfalfa meal.

STARTER NO. 7. KANSAS STATE COLLEGE,
Manhattan, Kansas

Ingredients	Pounds
Ground yellow corn or milo	30
Ground oats or barley	3
Wheat middlings	6
Dehydrated alfalfa meal 17% protein, 75000 or more I. U. of A per lb.	5
Soybean oil meal, 44% protein	35
Meat scrap, 50 or 55% protein	8
Fish meal, 60% or more protein	6
Dried whey or distillers sol	3
Calcium carbonate	1
Steamed bone meal	1.5
Salt	0.5
Vitamin fortifier (soybean meal premix)*	1
Manganese sulphate	25 grams
Total	100 lbs. 25 grams

Estimated protein content 28%; fiber, 5.27%; Ca., 2.05% and P., 1.18%.

***A VITAMIN MIX FOR TURKEYS**

(Calculated to give the essential levels)

	Per Ton
Vitamin A (10,000 I.U./gm.)	400 grams
Vitamin D_3 (3,000 I.C.U./gm.)	400 grams
Riboflavin mix (1 oz. = 1 gm. ribo)	100 grams
Choline chloride 25% mix	3,120 grams
Niacin, crystalline	40 grams
Pantothenic acid (calcium pantothenate)	20 grams
Vitamin B_{12}-antibiotic supplement	5,000 grams
(Example when the product contains: 1.8 mg. B_{12}, and 1.8 gm. antibiotic per lb. of supplement.)	
Total	9,080 grams (20 pounds)

Thoroughly mix 1 pound of this with each 100 pounds of mash. (1 ounce equals about 28 grams and 1 pound equals 454 grams.)

STARTER NO. 8. MICHIGAN STATE COLLEGE,
East Lansing, Michigan

	Pounds	Per cent
Ground yellow corn	303	30.07
Ground heavy oats	50	4.93
Wheat standard middlings	50	4.92
Alfalfa meal, dehydrated, 17% protein	50	4.92
Meat and bone scraps, 50% protein	75	7.40
Soybean oil meal	300	29.58
Fish meal[6]	75	7.40
Dried skimmilk	50	4.92
Dried yeast	25	2.46
Ground limestone	15	1.47
Steamed bone meal	10	0.98
Salt (iodized)	3.5	0.34
Vitamin A and D feeding oil, 3000-A, 600-D	4	0.15
Choline chloride, 25% dry mixture	3	0.34
Manganese sulfate, feeding grade	0.25	.024
Vitamin B_{12} supplement, 6 mg/lb	0.25	.024
D-activated animal sterol (2,000 D per gm.)	0.50	0.05
	1014	100.00

Analysis:

Protein	29.0%	Riboflavin	2.2 mg./lb.
Fat	3.6%	Niacin	18.0 mg./lb.
Fiber	5.2%	Pantothenic acid	6.0 mg./lb.
Calcium	2.3%	Choline	960.0 mg./lb.
Phosphorus	1.2%	Vitamin A	8900.0 USP units/lb.
Calories/lb.	750	Vitamin D	850.0 ICU units/lb.
		Vitamin B_{12}	5.6 micrograms/lb.

Alternate Feed Ingredients:

Optional additions: Antibiotic, arsenical and/or coccidiostat according to directions of manufacturer.

Ground wheat for the middlings.

Yellow hominy feed for ⅓ the corn.

Dried buttermilk for skimmilk.

Corn gluten meal for ¼ the soybean oil meal.

Dried whey, condensed fish solubles, distillers' grain solubles and/or other vitamin B-complex supplements for the milk or yeast.

Oyster shell flour for limestone.

Dicalcium phosphate or defluorinated phosphate for bone meal.

Manganese oxide for the sulfate.

Other sources of vitamin A and D on equivalent unit basis.

If corn is not available, oats and barley may be used, provided extra alfalfa meal is added.

Feeding Directions:

Feed all-mash to poults until they are 8 weeks of age after which change to turkey growing mash. The above formula is not recommended for the feeding of turkey broilers, because of the fish meal and cod liver oil.

STARTER NO. 9. UNIVERSITY OF MINNESOTA, St. Paul, Minnesota

Ingredients	Pounds
Ground yellow corn	37
Pulverized oats	5
Alfalfa meal, dehydrated, 17% protein	5
Meat and bone scrap, 50% protein	10
Fish meal	5
Soybean meal, solvent or expeller	35
Steamed bonemeal (or equivalent)	1
Ground oystershell (or limestone)	1
Iodized salt	0.50
Riboflavin concentrate, 227 mg. per lb.	0.50
Antibiotic B_{12} supplement with 12.5 mg. B_{12} per lb.[1]	0.25
Feeding oil 600 D - 3000 A	.30
Dry vitamin D, 680,000 I. C. U. per lb.	.05
Manganese sulphate, feed grade	.025
Choline chloride, 25% or equivalent	.300
Total	100.925

[1]Follow manufacturer's directions in regard to level of product used to supply an adequate level of antibiotic.

Estimated protein content about 28%.

STARTER NO. 10. UNIVERSITY OF NEBRASKA, Lincoln, Nebraska (Turkey starter No. 54-TS)

Ingredients	Per cent
Ground yellow corn	35
Wheat shorts	5
Alfalfa meal, 17% or more protein	5
Meat and bone scrap, 50% or more protein	25
Fish meal	2
Fish solubles blend	2
Soybean oil meal	20
Dried whey	2
Mineral mixture No. 59[1]	3
Vitamin concentrate No. 4[2]	1
Total	100
Calculated protein level	27.3%

[1]Mineral Mix No. 59

Limestone	41 lbs.
Salt, fine	16 lbs.
Steamed bonemeal (di-calcium phosphate equivalent)	41 lbs.
Trace Mix No. 1	2 lbs.
	100 lbs.

Trace Mix No. 1 to carry enough manganese, iodine, iron, cobalt, copper and other minor elements to supplement ingredients as called for in these formulas.

[2]Each pound of vitamin concentrate No. 4 to carry the following minimum values:

Riboflavin	0.2 grams	Vitamin D_3 powder 1500 I.U./gm.	.9 gm.
Calcium pantothenate	0.4 "	Vitamin B_{12}	1.25 milligrams
Niacin	0.6 "	Procaine penicillin	200 milligrams
Choline chloride	2.0 "	(or aureomycin, terramycin, or bacitracin, 1000 mg.)	

Such a mixture can be purchased or made up from the ingredients using soybean meal to make up to 1 pound.

STARTER NO. 11. ONTARIO AGRICULTURAL COLLEGE,
Guelph, Ontario, Canada

(For feeding poults as an all-mash diet to 4 weeks and with grain free choice from 5 to 8 weeks.

Ingredients	Per ton
	Pounds
Ground wheat	500
Ground yellow corn	300
Dehydrated alfalfa meal[1]	100
Fish meal (65% protein)	50
Meat meal (50% protein)	40
Dried buttermilk or skimmilk[2]	40
Soybean meal (44% protein)	840
Tallow or grease	20
Ground limestone (high calcium)	30
Steamed bonemeal or equivalent	65
Iodized salt	10
Vitamin A oil (10,000 I. U./gm.)	1
Dry vitamin D_3 (1,650 I.C.U./gm.)	2
DL methionine (feed grade)	0.5 lbs.
Riboflavin (pure)	3 grams (or equivalent in riboflavin supplement)
Vitamin B_{12}	3 milligrams
Niacin	15 gm.
Antibiotic	4 to 10 grams[3]
Manganese sulphate (technical grade)	0.3 lbs.
Arsonic acid	45–90 gms.
Alpha-tocopheryl acetate	5 gms.
	2002.8

Calculated analysis:

Minimum crude protein, %	27.1
Minimum crude fat, %	2.8
Maximum crude fiber, %	5.2
Calçium, %	1.9
Inorganic phosphorus, %	0.68

[1]Containing 100,000 I.U. of vitamin A per lb.

[2]May be replaced by 40 lbs. dried whey, adding 30 lbs. of soybean meal at the expense of the wheat.

[3]2 grams of procaine penicillin; 10 grams of aureomycin, bacitracin, or terramycin.

Grain: The grain mixture to be used with these mashes will depend to a considerable extent on conditions of price and local availability. For confined birds good quality oats is satisfactory as the sole grain during the growing period. However, for the last month before marketing in the case of confinement birds, and for breeding birds, at least half of the grain mixture should consist of wheat and/or corn. For birds reared on range at least one-third of the grain mixture during the growing period and one-half of the grain mixture during the finishing period should be wheat and/or corn.

Grit: Insoluble grit, such as granite, should be fed to turkeys of all ages with these diets (also Grower No. 17), using the size of grit appropriate to the size of the birds. The grit may be sprinkled on the mash at regular intervals (once or twice weekly) or it may be made freely accessible in separate hoppers.

STARTER NO. 12. PENNSYLVANIA STATE UNIVERSITY, University Park, Pennsylvania

Ingredients	No. 12 Pounds
Ground yellow corn	332.1
Ground wheat	50
Ground heavy oats	50
Alfalfa meal (low fiber, 50,000 A)	40
Soybean oil meal (44% pro.)	300
Fish meal ((Menhaden)	50
Meat scrap (50% or more protein)	60
Dried whey	40
Brewers' dried yeast	40
Ground limestone (high calcium)[1]	20
Steamed bonemeal (or dicalcium phosphate)	10
Iodized salt	3
Premix*	4.7
Anhydrous manganese sulphate ($MnSO_4$)[1]	0.2
Totals	1000
Calculated protein, percent	27.8
" fiber percent	4.5
" calcium (Ca) percent	2.1
" phosphorus (P) percent	1.0
" riboflavin mg. per lb.	4.1
" calorics per lb.	785

*Premix:

	Grams
Choline chloride, 25%	900
Riboflavin concentrate, 8 mg.	230
Vitamin D_3, 3,000 I.C.U./gm.	270
Vitamin A, 10,000 I.U./gm.	200
Antibiotic supplement	450
Niacin	15
Myvamix	80
(Approx. 4.7 pounds)	2,145

[1]The $MnSO_4$ and iodized salt should be premixed with the ground limestone and bonemeal.

STARTER NO. 13 and 14. PURDUE UNIVERSITY, Lafayette, Indiana

Ingredients	No. 13* Starter	No. 14* High Energy
	Pounds	Pounds
Ground yellow corn	388	814
Wheat bran	300	......
Wheat middlings	300	......
Meat and bone scraps (50% protein)	100	......
Soybean oil meal	620	930
Dried milk	100	......
Dehydrated alfalfa meal (17%)	100	100
Steamed bone meal	40	80
Ground limestone	20	20
Granite grit (fine)	20	20
Salt mixture[1]	10	10
Vitamin D_3 supplement (1500 I.C.U.)	2	2
Vitamin A supplement (5,000 A)	......	2
Fermentation by-product (500 units of riboflavin)	......	20
Vitamin B_{12} and antibiotic supplement[2]	......	2
Niacin	......	(1 oz. per ton)
Total	2000	2000
Percentage protein	24.93	25.01
Percentage calcium	1.76	1.72
Percentage phosphorus	1.16	0.97
Percentage fiber	6.71	5.03

*Complete all-mash to be fed first 8 weeks.

[1]Salt mixture: 9 lbs. iodized salt and 1 lb. manganese sulphate.

[2]Supplement containing 3 milligrams vitamin B_{12} and 2 grams penicillin per pound. If there is any reason to suspect a deterioration in the vitamin A content of the alfalfa and/or yellow corn, add from 1 to 2 lbs. of the "vitamin A supplement (5,000 A)" to Starter No. 13.

STARTER NO. 15. U. S. DEPARTMENT OF AGRICULTURE, Research Center, Beltsville, Maryland

Ingredients of Diet 225	Per cent
Ground yellow corn	26.50
Wheat middlings, standard or brown, with or without screenings	10
Pulverized oats	10
Alfalfa meal, dehydrated, 17% protein	5
Meat scrap, 55 (or 50) % protein	10
Fish meal	5
Soybean meal	30
Steamed bonemeal (or dicalcium phosphate)	0.50
Ground oyster shell or limestone	1.50
Fermentation byproduct riboflavin supplement, 225 milligrams per pound	0.50
600 D - 3000 A feeding oil	0.15
Dry vitamin D (1500 I. U. per gram)	.05
Manganized salt[1]	.55
Antibiotic B_{12} supplement (premix)[2]	.25
Total	100.00

Estimated protein, percent 28; estimated calcium, percent 2; estimated phosphorus, percent 1.1; estimated crude fiber, percent 5.8.

[1]A mixture of 1.75 lbs. of anhydrous manganese sulphate, $MnSO_4$, or 2.5 lbs. of the tetrahydrate, $MnSO4—4H_2O$, in 100 lbs. of common or iodized salt.

[2]Should be used according to manufacturers directions to provide about 0.50 gm. of aureomycin, terramycin, or bacitracin; or 0.15 gm. of stabilized penicillin per 100 pounds. Any one of these antibiotics may be used.

Remarks: This mash is suitable for the feeding of poults intended to be marketed as fryers or light roasters as well as those intended to be grown to regular market age. It also makes an excellent complete growing mash to be fed along with grain after 8 weeks, eliminating the antibiotic and A and D feeds for range birds. Complete mixing of the ingredients is essential in order to obtain even distribution of the several ingredients that are present in very small percentages.

STARTER NO. 16. CHARLES W. WAMPLER, Harrisonburg, Virginia

Ingredients	Pounds
Yellow corn meal	23.
Ground heavy oats	5.
Wheat middlings	10.
Alfalfa meal, 17%, dehydrated	5.
Corn gluten, soybean or peanut meal[1]	40.
Meat scrap, 55% protein	5.
Fish meal	5.
Dried whey	2.50
Ground oyster shell or limestone	1.00
Salt	0.50
300 D - 2250 A feeding oil	0.25
Dry vitamin D (1500 I. U. per gram)	0.05
Dried Distillers' Solubles	2.50
Steamed bone meal	1.00
Manganese sulfate	(4 oz. per Ton)

Estimated protein, 28%.

[1]Or a combination of two or three of these.

Remarks: Optional additions, coccidiostat and antibiotic supplement.

STARTERS[1] NO. 17, 18, 19, 20 and 21.
STATE COLLEGE OF WASHINGTON,
Pullman, Washington

(day-old to 8 weeks)

Ingredients	No. 17	No. 18	No. 19	No. 20	No. 21
	%	%	%	%	%
Ground yellow corn or milo[2]	40.2	37.2	37.2	43.7	41.5
Soybean oil meal (44% protein)	46.5	46.0	42.0	40.0	37.2
Fish meal (70% protein)	5.0	5.0	5.0	5.0	10.0
Meat and bone meal (50% protein)	....	....	....	5.0	....
Dehydrated alfalfa or grass (carotene–12 mg. /100 gms.)	2.5	2.5	2.5	2.5	2.5
Dried brewers' yeast	....	....	4.0	....	....
Dried whey	....	4.0	4.0	....	4.0
Salt (iodized, fine)	0.3	0.3	0.3	0.3	0.3
Limestone or shell flour	2.0	2.0	2.0	2.0	2.0
Steamed bone meal or deflourinated phosphate	3.5	3.0	3.0	11.5	2.5
ADD THE FOLLOWING TO EACH TON OF FEED					
Vitamin A, millions of I. U.[4]	3.5	3.5	3.5	3.5	3.5
Vitamin D_3, millions of I. C. U.[4]	1.2	1.2	1.2	1.2	1.2
Riboflavin, gms.	1.6	0.6	....	1.5	0.5
Calcium pantothenate, gms.	2.6	1.0	....	3.0	1.5
Choline chloride, lbs.	1.0	1.0	1.0	1.0	1.0
Methionine, lbs.	1.0	1.0	1.0	1.0	1.0
Manganese sulfate lbs. (100% pure)	0.4	0.4	0.4	0.4	0.4
Antibiotic[3]—See note 3					

Suggested analysis: Protein, 28.0%; calcium, 2.0%; phosphorus, 1.0%; vitamin A, 4000 I.U./lb.; vitamon D_3, 600 I.C.U./lb.; and riboflavin, 2 mg./lb.

[1]These mashes may also be used for turkey fryers. Suggested method of feeding: During the first 3-4 weeks feed as crushed pellets or crumbled feed. After this feed intermediate size (⅛") pellets. If feather picking develops, mix pellets and mash. After the poults are 6 weeks of age small amounts of cracked or whole grain may be placed with the feed to accustom the turkeys to grain. Feed granite or silica grit of the proper size one day each week.

[2]Up to 15.0% of ground wheat, ground barley or finely ground oats may be used to replace an equivalent amount of corn or milo, however, such substitutions will decrease feed efficiency. Up to 5% of molasses may be used to replace grains. Two pounds of molasses are approximately equal to one pound of grain.

[3]Diamine or procaine penicillin, use at rate of 3 gms./ton; if aureomycin, bacitracin or terramycin, use at rate of 10 gms./ton.

[4]If a fish-oil carrier is used, limit it to 0.15%.

GROWING DIETS

Greater latitude exists in the preparation of growing diets because turkeys, when they are grown to the age of about 8 weeks, are able to utilize a greater variety of feedstuffs and a greater variety of feedstuff combinations than they are during the first few weeks. The cost of the starting diet while seemingly high per pound is still low enough in total cost to make it economically sound to spare no reasonable expense in securing a diet adequate for normal growth and development. The cost of the growing diet represents a much larger share of the cost of raising a roasting turkey, consequently it is more important to pay more attention to the cost and try to obtain the most reasonably priced growing feed that will produce satisfactory results. Some diets are capable of producing a pound of gain on less feed than are other diets, because they contain a higher percentage of digestible nutrients.

From the following growing diets it ought to be possible to choose one or more, which if composed of good quality ingredients properly mixed, will produce turkeys efficiently and economically out of ingredients commonly available and reasonably priced in any section of the United States or Canada.

Growing diets may be classified as "simplified," "semi-complete" and "complete". The complete diets, when supplemented with water, insoluble grit, and direct sunshine, are expected to supply all the food elements necessary for growth and fattening. They may be used in any system of management but are especially adaptable for use in confinement rearing or in other situations where the birds obtain no green feed. The semi-complete diets may be fed to all turkeys except those raised in strict confinement without green feed of any kind. They are especially adapted to situations where the birds do not have continual access to succulent green feed or where the supply of green feed is limited or of doubtful quality. Such situations are the use of range areas of inadequate size; the typical fall shortage of green feed due to cessation of plant growth; limited hand feeding of green feed to confined birds; utilizing the dried-up summer range in arid or drought-stricken areas; ranging the birds on heavily-wooded or poor-soil areas that do not provide an adequate quantity of succulent grasses, legumes, or other edible plants; and the raising of turkeys in very large flocks resulting frequently in poor use of the range. The simplified diets are intended for use only when the birds are able to eat their fill constantly, of good quality succulent growing green feed. This is an extremely important point to consider when selecting a feed, whether home-mixed or commercial. The complete and semi-complete diets are easily recognized by the presence of dried alfalfa products, milk, and other vitamin-bearing ingredients which the simplified diets do not as a rule contain.

Growing diets should contain little or no fish oil because of the danger of causing off-flavor in the flesh. After the brooding period turkeys are almost always raised with continual access to direct sunlight and this will supply sufficient quantities of the vitamin D factor anywhere in the United States. In fact, it is not recommended that turkeys be raised without direct sunlight although it is possible to do so. Sunlight favors normal feather development in addition to supplying the vitamin D factor.

When continuous access to direct sunlight cannot be provided, the most practical method of supplying vitamin D with minimum danger of producing off-flavor in the flesh is to use a commercial dry non-fishy vitamin D concentrate such as activated animal sterol which usually furnishes vitamin D at less cost per unit than the fish oils. The vitamin A needed by growing turkeys can be supplied by yellow corn and alfalfa meal in the growing mash supplemented if necessary either by a non-fishy vitamin A supplement or by a quantity of high A potency feeding oil limited to about 0.20% of 18 to 28% protein growing mashes and to 0.25% in 30% or more protein concentrates. The quantities of fish meals and/or fish solubles should also be limited in growing feeds to not over about 8% in 18 to 28% protein growing mashes and to about 12% in 30% or more protein concentrates.

Growing mashes usually contain 18 to 22%, less commonly up to 28% protein; growing concentrates are mashes having 30% to 32% protein, less commonly up to 40%. The Washington, Michigan, and Missouri Stations have shown that (1) the higher the protein content of the mash the less mash, and consequently the more of grain, is eaten (if both are fed ad libitum); (2) when compared, mash and concentrate produce gains on about the same expenditure of feed per pound of gain; and (3) total percentage of protein consumed is higher with concentrates, resulting usually in higher over-all feed cost. Sometimes, however, the grain is low enough in relative cost to offset the cost of a higher protein level. Rate of consumption of mash and grain, availability of suitable feedstuffs and the price of mash and grain are points for consideration in planning for the most economical feeding program possible.

In the following list of diets there is no special reason for using the starter and the grower recommended by the same institution or person, though it often will be found economical to do so. Unless otherwise directed, all of the mashes listed are intended to be fed free choice with grain.

GROWERS NO. 1, 2, 3, 4, and 5.
THE UNIVERSITY OF CALIFORNIA,
Davis, California

(Growing mashes with about 21% protein)

Ingredients	No. 1	No. 2	No. 3	No. 4	No. 5
	Lbs.	Lbs.	Lbs.	Lbs.	Lbs.
Fish meal	100	—	50	100	—
Meat scrap	—	100	—	—	100
Soybean oil meal[1]	400	400	450	400	400
Alfalfa meal[2]	200	200	200	300	300
Wheat bran	300	300	300	300	300
Ground grains	897	907	849	779	789
Molasses	—	—	—	20	20
Bone meal, special steamed[3]	40	30	40	40	30
Ground limestone	40	40	40	40	40
Salt, iodized	20	20	20	20	20
Manganese sulphate	½	½	½	½	½
Dried whey	—	—	50	—	—
Vitamin D carrier (1500 D per gram)	1	1	1	1	1
Riboflavin supplement[4]	2	2	—	—	—
Approximate totals	2000	2000	2000	2000	2000

[1]Soybean oil meal may be partly replaced by cottonseed meal 10% (200 lbs.) or sesame seed oil meal 5% (100 lbs.)

[2]Should contain carotene equivalent to at least 67,000 I.U. of vitamin A per pound.

[3]Other products of equivalent calcium and phosphorus content may be used.

[4]Containing 225 milligrams of riboflavin per pound. A proportionate amount of a product of different potency may be used.

Remarks: Alfalfa hay and fresh greens are valuable supplements to these growing mashes.

GROWERS NO. 6, 7, and 8. THE UNIVERSITY OF CALIFORNIA, Davis, California

(Concentrates with about 30% protein)

Ingredients	No. 6	No. 7	No. 8
	Lbs.	Lbs.	Lbs.
(1) Fish meal	200	100	200
(2) Meat scrap	......	100	......
(3) Soybean oil meal[1]	800	800	800
(4) Alfalfa meal[2]	400	500	500
(5) Wheat bran	300	250	250
(6) Ground grains	94	68	......
(7) Molasses	......	......	50
(8) Bone meal, special steamed[3]	80	60	80
(9) Ground limestone	80	80	80
(10) Salt, iodized	40	40	40
(11) Manganese sulphate	¾	¾	¾
(12) Vitamin D carrier (1500 D per gm.[5])	2	2	2
(13) Riboflavin supplement[4]	4	......	......
Approximate totals	2000	2000	2000

[1]Cottonseed meal or sesame seed oil meal may be substituted for soybean oil meal in amounts up to 20 percent of the mash (400 pounds per ton).

[2]Should contain carotene equivalent to at least 67,000 I.U. of vitamin A per pound.

[3]Other products of equivalent calcium and phosphorus content may be used.

[4]Containing 225 milligrams per pound. A proportionate amount of a product of different potency may be used.

[5]Proportionate amounts of products of different potency may be used.

Remarks: Alfalfa hay or fresh greens are valuable supplements to these growing concentrates.

GROWER NO. 9. CLEMSON AGRICULTURAL COLLEGE, Clemson, South Carolina

Ingredients (Complete growing mash)	Pounds
Ground yellow corn	55.00
Wheat middlings	10.00
Alfalfa meal	5.00
Soybean oil meal	20.00
Fish meal	5.00
Dried whey	3.00
Steamed bone meal or equivalent	1.00
Calcium carbonate	1.00
Salt	0.50
Manganese sulphate	0.0125
Vitamin supplement (to supply 2 mgs. riboflavin, 4 mgs. pantothenic acid, 9 mgs. niacin and 90 mgs. choline per pound of mash)	0.10
Vitamin A and D concentrate—* (3000 A—600 D_3 per gram)	0.10
D-activated animal sterol—* (1500 D_3 per gram)	0.05
Total	100.7625

*Omit for poults on green range.

Estimated protein content, 19.2%; fiber, 4.0%; Calcium (Ca), 1.70%; P (phosphorus), 0.8%; vitamin A per lb., 9000 I.U., riboflavin, 2.3 mg. (2300 micrograms) per lb.; vitamin D_3 per lb., 600 ICU.; and manganese, 27 mg. per lb. (59 parts per million).

Remarks: Poults are changed to the growing mash, from the starter, at 8 weeks of age. Grains, as corn, oats, milo, and wheat are placed before the poults in hoppers and the birds allowed to consume, free choice after 16 weeks. Insoluble grit and oyster shell are kept in hoppers at all times.

GROWERS NO. 10 and 11. CORNELL UNIVERSITY, Ithaca, New York

(Complete turkey growing mashes*)

Ingredients	No. 10 Growing Mash 4–8 Weeks (all mash) Lbs./ton	No. 11 Growing Mash After 8 Weeks (with grain) Lbs./ton
Corn meal, yellow	628.5	669.5
Wheat middlings	300 (flour)	400 (standard)
Oats, pulverized	200	200
Soybean meal, solvent extracted	400	200
Fish meal**	100	—
Meat scrap	100	200
Yeast, dried brewers'	40	40
Distillers' dried solubles, dried whey or dried skim milk	60	60
Alfalfa meal***	100–	140–
Limestone ground	30	30
Steamed bone meal	30	40
Salt, iodized	10	20
Manganese sulphate (feeding grade 67% $MnSO_4$)	0.5	0.5
Choline chloride	1.0	——
Niacin supplement to supply	40 gms/ton	40 gms/ton
Alpha tocopheryl acetate supplement to supply	5 gms/ton	5 gms/ton
Riboflavin supplement to supply	1.0 gms/ton	4.0 gms/ton
High-potency vitamin A source to supply (units vitamin A/lb.)	——	(1000 units/lb.)
D-activated animal sterol to supply (units vitamin D/lb.)	(600 units/lb.)	(1200 units/lb.)
Antibiotic feed supplement to supply	(10 gms/ton)	(10 gms/ton)
Total	2000	2000
Approximate analyses:		
Protein, %	24.6	20.0
Fat, %	3.7	4.2
Fiber, %	5.1	5.7
Calcium, %	1.9	2.3
Phosphorus, %	1.0	1.1
Riboflavin, mg/lb.	2.3	3.8
Vitamin A, units/lb.	5750	8740
Vitamin D, units/lb.	600	1200

*Scratch grains may be fed free-choice starting at 12 weeks. It also would be advisable to feed extra oyster shell or some other source of calcium.

**Fish meal should be replaced by meat scrap if turkeys are to be marketed early as broilers so as to prevent off-flavor in the flesh.

***Formula calculations based on alfalfa containing 90,000 units of vitamin A activity per lb. If lower potency alfalfa is used, the required vitamin A level should be maintained by additional high-potency vitamin A supplement.

Growing Ration: Two growing rations are outlined in the foregoing table; one is designed for use as an all-mash ration for poults during the period four to eight weeks of age, and the other to be fed with grain for the rest of the growing period.

The major change involved in going from the starting mash to the first growing mash is a reduction in protein content. This ration is somewhat less critical with respect to available energy content, and more latitude in substitution of ingredients can be exercised. The turkey poult at this time is still in a period of rapid growth, and the statements made regarding the unidentified vitamins in the starting ration also apply to this preliminary growing mash. Attention is called to the fact that this preliminary growing ration, as formulated, contains fish meal. In areas where turkeys are marketed early as turkey broilers, the fish meal should be replaced by high quality meat scrap in order to prevent "fishy flavor" in the turkey broilers. Likewise fish-oil vitamins carriers should be limited to 0.15% in all-mash diets and 0.25% in the mash where grain is fed.

GROWER NO. 12. KANSAS STATE COLLEGE, Manhattan, Kansas

(Complete growing mash)

Ingredients	Pounds
Ground yellow corn or milo	30
Ground wheat or kafir	20
Ground oats or barley	10
Wheat bran	3
Dehydrated alfalfa meal, 75,000 I. U. of vitamin A per lb.	10
Soybean meal (soybean oil meal) 44% protein	15
Meat scrap, 50% or more protein	7
Calcium carbonate (or high calcium limestone or oyster shell), ground	2
Salt	1
Vitamin fortifier*	2
Manganese sulphate	0.033 (15 gms.)
Total	100.033

*A mixture of such potency that 2% in the mash supplies vitamins as follows, per pound of diet: dry vitamin A to supply 2000 I.U.; dry vitamin D to supply 900 I.U.; riboflavin to supply 2.3 mg.

Approximate analysis: Protein, 19.07%; fiber, 5.83%; Ca, 1.77%; P, 0.80%; vitamin A per lb., 9000 I.U.; vitamin D, 675 I.U. per lb.; vitamin G (riboflavin) 2500 mcgms. (2.5 mg.) per lb.; choline, 534 mg., per lb.; and niacin (nicotinic acid), 18.0 mg. per lb.

Remarks: If full access to good green range is given, the vitamin fortifier may be omitted from this diet. Milo alone is an economical grain to feed with this mash.

GROWER NO. 13. MICHIGAN STATE COLLEGE, East Lansing, Michigan

Complete Growing Mash	Mash to be fed with grains (B)	Mixing Mash (C)
	Pounds	Pounds
Ground yellow corn	300	......
Ground heavy oats	100	......
Wheat standard middlings	100	......
Wheat bran	100	200
Alfalfa meal, dehydrated, 17% protein	75	150
Soybean oil meal	225	450
Meat and bone scraps, 50% protein	25	50
Fish meal	25	50
Dried yeast	25	50
Dried whey	25	50
Ground limestone	25	50
Steamed bone meal	25	50
Salt (iodized)	7	14
Manganese sulfate, feeding grade	0.5	1
Dry Vitamin D_3, 1500/gm	2.5	5
Dry Vitamin A, 5000/gm	2	4
Choline Chloride, 25%	1	2
Riboflavin	1 (gram)	2 (gram)
	1063 lbs.	1126 lbs.

Optional Additions:

Antibiotic ....(according to directions of supplier)
Arsenical ....(" " " " ")
Coccidiostat ....(" " " " ")

	Calculated Composition		
	Units	B	C
Protein	%	22.0	33.8
Fat	%	2.8	2
Fiber	%	7.3	9
Calcium	%	2.2	4.4
Phosphorus	%	1.0	1.7
Energy	Cal/lb.	700	500
Riboflavin	mg/lb.	3	5.4
Niacin	mg/lb.	24	33
Pantothenic acid	mg/lb.	7	10
Choline	mg/lb.	730	1160
Vitamin A	USP/lb.	9340	10000
Vitamin D	ICU/lb.	1700	3400
Vitamin B_{12}	ug/lb.	1.3	2.7
Salt	%	0.35	0.7

(Continued next page)

GROWER NO. 13 (Continued)

Alternate Feed Ingredients:

Oats and barley for corn, but add 20 lbs. more alfalfa meal.

Ground wheat and wheat middlings for bran.

Yellow hominy feed for ⅓ the corn.

Corn gluten meal for ¼ the soybean oil meal.

Dried skimmilk or buttermilk, condensed fish solubles, distiller's grain solubles for the whey or yeast.

Oyster shell flour for limestone.

Dicalcium phosphate or defluorinated phosphate for bonemeal.

Manganese oxide for the sulfate.

Other non-fish-oil sources of vitamin A and D on equivalent unit basis.

Mixing and Feeding Directions:

Use grain mixture of 60% corn, 20% wheat and 20% oats with "B."

Mix equal amounts of ground grains (60% corn, 20% wheat and 20% oats) with "Mixing Mash" to make mash to be fed with grain.

Provide oyster shell or limestone and hard grit with all feeds.

Feed growing mash to birds, confined or on pasture, from 8-12 weeks of age to time of marketing.

For special finishing mash for birds after 20 weeks of age, replace fish meal and 100 lbs. soybean oil meal of feed "B" with corn.

GROWER NO. 14. THE UNIVERSITY OF MINNESOTA, St. Paul, Minnesota

(Semi-complete growing mash)

Ingredients	Pounds
Ground yellow corn	47
Oats, pulverized	5
Alfalfa meal, dehydrated	5
Meat and bone scraps, 50% protein	10
Soybean meal, solvent or expeller	30
Steamed bonemeal (or equivalent)	1
Oyster shell or limestone, ground	1
Iodized salt	0.50
Riboflavin concentrate (227 mg. per lb.)	0.50
Antibiotic B_{12} supplement (12.5 mg. of B_{12} per lb. with adequate antibiotic activity per manufacturer's recommendations)	0.05
Dry vitamin D (680,000 I. C. U. per lb.)	0.05
Manganese sulphate, feed grade	0.025
Total	100.125

Remarks: Turkeys should have range for best results with this mash. Feed grain free choice with mash after 8 weeks.

GROWERS NO. 15 and 16.
THE UNIVERSITY OF NEBRASKA,
Lincoln, Nebraska

Ingredients	No. 15 51-TG-1 (Complete)	No. 16 54-T6 (Simplified) For use with good pasture
	Pounds	Pounds
Corn, ground yellow	348	270
Wheat shorts	200	200
Whole or pulverized oats, whole preferred	100	200
Alfalfa meal, 17% protein	100	50
Meat and bone scrap, 50% protein	50	100
Fish meal	20	00
Soybean oil meal	150	150
Mineral Mixture No. 59[1]	30	30
Nebraska vitamin concentrate No. 4 or equivalent	2	0
Total	1000	1000
Calculated protein level	21.0%	21.0%

[1]See Starter No. 10 for formulas for mineral and vitamin mixes.

Remarks: Feed either mash free choice with grain after 8 weeks.

GROWER NO. 17. THE ONTARIO AGRICULTURAL COLLEGE, Guelph, Ontario, Canada

(Complete growing mashes and concentrate)

Ingredient	No. 17 Turkey Grower Mash (4-8 wk.)	No. 18 Turkey Grower Mash (8-24 wk.)	No. 19 Grower Concentrate (8-24 wk.)
	Pounds	Pounds	Pounds
Ground wheat	500	640	....
Ground yellow corn	200	200	....
Pulverized oats	100	200	....
Ground barley	100	200	....
Dehydrated alfalfa meal or cereal grass	100	120	270
Fish meal (65% protein)	40	20	70
Meat meal (50% protein)	40	40	100
Dried buttermilk or skimmilk	40	40	100
Soybean oil meal (44% protein)	750	425	1200
Tallow or grease	20	20	30
Ground limestone	30	30	70
Steamed bone meal or equivalent	65	50	125
Iodized salt	10	10	25
Vitamin A oil (10,000 I. U./gm.)	1	1	3
Dry vitamin D^3 (1,650 I. C. U./gm.)	2	2	5
Manganese sulfate (feed grade)	0.3	0.4	1.25
DL Methionine (feed grade)	0.5	....	....
	gms.	gms.	gms.
Riboflavin	3	3	8
Niacin	15	15	35
d-alpha-tocopheryl acetate	5	5	12
Antibiotic	4-10	4-10	10-35
Arsonic acid	45-90	45-90	110-220
	mgs.	mgs.	mgs.
Vitamin B^{12}	3	3	8
Calculated analysis	%	%	%
Min. crude protein	25.6	20.1	35.3
Min. crude fat	2.9	3.3	3.0
Max. crude fiber	5.7	6.0	7.6
Calcium	1.9	1.6	3.9
Total phosphorus	1.0	0.8	1.5
Inorganic phosphorus	0.67	0.54	1.24

Remarks: See footnote to Starter No. 11.

GROWERS NO. 20 and 21.
THE PENNSYLVANIA STATE UNIVERSITY
University Park, Pennsylvania

Complete Growing Mashes		No. 20 8–16 wk. Grower A	No. 21 16–24 wk. Grower B
		Pounds	Pounds
Ground yellow corn		520	608.1
Ground wheat		—	—
Ground heavy oats		100	100
Alfalfa meal, dehydrated		40	40
Soybean oil meal, 44%		171	100
Fish meal, menhaden		25	—
Meat scrap 50% protein		50	50
Dried whey		20	20
Brewers' dried yeast		30	30
Ground limestone		10	13
Steamed bone meal		25	30
Iodized salt		5	5
Manganese sulfate		0.2	0.2
		grams	*grams*
Choline chloride, 25%		(900)	(900)
Riboflavin concentrate, 8 mg.		(90)	(90)
D_3, 3000 ICU/gm.		(270)	(270)
A, 10,000 IU/gm.		(200)	(200)
Antibiotic supplement		(225)	(225)
Niacin		(15)	(15)
Myvamix		(40)	—
Total Premix in lb.		3.8	3.7
		1000	1000
Protein	%	20.6	16.5
Fiber	%	4.3	4.1
Calcium	%	1.9	2.0
Phosphorous	%	1.0	1.0
Riboflavin	mg/lb.	2.6	2.5
Calories	C./lb.	870	910

GROWER NO. 22. PURDUE UNIVERSITY,
Lafayette, Indiana

(To be fed free-choice from 8 weeks to market with corn and oats)

Ingredients	Grower No. 22A	High Energy No. 22B
	Pounds	Pounds
Ground yellow corn	340	936
Wheat bran	300	—
Wheat middlings	300	—
Meat and bone scraps	200	—
Soybean oil meal	500	620
Dehydrated alfalfa meal (17%)	200	200
Steamed bone meal	40	140
Ground limestone	60	40
Granite grit (coarse)	40	40
Salt mixture[1]	20	20
Vitamin B_{12} and antibiotic supplement[2]	—	4
Total	2000	2000
Percentage protein	23.84	19.74
Percentage calcium	3.07	2.99
Percentage phosphorus	1.32	1.28
Percentage fiber	7.81	5.64

[1]Salt mixture: 18 lbs. iodized salt and 2 lbs. manganese sulphate.
[2]Supplement containing 3 milligrams vitamin B_{12} and 2 grams penicillin per pound.

Remarks: Turkeys should have range for best results with these mashes.

GROWERS NO. 23, 24, 25, 26, and 27. THE U. S. DEPARTMENT OF AGRICULTURE
(Research Center, Beltsville, Maryland)

Growing Mashes	No. 23[1]	No. 24[1]	No. 25[2]	No. 26[3]	No. 27[9]
Ingredients	(Diet 205) Complete	(Diet 185) Complete	(Diet 184) Semi-Complete	(Diet 183) Simplified	(Diet 133) Simplified Concentrate
	%	%	%	%	%
Ground yellow corn	40	32	25	30	10
Wheat bran[4]	—	10	5	—	15
Wheat standard middlings[4]	15	15	15	—	20
Pulverized oats[5]	—	—	15	30	—
Alfalfa meal, dehyd. 17% protein	10	12	10	—	—
Soybean meal[6]	19	15	15	30	20
Fish meal	5	—	—	—	10
Meat and bone meal, 50% protein (or meat meal, 55% protein)	5	10	10	—	20
Riboflavin supplement[7]	0.50	0.5	—	—	—
Steamed bonemeal or dicalcium phosphate	1.50	3.5	2.5	7.0	—
Ground limestone (or oystershell)	2.80	1.0	2.0	2.0	3.5
Fine salt, manganized[8]	0.75	0.85	0.5	1.0	1.5
2250 A - 600 D feeding oil[10]	0.15	0.15	—	—	—
Antibiotic B_{12} Supplement[11]	0.25	——	——	——	——
Dry vitamin D	0.05	——	——	——	——
Totals	100.00	100.00	100.00	100.00	100.00
Estimated protein content	22.17	21.14	20.75	20.00	32.13
Estimated fiber content	5.88	6.35	6.90	5.70	4.62
Estimated calcium content	2.43	2.65	2.68	2.72	3.93
Estimated phosphorous content	0.97	1.40	1.26	1.29	1.78

[1]This mash is a complete diet suitable for turkeys raised in confinement with no supplement except grain, insoluble grit and water.

[2]When fed along with grain and access to a fair amount of green feed either hand-fed or obtained from range, this is a satisfactory and economical diet.

[3]When fed along with grain and full continuous access to good green range, this diet is excellent and economical.

[4]Wheat by-products may be replaced if desired by ground wheat or grain sorghum, or pulverized barley.

[5]The pulverized oats may be replaced by pulverized barley, or ground wheat, or grain sorghum.

[6]If desired, peanut meal, cottonseed meal, or corn gluten meal or a mixture of 2 or 3 may replace half the soybean meal in 205, 183, or 185, or all of it in 184 or 133.

[7]A fermentation by-product containing 500 micrograms of riboflavin per gram, or 227,000 per pound.

[8]Salt with which has been mixed 2% anhydrous manganese sulphate (MnO_4) or 4% of the tetrahydrate (MnO_4-$4H_2O$). Iodized salt may **be used** if desired.

[9]This mash (No. 27) must be accompanied by abundant fresh green feed, otherwise good results will not be obtained. Feed it free choice from 8 weeks to marketing along with any grain or grain mixture except with corn alone or oats alone.

[10]If obtainable, dry vitamin A (2000 A per gram) at 0.25% and dry vitamin D (1500 D per gram) at 0.10% would be excellent substitutes for the A and D oil. They are easier to mix (but must be premixed and the premix very thoroughly blended in the total mix) and produce no off-flavor in the flesh of turkeys.

[11]To supply 0.5 gm. of aureomycin, bacitracin or terramycin, or 0.15 gm. of stabilized penicillin per 100 lbs. of diet. The use of this supplement is optional (but recommended) for turkeys that do not have access to fresh green feed.

Remarks: These diets should be fed free choice with a grain or grain mixture selected so as to provide a variety of grains, counting those in the mash. As a single grain, barley or grain sorghum will produce excellent results with any of the mashes listed, like-

(Continued next page)

GROWERS NO. 23, 24, 25, 26 and 27. (Continued)

wise a mixture or free-choice offering of ½ oats and ½ yellow corn (cracked and cleaned to 18 weeks, whole if desired after 18 weeks) or ½ oats and ½ wheat, or ½ corn and ½ wheat. Barley may be used as part of mixture as well as by itself, but not free choice with other grains. Oats are excellent when fed free choice with other grain or as part of a mixture. They also can be fed as the only grain free choice with any of the mashes but not with the concentrate (No. 27). With the concentrate, two or more grains free choice or in a mixture should be fed.

GROWERS NO. 28, 29, and 30. U. S. DEPARTMENT OF AGRICULTURE

(Research Center, Beltsville, Maryland)

(Complete all-mash growing diets)

Ingredients	No. 28 8 to 16 weeks of age	No. 29 17 to 24 weeks of age	No. 30[5] 25 weeks on
	Per cent	Per cent	Per cent
Ground yellow corn or grain sorghum	12	20	27
Pulverized (or whole) oats	15	15	15
Wheat middlings, standard	15	15	15
Pulverized (or whole) barley	25	25	25
Fish meal[1]	4	3	2
Soybean meal	19	12	6
Alfalfa meal, 17% protein	5.20	5.20	5.20
Ground limestone	1.50	1.50	1.50
Steamed bonemeal	2.00	2.00	2.00
Riboflavin supplement[2]	0.25	0.25	0.25
Manganized salt[3]	0.50	0.50	0.50
Vitamin A supplement 2000 I. U. per gm.[4]	0.25	0.25	0.25
Vitamin D supplement 1500 I. U. per gm.[4]	0.05	0.05	0.05
Antibiotic B_{12} Supplement[6]	0.25	0.25	0.25
Totals	100	100	100
Estimated protein percent	20	17	14

[1]May be replaced by equal amount of 55% protein meat-and-bonemeal.
[2]Containing 500 micrograms per gram.
[3]A mixture of 97½ lbs. salt and 2½ lbs. anhydrous manganese sulphate or 4 lbs. of the tetrahydrate.
[4]The vitamin D supplement may be omitted if birds receive ample direct sunshine. These non-fishy vitamin A and D supplements are preferred to the feeding oils but if necessary the latter in 2250A—600 D at 0.15% may be used instead.
[5]This diet also may be used as a maintenance diet for feeding breeding stock from 28 weeks of age until breeder mash is fed.
[6]See footnote 11, Grower 23.

Remarks: These diets with no supplements except insoluble grit and water are suitable for turkeys raised in confinement or on poor range. If turkeys have access to good green pasture, the riboflavin supplement, alfalfa meal, antibiotic, and vitamins A and D supplements may be omitted, substituting for them 6% more oats.

GROWER NO. 31. CHARLES W. WAMPLER
Harrisonburg, Virginia

(Semi-complete growing mash)

Ingredients	Per cent
Yellow corn meal	35
Wheat middlings	20
Pulverized oats, heavy oats	5
Alfalfa leaf meal (or good quality alfalfa meal), dehydrated	5
Distillers' dried solubles	3
Meat scrap, 50% protein[1]	10
Crab meal	3
Oyster shell meal	2
Soybean meal or peanut meal	15
Defluorinated rock phosphate	1½
Salt	½
Manganese sulphate	4 oz. per ton
Total	100

[1]Half the meat scrap may be replaced by soybean or peanut meal if desired, to reduce cost of the diet.

Estimated crude protein, 21½%.

Remarks: This mash is intended for feeding turkeys on reasonably good range, free choice with grain. Antibiotic supplements may be added.

GROWERS[1] NO. 32, 33, 34, 35, and 36. STATE COLLEGE OF WASHINGTON
Pullman, Washington

(Semi-complete turkey developing mashes to be fed from 8 weeks to maturity)

Ingredients	No. 32	No. 33	No. 34	No. 35	No. 36
	%	%	%	%	%
Ground grains[2]	61.5	49.0	45.0	47.0	55.0
Wheat mixed feed (millrun)[2]	—	10.0	10.0	10.0	—
Fishmeal (70% protein)	5.0	2.0	—	—	5.0
Meat and bone meal (50% protein)	—	3.0	—	5.0	—
Soybean oil meal (44% protein)	23.0	20.0	28.0	22.0	14.0
Cottonseed meal (43% protein)	—	—	—	—	10.0
Dehydrated alfalfa or grass (carotene—12 mg./100gm.)	5.0	10.0	10.0	10.0	10.0
Limestone or shell flour	2.0	3.0	3.0	4.0	4.0
Steamed bone meal or defluorinated phosphate	3.5	2.0	3.0	1.0	1.0
Salt (iodized, fine)	1.0	1.0	1.0	1.0	1.0
ADD THE FOLLOWING TO EACH TON OF FEED					
Vitamin D[3], millions of I.C.U.	1.2	1.2	1.2	1.2	1.2
Riboflavin (grams)	1.7	1.0	1.0	1.0	1.0
Manganese sulfate, lbs., (100% pure)	0.4	0.4	0.4	0.4	0.4
Antibiotic	See Note 3				

Suggested analysis: Protein, 20.0%; calcium, 2.0%; phosphorus, 1.0%; vitamin A, 4000 I.U./lb.; vitamin D_3, 600 I.C.U./lb.; riboflavin, 2 mg./lb.

[1]Suggested method of feeding: feed the mash or pellets and whole grain free choice during the entire growing period. Feed granite or silica grit one day each week.

[2]Ground corn, milo, wheat, barley or oats may be used singly or in any combination with 0-10% millrun as long as the combined amounts of barley, oats and millrun do not exceed 35% of the mash. Molasses may be used at levels up to 5% replacing grain on a two for one basis.

[3]Diamine or procaine penicillin at rate of 3 gms./ton; aureomycin, bacitracin or terramycin at rate of 10 gms./ton.

BREEDER DIETS

For the production of large numbers of eggs of high hatchability, turkeys require a diet complete in the various nutrients with special attention paid to the vitamins. Good results may be obtained with a rather simple laying diet but only if it is supplemented by plenty of direct sunshine and green feed picked by the birds from the range. Such a diet may be compounded as follows:

MASH-GRAIN BREEDER DIET NO. 1
U. S. Dept. of Agriculture
(Simplified Breeder Diet)

Mash	Parts by weight	Grain	Parts by weight
Ground yellow corn	25	Yellow corn or grain sorghum	50
Bran	12	Heavy oats or barley	50
Middlings	15		
Pulverized oats	20		
Meat scrap (50% protein)	10		
Fish meal	10		
Ground oyster shell or limestone	7		
Salt	1		
Total	100	Total	100

This mash contains about 20% of crude protein. It should be kept before the birds at all times beginning about a month before eggs are expected. The grain mixture should be fed in separate troughs or on top of the mash at the rate of one-fifth pound (about 3 oz.) per day per bird in the mated flocks of large and medium types (2½ oz. for small types); the birds thus consuming about 60% mash and 40% scratch grain. However, unless the weather is mild and the birds are on good range this simplified diet will not be satisfactory. Gravel or other insoluble grit should be provided. No other supplements are necessary or desirable. However, if preferred, the oyster shell may be left out of the mash and fed separately in hoppers, ad libitum; increasing the corn to 32 parts.

As noted above, succulent green feed and direct sunshine, when obtained in abundance by the birds from range lots, supply the food elements necessary to render complete a combination of the common feedstuffs such as laying diet No. 1. However, in the management of turkey breeding stock, particularly in large flocks, conditions are frequently such that the birds cannot, or do not, obtain proper supplements for a simplified diet and in these cases a complete diet must be fed. Nearly all breeding stock kept in commercial quantities is in the latter category and is fed complete, well balanced diets even when given good range.

Breeder diets may be fed as (1) mash and grain, (2) in all-mash forms, or (3) as a mixture of grain and mash. The last two methods are recommended because they simplify management, save labor, and tend to insure that each bird receives a complete diet. Unless there is an economic reason for feeding grains separately, these breeder diets will be found best adapted to most breeding stock enterprises.

In the first method (grain and mash separately), the mash, either loose or pelleted, is fed free-choice and the grain is hand-fed in the desired proportion. However, it may be fed free-choice along with the mash. The latter plan often does not work too well as the turkeys tend to eat too much grain. Possibly the best method of avoiding this is to use a high-protein (28% or more) mash such as Breeder Diet No. 9, along with a not-too-palatable grain or a grain mixture; such as barley alone, grain sorghum alone, a mixture of both, or almost any mixture that contains not over 50% of oats. Breeder diets should contain a fair percentage of oats which can be obtained through the mash. To determine the correct proportion of mash to grain to obtain a total protein intake of 15 to 17%:—estimate the protein content of the mash and that of the grain fed and balance them so that the average protein intake is about 16%. For example, a 22-percent-protein mash or pellets fed 50-50 with a mixture of corn and oats equal parts, gives a protein intake of about 16% as follows:

	Protein
Mash	
50 lbs. × 0.22	11.00 lbs.
Grain	
Oats—25 lbs. × 0.11	2.75 lbs.
Corn—25 lbs. × 0.085	2.13 lbs.
Total	15.88 lbs.

The pounds of protein in 100 lbs. of feed consumed is obviously 15.88% protein.

In the second (all-mash) method of feeding breeding turkeys, the ingredients all are ground and are thoroughly combined into a homogeneous mixture which is fed ad libitum to both hens and toms. The latter appear to thrive on the all-mash breeder diet even though their need for protein may possibly be less than that of the hens. Also, there is some opinion that feed consumption by the hens may be reduced if an all-mash diet is fed. Observations by the authors indicate that this is not necessarily so with properly compounded diets fed correctly. To be palatable, all-mash diets should not be finely ground and should not be high in vegetable-protein supplements or alfalfa meal. Freshly mixed (bi-weekly) mash is more palatable and in addition, freshness helps to preserve the valuable vitamin A potency in the feeding oils which usually are included. Feeding the mash, by daily feeding, not permitting the accumulation of stale mash is helpful. In any event, feed should be ordered at least twice weekly and all stale mash should be eaten up before more is added. Hopper-type feeders should be checked to be sure no mash is by-passed and so allowed to become stale.

The third method of feeding breeding stock consists either (1) of mixing the mash and grain of a mash-grain diet together in the proper proportion and feeding the resulting mix as an all-mash diet, or (2) using whole or cracked grain in place of ground grain in an all-mash diet. For example, in Breeder Diet No. 13, the oats may be whole and the corn may be whole or cracked. When this type of mixture is used, the turkeys tend to pick out the grain and eat it first, hence care must be taken to see that the finely ground mash ingredients are not wasted or allowed to become stale. Possibly the best way to accomplish this result is to force the birds to empty the feeders daily or at least twice or three times per week, before fresh feed is added. In fact, this is good practice with any feeding program.

The Minnesota Station (1952) reported excellent results from feeding either a 28-percent-protein mash, such as Breeder Diet No. 9, free-choice with free-choice corn and heavy oats; or a mixture of breeder mash (Breeder Diet No. 8) and grain (corn and oats) equal parts. The latter method appeared to have a slight edge on the former.

If the breeders are to be confined to buildings, small bare yards, or to sunporches, it is advisable to use straw, hay, or other roughage for litter in the inside quarters building it up by regular additions. The litter provides roughage in lieu of that which the birds usually obtain when they have access to a lot containing vegetation and a suitable environment to favor normal mating and prevent foot and leg trouble.

Vitamin supplements based on condensed milk or whey which supply extra vitamins particularly A, D, and riboflavin are attaining some popularity. These products are fed in addition to the regular feed, usually by placing containers of them where the birds can eat at will. If the diet fed is complete in every respect this supplementary feeding should not be necessary. However, inadequacies do sometimes occur and many breeders consider it good insurance to feed one of these products, particularly where the cost per pound does not greatly exceed that of the mash.

Added vitamins (pantothenic acid, folic acid, and niacin) may be valuable in some breeder diets to reduce fluctuations in hatchability according to the Tennessee Station (Patrick, 1953). Such supplementation is included in some of the breeder diets listed. In other diets, the feed ingredients are expected to supply the necessary vitamins.

All feed for turkeys should be placed in feeders, never on the ground or in the litter. Feeders should be constructed so as to prevent waste and contamination from litter or droppings. One lineal foot of feeding space for 6 to 8 birds is advised for breeding stock.

To answer the question, "Is it better to restrict the feeding of breeders during the winter prior to the egg-laying season to prevent the birds becoming too fat?" the Kansas Station (1941) for two seasons compared the egg production and hatchability of breeding turkeys on full feed to those of other groups of turkeys on restricted feed. The results show that pens full-fed from market age to laying age produced 3.4 more eggs and gave 11.3% better hatchability than did hens on a ration amounting to 61% by weight of the normal intake; and 0.6 more eggs and 2.4% better hatchability than hens on a 78% normal ration. In all cases the birds received all the feed they would eat starting at the time egg production began. These results indicate that breeding birds should be full-fed throughout the fall and winter period prior to the start of egg production. It was noted that the hens in both groups lost weight during the breeding season.

The complete breeder diets listed in following pages are recommended by the various experiment stations. There are all-mash diets, as well as diets in which the grain portion is limited by hand feeding, and a concentrate-grain diet in which both the high-protein and high-vitamin mash and the grain both can be fed free choice.

BREEDER DIETS NO. 2, 3, 4, and 5, THE UNIVERSITY OF CALIFORNIA
Davis, California

Ingredients	No. 2	No. 3	No. 4	No. 5
	lbs.	lbs.	lbs.	lbs.
(1) Fish meal	100	200	50	100
(2) Meat scrap	100	—	150	—
(3) Soybean oil meal	200	200	250	350
(4) Alfalfa meal[1]	300	300	300	300
(5) Wheat bran or mill run	300	300	300	300
(6) Ground grains	810	850	820	780
(7) Bone meal special steamed[2]	20	30	10	50
(8) Ground limestone	40	40	40	40
(9) Salt, iodized	20	20	20	20
(10) Manganese sulphate	½	½	½	½
(11) Dried whey	100	50	50	50
(12) Vitamin A and D feeding oil (2250 A - 300 D per gram)[3]	6	6	6	6
(13) Vitamin D carrier (1500 D per gram)[3]	3	3	3	3
(14) Riboflavin	3g	3g	3g	3g
(15) Pantothenic acid	10g	12g	12g	12g
Approximate totals	2000	2000	2000	2000

[1]Should contain carotene equivalent to at least 67,000 I.U. of vitamin A per pound.
[2]Other products of equivalent calcium and phosphorus content may be used.
[3]Proportionate amounts of products of other potency may be used.

Protein content about 20% in each diet. To obtain percentages divide each figure by two and point off one place.

Remarks: Feed equal parts with a grain or grain mixture and keep water, oyster shell, and commercial soluble grit or ordinary gravel available to the turkeys at all times. This results in a protein consumption of about 15%.

BREEDER DIET NO. 6. CLEMSON AGRICULTURAL COLLEGE
Clemson, S. C.

(Complete turkey breeder all-mash diet)

Ibgredients	Pounds
Ground yellow corn	50.00
Ground oats	10.00
Wheat middlings or ground wheat	10.00
Alfalfa meal	5.00
Soybean oil meal	10.00
Meat scraps	2.50
Fish meal	2.50
Cond. fish solubles supplement (Supplies 1 mg. riboflavin; 12mgs. niacin; 110 mgs. choline and 0.005 mgs. vitamin B_{12} per pound of ration.)	2.50
Dried whey	3.00
Steamed bone meal or equivalent	3.00
Calcium carbonate	1.00
Salt	0.50
Manganese sulphate	0.0125
Vitamin A and D concentrate—if not on range (to supply 3400 IU of A and 700 D_3 ICU per pound of ration).	

Approximate analysis, %: Protein 16.0, fiber 4.2, calcium 1.73, phosphorus 0.88; vitamin A per lb., 5500 I.U.; riboflavin per lb., 2.40 milligrams; vitamin D_3 per lb., 700 I.C.U.; and manganese per lb., 24 mgs.

Remarks: Granite grit and oyster shell are kept in hoppers for turkey breeder hens. It is desirable that a succulent green range be provided for turkey breeders but this mash is adequate without supplementation of any kind except water, grit, and shell.

BREEDER DIET NO. 7. CORNELL UNIVERSITY
Ithaca, New York

(Complete breeder mash to be fed with grain)*

Ingredients	Pounds per ton
Corn meal, yellow	404.5
Wheat standard middlings	400
Oats, pulverized	200
Soybean meal, solvent extracted	240
Fish meal	100
Meat scrap	200
Dried skimmilk or dried whey	100
Distillers' dried solubles or dried brewers' yeast	100
Alfalfa meal, dehydrated	150**
Dicalcium phosphate	60
Ground limestone	25
Salt	20
Manganese sulphate (feeding grade, 67% $MnSO_4$)	0.75
Riboflavin supplement to supply	7.0 gms/ton
High-potency vitamin A source to supply	5000 units/lb.
D-activated animal sterol to supply (units/lb.)	1800 units/lb.
Alpha-tocopheryl acetate supplement to supply	5 gms./ton
Approximate Analyses:	
Protein, %	24.0
Calcium, %	2.8
Phosphorus, %	1.4
Fat, %	4.6
Fiber, %	5.8
Riboflavin, mg/lb.	6.1
Vitamin A, units/lb.	12950
Vitamin D, units/lb.	1800

*To be fed with scratch grain. Grain intake should not exceed 60% of total. Free-choice oyster shell or other calcium supplement should be provided.

**Formula calculations based on alfalfa containing 90,000 units of vitamin A activity per lb. If lower potency alfalfa is used, the required vitamin A level should be maintained by additional high-potency vitamin A supplement.

Remarks: The breeding period is comparable with the early growth period in that the highest requirements for necessary nutrients are encountered at these times. The discussion of unidentified vitamins in relation to the requirements of a starting mash applies equally well to breeding rations. The breeding mash, therefore, is similar to the starting mash in the types and variety of materials used in its formulation. It is somewhat less critical in regard to the need for high energy content, but the indiscriminate use of feedstuffs extremely high in fiber should be avoided.

BREEDER DIETS NO. 8 and 9.
THE UNIVERSITY OF MINNESOTA,
St. Paul, Minnesota

(Complete breeder mash to be fed with grain)

Ingredients	Diet 8 Breeder Mash[1]	Diet 9 Breeder Concentrate[2]
	Pounds	Pounds
Ground yellow corn	200	......
Wheat middlings	200	425
Pulverized oats	200	......
Alfalfa meal, 17% protein, dehydrated	75	100
Meat and bone scrap, 50% protein	100	150
Fish meal or fish solubles (equivalent)	50	75
Soybean meal, solvent or expeller	125	175
Steamed bone meal or equivalent	10	15
Oyster shell or limestone, ground	20	30
Iodized salt	10	15
Riboflavin concentrate (227 mg per lb.)	7	11
Vitamin B_{12} supplement (12.5 mg. B_{12} per lb. or equivalent)	¼	⅜
Feeding oil, 300 D - 2250 A, or equivalent (new units)	6	9
Dry vitamin D (680,000 I. C. U. per lb. or equivalent)	1	1½
Manganese sulphate, feed grade	½	¾
Total pounds	1004¾	1007⅝
Estimated protein content	22%	28%

[1]To be fed with equal parts of grain with oyster shell free choice .
[2]To be fed with corn and oats free choice with oyster shell free choice. The grain consumption should be limited to twice as much as the mash, i.e. 2:1. Free choice grain feeding of this mash with free-choice corn and heavy oats should result in satisfactory consumption and production.

[2]Each pound of Vitamin Concentrate No. 14 to carry the following minimum values:	
Alpha tocopherol acetate	2,000 Int. Units E per lb.
Riboflavin	0.2 gms.
Calcium pantothenate	0.4 gms.
Niacin	0.6 gms.
Choline chloride	2.0 gms.
Vitamin A	10,000 I. C. Units
D_3	135,000 I. C. Units
B_{12}	0.03 milligrams
Antibiotic (Procaine Penicillin)	200 Milligrams)
(Aureomycin) - 1 gm.	)Choice
(Terramycin) - 1 gm.	)

BREEDER DIET NO. 10. THE UNIVERSITY OF NEBRASKA
Lincoln, Nebraska

(Complete all-mash breeder diet)

Ingredients	Pounds	Per cent
Whole oats	200	20.0
Wheat bran	100	10.0
Wheat shorts	100	10
Ground yellow corn	230	23
Alfalfa meal, 17% protein	100	10.0
Meat and bone scrap, 50% protein	100	10
Fish meal	20	2.0
Soybean meal	100	10
Dried whey	20	2.0
Mineral mixture No. 56[1]	20	2.0
Nebraska vitamin concentrate No. 14[2]	10	1.0
Total	1000	100.0

Estimated protein content 18.7%.

Remarks: Supply fresh from 3 to 5 p.m. daily, and allow free access to road gravel or grit, oyster shell and good quality alfalfa hay.

Mineral Mix No. 56	Parts by Weight
Limestone	806
Salt	161
Trace Mix No. 1**	33
	1000

**Trace Mix No. 1 to carry enough manganese, iodine, iron, cobalt, copper, zinc, and other minor mineral elements when included at a level of 33 parts per 1,000 and the complete mineral supplement is used at a level of either 2 or 3 per cent to supplement ingredients as called for in these formulas.

BREEDER DIET NO. 11. OREGON STATE COLLEGE
Corvallis, Oregon

Ingredients	Per cent
Finely ground barley	5.0
Finely ground oats	10.0
Ground yellow corn	15.0
Millrun	13.5
Ground wheat	15.0
Meat meal	5.0
Soybean meal	10.0
Fish meal	7.5
Dried distillers' solubles	5.0
Alfalfa leaf meal or special meal, dehydrated	10.0
Steamed bonemeal	1.5
Ground limestone	1.5
Salt	1.0
300 D - 2250 A feeding oil	0.1
Irradiated animal sterol—1500D	.1
Vitamin mix—3000 mgs. riboflavin, 3000 mgs. pantothenic acid, and 300,000 mgs. choline per lb.	.05
Manganese sulphate, anhydrous ($MnSO_4$)	.025
Total	100.275

Calculated analysis:

Protein, %	20.9	Vitamin A, U.S.P. Units per lb.	7353
Fiber, %	6.2	Vitamin D, International Units per lb.	817
Calcium, %	1.8	Riboflavin, mcgms per lb.	3800
Phosphorus, %	1.0	Pantothenic acid, mcgms per lb.	8400
		Choline, mgs. per lb.	845

Remarks: Feed with mixed grains 50-50 along with free access to oyster shell, insoluble grit, and water.

BREEDER DIETS 12 and 13. U. S. DEPARTMENT OF AGRICULTURE
Research Center, Beltsville, Md.

Ingredients	No. 12 To be fed with grain (78-P)	No. 13 All-mash (78-O)
	(percent)	(percent)
Ground yellow corn	25	30
Wheat middlings, standard or brown	25	22
Wheat bran	10	5
Pulverized whole heavy oats	0	20[4]
Alfalfa meal, 17% protein, dehydrated preferred	10	5
Meat scrap, 55% protein (or 50% protein meat-and-bone meal)	8	4
Fish meal	8	4
Dried milk[1]	10	5
Riboflavin supplement[2]	0.50	0.25
Ground oyster shell or limestone	1	4
Steamed bonemeal (or dicalcium phosphate)	1	0
Vitamin A & D feeding oil, 2250 A - 600 D	0.50	0.25
Salt, preferably maganized[3]	1	0.50
Total	100	100
Estimated protein content, percent	22.6	16.5
Yellow corn (or grain sorghum or barley)	50	....
Heavy oats	50	....
Total	100	....
Estimated protein content of whole diet	16.1	16.5
Remarks	Mash and grain to be fed 50-50. Provide insoluble grit and keep granulated oystershell or limestone, and water before the birds.	Feed the mash without supplements except insoluble grit and water

[1]The dried milk, if too high priced, may be replaced, in Mash No. 12 by (1) 5% dried whey, and 5% more fish meal, or (2) by 6% soybean oil meal, and 4% dried whey; in Mash No. 13 by (1) 2½% dried whey and 2½% more fish meal, by (2) 2½% condensed fish solubles, 2% soybean oil meal, and ½% steamed bonemeal, or by (3) 3% soybean oil meal and 2% dried whey.

[2]A fortified fermentation by-product containing about 225,000 micrograms (225 milligrams) of riboflavin per pound.

[3]A mixture of 100 pounds of fine salt and 1¾ lbs. of anhydrous manganese sulphate ($MnSO_4$) or 2½ lbs. of tetrahydrate ($MnSO_4$-$4H_2O$).

[4]Whole oats may be used in place of pulverized oats.

FATTENING OR FINISHING DIETS

Finish in young turkeys normally comes only with maturity and attainment of maturity is hastened by the feeding of adequate quantities of protein, minerals, and vitamins. To provide a well balanced growing diet suitable to turkeys of different ages and sexes, the simplest and most efficient method is to feed the birds both dry mash or pellets and scratch grain, allowing them to choose the proportion of each they desire. It is a mistake to cut down on the sources of protein, minerals, and vitamins at any time during the growing period. It is best to let the birds decide how much they need. They will not overeat of mash, on the contrary they will eat less of it as they approach maturity. Any of the recommended growing diets (including their supplements) should serve to provide the turkeys with proper nourishment if they are mixed and fed as directed. In a high-quality roasting turkey the absence of short pinfeathers is as important as the presence of fat and flesh; and turkeys require protein and minerals to complete their feather growth. Uninterrupted, fast growth provides best assurance of good market quality. It seems certain that proper diet, favorable environment especially as related to uncrowded conditions, and freedom from diseases, parasites, and feather picking, will go far in promoting the normal growth of feathers within the usual time limit in any climate. Normally it is useless to expect the medium or large-sized strains of turkeys to be free of short pinfeathers at 24 weeks of age. Even under favorable conditions it requires about 26 weeks for the young hens and 28 weeks for the young toms. The Beltsville Whites (both sexes) mature for market four weeks earlier, on the average. Under ordinary conditions, turkeys attain a good finish at about the time the pinfeathers are grown out sufficiently to permit clean picking. This is the normal, at least for the strains of turkeys now generally grown. Prolonged hot weather appears to retard finishing, but does not prevent the production of acceptable market turkeys. In most northern states it is doubtful if there would be any retardation due to heat no matter when the turkeys are hatched.

Standard methods of feeding usually are more profitable than the special methods sometimes recommended. As stated, the use of well-balanced standard feeds is the best guarantee of a good market finish. The best chance of producing well finished roasting turkeys in the least possible time is to use modified high energy mashes with an antibiotic in the starter.

The high value of corn in fattening diets is generally recognized. Liberal portions of this grain will be found in most of the growing and finishing diets listed in the preceding pages. If corn is expensive, other grains, especially barley, may be substituted. It is believed, however, that there is no advantage to be gained, under most conditions, in radically changing the diet toward the end of the growing period. In fact, a radical change may do more harm than good. If changes are made they should be made gradually.

Iodine induces more rapid molting and recovery in chickens, according to some workers. Whether or not this mineral may have an effect on the feather growth of turkeys is not known. Iodine deficiency probably does not exist in the average poultry diet. However iodized salt may be used to avoid a possible deficiency (see page 251).

Milk as a fattening feed is widely recommended because of its supposed favorable effect on palatability and appearance of the flesh. However, in the experience of the authors, excellent flavor of meat, bleached skin and fat not too yellow can be produced without the use of milk. Kansas State College suggests the following methods of milk-fattening turkeys by (1) a mash, and (2) a condensed milk product. For the latter, use condensed skim or buttermilk spread over the feed, usually the grain, in the feeders; start the milk gradually and build up to about 3 lbs. per 100 turkeys per day over a 3- or

4-week period just before marketing. For the former the suggested mash is as follows:

White corn meal	300 lbs.
Oat groat meal or pulverized oats	200 "
Wheat shorts	100 "
Meat scrap	25 "
Alfalfa leaf meal	25 "
Total	650 lbs.

The mash should be moistened with milk to the consistency of paste. The birds should be fed all they will clean up in a half hour in shallow trays three times daily for 3 weeks before marketing. Liquid milk to drink also should be provided. Diets similar to the foregoing tend to produce a white finish especially if no fresh green feed is given. Unless a premium is obtained for birds so fed, special fattening methods have no practical advantage over continuing the growing diets through to marketing.

If a *yellow finish* is desired, yellow corn and fresh green feed should be supplied abundantly. Turkeys grown on grain, a large portion of which is yellow corn, and milk with green range, will have a yellow finish. Under most conditions 40% or more of yellow corn in the entire diet has a tendency to produce yellow finish and tender, juicy flesh in turkeys. The ability to assimilate yellow pigment, principally xanthophyll, and deposit it in the fat may be hereditary according to some observations. For example, certain strains of Bourbon Reds are said to be genetically of yellow skin. However, these authors have seen no evidence that could support such a belief. The skin of all turkeys, so far as is known, tends to be white, but a pale to deep yellow cast can be developed in the skin of females through feeding. The skin of males is less amenable, seldom deevloping more than a light yellow color.

If a *white finish* is desired, yellow corn should be fed in moderation so that it does not constitute more than about 30% of the total intake. When grain and mash are both fed, the yellow corn may well be confined to the mash of which it should comprise 25 to 40%. The scratch grain can be made up of barley alone or a mixture of grains excluding yellow corn. The vitamin A content should be carefully checked, especially when yellow corn is fed sparingly and succulent green feed is not provided. Green feed can be fed in sufficient quantity to supply all the vitamin A needed.

The *fattening* of mature stock appears to be impossible during the summer season. No matter how much is fed, the adult stock, hens and toms, will not become fat during the summer. However, in September, the birds will start to put on weight if they obtain sufficient feed. Access to growing green feed and grain affords the cheapest way to fatten adult stock although they fatten well on the regular mash and scratch feeds ordinarily fed to growing birds. If there is no objection on the market to yellow skin, a good share of the fattening feed may well be composed of yellow corn. If a white finish is desired, no yellow corn should be fed, using white corn or one of the other common grains instead. A mixture of equal parts of corn, wheat, oats, and barley makes a good fattening mixture for adult turkeys to supplement green pasture. If pasture is not available, these adult birds should be given access to direct sunshine and be fed one of the semi-complete growing diets. Since they are not growing at a rapid rate, any of the semi-complete diets should prove satisfactory without the addition of green feed. The Kansas milk-fattening method also should give excellent results especially if a premium can be obtained for white flesh of extra fine quality. It is possible, however, that turkeys will not always take readily to a radically different method of feeding. For all practical purposes the grain-pasture method is the easiest and cheapest method of fattening adult turkeys and will result in a carcase possessing an attractive appearance and fine table quality.

Flavor and palatability of flesh are important items in finishing turkeys. Meat scrap, dried skim and buttermilk, soybean meal, peanut meal, and corn gluten meal all produced palatable flesh at the Wyoming and Beltsville Stations. Cottonseed meal and linseed meal were somewhat inferior in this respect when 10 to 14% was included in the mash. Among the grains there were no demonstrable differences between corn, wheat, oats, barley, and rye. A variety of grains always gives good results, provided the diet as a whole is well balanced and no undesirable feedstuffs are included. As mentioned elsewhere, fish products and others containing unsaturated fatty acids should be avoided in finishing diets. Certain green feeds, such as wild onions and similar weeds are suspected by some growers of causing off-flavor in the flesh.

Rate of growth and deposition of fat in the body tissues and skin appear to be unaffected by the kind of grain fed, according to the South Dakota Station. Likewise, there is little difference between milk, fish meal, meat scrap, soybean meal, and peanut meal if diets are balanced. Corn gluten meal and cottonseed meal in limited quantities are good in finishing diets, but linseed meal is not recommended, in any amount.

CHAPTER 4

THE FEEDING STUFFS

OUTLINE

In selecting the ingredients of a diet, it is desirable to have as much information as possible about the individual feeding stuffs. Much information about the approximate chemical composition, the vitamin content, and the digestibility of a number of feeding stuffs used in feeding turkeys will be found in the tables near the end of this chapter. Other information about the more important feeding stuffs is given in the following pages.

FEEDS USED CHIEFLY AS SOURCES OF CARBOHYDRATES

Grains, Seeds and Their Byproducts

Wheat, corn, barley, oats, rye, and the grain sorghums, when shipped in interstate commerce, must be inspected and graded according to the Official Grain Standards of the United States. In buying grain for feeding, poultrymen should take advantage of this fact and buy according to grade. Annual yields for the U. S. are given in Table 4.1.

Yellow corn, either flint or dent type, is palatable, digestible, and readily available in most sections of the United States. Dent corn is the common Indian corn grown widely throughout the U. S. and Canada. Flint corn, an exceptionally hard type, is available chiefly in the coast states as the bulk of it is imported from South America. Corn is called maize in England, and other English speaking countries except Canada.

Table 4.1—U.S.D.A. Estimates of Production of Common Grains in the United States.

Thousand bushels[1]

Grain	1940 yield	1942-51 avg.	1953[2]
Corn	2,449,200	3,036,380	3,176,615
Oats	1,235,628	1,324,614	1,216,416
Wheat	816,698	1,088,548	1,168,536
Barley	309,235	295,299	241,015
Grain sorghums[3]	121,371	137,236	109,022
Rice	52,754	35,120	52,529
Rye	40.601	25,837	17,998
Buckwheat	6,350	6,370	3,193

[1]Add 000 to each number.
[2]Preliminary estimates.
[3]No. of 100 lb. bags; add 000.

Yellow corn is a good feed for turkeys of any age, is a fine fattening feed, and contains the vitamin A factor in quite large amount. Because of these and other characteristics it is a very desirable component of all turkey feeds. It tends to produce yellow fat and yellow skin in turkeys, but this effect is not noticeable until it comprises about 40% or more of the total diet. Disadvantages are that its protein is of rather poor quality and its effect on feather growth is not especially favorable.

Whenever the price is reasonable, yellow corn should constitute a large proportion of the diet. The percentage of corn is usually held to about 30% or less in starting diets since the use of higher percentages must necessarily result in the inclusion of smaller amounts of other desirable ingredients. In other diets corn may be used to the maximum extent of about 65% of total feed intake if the diet is properly balanced. However, because of the fact that a variety of 3 or more grains or their byproducts is generally more satisfactory than any one grain, the percentage usually varies from 30 to 50%, especially in diets fed to confined birds.

New corn properly dried so as to contain not more than 17½% of moisture (which is the maximum allowed for U. S. grade 3) is best for turkey feeding. Immature or poorly dried corn is considered to be undesirable as a poultry feed. Old corn more than a year old, if well preserved, is a good feed but is likely to contain less vitamin A than new corn since this vitamin is lost through aging and oxidation. Corn grading No. 1 or No. 2 is safest for turkey feeding, since its maximum moisture limits are 14 to 15.5%. If it is to be ground and stored for several weeks, No. 1 corn is preferred because it is least likely to heat. No. 3 corn usually is satisfactory where there is no danger of heating. The South Dakota Station found that low-test corn (50 lbs. per bu.) is a satisfactory turkey feed.

Whole corn is not eaten readily by turkeys under 18 to 20 weeks of age but can be handled by them when 12 to 16 weeks of age. When fed in the grain ration to birds younger than about 16 weeks, corn should be cracked and freed of meal. Cleaned cracked corn is good to use alone as scratch feed with dry mash containing up to 24% protein. Up to about 6 weeks of age, chick-size cracked corn should be used; from 6 to about 16, or preferably to about 18 weeks, ordinary cracked corn is best; after that, whole corn can be fed. The change should be made gradually, however, putting the whole grain in with the cracked. If the change is not made gradually, grain consumption will be reduced considerably over a period of a week or two.

Hybrid corn appears to be the equal nutritionally to open-pollinated corn (Wisconsin Station, 1947).

Corn chop, ground corn, or cracked corn is the entire product made by cutting or chopping the corn grain. It may be fine, medium or coarse and unscreened. It should contain not over 4 percent of foreign material if it is sold on the open market. If freshly prepared, this product is the equal, nutritionally, of the corn from which it was made. However, turkeys will not eat it eagerly if it is fed as a portion of the grain ration; its place being in the mash. It, and other corn products are likely to lose their vitamin A potency fairly rapidly if allowed to stand.

Screened corn chop, screened ground corn, or screened cracked corn is the coarse portion of the chopped, ground, or cracked corn from which most of the fine particles have been removed. It must not contain over 4 percent of foreign material if it is to be sold on the open market. These products often are sold as fine, medium, or coarse "scratch grain" and are readily eaten by turkeys since the meal has been removed.

Corn meal is finely ground unbolted corn. As sold commercially it may have the germ and bran removed and so may be lower than whole corn in vitamins A, B, E, and G. The term corn meal is popularly used to indicate ordinary ground or chopped corn.

Argentine corn is of the flint type. Most of it is yellow and is equal nutritionally to domestic corn.

White corn, except for its lack of the vitamin A factor, is as good a feed as yellow corn. When vitamin A is supplied in abundance by fresh or dried green feeds or fish-liver oil or both, the white corn can be used if it is cheaper. A special use for white corn is in fattening when a white finish is desired and when vitamin A is supplied by green feed, fresh or dried.

Variegated corn of the red, purple or other varieties frequently has a yellow endosperm (starchy portion) in which case it is equal in vitamin A content to yellow corn. The deeper the shade of yellow, the greater the vitamin A content. The feeding value of corn, except its vitamin

A content, is not related to the color of the grain. If the endosperm is white, vitamin A will be lacking, regardless of other surface colors.

Hominy feed (hominy meal or hominy chop) is a kiln-dried mixture of the mill-run bran coating, the mill-run germ (with or without a partial extraction of the oil), and a part of the starchy portion of the corn grain. It usually is made from yellow corn, but may be made from white corn. In general it is a dependable feed and if fresh appears to be equal or superior in poultry nutrition to the corn with which it was made. However, if dried at too high a temperature, reduction in quantity of vitamins A and B is likely to take place. If the germ has been removed, the quantity of certain vitamins will be lowered. The presence of the corn germ is an important point to consider because the germs of the grains and grain products are among the chief sources of vitamins B and E in poultry diets. Freedom from harmful chemicals should be demanded by purchasers.

Wheat is a valuable feed for turkeys but its byproducts are more generally used. It is a fattening feed and according to work at the Wyoming Station wheat-fattened birds compare favorably with those fattened on oats, rye, and barley in palatability. Wheat produces an attractive, hard white finish in turkeys. However if it is fed heavily to turkeys near market age which are allowed access to fresh green feed, there is a tendency for the skin color to become yellow. Wheat is digested as well by chickens as the other common grains, and this appears to be true also for turkeys. As far as energy value is concerned ground wheat is worth more per pound than its byproducts, bran and shorts, and except in starting the breeder mashes, may be used in place of middlings when it does not cost much more. The Nebraska Station recommends substituting whole wheat for bran and shorts in all but starting mashes. Wheat and its byproducts should not constitute more than from 55 to 60% of the total diet. Wheat ranks next to oats in palatability with turkeys. It is a good grain to use alone with growing mashes, but not with concentrates, as too much of it would be eaten.

New wheat is considered a dangerous feed by some, but available evidence indicates that new wheat if ripe and dry (containing not over 14% moisture) when fed is not harmful and is equal or superior in feeding value to older grain. However, there is a limit beyond which wheat should not be fed, probably around 60%. Noticeable ill-effects have been observed in adult turkeys summered on wheat as the only feed supplementing green range. New-wheat damage may be a wheat-variety problem. Marquis, Apex, and Reward varieties gave no bad results with chickens at the Saskatchewan Station. From the cost angle No. 3 wheat usually is best for poultry, but grades 4 and 5 usually are satisfactory, as is lightweight (46-lb. per bu.) wheat. Shrunken wheat testing 40 pounds per bushel was reported satisfactory as a poultry feed by the North Dakota Station. Grain of higher grade than 3 usually is too expensive to use as feed.

There are various types of wheat ranging from soft to very hard. Soft wheat has only about 9% protein while the hard western-grown wheats may have 14%. All types are rated as good feed for poultry.

Wheat screenings are an unreliable feed due to differences in composition. When composed principally of shrunken and broken kernels and free from injurious seeds they are equal to wheat in feeding value. They should never be used in starting feeds and should be used with caution in growing, laying, wintering, and finishing diets. They are commonly sold ground, in bran or millrun. The wheat millfeeds are bran, shorts, middlings, red dog, low-grade flour, and millrun. The U. S. annual production of these products from October 1949 through September 1950 was 4,573,100 tons.

Wheat bran is the coarse outer husk of the wheat grain and constitutes about 11% of the wheat grain. Although possessing rather low digestibility it is a valuable feed. It has slight laxative properties, adds bulk and improves the texture of mashes, is a good source of one of the anti-gizzard-erosion factors and of vitamins B and E, and contains some A and riboflavin. All-mash starting diets may contain 5% to 10% of bran; growing, wintering, and laying mashes when fed with scratch grain may contain about 10%; all-mash laying, growing, and wintering diets may carry about 5% even when ground wheat also is included.

Wheat middlings or shorts (called pollard in Australia and England) is that portion of the wheat grain between the bran and the flour. About 11 per cent of the wheat kernel is recovered as middlings or shorts during the milling process. Most of the germ goes into the shorts. There is considerable confusion in the classification and labeling of middlings. However, according to the U. S. feed control officials, there are 4 types of middlings, which are: (1) Wheat standard middlings consisting of fine particles of wheat bran, wheat germ, wheat flour and some of the offal from the "tail of the mill." This product shall be obtained in the usual process of commercial milling and shall contain not more than 9.5 per cent of crude fibre. (Adopted 1949); (2) Wheat brown shorts consisting of fine particles of wheat bran, wheat germ, and wheat flour and some of the offal from the "tail of the mill." This product shall be obtained in the usual process of commercial milling and shall contain not more than 7.5 per cent of crude fibre. (Adopted 1949); (3) Wheat gray shorts, wheat gray middlings, or wheat flour middlings consisting of fine particles of wheat bran, wheat germ, wheat flour, and the offal from the "tail of the mill." The product shall be obtained in the usual process of commercial milling and shall contain not more than 6 per cent of crude fibre. (Adopted 1949); (4) Wheat red dog, wheat white shorts, or wheat white middlings consists of the offal from the "tail of the mill" together with some fine particles of wheat bran, wheat germ, and wheat flour. This product shall be obtained in the usual process of commercial milling and shall contain not more than 4.0 per cent of crude fibre. (Adopted 1949). Some mills dispose of their screenings by adding them to the bran or middlings. When this is done the resulting product should be so labeled. Bran or middlings with screenings are not so desirable as the pure products, but in general give good satisfaction.

All middlings are excellent feeds, but like bran, should be used in moderation, probably not to exceed 25% of the mash in either mash-scratch, or all-mash diets. White middlings have a slightly higher percentage of total digestible nutrients than the other types and are especially valuable in finishing diets. In starting, wintering, and laying diets the other kinds, especially the standard and brown middlings, are better because they are less sticky when wet, and the mineral and vitamin values are higher. Degermed middlings are somewhat lower in vitamin B and considerably lower in E and G than those containing the full complement of germ. Middlings, especially standard and brown, are high in manganese, which is of value in preventing slipped tendon or perosis in chicks and turkeys. This property enhances the value of middlings as an ingredient of chick and turkey starting mashes. The Texas Station recommends 20% gray shorts in the diet of battery-brooded chicks to prevent slipped tendon. Manganese may be added, however, in the form of manganous sulphate ($MnSO_4$) to prevent slipped tendons.

Wheat red dog flour (or wheat red dog) is a byproduct consisting mainly of the aleurone layer with some bran, analyzing not more than 4% of crude fiber, and made from hard spring wheat. It is analogous to the white shorts obtained from winter wheat and is used for poultry only in fattening feeds. Its composition is similar to that of gray, flour, and white middlings and consequently it is worth no more than these products. Due to its sticky nature when wet it is used in comparatively small quantities. Wheat yields about 5% red dog flour during milling.

Low grade wheat flour is made from hard spring wheat

and contains not more than 1½% crude fiber. It is little used in poultry diets and only for fattening.

Wheat mixed feed (wheat mill feed, mill run, or shipstuff) consists of the coarse outer covering of the wheat kernel, fine particles of wheat bran, wheat germ, and wheat flour, and the offal from the "tail of the mill." This product shall be obtained in the usual process of commercial milling and shall contain not more than 9.5 per cent of crude fibre. (Adopted 1949). It can be used in place of both the bran and middlings in formulas calling for these products. It appears desirable to limit its use to a maximum of 20% of all-mash diets and 30% of the mash when scratch also is fed.

Oats are a valuable feed and should be included in all turkey feeds when the cost compares favorably with that of other common grains. The hulls appear to possess factors that increase growth, prevent gizzard erosion, and improve the feathering, according to the Iowa Station. Oats should be fed either whole or pulverized (in a hammer mill with 1/16 to 1/20th inch screen) to reduce the hulls to small particles that cannot harm the bird.

Crushed oats or coarsely ground oats should not be fed unless moistened. Coarsely ground oats were reported unsatisfactory in poult starting diets by the Texas Station (1947), while pulverized oats at the 20% level were very satisfactory. In turkey diets, oats are believed to favor feather growth and reduce feather picking. Good quality oats are fairly high in minerals, including manganese, and are highly regarded as a component of the diet of breeding stock. Oats are the best liked of all grains by turkeys and when free choice of oats and grain is allowed, turkeys usually eat too many oats, hence oats are not recommended as a *single* grain to be fed free-choice with growing mashes or concentrates. Combinations of oats with other grain, mixed or free choice are preferred. However, the South Dakota Station (1949) found that the weight of Broad Breasted Bronze turkeys at 28 weeks was not affected when oats at levels of 45, 50, 55 and 67.5%

were fed to turkeys on range starting with the 16th week through the remainder of the feeding period.

At the Michigan Station (1949) the mash-grain ratio during the period from 9 to 28 weeks was changed from 70:30 with corn alone as the grain to 47:53 when oats were added, using a 27% protein growing mash. Everything considered, oats are recommended as a desirable component of most turkey diets. As a general rule, from 5 to 15% of starting diets, from 10 to 50% of laying, wintering, growing and fattening diets, and 20 to 50% of the diet of confinement-reared turkeys may well consist of this grain. More would be tolerated but the high fiber content makes it inadvisable. However, entirely satisfactory growing and wintering diets can be formulated without including them. Standard No. 1 oats (32 to 34 lbs. per bushel) and heavy oats (35 lbs. or more per bu.) are more desirable than light oats but both are satisfactory. When light (No. 2 or 3) oats which weigh 27 to 31 lbs. per bu. are used, the lower percentages are suggested. Very light oats, less than 27 lbs. per bu., probably are not desirable feed for turkeys, except in small quantities. Hulled or germinated oats rarely justify the additional cost. Most commercial whole oats except very heavy ones (40 or more lbs. per bu.) are clipped. Clipping probably is of no value unless the oats are barbed or of very light weight. Oats can be harvested by turkeys when green but not when ripe without some impaction.

Oatmeal (rolled oats) and oat groats are the oat kernels divested of their hulls and fibrous jackets. They have high feeding value but since the oat hulls which have been removed have value, the extra cost of these products is seldom justified in turkey feeds. Indeed, in starting feeds these products are actually not so valuable as pulverized oats which contain the hulls. However, oatmeal may be scattered on the mash and on the water the first few days to aid in teaching the poults to eat and drink.

Oat mill feed (or oat feed) is a byproduct of the manufacture of oatmeal. It consists of the hulls, the loose fibrous covering of the oat kernel (the shorts), and particles from the kernel itself (the middlings). As usually sold, oat mill feed contains about 5½% protein, about 28½% crude fiber, and about 50% nitrogen-free extract. The highest grade has about twice as much protein and only two-thirds the fiber. Since it is a good stock feed it probably has value also in turkey diets. If used, the comparatively low nutritive value and high fiber content should be given due consideration.

Barley is an excellent turkey feed except in starting diets where its presence in excess of 10% is not generally recommended. It is not well liked by turkeys at first but they soon learn to like it, especially when it is given as the only grain. Its effect on feather growth is considered good. In growing, wintering, or laying diets it doubtless can be substituted for all or part of the yellow corn except in cases where the vitamin A in the yellow corn is needed. Where vitamin A is supplied by means of green feed or special carriers, barley is practically equal to yellow corn except in starting feeds. In growing and finishing diets, corn would be preferred to barley if its cost, based on comparative feeding values, does not significantly exceed that of barley. Where it does so exceed it, barley is preferred.

Consideration should be given to the fact that 40 per cent or more of yellow corn in the whole diet produces a yellowish skin, which may not be desired in certain markets. Barley and wheat tend to produce a white finish and flesh of good table quality. Together they should not constitute more than about 70% of the total diet of turkeys. If oats and barley are both fed, the two together should not constitute more than 60 to 65% of the diet. Barley, like oats, is best fed *either whole or pulverized,* as the broken hulls of coarsely ground or rolled barley or oats unless soaked may cause respiratory trouble. Ripe barley of the awned varieties cannot be harvested successfully by turkeys. Barley stubble fields likewise are dangerous.

Barley is a good grain to use alone as scratch feed. It can be fed from about six weeks up to market age and beyond. There is no tendency to over-eat of the grain diet when barley alone is used either with a growing mash or concentrate; in fact, with all but concentrates having over 28% protein, barley is the ideal single grain. Mixed feed barley (ground) should contain not less than 75% of barley and not more than 25% of other grains, seeds, etc. Barley grading No. 1, 2, or 3 is desirable for turkey feeding. Low-test barley (36 to 44 lbs. per bu.) is a satisfactory feed.

The Idaho Station (1952) found that barley, when fed as the only grain, after 8 weeks of age, was consumed at a satisfactory rate and reduced the final weights of the turkeys and the feed efficiency only to an insignificant degree. Broad Breasted Bronze and Beltsville Whites were used in the test, the former being marked at 28 weeks and the latter at 24 weeks. The finish was not affected to any noticeable degree. It was estimated that barley could be used when the cost is no greater than 80 percent of wheat.

The sorghums, such as milo, kafir, feterita, kaoliang, durra (Egyptian corn), shallu, hegari, and darso, all are considered good feeds for livestock and poultry. They may be harvested successfully by turkeys. Yellow milo ranks high among the grain sorghums for poultry feeding. Ground kafir and milo were found by the Kansas Station to be equal to corn and wheat when fed to chickens. The California Station found that the grain sorghums are excellent feed for turkeys and should be used if they can be obtained at reasonable prices. They can be fed whole in the scratch feed or ground in the mash. As with other grains it appears advisable to limit their use to about 60% or less of growing, wintering, or laying diets and to 15% or less of starting diets. Sorghums grading No. 1, 2, or 3 are best for poultry feeding. Shrunken sorghum seeds, especially hegari, when produced under drought conditions, are dangerous because they sometimes contain a poison, prussic acid. Mouldy sorghum is dangerous. Sorghums rate with cracked corn in palatability. The Kansas Station (1950) was able to reduce materially the cost of production by feeding grain sorghum in place of corn.

Sorghum gluten feed and sorghum gluten meal are new products obtained from the processing of grain sorghums for industrial uses. The former contains about 25 percent protein, 3.4 percent fat, 6.3% fiber, 48.4% nitrogen-free extract, 9.2% water, and 7.7% ash. The sorghum gluten meal contains 41%, 4.1%, 2.8%, 40.3%, 10.5%, and 0.72%, respectively. The amino acid balance of these feeds is poor, which limits their use in poultry diets. The Texas Station (1951) successfully substituted 10% sorghum gluten meal for 10% of the soybean oil meal in a 60% soybean oil meal—35% corn experimental diet for starting poults. Sorghum gluten feed is primarily a feed for ruminant animals but, as with corn gluten feed, could be worked into turkey diets as a partial grain substitute if relative costs favored its use. Sorghum gluten meal has about 1800 micrograms of riboflavin and 22,000 of nicotinic acid per pound while the sorghum gluten feed has about 5,000 micrograms of riboflavin and 45,000 of nicotinic acid. Both products are low in minerals.

Rye is not popular in poultry diets, but the Wyoming Station reported favorable results with it (as high as 45% of the diet) when fed to turkeys 8 to 26 weeks of age. Rye produced as economical gains and as palatable flesh as did wheat, oats, or barley. The Wisconsin Station found that ground rye could be used successfully in laying and growing mashes for chickens to the extent of about 30%, provided a like quantity of wheat, or wheat byproducts, or oats was also included. Rye is not recommended in starting diets or in breeder diets on the basis of the limited experimental data available.

Rye middlings are the combined byproducts of the milling of rye flour, these byproducts seldom being marketed separately as in the case of wheat. The term *rye feed* is practically synonymous with rye middlings although the former is not expected to contain any of the rye red dog. The feeding value for turkeys has not been reported. From its composition, it appears that it could be used in turkey mashes, to a limited extent, at least, probably not over 5 per cent in starting mashes and not over 10 per cent in mashes of other types. Like other less well-known feedstuffs the use of rye or its byproducts in turkey feeds is not recommended unless the cost is very low in comparison with standard products.

Emmer and spelt (used interchangeably in the U. S.) are two cereal grains closely related to each other and to wheat, but resembling oats in composition and appearance. They are not widely grown in the United States, but are common in Europe, where they are used as feed for animals and food for humans. Their use in turkey growing, fattening, and wintering diets in liberal quantities has been successfully demonstrated. Turkeys eat them readily, especially when fed whole in the grain portion of the diet. They may also be pulverized and used to replace up to half the corn or all the oats, wheat, or barley in mashes except starters.

Buckwheat is favorably regarded as a chicken feed if used in comparatively small quantities, 10% or less in the entire diet. In chickens it favors the production of white flesh of fine quality and may be fed whole or ground. Buckwheat middlings may be used in small quantities in chicken mashes. Although its use by turkeys has not been reported experimentally, small percentages in all except starting diets probably would be satisfactory. It is regarded as a good finishing feed by many growers and can be harvested by the turkeys when ripe.

Millet. Hog millet or proso is the only millet grown to any considerable extent as a grain crop in the United States. It resembles grain sorghum, but is not classed as such. It is rated 90% as valuable as corn when fed to poultry. It is considered somewhat more digestible than barley. The South Dakota Station found red proso a satisfactory turkey feed. The North Dakota Station found that ground proso (Early Fortune, a yellow variety) could satisfactorily replace all the yellow corn in a turkey starting mash similar to No. 13 and in a growing mash similar to No. 22 with which a scratch grain mixture consisting of 50% whole millet and 50% whole wheat was fed along with the mash, both free choice. The keeping qualities of millet are excellent and it is readily eaten. It is reported to favor the development of yellow skin color.

Miscellaneous grains and seeds such as alfalfa seed screenings, sunflower seed, and sweet sorghums probably can be used to a limited extent in growing, wintering, fattening, and laying diets but experimental evidence as to their value for turkeys is almost completely lacking. Consequently, their use is not recommended except for special purposes in limited quantity and then only when they are priced considerably under the well-known grains and byproducts. Chewings' fescue (grass) seed has given good results at a 30% level in chick starter.

Rice as it comes from the thresher includes the tough woody hull. It is sometimes called rough or paddy rice. Growing turkeys can utilize it as a portion of the grain diet. Hulled or brown rice is the rice grain with the hull and most of the bran layers removed. Cracked or whole brown rice may be used in grain mixtures for poults of all ages to the extent of about 25%. Cracked and ground rice have a tendency to become rancid, hence must be used with care. Rice, as far as known, has no growth-retarding properties when fed to poultry and is favorably regarded as a poultry feed when used in moderation. However, turkeys fed heavily on rice are said to possess a white, shiny, somewhat patchy appearance.

Rice bran can be used to the extent of 10 to 15% of the starting, growing, finishing, and wintering mashes to replace part of the wheat bran and middlings. It has been found to be very valuable in the diet at a level of 10 to 15% for the prevention of perosis in chickens and undoubtedly has a similar value in turkey diets. It is composed of the germ and most of the bran layers of the rice grain and should contain very little of the husk. Rancidity should be guarded against in the purchase and use of this byproduct.

Rice polish is a fine-grained byproduct analogous to wheat middlings and is sometimes called rice middlings. It is a good source of vitamin B_1, but is not generally available for poultry feeding. Where available it undoubtedly could be used successfully in turkey diets. An alcoholic water extract of rice called tiki-tiki is sometimes used in the treatment of beri-beri in humans and to furnish vitamin B_1 to animals. The U. S. total annual production of rice millfeeds (bran and polish) was 213,700 tons from October 1949 to September 1950.

Brewers' and distillers' grains. These products are fairly high in protein and often are reasonable in price. Little experimental evidence can be found relating to their use (either wet or dry) as turkey feeds, although some growers feed them successfully. In all probability the dried grains, sometimes called light distillers' grains, can be used in growing, wintering, or laying diets in quantities not to exceed 5% of the total diet. The wet grains and slop are perishable and hence available only in the vicinity of distilleries and breweries. Their use would be in the nature of an experiment and should be undertaken cautiously. These grains usually are labeled according to the grain that predominates. Distillers' grains usually contain corn, rye, and barley, while brewers' grain usually contains only barley. The Kentucky Station reported the successful use of thick distillery slop in separate troughs to replace as much as four-fifths of the corn in the mash of chicks. Thick slop was also mixed with the all-mash diet in the proportion of one of mash to two of slop. Thin slop was reported unfavorably.

When distillers' dried grains are mixed with dried solubles, often half and half, they are sometimes called dark grains and are good sources of B-G complex. They may be used in growing and starting mashes in quantities not exceeding 5% of total diet.

The annual production of dried brewers grains for the years 1945-49 averaged 225,000 tons; from October 1949 to September 1950 it was 233,000 tons. For distillers dried grains from October 1949 to September 1950 the production was 365,100 tons.

Peas. The Idaho Station suggests that ground peas may be used successfully in all turkey mashes. In the diets recommended by that station, the grain portion of the mashes consists of approximately equal quantities of ground yellow corn, ground peas, ground barley, bran, and ground oats. Peas, especially the green varieties, contain an appreciable quantity of vitamin A. Peas are low in minerals, hence diets containing them should be checked for calcium and phosphorus content. They are high in protein which is deficient in methionine. Additions of 0.25 to 0.75% of the latter in chicken diets resulted in good utilization of peas at the Idaho Station (1945). Cystine additions were of no value. Up to 12% of pea protein in the diet gave good results.

Beans like peas are a nourishing feed high in protein and low in minerals. They have been used successfully in chicken diets and in all probability can be worked into turkey diets when occasion demands. When this is done it is suggested that not over 8 to 10% of the entire diet consist of beans. Grinding them would be likely to increase the digestibility, although turkeys probably could make good use of whole beans in finishing and wintering diets. Cooking ordinary beans, as with soybeans, increase the digestibility but is impracticable under most conditions.

At the California Station pigeon peas and cull lima beans were well utilized by chicks up to 10 per cent of the diet. Cooking did not improve them.

Mung beans. (Once called chickasaw peas). Mung beans were found suitable as turkey feeds when cracked to replace up to 2/3 of vegetable proteins in turkey starting diets by the Oklahoma Station (1945). The mung bean is a good-sized, strong growing annual legume from Asia. It is grown to a limited extent in the Southwest. The beans do not have to be cooked before feeding. The plants make good poultry pasture.

Miscellaneous Carbohydrate Feeds

Cane molasses or blackstrap in the dehydrated form is recommended for use in turkey rations by the University of California to the extent of 1% of starting, growing and laying mashes to supply pantothenic acid. The Iowa Station found liquid molasses satisfactory in a milk-fattening diet for chickens, using it to moisten wet mash to the extent of 10% of the dry mixture before milk was added. The flavor of the chickens so fattened was reported as excellent. Ohio State University reported that liquid molasses could be used to replace cereal grains pound for pound up to 10% of the diet of chickens. The use of a power mixer was recommended. When little or no milk was fed in the diet, molasses was found to increase palatability, to increase water consumption, to serve as a mild laxative, and to keep the birds in a more healthy condition. The Louisiana Station reported favorably on the use of liquid molasses in chicken diets. On the other hand, the Dominion Experimental Station (Canada) found that 5% of liquid molasses gave poor results in chicken fattening diets. The Nebraska Station reported unfavorable results from the use of liquid blackstrap molasses in turkey diets and does not recommend it. The Pennsylvania Station obtained good results from 2, 4, and 6% liquid molasses in chick starting diets. Growth was excellent, but there was a slight lowering of feed efficiency in the 4 and 6% groups. It is useful as a binder in pelleted mashes, 2% sometimes being used.

Cocoanut oil meal or copra meal is too low in protein to be considered a protein supplement. It may be used to replace grains, being added to the mash (up to 10%).

The average amount of copra meal and cake available for animal feeding during the period 1945-49 was 154,000 tons and for the year October 1949 to September 1950 it was 135,900 tons.

Babassu oil meal was found by the California Station to be satisfactory in chick diets up to a 10% level. It is made from palm seed and contains only about 23% crude protein, 6% fat, and 12% fiber, hence must be used to replace the grains rather than as a protein supplement. It appears likely that it could be used satisfactorily in turkey growing diets and possibly in other turkey feeds.

Hempseed oil meal was found by the California Station to be acceptable in chick diets up to a level of 20% to replace grains. It has about 31% crude protein, 7% fat, and 23% fiber. No experiments with turkeys can be cited but it is not improbable that it could be used in turkey diets in limited amount—not over 5% of total feed.

Chocolate meal has little feeding value for poultry.

Dried fruits. There are possibilities in the use of surplus dried fruits in turkey feeding. The California Station mixed 2 parts barley and 1 part dried prunes and fed the mixture successfully to turkeys during the finishing period. The gains were better than those made on a control diet of the usual type.

Citrus meal. Citrus meals consist of processed peels and other residues of citrus fruits. The Florida Station reported "In general citrus meal (dried grapefruit and dried orange refuse) was not a satisfactory component of chick growing rations. However, up to 15 per cent could be fed to laying pullets without ill effect. Pullets 8 to 20 weeks utilized citrus meal fairly well." Orange peel and pulp meal was found by the California Station to be without value in chick diets. It has about 7% protein and 9% crude fiber.

Beet molasses is similar in composition to cane molasses, but is almost devoid of pantothenic acid and riboflavin. No reports can be found regarding its value in poultry diets. It is quite laxative.

Silage has been reported favorably as a turkey feed, and found valuable as a succulent feed for chickens by the Kansas Station.

Raisins. The California Station (1951) reported that when raisins were fed as 10% of the grain to turkeys from 18 to 30 weeks of age, 4.7% of the total feed consumed consisted of the raisins. When fed as 30% of the grain, the raisins constituted 14.3% of the feed consumed; and when fed free-choice, raisins made up 16% of the total diet. There were no differences in size or quality of the turkeys fed the varying quantities of raisins. Raisins contain 3 to 4 percent of crude protein and about 80 percent carbohydrates. Scattering the raisins on the feed at first, helps to attain good consumption. Cull raisins sometimes are available at a price low enough to justify their use in the turkey diet.

Potatoes are a possible source of carbohydrates. The California Station (1950) reported that sun-dried raw potatoes were unsatisfactory in turkey starting diets at 20 and 40% levels while oven-dried raw potatoes were unsatisfactory at 40%. Boiled sun-dried potatoes were satisfactory but expensive to prepare. The Oregon Station (1950) reported that cull potatoes fresh or cooked were not efficiently utilized by turkeys but that dehydrated cull potatoes had possibilities.

Apples. Unmarketable apples are used successfully in growing and wintering feeds for turkeys by some growers. They contain appreciable amounts of vitamins A, B, and riboflavin and consequently are of value when green feed is scarce. Turkeys eat them readily. Windfall apples as well as other fruit frequently contain the larvae (worms) of harmful insects which are destroyed when this low-grade fruit is eaten by turkeys.

FEEDS USED CHIEFLY AS SOURCES OF PROTEIN

All of the protein supplements should be purchased with a knowledge of their guaranteed protein content and their adaptability to turkeys so that the value per pound of protein may be calculated.

Animal Products

Meat scraps or meal is a satisfactory poultry feed of generally uniform quality. It is made from packing-house meat and bone residue without appreciable quantities of manure, stomach contents, hoofs, horns, blood, or "stick." As produced by the standard method of manufacture and properly preserved, it is a reliable source of protein. The Washington Experiment Station reported poor results when Argentine meat scrap (which may not have been manufactured by a standard American method) was used in a ration fed to growing chicks. Other observers have obtained very poor results from the use of domestic meat scrap when not produced by the standard process or when allowed to deteriorate. Although instances of this kind are not numerous, it is good practice to check the quality and growth-promoting ability of the meat scrap used, especially if it is produced by packers not equipped to manufacture a product that is fully up to standard. Free-fat-acid content of meat scrap and other animal protein supplements is not necessarily a gauge of the nutritional worth of these products.

When balanced properly with vitamins, meat scrap can be used with excellent success as the sole source of supplementary protein in growing, wintering, and finishing diets. It has shown a high supplementary relationship with corn and wheat middlings, altho not quite so high as soybean meal according to the West Virginia station (1945). In laying and starting diets it can be used with only fair success as the only source of supplementary protein. According to recent tests, better results are obtained when the meat scrap protein is supplemented with a good source of vitamin B_{12} such as milk or fish meal. So far as known, meat scrap produces no off-flavor in the flesh of turkeys and is very palatable to turkeys of all ages. Meat scrap is deficient in lysine and tryptophan.

When meat scrap contains bone in sufficient quantity to raise the percentage of phosphorus above 4.37%, it is required to be labeled as *meat-and-bone scrap.* If any one kind of meat is used in its preparation, the product may be named accordingly. Meat scrap usually is manufactured with a crude protein content of about 50% and this scrap has sufficient bone to be called meat-and-bone scrap. Meat scrap containing 55% or more protein when available, often is more economical than the 50% protein variety. If proper appraisal of the calcium and phosphorus content of the diet is made, there is no reason why 55% meat scrap, or even that with a higher percentage of protein, should not replace the 50% kind so commonly used.

Meat scrap production was 605,053 tons in 1950 (including meat-and-bone meal). In 1953 there were 782,457 tons produced.

Digester tankage (sometimes called meat meal tankage, feeding tankage, or tankage) is a product similar to meat scrap, but may contain considerable quantities of blood and "stick." Like meat scrap, the word "bone" is added to identify products containing more than 4.37% phosphorus. It is used mainly as a hog feed but can be used for chickens after they are 3 months old. Its digestibility is considered to be equal to that of meat scrap. Experimental data on its use by turkeys are lacking, but in the field it has been used successfully in place of meat scrap. However, caution is advised in its use, especially in starting, finishing, and laying diets, as it may possibly affect adversely the hatchability of the eggs, or the early growth rate. No data are available on these points, but it appears that high grade tankage has possibilities as a protein supplement if and when needed.

In 1953 in the U. S., 229,896 tons of tankage were estimated to have been produced.

Liver meal is a valuable feed, mostly imported. It is a fine source of pantothenic acid and riboflavin and has other vitamins, including A. When available it is popular with feed mixers.

Dried pork liver is an excellent source of riboflavin and a fair source of vitamin A.

Blood meal (or dried blood) is a high-protein product well utilized when fed at low levels along with corn gluten meal.

Pork cracklings are a high protein feed. Their use in turkey growing, wintering, and laying diets to supply a portion of the supplementary protein has been reported favorably by the Nebraska Station.

Chicken scrap is a byproduct of the poultry processing industry. It contains about 60% protein. In experiments at the Nebraska Station it proved satisfactory in starting diets for chicks.

Fish meal is defined by American Feed Control officials as clean, dried, ground tissues or undecomposed whole fish or fish cuttings, either or both, with or without the extraction of part of the oil. If it contains more than 3 per cent of salt (NaCl) the amount of salt must constitute a part of the brand name, provided that in no case shall the salt content of this product exceed 7 per cent. It is quite generally used to supply one-fourth to two-thirds of the supplementary protein in turkey starting and breeder diets. It also can be used as the sole source in these diets. However, because it sometimes causes an off-flavor in the fat and flesh of turkeys it should not be used during the last twelve weeks prior to slaughter except in quantities limited to 5% of all-mash diets, 10% of 18- to 27-% protein growing mashes fed with grain, and 15% of 28 (or higher) % protein concentrates fed with grain.

As nearly as can be ascertained from the evidence at hand, all fish meals do not always cause off-flavor under all conditions of feeding but apparently there are enough of them on the market that do cause it to warrant full precautions against the production of off-flavor which is exceedingly damaging to the turkey industry. (See remarks on fishy flavor near the end of this chapter.) However, high grade fish meal is an excellent feed and should be used as indicated elsewhere in starting and breeder diets.

Fish meals vary according to species of fish, parts of the fish used, and the method of processing. For the preservation of quality and general acceptability as a feed ingredient, fish meals should contain about 8% moisture with a minimum of 6% and a maximum of 10. Fat content should be about 8% with a minimum of 3% and a maximum of 10. In particle size all of the meal should pass a No. 7 U. S. standard or No. 7 Tyler standard screen and 98% should pass a No. 9 Tyler or No. 10 U. S. screen. Overheating during processing impairs protein quality. The content of common salt should be limited to 3%, preferably 1% and must be so labeled if over 3%.

Sardine (pilchard) fish meal has the highest net protein utilization, (84.5%) of any fish meal. Tuna meal was found to be somewhat inferior to sardine meal, and shark meal fair but unpalatable to chickens; according to the California Station. At the Beltsville Station, turkeys grew equally well on white, Alaska herring, and menhaden meals, and slightly better on sardine meals. White fish meal is made from trimmings (excluding entrails, except the liver) from non-oil fish such as cod, hake, haddock, and cusk. There appears to be little difference in the nutritive values of the various kinds of fish meals if they are of good quality and not allowed to deteriorate. Fish meals containing a large percentage of heads are of less feeding value. Good-quality sardine and tuna meals, in experiments so far reported, produce but little off-flavor in the flesh of turkeys, while the other common kinds, according to the evidence available produce considerable off-flavor. Nearly all fish meals are very palatable to turkeys.

The principal types of fish meal produced in the United States are menhaden and white fish meals on the Atlantic coast, and pilchard (sardine), tuna and mackeral on the

Pacific Coast. Table 4.2 gives the production of fish and shellfish meals for animal feeding in the United States. (U. S. Interior Dept., Div. of Coml. Fisheries: Annual Summary.)

Table 4.2—Production of Fish Meals for Feeding

Type	Where Produced	Tons Produced				
		1936	1940	1941	1944	1950
Pilchard (sardine)	Pacific coast[1]	121,739	69,975	97,979	86,196	43,009
Menhaden	Atlantic coast	42,411	56,249	75,316	69,170	103,365
Herring, Pacific	Pacific coast	14,193	7,694	11,764	8,122	12,313
Groundfish ("white fish")[3]	Atlantic coast	14,188	9,250	8,803	17,824	26,953
Tuna and mackerel	Pacific coast	11,847	14,955	8,693	9,648	25,377
Herring, Atlantic	Atlantic coast	2,587	1,522	756	6,090	5,484
Shrimp	Both coasts	2,133	1,716	1,798	950	1,394
Blue crab	Atlantic coast	2,015	2,554	....	1,600	8,631
Salmon	Pacific coast	1,657	969	1,892	625	1,401
Whale meat	Pacific coast	789	....	....	14	....
Miscellaneous[2]	Both coasts	....	....	14,850	9,986	11,786
Totals produced including miscellaneous kinds not listed above		220,312	176,141	221,851	210,225	239,713

[1]Includes Alaska.
[2]Includes miscellaneous fishmeals and fish scraps not listed separately.
[3]Includes ocean perch (rosefish).

Importations of fish meal for animal feeding during the years 1931-40 amounted to 20.3% of the total supplies; during 1941-47 it was 5.8%; and during 1948-50 it was 18.4%. In 1951, 40.3% (128,478 tons) of the total supply were imported. In 1952, 170,679 tons were imported. The domestic production was 189,998 tons in 1951 and 194,664 in 1952. Exports of fish meal by the United States never have been of sizeable proportions since 1935.

Since fish meals vary greatly in their chemical composition, the users of them should obtain all possible information concerning the type meal that is being used. If the species of fish can be ascertained, the data in the tables can be used to obtain the approximate chemical analysis and calcium and phosphorus content. If the type or species is not known, the guaranteed percentages of protein, fat, moisture, and ash can be used as a basis for calculation. The calcium and phosphorus content can be estimated from the guaranteed ash content by the use of conversion figures. The ash of fish meals, on the average contains 31.5% calcium and 18.0% phosphorus. The use of fish meals carrying no guarantee is not advised.

Starfish meal successfully replaced meat and fish meals in chicken diets at the Massachusetts and Rhode Island Stations.

Dogfish meal, carefully prepared, gave good results at the Washington Station with chickens, but its value as a turkey feed is unknown.

Crab meal gave good results when fed in an all-mash laying diet to chickens at the Beltsville Agricultural Research center. Egg production and hatchability compared favorably to that resulting from the feeding of diets containing the better known supplements. The Massachusetts Station also found crab meal to be a valuable protein supplement in chicken diets, especially breeder diets.

Crab meal is defined by American Feed Control Officials as the undecomposed ground dried waste of the crab and contains the shell, viscera and part or all of the flesh. It shall contain not less than 25 per cent of protein. If it contains more than 3 per cent of salt (NaCl)the amount of salt must constitute a part of the brand name, provided that in no case shall the salt content of this product exceed 7 per cent. Crab meal is high in iron, containing 1063 parts per million (p.p.m.).

Charles Wampler of Harrisonburg, Virginia, reports high quality crab meal to be a valuable ingredient in turkey feeds. Since some crab meals are unfit for animal feeding, care should be taken in the selection of them; and since they are a fish (shellfish) product, there is a possibility of producing off-flavor.

Shrimp meal. This product is often high in common salt and some other minerals (see analysis), hence the analysis and guarantee should be studied so as to make due allowance for the presence of these minerals. Sun-dried shrimp meal is not so good a product as steam-dried meal. According to the Louisiana Station, sun-dried meal is not desirable in chicken diets, but the steam-dried product may replace one-half to all of the meat scrap usually fed, which may run up to 18% of the mash to be fed along with grains. However, as the sole source of supplementary protein in chick diets it has given only fair results. Although no experimental evidence can be cited, it appears probable that good quality steam-dried shrimp meal could be used satisfactorily in turkey laying, wintering, and starting diets to supply a portion of the supplementary animal protein usually included. Its effect on the flavor of the flesh is not known, but it should be considered in the same category as fish meal in this respect.

Shrimp meal is defined by American Feed Control Officials as the undecomposed ground dried waste of shrimp and contains the head, hull or whole shrimp, either or in mixture. If it contains more than 3 per cent of salt (NaCl) the amount of salt must constitute a part of the brand name, provided that in no case shall the salt content of this product exceed 7 per cent.

Insects are rich in protein and minerals. The Kentucky Station reported that dried grasshoppers contained 75.28% of crude protein, 7.21% of fat and 5.61% of minerals. Probably one-fifth of the protein is indigestible so for all practical purposes grasshoppers (dried) contain about 60% available crude protein. Some disadvantages of insects and other small forms of animal life as an important component of the diet are: (1) they are intermediate hosts of tapeworms and other parasites of turkeys; (2) the supply and quality are not uniform, and (3) they may possess injurious physical or chemical properties. However, turkeys are well qualified to catch and digest insects and they naturally feed freely on them.

Turkeys have been used successfully to control Mormon crickets and grasshoppers in some sections of the West with no apparent injury to the birds. The turkeys are brooded for 7 to 10 weeks in the usual manner, then herded, or in some cases trucked long distances, to infected areas. Usually a special low-protein mash or regular growing mash and scratch grain are fed along with the insects and other feed obtained from the range. Although a heavy consumption of insects results in a saving in feed costs, whether or not the practice is economically sound has not been demonstrated experimentally. When insect control is actually effected, its worth should be considered when evaluating the project. An average consumption of 1000 immature grasshoppers per turkey per day, and later in the season, or 500 to 800 adult grasshoppers per bird per day has been reported by a county agricultural agent in Oregon.

When turkeys are raised on grasshopper range, the California Station recommends that the birds be fed all the grain they will eat along with some alfalfa meal or alfalfa hay. The alfalfa is included only when green feed is scarce. The grain may be ground and mixed with alfalfa and minerals according to the following formula:

MASH TO FEED ON GRASSHOPPER RANGE

Ground grains	80
Alfalfa meal	12
Steamed bone meal	4
Ground limestone or oyster shell	3
Salt	1
	100

There is but little experimental evidence relating to the use of insects and other small forms of life by turkeys. The Oklahoma Station in a small experiment, obtained good results from feeding dried grasshoppers to growing turkeys. Satisfactory results were obtained when 10% of the mash consisted of dried ground grasshoppers and also when the dried whole grasshoppers were offered free choice along with mash and scratch feed. In a previous experiment the grasshoppers were improperly processed, hence gave poor results.

Considering the obvious disadvantages and uncertainty of supply, it appears that unless insect control is an important factor, it is best as a general rule not to depend on insects for a very large share of the feed or to make special arrangements for supplying them as feed. Turkeys ranging in an apple or peach orchard have proven their ability to materially reduce the injury from codling moth and curculio. In consuming the windfalls of the insect-damaged fruit they break the life cycle of the insect larvae therein. If turkeys are raised with access to range they will obtain animal and plant life in greater or lesser quantities. If the grain and mash are both fed ad libitum (see Fig. 4.1) the turkeys will be able to balance their diet and will not be forced to eat excessive amounts of natural feeds.

Fig. 4.1. Wire platforms for feeders as well as water founts help to break the disease cycle.

Milk. The proteins, vitamins, and minerals of milk are very valuable in the nutrition of turkeys, and milk products are in general very palatable. The use of dried milk is recommended in starting and breeder diets especially when access to growing green feed is not permitted. It is a valuable source of the essential riboflavin factor, which also is found in green feed. Milk is thought by many to have a bleaching effect on the skin and fat and to increase the palatability of the flesh, but the Wyoming Station was not able to substantiate these beliefs in the case of turkeys.

Liquid buttermilk and skimmilk are not wholly satisfactory feeds for turkeys. Their use makes sanitation difficult, increases labor costs, tends to cause pendulous crops, and does not promote maximum growth. Consequently, milk in liquid form (skimmilk and buttermilk) is not recommended for turkeys of any age unless it is readily available at low cost. It probably is not economical to use when its cost or its sale value per pound exceeds 5% of the cost of dried milk. In most cases where milk is needed in the diet of turkeys it appears to be best to use dried or possibly condensed skim- or buttermilk unless the liquid milk may be secured much cheaper per pound of nutrients contained in it. Whole milk has no merit for poultry feeding over skimmilk and the butterfat is of too much value to consider as a poultry feed.

Condensed buttermilk* and skimmilk are excellent feeds. Whether or not they are used should depend mainly upon the cost per pound of protein as compared to the cost per pound of protein of dried milk. There is little evidence to show that condensed milk possesses any special qualities over and above those of an equivalent amount of milk solids in the dried form. Consequently, if it costs more per pound of protein than does dried milk, the dried milk is to be preferred. Dried milk contains about 90% solids, condensed milk about 30% solids, and liquid milk about 10% solids. The method of preparation of milk products does not appear to influence greatly their feeding value including the vitamin content. An apparent exception is sweet-cream dried buttermilk, which contains more riboflavin than the other types of dried milk. The solids of dried milk would be obtained at the same cost if the price of dried skimmilk or buttermilk were three times that of condensed buttermilk or nine times the value of liquid milk. However, many turkey raisers prefer

---

*Frequently sold as "Semi-Solid Buttermilk," a trade-marked name.

one form of milk over the other even though the cost of milk solids is greater. A practical way to compare the various sources of milk solids would be on a cost basis to the feeder; for example, would a dollar invested in dried skimmilk, dried buttermilk, condensed buttermilk or liquid milk produce the greatest net return? In such a comparison many factors other than cost must be considered such as (1) spoilage, (2) regularity of supply, (3) labor of hauling, feeding and cleaning of utensils, (4) relative growth rate, (5) finish and flavor of carcass, and (6) effect on appetite, health, and sanitation practices.

Condensed milk or condensed whey fortified with vitamins is a new product for which much is claimed. As with other products, the guaranteed values for protein, minerals, and vitamins should be studied and the worth of the product calculated for the conditions under which it is to be used.

The feeding of condensed milk at the rate of 3 lbs. per 100 turkeys per day on top of grain for 3 or 4 weeks before marketing has been tried with favorable results at the Kansas Station.

Dried milk appears to be desirable in all-mash starting diets to the extent of 5 to 10% by weight and in laying diets to the extent of about 5% of the total feed intake. It has a very favorable effect on early growth of poults and on hatchability of turkey eggs. However, proper use of other good protein bearing feeds along with adequate vitamin supplementation should eliminate the necessity for using milk. However, if the cost of a milk product compares favorably, its protein and vitamin content considered, with that of other good protein supplements, it should be used. In growing and wintering mashes, when they are fed along with scratch grain and without access to growing green feed, about 10% of dried milk is desirable. When access to growing green feed is allowed, milk is not necessary. In fattening feeds milk is desirable but, again, is not essential. Turkeys grown and fattened without milk are just as desirable as those fed milk, provided that products (such as fish oils and some fish meals) that may produce an undesirable flavor are not fed. Turkeys raised on diets containing only meat scrap or soybean oil meal (or both) as sources of supplementary protein are in the experience of the authors, fully the equal of milk-fed birds in flavor and tenderness of meat.

From the nutritional standpoint, milk as the sole source of supplementary protein should be satisfactory. In feeding practice, however, its high cost in dried and condensed form and its lack of adaptability in liquid form militate against its use as the sole source of supplementary protein. Furthermore the comparatively low protein content of milk in all its forms prevents its use as the only protein feed in turkey starting mashes because too much of it must be consumed. For example, even a 20% protein starting feed would require about 35% of dried milk if only milk were to be used to supply protein in addition to that supplied by the usual alfalfa products, grains, and grain byproducts in the mash. In growing, wintering, fattening, and laying diets this objection would not be quite so important, the main point to be considered being the cost in comparison with other sources of protein. Growing turkeys fed as high as 30 to 35% of dried milk in the all-mash starting feed tend to have watery droppings and do not develop properly at the start, but they will grow exceptionally well after about 12 weeks when fed scratch grain and an 18% protein growing mash containing about 28% of dried milk as the only protein supplement. However, cost usually makes such a practice unprofitable.

Dried skimmilk and dried buttermilk appear to be of equal feeding value, although the Washington Station reported that the riboflavin content of dried sweet-cream buttermilk was higher than that of dried skimmilk. The Purdue Station found no significant differences between dried skimmilk and dried buttermilk in chick starting diets. The authors have used both for turkeys with equal success.

Lactic acid is found in buttermilk in fairly large quantities but not in skimmilk. Lactic acid has no demonstrable value for poultry except the questionable claim of palatability.

In 1952 the United States production of dried skimmilk for animal feeding was only 25,306 tons whereas in 1941 at was 55,000 tons. In 1952 the production of dried buttermilk was 47,067 tons, most of which was used in animal feeding.

Casein is the product resulting from acid or rennet precipitation of skimmed milk and must contain at least 80% of crude protein. It is practically pure protein combined with a little water and containing very small percentages of carbohydrate, fat and mineral matter. Its chief use is to supply protein in experimental diets for research work rather in practical field trials.

Whey is discussed under vitamin-bearing feeds.

Eggs are a good source of protein as well as of vitamins A, B, C, D, E, and riboflavin. Although their use is not encouraged, hard boiled* chicken or turkey eggs ground or crushed, may be fed to poults when they are started in the brooding quarters, at the rate of one or two turkey eggs (3 chicken eggs equal 2 turkey eggs) per hundred poults about three times during the day. The crushed material, scattered lightly about on top of the mash or on boards or papers, is attractive to the poults and aids in teaching them to eat. Excessive quantities of eggs should not be fed, neither should they be fed raw because of danger of transmitting pullorum or other diseases. There is no objection to continuing the feeding of hard boiled eggs beyond the first two or three days in the brooder, but it is doubtful if the labor involved is justified except at the start unless special conditions exist, such as the availability of large quantities of eggs which have no other value. In the latter case the protein

*At least 10 minutes to insure killing any organism such as the pullorum bacteria that might be carried therein.

in the mash could be cut down in proportion to the quantity of eggs consumed, or scratch grain could be fed ad libitum, thus enabling the birds to eat more of it, and permitting them to adjust the protein intake according to their needs.

Plant Products

Soybean oil meal (soybean meal) is the ground residue left after the oil has been extracted from soybeans by three different processes; expeller, hydraulic, and solvent. According to the Cornell and Wisconsin Stations the feeding value of soybean meal depends mainly upon the temperature of processing and upon method of manufacture. Highest protein efficiency was obtained (1) in the expeller method by a processing temperature of 140° to 150° Centigrade (284° to 302° F.); (2) in the hydraulic method by a cooking temperature of 105° Centigrade (221° F.); and (3) in the solvent method by the moderate temperature (toasting) used. Meals produced as indicated in the first two processes were brown in color with a cooked or nutty odor and flavor. Meals produced by the solvent method were light in color, with no distinct odor or flavor but showed a protein efficiency ranking with that of the other meals. From these results it will be seen that color and flavor may not be a reliable criterion of feeding value, but that meals possessing a raw beany flavor are inferior because of insufficient heating. Solvent process meals are light colored but do not possess a raw flavor. However, like most other vegetable protein concentrates, it is not very palatable to turkeys. Soybean oil meals should be labeled according to the method of processing as "expeller" or "hydraulic" and "solvent" soybean oil meal or "extracted" soybean oil meal. Dehulled solvent extracted soybean oil meal must not contain more than 3% crude fiber. It usually contains about 50% protein.

Soybean oil meal of good quality is probably the best of the vegetable protein concentrates available for poultry feeding. It has a fairly good amino acid balance. When fed to turkeys it appears to favor feather growth and to produce a fine finish and excellent flavor of flesh. It is now commonly used to supply part of the supplementary protein in starting, breeding, and growing diets, and is known to have a good amino-acid balance.

The usual recommendation for the use of soybean meal in turkey starting and growing diets is 10 to 20% of the mash. It has been demonstrated, however, that this percentage can be raised considerably in starting and growing diets provided the proper mineral supplements (usually bone meal and oyster shell) are added to bring up the calcium and phosphorus content of the mash to the desired levels which vary with the ratios of mash to grain. The U. S. Dept. of Agriculture found that soybean meal of good quality could be used as the sole source of supplementary protein in turkey growing diets when the birds had access to good alfalfa or grass range. Soybean meal, however, is the only vegetable protein concentrate yet demonstrated to be successful as the sole source of supplementary protein, as far as the authors have been able to determine from the evidence available. Furthermore the only place in turkey feeding where an all-vegetable-protein diet is satisfactory appears to be during the growing and finishing period.

In turkey breeder diets, soybean meal and the other vegetable protein supplements have not been generally favored, although the Nebraska Station now includes 15% of soybean meal in the recommended all-mash breeder ration (see Breeder Diet No. 10). The Purdue Station has found that hatchability and egg production of turkey hens are not adversely affected by diets containing large percentages of corn and soybean meal provided the birds have access to bluegrass range. The Colorado Station reported that solubilized liver extract and fish solubles tend to correct the deficiencies in soybean meal that depresses hatchability. The unfavorable effects on hatchability obtained by various investigators indicate that in turkey breeding diets, the use of soybean meals in more than small amounts (in excess of 10% of the total diet) should be approached with caution, especially when the breeding stock does not have ample green range.

In 1941 in the U. S., 1,950,000 tons of soybean meal and cake were available for animal feeding. The yearly average, 1945-49, was 3,784,000 tons and for the year Oct. 1952-Sept. 1953, 5,551,300 tons. The availability of large quantities of soybean meal during and following the war period stimulated the increasing tendency to feed liberal amounts of this product in all turkey feeds except breeder diets.

Unprocessed soybeans are not generally favored as feed for poultry although it has been reported that in some sections they are used successfully as a small part of the growing and finishing diet of turkeys, allowing the birds to range in soybean fields in the fall when the plants are mature or nearly so. The Illinois Station reported good results when 5 to 10% of ground raw soybeans were used as a protein supplement along with meat scrap in growing diets fed to chicks. However, it has been reported by growers that heavy feeding of unprocessed soybeans or poor quality soybean oil meal produces soft oily fat in turkeys fattened in this way. Since this soft fat is very undesirable it appears best not to use raw soybeans in the turkey diet. If used, not over 5% of the total diet would be indicated, until or unless the successful use of a greater quantity is demonstrated.

Cottonseed meal can be used in turkey starting and growing feeds in limited quantities. As in other cases where plant protein supplements are fed, the calcium and phosphorus content of the diet must be properly adjusted and ample vitamin A provided. The Oklahoma Station (1954) suggests that 7½% of turkey starting and growing mashes may consist of cottonseed meal (preferably the prepress solvent type) if accompanied by 5% fish meal to supply lysine. Its use in turkey breeder diets is not recommended, although a small quantity, 3% or less of the total feed intake, probably could be used without ill effects. However, as with all other products that are not among the best feeds, its use is not advised except under special conditions. Lysine and methionine deficiency may explain the poor growth results often obtained from cottonseed meal, according to the Texas Station (1950).

Work at the Beltsville Station indicates that cottonseed meal may have a tendency to produce a rather strong, gamey flavor in turkey meat when fed at a level of about 18% of total feed. At the more common levels of 5 to 8% of total feed, no such effect was noted and growth was in no way impaired. Cottonseed meal is considered as interchangeable with soybean meal in chick diets up to 12% of the mash by the Texas Station.

About 1,900,000 tons of cottonseed meal and cake for animal feeding were produced in 1942, and about the same in 1944. The yearly average, 1945-49 was 1,888,000 tons. For the year October 1952 through September 1953 the production was 2,673,000 tons.

Corn gluten meal is a good source of vegetable protein and may be used to replace part of the animal products customarily used in diets for growing turkeys. Whether or not it should be used depends largely upon its protein content in relation to its cost, i.e., how much per pound does its protein cost as compared to other sources of protein? Its high vitamin A content, if made from yellow corn, is an advantage and should also be considered. Corn gluten meal is highly regarded as a fattening and weight-maintaining feed. These qualities point to its use in some turkey diets, especially those fed during the finishing period. It tends to produce palatable, yellow-fatted meat. The Beltsville and Wyoming Stations found that it produced excellent results in growing mashes up to 60% of total protein supplements. From 10 to 15% is sometimes recommended for use in starting, growing and finishing mashes. Gluten meal is low in arginine, lysine and tryptophan and has a high supplementary relationship with fish meal.

Corn gluten feed is not widely used for poultry. However, the Nebraska and Nevada Stations recommend its use in turkey mashes when its cost is not much higher than that of ground corn.

Peanut meal must not have more than 11% crude fiber and usually contains 48 to 50% protein. Peanut feed must not contain over 20% crude fiber and usually has from 41 to 45% protein. These products are good sources of protein in turkey diets. Experiments show peanut meal to be the equal of soybean oil meal as a partial source of protein for both turkeys and chickens. Many turkey feeders regard it highly and use it in the mash up to 15 or 25% by weight. Peanut-meal-fed turkeys possess good flavor which, however, does not resemble that of roasted peanuts. Since the keeping quality of peanut meal is rather poor and rancidity must be guarded against, feeds containing this product should be mixed frequently.

In 1941 in the U. S., 60,000 tons of peanut meal and cake were produced. The yearly average, 1945-49, was 85,000 tons; for the year October 1951 through September 1952 the production was 42,000 tons.

Linseed meal (or linseed oil meal, sometimes called simply "oil meal") can be used in turkey diets but its value is uncertain. Stimulation of feather growth and a beneficial effect on the digestive system sometimes are credited to it but no published experimental evidence is available so far as turkeys are concerned. Work at the Beltsville and Nebraska Stations has shown that linseed meal tends to depress the growth rate in turkeys, just as it does in chicks, when fed in excess of 5% of the total feed intake. In addition, it appeared to produce a strong, gamey flavor in the cooked flesh. Its laxative properties afford an additional reason for limiting its use to about 4% of the feed intake.

However, the California Station (1950) found that when one part of old-process linseed oil meal was mixed with 3 parts of water, allowed to stand for 24 hours, then dried at 65° C. (149° F.), and fed in poult diets it produced good livability and fast growth even at a level of

30%. Equal results were obtained when 30% of untreated linseed oil meal in the starting mash was supplemented by 0.002% of pyridoxine hydrochloride or when the untreated linseed meal was limited to a 10% level. No advantage was gained by supplementing water-treated linseed meal with pyridoxine. Linseed meal often is recommended as a conditioner to enhance the appearance of the feathers, but again there is no evidence to support this view. Old process meal (designated as O.P.) contains about 3% less protein and about 3% more fat than new process meal (designated as N.P.). The old process meal is favored by poultry raisers in preference to the new process or solvent meal. The authors do not recommend the use of linseed meal in turkey diets. The average annual commercial production of linseed meal and cake for animal feeding was 560,000 tons for the period 1945-49 and 457,700 tons for the year Oct. 1952 to Sept. 1953.

Rapeseed oil meal. This 37%-protein byproduct is produced in some quantity in Canada. The Saskatchewan Station (1948) found that it had goiterogenic properties even when iodine intake was adequate; that market-age turkeys fed it showed reduced resistance to cold; and that young poults on 4, 10, and 20% of it showed enlarged thyroid glands and slight retardation of growth. These results indicate that rapeseed oil meal has little value in turkey diets. When protamone (containing 3% of thyroxine) at 36 grams per 100 pounds of feed was added to the rapeseed oil meal diets, the thyroid glands of poults 9 weeks old were only one-tenth as large as those of the controls and feed efficiency was impaired.

Sesame meal is a product of the tropics. The California Station found it exceptionally good in combination with soybean meal in chicken diets. It contains about 40% protein, 13% fat, 6% fiber, 23% nitrogen-free extract, and 12% ash. It is high in methionine in which soybean meal is rather low; has 1400 units of riboflavin per pound.

Sunflower seed oil meal. This product satisfactorily replaced soybean meal in turkey starting diets (Ontario

Station 1949, 1950) at a 21% level but some achroma (white wing spotting) was produced due to lysine deficiency. At a 34% level supplemented by 0.4% L-lysine, growth and feathering of Broad Breasted Bronze poults were normal. In growing diets it appears that sunflower seed oil meal can replace soybean meal up to a level of about 20% with entire satisfaction and without special supplementation. South American sunflower seed oil meal may be a more valuable protein supplement. Destruction of lysine during expeller processing probably accounts for the deficiency of that amino acid. Solvent-process sunflower seed oil meal does not have this disadvantage. The methionine content is not affected by the method of processing.

Dried tomato pomace was found by the Massachusetts Station to be palatable when fed in a chick ration at a level of 11.6 per cent. The sample tested contained 4.28 per cent moisture, 24.08 per cent protein, 15.70 per cent fat, 32.11 per cent carbohydrates, 20.78 per cent fiber, and 3.05 per cent ash. The vitamin content per pound was as follows: A, 3080 international units; B 1810 international units; G 2500 chick units. It is estimated that about 12,000 tons of this dried product are available annually. Only a few feeding tests with this product have been reported and in none of these were turkeys used. Its high fiber content and reasonably high vitamin content suggests the possibility that it might have a place in turkey diets.

Tung oil meal, commercial type, is regarded as poisonous to livestock and poultry. According to the Florida Station solvent extracted tung oil meal appears to be non-toxic, unpalatable (to chickens) and of low feeding value. It has about 21.5% protein, 42.2% fiber, and 20.2% nitrogen-free extract.

Safflower seed oil meal contains up to 40% protein. It is a new product, sometimes available at reasonable prices. It has been used successfully up to 15% of laying diets for chickens along with fish meal. No reports have yet been given for its value in turkey feeding. It is low in methionine and lysine.

Miscellaneous Products

Cow manure has value in turkey diets as a partial substitute for alfalfa since 10% of it dried at 80-120° C. gave excellent results in starting diets at the Beltsville Station. It contained about 1800 units of riboflavin per pound, protein 10%, fiber 29%, Ca 1.6%, P 0.4%. In growing diets the Ontario Station (1949) obtained good results at the 5% level in an all-vegetable protein mash fed along with grain. Feathering was excellent.

Rumen contents dried at 47° C. had 6200 units of riboflavin per lb., 14% protein, 39% fiber, .79% Ca. and .67% P. This product was as effective as the manure in chick diets in the Beltsville experiments.

Prepared feed concentrates are fed primarily for their protein content, although the minerals and vitamin-bearing supplements which are usually included are equally important. When the ingredients are known and fully guaranteed, these concentrates are satisfactory and convenient. High-protein concentrates cause a higher percentage of protein to be consumed than do 18 to 22% protein growing mashes, but allow larger percentages of grain to be fed.

FEEDS USED CHIEFLY AS SOURCES OF VITAMINS

Plant Products

Green grasses and legumes *when eaten by the birds from the range in liberal quantities, along with the sunshine simultaneously obtained, will satisfy the needs of turkeys (except during the first few weeks of their lives) for all the essential vitamins and will make other important contributions in the form of minerals, carbohydrates, and proteins.* Because green feed and sunshine are cheaper than manufactured products it is good management to make use of them whenever possible and to eliminate from the diet, except in starting feeds, the ingredients (milk, alfalfa, A and D oils, riboflavin supplements) that are used to substitute for them. Maximum practical use of green range is obtained when a simplified growing diet is fed to turkeys on range after 8 weeks of age. Even confinement-reared turkeys can be fed freshly cut green feed with as good results as though the birds had pastured, though labor costs often limit the practice. The Wisconsin Station (1952) reported that "if turkeys are on a good ladino range they do not need antibiotics or B_{12} after the age of six weeks."

Alfalfa and ladino clover usually are rated excellent as pasture crops for turkeys. Bluegrass*, reed canary grass, Dutch clover, Sudan grass, Bermuda, vetch, cowpeas, fall rye, kudzu and rape, are all good forage plants. Korean and common lespedeza are unpalatable to turkeys but sericea, if of low tannic-acid content, is well liked. Soybeans in rows 20 inches apart provide good green feed and will do well on land too sour for clover or alfalfa. Orchard grass makes excellent forage and can be cropped for hay or seed when not needed for turkey pasture.

Kudzu, a rank growing perennial vine, will provide good pasture for turkeys in areas with mild climates; until late in the fall. It will grow in rough, out-of-the-way places that are not suitable for the growing of grasses and legumes and will maintain itself well against all kinds of competition, including weeds.

There have been some reports of prussic acid poisoning resulting from the consumption of drought-stricken sudan grass. The Nebraska Station, however, has fed sudan grass grown under all sorts of conditions and has never observed any prussic acid poisoning. With plenty of limestone or other source of calcium in the diet this poisoning is unlikely to occur.

While grasses and other green crops are at their best as sources of vitamins and proteins at an early stage in

*Local county agents and state agricultural authorities are best qualified to advise regarding pasture plants and management.

their development, experience has shown that they are valuable for this purpose at all stages even in a naturally cured condition such as prevails during summer and fall, especially in areas of low summer rainfall. If growing turkeys past the brooding stage are allowed free access to growing or to naturally cured green feed of almost any type that they will eat freely, little fear need be that their requirements for vitamins will not be satisfied. Even palatable weeds and pest grasses are satisfactory. Some turkey growers are successful in using turkeys to pasture off these plant pests and at the same time obtain a fairly adequate supply of vitamins.

Rotational grazing is necessary, not only from the angle of soil sanitation, but also from that of preservation and protection of the pasture. Frequent moving, weekly at least, of the turkeys and the rearing equipment accomplishes the same results as rotating from one field to another. Daily moving of all equipment is recommended by the Pennsylvania Station, in order to preserve the pasture plants and to provide the maximum in grazing and soil sanitation.

Alfalfa, ladino clover, orchard and brome grasses give summer green feed. Alfalfa and orchard grass make a good combination. Orchard grass can be seeded into thin alfalfa stands in early spring. In the midwest, alfalfa and brome grass make good turkey pasture. White clover and bluegrass are a common and successful mixture likewise Kentucky bluegrass and timothy where summers are cool with good rainfall. "Bromus enermus" a brome grass is recommended for all-year turkey pasture.

In corn growing regions seed 5 lbs. of rape and 10 to 15 lbs. of domestic ryegrass per acre immediately after last corn cultivation to furnish late summer and fall pasture. Rygrass alone, 15 to 20 lbs. per acre, is good. In fact most all grasses or palatable weeds are well utilized as pasture by turkeys. In areas where it will do well, ladino clover is considered by the Pennsylvania and Purdue Stations as perhaps the best forage crop for turkeys.

New growth is continually produced and if allowed to re-seed during the summer, stands of it will last for years. Fertile soil well-watered is required for this legume. The Pennsylvania Station suggests the following mixtures per acre for areas of climate and rainfall (35-45″ per year) similar to Pennsylvania.

	lbs.
Ladino clover	1
Alfalfa	6
Orchard grass	6
or	
Ladino clover	1
Red clover	4
Orchard grass	6
or	
Orchard grass	6
Alfalfa	6
Perennial ryegrass	5
Redtop	2
White clover (Louisiana white preferred)	1
Ladino clover	1

Birdsfoot trefoil, a drought-resistant long-lived legume has possibilities for turkey pasture, likewise alta fescue, a coarse hardy grass like orchard grass, and reed canary grass also coarse, hardy, and long-lived.

Sudan grass seeded 25 to 30 lbs. per acre or 1 bushel of soybeans and 15 lbs. sudan grass make good summer range pastured after the sudan is 16 to 18 inches high. It makes a thick stand with relatively fine stems. Another good method (Kansas Station, 1946) is to drill it in thickly in rows 3 feet apart to furnish shade and green feed. Successive plantings of rape may be supplemented by Sudan grass for hot-weather forage.

Seeding alternate strips, about 30-feet wide, to ladino or alfalfa and the following spring planting the bare spots to corn or sudan grass, has been found a satisfactory program in the Midwest.

Poisonous plants. There are only a few poisonous plants affecting poultry that are common in the United States. Crotalaria spectabilis, a legume sometimes grown for green manure in the South is considered poisonous to livestock and poultry; likewise the whorled milkweed.

"Short or cereal grasses." Oats, wheat, barley, or rye plants provide highly nutritious green feed. If pastured or cut and fed 18 to 20 days after sprouting, they are high in protein and in vitamins. At this stage they are sometimes cut, dried, and marketed as "short grasses" or mixed with milk products and fed that way. Some of these products carry a guaranteed vitamin A potency (furnished by carotene) equal to that of cod-liver oil so that as little as 2% will suffice to satisfy birds' requirements for that vitamin. In addition, vitamins B_1, E and riboflavin and some of the less well-known vitamin factors are supplied. The Wisconsin Station reported that alfalfa leaf meal carries some factor in addition to riboflavin which is essential for embryonic development and that this factor is supplied to a less extent by dried cereal grasses.

Yellow carrots, onion tops, tender alfalfa tips, swiss chard, green cabbage, celery and lettuce leaves, if fed crisp, are good to excellent sources of vitamins A, B_1, E, and riboflavin.

Rape and kale are good pasture crops for turkeys. The North Dakota Station found that rape did not cause off-flavor in the flesh of turkeys and was fully satisfactory in every way. The Oregon Station regards kale as an excellent green feed for poultry during late summer, fall, and winter where temperatures do not fall below 13° F.; and *vetch* as a high yielding green crop available during spring and early summer. Turkeys transferred from confinement to rape pasture may overeat of the rape, hence should be introduced to it gradually.

Lawn clippings are a good green feed but must be fed carefully and should not be fed at all to turkeys under 4 weeks of age because of danger of impaction of the digestive tract, unless they are freshly cut in short lengths and not allowed to remain before the birds in a wilted condition. The frequent timely mowing of a bluegrass or Dutch clover lawn by a mower equipped with a grass catcher is a good method of supplying green feed to turkeys of all ages. By mowing frequently, uniform clippings as short as 1/4 inch can be obtained and these are satisfactory for small turkeys if fed carefully, removing uneaten portions promptly. The more clover in the lawn the better since clover is more tender than the grass. An area of Dutch clover reserved especially for the purpose and kept well fertilized and watered is of great value.

The Pennsylvania Station found that confinement-reared White Holland turkey males, fed standard mashes and grain, averaged 18 pounds at 28 weeks of age while those handfed fresh lawn clippings in addition to the mash diet and grain (from 4 to 16 weeks) and then dehydrated, soaked lawn clippings from 17 to 28 weeks, weighed 19.2 pounds. Females fed the grass weighed 0.4 lb. more than the other lot. In addition, there were twice as many birds grading out as A in the grass fed lots for each sex and the skin contained considerably more yellowish pigment. Feed efficiency was not affected significantly by the feeding of green feed where the air dried weight of the green feed was added to that of the mash and grain consumed.

Sunflower and sudan grass provide both shade and green feed. Sunflowers produce in addition, a good crop of nutritious, gloss-producing seeds.

Sprouted grains (if fed before the sprouts become green) white carrots, pumpkins, inside cabbage, lettuce and celery leaves, and mangel wurzels appear to be of little value in poultry diets. In general they are low in vitamin A and have no qualities that are of special value in the nutrition of turkeys or other poultry. The New Jersey Station (1950) reported that celery fed to turkey broilers for six weeks before marketing had no effect on palatability of the carcass.

Alfalfa meal is a general term referring to ground alfalfa hay. It should be purchased with due attention to the fiber content, which should not be over about 30%; to the protein, which should not be less than 14%, and to the color, which should be a bright, natural green. Fresh, good-quality meal is a good source of betaine and of vitamins A, K, thiamin and riboflavin. To get the best quality, "U. S. extra green" should be specified along with freshness (not over one year old). Alfalfa meal also contains some of the less well known vitamin factors and is relatively rich in minerals. Alfalfa meals and leaf meals are primarily a substitute for fresh green feed and need not be fed when the latter is abundantly available except in starting feeds, which should contain either high quality alfalfa meal or alfalfa leaf meal. Properly dehydrated alfalfa products usually are richer in vitamin A than sun-cured products and consequently are worth more as vitamin A supplements. However, it is possible to obtain sun-cured meals that are as high in vitamin A as the dehydrated kind. On the average, the dehydrated products are to be preferred. Five percent in starters, up to 20 percent in high protein concentrates is commonly used and recommended.

Stemmy meal, no matter how well cured, is not a concentrated source of any of the vitamins. "Special alfalfa meal" contains at least 60% but less than 85% of leaf material and not less than 16% protein. Most of the alfalfa meal now produced corresponds to the special grade and carries a protein guarantee of 17% and a vitamin A guarantee of 67,000 or more International units per pound. It is widely used in poultry mashes. The so-called medium grind is probably the most satisfactory for use in mashes. Good quality leafy alfalfa hay ground in a hammer mill with a ⅛-inch screen is satisfactory for poultry mashes. The resulting meal is similar in quality to the special grade. Very fine grinding is sometimes used to conceal high fiber content and is unnecessary in meals intended for poultry feeding. Ordinary alfalfa meal may contain as low as 25% of leaf material and 12% of protein, but if less than about 14% it is not favorably regarded as a poultry feed. The special meal or a ground hay meeting the specifications for "special" meal in the medium grind probably is best for turkeys.

Alfalfa hays, meals, and leaf meals dehydrated or sun-cured have been shown to contain an unidentified growth-inhibiting factor which may affect poults and chicks. Some samples of alfalfa products carry the inhibitor while others do not and the reason for this difference is not known. This factor probably accounts for the fact that in some experiments alfalfa meals are well-tolerated in large amounts while in others a definite retardation of and growing diets exceed 10%.

The Utah Station in a series of tests (1942-48) obtained good results from feeding alfalfa meal to Broad Breasted Bronze turkeys at relatively high levels. In all-mash starters fed the first eight weeks, levels of 10, 13, 16 and 19% produced uniform and satisfactory growth results with no differences in feed conversion efficiency. Growth was retarded when 22, 25, 28 and 31% were fed. In growing mashes, alfalfa meal at levels of 15, 20, 30, 35 and 40% produced equal weights at 29 weeks of age, while at 50%, average weights of toms and hens were lowered about 1½ lbs.; from 16½ to about 15 lbs. These weights are not high for Broad Breasted Bronze turkeys, but the fact that the high levels of alfalfa meal gave as good results as the lower more nearly standard levels, indicates that alfalfa meal when cheap can be the means of making a substantial reduction in feed costs. Apparently the meals fed in these experiments carried but little of the inhibitory factor. Grain was fed with the growing mashes.

The Texas Station (1950) observed early growth depression, impaired feed efficiency, and increased mortality to 8 weeks when Broad Breasted Bronze poults were fed 8 and 20% of alfalfa meal known to carry the inhibitory factor. The Beltsville Station observed that Beltsville White turkeys fed 0, 4, 8 and 16% of dehydrated alfalfa meal in the mash (with grain free choice, starting at 8 weeks) made normal growth to 24 weeks, while in those receiving 24%, the males were retarded from 3 to 5% and the females up to 3%. Males receiving 32% alfalfa meal in the mash were retarded about 10% at 24 weeks and the females about 2½%. Apparently the inhibitory factor was present in these meals. In an all-mash breeder diet, 15% of alfalfa meal did not lower the egg production or hatch-ability of the Broad Breasted Bronze turkeys at the South Dakota Station (1949).

Vitamin A (carotene) retention in suncured alfalfa products is poor when stored at temperatures much above 50° F. (Texas Station). Dehydration tends to retard this loss of vitamin A potency and that of other vitamins in the alfalfa products. Freshness and good green color in these products are desirable whenever the vitamin A potency is important—as it usually is. The riboflavin content is little affected by storage or loss of high green color although a brown-colored meal may be low in all the vitamins due to leaching or other mistreatment.

For the year April 1950 through March, 1951, the U. S. output of suncured alfalfa meals was 233,800 tons; of dehydrated, 900,900 tons. The total production for 1952 was 1,056,700 tons.

Alfalfa leaf meal consists of 85% or more of leaf particles, not less than 19% of crude protein, and not over 18% of crude fiber. It usually is richer than alfalfa meal in the vitamins. To obtain the highest quality meal, "U. S. Extra Green" should be specified. For the main purpose for which it is fed (to supply vitamins), the leaf meal appears to be worth about twice as much as ordinary low protein alfalfa meal. The medium grind probably is best for poultry mashes. Alfalfa leaf meal no longer is produced in large quantities, instead, high-grade alfalfa meal with a 17% minimum protein guarantee predominates

Granulated alfalfa meal is a new (1952) free flowing dust-free product which improves palatability and pelleting qualities. The price is competetive with standard products.

Alfalfa stem meal is the ground product remaining after the leafy material is removed from the hay. It is of little value in turkey feeding except to boost the fiber content and this is seldom necessary in feeding practice.

Alfalfa hay (and hay made from other legumes such as clover and soybeans) have the same feeding value as the meals made from them but under most conditions turkeys will not eat enough of them to satisfy their needs for green feed, consequently it is necessary to incorporate them in the mash to be sure the birds eat adequate quantities. This is especially true in starting and laying diets. It is possible that growing birds will eat hay in sufficient amounts if it is kept before them in clean open troughs or clean areas on the range. Placing it in wire or slatted racks reduces the consumption. As with the meal, dehydrated hays usually but not always contain more vitamin A than sun-cured products. High-quality leafy hay when ground may meet the specifications for leaf meal or high-grade meal and often is more economical than commercially prepared products.

Lespedeza meals although of value are not generally available. The Kentucky Station found that the leaf meals of Korean lespedeza and lespedeza sericea were of approximately equal biological value in chick diets. No experiments with turkeys can be cited. In all probability lespedeza products if properly cured are equal in turkey diets to alfalfa products of similar quality.

Dehydrated pea vines were found by the Maryland Station to be a good substitute for alfalfa meal in chickens diets and were more palatable. They supplemented soybean meal (the only protein supplement) in a very efficient manner. The sample used contained 23.26% crude fiber and 77,000 I. U. of Vitamin A per pound.

Sudan grass meal prepared from grass 12 ″ to 18″ high and artificially dried was found to be a good substitute for alfalfa meal in chick rations at the Nebraska Station. It had 25% protein and about 117,000 units of vitamin A per pound when fresh.

Wheat and corn germ are concentrated sources of vitamins B_1 and E and *wheat and corn germ oils* are concentrated sources of vitamin E. However, these products are not needed in poultry feeds except under very special conditions. *Wheat germ meal* at a 5% level had no beneficial effect in the laying diet of chickens at the Michigan Station.

About 2% of the wheat grain is germ and 10% of the germ is oil, about 55% of which is retrieved by the hydraulic method of extraction.

Brewers' yeast is an excellent, although relatively expensive source of nearly all the vitamins of the B complex, including one or more unidentified factors. Its production for animal feeding purposes is increasing. It has special value as a supplement to use at levels of 3 to 6% as a preventive or corrective feed to supply thiamin and other members of the B complex.

The Cornell Station (1952) found it effective at the 5% level in turkey starting feeds, in preventing enlarged hock disease. This effect apparently was due to its high content of niacin and its antioxidant properties. The protein of brewers' yeast is of good quality containing all of the essential amino acids. It is a good source of the trace minerals copper, cobalt, iron and zinc and contains betaine. To provide nutrition insurance in turkey diets, dried brewers' yeast has been suggested by some nutritionists, at a level of 5% in the starting mash, 3% in the growing mash and 5% in the breeder mashes. Brewers' dried yeast is defined as the dried non-fermentive, non-extracted yeast resulting as a by-product from the brewing of beer and ale and should contain not less than 40% of crude protein on the moisture-free basis.

Cereal yeast feed, a brewery byproduct, has had a favorable reception as a supplement to supply B-complex vitamins in chicken and turkey diets. Cereal yeast feed contains about 30% protein, 4% fat, and 9% fiber. If of proved quality with a riboflavin content equal to or in excess of that of dried milk, and their cost somewhat less, they can be used to replace up to 100% of the dried milk generally included in turkey diets and are especially valuable in starting feeds and those fed to birds in strict confinement. However, there are some yeast feeds on the market the cost of which definitely exceeds their value for poultry feeding.

Yellow corn and variegated corn with yellow endosperm, while not fed primarily for the vitamin A content, must be considered as an important source of this vitamin in practical feeding operations.

Distillers' dried solubles:—a riboflavin concentrate with 7000-12000, often fortified to 113,000 or more units of riboflavin per pound. The Minnesota and Purdue Stations report that they are a substitute for dried milk in a ration moderately deficient in riboflavin and may be used up to a level of 7% of the total diet. Their protein is not of superior quality. During 1953 the estimates for use in livestock feeding indicate that 228,000 tons of brewers' dried grains and 244,000 tons of distillers dried grains were used. The 1954 estimates show 225,000 and 240,000 tons respectively.

Synthetic (dry) vitamin A is a new product which promises great help to the feed industry in providing a relatively stable and easily handled source of vitamin A free from any tendency to produce fishy flavor in the flesh of poultry to which it may be fed.

Animal Products

D-activated animal sterols (dry vitamin D) have proved effective in supplying vitamin D to turkeys as well as chickens. Their vitamin D, however, is readily destroyed if premixed with calcium carbonate but not with most other diet ingredients.

Fish oils are obtained from fish livers, from fish scraps, or from the whole bodies of fish. They are fed to poultry for the sake of the vitamins A and D contained in them. The livers of various species of fish vary considerably in their vitamin potency but most of them contain liberal amounts of vitamins A and D. They should be used only when guaranteed as to vitamin potency. Cod-liver oil and salmon-liver oil have proved satisfactory according to reports of various investigators and are used in poultry feeding for supplying vitamins A and D.

Sardine (pilchard) oil is obtained by extraction of part of the oil from the whole Pacific sardine (or pilchard) or from cannery refuse of this species. It appears to be satisfactory in most types of diets as a source of D but is only a fair source of A unless fortified. As with other fish oils, it is advisable to use sardine oil with due regard for its guaranteed vitamin potency.

Menhaden oil is made from the whole menhaden fish. It was reported by the U. S. Bureau of Fisheries to be 75% as efficient as cod-liver oil in providing the vitamin D factor. The Pennsylvania Station reported it to be a fair source of vitamin A. The Maryland Station reported that commercial menhaden oils collected from various sources "compare favorably with cod-liver oil as an antirachitic for poultry." The North Carolina Station reported that menhaden oil was equal to cod-liver oil (for vitamin D) in chick diets.

Salmon body oils vary considerably but in general are equal to cod-liver oil in vitamin D content but inferior to it in vitamin A unless fortified. They are made from salmon cannery refuse.

Tuna-liver oil generally is not rated so high as a source of vitamin D for poultry as some other fish oils, but from some species it rates high in vitamins A and D.

Shark-liver oil varies considerably but in general is an extremely potent source of A but only a fair source of D. Florida shark-liver oil, poultry grade, contains about 35 units of D and 9000 of A per gram.

Different grades and different lots of the same kind of oil vary greatly in vitamin potency and should be purchased on the basis of a guarantee of potency of International units determined by biological test. Only oils of guaranteed potency should be used.

Fish oils of standard potency (85 to 100-D) are now but little used at levels of 2% or less in the entire diet. Too much fish oil is toxic to poultry, more than about 2% of total feed intake being considered deleterious. The cause of the toxicity is not known, but it may be due to some factor which hinders the utilization of Vitamin E. Fish oils apparently decrease the creatine in the muscles and thus may aggravate hock disorder in turkeys.

Fish oils are commonly added to the mash by working them thoroughly into the bran or corn meal by hand, then mixing the resulting combination with the other mash ingredients. Another method is to spray the oil over mash or scratch grain while it is being mixed. Mixing can also be done by pouring the oil in a fine stream on the scratch grain while it is being stirred vigorously. It is not advisable to mix the oil with minerals or milk products on account of possible destruction of vitamins.

Condensed fish solubles are obtained by condensing the press water resulting from the hydraulic extraction of oil from fish. The product is available in semi-liquid form or as a dry product mixed with alfalfa or soybean meal. It is a good source of riboflavin, choline, B_{12}, and niacin and appears to carry one or more unidentified growth vitamins. At the 4% level it proved to be a good supplement in a corn-soybean meal starting diet for poults (Texas Station, 1948). Its protein is of good quality and contains methionine which helps to balance soybean protein. In the absence of reports dealing with the results of feeding fish solubles on the flavor of turkey flesh it seems advisable to consider it in the category of the average fish meal in this respect and limit alone or a combination of it with fish meal to 5% of the starting diet when fed to turkeys marketed as fryers or light roasters. In 1952 a total of 70,486.5 tons of fish solubles and homogenized condensed fish was produced in the U. S.

Cod-liver meal has not proved to be a reliable source of vitamins. Some investigators have found it satisfactory while others have found it unsatisfactory. The Nebraska Station found that 5% cod-liver meal did not prevent rickets in chicks as effectively as did 2% of cod-liver oil. Cod-liver meal contains about 50% crude protein, 30% fat, 1% fiber, 0.13% calcium, and 0.75% phosphorus.

Vitamin A and D feeding oils (fish oil concentrates or fortified oils) are blends of fish oils and fish liver oils with vitamin A concentrate or vitamin D concentrate, or both. Vitamin content per gram of oil is greatly increased by the addition of the dry A or dry D sources; so that much smaller quantities may be fed to guarantee the optimum vitamin level. They usually contain 300 to 800 International Units of vitamin D in 1,000 to 3,000 International Units of vitamin A per gram.These oils can be mixed by hand if care is taken to mix them thoroughly in a suitable quantity of bran or ground corn and then incorporate this mixture in a larger quantity of the bran or corn before it is finally mixed with the rest of the diet. Due to the smaller percentages used, A- and-D oils do not produce the degree of off-flavor produced by the straight oils but they do cause some fishy flavor at a level of ¼% of total diet of turkeys held to roasting age. At a level of ⅛%, the Oregon Station reported that no off-flavor was produced in roasting age turkeys.

While the vitamin A potency remains intact for long periods of time in fish oils kept in sealed containers, it diminishes quite rapidly when the oils are mixed in the feed or otherwise exposed to air. Consequently, when the vitamin A content is an important factor to be considered (as it usually is) it is important to mix the oils fresh every week to 10 days, or at the most every 14 days. Vitamin D is more stable, however. Investigation has shown that it persists for several months or even a year when the oil is mixed in the feed.

Vitamin A and D powder is a new product containing both vitamins supposedly in stable form even after mixing in the feed. Potencies from 1000A-300 D to 5000A-1500 D per gram are available.

Off-flavors in Turkey Meat

The Pennsylvania and Beltsville Stations have shown that the feeding of certain kinds of fish meal, or cod-liver oil, or both, will cause fishy flavor. This off-flavor remains in the bird's flesh for as long as 11 weeks after the fish products are removed from the diet. The Beltsville Station reported that fishy flavor persisted as long as 13 weeks after birds were removed from "off-flavor" starting diets and that other off-flavors persisted as long as 16 weeks even though no fishy products were fed after the turkeys were 8 weeks old. The California Station reported no ill effects from feeding 25 per cent good grade tuna or sardine meal in an all-mash diet provided the birds were properly fasted and chilled. The Oregon Station reported no unfavorable results from feeding Alaska herring meal up to 10 per cent of the total feed intake, but the Beltsville Station reported noticeable off-flavor in turkeys fed 10% of menhaden, sardine, white fish, or herring meal. Both Stations reported off-flavor from feeding as little as 0.25% (of total diet) of sardine or cod-liver oils, straight or fortified.

Although the number of birds tested has been small and the fish products used hardly constitute a fair sample of those found on the markets, the danger from feeding any kind of fish oil or fish meal during the finishing period is too great to justify their use, except at very low levels. Good quality west-coast sardine or tuna meal probably is the safest kind to use. At a level of 5% of the total feed intake, even if fish oil at a level of 0.15% or less is also used in addition to the meal, no ill effects should be produced. However, the use of non-fishy sources of vitamins A and D, which are now generally available, seem advisable in growing and finishing diets in preference to the sources from the various fish oils. Both dry A and dry D are now available on the market commercially. The dry D has been available for some length of time and is in a very favorable competitive price ratio to the oil sources of that vitamin. If such sources of vitamins A and D are fed, it will be safe to use up to 5% of the total full feed intake of any type of good quality fish meal. In starting mashes, including those fed to broiler turkeys, the suggested maximums are 5% of any good quality fish meal along with not over 0.15% of an A and D feeding oil with a fish-oil base. These maximums for fish meal include also any condensed fish solubles that may be used in the diet.

Other possible sources of fishy or other off-flavor in turkey flesh are unsaturated oils such as linseed and some other seed oils. These products usually are not fed to poultry but the corresponding meals sometimes are and since they contain some of the seed oil they are a potential cause of off-flavor if fed at high levels.

There probably is no way to remove the off-flavor in turkey meat, once the bird possessing it has been slaughtered. It is possible in some but not all cases to tell whether a dressed bird will have a strong fishy flavor when cooked. This is done by a "smell test" of the abdominal cavity when the blood-and-feather dressed bird is first opened for drawing. The viscera and body cavity sometimes, but not always, have a strong rank odor quite distinct from the ordinary abdominal odor found in the average dressed turkey that has been well handled. The adnormal odor cannot always be described as "fishy" but it is always rank and disagreeable.

In the case of turkeys fed fish products it is probable that the substance which is responsible for the fishy flavor of fish products is itself taken into the bird's system from the intestines. It is also possible that some trimethylamine oxide remains in the fish products and that it reacts with the fatty acids set free in the intestines during digestion to form the compounds responsible for fishy flavor. These compounds, whether obtained directly from the feed or produced in the intestines, presumably are carried in the blood stream to all parts of the body, and in addition probably are absorbed directly into the tissues from the body cavity. The off-flavor seems to be found almost exclusively in the fat of the bird. Therefore the skin is the worst offender. The gravy and the thigh meat, since both usually contain considerable fat, are next most likely to possess the off-flavor. The breast meat, since it contains the least fat of all, is least likely to be affected.

Fishy flavor in chicken or turkey eggs has not been reported even though the total diets fed have contained up to 1½% of cod-liver oil and 5% fish meal.

Whey has value as a feed mainly because of its riboflavin content. Dried whey is widely used as a riboflavin supplement. Its low protein content (only about 12%) lessens its feeding value so that it cannot be generally recommended for use under practical feeding conditions, as riboflavin usually can be supplied at less cost (the protein being considered) by green feed, distillers' solubles, or milk. Due to the laxative effect, not over 5%, preferably 3 or 4% of total feed intake should be used.

Dried whey product (delactosed whey) results from the partial removal of lactose from whey. It is higher in protein and vitamins than dried whey and sometimes is fortified with riboflavin. It should be superior to dried whey as a vitamin supplement if the cost is in line with nutritive value. Recent research indicates that dried whey and dried whey-products contain fair amounts of an unidentified factor probably necessary for poult growth. Both are good sources of pantothenic acid.

Poults tolerate the delactosed product better than the straight whey high in milk sugar (lactose).

Sunshine and ultra-violet light, though not classified as feeds, make possible the formation of vitamin D in the body which in turn permits the proper development of bones. Sunshine is free and consequently should be used whenever possible to supply the vitamin D factor in starting diets, however, it has been amply demonstrated in practice that vitamin D should be added to insure proper quantities of the factor when poults are raised artificially. The California Station, however, found that poults raised in a "sunshine brooder" were able to obtain enough of the vitamin D factor from sunshine without adding it to the feed. Added vitamin D also is necessary in breeder diets except under the most favorable range conditions where turkeys have continual access to succulent green feed and direct sunshine.

Artificially produced ultraviolet light is generally regarded as a satisfactory source of the vitamin D factor for poultry. The Ohio Station found the S-4 type of sunlamp to be an efficient and economical means of providing this important factor to growing turkeys, growing chicks, and (chicken) breeders. One hour exposure daily was sufficient. However, there was no saving in cost over the use of cod-liver oil to supply adequate vitamin D and there were practical disadvantages difficult to overcome.

FEEDS USED CHIEFLY AS SOURCES OF MINERALS

Common salt (sodium chloride, NaCl) is regarded as an essential factor in the diet of poultry. It now is fed usually as iodized salt (0.007% iodine) at a level of ½% of the total feed of all classes of poultry, including tur-

Fig. 4.2. Ample feeding and watering facilities have been provided for the Indiana flock. (Photo from Purdue University.)

keys. It should be fed in a finely granulated form, sifted to remove all lumps, and thoroughly mixed in the mash. One-half to 1½ per cent usually is added to the mash when grain also is fed, thus insuring a salt level of ¼ to ¾% of the total feed intake. In all-mash diets ½% is preferred; more than ¾% is undesirable. High levels of common salt, Glauber's salt (Na_2SO_4), or baking soda ($NaHCO_3$) have a laxative effect and may cause edematous abdomens ("water-belly"). The Pennsylvania Station (1951) reported that 0.9% to 3.6% of common salt added to normal starting diets produced considerable mortality from water-belly. A few animal products such as shrimp meal and some water used for drinking may contain rather high percentages of some of those salts which should be considered when planning the feeding.

Oyster shell is an excellent source of calcium. It has a fair value as a grinding agent and it contains iodine which may be valuable where iodine-deficient diets are fed. It also contains some of the other so-called trace elements. The inclusion of the proper quantity in ground or finely granulated form in the mash is the best way to feed oyster shell to turkeys of all ages, although some turkey raisers prefer to keep it before them in granulated form so that they may consume the calcium necessary to balance the phosphorus in their diet. The authors have obtained uniformly good success from feeding 3½ to 4% ground oyster shell in all-mash turkey laying diets. However, some evidence is accumulating to favor feeding calcium carbonate products in granulated form, such as chick-size grit.

Clam shell, which has some value as a grinding agent, has been regarded as inferior to oyster shell by poultrymen but according to recent work is of approximately the same value and same mineral content.

Limestone is a good source of calcium. It is fed in the same manner as oyster shell, that is, in ground or granulated form in the mash, or in granulated form (¼-inch chips) in hoppers. However, when oyster shell costs little or no more than limestone, it is advisable to use the former. Limestone of high calcium content is to be preferred although for turkey feeding the Nebraska Station obtained good results with low-calcium dolomitic limestone containing only about 50% calcium carbonate (of which about 40% is calcium) in granulated chick-grit size.

Gypsum (calcium sulphate) was reported by the Nebraska Station to be fully equal to calcium carbonate as a source of calcium in the diet of chicks, and slightly superior to it in the diet of turkey poults. However, it takes 1.75 lbs. of the sulphate to supply as much calcium as 1 lb. of the carbonate. If gypsum is considerably cheaper than oyster shell or limestone its use as a source of calcium for growing turkeys may therefore be indicated. However, because of the trace elements found in oyster shell its use is recommended by the authors.

Steamed bonemeal (sometimes called poultry bonemeal or 10-20% protein bonemeal) is made from bones cooked with steam under pressure. It is a good source of both calcium and phosphorus and contains considerable fat and protein. However, since the fish meal, meat scrap, and tankage, one or more of which usually is included in most poultry rations, all contain relatively large percentages of minerals, the use of bonemeal is not always necessary. In fact, it may be detrimental in that it may raise the phosphorus content of a diet to an excessive level. Its chief use is with diets containing much protein from vegetable sources (see recommended diets). When 10% or more of the entire diet of chicks consists of meat scrap, tankage, or fish meal, no additional calcium or phosphorus need be added. Although this recommendation may in a general way be applied to the feeding of growing turkeys, it is better practice to estimate the mineral content and balance the diet accordingly.

Special steamed bonemeal is similar to the above but has very little fat or protein and is somewhat higher in calcium and phosphorus.

Low fluorine rock phosphate must not contain over 0.5% fluorine. It can be substituted for steamed bone meal in turkey diets but defluorinated phosphate or dicalcium phosphate is preferred.

Defluorinated phosphate consists of either calcined, fused or precipitated calcium phosphate which must not contain over one part of fluorine to 100 parts of phosphorus. It is similar in composition to steamed bonemeal and can be used in place of it in all turkey diets.

Grit. There is no evidence to show that turkeys on finely ground all-mash diets need grit or are benefited significantly by it. Where whole or cracked grain or fibrous green feed is fed, however, the need for insoluble grit is well established. There is no known advantage in a sharp-edged grit since the sharp edges are soon worn off, the grit remaining available as a grinding agent. Readily soluble materials such as soft limestone are unsuitable as grit and are likely to be consumed in too-large quantities unless a good insoluble grit is also available to act as grinding material. Any grit that is easily broken up into small pieces is too friable to be valuable as grit and also is likely to be over-consumed.

Experience has shown that when turkeys have access to gravelly soil there is nothing to be gained by supplying them with additional insoluble grit such as is sold commercially. Small gravel such as is common in creek beds is quite as effective for grinding as most of the commercial grits. Birdseye or pea gravel is excellent. Excess feeding of insoluble grit to confinement-reared turkeys sometimes is dangerous. It may cause impaction and possibly stimulate feather-picking. Grit consumption by confined turkeys should be watched closely and, if excessive, should be cut down to desirable proportions, by removing the grit for several days or by hand feeding; sprinkling a sufficient amount on top of the mash.

MISCELLANEOUS FEED MATERIALS

Dried kelp is sometimes recommended as a source of minerals and vitamins for poultry. According to Silmore

(1939) Pacific Ocean kelp, Macrocystis pyrifera, which is native to South Pacific waters, has a much higher mineral content than common Atlantic or Pacific seaweed. One sample, dried, had a mineral content of 38.57% in contrast to 17.0% in a foreign dried kelp. The use of dried kelp as a supplement to normal, well-balanced poultry diets is not recommended. That it may have value as an ingredient of scientifically compounded feeds is entirely possible but its indiscriminate use by home-mixers is to be discouraged. The Washington and California Stations both found that common kelp, dried, had no value in chicken diets.

Sulfur. This element and supplements fed for their sulfur content are probably of no value in turkey diets therapeutically or nutritionally. Sulfur compounds may cause so-called "sulfur rickets" in turkeys.

Iodine. Potassium iodide (KI) in common salt at the rate of not less than 0.007% iodine or 0.01% KI is the form in which iodine is usually administered to animals. However, the need for iodine supplements in ordinary poultry diets has not been demonstrated. In spite of this, iodine supplementation of poultry feeds is advised as a safeguard (by many nutritionists) using iodized salt in lieu of plain salt. Since fish meal, meat scraps, milk, molasses and oyster shell all contain fair amounts of iodine, possible iodine deficiency could be guarded against by including one or more of these products in the feed. About 1 part per million in the feed is required for high hatchability in chickens, according to the Colorado Station. Deficiency of iodine in the feed causes goiter in turkeys as it does in other animals but there is no evidence to show that the goiter is harmful. On the contrary it may be an advantage in fattening and in preventing preseasonal moult in turkey toms.

To make a stable iodized salt, pulverize each of the following and mix in a ball mill: Common salt (NaCl) 100 lbs., potassium iodide (KI) 0.3 lb., sodium thiosulphate ($Na_2S_2O_3 \cdot 5H_2O$) 0.4 lb., calcium carbonate

($CaCO_3$) very fine or precipitated 3.3 lbs. and corn starch 6.0 lbs. One pound of this mixture in 100 lbs. of mixed feed should bring the iodine content up to about 21 parts per million. Commercial iodized salt must contain not less than 0.007% iodine, uniformly distributed.

Miscellaneous mineral elements and mineral compounds such as raw bone, rock phosphate,* iron, copper, and phosphatic limestones are unknown quantities as far as turkey feeding is concerned. For the feeding of chickens the U. S. Department of Agriculture makes the following statement: "Recent experimental results . . . indicate that it is feasible to add small quantities of manganous sulphate, ferrous sulphate, and copper sulphate to common salt and use the resulting mixture in place of common salt at the levels recommended for common salt." The following is suggested as a mixture:

	Parts by weight
Common salt	100.00
Manganous sulphate (anhydrous)	1.70
Ferrous sulphate	2.00
Copper sulphate	0.04
	103.74

However, as mentioned elsewhere, the need for using special mineral mixtures in poultry diets to supply the trace elements (except manganese) has not been demonstrated.

Mineral mixtures. Prepared mineral mixtures often are not necessary or economical. Feed mixers should make a study of the mineral requirements and the proper additions should be made in the form of standard mineral products of known content.

Charcoal. Ohio State University reports that "there is not sufficient experimental evidence to warrant its (charcoal) use in poultry rations." Charcoal has a destructive effect on vitamins A, K, riboflavin, and probably others and contains no available nutrients, except calcium and phosphorus in very small quantities.

*Defluorinated rock phosphate (superphosphate) is a good source of Ca and P.

Tonics and condiments. The best "tonic" is a well-balanced diet. Money spent for tonics would be better spent providing a complete diet. Most condiments are of the "shotgun" type; they have small amounts of many different minerals and other nutrients. Their apparent success in some cases may be due to the fact that the one essential element lacking in the diet was provided in the tonic. However, this method of approach is unscientific, expensive, and unsatisfactory.

DDPD (N^1, N^1-diphenyl-p-phenylene-diamine) is an antioxidant intended to protect the natural tocopherols (vitamin E) present in feeds and to act as a stabilizer for carotene and vitamin A. It sometime is used by commercial mixers at the rate of 0.0125% or ¼ lb. per ton of broiler (chicken) ration; and should be of equal value in deterring oxidation in turkey rations.

CHAPTER 5

NUTRIENT REQUIREMENTS

OUTLINE

The feed supplied to poultry may be broken down into the following general classes of chemical compounds:

(1) water, (2) proteins, (3) carbohydrates, (4) fats, (5) minerals, (6) vitamins and antibiotic, (7) extractives.

The following discussion is intended to give information insofar as it is available, of the uses and principal sources of the nutrients in the feed and the turkey's nutrient requirements.

WATER

Water makes possible the processes of digestion and absorption, transports waste products, softens feed, and cools the body by evaporation in air sacs and lungs. It makes up about 59% of the edible parts of dressed, fat turkey males and 54% of comparative females. All feed contains some water but the principal sources are fresh green feed, liquid milk, and the water supplied for drinking. Water should be supplied ad libitum to all classes of turkeys at all times except in some instances when liquid milk is fed ad libitum. Ample drinking space (see Chapter 2) should be provided, especially when starting poults in the brooding quarters since many young poults die

Fig. 5.1. Large water containers conserve labor and insure a constant water supply.

from water starvation at the start because they never learn to drink. Dirty, very cold or hot water is to be avoided and daily cleansing and rinsing of water containers is essential (see Fig. 5.1).

WATER CONSUMPTION

Turkeys need 2 pounds (1 quart) of fresh water for each pound of grain and mash eaten; according to results at the Purina Mills Experimental Farms. The water consumption of growing turkeys was carefully observed and measured by Morehouse (1949) ; the data being given in Table 5.1.

Table 5.1—Water Consumption of Broad Breasted Bronze Turkeys; Test Lot of 165 Birds Under Iowa Conditions

(Date of Morehouse, **Poul. Sci.** 28:151-153, 1949)

Week (age)	Water consumed per 100 turkeys	Week (age)	Water consumed per 100 turkeys
	gallons		gallons
1	8.34	14	105.58
2	11.65	15	117.36
3	17.56	16	123.55
4	25.53	17	112.63
5	34.08	18	125.48
6	42.01	19	118.46
7	58.89	20	116.75
8	55.50	21	104.46
9	62.22	22	96.48
10	68.43	23	96.62
11	81.17	24	96.24
12	93.90	25	102.10
13	100.05	26	103.10

No attempt was made to control or measure losses from evaporation but wastage was kept to a minimum. Most of the turkeys were hatched April 13 to 17, 1949. They were raised in confinement.

CARBOHYDRATES

The carbohydrates (starches, dextrin, sugars, pentosans, celluloses and certain gums) contain the elements carbon, hydrogen, and oxygen. The latter two elements are present in the same ratio as they are in water, two hydrogen and one oxygen (H_2O). In chemical analyses of feedstuffs, the carbohydrates are classified into two groups, nitrogen-free extract and crude fiber. During digestion the carbohydrates are broken down into simple sugars and are absorbed by the body in that form. They may be stored to a small extent as glycogen (or animal starch) in the liver and muscles.

Nitrogen-free extract on the whole is highly digestible by animals, including poultry. The starches and sugars in particular are highly digestible. They are readily oxidized (burned) in the bird's body to produce heat or are converted into fat and stored in that form. Because of these properties they are valuable in fattening diets. The pentosans are not readily digested by poultry and are, presumably, not valuable constituents of poultry feeds. The chief sources of starches and sugars in turkey feeds are green feeds (fresh or dried), grains, and grain byproducts. The birds' requirements for starches and sugars are well satisfied by the products generally fed.

Crude fiber represents a class of carbohydrates that apparently is not digested to any appreciable extent by chickens. It appears to have some value in the diet of turkeys although the nature of its use is not understood. Turkeys appear to require some fiber at all ages, possibly merely for mechanical purposes in the utilization of the more valuable nutrients. The principal sources of crude fiber in turkey feeds are oats, fresh green feed, and dried peanut meal, millet, wheat bran and wheat middlings, and rice bran. From available experimental results and from practical experience, the following recommendations are made.

For starting mashes fed under all conditions, practical limits are 3½ to 8% of crude fiber, 4 to 5½ the optimum; for growing and laying mashes 3½ to 7% where fibrous litter, high-fiber grain, or green feed is furnished, and 6 to 9% if roughage is not furnished. Crude fiber in the form of tough coarse material such as straw, hay, and fibrous green feed is quite detrimental to young poults, especially those under two weeks of age. Because of this, care must be taken in the management and feeding in order to prevent impaction of the digestive tract and other disturbances. For small poults the necessary roughage should be supplied by the crude fiber in the feedstuffs and in addition if desired, by tender, finely cut green feed or green feed picked by the birds from the yards. Barley and most types of soybean meal have a fair amount (around 6%) of fiber. The Kansas Station found that fiber supplied in the form of 10% of alfalfa stem meal or 5% of wood pulp when added to a starting diet already containing 4.6% crude fiber markedly improved growth in turkeys. On the other hand, Cornell University obtained excellent results with 3.5% fiber in a high-energy starting diet. Other experimental results have indicated that the lower percentages of fiber (3½ to 5½) are desirable during early growth stages. At other ages the fiber content of the feed may vary between limits of 4 to 9% of total feed. In pelleted 20% protein growing mashes Almquist (1953) suggested a fiber level of "6 percent or more" for ranged turkeys. He noted a feed cost per pound of gain 1 cent less on pellets containing 6.64% fiber than on those containing 5.5%, fed free choice with wheat (grain) which resulted in a fiber level of total feed of 4.96% and 4.67% respectively. The turkeys ate more of the low-fiber wheat with the high-fiber pellets which presumably accounted for the lower feed cost.

FATS

Fats are composed of the elements carbon, hydrogen, and oxygen, the latter in smaller quantity than in carbohydrates. Some fats also contain phosphorus. The common fats are chemical combinations of glycerin and fat acids, usually stearic, palmitic, and oleic. Fats provide heat and energy and are considered to be 2¼ (to 2½) times as potent as either proteins or carbohydrates for that purpose. However, they are digested slowly by poultry, and have a growth depressing effect when used in amounts much above 7% of the diet. However, the latter effect was relieved when 15 p.p.m. or procaine penicillin G was added, allowing the use of up to 10% corn oil added to a normal turkey starting mash (Ontario Station, 1952).

Fats are desirable in the diet to promote feather growth, egg production, and the production of prime market poultry. About 21% of the (dressed) edible portion of U. S. A A Grade turkey males and 28% of comparative females is fat. The bird or animal derives the fat used in its body processes from three sources; the fat, protein, and carbohydrates that are in the feeds eaten.

Fats occur in most of the body tissues and are carriers of the fat-soluble vitamins. Moreover, the substance formerly called vitamin F is composed of the fat acids linoleic and linolenic. Liberal deposits of fat in and under the skin, between the muscles, and around the internal organs are always found in high-quality roasting turkeys.

Principal sources of fats among the feedstuffs may be found in Table 5.6 (page 293). The turkey's requirement for fat is relatively low, presumably 3.5 to 4.5% of total diet. If this amount is not supplied by the feeds, the Cornell and Ontario Stations suggest adding enough stabilized grease or tallow to raise the fat content to 3½-4½%. The Beltsville Station (1952) showed that a fat content of 6 to 7% in the starting diet was conducive to increased tenderness and juiciness in frying-age turkeys. Fish oils, however, could not be used safely to bring up the fat content due to off-flavors produced. Feed products with large percentages of fat are subject to deterioration through age, heat and other factors, especially if the fats are unsaturated.

PROTEINS

The term crude protein, usually shortened to protein, includes all the nitrogen containing compounds in feeds. All proteins contain the elements carbon, hydrogen, oxygen, and nitrogen. Nearly all contain sulphur; many contain phosphorus; and a few contain iron, copper, iodine, manganese, or zinc. Simple proteins are combinations of the amino acids only. Examples are egg white and milk albumen. Conjugated proteins are those consisting of the amino acids combined with some non-protein compound such as nucleic acid, sugars, phosphoric acid, fats, hematin, or hemocyanin. Examples of conjugated proteins are casein and hemoglobin. When proteins are digested they are split up into their component parts, the amino acids, and when these are utilized by the animal body they are recombined to form animal proteins and other necessary substances.

About 35 *amino acids* have been reported and of these 23 are accepted as being generally present in proteins, though each protein does not necessarily have all 23, some proteins having all or nearly all while others are not so well supplied. Essential amino acids are those required by the animal but which cannot be made in its own body. They must be supplied in the feed. The best proteins are those containing all of the essential amino acids which, according to the most recent work are; (1) lysine, (2) tryptophan, (3) histidine, (4) phenylalanine (5) leucine, (6) isoleucine, (7) methionine, (8) valine, (9) threonine, and (10) arginine. For chickens glycine should be added as an essential amino acid. Turkeys, however, appear not to need glycine. They apparently are able to synthesize it.

Each amino acid is the same regardless of source, hence if the animal's need for these substances can be satisfied, the type of feed from which they are derived is no problem in itself, for example, animal feeds. The animal's need for each of the ten essential amino acids is critical—if any one is completely lacking, growth will stop and death will ensue eventually. As a rule, amino-acid deficiencies do not produce special symptoms. However, there are some exceptions. A deficiency of lysine, in turkeys may cause white wing bar or achroma. Tryptophan deficiency in chicks may cause enlargement and erosion of gall bladder, swelling of seminiferous tubules and degeneration of the spleen. Arginine deficiency reduces sperm count as well as motility and methionine is necessary for healthy livers in livestock. It furnishes methyl groups for choline synthesis. Glycine in excess of 1.5% of the diet will retard poult growth according to the California Station (1948).

While all ten of the essential amino acids must be supplied in certain amounts, it is considered advisable to supply five more of the non-essential ones which are commonly found in proteins: cystine, (contained in large amount in feathers), glycine, glutamic acid (important in flavor of meat), proline, and tyrosine. Turkeys do not store amino-acids hence all necessary ones must be on hand if the bird is to reconvert the proteins of its feed into body tissues. Surplus or unusable amino acids are deprived of their nitrogen in the liver, resulting in sugar residues which are stored as fat or burned for energy, an inefficient use for the normally expensive proteins.

The amino-acid requirements for growth and egg production theoretically are different because the bird's muscle and its eggs differ somewhat in amino-acid composition, but at present only one standard, that for growth, is utilized in formulating feed, the assumption being that the laying hen will be able to utilize a well-formulated growing type of feed effectively, so far as the amino-acid content is concerned. It appears, however, that turkey breeding birds need a more liberal supply of cystine and methionine (the sulphur containing amino acids) to manufacture egg protein and to molt, while the growing poult needs ample lysine to develop muscles and proper feather color and texture.

In amino-acid terminology the prefix L, as in L-lysine, usually means the naturally occurring form. Synthetic amino acid always is DL except in the case of glycine. The DL can be separated into D and L but this is an expensive procedure. Synthetic DL methionine is now being used commercially in broiler feeds. D methionine and

D leucine are utilized by chicks and presumably by poults. Most other D amino acids are not.

Amino acid requirements in general appear to go up as the percentage of protein in the diet increases. Lysine requirements go up when an antibiotic and/or B_{12} supplement is added (Ontario Station 1951). High-energy (low-fiber) feeds require more of the amino acids and some vitamins and other nutrient factors may increase these needs. In table 6 will be found some of the available data on amino-acid requirements of poultry. The Cornell Station (Scott, 1953) estimated the requirements for some of the amino-acids by turkeys fed a modern high-energy, 28 percent-protein starting diet containing an antibiotic. These estimates are shown in parentheses following the percentages required according to various workers using different types of diets. It was suggested by Scott that, in the absence of turkey reesarch data the chick requirement for an amino acid be increased by 40%. Although glycine, tyrosine, and cystine are not considered as strictly essential, it is desirable and easy to supply them

Table 5.2—Amino Acid Requirements of Poultry

Amino acid	Percent required in the diet for growth[1]	Species
Arginine	1.35 (1.6)	Turkeys (Calif.)
Cystine	0.3 (0.35)	Turkeys (Calif.)
Glycine	0.9 (1.1)	Turkeys (Calif.)
Histidine	0.3	Chicks
L-Isoleucine	0.8 (0.84)	Chicks
Leucine	1.4	Chicks
Lysine in 20% protein starter, no antibiotic	1.2 -1.3	Turkeys (Ont.)
Lysine in 24-30% protein starter, no antibiotic	1.3 -1.5	Turkeys (Calif.)
Lysine, with an antibiotic 28-29% pro. starter	1.8 -1.9	Turkeys (Ont.)
Methionine with cystine	0.42-0.53	Turkeys (Calif.)
Methionine without cystine	0.8	Turkeys
Phenylalanine without tyrosine	1.6	Chicks
Phenylalanine with tyrosine	0.8	Chicks
Threonine	0.6	Chicks
Tryptophan, DL	0.26 (0.28)	Poults (Calif.)
Tryptophan, DL	0.41	Young poults (Conn.)
Tyrosine	0.8	Chicks
Valine	0.8	Chicks

The use of the data in the foregoing table serves as a guide to the requirements, along with the information in Table 5.12 (the amino-acid content of feedstuffs, page 305 makes it possible to speculate on the amino acid values of most turkey diets likely to be formulated. However, the present incomplete knowledge of amino-acid requirements of livestock and the amino-acid constitution of feedstuffs does not allow great dependence to be placed upon these standards.

[1]Figures in parenthesis are Cornell estimates for 28% protein, **high-energy** diet with an antibiotic.

in the proportions indicated as required in Table 5.2. Cystine is not required if there is enough methionine to supply the need for both since methionine can be converted to cystine (but not vice versa). Likewise tyrosine is not needed if there is enough phenylalanine to supply the needs for both, although glycine apparently can be synthesized by turkeys, enough of it as well as tyrosine and glycine usually is supplied by the feedstuffs comprising the average turkey diet.

In contrast to all the accumulated evidence indicating that the protein requirements of turkeys are relatively high as reflected in the recommendations in this text, comes a news report from the Stine Laboratory (Baldini, *et al,* 1953). In the work reported, Jersey Buff poults made as good growth on a practical type 20-percent protein starter supplemented with small amounts of L-lysine hydrochloride and synthetic D-L methionine as they did on a typical 28% protein starter. Both diets contained procaine penicillin. This finding points toward a possible saving in feed costs through the use of synthetic amino acids which may prove to be a desirable and economical practice if and when these products are made available at reasonable cost. So far only synthetic methionine has been offered as a commercial feed supplement and its use in poult starting diets at levels of ½ to 1½ pounds per ton has been reported favorably by several experiment stations. In nearly all cases methionine at 1 pound per ton was shown to be economical; the methionine usually produced faster growth and always improved the feed efficiency. All-vegetable-protein diets showed greatest improvement from the use of methionine. At the Utah Station (1953) it was shown that Austrian and Alaska peas could be used as a protein supplement in poultry starters if methionine is added. The synthetic D-L methionine of commerce is reported to have good keeping qualities if kept well-packaged.

Antibiotics do not in general increase the amino acid requirements; on the contrary the effect tends to be in the opposite direction (Almquist, 1953). However, an antibiotic is not required in order to obtain good results from methionine. In some experiments the addition of 1 lb. per ton of methionine to a non-antibiotic basal diet produced growth increases and feed efficiency equal to those obtained from an antibiotic and sometimes when both were used, growth increases due to methionine were 10 to 34% over those obtained from the antibiotic alone (Texas Station, 1953). It also was reported that the methionine (1 lb. per ton in the starting feed) produced 68 to 80 lbs. more of turkey meat for each ton of feed at 8 weeks of age. The full explanation for the favorable effect of the free D-L methionine in the synthetic product is not yet available.

Quality of proteins depends fundamentally upon the number and kinds of amino-acids composing them. Quality in a single protein or protein-bearing feed is linked for all practical purposes, to the supplementary values of other proteins that may be fed with it. Various Experiment Stations have devised means of measuring the feeding value of the common protein supplements. These values have been variously termed "biological value," "protein efficiency," "protein quality" and "protein value." The results with various feedstuffs vary considerably but they all agree in giving milk and steam- and vacuum-dried fishmeals a high rating as poultry feeds. This high rating is in a large measure due to the vitamins and trace elements which milk and fishmeals carry. Recent research indicates that when the proteins of soybean meal and peanut meal are properly supplemented virtually equal results are secured as when animal protein feeds are fed. Meat scrap in all cases stood midway between a high and a low rating. The Washington Station calculated that at prevailing prices the available protein in meat scrap cost almost twice as much as that of herring meal. These findings are most significant and render necessary a revision of the current plan of evaluating the common protein supplements exclusively on the cost per pound of their crude protein. Until the picture is clarified, however, it seems desirable to make no drastic changes in evaluating feedstuffs except that of lowering the value of meat scrap and raising that of vacuum- and steam-dried fishmeals.

The protein of corn is of low quality because it contains no lysine and only a small quantity of tryptophan. However, properly supplemented it produces excellent results. The proteins of milk, oats, wheat, and barley are of better quality because they contain three of the more important essential amino acids, lysine, tryptophan and histidine.

Proteins are used in the bird's body to build muscle, connective tissue, skin, feathers, internal organs, and eggs. They furnish energy when supplied in excess of building needs. Protein comprises about 19% of total (dressed) edible parts of fat (U. S. AA Grade) turkey toms and 17% of comparable hens (toms 18 lbs., hen 10.8 lbs., live wt.).

Concentrated sources of proteins in feedstuffs of animal origin are milk, fish meal, meat scrap, and tankage; of plant origin the most common concentrated sources are soybean oil meal, corn gluten meal, distillers dried grains and solubles, linseed meal, cottonseed meal, and peanut meal.

The protein requirements of poultry usually are expressed in terms of percentage of crude protein in the diet. When turkeys are growing fast or laying at a heavy rate the percentage of crude protein required is higher than it is when growth is slow or no eggs are being laid.

From the standpoint of economical production, the optimum quantitative crude protein requirements of turkeys as nearly as can be determined from the experimental evidence available are as shown in Table 5.3. The figures on protein requirement by periods are based on data from medium and large-sized strains of turkeys in mixed flocks, half males and half females. Limited data on small-type turkeys indicate that their protein requirement curve descends a little more rapidly due to the fact that they mature faster than the larger strains, reaching Grade A roasting condition about 4 weeks earlier, 22 to 24 weeks instead of 26 to 28 weeks. The slight differences are indicated in the week-by-week data in Table 5.3.

The Borden Laboratory (1947) reported that the protein, calcium requirements of standardbred Bronze poults apparently were similar to those of the Broad Breasted Bronze variety and (1949) that starters with 24 and 28% protein gave best results in growth and feed efficiency; those with 20% or 32% protein producing slower growth and those with 36% protein from meat scrap and soybean meal producing still slower growth and swelled feet. Other recent (1951-53) reports have favored starting diets, usually fed without grain, containing 28 percent protein and about 4 to 5½ percent fiber for both Broad

Table 5.3—Crude Protein Requirements of Turkeys (estimates)

By periods	Percentage protein in total feed consumed	
	Optimum	Acceptable range
1 to 4 weeks incl., both sexes	28	23 - 30
5 to 8 weeks incl., both sexes	24	21 - 30
9 to 12 weeks incl., both sexes	21	18 - 26
13 to 16 weeks incl., both sexes	19	16 - 24
17 to 20 weeks incl., both sexes	17	15 - 21
21 to 24 weeks incl., both sexes	15	13 - 19
25 to 28 weeks incl., both sexes	14	12 - 18
Weighted average, 1 to 28 weeks, incl.*	17.28	14.8 - 19.4
29 weeks to sexual maturity	13	11 - 17
Breeding males	13	11 - 17
Breeding females	16	14 - 20

*Based on pounds of feed eaten per bird during each period by Broad Breasted Bronze turkeys of both sexes according to Table 5.23.

Table 5.3—Crude Protein Requirements of Turkeys (estimates) (Continued)

By weeks	Optimum crude protein requirement as per cent of total diet*	
	Large and medium-sized strains	Beltsville Small White
0 (at hatching)	28.0	28.0
1	27.1	27.1
2	26.3	26.3
3	25.5	25.5
4	24.8	24.8
5	24.1	24.0
6	23.4	23.3
7	22.7	22.6
8	22.0	22.0
9	21.4	21.3
10	20.8	20.6
11	20.2	20.0
12	19.6	19.3
13	19.0	18.6
14	18.5	17.9
15	18.0	17.3
16	17.5	16.7
17	17.0	16.1
18	16.5	15.6
19	16.1	15.1
20	15.7	14.7
21	15.3	14.3
22	14.9	13.9
23	14.5	13.5
24	14.2	13.2
25	13.9	13.0
26	13.6	13.0
27	13.3	13.0
28	13.1	13.0
29 + (maintenance diet)	13.0	13.0
During the breeding season	16.0	16.0

*Turkeys will tolerate variations in protein intake on the order of 5 per cent below to 2 per cent above these figures from hatching through 4 weeks; 3 per cent below to 6 per cent above from 5 through 8 weeks; 3 per cent below to 5 per cent above from 9 through 16 weeks; and 2 per cent below to 4 per cent above from 17 weeks on, including the breeding season.

Breasted Bronze and Beltsville White turkeys either with or without an antibiotic. As is indicated in Table 5.3, protein requirements decrease rapidly so that at 4 weeks of age the optimum is around 25%. Growing mashes fed with grain starting at 8 or 9 weeks popularly contain around 21 percent which is reduced gradually as the turkeys consume more grain. However turkeys grow well after 8 weeks on growing mashes containing all the way from 18 to 36 percent protein as long as grains are fed free choice.

Sources of protein. In the average turkey diet from 30 to 70% of the protein contained is supplied by the grains, mill byproducts, and dried alfalfa products. The remainder is supplied by concentrates (or protein supplements) either of plant or animal origin or both. The higher the percentage of total protein the higher is the percentage that must be supplied by the protein concentrates. For best results the grain portion of growing, finishing and wintering diets should consist of 2 or more grains or products derived from them, while in starting and breeder diets, it seems best to use three or more grains. The supplementary protein should be supplied from two or more sources in starting and laying diets, while in growing, finishing, and wintering diets only one source need be used if access to green range is provided. If green range is not provided, the supplementary protein should come from two or more sources.

Adjustment of protein intake. Since it usually is impracticable or inconvenient to alter the protein content of starting and growing diets every few weeks to fit the changing needs of both sexes of growing birds, some other method must be found to do so. Fortunately, growing turkeys have demonstrated their ability to make their own adjustments in protein intake if they are given access to a suitable mash and a grain or grains.* Experiments have shown that the most satisfactory plan of feeding is

*See pages 324 for mash-grain consumption, and 308-315 for information on palatability of feedstuffs, and planning of diets.

that of giving the poults about a 28% protein starter the first 8 weeks, adding scratch grain by the end of the sixth to eighth week and keeping both mash and grain before the birds from then on and changing to the 18 to 22 per cent grower by the beginning of the 9th week. Growing mashes with 20 to 22% crude protein are favored, except in cases where it is desirable to make the best possible use of home-grown whole grains. In these cases, growing concentrates with about 30% protein are best.

In cases where it is desirable for experimental or other reasons to feed all-mash diets to growing turkeys throughout their lives, the data in Table 5.3 can be used as a guide to the formulating of a series of growing diets containing proper percentages of protein. Such changes could be made at weekly or greater intervals throughout the growing period. In 1953 at the Beltsville Station, Beltsville White turkeys of both sexes were economically and successfully grown on a series of three all-mash diets containing an aureomycin-B_{12} supplement as follows: hatching to 10 weeks inclusive 23 percent protein; 1 to 17 weeks inclusive, 20 percent protein, and 18 to 24 weeks inclusive, 17 percent protein. These diets were modeled after starting diet No. 15 (USDA starter).

Experience and observation indicate that laying turkeys do not adjust their protein intake as well as growing turkeys, hence it is best to feed either a 15.5 to 17% protein all-mash laying diet ad libitum or a 22 per cent protein laying mash and a grain mixture limited to one-half the total feed intake, including that of the males, in the mated pens. The grain in this case may be mixed with the mash on a 50-50 basis and the resulting mix fed in the manner of an all-mash diet. Likewise a strictly all-mash diet may be altered by feeding in whole or cracked form all or part of the grains normally fed ground. The resulting mix is believed by some to be more palatable than an all-ground mixture. The pelleting of mashes generally is satisfactory but, unless wind wastage is a problem, there probably is no advantage in pelleted mash over loose mash for most classes of turkeys.

MINERALS

The term mineral refers, in feeding, to the inorganic elements found in the ash after a feedstuff is burned. These elements are found in all body tissues and some of them are necessary to normal functioning of the body organs. Through them the body maintains its acid-base balance and the correct surface tension and osmotic pressure relationships. Egg shells are largely calcium and this element is needed in the clotting of blood; bones are largely calcium and phosphorus; calcium is needed for the efficient elimination of excess phosphorus in the diet; copper and iron are essential blood elements—deficiency causes anemia; iodine is necessary in the functioning of the thyroid gland; sodium and potassium are required in maintaining an acid-base equilibrium of the body; chlorine is important in the digestive process; and manganese is essential in the blood and for maintenance, growth, hatchability, normal bone growth, and the prevention of perosis. About 1.0% of the (dressed) edible portion of U. S. Special turkey toms (32 weeks old) is ash; 0.93% for comparable females.

Calcium, phosphorus, and manganese are the inorganic elements with which the turkey feeder is most concerned. Improperly developed bones usually are caused by a deficiency or improper balance of these three elements, sometimes combined with vitamin deficiencies. The other mineral elements are believed to be supplied in sufficient quantity in the feeds generally given to poultry. Not only the net quantities of calcium* (Ca) and phosphorus (P) but relative quantities of each, (the calcium-phosphorus ratio) are important in poultry nutrition. Calcium carbonate ($CaCo_3$), the form of calcium found in limestone and oyster shell, is believed to be a better

*Calcium is the important element in lime, hence the two terms are sometimes used interchangeably in feeding parlance.

source of calcium than calcium chloride, or calcium phosphate. High-calcium limestone containing about 98% of $CaCo_3$ is as effective as chemically pure calcium carbonate. Gypsum and low-calcium dolomitic limestone containing only about 50% $CaCo_3$ were both found to be satisfactory at the Nebraska Station. However, dolomite has been reported as unsatisfactory in chicken diets by other investigators. This variation in results is probably due to variations in the composition of dolomite, as well as a species difference. Oyster shell, calcite and steamed bonemeal are good sources of calcium.

Concentrated sources of phosphorus are steamed bonemeal, dicalcium phosphate, defluorinated rock phosphate and defluorinated superphosphate. Good sources are meat scrap and fish meal. That the phosphorus in plant products (phytin-P) appears to be well utilized by poults was shown by the Borden Laboratory (1947) and the Connecticut Station (1950); especially in the presence of D-activated animal sterol so that in general plant phosphorus may be counted on when formulating turkey diets. Excess phosphorus in turkey diets is undesirable, hence phosphorus containing supplements never should be fed free choice. If excess P has to be fed, sufficient Ca also should be provided to keep the Ca-P ratio within recommended limits. With or without an antibiotic a level of Ca of 2.0 was found to be most effective for poults along with 0.55% of inorganic phosphorus. In the same experiment (Ontario Station 1952), 0.4% of inorganic phosphorus gave inferior results while 0.85% inorganic P was no better than 0.55%.

As a result of these and other experiments, the following recommendations concerning calcium and phosphorus in the diet of turkeys are made: (1) In starting and growing diets, all-mash basis, 1.0 to 3.0 per cent of calcium, from 1.9 to 2.1 per cent preferred; phosphorus 0.6 to 1.8 per cent, about 1.0 per cent preferred about 0.55% from inorganic sources. (2) In wintering (maintenance) diets, .5 to 3.0% calcium, 1.0% preferred;

and .4 to 1.8% of phosphorus, .6% preferred. These figures apply to starting mashes fed with or without scratch grain, and to all-mash growing and wintering diets. In the case of 20% protein growing and wintering mashes fed with scratch grain, the mash should contain 2.25 to 3% calcium, about 2.7% preferred; and 1.0 to 1.8% phosphorus, 1.5% preferred. The calcium-phosphorus ratio of the mash in this case may range from 1.7:1 to 2.5:1, 1.8:1 preferred. (3) In laying or breeder diets, all-mash basis, calcium 2.0 to 2.5 per cent, 2.3 per cent preferred; phosphorus 0.7 to 1.0 per cent, .8 per cent preferred; Ca/P ratio 2.3:1 to 3.1:1, 2.9:1 preferred. If scratch grain is fed, the percentage should be doubled or a more accurate adjustment made. For all turkeys raised in confinement, the percentages indicated as preferred should be supplied. Whenever the diet can be so planned, the calcium-phosphorus ratio should always be considered. If it is necessary to include larger quantities of phosphorus than needed, it should be balanced suitably with calcium and manganese, increasing these elements as suggested.

Manganese deficiency in the diet of breeding stock produces high embryo mortality and in the diet of poults and chicks it causes slipped tendon. Results at the Kentucky Station show that manganese is essential in the turkey diet; and that manganese deficiency produces low hatchability in breeders, and *slipped tendon (or perosis)* in turkeys under 8 weeks. It probably is a causative factor in the production of crooked breast bones. Fortunately it is found in quite liberal quantities in rice bran, wheat, wheat bran, wheat middlings, and green feed. Orchard grass was found by the Idaho Station (1938) to be very high in manganese, containing about 4 times as much as alfalfa.

Recent research has indicated that the perosis-preventing properties of manganese are nullified to some extent by a high percentage of inorganic phosphorus, and possibly of calcium, in the diet. It also has been found that there is an organic factor that must be supplied if the perosis-preventing properties of manganese are to be expressed. This factor is *choline,* which is fairly well distributed in feedstuffs. Fair to good sources are dried whey, cottonseed meal, sardine meal, herring meal, meat scrap, soybean meal, and barley. Poor sources are wheat bran, rice bran, corn, wheat, oats, alfalfa meal, and dried skimmilk. Starting diets such as those listed (Chapter 3) should prevent perosis in turkeys, as they nearly always will carry in the recommended feedstuffs sufficient manganese and choline to be well above the minimum level. However, in most cases additional manganese is recommended. Synthetic choline chloride is available for inclusion in feeds and is included in some of the recommended diets.

Where the percentage of phosphorus exceeds about 1.5% of the total feed intake, 50 to 60 parts of manganese are recommended. To obtain this quantity it is sometimes necessary to add manganese in the form of manganous sulphate, a pinkish material which is now available on the market as the tetrahydrate ($MnSO_4$-$4H_2O$) and in the anhydrous form ($MnSO_4$). At the rate of 4 ounces per ton of feed the tetrahydrate will supply about 31 parts per million of manganese while the anhydrous form will supply about 45.5 parts. These quantities when added to the quantity already present in the average practical diet insure an adequate supply of manganese without raising it anywhere near the dangerous level, and without measurable loss of vitamin D. A practicable method of adding manganese is to thoroughly mix 1¾ lbs. of anhydrous $MnSO_4$ (or 2½ lbs. of the tetrahydrate) in 100 lbs. of common salt. The resulting mixture is then used like common salt, ½% of it in the all-mash diet supplying about 31 parts per million of manganese. Dissolving the manganese sulphate in water and mixing the solution in the feed is recommended by the California Station. Green manganous oxide (MnO) containing about 70% manganese and precipitated manganese carbonate also are satisfactory sources of manganese but natural $MnCO_3$ (rhodochrosite) is not. (See Table 5.7 for per cent Mn in the feedstuffs and the sulphates.)

In turkey growing diets the authors have found that 30 to 35 parts per million of manganese are sufficient, even for birds grown in confinement, provided the percentage of phosphorus is kept down to the recommended amounts. However, it probably would be advisable to raise the level to 40 or 50 parts when practicable to do so. For breeders not given range it is important that 50 to 60 parts per million be provided. The Nebraska Station found that up to 385 parts per million of Mn were tolerated by turkey poults. The Wisconsin Station found that sunlight (but not vitamin D in feeds) had a sparing effect on manganese which means that birds exposed to sunlight have less rigid manganese requirements. The interdependence of manganese and the B-complex vitamins has been demonstrated, indicating that a liberal supply of 5 of them, including B_1, is necessary for complete protection from perosis, according to the California Station.

Table 5.4—Blood Levels of Ca, Na, and K in Turkey Hens
South Dakota Station (1948)
Milligrams Per 100 Milliliters

	Plasma	Whole blood	
	Ca	Na	K
Before egg production, average	12.68	203.6	209.9
At peak of egg production, average	19.76	213.2	153.3

(See Appendix for table of elements, symbols, and atomic weights.)

Careful attention should be given to the mineral content of diets fed to birds raised in confinement. When on range, poultry can tolerate much greater variations in this nutritional factor.

To summarize, about fourteen inorganic elements are needed in the diet of poultry. All of these elements are required in small quantities, some of them in minute quantities. These elements are calcium (Ca), sodium (Na), phosphorus (P), chlorine (Cl), sulphur (S), iodine (I), potassium (K), magnesium (Mg), manganese (Mn), zinc (Zn), copper (Cu), iron (Fe), and silicon (Si). Cobalt (Co) apparently is required. It is a constituent of vitamin B_{12}. Some other elements, among them arsenic (As), Nickel (Ni), bromine (Br), aluminum (Al), and boron (B), have been found in animal tissues. Aside from Ca, P, and Mn and possibly iodine it appears unnecessary to add any other trace elements since they are believed to be present in sufficient quantities in the feedstuffs normally provided. Excesses of the so-called trace elements may react unfavorably hence the "shotgun" feeding of these minerals is not advised. Furthermore, deficiencies of any of the trace elements sulphur, potassium, magnesium, zinc, copper, iron, silicon and cobalt are extremely unlikely when good standard diets are used. Sodium and chlorine are supplied in the salt (NaCl) normally included in poultry feeds; calcium, phosphorus, and manganese are in the standard supplements in common usage.

Data on the need for iodine by turkeys apparently are non-existent. It is well established that chickens and presumably other kinds of poultry as well, need some iodine to prevent goiter, or enlargement of the thyroid gland. How much iodine is needed has not been established. Possibly 0.5 to 1.0 p.p.m. is sufficient and this much would be supplied by using 0.5% of standard iodized salt (containing 70 p.p.m. of iodine) in the average practical diet. However, Titus (1949) recommends an allowance of 2.0 to 5.0 p.p.m. of iodine in chicken diets as a safety measure. About 2 p.p.m. could be provided by using 0.5% of a special iodized salt containing 400 p.p.m. of iodine. This much iodine added to the 0.25 to 1.0 p.p.m. contained in practical diets, brings the total iodine within the minimum allowance suggested.

The Mg requirement of poultry is believed to be about .04% of the diet and the potassium requirement about

.17%. There is a little danger of magnesium deficiency in ordinary poultry diets. Sulphur comprises 0.2% or so of eggs and 2.0% of feathers so sulphur-containing compounds are obviously necessary for normal functioning. Iron deficiency in turkeys is believed to produce wart-like growths on the feet and legs.

The element fluorine (*Fl*) when present even in minute amounts is toxic to chickens and probably also to turkeys. Rock phosphate and some other minerals which contain fluorine are likely to produce detrimental effects unless defluorinated rock phosphate is used.

Selenium poisoning or alkali disease was first noted in South Dakota where the Experiment Station found that the non-metallic element selenium (*Se*) when present in the feed of chicken hens in quantities exceeding 5 parts per million (p.p.m.) caused low hatchability and the chicks were weak, deformed, and had wiry down. Selenium has been reported in the soil of eleven western states: Montana, N. Dakota, S. Dakota, Wyoming, Utah, Colorado, Nebraska, Kansas, Oklahoma, New Mexico and Arizona. From the soil selenium is taken up by some forage plants and grains. In affected grains the selenium is found in the protein fraction. The South Dakota Station (Ann. Report, 1949) found that with selenium at a level of 15 p.p.m. in the breeder diet of Broad Breasted Bronze turkeys hatchability was zero. At 9 p.p.m. hatchability, egg production, and fertility were normal but the resulting poults had defective down. Seleniferous grains if eaten in large quantities, 65% or so of the diet, can provide enough selenium to cause trouble in both breeding stock and poults. The effect of selenium can be counteracted by adding very small quantities of arsenic to the salt or drinking water (Moxon 1952).

Principal Sources of Mineral Elements

In general, the free-choice feeding of minerals is to be discouraged because birds may eat too much of them.

However, the Pennsylvania Station suggests feeding limestone or oyster shell free choice to turkeys during the growing period in order to insure sufficient amounts of calcium to prevent a deficiency during the finishing period when the mash-grain ratio is decreasing rapidly. The authors, however, prefer to add the desired quantities of minerals to the feeds rather than feed them free choice.

Meat scrap, tankage, and bonemeal all are high in calcium, phosphorus, and magnesium; fish meal is a good source of calcium, phosphorus, and magnesium and contains iodine; oyster shell is a good source of calcium, and contains magnesium and iodine; limestone is high in calcium and contains magnesium; wheat, wheat byproducts, rice bran, and green feed contain manganese; crab meal is high in iron and iodine; salt is sodium and chlorine; grain, grain byproducts, green feed, and soil contain a variety of minerals and in all probability the requirements of turkeys for the rarer elements such as sulphur, iodine, copper, iron, and others are satisfied when the diet contains a sufficient variety of feedstuffs combined in a suitable manner. The inclusion of one-half per cent of common *salt* in the total diet satisfies the requirements for sodium and chlorine. The Washington Station reported that *defluorinated rock phosphate* may replace bonemeal pound for pound (if necessary) to supply phosphorus in chicken diets. *Dicalcium phosphate,* $Ca_2H_2(PO_4)_2$, if defluorinated, may be used as a substitute for bonemeal.

VITAMINS

Vitamins are accessory feed constituents which are necessary for the vital processes and as such are of great practical importance in feeding for growth, egg production, and hatchability. Minute quantities are required.

The vitamins are designated by letters, by descriptive names, or by chemical names when they are known, but the exact number of different and distinct vitamins is not known. There are many vitamins or vitamin complexes*

*A complex is a closely associated group of vitamins.

generally recognized and the existence of several more is suspected. The generally recognized vitamins, some of which may be vitamin complexes, are designated as A, B or B_1), C, D, E, riboflavin, K, B_6 (pyridoxine), biotin, choline, niacin, pantothenic acid, para-amino-benzoic acid, B_{12}, folic or pteroylglutamic acid and the "grass juice" factor. There are several other factors believed to be dietary essentials.

A diseased condition resulting from the lack or deficiency of a vitamin is called an avitaminosis although strictly speaking an avitaminosis is the result of a complete lack of a vitamin and a hypovitaminosis is the result of a partial lack (or deficiency) of a vitamin. A condition resulting from an excess is a hypervitaminosis.

The vitamin theory was postulated by Funk in 1912. All of the vitamins now known have been "discovered" since then although the foundation for their discovery was laid by the classical work of Eijkmann at the close of the nineteenth century. Hippocrates in 500 B. C. knew that night blindness, a form of vitamin A deficiency, could be cured in humans by eating ox liver. Much research work has been done recently on vitamins.

VITAMIN A

Vitamin A has been synthesized and given the chemical formula $C_{20}H_{30}O$. It is now available for feeding animals. Pure vitamin A is a viscuous, nearly colorless oil which is readily destroyed by heat in the presence of oxygen but not by heat in the absence of oxygen. Aging, light, and exposure to air under ordinary temperatures have a destructive influence. In poultry, vitamin A is necessary for growth, reproduction, hatchability, and health. Vitamin A is measured by the International or U. S. P. unit, and by the Sherman and A.D.M.A. units which equal approximately 1.4 International units. Because of an improved method of analysis (announced in 1950) for true vitamin A potency in oils, the guaranteed analyses of feeding oils have been reduced by one-fourth, although the actual content of the oils (of vitamin A) remains the same. The International unit is the activity of .0006 milligram of pure crystalline beta-carotene, deep red crystals which in solution are intensely yellow. When the vitamin A content is expressed in milligrams of beta-carotene, conversion to International units is easily done by dividing the number of milligrams of carotene by .0006 (or dividing by 6 and multiplying the result by 10,000). Vitamin A exists in 2 forms: A_1 found in the livers of salt-water fish, and A_2 in fresh-water fish. They differ physically and chemically but have similar biological effects.

Carotene ($C_{40}H_{56}$), a yellow plant pigment found in green parts of plants and in certain yellow parts of plants (yellow corn and yellow carrots) acts as a *precursor* of vitamin A in rats and birds, its conversion taking place in the liver. Thus for practical purposes in poultry feeding, carotene (sometimes called provitamin A) takes the place of the true vitamin. Carotene, however, is a substance distinct from vitamin A and like the latter is readily destroyed by heat in the presence of oxygen although not by aging. Temperatures over 120° F. are very destructive, while those under 50° cause but little destruction of carotene in alfalfa products. Ordinary summer temperatures are quite destructive, according to the Texas Station, while winter temperatures are not. After a year's storage, under average conditions, alfalfa meals may be expected to lose 50 to 75% of their original potency when freshly cured. Dehydrated alfalfa meals however, tend to retain their vitamin A potency very well due to the inactivation of the enzyme lipoxidase which causes the deterioration.

Yellow corn stored three years under favorable conditions loses about half of its vitamin A. Mixtures of carotene and vegetable oils are sometimes used to bring up the vitamin A content of commercial feeds and to preserve a good share of it until feed can be distributed and consumed. Carotene occurs in four forms: alpha, beta, gamma, and delta. Upon conversion, beta-carotene yields twice as much vitamin A as the other types. Cryptoxanthin, probably zeaxanthin, and some other substances also act as precursors of vitamin A.

Vitamin A and its precursors are unstable under average holding conditions. Feedstuffs containing them should be protected as much as possible from the air and from heat. The new dry vitamin A supplements are quite stable when mixed in feeds. The vitamin A in fish oils while quite stable as long as the oil is kept in sealed-in, darkened containers, is destroyed fairly quickly when the oil is mixed in feeds. The Wisconsin Station (1951) found that a good share (60%) of the added vitamin A (from cod-liver oil) remained in a feed mixture after 2½ to 3 months' storage at 91-97° F. When trace minerals (iron, copper, cobalt, and manganese) were included in the mixed feed in free form, the vitamin A was destroyed rapidly but when included as a gelatin-mineral mixture this destruction was largely prevented.

Other tests have shown that the vitamin A from fish oils which were first mixed with cottonseed oil and then mixed in feed and stored under average conditions was very well preserved during 8 weeks of summer temperatures and was moderately well preserved for as long as 21 weeks. The New Jersey Station made the following statement ". . . in mixed feeds . . . the vitamin A of fish liver oils . . . does not decrease in potency in significant amounts during a month and probably not during two months." Alfalfa and grass silages well prepared and preserved, are high in Vitamin A.

Different feedstuffs vary greatly in the power to retain vitamin A and its precursors. The estimates of vitamin A content of various feeds found in Table 5.9 are conservative and based on what may be expected under average conditions when the feedstuffs are held for normal lengths of time and under normal storage conditions before being fed. The estimates do not, however, make any allowance for loss of vitamin A in fish oils after they are mixed in feeds.

In turkeys, deficiency of vitamin A causes some of the symptoms listed for chickens. The Kansas Station found that it caused hemorrhagic enteritis in young poults and impaired the functions of the oviduct in mature turkeys. The California Station found in young poults, xerophthalmia* or watery eyes (without hardened pus), tracheitis, sinusitis, and whitish lesions in mouth, crop, esophagus, and the bursa. Deposits of urates in the kidneys or ureters and on the heart and other viscera were rare as was hemorrhagic enteritis. In adults, whitish-yellow, caseated pus collected in the eyes. The Connecticut Station (1949) reported ataxia (muscular incoordination) as a fairly constant symptom but found very little of the characteristic A-deficiency lesions.

Concentrated sources of vitamin A are dry vitamin A, cod-liver oil, halibut-liver oil, shark-liver oil, salmon-liver oil, and some other fish oils. Pilchard (sardine) oil, unless fortified with vitamin A, is only a fair source, since it normally contains only 12½ to 20% as much A as cod-liver oil. Excellent sources of carotene (provitamin A) are yellow carrots, succulent green feed, and properly cured, reasonably fresh alfalfa, other legumes and grasses. Palm oil has been reported as a good source, likewise red clover hay and yellow corn gluten meal. Yellow corn is rated a fair source. According to Cornell University, yellow corn contains about 7 U.S.P. units of vitamin A per gram while corn gluten meal contains 7 to 25 units. About 15 units per gram or 7,000 per pound is considered average. Standard poultry cod-liver oil should contain not less than 600 U.S.P. units of vitamin A per gram. The average is about 750 units per gram or 340,000 per pound. Shark-liver oil, a relatively new product, contains as high as 2,700,000 units per pound. Other fish oils sold as a source of vitamin A should contain at least 750

*Inflammation with thickening of the membrane lining the eyes.

units per gram. The fish oils sold as animal feed vary in potency from 600 to 3000 units per gram. Synthetic vitamin A in dry, stable form is now available in potencies ranging from 2000 to 250,000 I.U. per gram.

Dehydrated alfalfa leaf meal may be counted upon to supply about 264 units per gram or 120,000 per pound. Sun-cured meal may be practically equal in vitamin A potency to machine-cured products if hay-making weather is ideal. On the average, however, the dehydrated meal is rated considerably higher than average green-colored, field-cured meal. Improved methods of machine curing have resulted in increased vitamin A values in recent years. Brown-colored alfalfa products are never reliable sources of vitamin A, and, since carotene loss is faster than color loss, green color alone is no guarantee of high vitamin potency, though it is a prerequisite of it.

Fresh green alfalfa and grasses may contain 200 to 400 units and specially cured "short grasses" may contain up to 1000 units per gram. The products listed, plus other types of fresh green feed are the only dependable sources of vitamin A, or its chief precursor carotene, that are generally available to the poultry feeder. Some other feeds contain it but they are not generally available or the quantities of the vitamin contained are small or are too variable to be dependable.

Fresh green feed of almost any ordinary kind, if constantly accessible to the turkeys so they can pick it from the growing plants, will supply all the vitamin A necessary in growing, wintering, and laying diets. In starting diets some other source is desirable. The fish-liver oil and alfalfa products generally included in starting feeds will, if of normal potency when eaten, supply ample quantities of vitamin A. Dry vitamin A along with dry vitamin D makes it possible to eliminate, where desirable, the fish oils from the diet or reduce them to safe levels.

Turkeys of all ages require relatively large quantities of vitamin A or its precursor carotene. The California Station reported that Bronze turkeys required a ration including 8% of dehydrated alfalfa leaf meal containing approximately 130 gamma* of beta carotene (equal to about 217 U.S.P. units of vitamin A) per gram for normal growth to 30 weeks of age. This would supply 1736 U.S.P. units per 100 grams, or 7874 units per pound, of the feed mixture. The diet contained no source of vitamin A other than the alfalfa. The Connecticut Station (1949) reported a minimum requirement of Broad Breasted Bronze poults of 5440 I. U. per pound of feed and an optimum of 7700 I.U. The Colorado Station reported that poults required 4 times as much vitamin A (supplied by carotene) as chicks, that is, approximately 2700 units per pound. The Kansas Station reported that the vitamin A requirements of poults up to 11 weeks of age were satisfied by feeding 50 to 60% of yellow corn in the diet, no other source of the vitamin being available. This would total 350-375 U.S.P. units per 100 grams, or 1560 to 1625 per pound of feed. These results are in great disagreement since the California and Connecticut Station findings indicate a requirement about 5 times as great as that indicated by the Kansas experiments, while the Colorado results are in between. The difference may be explainable by the fact that the carotenoid pigments in alfalfa appear to be less efficiently utilized by poultry as vitamin A than are the precursors of the vitamin A that are found in yellow corn or vitamin A itself.** Therefore, if carotene is depended upon as the main source of vitamin A, the requirements might be very high (1700 units per 100 gms. feed) while if vitamin A is supplied mainly by yellow corn or fish oils or both, the requirement would be considerably less. The New Jersey Station found that in low-fat diets, carotene was less efficiently utilized than in normal poultry diets containing the usual 3 to 4 per cent of fat, and that in normal poult diets (1951) at least 2270 I.U. per pound of feed were required by young poults to prevent signs of deficiency. The National Research Council

*A gamma is one-millionth of a gram, or one microgram.

**Scientific evidence on these points is conflicting but the authors believe that the best evidence supports the above viewpoint.

(1950) recommends a minimum of 4000 I.U. per pound of feed for all classes of turkeys. The Borden Laboratory (1950) obtained optimum growth of poults on 2000 U.S.P. units of vitamin A per pound of diet regardless of source whether vitamin A alcohol, natural esters, or synthetic vitamin A palmitate. The poults utilized carotene from alfalfa efficiently when ample tocopherol was supplied but used it inefficiently in the absence of tocopherol (vitamin E). Some of the conflicting results in vitamin A levels may be due to the low protein or restricted vitamin content of the control rations used, resulting in retarded growth irrespective of the vitamin A level.

In practice, vitamin A usually is supplied in starting mashes well in excess of the Colorado figure and equal to three-fourths of the California recommendations. The high figure is desirable because of the instability of vitamin A and the great variation in the vitamin A content of feedstuffs. The recommendations of the authors when the accompanying tables are used, are 6000 units per pound of starting feed, 3500 per pound of growing feed, 2500 per pound of maintenance feed, and 5000 units per pound of laying feed. Where mash and grains are both fed, the vitamin A unitage (as well as the other factors) should be calculated with due regard to the consumption of both. In starting diets, however, only the mash need be considered because under most conditions the birds do not overeat of scratch grain even when it is fed, throughout the brooding period.

The Texas Station (1953) recommends 7000 units (I.U.) of vitamin A per pound of feed for growing turkeys and 5000 units for breeders.

Some general suggestions for supplying vitamin A to turkeys are: (1) In starting diets use about 0.15% of vitamin A (and D) fortified oils containing 2250 to 3000 units of A per gram or dry vitamin A to supply 1500 units of A, freshly mixed plus 5 to 10% of high quality alfalfa meal or leaf meal, plus 15 to 30% of yellow corn. (2) In all-mash laying diets use 0.2% of 2250 A-600 D oil mixed every 2 weeks, plus 20% or more of yellow corn, plus 4 to 8% of alfalfa leaf meal or special alfalfa meal, dehydrated preferred; otherwise provide access to growing green feed. (3) In growing diets, access to growing green feed usually is the most practical method. (4) If confinement rearing is practiced and no green feed given, use 30 percent yellow corn in the entire growing diet and 5 to 10% dehydrated, or 10 to 15% of good quality suncured, 17% protein alfalfa meal. If yellow corn gluten meal is fed, the percentage of corn may be reduced accordingly. (5) For the maintenance diets in the absence of green feed, use 30% yellow corn in the entire diet and 7% of good quality alfalfa meal (or 10% alfalfa, no corn) in the mash. Since fresh green feed supplies vitamin A (along with other vitamin factors) in large but unmeasurable quantities, the addition of it to a diet complicates the picture from a theoretical standpoint but simplifies it from the feeder's standpoint because accurate estimates of vitamin content need not then be made. In any case, fresh green feed no matter what the source should be reckoned with as an excellent source of vitamin A.

To prevent loss of vitamins, oil containing vitamins A or D should not be pre-mixed with certain substances, namely those mentioned under Vitamin D.

THE B COMPLEX

(1) *Vitamin B_1* formerly designated as B is soluble in water and alcohol and probably slightly soluble in fat. It is now available in pure form under the name of *thiamin*, thiamin chloride, or aneurin. The chemical formula for it is $C_{12}H_{17}N_4OSCl$. Vitamin B_1 is destroyed by heat and alkalinity, but not by oxidation, drying, or aging. It is necessary for growth and proper functioning of the nervous system. It is important in the utilization of carbohydrates by animals, being necessary to the production of the metabolic enzyme cocarboxylase.

The International Unit (I.U.) of vitamin B_1 is the activity of 10 milligrams of a standard adsorption concentrate of the vitamin on fuller's earth or 3 micrograms of reference standard thiamin hydrochloride. One gram of crystalline B_1 contains about 200,000 I.U. of the vitamin. It can be dissolved in water or water and alcohol, and sprayed on, or otherwise incorporated in, the feed of animals and the food of humans.

In humans, vitamin B_1 prevents beri-beri and in some cases stimulates the appetite.

Deficiency of vitamin B_1 in the diet of chickens results in the following symptoms; loss of appetite, retarded growth, emaciation, general weakness, inability to stand or exercise in a normal manner, paralysis, convulsions, head retraction, cartwheeling (a condition in which a bird lies on its side and kicks, resulting in a circular motion), impaired digestion and reproduction, decelerated heart action, subnormal temperature, and impaired internal secretion. The nervous symptoms described are collectively termed polyneuritis. Abrupt failure of appetite is characteristic. Postmortem findings are not marked although reduction in weight of testes, increase in the weight of the adrenal glands, and inflammation of the brain have been reported.

In the case of turkeys, no investigational work dealing with vitamin B_1 deficiency can be cited. The authors, however, have observed what appeared to be vitamin B_1 deficiency in poults. The following symptoms were noted; loss of appetite, retarded growth, emaciation, weakness, inability to stand and walk properly, convulsions, head retraction, and cartwheeling. Upon postmortem, inflammation of the large lobes of the brain was noted in a few cases. These symptoms resemble in a general way those of the anti-encephalomalacia factor (vitamin E) deficiency.

Vitamin B_1 is found in yeast, egg yolk, green feeds, oilcake meals, mill feeds and grains. Under ordinary conditions of feeding, no deficiency occurs. Rough calculations indicate that average rations contain two or three times the requirements for B_1. However, it is not impossible that the extensive use of degermed grain products may result in a deficiency. See Table 5.10 for thiamin content of feedstuffs and Table 5.15 for thiamin requirement of turkeys.

Russian workers reported that 20% of millet and 15% of wheat bran protected chickens fully against polyneuritis and permitted normal growth.

(2) *Pyridoxine* ($C_8H_{11}NO_3$), formerly was designated as B_6. Deficiency of it in poults causes slow growth, depraved appetite, poor feed utilization, hyper-excitability when disturbed, weakness, convulsions, and death. These symptoms may be prevented by the administration of 0.3 mg. of synthetic pyridoxine per 100 grams (1.35 per pound) of diet. Hatchability and egg production both drop when the diet is deficient. Good sources of this factor are cane molasses (but not beet), grains, meat products, grass, alfalfa meal, yeast, kidney and milk. With the common poultry feeds a deficiency of it is unlikely to occur. It is soluble in water and is destroyed by rancidity and radiation, ultra-violet and visible. Poults require, including a safety factor, about 2,000 micrograms of pyridoxine per pound of diet, according to workers at the Cornell (N. Y.) and California Stations. It is now obtainable in crystalline form.

In milligrams per pound, the average pyridoxine content of dehydrated special alfalfa meal is about 6; of dried brewers' yeast (feed grade) about 22.6.

(3) Nicotinic acid, or niacin ($C_6H_5O_2N$) or its amide, nicotinamide ($C_6H_6ON_2$) prevents black-tongue in dogs, pellagra in chicks and humans, and perosis in poults. The unit of measurement is the microgram. Best sources are wheat bran, rice bran, rice polishings, condensed fish solubles, dried pork liver, and dried yeast. Turkey embryos synthesize it. It is relatively stable. See Table 5.10 for niacin content of feedstuffs.

The Wisconsin Station (1947) showed that poults require nicotinic acid (niacin) for normal growth. Ten milligrams of niacin per 100 grams of diet (45 mg. per pound) stimulated growth by ⅓ over a low-niacin diet. Tryptophan, one of the essential amino acids, appeared to be the precursor from which turkey and chick embryos synthesized nicotinic acid.

Nicotinic acid deficiency in poults (Maryland Station, 1946) is indicated by poor growth, inflammation of the mouth, diarrhea, and perosis. Apparently 3 to 5 mg. per 100 grams of diet (13.5 to 22.5 mg. per pound) are required and this amount normally would be provided by the feedstuffs used in well-planned diets. Data from the Ontario Station (1953) indicates that the niacin requirement of young poults for growth and the prevention of perosis was 16.7 to 21.7 mg. per pound of feed with an antibiotic (penicillin) but was considerably higher without it. However, the Cornell Station (1952) reported that a total of about 32.5 mg. of niacin per pound of diet along with 20 mg. of alphatocopheryl acetate (a vitamin E supplement) or 2.5 mg. of ATA plus 1% dried brewers' yeast was necessary for the complete prevention of enlarged hock disorder (different from perosis) in poults 8 to 20 weeks old. In breeder diets the Cornell Station (1953) reported no need for niacin supplementation.

(4) The *filtrate* or chick *antidermatosis* or anti-pellagric factor has been identified as pantothenic acid, a compound known for several years. It is stable to moist heat and oxidation but is destroyed by prolonged dry heat. Its chemical formula is $C_9H_{17}O_5$. Liver and yeast are excellent sources. Very good sources are cane molasses, wheat bran, rice bran, peanut meal, dried milk, dried whey, dried fermentation solubles, condensed fish solubles, good quality alfalfa meals. Fair sources are fresh greens, wheat middlings, whole soybeans, soybean meal, peas, cottonseed meal, and all grains. (See Table 5.10 for pantothenic acid in feeds). From the experimental work so far done, the authors conclude that the feeds ordinar-

ily supplied in turkey and chicken diets contain enough of this factor to promote normal growth and well-being. However, it now is commercially available as calcium pantothenate for inclusion in mixed feeds and some nutritionists recommend that it be added to turkey mashes at the rate of 8 to 10 grams per ton. Pantothenic acid is necessary to the maintenance of the skin and spinal cord in chicks, and probably poults also. In poults the symptoms of deficiency are slow growth, quick, high mortality, broken feathers, and perosis. Dermatosis in the form of lesions in corners of the mouth and "stuck" eyelids are sometimes present, but often the poults die before showing these symptoms. Poults fed diets markedly deficient in this factor were almost all dead in three weeks in experiments at the California Station.

The Jukes-Lepkovsky unit of pantothenic acid, when multiplied by 635, results in conversion to milligrams of pantothenic acid per pound. The accepted unit of measurement is the microgram. Growing turkeys require about 6,000 mcgms. per pound of feed, breeding turkeys 6,000 to 7,500. The latter amount is suggested for sustained high uniform hatchability. The average pantothenic acid content of many feedstuffs will be found in Table 5.10.

(5) *Riboflavin* is discussed on page 273.

VITAMIN C

Vitamin C prevents scurvy in man, monkeys, and guinea pigs. It is essential to digestion and aids mineral assimilation. It builds up resistance to colds and to toxins and poisons of bacterial origin. It is water-soluble and is destroyed by exposure to oxygen, air, heat, age, and alkaline substances. It has been isolated in pure form under the name of ascorbic acid (cevitamic acid) which has the chemical formula $C_6H_8O_6$. The International unit of vitamin C is .05 milligram of ascorbic acid. Vitamin C is obtained almost solely from fruits, vegetables, and green plants.

Birds apparently are able to manufacture vitamin C from the feeds supplied to them, hence this vitamin appears not to be a problem in poultry nutrition.

VITAMIN D

Vitamin D ($C_{28}H_{44}O$) is known as calciferol, the sunshine vitamin, the antirachitic vitamin, and the rickets-preventing vitamin. When acted upon by ultra-violet rays (invisible rays) produced by the sun or by sun lamps, animal sterols, either in the skin or isolated in pure form, are converted into vitamin D which has the same chemical formula as cholesterol but a slightly different chemical structure or arrangement of the atoms within the molecule. The pro-vitamin D found in the skin of poultry is 7-dehydrocholesterol. University of Chicago workers reported that feathers and the preen gland of chickens contained no cholesterol and that the skin of the legs and feet had 8 times as much of it as the body skin. The Wisconsin Station reported that heads and feet of chickens had much more cholesterol than the body skin.

The beneficial effect of sunlight in preventing rickets is explained by the plant and animal sterol theory. That vitamin D exists in several forms has now been established. The first form of vitamin D to be isolated was named calciferol or vitamin D_2. It was obtained by a complicated chemical process from irradiated ergosterol. Irradiated ergosterol itself is a very effective antirachitic substance when fed to rats but is much less effective when fed to poultry, chicks requiring about 50 times as many vitamin D units as of irradiated 7-dehydrocholesterol or crystalline vitamin D_3 which is more stable than D_2. Irradiated animal sterols have become one of the most important sources of vitamin D available for poultry feeding. These sterols are now available principally in concentrations of 680,000 or 900,000 International units per pound or 1500 or 2000 respectively, per gram.

Vitamin D makes possible the assimilation of the minerals, calcium and phosphorus from the feed. Without vitamin D, calcification of the bones cannot take place and conversely vitamin D itself cannot function properly in promoting mineral assimilation if there is (1) too little calcium, (2) insufficient phosphorus, or (3), a calcium-phosphorus imbalance.

Vitamin D is fat-soluble, stable to heat, and reasonably stable to oxidation. Long storage, however, results in some destruction unless in air-tight containers. Mixing either fish oils or irradiated cholesterol with dried milk, dried whey, lactose, sucrose, oyster shell, manganese sulphate, mineral mixtures, and sand, and then mixing with feed and storing, resulted in a considerable loss of vitamin D in 2 months' time. Similar results were obtained when pre-mixed with meat scrap.

The International chick unit of vitamin D is the activity of 0.025 microgram (0.025 gamma) of crystalline vitamin D_3 (activated 7-dehydrocholesterol). The A.O.A.C.* chick unit is equivalent to one International unit of the kind of vitamin D supplied by cod-liver oil. Until January 1, 1951 it was the recognized unit for expressing the vitamin D content of all carriers intended to supply this vitamin to poultry but has been replaced by the International chick unit, the standard for which is a solution of crystalline D_3 in cottonseed oil containing 400 International units per gram. To convert A.O.A.C. units to International chick units, subtract 25%. Thus the recommended allowances for vitamin D are now 25% lower than previously, if based on statements published prior to 1951.

Turkey poults show in general the same rachitic symptoms and lesions that are found in chicks but may develop them in earlier life. Characteristic rachitic leg weakness in turkey poults 17 days of age has been observed. The Kansas Station observed the characteristic symptoms 18 days after poults (then 25 days old) were placed on rickets-producing diets. Upon autopsy these poults showed

*Association of Official Agricultural Chemists.

the beaded-rib condition. At six weeks of age the ash content of the femur (thigh) and tibia (drumstick) leg bones was 28.95% as against 48.55% for the bones of irradiated poults of the same age. The University of Nebraska found that the ash in the tibia of normal poults varied from 50% to 57% at 8 weeks of age and that in Bronze poults deprived of the D factor, rickets appeared as early as the 10th day of age. At 28 days of age the ash in the tibia varied from 24.7 to 28.5%. The California Station reported that crooked breastbones could be caused by vitamin D deficiency. Vitamin D deficiency in the diet of turkey breeding stock results in low hatchability, the embryos dying around the 22nd and 23rd days of incubation according experiments at the Kentucky Station.

The Pennsylvania Station (1950) found that egg production in White Holland turkeys was not measurably affected by variations of from 908 to 7264 A.O.A.C. units (680 to 5,448 I. U.) of vitamin D per pound of feed furnished by cholesterol or from 1,030 to 8,240 A.O.A.C. (770 to 6,180 International) units furnished by tuna-liver oil. Without any vitamin D, egg production was good, presumably due to stored vitamin D and soft-shelled eggs were not laid in unusual numbers. However, hatchability of fertile eggs was lowered drastically after about three weeks on the D-deficient diet but was not affected by the variations listed. The percentage of ash in the toes of 2-week old poults was directly related to the lack of vitamin D in the diets of the parent hens when the poults received 91 to 204 A.O.A.C. units (68 to 153 I.U.) per pound of diet. When the poults received 306 or 454 A.O.A.C. (230-340 International) units no differences due to nutrition of the dams were observed. Transmission of D from hen to poult through the egg yolk was somewhat better with cholesterol as the source than with tuna-liver oil. Poults fed adequate quantities of Ca and P utilized the D from both cholesterol and the oil equally well.

The invisible ultraviolet rays of the sun are the cheapest source of the vitamin D factor. Artificial ultra-violet radiation is satisfactory; likewise are fish-liver oils, irradiated cholesterol, and various blends of the last two into "A and D feeding oils." Irradiated cholesterol, being non-fishy, is a means of supplying vitamin D without danger of tainting the flesh. Although the antirachitic effect of ultra-violet irradiation is well known, only recently has it been shown to have other effects. In one report from the Connecticut Station (1953) Broad Breasted Bronze hens under ultraviolet light in the 2800-3200 Angstrom-unit range combined with ordinary incandescent lighting (4 a.m. to 6 p.m.) laid 15% more eggs than those under incandescent lights alone. Previously similar results with chickens had been reported from Beltsville. In neither case could the results be explained.

Sunshine varies in ultraviolet ray potency with the seasons and with atmospheric conditions, being much more potent when the sun is high in the heavens (April to September north of the equator) than other times of the year. Cloudiness and fog reduce the potency greatly. Very young turkeys do not have the opportunity of older birds to utilize the sun's rays effectively in obtaining the vitamin D factor. Because of these facts and circumstances, vitamin D should be supplied in some way in addition to the sunshine, to turkeys during the brooding period when they are grown in commercial numbers under artificial conditions. If this is not done, rickets may develop and irreparable damage be done the poults. If at all possible, direct sunshine should be provided in addition to the vitamin D in the feed as it appears to provide some factor not found in the vitamin-bearing feeds. This factor seems to aid in feather development. Good poults can be grown without sunshine but they develop better plumage with it. Breeding turkeys also are benefited by sunshine even though they may be receiving plenty of vitamin D in the feed. The Washington Station found that poults from dams receiving sunshine and fish oil weighed more at 2 weeks of age than did poults from dams receiving only the fish oil.

Laying turkeys if forced for egg production in winter or early spring should also receive vitamin D supplements in addition to the sunshine normally obtained. During the usual laying season for turkeys of February through June in the South, March through June in the northern States, no vitamin D supplements appear to be necessary provided the birds have the freedom of outside yards and the weather is warm and sunny. However, since conditions seldom are ideal, it is considered good practice to feed vitamin D to turkey breeding stock whenever kept under commercial conditions.

Growing turkeys past the early brooding stage appear to be able to utilize the sun's rays efficiently so that when they are given access to direct sunshine in sun porches or yards, no vitamin D need be supplied in any other way under normal climatic conditions in North America. If out-of-season turkeys are raised or if conditions are such that the birds are not exposed to normal summer and early fall weather conditions while growing rapidly, vitamin D should be supplied in the feed or possibly by irradiation. Since fish oils, fish-liver oils, and fish-liver oil concentrates cause off-flavor in the flesh, they should not be fed in quantities exceeding 0.15% of total feed intake within 12 weeks of marketing. Irradiated cholesterol offers a desirable source of vitamin D in finishing diets for turkeys, when sunshine cannot be utilized.

Since ordinary window glass filters out almost all of the ultraviolet rays of the sun, the sunshine must be supplied directly to the birds or through some reliable glass substitute or special glass that will permit the ultraviolet rays to pass through. However, the use of these substitutes is seldom desirable in turkey raising because vitamin D can easily be supplied in the feed or else it is practicable to utilize the direct rays of the sun as is done commercially during the growing and finishing periods.

As a rule 0.15% of 600-D feeding oil plus 0.05 percent of 1500-D (per gm) cholesterol, are desirable in starting diets; the same or 0.25% of the oil alone in laying diets to insure sufficient quantities of vitamin D for high hatchability. These recommendations are based on total feed intake as are those in Table 5.15. The Beltsville Station, using small-type turkeys, found that 360 I.U. were sufficient when supplied by refined cod-liver oils, by A and D feeding oil, or by activated animal sterol. The Washington Station, using Broad Breasted Bronze, obtained similar results. If thoroughly dependable products are used 0.125% of 600 D oil (or its equivalent in other guaranteed products) is theoretically sufficient to protect turkey poults against rickets. However, the National Research Council in 1950 recommended 600 I.U. of D per pound as desirable to allow an adequate margin of safety and the Texas Station in 1953 advised 1000 to 1200 I.U. in poult starting diets.

Turkey breeders require 350 units per pound (80 per 100 grams). The authors recommend 600 units per pound in turkey breeder diets when the birds obtain little or no direct sunshine and 300 units when a moderate amount of sunshine is obtained. Only when the birds are permitted continuous access to green pasture in sunny weather can vitamin D supplements safely be omitted.

VITAMIN E

This is the *antisterility* or *reproduction vitamin.* It is fat-soluble (and probably to some extent water-soluble), and stable to heat, light, and some types of mild oxidation. It is readily destroyed by rancid fats and by contact with iron (ferric) salts. Chemically it is a high molecular alcohol, alpha tocopherol ($C_{29}H_{50}O_2$). Beta and gamma tocopherols also show vitamin E activity but are less potent. The I. U. for vitamin E is the activity of one mg. of synthetic racemic alpha-tocopheryl acetate ($C_{31}H_{52}O_3$). The acetate is now commercially available as a source of vitamin E.

Lack of vitamin E causes a definite type of sterility in animals. In chickens vitamin E apparently is necessary for the production of hatchable eggs and to proper growth of chicks. Lack of it in breeder diets causes sterility in males and low hatchability of eggs. Lack of it in chick diets causes slow growth, a characteristic unsteady gait during the 4th to 6th weeks of life, termed encephalomalacia, and a rather high mortality.

In turkeys deficiency of vitamin E produced experimentally caused degenerative changes in the gizzard musculature which took the form of grayish patches that could often be seen on the outside wall of the organ upon post-mortem. Crazy-chick disease does not occur in poults.

Wheat-germ oil and cottonseed oil are excellent sources of vitamin E. Wheat germ and dehydrated alfalfa meal are very good sources. All grains are fair to good sources, as are green feed (fresh and dried), wheat bran, wheat shorts (if the germ is not removed), peanut meal, and hominy feed. Because of this wide distribution in the common poultry feedstuffs, there appears to be no need for making additions of vitamin E concentrates to the diets of chickens or turkeys.

Recently, however, the Cornell Station (1953) obtained a marked improvement in hatchability when 20 milligrams of alpha-tocopheryl acetate per pound of feed was added to a practical all-mash turkey breeder diet. It was suggested that the ground grains in the diet might have lost much of their vitamin E activity thus producing a deficiency. Since dehydrated alfalfa, a good source of E, was not contained in the diet used, the ground grains would be the only source of this vitamin. This work suggests that supplementing turkey breeder diets, as noted above, may be desirable as a safety factor in breeder nutrition.

The requirements of turkeys for vitamin E have not been reported, but these requirements appear to be adequately met on the whole, since deficiency of this vitamin has apparently not occurred in practice. The fact that vitamin E is found along with vitamin B_1 in the same feedstuffs simplifies the problem considerably because, by providing adequately for vitamin B_1, the vitamin E requirements will, as far as is known, be satisfied. The only probability of a deficiency is, as with vitamin B_1, through the wide use of degermed grain products. Degerming of grain products affects the vitamin E content more than it does the B_1, however, so that the extensive use of degermed products is more likely to cause a deficiency of the former. Vitamin-E loss in stored whole grains is very slight; in ground grains it is considerable (about 30% in 5 months).

RIBOFLAVIN (VITAMIN G)

After Mitchell in 1919 suggested that the then "water soluble B" was composed of two factors, one antineuritic and the other growth-promoting, the term vitamin B soon came to be used in America to designate the antineuritic, heat-labile factor while the heat-stable, growth-promoting factor was called vitamin G. British workers called the antineuritic factor B_1, and gave the growth-promoting factor the designation B_2. However, early in the history of investigation of the growth factor (G or B_2) it became apparent that a complex of perhaps 8 vitamins rather than a single one was involved. These factors usually are found associated with each other in nature although a feedstuff containing members of the complex may contain relatively large quantities of one or more factors and only relatively small quantities of one or more of the other factors. The term "G complex" sometimes is used to designate all of the B vitamins except B_1. These factors are all more or less soluble in water and stable to heat. In feeding, the most important of these is riboflavin which is growth-promoting in all animals so far tested. In chickens, and in turkeys, this factor is necessary for hatchability as well as growth, and also for the normal functioning and maintenance of the nervous system. Deficiency in the diet of breeding stock causes embryos to die at the 4th and 5th and the 12th to 15th days of incubation.

Riboflavin is now available in pure form, its chemical formula being $C_{17}H_{20}N_4O_6$. It is a deep yellow pigment with a greenish flourescence and a very bitter taste. It is stable to heat and moderately stable to acids, alkalis, and oxidation. It is soluble in water and is destroyed by light. This factor constitutes the main growth- and hatchability-promoting fraction of the vitamin B-G complex. According to various investigators, lack of riboflavin in poult diets results in perosis, dermatitis, lowered efficiency of feed utilization, slow growth, emaciation, poor feathering, and high mortality. Lack of riboflavin in breeder diets appeared to make more probable the occurrence of dermatitis in the poult offspring since apparently poults do not synthesize it in the intestines (Washington Sta. 1947). In poults from which feed and water were withheld longer than normal, livability was very poor in the presence of a riboflavin deficiency (Tenn. Sta., 1953).

As indicated by the data in Table 5.10, excellent sources of riboflavin are liver meal, dried milks, dried whey, condensed fish solubles, distillers' dried solubles, dried yeast, alfalfa meals, clover hay, and dried cereal grasses. Fair to good sources are bran, middlings, peanuts, all the oil-cake meals, liquid milk, cane molasses, rice bran, fresh greens, soybeans, casein, fish meal, meat scrap, and tankage. Fermentation and distillery by-products are important sources of riboflavin. They should be purchased on the basis of guaranteed unitage of riboflavin per pound. They are not practical sources of protein. Fish meals vary considerably in riboflavin content. The Ohio Station found that cod and haddock meals (white fish of North Atlantic) were of high potency; shrimp, sardine, and salmon meals of moderate potency; and tuna and menhaden meals of rather low potency. Cow manure was found to be a source of riboflavin for poultry by the Beltsville Station, which found that poultry manure also contains riboflavin. However, manure of any kind is not recommended for use as feed.

Due to the relatively small percentages of the protein concentrates normally used in poultry diets, only those classified as excellent (5 + mg. per lb.) sources contain enough of the vitamin to be of great value in supplying it. In the diets usually fed the whey, distillery and fermentation solubles, and green feeds are the main sources of riboflavin and should be included in sufficient quantity to supply a total of about ⅔ or more of the birds' estimated minimum requirements as listed in the paragraphs following. The other constituents of the diet may then be counted on to supply the remaining ⅓ and any excess that may be included.

The results of experiments at the Ohio Station indicate that either dried skimmilk and alfalfa leaf meal together; or the latter alone, gave better results in hatchability of chicken eggs than the dried milk alone. Presumably the milk alone did not supply in sufficient quantity all members of the vitamin G complex. In these experiments the best results were obtained when 5% alfalfa leaf meal and 5% dried skimmilk were included in the all-mash diet.

The chick unit is based on the activity of one microgram of flavin which equals approximately 2.1 Sherman-Bourquin units of vitamin G. In spite of the extensive literature on the other B-G vitamins, it appears that they are contained in adequate quantities in the diets generally used in feeding poultry and hence need not be considered separately when planning diets, provided ample quantities of riboflavin from natural sources are supplied.

In terms of chick units per pound, the vitamin table gives the riboflavin potency of the common feedstuffs. Dried pork liver is the basis for calculation, this product containing, on the average, 100 units per gram, or 45,359 units per pound.

During the first few weeks of life, turkey poults were found by the Cornell Station to require about 320 units per 100 grams of feed (or 1452 per pound) during the first 4 weeks of life, about 200 units during the second four weeks, and about 140 units after 8 weeks of age. The California Station (1946) reported that poults required 325 to 375 micrograms per 100 grams of diet (1475 to 1700 per pound) during the first four weeks of life. The Lederle Laboratory (1947) reported a requirement of 1370 to 1600 per pound for the first 6 weeks. From a study of various diets fed to turkeys, the authors recommend for starting diets a minimum of 2,000 units per pound, for growing diets 1500 per pound, and for breeder diets 1,800 per pound, be used as standards when the accompanying tables are used to estimate the vitamin content.

OTHER CLAIMED COMPONENTS OF THE VITAMIN B-G COMPLEX

Folic or pteroylglutamic acid is found in leaves and some animal tissues. It is acidic in nature, insoluble in common organic solvents, and is not easily destroyed by oxidation. Green leaves, liver, and yeast are excellent sources; alfalfa meals and soybean meal are good sources; linseed meal, cottonseed meal, and wheat bran are fair sources; the grains, milk, fish and meat products are poor sources. Folic acid is believed to be an essential vitamin of the B-complex and to replace factors U, R, B_{11}, B_c and vitamin M. Symptoms of deficiency in poults are slow growth, neck (cervical) paralysis, high mortality, and broken feathers. Breeding stock need it to promote high survival rate in the poults. The Cornell Station (1948) reported that about 360 micrograms per pound of feed were required for optimum results in poult starting diets while the California Station (1949) says 500 mcgms and the Texas and Missouri Stations (1948) reported that about 900 micrograms are needed. A sufficient amount is believed to be supplied in practical turkey diets without special precautions. Folic acid apparently occurs in nature as part of the so-called citrovorum factor from which it may be extracted by chemical process.

Dehydrated alfalfa meal 17% protein, has an average (Ewing, 1951) of 5 mg. per pound of folic acid; dried brewers' yeast, feed grade, 22 mg.; barley 0.27; corn 0.14; oats 0.09; rye 0.75; wheat 0.18; wheat bran 0.90; middlings 0.40; soybean meal 4.50; cottonseed meal, 1.04; corn gluten meal 0.091; dried skim milk 0.27; fermentation solubles 2.32; dried whey 0.40; tankage 0.70.

Factor S of Cornell Station workers is a water-soluble, growth, and reproduction factor as yet chemically unidentified, but is believed to be necessary for poultry. It is destroyed by prolonged dry heat and absorbed in fullers' earth. Cornell Factor S is found in wheat by-products, milk, yeast, alfalfa meal, liver, and fresh green feed. Factor S may be *strepogenin,* a microbiological growth factor with anti-anemic activity which may be required by turkeys (Cornell Station, 1947) in relatively large quantities. Good sources are dried whey, dried skimmilk, soybean meal, and sunflower seed oil meal. Fair sources are oats, wheat products, fish meal and meat scrap.

The fish factor (Tex. Sta. 1955), believed necessary for maximum turkey hatchability, usually is supplied adequately by 5% of fish meal. If used, it should be accompanied by 5% alfalfa meal and 2 mg. per pound of alpha-tocopheryl acetate.

Vitamin B_{12} (or cobalamin) formerly the animal protein factor (APF) is found in most animal products. Some of the best sources are fish solubles, liver meal, fish meal, meat scrap, and milk. This vitamin, dark red in color in the crystalline form, is necessary for hatchability and possibly for optimal growth in turkeys. It can be obtained in pure crystalline form or, more commonly as a B_{12} supplement (which must contain at least 1.5 mg. of B_{12} per pound) alone or in combination with an antibiotic. It may occur in a number of slightly different forms, most of which appear to be equally effective in poultry diets. The B_{12} requirements of poults have not been determined. Several Stations (Tennessee and Texas, 1950) have reported no response from B_{12} additions to all-vegetable-protein starting diets without an antibiotic. The authors have secured normal growth for many years in range-grown turkeys fed all-vegetable protein growing diets after 8 weeks of age without B_{12} additions. The

Table 5.5—Vitamin B_{12} Potency of Some Feedstuffs and Concentrates Obtained by Microbiological Assays (Lillie, et al A.R.C.)

Material assayed	Vitamin B_{12} potency by Microbiological (4) Assay Micrograms per gram
Crab meal	0.45
King crab meal	0.62
Herring fish meal (2 samples)	0.23
Salmon fish meal No. 2	0.14
Tuna fish meal No. 5 (2 samples)	0.22
Sardine fish meal No. 9	0.17
Steam fish meal No. 10	0.16
Whale solubles No. 2	0.04
Liver substance defatted	0.90
Broiler viscera meal	0.20
R-1[1]	0.61
Commercial B_{12} concentrate E	3.30
Acid precipitate[2]	6.20
ERRL P90-79-1[3]	2.40
BRRL P90-79 2[3]	1.90
Feather protein	0.02
Skim milk, dried	0.03
Whey,dried,sample (2 samples)	0.02

[1]A fermentation by-product of vitamin B_{12} manufacture.
[2]Acid precipitate of water extract of cow manure.
[3]Sewage sludge preparations.
[4]Each value represents an average of 3 cr more trials.

poult's requirement for B_{12} appears to be less per unit of feed during the later growth stages than during the early stages.

In poult starting diets the Washington Station (1951) found that 20 micrograms of B_{12} per kilogram of feed (18 mg. per ton or 9 mcgms. per pound) were needed for maximum growth when an antibiotic was included in the diet. The Texas Station (1950) obtained similar results. This much B_{12} presumably would not be supplied by the average starting feed. Therefore a B_{12} supplement would be needed, and the most practical solution seems to be to use an "antibiotic B_{12}" supplement to supply the needed B_{12} along with the desired antibiotic.

There is some support for the view (Tennessee and Texas Stations 1950, 1951) that B_{12} is not required by poults in amounts over those supplied by ordinary plant feedstuffs but that there is an unknown factor contained in fish meal, fish solubles, and casein which is required by poults for optimum growth. On the other hand the Minnesota Station (1951) obtained good growth response in 4-week Broad Breasted Bronze poults by adding 10 or 20 micrograms of crystalline B_{12} per kilogram of an all vegetable starting diet without an antibiotic and suggested that the optimum level of B_{12} for young poults is 5 to 10 micrograms per kilogram or 2.3 to 5 micrograms per pound. The Ontario Station (1951) by adding 22 micrograms of crystalline B_{12} per kilogram of an all vegetable starting diet (10 mcgm. per lb.) without an antibiotic obtained a 12% growth stimulation at 4 weeks and 17% at 8 weeks, in Broad Breasted Bronze poults. The California Station (1953) concluded that it is difficult to show need for B_{12} in poult diets in which choline is in adequate supply unless the poults were from B_{12} depleted hens. B_{12} can reduce the amount of choline needed for growth but not the amount needed for the prevention of perosis. Apparently some B_{12} is needed for proper utilization of methionine. Feed efficiency seems not to be affected by B_{12} (Minn. Sta. 1952).

B_{12} seems necessary in turkey breeder diets. A combination of 4% fish meal, 4% meat scrap, and 5% dried milk on an all-mash basis will supply a sufficient amount. In lieu of the 5% milk an extra 3% of meat scrap should be effective, likewise 3% of either dried fish solubles or condensed fish solubles, or an extra 2½% of fish meal. Estimates from various sources indicate that, per pound, 17% of protein dehydrated alfalfa meal averages about 16 micrograms of B_{12}; meat scrap about 60; fish meal 100;

fish solubles dried or condensed 250; dried milk 80; the grains 3 to 5; soybean meal 10; dried yeast 4; casein 50 (see Table 5.5).

Biotin, or **vitamin H,** is found in grains, liver, kidney, wheat bran, fish meal, meat scrap, milk, soybean meal, yeast, cane molasses, alfalfa meal, fresh green feed, and dried distillers' solubles. All of these contain riboflavin, so that as a general rule it can be assumed that to supply enough riboflavin from natural sources is to supply sufficient biotin. However biotin is available in crystalline form and its inclusion in poult starting diets is recommended by the Alberta Station (1950), at one gram per ton of diet along with calcium pantothenate at 10 gm. per ton to prevent high mortality, perosis, broken feathers, and dermatitis in late-hatched poults. Biotin in dried brewers' yeast, feed grade, averages in mg. per pound 0.5; barley 0.17; wheat bran 0.10; wheat 0.05; corn 0.036; oats 0.12; meat scrap 0.027 and soybeans 0.32. How much of this biotin in feedstuffs is available to poultry has not been determined. For chicks the National Research Council recommends 0.045 mg. per pound of feed; for breeding chickens 0.07 mg. No recommendations are made concerning biotin allowances for turkeys but it can safely be assumed that turkeys require at least as much as chickens.

Biotin deficiency in poults has not been widely investigated but the California Station (1946) reported that a daily calculated biotin intake of 2.5 to 9.6 micrograms per poult prevented dermatitis and mortality. On the other hand the Washington Station (1947) reported that the lack of biotin did not cause slow growth or dermatitis but did cause perosis under certain conditions. It appears to be necessary for high hatchability.

The grass juice factor is believed essential for turkeys for optimum growth by Cornell workers (1951). It may be identical with the grass juice factor required by rats and guinea pigs and may be the same as Cornell factor S. It is a growth factor, producing at the 5% level, a 15%

weight increase at 4 weeks, over that obtained on a good, modified high-energy starting feed. The juices of cereal grasses, young white clover, pea shells, cabbage, turnip tops, spinach, and alfalfa all are excellent sources of the factor. Alfalfa juice at a 10% level produced less growth than the control diet and much less than it did at the 5% level due to the growth inhibiting factor that it carried. Rice bran, brewers' yeast, and freshly prepared dried skimmilk are fair sources. Animal products are poor sources. In growing plants the concentration of the factor is high but in mature and old plants it is considerably lower. Oxidation and aging apparently destroy it.

Inositol, a type of saturated sugar, is considered by some to be a vitamin and necessary for normal growth and prevention of anemia in poults (Lederle Laboratory, 1947). The Missouri Station (1948), however, reported that it apparently was not necessary in turkey diets. It is widely distributed in natural feeds so would normally be supplied in goodly quantities. Dried brewers' yeast, feed grade, has an average of about 2250 mg. of inositol per pound.

Choline ($C_5H_{15}O_2N$), considered by some to be a B-vitamin, is discussed in connection with calcium and phosphorus requirements. It is a strong base essential in the diet of poultry for egg production, growth, normal fat metabolism, and normal bone development. It is relatively stable to heat and oxidation. Deficiency results in perosis or slipped tendon. Two tenths per cent (900 milligrams of choline per pound of feed) is recommended by the National Research Council (1950). However, the Connecticut Station (1950) obtained normal growth of Broad Breasted Bronze poults on 400 to 450 mg. per pound and obtained exceptionally good growth on 0.54% of an APF supplement along with 400 to 450 mg. of choline, indicating a possible interaction between choline and one or more unidentified factors in the A.P.F. Choline at 0.18% was satisfactory in growing mashes fed with grain at the Ontario Station (1946).

The Beltsville Station (1948) found that 0.14% (635 mg. per pound) of choline in the diet of Beltsville White poults produced normal poults with no perosis. Additions of 0.10 or 0.15% choline chloride made no improvement in starting diets. Breeding turkeys at the Connecticut Station (1950) receiving about 570, 770, or 1170 mg. of choline per pound of feed showed no better hatchability than those receiving only 370 mg. but poults from hens receiving 770 or 1170 mg. were consistently larger at 8 weeks than those from hens receiving 370 mg. regardless of the level of choline in the starting diet whether 400, 600, 800, or 1200 mg.

Betaine is a "choline equivalent" and can replace a large part of the choline and methionine required in chicken diets. It is obtained commercially from sugar beet juices and this crude betaine carries water-soluble vitamins. It provides methyl groups (CH_3) which spare methionine and choline and is useful as a feed supplement at a level of 0.4 to 0.5%, in chicken diets. However, there is a possibility that betaine may be involved in the development of fishy flavor through combining with free fatty acids hence it would not be advisable to use any more of it than is supplied by the feedstuffs ordinarily used in turkey diets. Furthermore the California Station (1952) reported that betaine did not substitute well for choline in turkey diets.

ANTI-GIZZARD-EROSION FACTORS

The existence of a factor or factors preventing gizzard erosion is doubtful. Diets containing ordinary percentages of green feed (fresh or dried), oats with hulls, soybean meal, and bran apparently prevent gizzard erosions and lesions in chicks and poults. The Nebraska Station suggested that gizzard erosions might be prevented by 5% of high quality dehydrated alfalfa meal. For normal gizzard-lining development, the diet should be reasonably coarsely ground or insoluble grit should be provided. Gizzard erosions are crater-like lesions with darkened, roughened surfaces occurring in the lining of the gizzard. True gizzard erosions are not associated with hemorrhages in the gizzard lining and there appears to be no link between vitamin K and bile deficiencies or slimy gizzard linings and gizzard erosions.

VITAMIN K

This is known as the *anti-hemorrhagic* vitamin. It is a fat-soluble dietary factor which has been found in grass, alfalfa meals, egg yolk, cabbage, kale, tomatoes, hempseed, the fat of hog liver, putrifying fish, and some cereal grains. Lack of it in the diet of chicks, ducks, and geese causes delayed blood clotting and hemorrhages under the skin, within the muscles, and in the abdominal cavity. A water-soluble form is available to inject into the blood. This vitamin is destroyed by alkali but is quite stable to heat, to visible light, and to aging. There is no record of experiments on the vitamin K requirements of turkeys, but undoubtedly it is required by them. It is believed to be supplied in sufficient quantities by the green feed (fresh or dried) generally included in turkey diets. The grains tested were found to be rather poor sources, corn being less potent than oats and barley. Five percent of dehydrated alfalfa meal in poult and breeder diets should supply ample vitamin K.

Vitamin K_1, or phylloquinone, is a light yellow oil isolated from alfalfa. It has about twice the potency of vitamin K_2, a light yellow crystalline solid isolated from putrifying sardine meal. The various preparations of vitamin K are derivatives of 1, 4-naphthoquinone ($C_{23}H_{48}O_2$). Potency of K_1 is 1000 units per milligram; of K_2, 660 units. At least 0.17 g. of K per 100 gm. of feed is required for maximum growth of chicks, according to the California Station.

PARA-AMINO-BENZOIC ACID

Para-amino-benzoic acid ($C_7H_7O_2N$) is the anti-gray-hair factor. It is essential for growth in rats. Little is known of its role in poultry nutrition, but it is believed to act indirectly by stimulating intestinal micro-organisms to produce various essential factors, one of which is folic acid. It can be obtained in pure crystalline form. It is believed to be a B vitamin but not a specific growth factor for poultry.

THE ANTIBIOTICS

The discovery in 1950 that some of the antibiotics stimulate growth, when fed to chickens, turkeys and swine rates as one of the most spectacular discoveries in the science of nutrition. The discovery came as a result of experimentation with antibiotic sources of vitamin B_{12}. Swine show the greatest response, turkeys following, with reported early growth stimulation of 10 to 30 averaging perhaps 20%. Chickens show less, averaging perhaps 10%. Growth of Broad Breasted Bronze poults to 8 weeks was increased by 20 to 25% and feed required to produce a pound of gain to 8 weeks of age was reduced by 20%, when an antibiotic-B_{12} supplement was fed at the Texas Station. At the Ontario Station (1951) growth, uniformity of growth, feed efficiency; and freedom from leg deformities at 4 weeks were markedly improved with Broad Breasted Bronze poults by the addition of an antibiotic B_{12} supplement containing 2.5 gm. of procaine penicillin per pound regardless of amount of common salt (0 to 1%) or amount of manganese sulphate (0, 4, or 8 oz. technical grade per ton) added. Without the antibiotic the 8 oz. per ton of $MnSO_4$ caused growth depression and 1% of common salt apparently caused leg disorders which were not reduced by the added manganese.

Antibiotics also apparently increased turkey's need for the amino acid lysine to prevent achroma or white wing spotting. About 1.9% lysine was required with an antibiotic and 1.3% without it, in 28 percent protein starting feeds. The antibiotics apparently do not decrease the poults' protein requirement but probably do increase their tolerance for fat in the diet. Penicillin, at least, has a riboflavin-sparing effect.

The well-known antibiotics (there are over 1200 antibiotics known) stimulate growth in turkeys; the best so far known being penicillin followed by bacitracin, terramycin, and aureomycin with little difference between the last three. About 9 or 10 grams per ton of feed (4½ to 5 milligrams per pound or 10 to 11 parts per million (*p.p.m.*) are widely recommended for use in turkey starting diets. Evidence at present does not indicate any need for an antibiotic in breeder diets. The South Dakota Station (1953) reported that hatchability of Broad Breasted Bronze breeding stock was not improved by feeding an antibiotic B_{12} supplement but that poults from stock so fed grew a little faster with less mortality when fed a starting diet low (0.1%) in the antibiotic B_{12} supplement than did poults from parent stock not fed the supplement. However, when the antibiotic B_{12} supplement was increased to a suitable level in the starting diet differences due to source of poults disappeared. Crystalline aureomycin in a breeder diet was found to reduce hatchability at the South Dakota Station (1952). Evidence against the use of antibiotics in breeder diets was obtained at the Montana Station (1953) where an aureomycin-B_{12} supplement reduced hatchability 6% and egg production 12%. At the Ontario Station (1953) penicillin reduced hatchability about 12%.

The Washington Station (1950, 1951) obtained an average of about 25% growth response in Broad Breasted Bronze poults 4 weeks of age over that of the control diet containing 10% herring meal by adding 10 parts per million (ppm) of terramycin, penicillin G (potassium salt), or procaine penicillin. Five *p.p.m.* of bacitracin had the same effect and 10 *p.p.m.* increased growth by 30% while 10 *p.p.m.* of streptomycin increased it an average of only about 10%. With Beltsville Whites these gains were about half as great as with the Bronze. The Tennessee Station (1951) obtained results showing great variations in growth response and as a result concluded that growth stimulation of young turkeys by the antibiotics can result, but that the extent of the stimulation can not be predicted. Since the antibiotics appear to increase the need of poults for B_{12} (Texas Station, 1951) it seems advisable to use a diet with a high content of B_{12} or use an antibiotic B_{12} supplement to supply both the antibiotic and the B_{12}.

Fish meal increases the favorable effect of antibiotics on growth rate and feed efficiency and 5% of it is enough in a well designed practical diet. The B_{12} in the fish meal seems not to be the factor responsible for this effect—the possession of an unidentified faction was suggested by the Washington Station (1953). The Cornell Station (1953) suggested that turkeys require two unidentified growth factors, (1) found in animal products such as meat scrap and fish meal and (2) found in casein, yeast, dried skim milk, and grass juice. Antibiotics appear to have a sparing effect on (2) but not on (1).

The removal of antibiotics from turkey diets fails to produce setbacks to the poults. Although it was first thought that antibiotic feeding during the growing period (after the first 8 to 12 weeks) would be unnecessary and uneconomical, evidence is accumulating which indicates that continuation of the antibiotic to the end of the growing period is advisable and economical. In 1952 the Cornell and Texas Stations recommended this and in 1953 the Washington Station reported that the continuation of 3 p.p.m. of diamine penicillin in the growing mash fed with grain resulted in significant improvement in body weight and feed efficiency of Broad Breasted Bronze turkeys up to the age of 24 weeks. Fat deposition in hens also was improved. The Cornell University Station (1952) found an advantage in adding an antibiotic at 8 weeks but preferred it throughout the poults' growing period starting at day old.

The degree of toxicity of antibiotics is slight, they

do not accumulate significantly in the tissues of poultry and do not have any effect on palatability of the meat.

That the antibiotics reduce mortality and improve uniformity of early growth seems to be well accepted; in addition they are believed by some to eliminate vent inflammation and the picking that frequently goes with it. The Oregon Station (1954) reported that with 4 grams of crystalline procaine penicillin in the starting feed of BBB turkeys, mortality was markedly reduced especially in poults from which feed and water were withheld for 24 to 48 hours.

The antibiotics are thought to act as drugs, not as vitamins. They are ineffective when injected into the breast muscles. Their exact mode of action in the digestive tract is not understood but the bacterial flora of that tract seems to be affected in a beneficial manner. It may be that the antibiotics have in themselves no growth stimulating properties but merely prevent the retardation of growth more or less normally caused by harmful bacteria in the digestive tract. The competition theory is that certain germs in the intestinal tract compete with the body of the animal for the food and when the organisms are killed or inhibited by the antibiotic, greater growth results. The "factor synthesis" theory suggests that the antibiotic inhibits certain bacteria of the intestinal tract, allowing a type of growth which synthesises an unknown (as yet) growth factor. The "nutrient" theory proposes that a part of the antibiotic molecule may act as a nutrient or factor for growth.

Of the various theories proposed to explain the action of antibiotics in the feed at low levels the first one mentioned is receiving greatest support. At Lobund Institute (Notre Dame University, 1952) it was shown that germ-free poults grew faster than conventionally reared poults and showed no response to antibiotic feeding, indicating that where there were no injurious microorganisms present in the digestive tract, the antibiotics had no effect on growth. It also has been shown that normal healthy chicks and poults placed in a new, clean, relatively germ-free environment made excellent growth which was not increased by feeding antibiotics. These experiments point to the desirability of cleanliness in the brooding quarters but should not be interpreted to mean that the feeding of antibiotics can be substituted for sanitation. However, absolute sanitation in the brooding quarters seldom can be accomplished under practical conditions hence the place of antibiotic feeding to growing turkeys seems assured.

Wire-brooded poults should be less responsive to antibiotic feeding because their environment theoretically is more sanitary than it is when litter is used. The Washington Station (1952) reported that wire-brooded poults were less responsive but not significantly so. At Beltsville, wire-brooded poults have given, on the average, good responses to the antibiotics but these responses are highly variable. Early in the season, in a thoroughly cleaned and disinfected brooder house, growth rate without an antibiotic usually is excellent with the result that response to an antibiotic is relatively small, more on the order of 10% rather than the 15 to 35% increases obtained in so many experiments.

For growing turkeys the Washington Station (1952) rated diamine and procaine penicillin as equal with bacitracin intermediate between them and terramycin. Most effective levels, in parts per million (essentially the same as grams per ton) were feeding grade bacitracin, 5 to 10, procain penicillin 5, diamine penicillin 2 and 10, and pharmaceutical grade bacitracin, 5. Aureomycin (chlortetracycline) and terramycin (oxytetracycline) were not rated but seem as effective as bacitracin at the 10 p.p.m. level. Streptomycin is not often used in poultry nutrition though it has value in disease control.

Aureomycin hydrochloride contains about 93% of the pure antibiotic, chlortetracycline. Penicillin G potassium contains about 90% of the pure antibiotic; streptomycin sulfate contains about 86% of the pure antibiotic; terramycin hydrochloride contains about 93% of the pure antibiotic; bacitracin, pharmaceutical grade is the pure antibiotic. Diamine penicillin is a new stable form which is regarded by many as the best form so far made available for the feeding of poultry. It showed 90% retention in pelleting compared to 52% for procaine penicillin, 79% for aureomycin, and 51% for bacitracin (Pfizer Co., 1952.)

The antibiotics are commercially available as dilute premixes (sometimes combined with other nutritional factors) suitable for incorporation in feeds by ordinary mixing methods. If the pure antibiotic is used, a premix should be prepared so that at least one pound per ton is used to supply the antibiotic in proper amount. Soybean meal is a suitable diluent in preparing a premix.

The Washington Station (1953) reported that the pelleting of feeds did not affect the value of antibiotics and that antibiotics contained in commercial antibiotic supplements were as effective as when supplied by pharmaceutical-grade antibiotics. Pelleting at a maximum temperature of 55° C. (131° F.) and regrinding the pellets to produce crumbles had no effect on the favorable action of procaine penicillin in poult starting feeds at the Dominion (Swift Current, Sask.) Experiment Station (1952).

Although evidence is somewhat scanty it appears that combinations of two or more antibiotics usually are no more effective than one antibiotic. However, Almquist and Merritt (1953) suggested a combination of 5 p.p.m. of aureomycin and 2 p.p.m. of diamine penicillin as being consistently beneficial to Beltsville White poults up to 16 weeks of age and the Texas Station (1953) suggested 5 p.p.m. of bacitracin and 4 p.p.m. of penicillin as being "foolproof" in turkey diets.

High-level feeding of antibiotics came about through the demands of producers for a shotgun prescription to control various infections of both high and low virulence.

Although not as yet backed up to any great extent by research, the practice is in wide use by chicken broiler producers and to a lesser extent by turkey growers to prevent and control chronic respiratory disease, enteritis, sinusitis, mud fever (blue comb or monoytosis), and early mortality.

The Delaware Station (1950) reported that 100 grams of aureomycin per ton prevented infectious sinusitis from breaking out in exposed 5-week-old turkeys. So far, harmful effects of feeding up to 500 grams per ton of aureomycin, bacitracin, or terramycin have not shown up. Aureomycin and terramycin (the wide-spectrum antibiotics) often are used separately or in combination at 50 grams per ton of feed for each or 100 grams if one only is used as a preventive of early mortality from sinusitis and air-sac disease; at 200 grams per ton as a treatment after the disease has appeared. For prevention of mud fever in turkeys 30 to 40 grams of terramycin per ton of feed; for treatment, 75 to 100 grams per ton are suggested by The Pfizer Co. (1954). Terramycin is claimed to have value in the prevention of and treatment for paratyphoid and hexamitiasis.

Sulfaquinoxaline and **nitrophenide** appear to have some growth promoting ability in addition to their value in the control of coccidiosis but this effect which appears to be similar to that of the antibiotics is not striking nor uniform. Vitamin C and sulfasuxidine at the 0.1% level each has been reported to have some growth promoting ability.

ARSENICALS

The arsonic acid derivative 3-nitro-4 hydroxyphenyl arsonic acid at 50 p.p.m. in the feed has in many cases caused poult growth increases equalling those caused by antibiotics, but when fed along with an antibiotic appears not to product any extra effect. The same may be said for arsanilic (para amino phenylarsonic) acid at .01% (Borden Lab., 1951) in a good starting diet. Feeding these organic arsenic compounds constitutes medication but so far there is no indication of a significant accumulation of arsenic in the flesh of poultry, when fed these compounds which are toxic to chicks if fed at 4 times the recommended level. Three-nitro-4 hydroxyphenyl-arsonic acid is more toxic to chicks than sodium arsenilate. The compound 4- amino phenylarsonic acid (34% elemental arsenic) and some other arsonic acid derivatives are now available as commercial feed supplements but their use is not recommended for small feed mixers, antibiotic supplementation being preferred. If fed, arsenicals must be removed from the feed of poultry 5 days before slaughter according to U. S. Food and Drug rules.

DETERGENTS

Certain *surfactants* (surface active agents) including some commercial preparations used for home laundry and dishwashing have been found to increase growth in 10 to 12-week old chicks, up to 12% above control lots. No turkey work has as yet been reported and the work on chicks still is in the laboratory stage. There appears to be no interaction between the surfactants and B_{12} antibiotic supplements. The unusual wetting characteristics of these so-called "soapless" soaps may make it easier for the intestines to absorb nourishment, particularly fats and fat solubles, from the liquified food that flows through that tract. The bacteria of the intestinal tract may be affected or peristaltic action stimulated. It is not known whether or not the surfactant, if added to a diet containing an antibiotic, would result in further growth stimulation or if a surfactant could be used in place of an antibiotic to produce increased growth at less cost.

Most surfactants or detergents are classified as emulsifying agents, or cleansers, but some of them are primarily of a germicidal nature. All of them have germicidal properties however, as indicated in Chapter 15. There is little evidence as yet to justify the use of surfactants in chicken or turkey diets, except on an experimental basis.

EXTRACTIVES

This class of nutrients is relatively unimportant in feeding. It consists of organic acids, amids, and other organic compounds not classified as proteins, carbohydrates, fats, minerals, or vitamins. The extractives, as far as is known, need not be considered in the preparation of poultry diets.

THE CHEMICAL COMPOSITION, VITAMIN AND MINERAL CONTENT, AND DIGESTIBILITY OF FEEDS

In preparing diets for poultry the five factors of greatest concern are the chemical composition, palatability, digestibility by poultry, adaptability, and availability. Fairly reliable analyses giving the average chemical composition of the common feeds are available. The following tables were compiled from averages of available figures, and they are dependable to the extent that any average is reliable. A complete analysis is not necessary. For practical purposes the following information relating to chemical composition is useful; (1) percentage of moisture, (2) percentage of ash, (3) percentage of crude protein, (4) percentage of crude fiber, (5) percentage of nitrogen-free extract (sugars, starches, pentosans), (6) percentage of fat, (7) percentage of calcium, (8) percentage of phosphorus, (9) quantity of manganese, (10) amino-acid composition, and (11) vitamin content. These figures are listed in the following tables. The iron, copper and iodine content of certain feedstuffs is also given.

Table 5.6—The Average Composition of Some Feedstuffs Used in Poultry Diets.*

Feedstuff	Moisture	Ash	Crude protein	Carbohydrates		Fat or ether extract
				Crude fiber	Nitrogen-free extract	
Grains and seeds, and their by-products	Per cent	Per cent	Per cent	Per cent	Per cent	Per cent
Acorns of the willow oak	11.3	1.6	4.4	16.5	46.8	19.4
Babassu meal	9.0	5.0	23.0	12.0	44.0	7.0
Barley (except Pacific Coast)	11.0	2.9	11.0	6.0	67.0	2.1
Barley (Pacific Coast States)	10.1	2.6	8.7	5.7	71.0	1.9
Barley, hull-less	10.0	2.0	11.5	2.5	72.0	2.0
Beans, mung	11.3	3.0	21.7	4.0	59.4	.6
Beans, navy	13.4	3.6	22.7	5.8	53.0	1.5
Beans, pinto	9.1	4.5	22.7	4.5	58.0	1.2
Bread	33.8	1.5	7.9	.7	55.4	.7
Bread, kiln dried	10.5	2.2	12.5	.5	72.7	1.6
Brewers' grain, dried	7.5	3.9	26.0	16.0	39.9	6.7
Broom corn	11.6	3.1	10.5	8.3	63.0	3.5
Buckwheat	11.9	1.8	10.1	10.4	63.5	2.3
Buckwheat middlings	11.4	4.7	28.6	6.0	42.2	7.1
Coconut meal, old process	9.3	6.1	20.5	11.0	44.8	8.3
Coconut meal, solvent extracted	9.5	6.6	22.0	12.0	47.4	2.5
Corn or ground corn, dent	12.5	1.3	8.3	2.2	71.7	4.0
Corn, Argentine	11.0	1.7	11.0	1.8	68.8	5.7
Corn, flint or Indian	11.3	1.5	10.0	2.0	70.9	4.3
Corn-and-cob meal	11.5	2.1	8.2	11.0	64.2	3.0
Corn bran	10.0	2.4	8.6	10.3	63.7	5.0
Corn feed meal	11.0	1.8	9.0	3.0	71.2	4.0
Corn germ meal	9.0	3.0	20.0	9.0	51.0	8.0
Corn gluten feed	9.5	6.0	25.0	7.5	49.0	3.0
Corn gluten meal	9.0	2.0	41.0	3.0	43.0	2.0
Corn oil-cake meal	9.0	2.2	22.0	10.5	49.0	7.3
Cottonseed meal (43%)	7.3	5.9	43.0	12.0	25.8	6.0
Cottonseed meal (41%)	7.5	6.0	41.0	13.0	27.0	5.5
Cowpeas	11.1	3.5	23.5	4.1	56.3	1.5
Durra	10.0	2.0	10.2	1.7	72.6	3.5
Emmer	10.5	3.0	11.5	10.0	63.0	2.0
Feterita	10.2	1.6	13.2	1.8	70.2	3.0
Field peas	9.3	3.3	23.3	5.9	57.0	1.2
Flaxseed	11.5	4.5	22.0	7.0	23.0	32.0
Garden peas, dried	11.8	3.0	25.6	4.4	53.6	1.6
Hegari	10.5	2.0	9.5	2.5	73.0	2.5
Hempseed	7.5	5.9	23.0	18.6	18.4	26.6
Hempseed meal	7.3	7.8	31.7	24.0	25.2	4.0
Hominy	11.8	.7	7.4	.6	77.6	1.9
Hominy feed	8.5	3.0	11.0	6.0	65.0	6.5
Kafir	12.0	1.6	11.0	2.4	70.0	3.0
Linseed meal, O. P.	9.7	5.8	34.0	8.5	36.0	6.0
Malt sprouts, dried	7.9	5.9	26.0	13.0	45.7	1.5
Millet (proso)	9.6	3.4	11.6	8.7	63.1	3.6
Milo	11.5	2.0	10.5	2.5	70.5	3.0
Oats, avg. (not incl. Pacific Coast)	9.0	3.5	12.0	11.0	60.0	4.5
Oats, (Pacific Coast)	10.0	3.4	9.0	11.0	61.2	5.4
Oat meal, groats or rolled oats	8.3	2.2	15.0	2.0	66.0	6.5
Oat mill feed	7.0	5.7	6.0	27.7	51.5	2.1
Peanuts (hulls on)	6.0	2.8	24.8	17.8	14.0	34.6
Peanut kernels	5.4	2.3	30.4	2.7	11.6	47.6
Peanut meal (43%)	9.0	5.5	43.0	11.5	23.5	7.5
Peanut meal (41%)	8.8	5.7	41.0	12.5	24.0	8.0
Rice, rough	10.5	5.0	8.0	9.0	65.5	2.0
Rice, polished	11.8	.5	7.5	0.4	79.4	0.4
Rice, brown or hulled	9.8	1.1	9.0	1.0	77.1	2.0

*Reprinted with some changes and additions from "Scientific Feeding of Chickens" (1949, 2nd Ed.) by Harry W. Titus, with permission.

Table 5.6—(Continued)

Feedstuff	Moisture	Ash	Crude protein	Carbohydrates: Crude fiber	Carbohydrates: Nitrogen-free extract	Fat or ether extract
	Per cent	Per cent	Per cent	Per cent	Per cent	Per cent
Grains and seeds, and their by-products						
Rice bran	8.8	10.9	13.0	12.5	41.1	13.7
Rice polishings or middlings	9.5	7.0	12.5	3.5	56.5	11.0
Rye	10.7	2.0	11.5	2.1	72.0	1.7
Rye middlings	9.8	3.8	16.5	5.0	61.5	3.4
Sorghum grain, sweet	12.0	2.0	9.5	2.4	70.7	3.4
Sesame seed meal	8.5	10.0	41.0	6.0	22.5	12.0
Shallu	10.0	1.8	13.0	2.0	70.0	3.2
Soybeans	8.8	4.6	38.0	5.0	26.6	17.0
Soybean meal, avg. hydraulic and expeller	9.0	5.6	44.0	6.0	30.0	5.4
Soybean meal, solvent high protein	10.0	5.0	52.0	3.0	29.0	1.0
Soybean meal (43%)	10.0	5.6	43.0	5.2	30.2	6.0
Soybean meal (41%)	9.0	6.5	41.0	7.0	31.5	5.0
Soybean meal, solvent extracted (45%)	10.0	6.0	46.0	6.0	31.0	1.0
Sunflower seed	7.4	3.4	16.0	28.6	21.4	23.2
Sunflower seeds (hulled)	5.0	3.8	28.0	6.0	16.2	41.0
Sunflower seed, Hungarian striped	6.7	3.7	18.0	26.0	19.2	26.4
Sunflower seed meal (with hulls)	10.5	5.5	21.0	34.0	27.5	1.5
Sunflower seed meal (hulled)	10.0	6.2	36.0	14.0	20.3	13.5
Velvet beans	10.0	3.0	24.8	6.2	50.8	5.2
Wheat, avg. all types	12.0	1.7	12.0	2.4	70.0	1.9
Wheat, hard	10.5	1.8	13.5	2.4	69.8	2.0
Wheat, soft (Pacific Coast)	11.0	1.8	10.0	2.7	72.5	2.0
Wheat mixed feed (shipstuff or millrun)	10.3	4.8	16.7	7.0	56.8	4.4
Wheat bran, avg.	10.2	5.9	16.0	9.0	54.7	4.2
Wheat flour	12.9	.4	10.7	.4	74.2	1.4
Wheat flour middlings	10.5	3.5	17.0	5.0	59.4	4.6
Wheat germ meal	10.5	4.5	28.5	2.5	44.5	9.5
Wheat middlings, standard	11.0	4.5	17.0	7.0	56.0	4.5
Wheat red dog flour	10.2	2.7	17.0	3.0	62.6	4.5
Wheat shorts (brown)	10.5	4.0	17.5	7.5	56.0	4.5
Wheat shorts (gray)	10.5	4.1	17.7	5.6	57.5	4.6
Wheat shorts (white)	10.5	2.0	16.0	3.0	65.3	3.2
Products of animal origin						
Beef scrap	6.5	21.5	58.0	2.2	0.8	11.0
Blood meal	13.2	3.5	80.0	.7	1.8	.8
Bone meal, steamed	3.1	73.8	13.0	.8	2.8	6.5
Bone meal, special steamed	3.1	85.1	6.5	2.6	2.1	.6
Buttermilk, liquid	90.8	.8	3.2	.0	4.6	.6
Buttermilk, condensed	71.6	3.5	10.6	.0	12.2	2.1
Buttermilk, dried	8.0	10.0	32.5	.2	44.0	5.3
Casein	7.1	3.0	85.0	.1	4.0	.8
Crab meal	8.0	39.0	36.0	8.0	6.5	2.5
Eggs (contents), liquid	73.1	1.0	13.3	0.0	1.1	11.5
Eggs (contents), dried	5.0	3.5	47.0	0.0	3.9	40.6
Fish meal, unidentified	10.0	20.0	60.0	.7	2.5	6.8
Fish meal, Alaska herring	6.3	11.0	71.0	0.6	3.1	8.0
Fish meal, menhaden	7.0	19.0	60.0	1.0	4.0	9.0
Fish meal, white (high ash)	7.8	26.0	62.0	.0	1.2	3.0
Fish meal, white (low ash)	12.0	17.6	61.0	.6	.1	8.7
Fish meal, sardine	8.0	15.5	65.0	.5	4.5	6.5
Fish meal, tuna	8.0	20.0	59.0	.5	4.5	8.0
Fish meal, salmon	7.2	19.0	60.0	....	3.8	10.0
Fish solubles, dried	....	....	56.0	....	....	....
Fish solubles, condensed	50.0	10.0	33.0	.0	3.0	4.0
Gelatin	2.9	2.0	95.0	.0	.0	0.1
Liver meal, commercial	5.0	5.1	65.0	.8	9.9	14.2
Liver-and-gland meal	7.5	6.0	65.0	1.5	4.5	15.5
Meat scrap (55%)	7.5	24.3	55.0	2.2	1.0	10.0
Meat-and-bone scrap (50%)	7.5	29.0	50.0	2.5	1.5	9.5
Meat-and-bone scrap (45%)	7.5	32.5	45.0	2.5	2.0	10.5

Table 5.6—(Continued)

Feedstuff	Moisture	Ash	Crude protein	Carbohydrates: Crude fiber	Carbohydrates: Nitrogen-free extract	Fat or ether extract
	Per cent	Per cent	Per cent	Per cent	Per cent	Per cent
Products of animal origin						
Milk, whole liquid	87.0	.7	3.5	.0	4.9	3.9
Pork liver, dried	5.0	5.5	63.5	.5	15.0	10.5
Pork cracklings	5.0	2.5	56.5	.0	4.0	32.0
Shrimp meal (or bran)	11.0	34.0	42.0	9.5	1.5	2.0
Skim milk, liquid	90.5	.7	3.5	.0	5.1	.2
Skim milk, dried	6.6	8.0	34.0	.0	50.5	.9
Tankage (60%)	8.0	19.5	60.0	2.7	1.8	8.0
Whey, liquid	93.7	.6	.8	.0	4.9	.0
Whey, dried	6.3	8.5	12.5	.3	71.7	.7
Whey product, dried	4.5	13.2	17.5	0.3	63.0	1.5
Vitamin supplements and green feed						
Alfalfa, fresh, green	73.8	2.5	4.6	7.5	10.7	.9
Alfalfa leaf meal (20%)	8.2	12.2	20.0	17.0	40.0	2.6
Alfalfa meal (17%)	8.3	8.8	17.0	26.0	37.4	2.5
Alfalfa meal (13%)	9.0	7.5	13.0	34.0	35.0	1.5
Alfalfa stem meal	8.0	6.8	10.0	40.0	34.0	1.2
Apple pomace, dried	10.7	3.5	4.5	15.5	61.0	4.8
Beet molasses, liquid	21.5	10.5	6.5	.0	61.5	.0
Beet pulp, dried	10.5	4.9	8.5	18.5	57.0	.6
Cabbage	90.8	.8	1.8	1.1	5.2	.3
Cane molasses, liquid	25.0	7.0	3.0	.0	65.0	.0
Cane molasses, dried	5.0	10.0	5.0	.1	79.4	.5
Cereal grass, dehydrated	12.0	11.0	23.0	13.5	35.0	5.5
Carrots	20.0	2.5	.3	.0	77.2	.0
Corn molasses	88.4	1.1	1.1	1.2	7.9	.3
Distillers' grains, corn, dried	11.0	2.0	30.0	11.0	35.0	11.0
Distillers' grains, rye, dried	11.0	2.0	22.0	15.0	44.0	6.0
Distillers' grains, mixed, dried[1]	11.0	2.0	28.0	12.0	37.0	10.0
Distillers' grain, corn, dried, with solubles[2]	11.0	6.0	28.0	7.0	39.0	9.0
Distillers' grains, rye, dried with solubles[2]	11.0	6.0	29.0	8.0	43.0	3.0
Distillers' grains, mixed, dried, with solubles[2]	11.0	6.0	28.0	7.0	40.0	8.0
Distillers' solubles, corn, dried	5.0	8.5	25.0	2.0	53.5	6.0
Distillers' solubles, rye, dried	5.0	8.5	37.0	2.0	46.5	1.0
Distillers' solubles, mixed, dried[1]	5.0	8.5	27.0	2.0	52.5	5.0
Fermentation byproduct riboflavin supplement, 500 mcgms. per gm.	6.7	4.8	33.0	5.2	43.5	6.8
Grapefruit residue, dried	9.3	4.3	4.8	11.6	68.7	1.3
Kale	88.4	1.9	2.4	1.5	5.3	.5
Kelp, dried	8.4	37.0	6.7	8.0	39.2	.7
Mangels	90.1	1.1	1.5	.8	6.4	.1
Orange peel, dried	14.0	4.1	5.8	10.6	64.8	.7
Orange pulp meal	10.8	3.4	7.5	8.9	67.9	1.5
Potatoes	78.8	.9	2.0	.5	17.7	.1
Rape	84.6	2.2	2.6	2.4	7.6	.6
Red-clover hay	12.3	6.7	12.7	25.7	39.6	3.0
Rutabagas	88.8	1.0	1.2	1.5	7.3	.2
Sweet potatoes	72.0	1.0	1.5	1.3	23.7	.5
Turnips	90.6	.8	1.3	1.1	6.0	.2
Yeast, brewers', dried, feed grade	7.0	7.0	45.0	1.5	38.0	1.5

[1]Typical mash of 70% corn, 20% rye and 10% barley malt.
[2]Sometimes called distillers' dark grains.

Table 5.7—The Average Calcium, Phosphorus, and Manganese Content of Some Feedstuffs Used in the Feeding of Poultry

Feedstuff	Calcium (Ca)	Phosphorus (P)	Manganese (Mn)
	Per cent	Per cent	Parts per million (p.p.m.)
Grains and seeds, and their by-products			
Acorns of the willow oak	0.26	0.07	?
Babassu meal	.30	.67	?
Barley, except Pacific Coast	.07	.40	16
Barley, Pacific Coast	.05	.40	16
Beans, navy	.16	.45	13
Bread	.03	.10	4
Brewers' grains, dried	.20	.46	20
Buckwheat	.06	.35	80
Coconut meal, old process	.28	.60	85
Corn or ground corn, dent	.02	.27	5
Corn, Argentine	.02	.30	12
Corn-and-cob meal	.01	.22	?
Corn bran	.03	.25	16
Corn feed meal	.03	.36	?
Corn germ meal	.04	.58	?
Corn gluten feed	.12	.66	24
Corn gluten meal	.07	.37	4
Corn oil-cake meal	.06	.57	?
Cottonseed meal (43% protein)	.24	1.15	18
Cottonseed meal (41% protein)	.25	1.18	18
Cowpeas	.10	.46	30
Feterita	.02	.32	?
Field peas	.08	.40	30
Flax seed	.25	.66	35
Garden peas	.08	.40	25
Hegari	.03	.31	?
Hempseed	.22	.87	10
Hominy	.01	.08	2
Hominy feed	.03	.50	16
Kafir	.03	.35	16
Linseed meal (o. p.)	.33	.74	40
Malt sprouts, dried	.22	.70	35
Millet (proso)	.01	.33	35
Milo	.04	.32	15
Oats	.10	.36	34
Oat meal or groats	.08	.44	25
Oat mill feed	.24	.20	?
Peanut kernels	.07	.39	16
Peanut meal (43% protein)	.18	.56	18
Peanut meal (41% protein)	.16	.52	?
Rice, rough	.07	.21	?
Rice, polished	.01	.09	12
Rice bran	.10	1.84	280
Rice polishings	.04	1.10	?
Rye	.05	.36	35
Rye middlings	.10	.44	*
Sesame seed meal	2.00	1.60	?
Soybeans	.21	.60	31
Soybean meal (43% to 45% protein)	.30	.70	30
Soybean meal (solvent, high protein)	.30	.66	**
Sunflower seed	.41	.70	?
Sunflower seed meal (hulled)	.43	1.00	?
Wheat, average all types	.04	.40	39
Wheat bran	.11	1.20	119
Wheat flour	.02	.11	4
Wheat flour middlings	.08	.73	113
Wheat germ meal	.07	1.00	160
Wheat middlings, standard	.10	.90	119
Wheat mixed feed	.10	1.00	119
Wheat, soft (Pacific Coast)	.04	.29	50

*Highly variable.
**Use next preceding value (30) as an estimate of the quantity.

Table 5.7—(Continued)

Feedstuff	Calcium (Ca)	Phosphorus (P)	Manganese (Mn)
	Per cent	Per cent	Parts per million (p.p.m.)
Grains and seeds, and their by-products			
Wheat red dog flour	.07	.60	35
Wheat shorts	.10	.90	60
Products of animal origin			
Beef scrap	7.23	3.73	5
Blood meal	.34	.25	1
Buttermilk, liquid	.18	.10	0.06
Buttermilk, condensed	.56	.33	0.2
Buttermilk, dried	1.50	1.00	0.4
Casein	0.28*	0.75	2
Cod-liver meal	.27	.34	?
Crab meal	13.00	.80	150
Eggs (contents), liquid	.06	.22	4
Eggs (contents), dried	.19	.78	14
Fish meal, unidentified, avg	5.00	3.00	45
Fish meal, Alaska herring	3.00	2.00	?
Fish meal,menhaden	6.00	3.60	25
Fish meal, salmon	5.00	3.50	8
Fish meal, white (high ash)	9.10	4.70	11
Fish meal, white (low ash)	5.80	3.00	40
Fish meal, sardine	4.70	2.70	40
Fish meal, tuna	6.00	3.40	?
Fish solubles, condensed	.08	.80	4
Liver meal	.10	.90	4
Liver-and-gland meal	.35	.90	5
Meat scrap (55%)	8.00	4.00	18
Meat-and-bone scrap (50%)	10.00	4.50	12
Meat-and-bone scrap (45%)	11.00	5.00	10
Milk, whole liquid	.12	.09	0.06
Pork cracklings	.82	.40	20
Pork liver, dried	.06	1.12	4
Skim milk, liquid	.13	.11	0.06
Skim milk, dried	1.30	1.00	0.6
Tankage (60%)	7.00	3.50	14
Whey, liquid	.05	.04	0.1
Whey, dried	.85	.70	1
Whey product, dried	1.50	1.00	?
Vitamin supplements and green feed			
Alfalfa, fresh, green	.42	.07	7
Alfalfa leaf meal (20%)	1.90	.25	30
Alfalfa meal (17%)	1.40	.20	26
Alfalfa meal (13%)	1.20	.20	?
Apple pomace	.10	.10	?
Beet molasses	.04	.02	?
Beet pulp, dried	.70	.07	23
Cabbage	.07	.04	2
Cane molasses, liquid	.60	.06	?
Cane molasses, dried	1.10	.08	?
Carrots	.05	.05	3
Distillers' grains, mixed, dried	.04	.21	20
Distillers' grains, mixed, dried with solubles	.2	.8	60
Distillers' solubles, mixed, dried	.4	1.6	100
Fermentation byproduct (500 riboflavin per gm.)	.5	.6	*
Grapefruit residue, dried	.74	.10	3
Kale	.18	.07	6
Mangels	.02	.04	?
Orange peel, dried	.73	.11	8
Orange pulp meal	.64	.10	?
Potatoes	.02	.06	2

*Highly variable.

Table 5.7—(Continued)

Feedstuff	Calcium (Ca)	Phosphorus (P)	Manganese (Mn)
	Per cent	Per cent	Parts per million (p.p.m.)
Vitamin supplements and green feed			
Rape	.34	.07	50
Red clover hay	1.17	.18	40
Rutabagas	.06	.04	16
Sweet potatoes	.04	.06	4
Turnips	.05	.05	1
Yeast, brewers' dried, feed grade	0.12	1.50	5
Calcium, phosphorus, and manganese supplements			
Bone meal, steamed	28.00	13.00	25
Bone meal, special steamed	31.00	14.00	10
Calcite, high calcium	39.00	0.00	200
Calcite	35.00	Trace	300
Charcoal	1.73	.03	?
Crabshell	23.75	2.55	300
Defluorinated phosphate (rock), high phosphorus	28.00	13.50	40
Defluorinated phosphate (rock), low phosphorus	21.00	8.50	?
Defluorinated superphosphate	26.50	12.30	?
Di-calcium phosphate	27.00	20.00	?
Gypsum	25.00	Trace	?
Kelp, dried	3.42	.34	12
Limestone, high calcium	39.00	0.00	100
Limestone	35.00	Trace	200
Mica grit	.59	.08	?
Oystershell, washed	38.00	Trace	100
			Per cent
Manganous sulphate, anhydrous, C. P.	.00	.00	36.3
Manganous sulphate, tetrahydrate, C. P.	.00	.00	24.6
Manganous sulphate, feeding grade (65% $MnSO_4$)	Trace	Trace	23.7

Table 5.8—The Average Iron, Iodine, and Copper Content of Some Feedstuffs Used in the Feeding of Poultry.*

Feedstuff	Iron (Fe)	Iodine (I)	Copper (Cu)
Grain and seeds and their by-products	Parts per million (p.p.m.)	Parts per million (p.p.m.)	Parts per million (p.p.m.)
Barley	50	0.05	7.5
Beans, navy	85	.05	9.5
Bread	9	?	2.0
Buckwheat	?	.08	?
Coconut meal	40	?	?
Corn or ground corn	35	.05	4.5
Cottonseed meal	600	.12	17.0
Cowpeas	10	?	?
Distillers' grains with solubles	300	?	70.0
Garden peas, dried	55	?	8.0
Hominy	9	?	?
Linseed meal	250	.07	26.0
Millet	?	.01	?
Milo	50	.02	12.0
Oats	70	.06	8.0
Oatmeal or groats	45	.05	7.0
Peanut kernels	20	?	?
Rice, rough	20	?	?
Rice, polished	8	?	1.5
Rye	45	.05	6.5
Soybeans	25	.09	?
Soybean meal	30	.13	17.0
Wheat	50	.04	7.8
Wheat bran	75	.07	10.0
Wheat flour	10	?	?
Wheat germ meal	75	?	?
Wheat middlings standard	100	?	20.0
Wheat shorts (gray)	?	.11	?
Products of animal origin			
Bone meal, steamed	800	.00	19.0
Buttermilk, dried	30	.70	?
Casein	?	.30	?
Eggs (contents), liquid	30	.08	2.5
Fish meal, sardine	?	1.05	20.0
Meat-and-bone scrap (50%)	?	1.30	11.0
Milk, whole liquid	2	.04	.3
Skim milk, dried	30	?	3.0
Whey, dried	7	?	50.0
Green feeds, etc.			
Alfalfa, fresh, green	15	0.15	2.5
Alfalfa meal (17%)	60	.55	10.0
Beet molasses, liquid	?	1.60	?
Cabbage	10	.01	1.0
Cane molasses (liquid)	95	1.60	?
Carrots	6	.01	1.1
Cereal grass, dehydrated	?	.20	?
Corn molasses (liquid)	15	3.50	?
Distillers' solubles, dried	500	?	80.0
Kelp meal	?	1920.0	?
Potatoes	10	.01	1.6
Rutabagas	6	?	?
Sweet potatoes	10	?	1.8
Turnips	6	?	.8
Yeast, brewers', dried, feed grade	200	.01	35.0

*The range of the content of these three elements in the feedstuffs listed above is so great that the minimum value often is only 5 to 25 per cent of the average.

Note: Potassium iodide, C.P., contains 76.45 per cent, or 764,500 parts per million, of iodine.

Stabilized potassium iodide (90% KI) contains 68.8 per cent or 688,000 parts per million, of iodine.

Table 5.9—The Average Vitamin Content of Some Feedstuffs Used in the Feeding of Poultry: Vitamin A, Vitamin D, and Vitamin E

Feedstuff	Vitamin A per pound	Vitamin D per pound	Vitamin E per pound
Grains and seeds, and their by-products	International Units (I. U.)	International Units (I. U.)	Milligrams
Acorns of the willow oak	80,000	—	?
Barley	320	Trace	20
Bread	40	?	?
Brewers' grains, dried	400	—	20
Corn, white	Trace	—	11
Corn, yellow, average	2,300	—	14
Corn, yellow, new	3,500	—	**
Corn, one year old	2,500	—	**
Corn, two years old	1,800	—	**
Corn, three years old	1,300	—	**
Corn, gluten feed (yellow)	1,000	—	18
Corn, gluten meal (yellow)	7,000	—	10
Cottonseed meal	150	—	12
Cowpeas	220	—	?
Distillers' grains (corn), dried with solubles	1,100	—	21
Feterita	230	—	?
Hegari	300	—	?
Hominy feed (yellow)	800	—	15
Kafir	280	—	?
Linseed meal	200	—	12
Malt sprouts	?	—	18
Milo	370	—	12
Oats	250	—	23
Peanut kernels	120	—	?
Peanut meal	150	—	?
Rye	70	—	?
Soybeans	400	—	17
Soybean meal	170	Trace	10
Wheat, average	125	—	17
Wheat, new	200	—	**
Wheat bran	175	—	1
Wheat flour middlings	160	—	?
Wheat germ meal	190	—	72
Wheat germ oil	?	—	860
Wheat middlings, standard	225	—	10
Wheat red dog flour	100	—	26
Wheat shorts (gray)	175	—	14
Products of animal origin			
Buttermilk, liquid	25	Trace	—
Buttermilk, dried	200	5	—
Cod-liver oil (850 A—85 D)	385,500	38,550	—
D-activated animal sterol (1,500 D)	none	680,000	none
D-activated animal sterol (2,000 D)	—	900,000	—
Eggs (contents), liquid	6,000	56	Trace
Fish solubles, condensed	1,800	900	—
Liver, pork, dried	45,000	200	?
Milk, whole, liquid	1,080	?	—
Sardine oil	52,000	38,550	—
Skim milk, liquid	15	—	—
Skim milk, dried	130	—	—
Vitamin-A-and-D feeding oil			
(3000 A - 600 D)	1,360,000	272,000	—
(3000 A - 300 D)	1,360,000	136,000	—
(2250 A - 600 D)	1,000,000	272,000	—
(2250 A - 300 D)	1,000,000	136,000	—
(1875 A - 300 D)	850,000	136,000	—
(1500 A - 300 D)	680,000	136,000	—
(1125 A - 300 D)	500,000	136,000	—
(750 A - 300 D)	340,000	136,000	—

****Use next preceding value (above) as an estimate of the quality.**

Table 5.9—(Continued)

Feedstuff	Vitamin A per pound	Vitamin D per pound	Vitamin E per pound
Green feeds, etc.	International Units (I. U.)	International Units (I. U.)	Milligrams
Alfalfa, fresh, green	56,000	?	69
Alfalfa hay, dehydrated	90,000	100	?
Alfalfa hay. suncured	20,000	400	12
Alfalfa meal, dehydrated, average	75,000	?	120
Alfalfa leaf meal, dehydrated (20%)	120,000	?	174
Alfalfa meal, dehydrated (17%)	90,000	?	120
Alfalfa meal, dehydrated (13%)	50,000	?	?
Alfalfa meal, sun-cured, average	25,000	500	12
Alfalfa leaf meal, sun-cured, (20%)	50,000	?	20
Alfalfa meal, sun-cured, (17%)	30,000	?	12
Alfalfa meal, sun-cured, (13%)	15,000	?	?
Cabbage	450	—	?
Carrots	18,000	—	?
Grass oats, young dehydrated	200,000	—	?
Kale	45,000	—	?
Potatoes	220	—	?
Red-clover hay	9,000	?	30
Silage, alfalfa, moist	25,000	?	?
Silage, grass, moist	60,000	?	?
Sweet potatoes	10,000	?	?
Yeast, irradiated (6,750 D)	—	3,000,000	—

Table 5.10—The Average Vitamin Content of Some Feedstuffs Used in the Feeding of Poultry: Riboflavin, Pantothenic Acid, Choline, Niacin, and Thiamin

Feedstuff	Riboflavin per pound	Pantothenic acid per pound	Choline per pound	Niacin per pound	Thiamin per pound
Grains and seeds, and their by-products	Mg.[1]	Mg.	Mg	Mg.	Mg.
Babassu meal	0.58	?	?	?	?
Barley	0.60	3.00	500	31.10	2.50
Beans, mung	?	?	?	10.50	1.80
Beans, navy	1.40	.60	?	13.00	2.40
Beans, pinto	3.60	?	?	?	7.00
Bread	.45	?	?	?	.15
Brewers' grains, dried	.40	1.10	?	?	?
Buckwheat	.25	.65	?	20.00	2.30
Coconut meal	2.30	.40	?	?	?
Corn, white	.55	3.00	200	7.30	2.30
Corn, yellow	.50	2.70	200	7.50	2.10
Corn bran	.75	?	?	?	2.00
Corn feed meal	.54	1.65	?	?	?
Corn germ meal	1.55	6.00	?	13.90	9.90
Corn gluten feed	1.50	4.50	350	47.20	?
Corn gluten meal	1.00	6.00	750	13.60	?
Corn oil-cake meal	3.00	?	?	?	4.50
Cottonseed meal	1.50	5.00	1,200	20.00	5.80
Cowpeas	1.25	8.00	1,000	10.80	3.90
Distillers' dark grains (corn), dried[2]	3.45	5.00	1,800	38.00	2.00
Distillers' dark grains, dried[2]	3.45	5.30	1,800	37.00	1.80
Durra	0.40	4.45	?	24.00	?
Garden peas	.78	7.00	1,000	?	2.70
Hegari	0.50	3.95	?	30.00	?
Hempseed meal	1.25	3.00	?	?	?
Hominy	.20	?	?	4.00	.35
Hominy feed	.90	3.3	?	9.00	5.0
Kafir	.50	2.80	?	18.00	2.30
Linseed meal	1.15	3.60	750	22.00	5.90
Malt sprouts	4.30	?	?	25.00	?
Millet	?	?	?	10.00	3.20
Milo	.50	4.0	250	33.00	1.8
Oats	.60	4.65	400	6.60	3.45
Oat meal or groats	.50	4.70	600	7.80	2.50
Oat mill feed	1.25	?	?	?	?
Peanut kernels	1.60	20.00	?	58.00	3.70
Peanut meal	2.20	24.00	900	75.00	3.00
Rice, rough	.60	?	?	13.00	1.20
Rice, polished	.20	1.80	350	6.90	.25
Rice bran	1.20	10.00	?	125.00	9.00
Rice polishings	1.00	5.0	500	300.00	8.90
Rye	.68	4.60	?	6.80	2.00
Sesame seed meal	1.45	2.50	680	?	?
Soybeans	1.20	5.90	850	16.50	5.50
Soybean meal (43% protein)	1.50	6.0	1,500	18.50	4.00
Soybean meal (solvent, 47% protein)	1.80	6.00	1,200	20.00	1.0
Sunflower seed	?	?	?	?	.25
Sunflower seed meal (with hulls)	1.35	19.00	?	113.00	?
Wheat	.55	5.00	330	27.50	2.30
Wheat bran	1.30	11.40	400	139.00	3.50
Wheat flour	.18	?	190	?	.20
Wheat flour middlings	.85	4.60	?	44.00	4.00
Wheat germ meal	2.10	3.90	1,200	23.70	9.00
Wheat middlings, standard	1.10	7.20	500	56.50	5.00
Wheat red dog flour	1.25	6.20	325	21.00	9.00
Wheat shorts (brown)	1.35	6.00	530	43.00	6.00
Wheat shorts (gray)	1.30	5.00	?	?	?
Wheat shorts (white)	?	?	?	40.00	?

[1]Multiply by 1,000 to obtain micrograms (mcgm.) or gamma.
[2]Distillers' dark grains=distillers' grains with solubles.

Table 5.10—(Continued)

Feedstuff	Riboflavin per pound	Panto-thenic acid per pound	Choline per pound	Niacin per pound	Thiamin per pound
Products of Animal Origin	Mg.	Mg.	Mg.	Mg.	Mg.
Blood meal	.85	.70	?	12.00	?
Buttermilk, liquid	1.25	2.00	?	.60	.20
Buttermilk, dried	13.00	19.50	?	7.00	1.30
Buttermilk, sweet cream, dried	15.00	20.00	?	?	1.40
Casein	2.50	?	100	?	?
Crab meal	2.10	3.0	800	?	?
Eggs (contents), liquid	1.40	12.20	?	?	.85
Fish meal, average	3.00	3.70	1,400	?	.45
Fish meal, Alaska herring	3.10	4.00	2,000	?	.45
Fish meal, menhaden	2.20	?	1,500	25.0	.20
Fish meal, sardine	3.00	3.50	1,300	?	.45
Fish meal, white	3.80	?	?	36.00	.45
Fish solubles, condensed	9.00	16.30	1,200	145.00	2.00
Liver meal	20.00	40.00	4,700	80.00	1.00
Meat scrap (55% protein)	2.60	4.00	1,000	30.00	.55
Meat-and-bone scrap (50%)	2.40	2.00	1,000	25.00	.50
Meat-and-bone scrap (45%)	2.30	1.50	750	22.00	?
Milk, whole, liquid	.85	1.40	390	.75	.20
Pork liver, dried	40.00	80.00	?	450.00	7.00
Shrimp meal	1.00	?	?	?	?
Skim milk, liquid	.90	1.70	?	.50	.20
Skim milk, dried	8.50	15.00	500	7.00	1.55
Tankage (60%)	.80	1.00	1,000	30.00	.45
Whey, liquid	.75	2.40	?	.40	?
Whey, dried	11.00	21.00	700	8.50	1.00
Whey product, dried	13.50	34.30	900	7.00	2.00
Green feeds etc.					
Alfalfa, fresh, green	2.20	5.00	?	8.20	.70
Alfalfa leaf meal, dehydrated (20% pro.)	8.40	17.00	600	24.00	3.00
Alfalfa meal, dehydrated (17% pro.)	6.70	16.00	500	22.00	2.00
Alfalfa meal, dehydrated (13% pro.)	6.00	?	250	18.00	?
Alfalfa leaf meal, sun-cured (20% pro.)	7.30	12.00	600	22.00	1.60
Alfalfa meal, sun-cured (17% pro.)	5.00	10.00	500	21.00	1.30
Alfalfa meal, sun-cured (13% pro.)	4.20	?	250	17.00	?
Apple pomace, dried	.50	?	?	?	?
Beet pulp, dried	.65	?	?	11.50	?
Cabbage	.25	?	900	1.30	.33
Cane molasses, liquid	1.00	15.00*	300	21.00	.45
Carrots	.30	1.00	325	6.70	.30
Cereal grass, dried	7.00	?	?	33.50	?
Corn molasses	.25	?	?	?	?
Distillers' solubles (corn), dried	7.50	10.40	2,700	70.00	4.00
Distillers' solubles, dried	7.00	11.00	?	68.00	3.20
Kale	2.25	2.00	?	?	.75
Potatoes	.15	1.50	400	5.00	.65
Red-clover hay	8.00	?	?	16.60	1.25
Sweet potatoes	0.25	5.00	130	6.00	0.45
Turnips	.20	1.50	320	?	.35
Yeast, brewers', dried, feed grade	16.00	55.00**	2,000	225.00	55.0

*Highly variably ranging from 5 to 30 mg. per lb.
**Very highly variable, ranging from 10 to 150 mg. per lb.

Note:

1 pound of riboflavin = 453,590 mg. of riboflavin or 100%
1 pound of calcium (d,1) pantothenate = 220,500 mg. of pantothenic acid or 48.612%
1 pound of calcium (d) pantothenate = 417,390 mg. of pantothenic acid or 92.019%
1 pound of choline chloride = 393,650 mg. of choline or 86.785%
1 pound of niacin = 453,590 mg. of niacin or 100%
1 pound of thiamin = 453,590 mg. of thiamin or 100%

Table 5.11—The Average Amino Acid Content of Some Feedstuffs Used in the Feeding of Poultry: Arginine, Lysine, Methionine, Cystine, Tryptophan, and Glycine

Feedstuff	Protein (N x 6.25)	Arginine	Lysine	Methionine	Cystine	Tryptophan	Glycine
Grains and seeds and their by-products	Per cent	Per cent	Per cent	Per cent	Per cent	Per cent	Per cent
Babassu kernel oil meal	23.4	3.33	0.9	0.38	—	0.18	?
Barley	11.0	0.46	0.21	0.22	0.20	0.13	?
Coconut oil meal	22.0	2.35	0.65	0.33	—	0.21	?
Corn	9.0	.36	.22	.21	.12	.06	0.38
Corn, high protein[2]	11.4	.61	.27	.22	.22	.113	.33
Corn, low protein[2]	7.6	.46	.23	.17	.14	.09	.24
Corn germ meal	20.0	1.5	1.2	.31	.32	.26	?
Corn gluten feed	25.0	.61	.68	.25	?	.14	?
Corn gluten meal	41.0	1.3	.68	.99	.58	.24	1.7
Cottonseed meal	41.0	3.1	1.2	.72	.84	.47	2.2
Distillers' dark grains (corn), dried	28.0	.77	.60	.43	?	.10	?
Distillers' dark grains, dried	28.0	.89	.71	.45	?	.17	?
Flaxseed	22.0	1.5	.44	.51	.42	.37	?
Garden peas, dried	25.6	2.3	1.3	.20	.31	.18	?
Hominy feed	12.0	0.53	.48	.10	—	.10	?
Linseed meal	34.0	2.5	.94	.68	.65	.54	?
Milo	10.5	.36	.27	.16	?	.11	?
Oats	11.0	.55	.37	.24	.18	.11	?
Oatmeal or rolled oats	15.0	.89	.50	.29	.26	.20	?
Oat mill feed	6.6	.27	.21	.03	—	.05	?
Peanut meal	41.0	4.1	1.2	.49	.66	.39	2.3
Rice, polished	7.5	.54	.24	.24	.11	.10	.77
Rye	11.5	.49	.46	.13	?	.13	?
Sesame seed meal	41.0	3.7	1.2	1.3	.53	.70	3.8
Soybean meal	41.0	2.5	2.4	.70	.72	.51	6.6
Sunflower seed meal	21.0	1.7	.87	.76	.32	.27	.84
Tomato pomace, dried, ground	22.0	1.2	1.63	.05	—	.21	?
Wheat	12.0	.43	.32	.23	.20	.14	.87
Wheat bran	15.6	.90	.51	.28	.22	.23	?
Wheat flour	10.7	.36	.22	.11	.20	.09	.77
Wheat germ meal	28.5	1.7	1.6	.50	.21	.29	?
Wheat middlings, standard	16.0	.77	.43	.27	.18	.18	.45
Wheat red dog flour	16.9	.64	.50	.25	?	.14	?
Products of animal origin							
Blood meal	80.0	3.3	6.9	1.0	1.5	1.0	?
Buttermilk, dried	32.5	1.0	2.2	.66	?	.42	?
Casein	85.0	3.5	6.7	2.7	.28	1.1	1.8
Eggs (contents), liquid	13.3	.86	.80	.64	.33	.20	.32
Fish meal, menhaden	62.0	4.01	5.42	1.87	—	.66	?
Fish meal, red	71.5	5.10	7.85	2.41	—	.69	?
Fish meal, red	62.0	3.72	6.49	1.68	—	.69	?
Fish meal (unidentified)	60.0	3.8	4.0	1.8	.63	.69	3.3
Fish meal, sardine	65.0	4.8	4.2	2.2	.75	.81	3.6
Fish solubles, condensed	33.5	1.5	1.5	.49	.20	.20	?
Fish solubles, dried	56.0	2.73	4.4	1.25	—	.33	?
Gelatin	95.0	7.6	4.4	.74	.10	.00	22.4
Liver meal	65.0	3.5	4.1	1.5	.91	.78	?
Meat scrap	55.0	3.7	3.1	1.2	.58	.35	2.5
Milk, whole	3.5	.11	.28	.10	.04	.05	.08
Skim milk, dried	34.0	1.2	2.5	.87	.41	.41	.19
Tankage	60.0	3.6	3.9	.64	?	.45	?
Whey, dried	12.5	.39	.88	.29	.38	.21	.00
Green feeds, etc.							
Alfalfa meal (20% pro.)	20.0	.86	.94	.30	.32	.28	?
Alfalfa meal (17% pro.)	17.0	.80	.80	.27	.28	.26	?
Beet pulp, dried	10.0	.33	.6	.01	—	.1	?
Brewers' dried grains	30.0	1.67	1.2	.43	—	.40	?
Carrots	1.1	.007	.013	.006	?	.003	?
Distillers' solubles (corn), dried	25.0	.56	.65	.38	?	.09	?
Distillers' solubles, dried	27.0	.85	.66	.38	?	.09	?
Yeast, brewers', dried	45.0	2.2	3.4	1.00	.60	.80	3.4

[1]Distillers' dark grains=distillers' grains with solubles.
[2]From Sauberlich, "J. Nutr." 51:241-250, 1953.

Table 5.12—The Average Amino Acid Content of Some Feedstuffs Used in the Feeding of Poultry: Isoleucine, Leucine, Phenylalanine, Threonine, Valine, Histidine, and Tyrosine

Feedstuff	Protein (N x 6.25)	Isoleucine	Leucine	Phenylalanine	Threonine	Valine	Histidine	Tyrosine
	Per cent	Per cent	Per cent	Per cent	Per cent	Per cent	Per cent	Per cent
Grains and seeds and their by-products								
Barley	11.0	0.43	0.61	0.44	0.25	.044	0.18	0.12
Babassu kernel oil meal	23.4	0.86	1.44	0.96	0.65	1.15	0.38	0.45
Coconut oil meal	22.0	0.78	1.86	0.86	.064	1.06	0.34	0.41
Corn	9.0	.41	1.4	.44	.32	.46	.22	.54
Corn, high protein[2]	11.4	.44	1.43	.55	.53	.81	.25	.47
Corn, low protein[2]	7.6	.27	.71	.31	.38	.57	.16	.28
Corn germ meal	20.0	.76	2.8	1.1	.92	1.2	.58	1.2
Corn gluten feed	25.0	1.1	2.8	.90	.88	1.3	.63	.38
Corn gluten meal	41.0	2.1	7.7	2.4	1.5	2.1	.76	2.3
Cottonseed meal	41.0	1.6	2.1	2.2	1.2	1.8	1.1	1.2
Distillers' dark grains (corn), dried[1]	28.0	1.7	2.1	1.8	.94	1.5	.69	.60
Distillers' dark grains, dried	28.0	1.7	2.0	1.7	.98	1.6	.43	.53
Flax seed	22.0	.66	1.3	1.3	.88	1.6	.42	.99
Garden peas, dried	25.6	1.0	1.6	1.2	1.0	1.0	.31	?
Hominy feed	12.0	.35	1.0	.34	.35	.47	.24	.46
Linseed meal	34.0	1.4	2.1	1.7	1.6	1.9	.51	1.7
Milo	10.5	.53	1.4	.48	.27	.57	.27	.18
Oats	11.0	.50	.72	.52	.33	.55	.29	.33
Oatmeal or rolled oats	15.0	.83	1.2	1.0	.53	.96	.33	.69
Oat mill feed	6.6	.14	.4	.16	.15	.25	.06	.19
Peanut meal	41.0	1.4	2.4	2.2	.62	2.3	.86	1.8
Rice, polished	7.5	.39	.65	.50	.30	.47	.10	.42
Rye	11.5	.46	.68	.59	.32	.55	.20	.14
Sesame seed meal	41.0	2.0	3.1	3.4	1.5	2.1	.62	1.8
Soybean meal	41.0	2.1	3.0	2.1	1.7	2.0	1.0	1.4
Sunflower seed meal	21.0	1.1	1.3	1.1	.84	1.1	.36	.55
Tomato pomace, dried gr.	22.0	.74	1.7	.86	.76	1.0	.42	.93
Wheat	12.0	.44	.73	.61	.35	.52	.26	.49
Wheat bran	15.6	.60	.86	.51	.33	.77	.42	.17
Wheat flour	10.7	.42	.78	.57	.27	.42	.21	.35
Wheat germ meal	28.5	.86	2.0	1.2	1.1	1.2	.71	1.1
Wheat middlings, standard	16.0	.76	.90	.69	.41	.83	.41	.20
Wheat red dog flour	16.9	.71	.92	.36	.43	.85	.43	.14
Products of animal origin								
Blood meal	80.0	.92	9.4	5.5	4.4	6.2	4.2	3.0
Buttermilk, dried	32.5	2.3	3.1	1.3	1.4	2.5	.84	.92
Casein	85.0	5.7	8.9	4.6	3.7	6.0	2.7	5.4
Eggs (content), liquid	13.3	1.1	1.2	.80	.61	.97	.28	.63
Fish meal, menhaden	62.0	2.86	5.07	2.68	3.00	3.12	1.63	1.61
Fish meal, red	71.5	3.67	5.05	2.83	2.86	3.68	1.47	1.88
Fish meal, red	62.0	3.17	6.16	2.71	2.74	3.57	1.39	1.89
Fish meal, (unidentified)	60.0	3.2	4.9	2.7	2.6	3.1	1.5	2.0
Fish meal, sardine	65.0	3.6	4.6	2.5	2.9	3.9	1.6	2.9
Fish solubles, condensed	33.5	.50	.77	.62	.70	.87	1.2	.25
Fish solubles, dried	56.0	2.14	4.25	1.69	1.43	2.67	.71	1.05
Gelatin	95.0	1.8	3.4	2.2	1.7	2.8	.95	.19
Liver meal	65.0	3.1	5.4	3.1	2.5	3.7	1.6	2.3
Meat scrap	55.0	2.3	4.7	2.4	2.1	2.7	1.4	1.7
Milk, whole	3.5	.24	.37	.19	.15	.26	.09	.19
Skim milk, dried	34.0	2.2	3.2	1.6	1.4	2.4	1.1	.77
Tankage	60.0	1.9	5.0	2.7	1.9	3.3	1.6	1.5
Whey, dried	12.5	.78	1.1	.44	.63	.73	.19	.25
Green feeds, etc.								
Alfalfa leaf meal	17.0	.74	1.1	.74	.58	.78	.31	.61
Alfalfa meal	17.0	.74	1.1	.74	.58	.78	.31	.61
Beet pulp, dried	10.0	.30	.70	.30	.40	.40	.20	.50
Brewers' dried grains	30.0	1.2	2.2	1.3	1.0	2.0	.53	1.3
Carrots	1.1	.032	.053	.031	.030	.037	.008	?
Distillers' solubles (corn), dried	25.0	1.4	1.3	1.6	.84	1.3	.65	.56
Distillers' solubles, dried	27.0	1.6	1.5	1.6	.95	1.5	.66	.57
Yeast, brewers', dried	45.0	2.7	3.3	1.8	2.5	2.4	1.3	1.9

[1]Distillers' dark grains=distillers' grains with solubles.
[2]From Sauberlich, "J. Nutr." 51:241-250, 1953.

Table 5.13—Average Digestibility in the Chicken of Some Feedstuffs Used in the Feeding of Poultry

Feedstuff	Organic matter	Crude Protein	Crude fiber	Nitrogen-free extract	Fat, or ether extract	Total digestible nutrients
Grains and seeds, and their by-products	Per cent	Per cent	Per cent	Per cent	Per cent	Per cent
Barley	76	74	9	82	69	67
Beans, navy	40	42	16	41	64	34
Beans, pinto	40	43	28	39	96	36
Brewers' grains, dried	65	80	12	78	60	63
Broom corn	71	46	13	82	91	65
Buckwheat	73	60	9	85	87	65
Coconut meal	41	56	15	32	92	45
Corn, whole or cracked	87	76	12	90	86	79
Corn, Argentine, whole or cracked	88	84	16	91	81	83
Corn, ground	88	79	6	92	88	81
Corn, Argentine, ground	88	83	13	91	91	84
Corn bran	35	54	7	33	89	36
Corn feed meal	80	85	5	82	82	74
Corn gluten feed	47	62	4	44	65	42
Corn gluten meal	66	81	11	57	55	61
Cottonseed meal	70	74	11	86	90	67
Cowpeas	75	55	12	87	89	65
Feterita	89	88	33	91	81	82
Field peas	74	76	12	80	80	66
Flax seed	76	90	6	60	93	101
Garden peas, dried	83	88	10	86	86	72
Hegari	92	86	42	95	77	83
Hempseed	61	75	?	65	90	83
Kafir	89	82	15	93	80	80
Linseed meal	41	62	9	24	76	41
Millet (proso)	74	76	17	89	82	73
Milo	91	85	26	94	82	81
Mixed feed (laying mash)	85	85	44	88	92	83
Oats	65	73	13	72	82	61
Oat meal or groats	88	82	16	91	89	86
Peanut kernels	77	80	4	84	78	118
Peanut meal	65	75	6	70	80	63
Rice, rough	74	74	5	84	72	65
Rice, polished	95	90	50	96	75	84
Rice bran	55	59	4	60	80	57
Rice polishings	84	79	9	88	93	83
Rye	79	67	13	84	42	70
Shallu	90	78	39	94	85	83
Soybeans	76	74	28	80	90	85
Soybean meal	76	81	3	82	80	70
Soybean meal, solvent extracted	74	80	1	80	38	62
Sunflower seed	44	66	11	16	95	67
Sunflower seed, Hungarian striped	49	84	4	17	94	75
Sunflower seed meal (with hulls)	50	77	3	86	82	44
Wheat	84	79	9	88	50	73
Wheat bran	45	61	8	46	58	41
Wheat flour	94	85	81	95	96	83
Wheat middlings, standard	62	72	7	67	60	56
Wheat shorts (gray)	65	69	10	67	85	60
Products of animal origin						
Beef scrap	89	92	—	—	94	77
Blood meal	87	90	—	—	90	74
Bone meal, steamed	76	90	—	—	90	25
Buttermilk, dried	81	81	—	81	86	72
Casein	80	85	—	?	?	72
Fish meal, unidentified	88	90	—	50	91	69
Gelatin	75	75	—	—	—	71
Liver meal	66	65	—	45	91	76

Table 5.13—(Continued)

Feedstuff	Organic matter	Crude Protein	Crude fiber	Nitrogen-free extract	Fat, or ether extract	Total digestible nutrients
Products of animal origin	Per cent	Per cent	Per cent	Per cent	Per cent	Per cent
Meat scrap (55%)	85	90	—	—	90	70
Meat-and-bone scrap (50%)	84	90	—	—	90	64
Meat-and-bone-scrap (45%)	83	90	—	—	90	62
Milk, whole, liquid	94	95	—	94	92	16
Shrimp meal, (or bran)	53	60	17	55	86	32
Skim milk, liquid	94	93	—	94	92	8
Skim milk, dried	83	81	—	85	85	72
Tankage (60%)	80	85	—	—	90	67
Green feeds, etc.						
Alfalfa, fresh, green	56	65	44	61	55	14
Alfalfa meal (20%)	44	85	4	40	40	36
A·falfa meal (17%)	23	65	—	20	25	20
Beet pulp, dried	18	37	4	20	50	16
Cabbage	76	72	71	80	57	7
Carrots	85	68	54	93	64	9
Potatoes	83	63	13	88	41	17
Red clover hay	35	65	10	40	50	30
Rutabagas	84	75	65	90	75	9
Sweet potatoes	82	40	20	88	70	23
Turnips	79	68	35	90	74	7
Yeast, brewers', dried	71	75	5	70	60	63

An example of the use of digestibility data will be found in Table 5.14 in which barley and corn are compared.

Table 5.14—Computation of Digestible Nutrients Per 100 Pounds of Whole Corn and Barley

	Corn	Barley
Pounds of crude protein contained	8.5	11.0
Percent of protein digestible	76.0	74.0
(1) Pounds of digestible protein contained	6.5	8.1
Pounds of nitrogen-free extract contained	71.5	67.0
Percent of nitrogen-free extract digestible	90.0	82.0
(2) Pounds of digestible nitrogen free extract contained (which supplies heat and energy)	64.4	54.9
Pounds of fat contained	4.0	2.1
Per cent of fat digestible	86.0	69.0
Pounds of digestible fat contained	3.4	1.4
(3) Pounds of digestible fat x 2¼	7.9	3.2
(The factor 2¼ is used because fat supplies approximately 2¼ times as much energy as an equal weight of carbohydrate)		
Pounds of fiber contained	2.0	6.0
Percent of fiber digestible	12.0	9.0
(4) Pounds of digestible fiber contained	0.2	0.5
Totals of (1), (2), (3), and (4)	79.0	66.7

Through this comparison it will be seen that as an energy feed barley is worth approximately 85% as much as corn. At $3.60 per cwt. for corn, barley would be worth about 3.06 per cwt.

PLANNING AND BALANCING DIETS

A study of the diets recommended in Chapter 3 gives the reader an idea of how the various classes of diets are compounded and offers a fairly wide selection to suit a variety of conditions. In general it is advisable to select a tried and tested formula, obtain good quality ingredients, and have them thoroughly mixed in exactly the proportion recommended. However, conditions often demand special treatment and so in planning diets to suit special conditions it is well to determine first the nutritive requirements of the classes of turkeys to be fed, build the diet around one of the recommended diets and consider the 5 factors of (1) chemical composition, including vitamin content, (2) digestibility, (3) adaptability, (4) palatability, and (5) availability (at reasonable cost) of the ingredients under consideration.

Palatability. In general animal protein supplements are palatable and tend to make a mash containing them palatable. Vegetable protein feeds and alfalfa products have the opposite effect. Ground grains and fish oils are moderately palatable. Of the grains, unground, oats rank first, then wheat and spelt, followed by sorghum grains, proso, rye, and cracked corn (if cleaned). Whole corn is not well liked by turkeys under 18 weeks of age but is moderately palatable to older birds. Barley is least well liked of all common grains. However any grain if fed singly along with dry mash will be eaten liberally.

Any of the common grains may be fed as the only grain with mashes containing 24% or less protein than any single grain except wheat and possibly oats may be fed as the only grain free choice with mashes containing more than 24% protein. Heavy oats are so palatable that too high a percentage may be eaten at the expense of the mash. However the South Dakota Station (1951) reported that whole heavy oats could be fed to turkeys free choice with a growing mash between 12 and 28 weeks of age to the extent of 30 to 67% of total diet without decreasing the weight of the turkeys at 28 weeks. In all-mash growing diets, finely ground heavy oats up to 80% of the diet were consumed without ill effects. Lightweight oats are not so palatable as heavy oats so are less likely to be eaten in quantities sufficient to be damaging. On account of their high fiber content their consumption in very large quantities would seem inadvisable. Along with another palatable grain or grains, free choice or in a mixture, both types of oats are very satisfactory. Wheat in very large quantities, as would be eaten along with a high-protein mash both free choice usually produces good results, but occasionally does not hence is not recommended as the only grain with concentrate mashes. A combination of wheat and oats half and half made best and cheapest grains in finishing turkeys at the Manitoba Station (1934). In the same experiment it was noted that wheat was eaten best when coarsely ground or whole; oats when rolled or whole and barley when rolled or whole—in that order.

If the mash and the grain both are palatable, or both unpalatable, proper amounts will be consumed; but if one is palatable and the other not, an undesirable ratio of grain to mash may result. Free-choice feeding of 2 or more grains usually is more convenient than mixing them but both methods are good. Any 2 or 3 grains (excluding barley) free choice make a good combination with mash concentrates or with ordinary growing mashes containing 28% or more protein. Barley as the only grain also is good with any type mash. However when offered free choice with any other grain (except whole corn to turkeys under 18 weeks old) barley will not be eaten liberally.

Table 5.15—Summary of Certain Nutritional Standards for Turkey Diets Based on Total Feed Intake

	Crude Protein	Crude Fiber	Calcium	Phosphorus	Manganese	Vitamin A	Thiamin or Vitamin B[1]	Vitamin D	Riboflavin	Pantothenic acid
	Per cent	Per cent	Per cent	Per cent	Parts per Million[1]	International units per pound of feed	International units per pound of feed	International units per pound of feed	Units (micrograms) per pound of feed	Units (micrograms) per pound of feed
Starting mash with or without scratch grain[8]	22 to 28 (28 preferred)	3.5 to 8 (4 to 5½ preferred)	1.0 to 3.0 (2.0 preferred)	.6 to 1.8 (1.0 preferred)[6]	40 to 60 (50 to 60 preferred)	6000[2]	250	340 to 1000 (600 preferred)	2000	6000
Growing diet[9]	14 to 22[7]	4 to 9.0 (5 to 7 preferred)[11]	1.0 to 3.0 (2.0 preferred)	.6 to 1.8 (1.0 preferred)	30 to 60 (40 to 50 preferred)	3500[7]	225	225 to 340[4] (340 preferred)[4]	1500	5000
Maintenance diet[9] [10]	13 to 14	4 to 9.0 (5 to 7 preferred)[11]	.5 to 3.0 (1.0 preferred)	.4 to 1.8 (.6 preferred)	20 to 40 (30 to 40 preferred)	2500	225	126	1000	4000
Breeder diet[9]	15 to 17 (16 preferred)	4 to 9.0 (5 to 7 preferred)[11]	2.0 to 2.5 (2.3 prefer ed)	.7 to 1.0 (.8 preferred)	40 to 60 (50 to 60 preferred)	5000[3]	250	340 to 600[5] (600 preferred)	1800	6000 to 7500 (7500 preferred)

[1]Or milligrams per kilogram.

[2]In growing diets nearly all the vitamin A usually will come from plant sources. Only 1000 units may be supplied by fish oils according to the 1942 regulations. In confinement it is desirable that yellow corn products furnish 1,000 or more units.

[3]Of this amount about one-fourth should be true vitamin A from fish oil or a special vitamin A supplement.

[4]Preferably supplied by exposure to direct sunshine, irradition, or irradiated cholesterol.

[5]If considerable direct sunshine is obtained, the smaller quantity will suffice; if little or no sunshine is obtained, the larger quantity is desirable.

[6]Due to the high calcium and phosphorus content of some of the feedstuffs found in some turkey diets, it is often impracticable to lower the percentages to the optimum values given.

[7]The requirement depends upon age. The usual method of feeding during the growing period is to offer mash and scratch grain both ad libitum. The mash should contain about 20% protein for the most economical utilization of it.

[8]The scratch feed intake is small enough to be disregarded in the case of starting diets unless the poults are forced into abnormally heavy grain consumption. Starting diets are fed during the first 8 weeks.

[9]If mash and scratch grain both are fed, the approximate percentages (of total feed intake) of each must be known or estimated. Growing diets are fed starting at the end of the eighth week.

[10]Summer maintenance diets for adult stock and wintering diets for both young and adult stock are included.

[11]If green feed, or access to fibrous litter is provided, 5 to 5.5 % fiber is preferred, without it, 6 to 7% is preferred.

Confinement Growing Diets. Suggestions regarding diets for poults raised on wire floors and sunporches (without access to any pasture or other source of green feed) are given herewith and in Chapter 3.

First the nutritive requirements are determined by consulting Table 5.15 and suggestions dealing with the growing turkey's requirements for protein, fiber, minerals, and vitamins. Mash and scratch grain both fed ad libitum are recommended, in preference to all-mash or to restricted grain and mash. A growing mash for best results should contain from 20 to 22% of crude protein. Where turkeys have access to green range the grain may consist of corn or wheat or barley, alone, or any mixture of common grains, but when raised in strict confinement yellow corn and oats free choice or in a 50-50 mixture are preferable. The desired fiber content of the mash, as birds will not have access to growing green feed or other roughage, lies between 6 and 9%; preferably about 6 to 7%. The desired calcium content is about 2.8% and the desired phosphorus content about 1.4% with a calcium-phosphorus ratio (the percent of calcium divided by the percent of phosphorus) falling between 1.3:1 and 2.5:1, about 2:1 preferred. The fat content should be 3.5 to 4.5% and an excess of unsaturated fats should be avoided because of danger of off-flavor in the meat. The minimum vitamin needs (on total feed basis) are: A, 3,500 units; B, 225 units; and riboflavin, 1,500 units. An antibiotic should be added; the D factor will be furnished by direct sunshine obtained on the porches. Assuming it has been decided to use a 20%-protein growing mash, the approximate mash-scratch consumption, according to Table 5.24, is approximately 60 parts mash to 40 of scratch feed from 9 to 28 weeks inclusive. Thus in determining the total intake of proteins, vitamins, et cetera, the quantities of each are estimated for the mash and the scratch separately. The estimate for the mash is then multiplied by .60 and that for the scratch by .40.

The sum of the two represents that estimate for the total feed intake.

In order to estimate chemical composition and vitamin content of a diet the formula must be expressed in percentages. It will be noted that many formulas listed are expressed in percentage, that is, pounds of each ingredient in a hundredweight of the mixture. If the diet is all-mash the process is simple for only the mash need be considered; if it is mash-grain then either the quantities of each to be fed must be known, or estimated as noted above. Formulas expressed in pounds per ton must be converted to percent by dividing by 20. Thus 200 lbs. per ton is 10%. The same effect is obtained by first multiplying the percentage of each nutrient in each ingredient by the number of pounds per ton listed, adding up the results and dividing the total by 2000 or by whatever is the total poundage. Formulas adding up to 1000 pounds are converted to percent merely by pointing off one place.

Table 5.16 illustrates one method of making a complete estimate of all the nutritive factors now known to be necessary in planning poultry diets. Some other nutritive factors are considered necessary or desirable but these are handled by adding small, sometimes very small, amounts in the form of special premixes or concentrates or else they are considered to be furnished in adequate quantities by the ingredients normally included. Examples are pantothenic acid, choline and niacin.

If actual chemical analyses and vitamin assays of fair samples of the ingredients to be used could be obtained (which is rarely possible in practice) then the diet could be accurately compounded. However, in most cases only the averages are available and the use of these averages is the only method possible for the average turkey grower and indeed for many individuals and organizations that mix poultry feed. Quite often special information is available on the chemical composition or vitamin content of certain feedstuffs that are to be used. This information is valuable and should be utilized in preference to the

Table 5.16—An Estimate of the Nutritive Factors in 100 lbs. of a Turkey Growing Diet

(Division by 100 reduces the vitamin figures to a 1-lb. or percentage basis)

Ingredients	Units Supplied								
	Pounds in the mixture	Lbs. crude protein	Lbs. crude fiber	Lbs. calcium (Ca)	Lbs. phosphorus (P)	Parts per million of manganese	Intl. or U.S.P. units of vit. A	Intl. or U.S.P. units of vit. B	Chick units of riboflavin
Mash									
Ground yellow corn	15	1.40	.32	.0015	.0435	.8	34,500	9,750	7,500
Wheat bran	9	1.40	.81	.0099	.1089	10.8	5,400	7,560	10,800
Wheat middlings standard	15	1.54	.99	.0120	.1395	18.0	15,000	15,000	10,500
Finely ground oats	15	1.68	1.70	.0150	.0540	5.1	3,000	17,250	6,000
Corn gluten meal	10	4.30	.26	.0060	.0400	.4	100,000	....	8,000
Alfalfa leaf meal dehydrated	10	2.04	1.71	.1900	.0220	4.0	950,000	4,500	80,000
Distillers' dried solubles	10	2.68	.34	.0280	.1340	5.4	....	12,100	100,000
Meat-and-bone meal (50% protein)	10	5.00	.22	1.0200	.4910	1.0	....	1,800	24,000
Steamed bonemeal	3	.39	.03	.7800	.3798	.3	....	....	....
Ground oystershell	2	....	....	.7600	....	2.0	....	....	....
Salt (fine sifted)	1	....	....	....	....	....	....	....	....
Totals for the mash	100	20.43	6.38	2.8224	1.4127	47.8	1,107,900	67,860	246,800
Scratch grain									
Oats	50	5.60	5.65	.0500	.1800	17.0	10,000	57,500	20,000
Yellow corn	50	4.65	1.05	.0050	.1450	2.5	115,000	32,500	25,000
Totals for the grain	100	10.25	6.70	.0550	.3250	19.5	125,000	90,000	45,000
Entire feed intake									
Mash	60	12.26	3.83	1.6934	.8476	28.7	664,740	40,716	148,080
Grain	40	4.10	2.68	.0220	.1300	7.8	50,000	36,000	18,000
Totals for entire diet	100	16.36	6.51	1.7154	.9776	36.5	714,740	76,716	166,080

averages. For example, soft wheat known to contain only about 9% protein, as well as hard wheat, analyzing 14% protein (two extremes), are often available. Special meat and fish products that contain by guarantee as high as 70% protein and have a mineral content lower than average are sometimes purchasable. Many times the manufacturers or vendors of a product can furnish analyses and these analyses should always be used whenever they can be obtained. Protein supplements especially should be purchased on the basis of guaranteed protein content. In many States the composition of feeds must be stated on the feed-control tag attached to each container and this information is valuable to the purchaser. Guarantees of vitamin potency are most valuable and should always be insisted upon when purchasing feedstuffs that are to be fed for the vitamins they contain.

Starting diets are difficult to plan because the poults' requirements are much more exacting during the first few weeks of their lives than they are at any other time. Consequently it is well to use either a very reliable commercial starting mash or a tried and tested starting formula, mix it exactly as directed, and make every effort to obtain high-grade ingredients. However, when conditions demand that substitutions be made, the data regarding feedstuffs and poult requirements should be carefully studied and the diet planned accordingly. An advance test with poults brooded under favorable conditions will soon determine whether or not the diet is satisfactory. The vitamin requirements are especially important, vitamins A, B, D and riboflavin being required in liberal quantities.

Providing the poults with access to green pasture, or in lieu of that, chopped tender green feed and direct sunshine, will go a long way toward insuring the success of a feeding program. The fact that many mill byproducts are now degermed increases the desirability of supplying fresh green feed with home-mixed starting diets or else being certain that all the known vitamins are provided in liberal quantities. The protein should come from 2 or preferably 3 or 4 sources (meat scrap, fish meal, milk, vegetable protein concentrates) one of which should be fish meal. Dried milk is desirable but not necessary, since substitutes can be used.

Laying (or breeder) diets. For feeding breeders all-mash diets appear to be the most satisfactory, all factors considered, because the content of vitamins, minerals and proteins is under better control and less labor is required. When mash and grain are both fed, the latter should be limited in such a manner that the total protein consumption is 15% to 17%, preferably about 16%. If a 22% protein breeder mash is fed 50-50 with grain averaging about 11% protein, the total protein intake will be about 16½% (50% of 22 plus 50% of 11). If a 20% protein breeder mash is used, the grain should be limited to 40% (60% of 20 plus 40% of 11).

All of the cereal products used in laying diets should contain normal quantities of the germs of the grains in order to insure a sufficiency of vitamins B and E. For best results the protein supplements in breeder mashes should, under most conditions, be made up of three types: (1) a meat product, (2) a marine product, and (3) a milk product, omitting vegetable-protein supplements, or limiting them to soybean oil meal at a level of not over 15% of the total diet. (See breeder diets Chap. 3.)

GROWTH STANDARDS

The size attained by turkeys and the rate at which gains are made are chiefly determined by breeding, feeding, and management. Proper feeding and management permit the birds to develop to the full extent of their hereditary possibilities. Growth standards for the different varieties are difficult to construct due to the variation in the environment and in the different strains within each variety. For example, the weights shown for Broad Breasted Bronze at 26 weeks in Table 5.17 may be exceeded by 2½ pounds for the hens and 5 pounds for toms in some strains. However, it is believed that the data presented are fairly typical of the results that may be expected from representative strains of these varieties when raised under favorable conditions. It will be noted that "Standard" weight is attained at the age of about 36 weeks by stock of medium to large size and at 30 weeks by Beltsville Small White turkeys. Increase in weight usually continues well beyond that, provided turkeys are well fed and reproduction does not commence. The birds attain full growth at 7 to 8 months, after which time increases in weight are due largely to the deposition of fat.

The weights listed in Table 5.17 were obtained from smoothed growth curves of the three most popular varieties of turkeys fed high-energy diets containing an antibiotic at practical growth-promoting levels. They represent increases above those obtained on standard diets used prior to 1950, averaging about 15% the first 8 weeks; 12%, 9 through 16 weeks; 7%, 17 through 24 weeks; and 2½%, 25 through 34 weeks after which increased growth due to the antibiotic and/or the high-energy content of the diet ceases to be in evidence. For comparison with the older figures see Tables 5.20 to 5.23 inclusive. The weights in Table 5.18 were obtained from smoothed growth curves of stock fed standard diets. To convert to averages expected when low-fiber antibiotic diets are fed, use the figures suggested above.

FEED CONSUMPTION

The quantity of feed (hopper and hand fed) consumed by turkeys for each pound of live weight at any given age depends upon a number of factors, chiefly the composition of the diet fed, and the quantity of feed obtained by the birds from the range. In actual practice wastage and mortality are important factors, especially when the poults die after they have attained considerable size. It is probable that on the average there is little difference in feed consumption (per pound of gain at comparable stages of growth) that can be attributed to the influence

Table 5.17—Average Live Weight in Pounds at Weekly Intervals. Turkeys of Three Varieties Fed High Energy Diets Containing an Antibiotic Throughout the Entire Growing Period[1]

Age in Weeks	Broad Breasted Bronze			White Holland			Beltsville Small White			Age in Weeks
	Males	Females	Both Sexes	Males	Females	Both Sexes	Males	Females	Both Sexes	
1/7 (one day)	0.134	0.134	0.134	0.120	0.120	0.120	0.115	0.115	0.115	1/7 (one day)
1	.26	.24	.25	.21	.20	.21	.20	.19	.20	1
2	.48	.42	.45	.39	.36	.38	.38	.34	.36	2
3	.85	.75	.80	.68	.55	.62	.65	.53	.59	3
4	1.27	1.12	1.20	1.01	.86	.94	.96	.83	.90	4
5	1.85	1.60	1.73	1.50	1.25	1.38	1.43	1.15	1.29	5
6	2.60	2.21	2.41	2.10	1.70	1.90	2.00	1.56	1.78	6
7	3.45	2.85	3.15	2.73	2.22	2.48	2.60	2.00	2.35	7
8(2)	4.26	3.57	3.92	3.50	2.76	3.13	3.25	2.50	2.88	8
9	5.35	4.40	4.88	4.28	3.40	3.84	3.95	3.00	3.48	9
10	6.50	5.25	5.88	5.05	4.00	4.53	4.65	3.50	4.08	10
11	7.60	6.10	6.85	5.85	4.70	5.28	5.35	4.00	4.68	11
12	8.78	6.98	7.88	6.75	5.30	6.03	6.10	4.50	5.30	12
13	10.00	7.80	8.90	7.65	5.90	6.78	6.85	5.00	5.93	13
14(3)	11.25	8.60	9.93	8.50	6.50	7.50	7.60	5.45	6.53	14
15(3)	12.60	9.45	11.03	9.40	7.05	8.26	8.35	5.90	7.13	15
16(3)	14.08	10.34	12.21	10.23	7.65	8.94	9.10	6.35	7.73	16
17(3)	15.20	11.20	13.20	11.10	8.15	9.63	9.85	6.75	8.30	17
18(3)	16.40	11.80	14.10	12.00	8.65	10.33	10.60	7.15	8.88	18
19	17.50	12.40	14.95	12.85	9.10	10.98	11.35	7.55	9.45	19
20	18.60	13.00	15.80	13.75	9.60	11.68	12.10	7.95	10.03	20
21(4)	19.75	13.50	16.63	14.55	10.00	12.28	12.85	8.30	10.58	21
22(4)	20.80	14.00	17.40	15.50	10.40	12.95	13.60	8.65	11.13	22
23(4)	22.00	14.50	18.25	16.40	10.80	13.60	14.35	9.00	11.68	23
24(4,5)	23.10	14.90	19.00	17.20	11.20	14.20	15.10	9.35	12.23	24
25(5)	24.20	15.25	19.73	18.00	11.50	14.75	15.85	9.70	12.78	25
26(5)	25.20	15.65	20.43	18.75	11.80	15.28	16.60	10.00	13.30	26
27(5)	26.00	16.00	21.00	19.60	12.10	15.85	17.35	10.30	13.83	27
28(5)	26.80	16.35	21.58	20.30	12.40	16.35	18.10	10.55	14.33	28
29	27.60	16.65	22.13	21.05	12.70	16.88	18.80	10.80	14.80	29
30(6)	28.40	16.95	22.68	21.87	13.00	17.44	19.40	11.00	15.20	30
31	29.10	17.20	23.15	22.50	13.25	17.88	19.80	11.15	15.48	31
32	29.75	17.45	23.60	23.20	13.45	18.33	20.20	11.30	15.75	32
33	30.40	17.70	24.05	23.75	13.70	18.73	20.60	11.40	16.00	33
34	31.00	18.00	24.50	24.35	13.90	19.13	21.00	11.50	16.25	34
35	31.50	18.25	24.88	24.65	14.00	19.33	21.20	11.60	16.40	35
36(7)	32.00	18.50	25.25	25.00	14.10	19.55	21.40	11.70	16.55	36
37	32.50	18.70	25.60	25.25	14.18	19.73	21.60	11.75	16.68	37
38	33.00	18.85	25.93	25.50	14.25	19.88	21.75	11.80	16.78	38

[1]Data from USDA Research Center, Beltsville, Md.
[2]End of normal brooding period.
[3]Fryer-light roaster age for Beltsville White turkeys.
[4]Roaster age for Beltsville White turkeys.
[5]Roaster age for large and medium-sized varieties and strains.
[6]Age at which Beltsville White young hens and toms attain standard weight.
[7]Age at which Broad Breasted Bronze and White Holland young toms and hens attain standard weight.

Table 5.18—Average Live Weight in Pounds at Two-Week Intervals of Turkeys of Three Varieties Fed Standard Diets Without an Antibiotic

Age	Standardbred Bronze[1]			Jersey Buff[3]			Wild[4]		
	Males	Females	Both Sexes[2]	Males	Females	Both Sexes	Males	Females	Both Sexes
	(Lbs.)	(Lbs.)	(Lbs.)	(Lbs.)	(Lbs.)	(Lbs.)	(Lbs.)	(Lbs.)	(Lbs.)
One day	0.125	0.125	0.125	0.120	0.120	0.120	0.110	0.110	0.110
2 weeks	.36	.34	.35	.34	.32	.33	.28	.28	.28
4 "	.98	.84	.91	.85	.75	.80	.62	.60	.61
6 "	1.94	1.60	1.77	1.70	1.30	1.50	1.27	1.19	1.23
8 "	3.13	2.68	2.91	3.01	2.11	2.60	2.06	1.84	1.95
10 "	4.83	4.00	4.42	4.32	3.14	3.73	2.90	2.46	2.68
12 "	6.54	5.28	5.91	5.75	4.13	4.94	3.72	3.16	3.44
14 "	8.42	6.55	7.49	7.20	5.15	6.18	4.77	4.02	4.40
16 "	10.35	7.67	9.01	9.00	6.10	7.55	5.95	4.95	5.45
18 "	12.40	8.70	10.55	10.72	7.01	8.86	7.40	5.96	6.68
20 "	14.47	9.67	12.07	12.44	7.73	10.09	8.94	6.82	7.88
22 "	16.38	10.50	13.44	13.98	8.50	11.24	10.10	7.30	8.70
24 "	18.23	11.30	14.77	15.54	9.17	12.36	11.13	7.65	9.39
26 "	20.00	12.04	16.02	17.34	9.90	13.62	12.16	8.00	10.08
28 "	21.60	12.68	17.14	18.88	10.55	14.72	12.58	8.34	10.46
30 "	22.90	13.28	18.09	20.41	11.20	15.82	13.70	8.70	11.20
32 "	23.95	13.70	18.83	21.80	11.71	16.76	14.20	9.05	11.63
34 "	25.00	14.20	19.60	22.50	12.10	17.30	14.60	9.50	12.05
36 "	26.00	14.50	20.25	23.00	12.40	17.70	—	—	—
38 "	26.80	14.80	20.80	23.50	12.60	18.05	—	—	—

[1]Data from U. S. Dept. of Agriculture, Beltsville, Md.
[2]Unweighted averages assuming that half are males and half females.
[3]Data from New Jersey State Turkey Expt. Station, Millville, N. J.
[4]Game-farm stock reared on full feed and limited range (USDA, Beltsville, Md., 1936.)

of variety or strain, except that the large broad-breasted strains appear to be slightly more efficient in the use of feed than are the small and medium-sized strains during the last half of the growing period.

The Minnesota Station (1950) reported that Broad Breasted Bronze turkeys started in batteries, electrically brooded to 8 weeks, then placed on millet-rape range were, at 24 weeks, produced on 4.42 lbs. of feed per pound of live market turkey. The feed cost was 12.4c per pound and the diet used was a standard 26-per cent protein growing mash with corn and oats free choice after 8 weeks. On a modified high-energy growing mash (similar to Grower No. 14) the feed consumed was 3.96 lbs.,—the cost was 12.4c, and the birds averaged about a pound heavier at 24 weeks. With a group on an all-plant-protein, modified high-energy mash the amount was 4.10 lbs., and the cost was 12.5c. The addition of 3% dried yeast to the plant protein diet yielded no advantage in a fourth group.

FEED EFFICIENCY

Almquist (1951) accumulated highly valuable field data gathered from flocks of several thousand B. B. Bronze and one flock of several thousand Beltsville Small White turkeys raised in 1949 and 1950 on varied diets under California conditions. His feed consumption, growth rate and feed efficiency data are given in Table 5.19.

Table 5.19—Growth Rate and Feed Efficiency*

Broad Breasted Bronze

(Average weight in lbs. for mixed sexes)

Age in Weeks	Weight	Mash to date	Grain to date	Total Feed	Feed Efficiency
2	0.49	0.7	0.0	0.7	1.42
4	1.15	2.5	0.0	2.5	2.18
6	2.15	4.7	0.3	5.0	2.33
8	3.70	8.1	1.1	9.2	2.49
10	5.45	11.8	2.3	14.1	2.58
12	7.40	16.3	3.8	20.1	2.71
14	9.40	21.6	5.3	26.9	2.86
16	11.42	27.2	7.7	34.9	3.06
18	13.38	33.6	10.2	43.8	3.27
20	15.35	40.7	13.2	53.9	3.51
22	17.30	48.0	16.2	64.2	3.71
24	19.20	55.4	22.0	77.4	4.03
26	21.00	62.5	29.0	91.5	4.36

*Field data of H. J. Almquist, Modesto, Calif.; reprinted from World's Poul. Sci. Jour., Apr.-June, 1951.

Table 5.19—(Continued)

Beltsville Small White

(Average weight in lbs. for mixed sexes)

Age in Weeks	Weight	Mash to date	Grain to date	Total Feed	Feed Efficiency
2	0.40	0.6	0.0	0.6	1.50
4	0.85	1.8	0.1	1.9	2.24
6	1.58	3.3	0.5	3.8	2.40
8	2.60	5.3	1.1	6.4	2.46
10	3.82	7.9	1.9	9.8	2.56
12	5.13	10.9	2.9	13.8	2.69
14	6.50	14.3	4.3	18.6	2.86
16	7.72	18.7	5.7	24.4	3.16
18	8.80	23.4	7.7	31.1	3.53
20	9.66	27.6	9.9	37.5	3.88

A comparative study of two strains of Bronze, the Bourbon Red and the Beltsville Small White varieties was made at the Dominion Experimental Farms at Ottawa, Ontario (Canada). The results are given for males in Table 5.20, for females in Table 5.21 and for both sexes (weighted on basis of equal numbers) in Tables 5.22 and 5.23.

These turkeys were fed well-balanced diets and were reared under near-ideal range conditions resulting in better-than-average growth rate and feed efficiency which compare favorably with the results of feeding modern antibiotic- high energy diets. The stock was typical of modern strains of the four varieties used.

Table 5.20—Body Weights and Cumulative Feed Consumption[1]
(Males raised on range—1942)

	BROAD BREASTED BRONZE			STANDARD BRONZE[2]			BOURBON RED[3]			BELTSVILLE SMALL WHITE		
Age in wks.	Av. body wt. lb.	Av. feed cons. per bird lb.	Lb. feed per lb. body wt.	Av. body wt. lb.	Av. feed cons. per bird lb.	Lb. feed per lb. body wt.	Av. body wt. lb.	Av. feed cons. per bird lb.	Lb. feed per lb. body wt.	Av. body wt. lb	Av. feed cons. per bird lb.	Lb. feed per lb body wt.
0	.12			.12			.13			.11		
1	.24	.15	.63	.22	.15	.68	.22	.16	.73	.18	.21	1.17
2	.44	.51	1.16	.39	.50	1.28	.40	.51	1.28	.35	.55	1.57
3	.69	1.05	1.52	.62	1.03	1.66	.64	1.13	1.77	.56	1.02	1.82
4	1.02	1.86	1.82	.85	1.82	2.14	.88	1.93	2.19	.79	1.65	2.09
5	1.54	2.94	1.91	1.33	2.85	2.14	1.24	2.93	2.36	1.16	2.44	2.10
6	2.21	4.35	1.97	1.86	4.16	2.24	1.77	4.31	2.44	1.56	3.47	2.22
7	2.98	5.99	2.01	2.57	5.82	2.26	2.43	5.97	2.46	2.16	4.60	2.13
8	3.77	8.28	2.20	3.19	8.04	2.52	3.09	7.97	2.58	2.69	6.26	2.33
9	4.77	10.61	2.22	4.14	10.61	2.56	3.86	10.19	2.64	3.38	7.97	2.36
10	5.80	13.47	2.32	5.12	13.78	2.69	4.65	12.98	2.79	4.10	10.02	2.44
11	6.96	16.77	2.41	6.13	17.62	2.87	5.47	16.25	2.97	4.82	12.45	2.58
12	8.01	19.99	2.50	7.16	21.75	3.04	6.26	18.59	2.97	5.64	14.82	2.63
13	9.14	23.60	2.58	8.08	24.76	3.06	6.91	21.50	3.11	6.24	17.85	2.86
14	10.17	26.98	2.65	9.30	28.25	3.04	8.05	24.45	3.04	7.17	20.48	2.86
15	11.52	30.57	2.65	10.55	32.66	3.10	9.00	27.65	3.07	8.11	23.38	2.88
16	12.71	35.99	2.83	11.64	37.82	3.25	9.99	30.96	3.10	8.88	27.60	3.11
17	13.90	40.99	2.95	12.92	42.34	3.28	11.19	35.60	3.18	9.86	32.06	3.25
18	15.46	45.05	2.91	14.21	46.47	3.27	12.32	38.68	3.14	10.91	34.61	3.17
19	16.80	49.80	2.96	15.28	50.20	3.29	13.21	42.25	3.20	11.84	38.59	3.26
20	18.09	55.53	3.07	16.25	55.27	3.40	13.63	46.24	3.39	12.49	42.75	3.42
21	19.18	62.27	3.25	17.41	60.46	3.47	14.80	50.53	3.41	13.61	46.73	3.43
22	20.58	67.33	3.27	18.22	65.37	3.59	15.16	53.42	3.52	14.41	50.32	3.49
23	20.89	74.24	3.55	18.83	69.57	3.69	15.83	58.25	3.68	14.58	54.81	3.76
24	22.46	81.10	3.61	19.81	74.01	3.74	16.77	63.26	3.77	15.72	60.05	3.82
25	23.23	88.64	3.82	20.11	80.40	4.00	16.91	68.79	4.07	16.36	65.15	3.98
26	25.04	95.63	3.82	21.33	85.87	4.03	18.58	74.39	4.00	17.54	69.90	3.99
27	25.80	102.02	3.95	21.48	92.50	4.31	18.40	80.15	4.36	17.70	74.74	4.22
28	25.20	109.09	4.33	21.70	98.48	4.54	18.26	85.52	4.68	17.39	79.58	4.58
29	25.61	116.69	4.56	22.65	106.28	4.69	18.66	91.16	4.89	17.79	85.41	4.80

[1]Data from Poultry Division, Dominion Experimental Farms, Ottawa, Ontario

[2]These Standard Bronze data are about the same as those from strains of White Hollands of larger than standard size.

[3]These Bourbon Red data are approximately equal to those obtained from White Hollands of standard size.

Table 5.21—Body Weight and Cumulative Feed Consumption[1]

(Females raised on range—1942)

	BROAD BREASTED BRONZE			STANDARD BRONZE			BOURBON RED			BELTSVILLE SMALL WHITE		
Age in wks.	Av. body wt. lb.	Av. feed cons. per bird lb.	Lb. feed per lb. body wt.	Av. body wt. lb.	Av. feed cons. per bird lb.	Lb. feed per lb. body wt.	Av. body wt. lb.	Av. feed cons. per bird lb.	Lb. feed per lb. body wt.	Av. body wt. lb.	Av. feed cons. per bird lb.	Lb. feed per lb. body wt.
0	.12			.12			.12			.11		
1	.22	.14	.64	.21	.13	.62	.20	.14	.70	.17	.21	1.24
2	.39	.49	1.26	.38	.44	1.16	.34	.44	1.29	.31	.50	1.61
3	.61	.97	1.59	.57	.91	1.60	.52	.92	1.77	.48	.92	1.92
4	.85	1.64	1.93	.79	1.57	1.99	.72	1.56	2.17	.67	1.47	2.19
5	1.31	2.55	1.95	1.22	2.47	2.02	1.09	2.43	2.23	.98	2.20	2.24
6	1.79	3.80	2.12	1.64	3.59	2.19	1.52	3.56	2.34	1.33	3.15	2.37
7	2.38	5.15	2.16	2.22	5.04	2.27	2.11	4.89	2.32	1.81	4.18	2.31
8	3.02	6.97	2.31	2.82	6.67	2.37	2.60	6.61	2.54	2.26	5.57	2.46
9	3.73	8.88	2.38	3.50	8.40	2.40	3.22	8.48	2.63	2.76	7.01	2.54
10	4.53	11.26	2.49	4.27	10.73	2.51	3.83	10.69	2.79	3.27	8.90	2.72
11	5.36	14.03	2.62	5.05	13.23	2.62	4.43	13.38	3.02	3.82	10.80	2.83
12	6.12	16.28	2.66	5.80	15.53	2.68	5.06	15.79	3.12	4.36	12.65	2.90
13	6.71	18.80	2.80	6.41	17.80	2.78	5.58	18.02	3.23	4.73	14.66	3.10
14	7.55	21.26	2.82	7.14	20.27	2.84	6.36	20.25	3.18	5.36	16.43	3.07
15	8.38	24.27	2.90	7.93	23.15	2.92	6.97	22.62	3.25	5.72	17.91	3.13
16	9.29	27.99	3.01	8.77	27.51	3.14	7.65	26.33	3.44	6.28	20.47	3.26
17	10.09	31.65	3.14	9.39	31.36	3.34	8.14	29.22	3.59	6.83	23.27	3.41
18	11.06	34.52	3.12	10.21	34.22	3.35	8.80	31.51	3.58	7.33	25.08	3.42
19	11.76	36.77	3.13	10.94	37.24	3.40	9.29	33.93	3.65	7.73	27.48	3.55
20	12.31	40.59	3.30	11.13	40.45	3.63	9.50	36.56	3.85	7.97	31.23	3.92
21	12.94	44.14	3.41	11.71	43.68	3.73	10.27	39.28	3.82	8.47	33.60	3.97
22	13.49	46.94	3.48	12.64	46.51	3.68	10.77	41.71	3.87	8.65	36.66	4.24
23	13.47	50.59	3.76	12.37	49.96	4.04	10.62	43.63	4.11	8.88	39.11	4.40
24	14.36	54.51	3.80	12.87	54.54	4.24	11.02	45.97	4.17	9.44	41.82	4.43
25	14.46	58.41	4.04	13.27	57.76	4.35	11.30	50.01	4.43	9.51	44.48	4.68
26	15.37	62.12	4.04	13.98	61.16	4.37	12.01	53.20	4.43	9.91	47.13	4.76
27	15.54	66.06	4.25	14.01	65.49	4.67	12.10	57.67	4.77	9.97	49.72	4.99
28	15.52	69.92	4.51	14.00	68.64	4.90	12.26	60.53	4.94	9.88	52.33	5.30
29	15.73	74.06	4.71	14.34	72.96	5.09	12.58	64.25	5.11	9.99	55.22	5.53

[1]See footnotes preceding table.

Table 5.22—Body Weights and Cumulative Feed Consumption[1]

(Both sexes—equal numbers—range reared—1942)

	BROAD BREASTED BRONZE			STANDARD BRONZE[2]			BOURBON RED[3]			BELTSVILLE SMALL WHITE		
Age in wks.	Av. body wt. lb.	Av. feed cons. per bird lb.	Lb. feed per lb. body wt.	Av. body wt. lb.	Av. feed cons. per bird lb	Lb. feed per lb. body wt.	Av. body wt. lb.	Av. feed cons. per bird lb	Lb. feed per lb. body wt.	Av. body wt. lb.	Av. feed cons. per bird lb.	Lb. feed per lb. body wt.
0	.12			.12			.12			.09		
1	.23	.14	.63	.22	.14	.64	.21	.15	.71	.18	.21	1.17
2	.42	.50	1.19	.39	.47	1.21	.37	.47	1.28	.33	.53	1.61
3	.65	1.01	1.55	.60	.97	1.62	.58	1.02	1.77	.52	.98	1.88
4	.94	1.75	1.86	.82	1.70	2.07	.80	1.74	2.18	.73	1.57	2.15
5	1.43	2.74	1.92	1.28	2.66	2.08	1.17	2.68	2.29	1.07	2.33	2.18
6	2.00	4.07	2.04	1.75	3.88	2.22	1.65	3.93	2.38	1.45	3.32	2.29
7	2.68	5.57	2.08	2.40	5.43	2.26	2.27	5.43	2.39	1.99	4.40	2.21
8	3.40	7.62	2.24	3.01	7.36	2.45	2.85	7.29	2.56	2.48	5.93	2.39
9	4.25	9.74	2.29	3.82	9.51	2.49	3.54	9.33	2.64	3.07	7.51	2.45
10	5.17	12.36	2.39	4.70	12.26	2.61	4.24	11.83	2.79	3.69	9.49	2.57
11	6.16	15.40	2.50	5.59	15.42	2.76	4.95	14.81	2.99	4.32	11.66	2.70
12	7.06	18.13	2.57	6.48	18.64	2.88	5.66	17.19	3.04	5.00	13.77	2.75
13	7.93	21.20	2.67	7.25	21.28	2.94	6.25	19.76	3.16	5.49	16.29	2.97
14	8.86	24.12	2.72	8.22	24.26	2.95	7.21	22.35	3.10	6.27	18.50	2.95
15	9.95	27.42	2.76	9.24	27.90	3.02	7.99	25.13	3.15	6.97	20.70	2.97
16	11.00	31.99	2.91	10.21	32.67	3.20	8.82	28.64	3.25	7.58	24.09	3.18
17	12.00	36.32	3.03	11.16	36.85	3.30	9.67	32.41	3.35	8.35	27.73	3.32
18	13.26	39.78	3.00	12.21	40.35	3.30	10.56	35.09	3.32	9.12	29.92	3.28
19	14.28	43.27	3.03	13.11	43.72	3.33	11.25	38.09	3.39	9.79	33.11	3.38
20	15.20	48.06	3.16	13.69	47.86	3.50	11.57	41.40	3.58	10.23	37.07	3.62
21	16.06	53.20	3.31	14.56	52.07	3.58	12.54	44.90	3.58	11.04	40.25	3.65
22	17.04	57.13	3.35	15.43	55.94	3.63	12.97	47.56	3.67	11.53	43.58	3.78
23	17.18	62.41	3.63	15.60	59.77	3.83	13.23	50.94	3.85	11.73	47.05	4.01
24	18.41	67.80	3.68	16.34	64.28	3.93	13.90	54.61	3.93	12.58	51.03	4.06
25	18.85	73.52	3.90	16.69	69.08	4.14	14.11	59.40	4.21	12.94	54.92	4.24
26	20.20	78.87	3.90	17.65	73.52	4.17	15.30	63.79	4.17	13.73	58.63	4.27
27	20.67	84.04	4.07	17.75	79.00	4.45	15.25	68.91	4.52	13.84	62.35	4.51
28	20.36	89.50	4.40	17.85	83.56	4.68	15.26	73.02	4.79	13.64	66.08	4.84
29	20.67	95.37	4.61	18.45	89.62	4.86	15.62	77.70	4.97	13.89	70.44	5.07

[1]Data from Poultry Division, Dominion Experimental Farms, Ottawa, Ontario.
[2]See footnotes table 5.20.
[3]See footnotes table 5.20.

Table 5.23—Body Weights and Cumulative Feed Consumption[1]

(Equal numbers males and females)

	BROAD BREASTED BRONZE			STANDARD BRONZE[2]			BOURBON REDS[3]			BELTSVILLE SMALL WHITES		
Age in wks.	Av. body wt. lbs.	Av. feed cons. per bird lb.	Lb. feed per lb. body wt.	Av. body wt. lb.	Av. feed cons. per bird lb.	Lb. feed per lb. body wt.	Av. body wt. lb.	Av. feed cons. per bird lb.	Lb. feed per lb. body wt	Av. body wt. lb.	Av. feed cons. per bird lb.	Lb. feed per lb. body wt.
Raised on Wire												
	1943			1943			1942			1943		
0	0.12			0.13			0.14			0.11		
4	1.00	1.87	1.87	0.86	2.06	2.40	1.05	1.99	1.90	0.78	1.77	2.28
8	3.14	8.40	2.68	2.64	7.61	2.88	2.97	7.35	2.48	2.23	6.93	3.11
12	6.29	20.25	3.22	5.51	17.49	3.17	5.78	16.40	2.84	4.40	15.01	3.41
16	10.52	36.70	3.49	9.51	33.69	3.54	8.19	28.58	3.49	7.25	27.71	3.82
20	14.43	56.54	3.92	12.58	51.64	4.10	9.51	41.70	4.39	9.65	42.60	4.41
24	17.22	75.82	4.40	15.28	69.92	4.58	11.88	58.82	4.95	11.94	57.20	4.79
28	20.05	98.29	4.90	17.06	90.71	5.32	14.10	78.15	5.54	13.66	73.65	5.39

[1]Data from Poultry Division, Dominion Experimental Farms, Ottawa, Ontario.

[2]These Standard Bronze data are about the same as those from strains of White Hollands of larger than standard size.

[3]These Bourbon Red data are approximately equal to those obtained from White Hollands of standard size.

RATIO OF MASH TO SCRATCH FEED EATEN BY GROWING TURKEYS WHEN FED AD LIBITUM

That growing turkeys are able to balance their diet if a palatable mash and scratch grain are both fed ad libitum has been definitely proved. Table 5.24 shows the ratio of mash to scratch feed consumed by turkeys fed mashes containing eight different levels of protein. These data are useful in estimating feed requirements and in planning diets so that adequate quantities of the various nutrient factors may be available to turkeys. It must be stressed, however, that these figures were obtained when certain types of diets were fed and might not be representative if, for instance, a mash or a scratch ration of low palatability were to be fed. The mashes used contained bran, middlings, corn, oats, alfalfa meal, meat scrap, dried milk, and salt. The grain fed from 2 to 12 weeks consisted of cracked wheat and chick-size cracked

Table 5.24—Ratio of Mash Consumption to Grain Consumed by Growing Turkeys[1]

Approximate percentage of crude protein in the mash	Proportion of mash to grain consumed during the brooding and rearing period.											
	1—4 weeks	5—8 weeks	9—12 weeks	13—16 weeks	17—20 weeks	21—24 weeks	25—28 weeks	29—32 weeks	1—24 weeks	1—28 weeks	1—32 weeks	9—28 weeks
18	96:4	97:3	90:10	97:3	90:10	67:33	39:61	....	85:15	74:26	....	72:28
20[2]	94:6	97:3	87:13	87:13	71:29	46:54	30:70	....	72:28	61:39	....	60:40
22	92:8	94:6	85:15	89:11	67:33	47:53	30:70	24:76	72:28	62:38	54:46	58:42
24	84:16	88:12	74:26	77:23	65:35	48:52	30:70	....	67:33	58:42	....	55:45
26	91:9	86:14	79:21	77:23	61:39	39:61	27:73	....	63:37	54:46	....	50:50
28[4]	87:13	82:18	72:28	70:30	54:46	33:67	25:75	....	57:43	49:51	....	46:54
30[2] [4]	84:16	81:19	70:30	63:37	48.52	28:72	26:74	....	53:47	46:54	....	43:57
32[3] [4]	....	....	72:28	54:46	41:59	35:65	23:77	....	....	....	....	41:59

[1]From Hammond and Marsden. Note the wide variation of proportion of mash to grain the first 24 weeks; 85:15 (5.7 mash: 1 grain) with low protein and 53:47 (1.1 to 1) with high protein (Poul. Sci. 39:11, 1939). Data from U. S. Exp. Sta., Miles City, Mont. (1934).

[2]Growing diet No. 23, Chap. 10, 5th Edition T. Mgt.

[3]These data were obtained from a group of turkeys fed from 9 to 28 weeks on a 32 per cent protein simplified growing concentrate containing 22 per cent soybean meal and 33 per cent meat scraps. The birds had access to a green range and a grain mixture composed of ⅓ oats, ⅓ whole yellow corn, and ⅓ barley.

[4]These may be classed as "concentrates."

corn in equal parts; after 12 weeks it consisted of whole yellow corn 40, hard wheat 40, and oats 20. The birds had access to dry-land range supplying green feed of rather poor quality, mainly cured grasses.

THE EFFECTS OF SEASON ON GROWTH

An analysis of weather and range conditions indicates that the turkey growing and hatching season can be extended by the wider use of early- and late-hatched poults. In the northern States, early-hatched (January to March) turkeys are practical because they mature for market during the cool summer and early fall when range conditions are good. In the southern States, late-hatched (June to August) turkeys are practical because they mature in the not-too-cold fall and early winter when the ranges are good. Confinement rearing likewise is at its best at these times. Winter brooding is highly successful with good equipment and the birds go on range at the ideal time. Summer brooding, although somewhat more difficult, is entirely practicable if good sanitation and fly control and an antibiotic in the starting diet are provided.

The Kansas Station reported that turkeys hatched March 10 and marketed in September were as fat and free from pinfeathers as were later hatched turkeys which matured for market after cool weather had set in. The Oklahoma Station, however, was not able to obtain good finish in turkeys marketed in mid-summer. The authors have observed that in Kentucky and Maryland, turkeys hatched before about April 15 are noticeably slower than normal in developing plumpness, while those hatched in June and July are slightly more rapid in finishing.

In hot climates, early-hatched turkeys reaching roaster market age in the hot summer weather appear not to attain a satisfactory fleshing condition though in finish they may be acceptable. Broad-breasted turkeys reared under these conditions may not resemble their parents in conformation until or unless they are carried to a much more mature age or the weather becomes at least moderately cool. In turkey broilers, especially white varieties, these differences are not striking as the early growth and development seems not to be inhibited by hot weather provided shade and good ventilation are provided in the rearing quarters. Recent research indicates that the shorter nights of mid-summer may be more of a factor than the summer heat.

THE INFLUENCE OF TIME OF YEAR ON THE FEED REQUIREMENTS OF TURKEYS [1]

During the 12- to 18-week period for Beltsville Whites and the 20- to 24-week period for Broad Breasted Bronze, the turkeys ate much better and gained weight more rapidly as the weather became cooler. However, BBB poults 14- to 20-weeks old showed a reduction in rate of weight gains during cooler weather. Grain intake showed up to 300% variation being at a minimum in July, August and September (the three hot months) followed by a quick rise when the weather changed to cool nights and mornings.

Feed efficiency of BSW hens was about 85% that of toms during cool weather but was 91 to 92% as much during hot weather. In the younger BBB the hens were 79 to 89% as efficient as the toms, in the older BBB, 69 to 72%; and the change in weather did not affect this relationship.

In the BSW, feed cost per pound of gain changed from 2.0% less for toms than for hens in cool weather to practically no difference in hot weather. In the younger BBB the change was for 5.2c less for the toms than for hens in cool weather to 1.4c less in hot weather. In the older BBB, feed cost of gains in hens was uniformly about 10c per pound more than in toms regardless of the weather.

[1]From H. J. Almquist, 1953 (Grange Co., Modesto, Calif.)

These data point to the probability that when marketed as turkey broilers or light roasters (at 12 to 18 weeks of age), the hens are about as economical as the toms, so far as feed cost is concerned.

THE INFLUENCE OF SEX ON FEED CONSUMPTION

The available evidence indicates that during the late growing period, male turkeys are more efficient in the utilization of feed than are female turkeys. It appears probable, moreover, that this difference manifests itself earlier in the growing period than has been shown in the experiments which have dealt only with turkeys 18 weeks of age and older. Since methods have only very recently been developed for determining the sex of turkeys at hatching or during the early growth stages, most of the published material dealing with differences in feed utilization due to sex deals only with birds mature enough to determine sex by the secondary sexual characters such as size, plumage, body type, and headgear.

The Beacon Mills Research Farm (1947, 1950) using day-old commercially sexed poults of three varieties, reported as follows on feed consumption of males and females raised separately from hatching to market age in confinement:

Table 5.25—Feed Efficiency and Cost

(Beacon Mills data)

Data at 26 weeks	Medium-sized White Hollands Hatched April 28, 1947				Narragansetts Hatched June 3, 1947			
	Mash and grain		Pellets and grain		Mash and grain		Pellets and grain	
	100% hens	100% toms	98% hens	97% toms	86% hens	100% toms	94% hens	100% toms
Av. live wt.....lbs.	11.38	19.17	11.85	18.65	13.88	22.31	14.14	21.34
Lbs. feed per lb. live weight	5.36	4.86	5.07	4.53	5.23	4.61	5.24	5.12
Feed cost* per lb. live wt.....cents	27	25	26	23	27	23	27	26

*Based on retail feed costs at Cayuga, New York; at the dates given.

Table 5.26—Feed Consumption and Growth

(Beacon Mills data)

	Large-type, broad breasted White Hollands Hatched May 23, 1950		Large-type, broad breasted White Hollands hatched June 24, 1950	
1-8 Weeks	98% hens 2% toms	100% toms	100% hens	96% toms 4% hens
Starting mash, lbs. per bird	6.59	7.76	6.39	7.97
Flushing mash, lbs. per bird	.60	.61	.76	.91
Average weights, both sexes, lbs.	3.14	3.82	3.12	3.89
9-24 Weeks				
Growing mash, lbs. per bird	26.65	41.73	23.78	38.61
Flushing mash, lbs. per bird	.72	.75	1.08	1.29
Fitting feed, (grains) lbs. per bird	36.14	46.51	40.19	53.53
1-24 Weeks				
Total feed consumed, lbs. per bird	70.70	97.36	72.29	102.31
24th Week				
Avg. live weight of toms, lbs.	26.00	23.72		22.96
Avg. live weight of hens, lbs.	14.95		14.23	17.50
Feed eaten per lb. live wt., lbs.	4.67	4.10	5.08	4.48
Feed cost* per lb. live wt., cents	22	19	24	21
Total feed cost per bird, dollars	3.28	4.54	3.39	4.82

*Based on retail feed costs at Cayuga, New York; at the dates given.

These data show the relatively small differences in feed efficiency between the two sexes and also the still smaller differences in feed cost. Feed cost varied from 1 to 4 cents, with an average of about 3c per pound of live weight, in favor of the toms. There was more of a difference in the feed efficiencies of the two sexes, an average of ½ lb. of feed per pound of live weight of turkey produced, but this was partially offset by the relatively higher consumption of grain (with lower cost) by the hens. As to pellets versus mash, the former showed better feed efficiency in the Whites and poorer feed efficiency in the Narragansetts—the average would give a slight edge to the pellets. There were no significant variety differences.

The Oregon Station found marked differences in feed utilization of male and female Bronze turkeys from approximately 20 weeks to 32 weeks. The weight at 20 weeks on October 1st was approximately the same for all lots. Four diets were fed. Each lot, in addition to the regular feed, had free access to alfalfa range, alfalfa hay, insoluble grit, oyster shell, and water. The grain was hand-fed twice daily and consisted of a mixture of 60 parts wheat and 40 parts cracked yellow corn. The results of this experiment are summarized in Table 5.27.

Table 5.27—Feed Consumption in Relation to Sex

(Data from Oregon Station—Circ. Inf. 124; 1935)

Lot No.	Feed	Number and sex	Average wt. at marketing	Total feed cost per lb. of gain	Feed including dry matter in the milk consumed per lb. of gain
			lbs.	cents	lbs.
1	An 18 percent protein growing mash with grain, and liquid milk.	15 males 20 females	24.10 14.50	14.40 20.40	7.79 10.99
	Average of 35 birds.			17.40	8.89
2	A 14 percent protein growing mash with grain.	15 males 20 females	22.50 14.10	13.94 21.28	7.25 11.34
	Average of 35 birds.			17.61	8.81
3	Grain and liquid milk.	15 males 20 females	22.30 14.30	12.93 16.92	6.94 9.04
	Average of 35 birds.			14.93	7.99
4	Grain only.	15 males 20 females	20.60 13.20	17.04 17.06	9.26 9.27
	Average of 35 birds			17.05	9.26

These figures show clearly the differences due to sex when the turkeys were fed adequate rations. When fed an inadequate diet (lot 4) the males were no more efficient than the females, the efficiency of the males being impaired by the low-protein diet. The birds in all lots graded equally well when marketed at 32 weeks, although those of lot 4 were smaller and matured later.

In perhaps the first test of the feed efficiency of turkeys in which the sexes were separated at hatching time the Missouri Station reported (1943) that male Bronze turkeys produced more economical gains than the females. The difference was not significant until the time the turkeys reached 16 weeks of age, then the males became increasingly more efficient with their feed as measured in terms of gains from 16 weeks to market age (26 to 28 weeks). Other interesting conclusions were:

Male turkeys required less protein per pound of gain from 17 to 30 weeks of age.

Male turkeys utilized the carbohydrates in the feed more efficiently than the females.

Male turkeys consumed less grain in proportion to mash than the females.

Males grew at a more rapid rate than did the females; and efficiency of gain was closely related to rate of growth.

Males gained in weight at less feed cost per pound than females.

SEPARATING THE SEXES

Young male turkeys do not interfere seriously with the growth and development of their female companions as do chicken males. For this reason there is no great incentive to separate the sexes so far as commercial turkey raising is concerned. Some growers however prefer to do so at 12 to 16 weeks. At this age sexing can be done with fair accuracy by external characteristics or it can be done at any age from hatching on, with fair accuracy (95% or so) by vent examination. See Chapter 13 for sexing methods. Sexed poults now are available at hatcheries but the demand at present is limited.

Although some advantages are claimed for separate rearing of the males and females, practically no experimental evidence on the subject has so far appeared. Special situations rather than any important advantages are likely to cause growers to rear the sexes separately. One advantage is important with early-hatched poults—

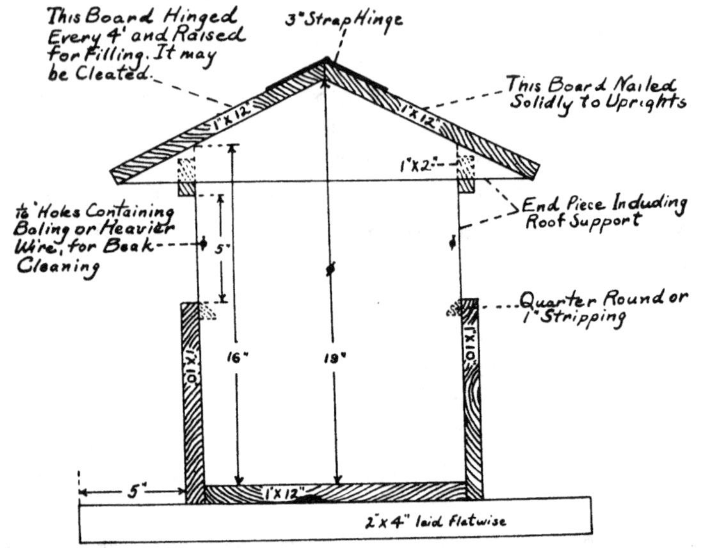

Fig. 5.2. Diagram of end of outdoor turkey feeder. One or more sliding partitions may be put in each feeder to permit adjustment of feeding space for mash, scratch, and other components of the diet. The feeder may be 3 to 20 feet in length with intermediate supports (patterned after the end pieces) placed 4 to 6 feet apart. Handles may be placed on each end and (or) runners underneath, for moving. A 2" x 4", a 2" x 6", or a pole may be laid across the bottom pieces so that small birds can reach the feed. If this feeder is not filled more than two-thirds or three-fourths full, very little wastage will occur from any cause.

treading, with resultant injury to females, is avoided. Normal season turkeys however seldom become sexually active before they are marketed. In confinement operations with heavy turkeys some growers like to separate the sexes at about 4 months of age, transferring the toms to range or to littered quarters to avoid leg and foot trouble, but continuing the hens on the wire or slat floors.

REDUCING FEED COSTS

The Kansas Station (1950) demonstrated the value of home-grown sorghum grains, economical mashes, and range in reducing the cost of production. Broad Breasted Bronze and Beltsville Small White turkeys were fed standard home-mixed diets; starters for 8 weeks, grow-

Table 5.28—Comparative Cost of Gains with Varying Supplements to Simple Diets

(Kansas Station data)

	Lot 1	Lot 2	Lot 3	Lot 4	Lot 5	Lot 6
Variety (Bronze or Whites)	BBB	BBB	BSW	BBB	BBB	BBB
Range	Western wheat grass, abundant	Western wheat grass, abundant	Western wheat grass, abundant	Western wheat grass, abundant	Sudan grass, abundant	Western wheat grass, deficient
Starting feed, 1-8 wks.	Corn and wheat base, 6.4% fiber	Milo base, 4.3% fiber	Milo base, 6.2% fiber	Milo base, 6.2% fiber	Milo base, 6.2% fiber	Milo base, 6.9% fiber
Merck's APF supplement, %	.05	.05	.05	.05	.05	.1
Growing feed, 9-20 wks.	Corn, wheat, oats base, 6.6% fiber	Milo base, 5.5% fiber	Milo base, 6.4% fiber	Milo base, 6.4% fiber	Milo base, 6.4% fiber	Milo base, 6.8% fiber
Merck's APF supplement, %	.05	.05	.05	.05	.05	.1
Finishing feed, 21-28 wks.	Corn, wheat, oats base, 6.4% fiber	Milo base, 5.2% fiber	Milo base, 6.2% fiber	Milo base, 6.2% fiber	No mash fed	Milo base, 6.6% fiber
Sources of protein in mashes	Animal and vegetable	Animal and vegetable	Animal and vegetable	Animal and vegetable	Animal and vegetable	Vegetable only
Merck's APF supplement, %	None	None	None	None	None	None
Grain, fed 9-28 weeks	40% yellow corn, 20% oats, 40% wheat	Milo	Milo	Milo	Milo	Milo
Av. live wt. of turkeys at 28 wks., lbs.	19.18	17.71	12.37	18.05	17.69	17.29
Feed cost* per pound of gain, cents	15.83	11.93	13.52	12.17	11.33	12.28
Feed and poult cost* per lb. of gain, cents	20.5	17.0	20.0	17.1	16.4	17.6
Total estimated cost* (feed, poult, labor and overhead) per lb. gain, cents	24.0	21.5	24.7	21.5	20.9	22.1
Profit per bird above total estimated costs, dollars ($)	.50	.90	.24	.92	1.01	.78

*At retail feed prices in western Kansas the 27 to 28% protein starting mash, corn and wheat base, cost $4.46 per cwt., while the same mash with milo substituted for the corn and wheat cost $4.00 to $4.17. The 20% protein growing mash with corn, oats, and wheat cost $3.88 to $4.26, while the growing mashes with milo substituted cost $3.20 to $3.68. The 16% protein finishing mash with corn, oats, and wheat cost $3.77 to $3.86 while those with milo substituted cost $3.03 to $3.24. The corn cost $2.75 to $3.20; the wheat $3.75 to $3.90; the oats $3.20; and the milo $2.10 to $2.30

ers from 9 to 20 weeks, and finishers from 21 to 28 weeks. The poults were brooded in lots of 200 in a standard manner in wire-floored brooder houses with attached gravel-covered yards. The wire floors were covered with crinkled paper for the first 4 weeks. At 8 weeks the poults were moved to 3-acre dry-land range lots. After 16 weeks the sexes were separated and the groups moved every 2 weeks to clean areas. Portable shelters provided shade. Table 5.28 gives the results in tabular form.

Even at the low price received of 23c per pound for toms and 32c for hens (which on basis of relative weights of the two sexes amounts to an average of 26.6c) it will be seen that the groups of Bronze turkeys fed milo, particularly the one fed no mash after 20 weeks (Lot 5) returned a fair profit above the estimated total cost. This profit ranged from $1.01 per bird in Lot 5 to $.50 in Lot 1. The Beltsville White turkeys even when carried to 28 weeks, 4 to 6 weeks longer than necessary, and sold at Bronze prices, returned a profit of 24c per bird. Had they been sold at 24 weeks and both hens and toms at Bronze hen price, the profit would have been about 91c per bird, about equal to that of the corresponding lot of Bronze, No. 4. The good feeding value of milo (a grain sorghum) even when it comprised up to 75% of the total diet is shown by these results. The other common grain sorghums such as kafir, durra, darso, feterita, shallu, kaoliang, and hegari, unless drought-damaged, should be nutritionally equal to milo.

REFERENCES

A more complete bibliography will be found in the 1st Edition (1939) of this book, pages 474-488, inclusive. Only a few recent publications are given herewith.

Almquist, H. J., 1951. Comparative mash consumption, grain consumption and feed efficiency of B. B. Bronze and Beltsville Small White turkeys. World's Poul. Sci. Jour. 7(2):103-109.

Assoc. of American Feed Control Officials, 1950. Official publication. L. E. Bopst, Secy., College Park, Md.

Ewing, W. Ray, 1951. Handbook of Poultry Nutrition. Pub. by W. R. Ewing, Pub., So. Pasadena, Calif.

Marsden, et al, 1952. Starting diets as a factor in edible quality of turkey. Poul. Sci. 31:451-458. (Beltsville Station).

Morrison, F. B., 1948. Feeds and Feeding. Published by Morrison Pub. Co., Ithaca, N. Y.

Scott, M. L., 1951. The grass juice factor in turkey nutrition. Poul. Sci. 30(2):293-297.

Titus, Harry W., 1949. The Scientific Feeding of Poultry. The Interstate, Danville, Ill. 2nd Edition, 253 pp.

CHAPTER 6

PREPARATION FOR MARKET

OUTLINE

Turkeys marketed ready-to-cook in 1954 cost the city consumer from three dollars for a small "broiler" to fifteen dollars for a large roasting tom. The average probably is between seven and eight dollars. With this much investment in one article of food, satisfaction is demanded. Responsibility for the consumer's satisfaction in the final analysis rests jointly upon the grower of the turkey, the processor who kills and dresses the bird, the transportation agency and finally the retailer. The cook is responsible to a large extent for ultimate satisfaction. A major portion of the turkey crop still is consumed during the Thanksgiving, Christmas, and New Year holidays; although during the period of 1940 to 1950 there was an increased tendency to spread the marketing of turkeys through many more months. During October 1948 a larger per cent of turkeys found their way to market than in any preceding October. However, still well over half of the birds produced were marketed that year during the Thanksgiving and Christmas Holidays. By 1954 approximately half (50.2%) of the turkeys were marketed at other than the Thanksgiving and Christmas seasons; as indicated in Table 6.1.

Table 6.1—Changes in Turkey Marketing 1940-1954

(Data by Kohls, from USDA)

Area	Before Aug. 1	Aug.	Sept.	Oct.	Nov.	Dec.	Jan., later
	Percent of total						
United States, All....							
1940	←8.8→				42.1	36.9	12.2
1954	9.9	6.8	11.1	17.3	30.2	19.6	5.1
United States, by breeds 1954:							
Heavy	4.6	6.4	11.5	17.5	32.9	21.1	5.6
Light	23.8	7.8	10.4	15.7	22.8	15.8	3.7
All, by Regions 1954:							
East North Central	4.1	6.1	9.2	16.8	37.0	23.2	3.6
West North Central	8.9	8.6	13.1	23.8	27.4	14.9	3.3
Western	9.4	7.5	11.7	18.7	28.1	18.4	6.2

LIVE VS. DRESSED MARKETING

One of the first problems confronting the turkey raiser when planning to market turkeys is whether he shall market them alive or dressed. His choice as to the methods depends to a great extent upon (1) his proximity to the market, (2) the opportunities for sales, (3) whether or not adequate refrigeration facilities are available or can be made so, (4) the comparative net price received for live and dressed birds, (5) the preferences of the market upon which he expects to sell his birds, and (6) his own preferences and abilities. In some situations it is almost impossible to market turkeys alive, while in others dressed or eviscerated marketing is impracticable or impossible.

In comparing the relative prices offered for one's turkeys, it should be kept in mind that there is a loss due to fasting or due to transportation and always a further loss due to dressing which should be checked against the live bird price and the dressed bird price. Many turkeys are dressed only by bleeding after overnight starving, sticking, and then picking. This method is referred to as blood-and-feather dress, New York dress, or simply dress. The loss of feathers and blood causes a shrinkage of from 8 to 12 per cent of the live weight before or after fasting (see Table 6.2). Overnight fasting (feed withheld, water freely given) which, in the customary handling, results in the turkeys being off feed for an average of about 18 hours, causes a loss of about 3.33% of the normal field weight. A loss, usually smaller, may be experienced during live bird transportation. However, fasted turkeys seldom are transported very far alive; they usually are loaded directly from full feed. These weight losses increase the amount per pound that one must get for the turkey in order to receive the same money he would have received for the live bird in the field.

Many commercial concerns buy on field weight and do not starve the turkeys before killing, depending on immediate evisceration or the water-pressure removal of crop contents. In this case the fasting shrinkage may be disregarded and the live field weight used as the basis for computation. A prime young Broad Breasted Bronze turkey hen weighing 16 pounds in the field should on the average weigh 15.47 pounds after overnight fasting or long live hauling; 14.11 pounds after being bled, picked and chilled; 12.28 pounds, including giblets and neck, after full drawing. Expressed in percentage of field weight this is (1) field weight 100%, (2) fasted

Table 6.2—Live-Dressed-Drawn Price-Per-Pound Relationships of Turkeys, 1951[1]

Class of turkeys	Average dressing loss as per cent of live field weight[2]	Average drawing loss as per cent of dressed weight	Per cent live field price is of drawn price	Per cent live field price is of dressed price	Per cent dressed price is of drawn price
Standardbred turkeys—Bronze, White Holland, Narragansett, Black, Bourbon Red, and Slate—5½ to 7 months old, both sexes	11.0 [3](Live price 40.00c)	15.0	60.71 (Drawn price 65.89c)	78.52 (Dressed price 50.94c)	77.31
Broad-breasted turkeys, medium to large size 5½ to 7 months old, both sexes	10.0 (Live price 35.00c)	12.5	61.08 (Drawn price 57.30c)	77.97 (Dressed price 44.89c)	78.34
Beltsville Small White and other small varieties 4½ to 6 months old, both sexes	12.5 (Live price 45.00c)	13.5	62.20 (Drawn price 72.35c)	78.36 (Dressed price 57.43c)	79.38
Breeder turkeys, all varieties, both sexes	9.0 (Live price 30.00c)	17.0	56.64 (Drawn price 52.97c)	76.98 (Dressed price 38.97c)	73.57
Unweighted averages for all four classes, both sexes	10.63c (Live price 37.50c)	14.5	60.16 (Drawn price 62.13c)	77.95 (Dressed price 48.05c)	77.15
Beltsville White fryer turkeys 12 to 16 weeks old, average 14 weeks old, both sexes	12.0 (Live price 45.00c)	21.2	57.38 (Drawn price 78.43c)	78.75 (Dressed price 57.14c)	72.85

[1]Based on an estimated charge of 6c per pound of dressed weight for dressing (5c) and grading including down grading allowance (1c); and an estimated charge of 6c per pound of drawn weight, including neck and giblets, for eviscerating and packaging (5c) and first month's storage (1c). The storage charge sometimes need not be considered. If individually cartoned, an extra charge of ½c per pound would have to be added.

[2]The percentages in this column include a 1% loss due to trucking to plant but no allowance for loss during overnight fasting.

[3]Figures in parenthesis are examples to illustrate computations. "Dressed weight" and "dressed price" refer to N. Y. dressed (bled and picked).

overnight 96.67%, (3) bled, picked and chilled 88.19% and (4) full drawn 76.75%. A simple rough device for calculating average fasting, dressing and drawing loss is the rule of 13; 13% of the live weight in the field is lost in overnight fasting, and/or live transportation, bleeding, and picking; and another 13% of live weight is lost in full drawing.

A useful schedule allowing for processing costs and weight losses is given in Table 6.2.

A very large proportion of U. S. turkeys is sold by the producer on a live basis but most of them reach the ultimate consumer New York dressed or eviscerated—relatively few reach market alive, mostly for the orthodox Hebrew trade.

Turkeys sometimes are sold at a flat rate on the basis of field inspection; sometimes on basis of the dressed grades after processing. U. S. grades for live poultry including turkeys have been established by the U. S. Department of Agriculture. The live grades are A, B, and C, and they correspond to the A, B, and C grades of the dressed turkey grading system described later. Information concerning live and dressed grades may be obtained from the Poultry Division of the U. S. Agricultural Marketing Service, Washington 25, D. C., in the publication entitled "Regulations governing the grading and inspection of poultry and domestic rabbits and edible products thereof, and the U. S. specifications for classes, standards, and grades with respect thereto."

WHEN TO MARKET

Turkeys intended for normal roasting purposes usually reach market maturity sometime between 22 and 30 weeks, the age depending on the variety and conditions under which they are grown. They are ready for market at the time they accumulate a reasonably good covering of fat in the skin over the breast and the feathers in general have grown past the short pinfeather stage. At this time if the birds are dressed there will be no pinfeathers to discolor the carcass and detract from the appearance of the bird. Under normal conditions, date of market maturity can be indicated with a fair degree of accuracy by age so that advance arrangements for marketing can be made. The turkey "broiler" or light roaster is an immature bird marketed at 12 to 18 weeks of age.

The practice of feeding a special fattening ration toward the end of the growing period is no longer common. It has been found that a turkey is in excellent condition for dressing when it has been growing rapidly on a well balanced ration. If the turkey grower makes it a practice to market the birds as they finish, he will avoid the heavy feed cost involved in carrying turkeys beyond the point at which they first reach market maturity. By staggering the dates upon which poults are obtained it is possible to place turkeys on the market over a long season. Storage facilities the use of which costs less than that of feeding beyond the normal market age often can be utilized to hold turkeys for a more favorable market.

SELECTION OF LIVE TURKEYS FOR MARKETING

In order to market stock that will when dressed be of U. S. Grade A quality, the turkey raiser should learn to gauge conditions in the live birds. Likewise the buyer of live turkeys must also be able to tell when birds are in good market condition. Such a rough grading for which speed is usually a necessity is best accomplished by looking at the underside or ventral surface of the wing web. If the scattered feathers of the ventral surface of the wing in general are well grown out so they can be picked cleanly, and if the skin of the web appears slightly creamy or waxy, the bird will in all probability grade U. S. Grade A or better. If, on the other hand, a fair share of the feathers are still in the short-pinfeathered stage and the skin of the web appears thin and reddish, the bird will grade U. S. "B" or "C." In this connection it should be noted that the skin over the fleshy and bony portions of the ventral (*under-wing*) surfaces always is thin in all but extremely fat birds (as will be noted in succeeding paragraphs.) Inspection of the bases of the main-tail feathers to determine how much pulp ("ink" in dark-colored feathers) they contain is sometimes recommended as an aid in quickly determining the state of finish. This method is inaccurate, however, since it is only in extremely well-matured specimens that the "ink"

Fig. 6.1. A flock rapidly approaching market age. Note roosts on top of range shelters.

is nearly or completely gone. It is true that a turkey whose main-tail feathers are nearly or completely free of ink will practically always grade "A" or higher. It is also true that a large percentage of A-grade turkeys carry considerable "ink," hence these high quality birds would be turned back if such a rough grading plan were to be used. A similar situation exists, though to a smaller degree, when wing-web inspection is used; that is, many Grade A birds will be turned back because their wing webs do not show noticeable fat even though they may carry enough fat on the breast to qualify as Grade A.

The grower or buyer who knows the age of the birds and the care they have received has a great advantage in grading because, as a rule, the birds in a well-cared-for, disease-free flock of both sexes of medium- or large-type turkeys fed standard diets mature evenly and can be relied on to grade 90 per cent or more U. S. Grade A at 28 weeks*. For a high Grade A to U. S. Special (AA) finish, 30 weeks* must be allowed, and for a high Special (numerical grade 1) finish, 32 to 34 weeks* are necessary. Beltsville Whites should reach these grades 4 weeks earlier. However, the strain, the feed, the management, the weather, the date of hatch, diseases, parasites, and other factors may operate to cause a change in this timetable in either direction. Feeding well-balanced, high-energy feeds and/or antibiotics may step up the schedules by about two weeks. If the grower will handle and carefully inspect a few representative specimens he should be able to determine the condition of the flock. Then, to avoid two handlings, each bird can be inspected as it is caught for slaughtering or for live sale, and unfinished birds then turned back. The inspection of the underside of the wing web, as already described, will suffice to separate the very well-finished from the moderately and poorly finished birds. However, as noted above, this method is not accurate because a good many Grade A birds will be turned back as unfinished. To obtain greater accuracy, the skin of the breast of doubtful birds may be inspected in addition to that of the wing. To do this (see page 348) pluck a few feathers *at a point halfway down the breast* just below the front point of the keel and in the sparsely feathered area between the two feather tracts of the breast. This point is just to rear of an imaginary line drawn between the shoulder joint and the front end of the keel. (It is not up on the side of the bird.) Then, when the bird is held by an assistant so its skin is relaxed, take a fold of the skin between the thumb and forefinger of each hand (Fig. 6.3) and feel and look

*Compare with the table on page 476 which represents the ideal rather than the average.

at it, using the diagrams on page 349 as a guide in classifying specimens according to the various subgrades based on the original four U. S. grades. For this test *the skin at the point specified* must be used. No other point is satisfactory. The condition of the feathering of underside of wing, the shoulder joint, the drumsticks, and the breast should be determined, using the diagram and descriptive material on pages 351-353. Noticeable numbers of pinfeathers too short to be cleanly removed will cause the bird to grade below U. S. Grade A if the pinfeathers are dark-colored. If they are light-colored so as to be invisible in the dressed bird, greater latitude is allowable in actual market grading. However in grading turkeys on ability to mature quickly, the system as described should be adhered to with respect to pinfeathers regardless of color of plumage.

THE GRADING OF LIVE TURKEYS FOR BREEDING, SHOWING, AND EXPERIMENTAL PURPOSES

When live turkeys are to be classified according to market quality in experimental work, for showing, when judging market classes, or in connection with breeding operations it is desirable to use a grading plan involving more than the present three U. S. Grades "A," "B," and "C." Accordingly the authors have worked out a system of numerical grades based on the original four U. S. Grades for dressed turkeys:—Special, AA; Prime, or A; Choice, or B; and Commercial, or C. The live grading is essentially a process of determining how the live turkey, if properly dressed, would rate according to the U. S. Standards for dressed turkeys. The reason for bypassing the live grading system is that it is impracticable to learn turkey grading from live birds without a basic knowledge obtained from the dressed ones. The system to be described has proved to be practicable and has the advantage of requiring the mastery of only one set of rules and terms.

Market quality in the live or dressed turkey depends upon six factors, each of which is influenced, or can be influenced, by heredity and hence by selection. The six grading factors are as follows:

(1) The shape and quantity of breast fleshing.

(2) The finish, that is, the amount and distribution of fat in the skin.

(3) The feather development insofar as it affects the clean removal of dark-colored pinfeathers.

(4) The shape and quantity of lower thigh (drumstick) fleshing.

(5) Skeletal measurements, including length of keel, length of shank, and body depth.

(6) Deformities of bone, skin, and crop.

The first four grading factors may be expressed by numerical grades which are added and averaged to obtain a composite market grade for the specimen. The measurements (No. 5 above) are used in addition to the composite market grade in selecting breeders or show birds. As a guide for making selections based on measurements, see the tables in Chapter 11. Selection toward these measurements will aid (1) in developing a strain possessing a desirable and symmetrical body and (2) in avoiding extremes in leg or body length and in body depth. Deformities of bone, skin, and crop are to be avoided in the selection of breeding stock and of high-quality birds for slaughtering or showing. The presence of these deformities is recorded on the breeding records and the data taken into consideration in the selection of breeders or market-type show birds.

The other factors contributing to desirable market quality in dressed turkeys are associated with their preparation for marketing, that is, the handling and dressing, and hence are not of direct concern to the breeder or the exhibitor when selecting for breeding or live showing.

Breast Type or Shape. The variations in breast fleshing that occur naturally in healthy, well-managed turkey flocks are illustrated by the diagrams in Fig. 6.2, representing nine grades which can be differentiated with an acceptable degree of accuracy by parting the breast feathers and looking at and feeling of the breast over its entire length. A tenth grade "O" may be added to take in extremely broad, "dimple-breasted" specimens. To permit adequate examination, an assistant should hold the bird by its legs, breast toward the grader. The diagrams may be used as a guide to grading for breast type as the grader sees it from a head-on viewpoint. If the breast width carries out in a normal manner to the rear of the keel, the grade as determined from the front part of the breast is assigned the specimen. However, if the fleshing tapers off markedly toward the rear, the breast-type rating is reduced one or two grades, depending on how much it tapers. Grading or judging breast type should be done while the bird is suspended freely by the legs, *not lying on its back.* When in the latter position the breast width is accentuated, causing the bird to be graded higher than it deserves.

Knobbiness also affects the breast-type grade. For example, if a bird had a breast width grade of 4 but possesses a very prominent or bad knob, the grade would be reduced to 7; for a moderate knob to 6; and for a slight knob, to 5. Failure of the flesh of the breast to surround the forward tip of the keel results in a prominence or knob. If the bird also has a roost-dent, the knob is accentuated in appearance. The diagrams for toms, it will be noted, make a reasonable allowance for sex differences in conformation, but in general, grades for breast type are not markedly different for toms and hens. Grade No. 5 is considered as "average" or "normal" for ordinary well-fleshed non-broad-breasted turkeys. For broad-breasted types, breast grade 4 would be minimum for qualification as broad-breasted. To obtain a clear conception of this "average" breast conformation is the first step in learning to classify variations in breast type.

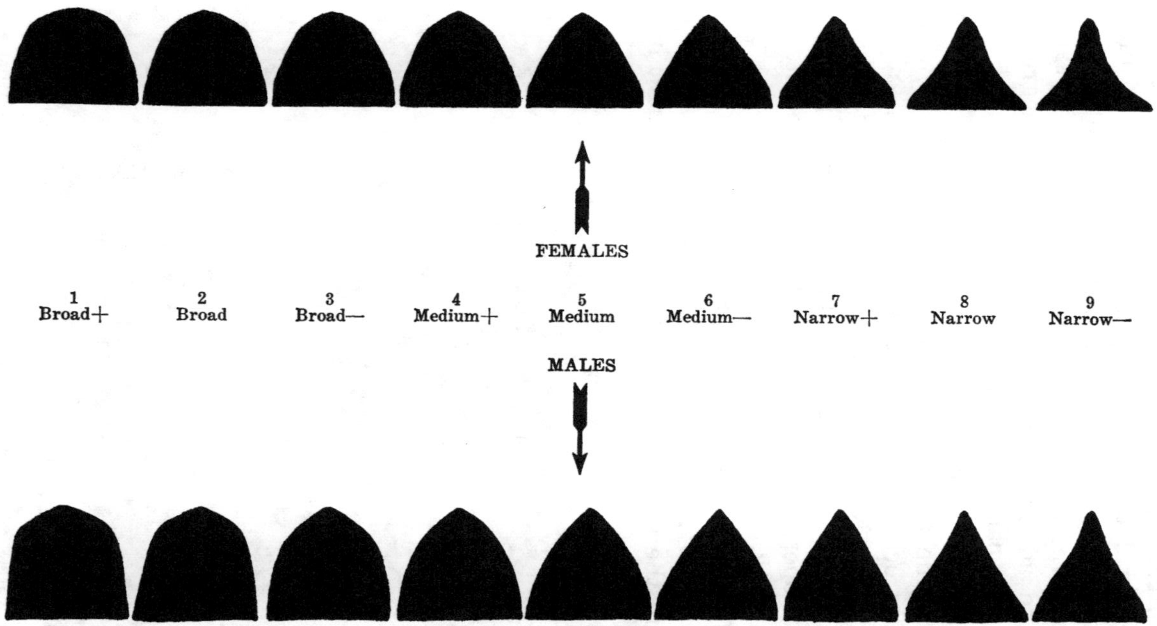

Fig. 6.2. Diagrams of cross-sections through the breast near the front, showing variations in shape and quality of musculature or breast type, of fat roasting turkeys, females at top, males at bottom. Number 1, 2, 3, and 4 are desirable: numbers 5 and 6 are acceptable: number 7, 8, and 9 are undesirable. While differences in breast type are sometimes due to condition, the variations here illustrated may also be due to genetic factors governing development of breast musculature which express themselves independently of condition. Due to the heritable nature of breast type, proper selection for it is important in turkey breeding.

Gauging breast type by molding solder- or fuse-wire over the breast has some merit and some special uses but for general use in grading large numbers of birds the authors have found it too slow and too difficult to apply. The grading plan described here can be readily learned and administered with a fairly high degree of accuracy by a careful operator.

Extremes of breast type narrower than No. 9 occur. To classify the excessively narrow-breasted birds which occasionally occur, 3 more grades, Nos. 10, 11, and 12, have been used, No. 12 being a bird possessing no musculature on the keel except at its base where it joins the sternum proper. Grades 10 and 11 are intermediate between 9 and 12.

The effect of breast type on the four U. S. Grades is as follows: To grade U. S. "AA" (a grade no longer included in the U. S. Grading System) a bird must have

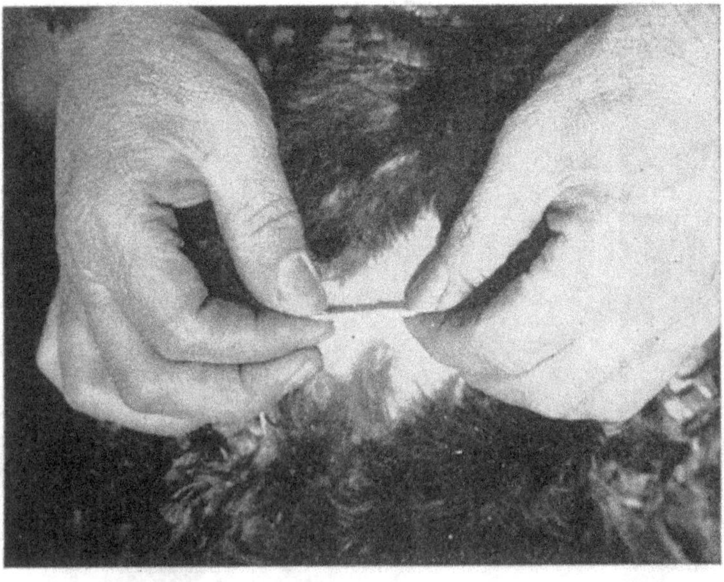

Fig. 6.3. Estimating the thickness of the skin of the breast to determine the degree of finish (Fat Grade) of a turkey. The fold of skin shown indicates a No. 5 grade of finish. (Photo by U. S. Dept. of Agr.)

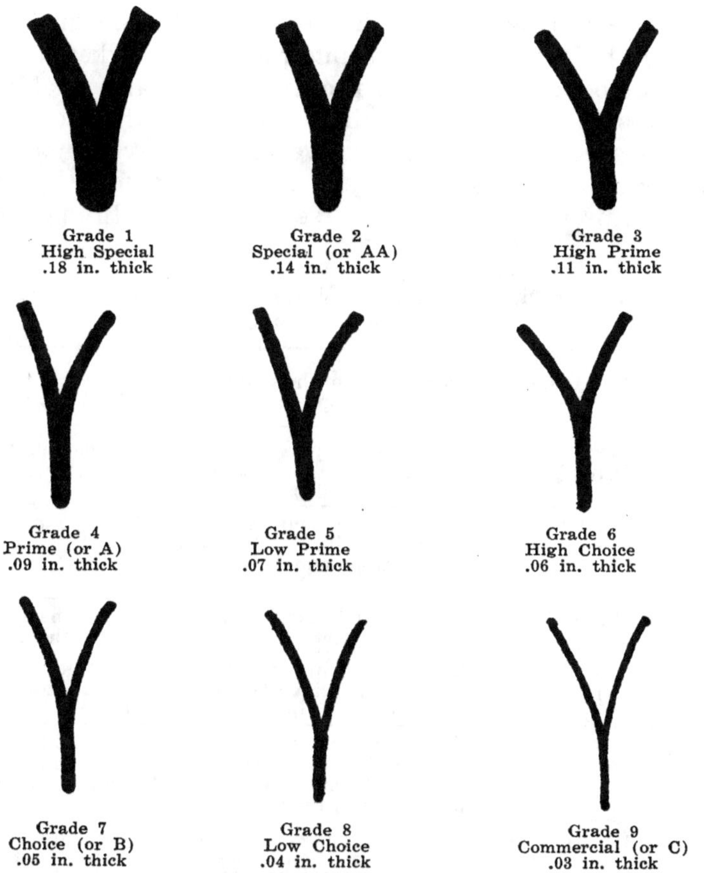

Fig. 6.4. Diagrams showing the relative thickness of the skin midway between the two feather tracts of the breast and slightly to the rear of a line extending between the shoulder joint and the forward tip of the keel. The figures are actual size and represent a cross-sectional view of a single fold of skin held between the thumbs and forefingers of both hands while the bird, with skin fully relaxed, is held by an assistant. A few feathers are first plucked.

a No. 5, or better, breast; to grade "A," a bird must be No. 7 or better; to grade "B" it must be No. 8 or better. No. 9 grade birds must be graded "C" regardless of other qualities. Birds grading below No. 9 in breast type must be considered as undergrade because they possess practically no breast meat.

Fat Grade or Finish. The quantity of external fat carried by turkeys when slaughtered for market is important because it affects the appearance and salability of dressed birds and their palatability when cooked. The degree of finish in the live turkey may be gauged and graded in a relatively simple manner by the method previously described (see page 343). In applying

Table 6.3—Grading Plan for Classifying Live Turkeys According to Finish—to be Used with Fig. 6.4

Grade No.	Grade Name	Description of the skin between the feather tracts at the point specified (pages 351-353.)
1	Special Plus (AA+)	Very creamy and extremely thick, equalling or exceeding the thickness indicated (.18 in. for the fold of skin.)
2	Special (AA)	Very creamy and very thick, the thickness of the fat coating at the specified point being approximately equal to that indicated. (.14 inch for the fold of skin).
3	Prime Plus (A+)	Creamy and thick, equalling the thickness specified for High Prime. (.11 inch for the fold.)
4	Prime (A)	Creamy and moderately thick, the thickness at the point specified being equal to that indicated for Prime. (.09 inch for the fold.)
5	Prime Minus (A−)	Slightly creamy with no readily discernible pinkness, and a thickness equalling that indicated for Low Prime and illustrated in Figs. 32 and 33. (.07 inch for the fold.)
6	Choice Plus (B+)	Creamy pink with a thickness equalling that specified for High Choice. (.06 inch for the fold.)
7	Choice (B)	Decidedly pink with only slight creaminess and a thickness equal to that specified for Choice. (.05 inch for the fold.)
8	Choice Minus (B−)	A thickness equalling that specified for Low Choice. The skin though decidedly pinkish in appearance with only a very slight creaminess, has definite "body" and is not paper-thin as in Commercial. (.04 inch for the fold.)
9	Commercial Plus (C+)	The skin appears nearly devoid of fat so that it is drawing-paper thin. (.03 inch for the fold)
10	Commercial (C)	Very nearly devoid of fat so that it is bond-paper-thin (0.2 inch for the fold.)
11	Commercial Minus (C−)	Devoid of fat so that it is tissue-paper thin (0.1 inch for the fold.)

this method the position on the side of the forward portion of the bird's breast between the feather tracts must be located and carefully inspected, paying no attention to the fat coating anywhere else. Reference to the diagrams and the written description, page 350, combined with practice on turkeys that can be live-graded and then dressed and chilled will serve to master this useful grading and selecting plan.

Eleven grades of finish in live turkeys are described in Table 6.3, but classifying of live birds into subgrades of the U. S. "C" grade of finish is not practicable. In dressed turkeys these "C" subgrades (9, 10, and 11) can be differentiated. They occur mainly in fryer-roaster turkeys.

Feathering. Grading live turkeys on state of feather development is important in the selection of market, show, and breeding birds since comparative freedom from pinfeathers too short to be cleanly plucked is necessary to rate a high U. S. grade in dark-plumaged turkeys. Fig. 6.5 illustrates differences between short (non-pickable) and long (pickable) pinfeathers. The following

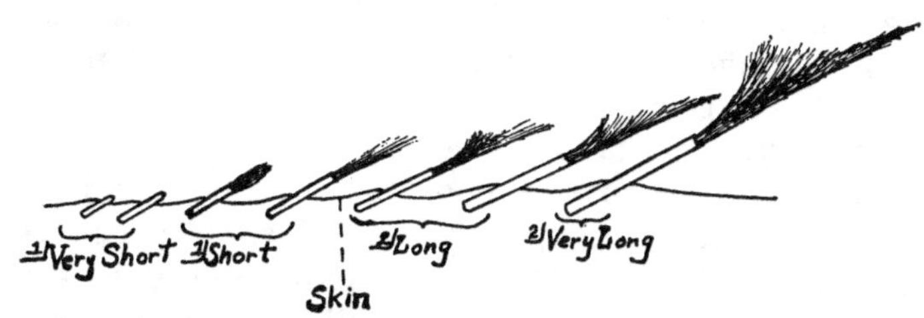

Fig. 6.5. Diagrams illustrating feather growth from the very short to very long pinfeather stages. The feathers to the left of the vertical broken line are so short that they cannot be picked clean; those to the right are sufficiently long to permit clean picking.

[1]Part of the feather pigment often remains in the skin when these feathers are pulled. In white birds this is not so serious as it is in birds with dark colored plumage.

[2]These feathers can almost always be pulled without leaving a residue of feather pigment in the skin.

grading plan is based on number, location, and length of pinfeathers found on the live turkey at the time of examination. Seven **feathering grades** numbered 1 to 7, inclusive, are named and described in Table 6.4.

To classify live turkeys according to this plan, an assistant holds the bird on its back, breast toward the grader, who first spreads one wing and examines the scattered feathers on its underside and at the shoulder joint where the wing joins the breast. Then the feathering in and between the two feather tracts of the breast, and finally the drumstick feathering are examined. The condition of the feathering of the drumstick and underside of the wing is a good index of its condition over back, sides, neck, abdomen, and upper thighs. Condition of feathering of the breast region which includes the shoulder joint affects the grade more than that of other sections as is indicated in the specifications of U. S. grades. For this reason special notice is taken of these parts. A fairly high percentage of these short pinfeathers can be removed, hence the specifications for the live grades are slightly more liberal than for the dressed grades. The effect of feathering on U. S. grades is shown in Table 6.4.

The state of the feathering is one of the two important measurements of rate of maturity in market turkeys (the other measurement being that of the fat in the skin.) The age of the turkeys to be examined must be known if selection is to be made for quick maturity. It is also necessary to know what to expect in turkeys of various ages and to this end the table on judging for rate of market maturity on page 476 should serve as a guide. These high requirements express the ideal; an objective for the breeder and exhibitor. In judging for rate of maturity the state of feathering and the quantity and distribution of fat in the skin are of primary importance, the other market-quality factors such as breast type, leg type, skeletal measurements, and deformities being of little significance. That environment and heredity both affect the rate of maturity, including feather development, is well known. However, the standards for rate of maturity herein described are based on flocks of turkeys raised in a standard manner and may be used

Table 6.4—Feathering Grades for Live Turkeys

Grade No.	Grade Name	Description
1	Excellent	No pinfeathers. High Special or AA+ (subgrade 1) as far as pinfeathers are concerned when picked.
2	Good	A few long pinfeathers; no short ones. High Special (subgrade 1) when picked.
3	Good minus	A moderate number of long pinfeathers; a few short ones, practically none on the breast. Special or High Prime (subgrades 2 or 3) when picked.
4	Medium plus	Many long pinfeathers; a few scattered short ones, very few on the breast. Prime or Low Prime (subgrades 4 and 5) when picked.
5	Medium	Many long pinfeathers; a moderate number of short pinfeathers scattered generally over the body but relatively less numerous on breast. High Choice (subgrade 6) when picked.
6	Medium minus	Many long pinfeathers; a moderate number of short ones scattered generally over the body, including the breast. Pickable to Choice or Low Choice (subgrades 7 and 8) when picked.
7	Poor	Many long and short pinfeathers scattered over the body in general. Birds as pinny as this cannot be graded above U. S. Commercial (subgrade 9, 10, and 11).

as a yardstick in judging and in measuring progress in breeding and success in management and feeding. However, improved methods of feeding may necessitate a change in these standards as indicated on page 476.

Furthermore, with modern scald-pick methods and white or light-plumaged turkeys, the state of the feathering is of considerably less importance than it is with dark-plumaged turkeys and dry-pick methods. A good finish in well-matured roasting turkeys should, but not necessarily does, go hand-in-hand with relative freedom from short pinfeathers since both qualities are desirable in roasting age turkeys of any plumage color. In turkey fryer-roasters it is useless to attempt to attain freedom from short pinfeathers and for this reason light plumaged varieties are to be favored for the production of this class of turkey meat.

Leg fleshing. The U. S. grading system makes no special reference to leg fleshing, although it is included in the general requirements for fleshing. Well-fleshed legs are essential to symmetry and to a high percentage of edible meat in the dressed carcass, hence are an important factor in judging and breeding of turkeys for market quality. In the commercial grading of dressed turkeys, leg fleshing where noticeably deficient (grade 6) serves to keep specimens from grading U. S. "AA" or its equivalent and where it is very decidedly deficient (grade 7) the bird cannot be graded "A". It is in breeding and showing turkeys that attention should be paid to leg fleshing because hereditary deficiency in this respect is rather common and results in a poorly balanced (unsymmetrical) market carcass. Examining the drumstick for quality and shape of fleshing is a practicable method of grading the bird on total leg fleshing. Breadth of rear half of the back is due partly to the flesh of the upper thigh and can be used to supplement the drumstick grading in gauging total leg fleshing. The diagrams in Fig. 6.6 illustrate seven grades of drumstick fleshing as seen in dressed carcasses. Grade No. 4 is the average, well-fleshed drumstick of normal size. The recognition of the average grade is the first step in learning to grade the variations in leg type and size. Examination of the

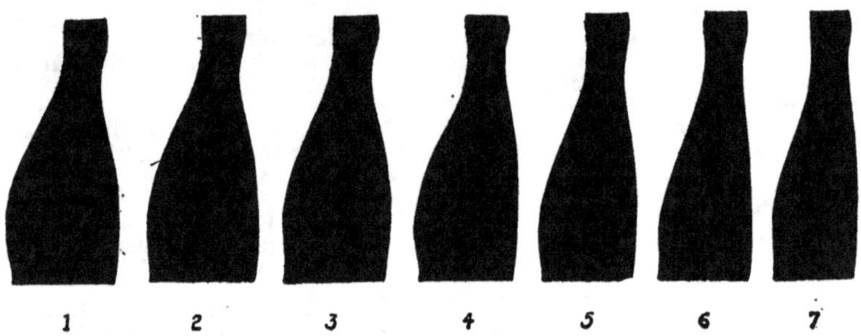

Fig. 6.6. Diagrams illustrating the plan for grading variations in the drumstick fleshing of male and female turkeys. The outline of the bird's left leg is shown as the grader sees it when viewing the hanging bird straight from the front. Nos. 1 to 3 are desirable. No. 4 is acceptable. Nos. 5, 6, and 7 are undesirable. Note the resemblance to a "ten pin" in grades 1, 2 and 3.

Table 6.5—Leg-fleshing Grades for Live and Dressed Turkeys—to be Used with Fig. 6.6

Grade No.	Grade Name	Description
1	Perfect or extremely plump	Appears extremely well rounded and in decidedly excess proportion to the remainder of the carcass. The fleshing extends well down toward the hock.
2	Very plump	Appears very well rounded and of noticeably greater than normal size.
3	Plump	Appears well rounded and of slightly greater than normal size.
4	Moderately plump or average	Appears fairly well rounded and of normal size.
5	Rather slender	Appears somewhat straight-sided and/or slightly smaller than normal.
6	Slender	Appears decidedly straight-sided and/or decidedly smaller than normal.
7	Very slender	Appears elongated and stringy and/or very much smaller than normal.

In some birds the legs are of good shape but are undersized in comparison to the remainder of the body. This fact must be considered when grading for leg type.

drumsticks of the live bird must be mainly by feel; of the dressed carcass by appearance.

The seven leg grades are described in Table 6.5 and illustrated in Fig. 6.6.

CAUSES OF LOW GRADES

There are a number of causes of low-grade turkeys, many of which may be avoided by careful growers. The most important causes are lack of fat in the skin and the presence of dark-colored pinfeathers too short to be picked cleanly without leaving a dark pigmented spot in the skin. Proper feeding methods must be used and sufficient time allowed for fat to be laid down in the skin and for the pinfeathers to grow out, except in turkey broilers or light roasters which are marketed in an immature state and of necessity with relatively poor finish and considerable numbers of pinfeathers. Varieties with inconspicuous pin feathers are best suited to turkey "broiler" production.

Abrasions and bruises frequently cause birds to be graded down. These spots can be avoided by due care in catching and handling the birds and bringing them to market. Care in picking is equally necessary. Grabbing birds from underneath the roosts frightens them and as they struggle and while one is handling them they may tear or scar the flesh and cause bruises and injuries. Unless subjected to federal grading, tears in the skin of dressed birds should be sewed to protect the flesh that would otherwise be exposed. Any strong white sewing thread or doubled silk thread may be used with a strong needle, bringing the edges of the torn skin together with an over and under or "baseball" stitch. A surgeon's needle held at the eye-end affords the best method. Such needles may be secured from the drug store. Sewn tears, however, are not permitted in U. S. Grade A birds.

Another cause of low grade or rather, no grade, is the presence of feed in the crop at the time the bird is dressed by the producer for selling on a New York dressed basis. It is a wise practice for the grower not to feed his turkeys grains during the twenty-four hours nor mash for fourteen hours preceding the time they will be slaughtered. A simple procedure is to feed only mash the last day or two before slaughtering and take it, but not the water, away at sundown of the day before slaughter. When necessary however, feed in the crop can be removed by the water-pressure decropper.

Poor dressing is another cause of birds being graded down. When the birds are sold alive this is out of the control of the turkey producer. However, if one dresses his own birds he can be very careful that the flesh is not scratched, rubbed, bruised, or torn, and that the bleeding is complete. Bruises which frequently result at the time of killing can be avoided by handling the birds carefully as they are hung up for killing. To prevent bruises during the first few flops that a bird gives immediately after it is killed, the picker should allow the bird to flop free of interference by standing back well out of reach of the wings. If two people are working together, one may embrace the bird firmly while the other bleeds and sticks it. Metal killing funnels or funnel-shaped boxes are very useful when birds are to be scald-picked. The turkeys' head protrudes from below and can be grasped for killing. The bird remains in the funnel for a few minutes until bleeding is complete, then is scalded. For birds below and over 17 pounds respectively, sizes of funnels desired are: length, 22½″ and 30″; top diameter, 12″ and 15″; bottom diameter, 5″ and 6″.

Pink-fleshed, off-flavored turkeys occasionally are found. The cause is obscure but it has been suggested that birds recovered from a cholera or swine erysipelas infection may show this condition.

CATCHING AND HANDLING LIVE BIRDS

Turkeys possess the herding instinct to a marked degree and in contrast to chickens, can easily be driven from place to place in large flocks. For catching they can be driven into a building, chute, or corral (see pages 359 to 362) and handled in groups of perhaps 100 to 150. The best method of catching is to drive the birds into a narrow runway, or catching chute (described below.) The turkeys arc removed by the legs. Since their heads are up they cannot see the operator approaching and can be caught with a minimum of piling and disturbance. When turkeys are caught off the floor or ground in buildings or corrals, care must be taken not to drive too may in at one time and to stir them up frequently to avoid losses from piling and smothering. In these operations some catching hooks about three feet long are indispensable. A good hook can be made as shown in Fig. 6.7, either all metal or a metal hook fastened to a strong stick.

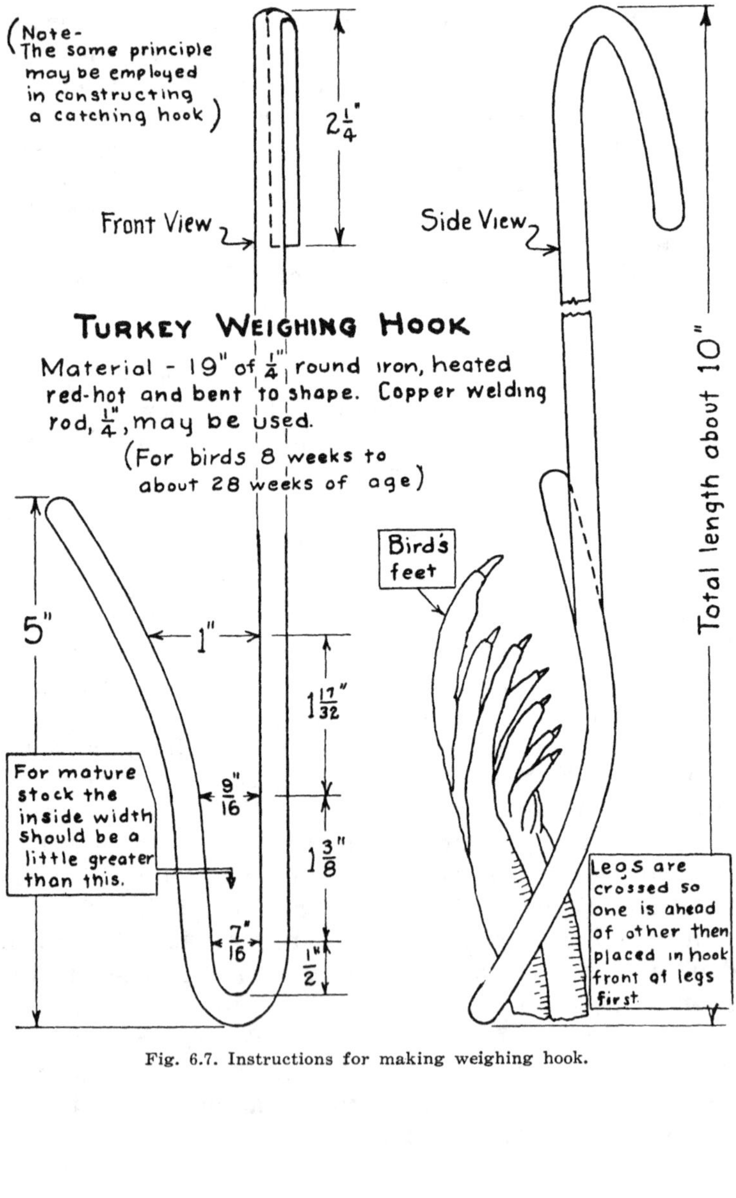

Fig. 6.7. Instructions for making weighing hook.

One type of catching chute is 8 feet or more long by 24 to 30 inches wide by 3 feet high. The top is wire- or slat-enclosed and there may or may not be a floor. There is a sliding gate in each end and a set of fencing wings for one end. The sides consist of a framework covered by slats or wire and must be relatively open as turkeys will not go into a darkened chute unless they have become accustomed to it over a period of time. On one or both sides, the framework comes to within about 16 inches of the bottom leaving an opening which should be covered with a sliding or hinged shattered panel to close while the

Fig. 6.8. A portable turkey grading or culling chute. Note flared portion at right to fit gate. (Photo from Dwight Rutherford.)

birds are being driven in. The whole side may consist of sliding slatted panels as in Fig. 6.8. Turkeys are driven into the chute via the flared wings and are easily removed by the legs through the bottom opening.

Another type is set on wheels and has about the same dimensions. The sides are constructed as directed above. The bottom is solid but littered or padded (burlap and straw.) There is a removable incline and removable wings, and the whole device is intended to be towed by or carried on a truck or wagon. A wheeled chute is shown in Fig. 6.8.

CATCHING AND SELECTING EQUIPMENT

There is need on every turkey farm for facilities for catching and sorting the birds. Permanently located equipment on well-drained ground usually is preferred to the temporary type. Turkeys may be driven to it from relatively long distances with little trouble. The best catching device is a chute preferably on wheels (Fig. 6.8). A corral or U-shaped enclosure 15′ to 20′ wide and 15′ to 20′ deep open at one end with sides well supported with heavy gauge poultry wire, preferably 6 ft. high will do. At least the rear half should be covered with poultry wire and provided with a gate so birds may be securely confined therein as is often desired. Driving wings are often attached to the open end. Turkeys in groups of 75 to 100 are driven into the corral and caught quickly with wire hooks. Losses from piling and smothering may be avoided by "stirring" the birds, or better by the use of slide gates or partitions in the chute for holding smaller groups; darkening the chute if possible.

When only marketing operations are carried on, two holding pens are needed, one for market birds and one for market rejects. When selection of breeding stock is carried on, there should be several holding yards and a "typing" walkway. With this plan the flock is driven into the large holding yard then, in small groups, into the driving funnel and thence to chute or corral from which they are removed, one at a time to the work area where they are handled, and examined for body type, balance, color and deformities. Each bird is then tentatively classified, usually the toms, and frequently both hens and toms, into 3 groups as follows; (1) first-choice breeders, (2) second-choice breeders, (3) market birds which are at once placed in the market enclosure. Those picked as breeders are then placed in the near end of their respective walkway and, as they walk along it to their holding pen, are observed for carriage, or balance, for ability

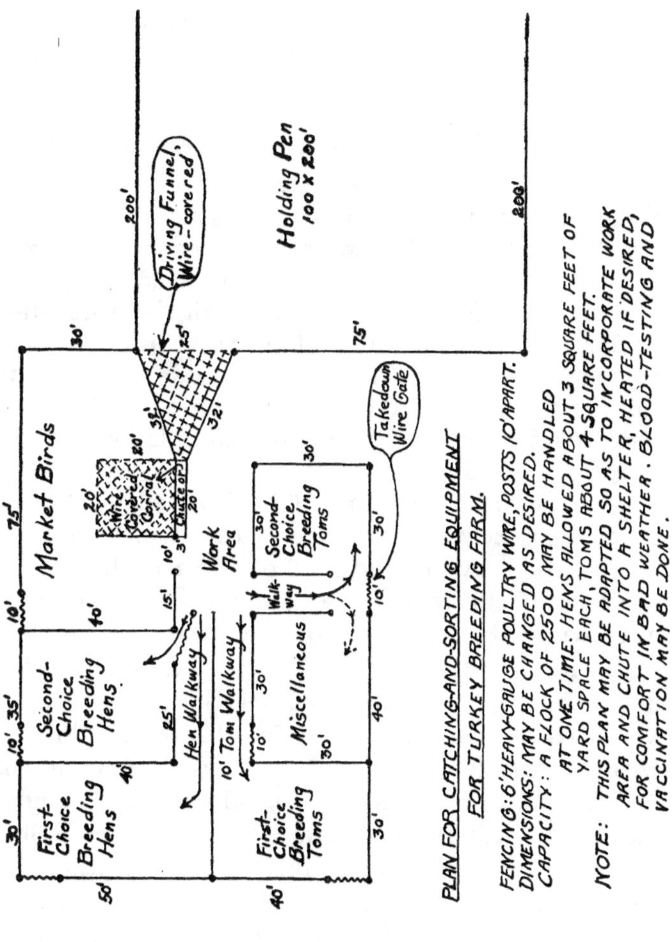

Fig. 6.9. Sorting arrangement.

to walk normally, and for possible leg deformity or weakness which frequently cannot be detected during handling. If defects are thus noted the bird is caught (before it reaches the holding pen) and placed in the market pen or it may be shunted into the market pen by a swing gate. A suggested arrangement for sorting yards is shown in Fig. 6.9.

PROCESSING FOR MARKET

Dry Picking. The bird is hung by its feet and its head held in one hand of the operator by pressure on the bird's skull at the angle of the jaw, not the neck, since compressing the neck retards the flow of blood. With this hand the mouth is forced opened and with the other hand the jugular veins are cut from left to right at a 45° angle far back in the throat just below the base of the skull. For this purpose a knife with a pointed, narrow 5″ blade with perfectly straight cutting edge is best. Immediate profuse bleeding indicates that the correct cut has been made. The bird should be brained or "stuck" as soon as this occurs. The knife is reversed and thrust up though the groove in the roof of the mouth and into the rear lobe at the rear of the skull, then the point of the knife wiggled a little to destroy brain tissue (Fig. 6.10). Learning the "stick" is rather difficult. A stick that is too short is ineffective and one that is too long, or which penetrates the middle or fore-brain will kill the bird without first loosening the feathers. When the correct "stick" is obtained the bird usually gives a peculiar squawk and its tail feathers spread out. Electric killing promotes fast and complete bleeding and the feathers are loosened without the customary convulsions so that a blood cup is unnecessary. Suitable "electronic" equipment may be purchased. A new method of immobilizing the bird prior to killing is to introduce the bird into a carbon dioxide chamber which renders it unconscious and promotes free bleeding without any struggling following the cutting of the veins in the neck as soon as the bird

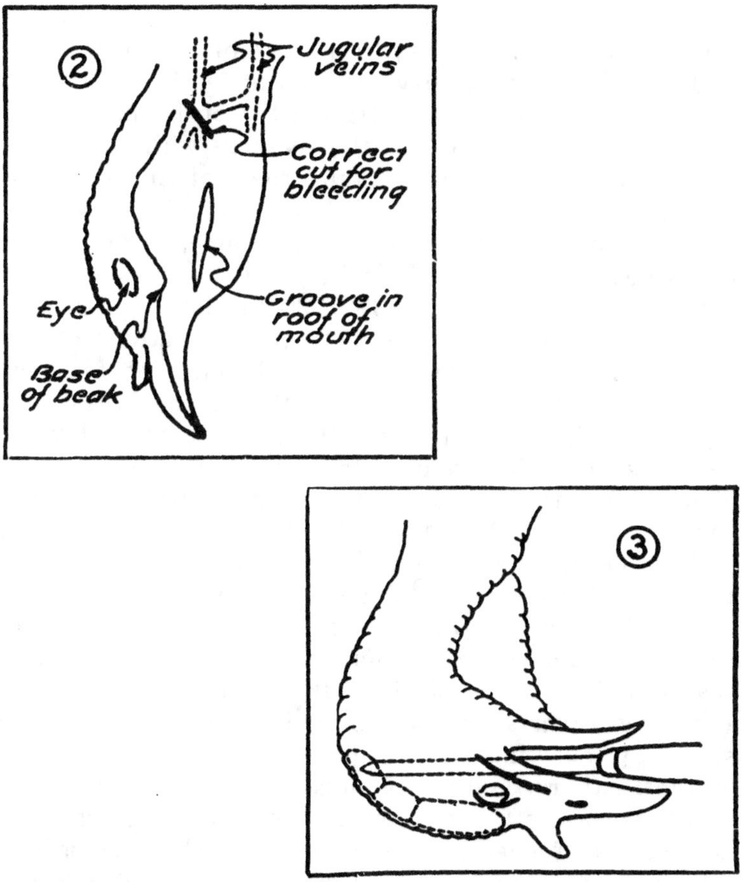

Fig. 6.10. Sticking (2) and braining (3) for dry picking. (Photo from Turkey World.)

collapses. (Davis and Coe, 1952). Its economic value is yet to be proved.

Whether or not one has stuck or "brained" a bird correctly can be told by the looseness of the wing and back feathers. After sticking, the bird's head remains in the hand until a blood cup has been fastened into the lower jaw. The bird's wings should not be locked as this often results in a broken wing. By stretching the bird's neck well out, the operator can stoop clear of the flapping wings, thus avoiding injury to the bird or operator. If the bird is to be properly dry picked the feathers must be plucked immediately after sticking the bird. The picking should be done by grasping the feathers while the hand is in a sweeping motion. Holding the feathers and then pulling often results in torn skin. A definite order for the removal of the feathers is desirable, such as the following: (1) main-tail feathers, (2) primaries and secondaries of the wings, (3) back, (4) legs, (5) breast, (6) area around neck, vent, (7) contour feathers of the wings, and (8) the neck. A small handful of feathers should be pulled at a time, especially on the breast and thighs which are easily torn. If the feathers of breast, neck, and wing tips are plucked first there is more chance of producing the red pin spots caused by plucking while there is still blood in the skin. These pin spots are a serious defect, especially if occurring on the breast. Removing feathers by rubbing will invariably injure the skin and lower the grade. Pliers are useful in removing the short primaries, the primary coverts, and the pinion feathers. The skin at the base of the fan feathers (primary coverts) should first be loosened from the shaft by exerting strong pressure toward the base of the feathers. Poultry pinning knives and strawberry hullers are useful in removing ordinary pinfeathers. Likewise a crochet hook for removing short embedded feathers. If a pinfeather cannot be removed without cutting the skin, it should be allowed to remain because a skin injury is worse than a pinfeather. All turkeys should be picked clean. Turkeys to be marketed New York dressed should have the head "snapped" to remove clotted blood in the mouth, and dirty feet should be scrubbed when the birds are killed, and the vent area squeezed to remove any feces from the cloaca.

The slack or semi-scald. After sticking, and bleeding for 1½ minutes, the birds are moved about for about 30 to 50 seconds in water kept at 126 to 130° F. or about

75 seconds at 124° F. and then picked by hand, machine, or with wax. Some plants spray the birds for 80 seconds in 130° water. In the slack-scald, soft water is better than hard water for loosening feathers. To soften hard water, add 1 pound of sodium bicarbonate to each 300 gallons of water. Bleeding by cutting the throat from the outside at a 45° angle from left to right with a very sharp knife is preferred by many slack-scald operators, who do not stick the brain. A new plan for quick and easy removal of all feathers including pinfeathers is to use the slack-scald at 130° F., for about 30 seconds and ice the turkeys right after rough picking and while being pinned. The chilling makes feather removal much easier at the same time preserves intact the outer skin layer, provided the rough picking is done carefully. Small operators sometimes use both dry picking and semi-scalding. After sticking and before immersing, the feathers desired for use or market are removed dry and then the bird is semi-scalded to loosen the remaining feathers.

The subscald (medium) and hard (full) scald. If the birds are to be marketed iced immediately or are to be eviscerated and placed in moisture-proof wrapper, subscald at 138 to 142° F. for 25 to 30 seconds may be practiced. There is a little place for the old-fashioned hard scald with water at 160 to 180° (preferably 150-160) as this process invariably injures and discolors the flesh. Hard-scald birds even if properly handled do not promote the turkey business. Singeing may be done with a blow torch or gas flame, but carefully; also an alcohol wick lamp is excellent. The subscald is the best method available for removing feathers from turkey broilers. However, subscalded turkeys, particularly immature birds, tend to develop a reddish appearance unless they are frozen quickly at minus 20° F. or lower, on open trays. Wrapping skin and air-tight helps. Subscalding also results in the development during ordinary roasting of a much tougher and somewhat wrinkled skin (U. S. West. Util. Res. Br., Albany, Calif.).

The subscald results in the reasonably complete removal of the thin epidermal layer of skin (the cuticle or bloom) which contains the yellow skin pigment, if any. Therefore subscalded turkeys, regardless of their surface skin color, always appear white-skinned. For easy removal of cuticle, some operators subscald at 142° F., pick, drop the picked turkeys into warm water and sponge off the cuticle with a celanese sponge. Recommendations for subscald temperatures vary from 135 to 145° F. The U.S.D.A. (1953) recommends 138 to 140° for 25 to 30 seconds depending on kind, class, and age of the poultry. Cartoning or individual air-tight wrapping are necessary for the preservation of subscalded turkeys unless they are kept in crushed ice until they reach the consumer. If stored (frozen) without individual wrapping, the semi-scald is preferred to the subscald.

Information on methods of scalding poultry may be obtained from the Poultry Division, Agricultural Marketing Service, USDA, Washington 25, D. C.

Cooling

The carcass should be cooled to an internal temperature of 32° to 36° as soon as possible after killing. Temperatures can be taken with a thermometer thrust about four inches into the vent so as to reach the center of the body. An ordinary dairy thermometer is satisfactory for this purpose. Delay of cooling for more than a very few hours may start decomposition. Freezing of the neck and wings before the interior is cooled causes discoloration and should be avoided as it reduces market value. A bird can be frozen on the outside and yet be relatively warm on the interior.

Cooling in water is the usual practice in connection with semi- and medium-scalding. Water-cooled birds are weighted down in ice water sometimes after being precooled in ordinary cold tap water for 1 or 2 hours until the internal temperature reaches 40° or less. Cooling to internal temperatures as high as 45° is satisfactory in many instances although cooperative pools and commercial packing plants usually require a maximum internal temperature of about 36°. The body heat should be gone in 12 hours or a little less in air and 3 or 4 hours in water provided the birds are not over-crowded in the cooling medium. Freezing of the turkey carcass starts at about 28° F.

Greenstruck turkeys are birds showing a green discoloration around the crop, ribs, back, or vent. Insufficient or improper cooling is considered to be the chief cause. Consequently, an internal temperature of 36° F. or less is desirable before birds are packed and birds should not be allowed to freeze on the surface before the interior is thoroughly cooled. Maintenance of proper temperature until the birds reach the consumer is of course essential. Bruises and scratches sometimes turn green but these do not properly cause the bird to be considered greenstruck. Removal of cloacal droppings by squeezing the abdomen and elimination of feed in the crop will contribute to the prevention of greenstruck dressed birds. Crop removal in which the crop connective tissue is not permitted to remain attached to the skin may cause discoloration and consequently lowering of grade. This type of discoloration, along with greenish colored bruises and scratches, and the bluish-black discoloration known as blue-back should not be confused with the true greenstruck condition which causes any bird possessing it to be classified as undergrade, or below-grade.

Wax Picking

Wax picking of turkeys is no longer considered practical. It has been replaced largely by mechanical pickers of the revolving drum rubber-fingered type followed by hand "pinning."

Wax picking is most valuable in the dressing of pinny birds. If pinfeathers are properly grown out before killing, wax picking, although faster than dry picking, loses its greatest asset—the quick and easy removal of pinfeathers. Even wax, however, will not remove pinfeathers that protrude only slightly, or not at all, from the skin. Birds possessing many such extremely short pinfeathers should not be slaughtered.

Headwrapping

Headwrapping adds much to the appearance of dressed turkeys and in addition prevents blood from dripping onto other birds during handling. All well-packed New York dressed turkeys are headwrapped. Headwraps can be pur-

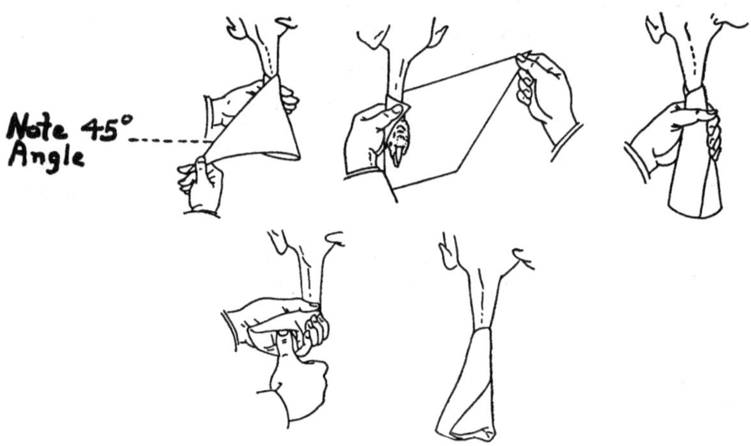

Fig. 6.11. A quick method of head wrapping.

chased or made at a very small cost. Wraps made of heavy white waxed paper present the best appearance and are easiest to put on (see Fig. 6.11 for detailed instructions). The following sizes are needed: (1) for large toms: 22″ long side x 11″ short side x 11″ end x 15½″ diagonal; (2) for medium toms: 20″ x 10″ x 10″ x 14″, respectively; (3) for small toms and medium and large hens: 18″ x 9″ x 9″ x 12½″; (4) for small hens: 15″ x 7½″ x 7½″ x 10½″. Frequently only the first and third sizes are used. Headwraps are somewhat difficult to apply, but once learned, the wrapping is quickly and easily done (see Fig. 6.11).

Evisceration Before Cooling

With the advent of modern refrigeration practices and the merchandising of turkeys "oven ready", evisceration prior to cooling has become the common practice. The merchandising of New York dressed turkeys is gradually becoming obsolete. Evisceration prior to cooling has distinct advantages in both labor saving and efficiency. The warm carcass is easier to handle and evisceration is simpler before cooling. In most cases of commercial operations, the bird stays on the same line and passes from the last washing to the evisceration line. Removal of viscera at this time also has the advantage of saving on the cost of cooling since the discarded portions are not involved in the cooling process, and the opening of the body cavity makes cooling more efficient. Since cooling subsequent to evisceration is accomplished more quickly, this tends to hold the quality of the dressed bird better and thus increases the "shelf life" of the dressed product.

Packaging

The principle purposes of a package or a wrap for the dressed turkey are the protection of the carcass from contamination, and the prevention of the loss of moisture from the dressed bird. Consequently, a moisture vapor barrier type of wrapping material is to be desired. If that material is transparent, it has certain advantages in merchandising although form-fitting opaque wraps do not detract from the appearance of the bird.

The old fashioned parchment paper, referred to as "butcher paper" may still be used where the birds are sold direct to the consumer shortly after dressing. However, if the bird is to be handled through the channels of trade, or a day or more would elapse before marketing, it is wise to place the bird in some type of container such as a pliofilm bag.

If it is desired to increase the attractiveness of the pack and thereby promote sales, some type of form-fitting wrap should be adopted; or a polyethylene bag may be used. Polyethylene is a high-molecular-weight wax which cracks much less than cellophane in zero and sub-zero temperatures. However, most all wrapping materials, except aluminum foil, tend to become brittle and crack on rough handling at sub-zero temperatures. Pliofilm and Cry-o-vac are less transparent, but will withstand lower temperatures without cracking; and are quite comparable to polyethylene. Any form-fitting wrap has the advantage of reducing the surface of the bird that is directly exposed to air and thereby reducing moisture loss from the carcass.

Attractive and appropriately wrapped eviscerated turkeys will not retain their consumer appeal unless placed in an appropriate carton or box. If such a box is wax treated on the interior, the necessity of heavy waxed papers for lining the boxes is eliminated.

Crop Flushing

Birds with crops containing solid material should not be slaughtered if intended for dressed marketing, but if they are slaughtered, the crops should be emptied by the water-pressure decropper or as a last resort, removed along with their contents. To remove, cut a slit 2 or 3 inches long in the skin along the right side of the neck where it joins the body. With the fingers push aside, but do not remove, the crop fat. With 2 or 3 fingers, or the whole hand if the bird is large, remove with care the crop and its contents, cutting off the esophagus well to rear of the crop and well up into the neck so that the contents are not spilled. The skin of the neck may be sewed up with white thread, unless the bird is to be federally graded when it should be left unsewn. If the cut has been made at the side of the neck and is neatly sewed up, the specimen if otherwise worthy should be readily saleable at no price handicap. If the cut is made on the front of the bird, its appearance is spoiled and the grade will be lowered.

Crops containing only liquid matter can be emptied by "milking" before bird cools off, or before killing. Removing crop contents by water pressure is the usual method in processing plants. Turkeys intended for evisceration often are killed with feed in the crop but this feed usually is removed by water pressure at the end of the picking line unless evisceration is to be done immediately.

Home Dressing Outline

Turkeys should be handled carefully at all stages in the marketing process to prevent damage and deterioration. A few rules to follow in dressing turkeys for marketing New York dressed are given below. If birds are to be drawn at once, certain obvious procedures in this outline may be omitted.

1. Select for marketing only birds of U. S. Grade A quality or better. Under favorable conditions 90% or more of a flock will be "ripe" for slaughter at 28 weeks of age, for all but small-type turkeys which are ready at 24 weeks. There is a modern tendency, however, to market turkeys earlier than this, 24 to 26 weeks for medium and large-type, 21 to 22 weeks for small type. Improved feeding and processing white or light-colored plumage, and the tendency to accept birds with finish in the low A (subgrade 5) and high B (subgrade 6) classifications make this possible.
2. Remove ground feed at sundown, or scratch grains in the early morning, of the day before slaughter. It is preferred to feed only ground feed on the day before slaughter, especially to old stock. Provide water to drink at all times.
3. Before slaughtering, examine each bird for market quality and solid feed in the crop. Turn back birds that are not finished or have solid matter in the crop; unless the crops are to be emptied by the water-pressure decropper, or evisceration is to be done promptly after picking.
4. Wash the bird's feet with a scrub brush if feet

are dirty (as they usually are.) Confining to clean litter or on wire helps to keep feet clean.

5. Place bird head down in a killing funnel (when using slack or medium scald) or hang it by the legs on a metal shackle or strong cord knotted at the end and passed through a hole in a wooden block 2 inches square.
6. Cut birds' jugular veins which are far back in and on the (bird's) upper surface of the throat. Make 2 criss-cross cuts and do not give up until the blood gushes from the throat; or, if birds are to be eviscerated and scald-picked (slack or medium), the crosswise, or "Kosher," cut from the outside may be used.
7. Brain the bird by thrusting the knife upward through the slit in the roof of the mouth at a small angle (Fig. 6.10) with the long axis of the head. Do not give up until the bird emits the characteristic squawk or releases its feathers as determined by pulling out a few on the back. The end of the knife blade should be twisted slightly in order to destroy some brain tissue. Insert blood cup which for large birds should weigh 7 pounds, for small ones, 5.
8. Pull large wing and tail feathers, the former with quick, strong, slightly rotary jerks, several feathers at a time, the latter with a single strong rotary motion.
9. For slack scalding move the bird about in water at 128° F., for about 40 seconds or, for less skin damage, for about 1¼ minutes at 124° F. For the medium scald, dip 25-30 seconds at 138 to 140° F.
10. If the bird is to be dry-picked, proceed with the picking of the feathers in the order indicated on page 364. When feathers do not pull out easily, remove only a few feathers at a time to avoid tearing the skin.

11. Remove all feathers except those found among the caruncles and on top of the head. Remove all pinfeathers that will come out by pulling—do not attempt to dig them out except with a crochet hook and do not pull out the beard, which should

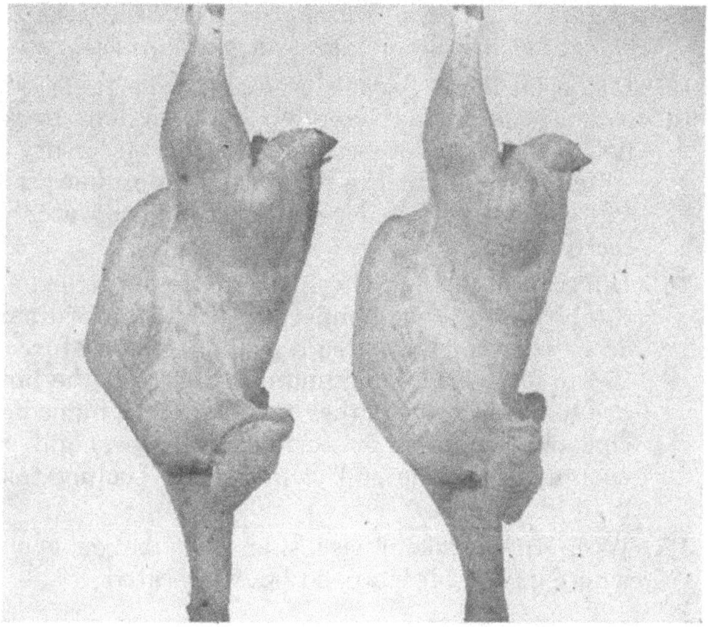

Fig. 6.12. Good (left) and poor (right) balance in Broad Breasted Bronze hens. Note absence of knob and that back line and breast are parallel in hen at left. (Photo from U. S. Dept. of Agr.)

first be clipped close to the flesh with scissors if bird is to be wax-picked. All U. S. Grades require clean picking.

12. If wax picking is practiced, follow directions of the manufacturer of wax or wax-picking equipment.

13. After picking is completed, snap blood out of bird's mouth and squeeze the abdomen at the vent to remove any dropping that may be in the cloaca.

14. Sew up any tears (unless birds are to be federally graded*) and remove any crops that contain solid matter. Crops containing only liquid matter sometimes can and should be "milked" before the bird cools, or preferably before it is killed. Normal crop contents can be removed very quickly and efficiently by the water-pressure decropper, an elongated narrow nozzle on a garden hose.
15. If the birds are bloody or dirty, wipe them with a damp cloth and hang up by both legs or lay flat on back, for air-cooling. If it is necessary to water-cool, weight the birds down in ordinary cold tap water, then in ice-water, or in ice-water directly.
16. During air cooling, check the air temperature and the bird's internal temperature with an ordinary dairy thermometer. Use a cooling temperature of 29° to 36° F. The maximum is 40° F. if the birds are to be shipped. If they are to go into immediate consumption, 45° is permissible. Overnight air cooling usually is sufficient; water cooling takes 3 to 4 hours.
17. Wrap the heads unless this is to be done elsewhere or the birds are to be eviscerated.

COMMERCIAL DRESSING METHODS

The large-scale operator accomplishes a major portion of the dressing process with specially designed machines of which there are several makes, both small and large, for the mechanical removal of feathers. Their practical and economical use is dependent upon the volume of birds to be processed. Most smaller processors use only one machine, a "rougher" for mechanically removing the heavier feathers left after the first rough plucking, and then finish the bird by removing pin and short feathers by hand

---

*U. S. Grade A bird may not have sewn tears.

The larger operator processes his volume of turkeys "on the line," that is, the bird is hung by the feet in a specially designed shackle affixed to a moving conveyor. The speed of the chain is adjustable to accommodate various sizes of birds, and the consequent speed at which the machines and people can clean them. At the start, the killer makes the external throat cut on birds about 8 pounds and under or the internal throat cut and brain stick on larger birds. An operator standing close by removes the tail feathers, and usually the large wing feathers as the bird is in its death struggle. Frequently a quill pulling machine is used which will operate very well on shorter length feathers, so frequently the large wing feathers are clipped about half off at the time the bird is first hung on the shackle to reduce the load on this machine. Wing feathers may be removed on a wing stripper after scalding.

From the killing point, the birds pass through a scalding machine or tank. Some machines are box-like equipped with nozzles for spraying 128 to 130° F. water on the birds for about 80 seconds; other machines are long tanks with water 128 to 140 degrees, the water constantly agitated by propellors, so that the bird is completely wetted. From the scalder, the birds pass through the first or "rough picker," which consists of two large cylinders, fitted with long, whiplash fingers that are covered with corrugations. The speed of this drum causes these fingers to lengthen out, so they do not strike the bird with their ends, but with some of the side corrugations which pull the feathers loose. The quill-feather puller if used, is next in line. From this machine the bird travels through several finishing machines, which consist of revolving drums filled with broad flat "fingers" of soft pliable rubber. Portions of the bird are dropped between or held against these rollers whose action is "massage" the small feathers and pins from the carcass. A common practice is to have separate neck and leg pickers. The turkey then passes through another machine, similar to the first rough picker, where, under water, these long fingers pull out any remaining pinfeathers. At this point helpers walk along the moving line carefully going over each bird to remove with a special dull-edged knife any pins left on by the machines.

With running water the fecal matter and the crop contents are also removed. Sometimes the crop is emptied by a small tube inserted in the crop through the mouth, the water turned on, and the feed washed out. At other times the skin at the back of the neck is opened and the crop and its contents removed. If the bird is to be made ready to cook, the decropping is performed at the time the bird is eviscerated. After cropping and venting, the bird passes through a washer which consists of two water-flushed revolving drums, with long soft rubber fingers that scrub away any dirt and loose feathers on the carcass.

TENDON PULLING, DRAWING, AND BONING

Starting with the dressed bird, the hair feathers first should be singed, by holding the bird over a hot flame. An alcohol lamp is excellent for singeing, likewise a gas flame or blowtorch. Care should be taken not to burn the skin. If the bird has been cooled, the body will be stiff, and it will be necessary to flex the legs, grasping them firmly, and bending the legs against the body.

Removing the tough tendons from the legs will add to consumer satisfaction. This may be done by breaking the leg about two inches below the hock joint, placing the foot in a bracket that holds it firmly, grasping the thigh, and pulling downward with the hands until the tendons pull out. Commercial tendon pullers do the job more quickly and easily.* Another method of pulling tendons is as follows: Cut skin along side of shank to expose the main tendons, then cut skin in front of hock to expose

*The All-Poultry Tendon Puller Co., (826 Pico Blvd., Santa Monica, Calif.) makes a puller that handles both legs at once.

tendons on front of leg. Tendons then can be removed one by one with a steel skewer or tendon puller.

DRAWING

Slit the skin on the back of the neck from the shoulders to the head. The crop can then be reached. Work it loose with the hands and pull it out, bringing with it the windpipe and esophagus, all of which remain attached to the head. Now disjoint the head at its junction with the neck and remove it with the attached crop, windpipe, and esophagus. Cut through the neck muscles at the shoulders, and the neck may be twisted off easily. There are three methods of removing the entrails. The method in most common use is to make an incision crosswise of the abdomen about midway between the vent and the point of the breastbone. Some prefer to make the incision about one and one-half inches to two inches from the vent. The hand is then thrust through this incision and the entrails loosened and carefully pulled out, being especially careful not to break the gall-blader. Cut the skin entirely around and slightly away from the vent and remove it with the entrails.

The second method of removing the entrails is to cut around the vent first and remove the entrails through this opening without making a second incision.

The third method is to use a longitudinal incision from end of breastbone to vent. Carefully remove the gall bladder from the liver. Open the gizzard and remove the horny lining and contents. Separate the heart from the other organs. Wipe the carcass with a clean cloth. If it is to be prepared ready for the oven it should be washed. Place the neck, liver, gizzard, and the heart inside of the carcass. They may be wrapped in waxed paper. The skin of the neck should be folded under.

If the first method of drawing is used, there will be a strip of skin and flesh between the point where the first incision was made and where the vent was removed. The legs then can be thrust through this fold of skin, which helps to hold the desired shape. Clean butcher's twine tied around the legs and tightly to the tail will serve the same purpose with birds drawn according to the second or third method. Sometimes the legs are allowed to remain a natural position and the incision sewed up.

BONING

The process of boning a turkey is comparatively simple, but rather difficult to describe.[1] First, remove a part of the wings at the second joint. Slit the skin from neck to vent along a center line over the breast bone. Remove the skin in one piece by pulling from the flesh. Loosen with the knife where necessary. After the bird has been skinned, remove the legs by disjointing at the "hip" joint. Also remove the wings. Remove the flesh from the carcass. Start at the end of the breast and pull and work flesh from the bone. Remove the remainder of the flesh from the carcass.

U. S. GRADING SYSTEM

The U. S. Grading system for dressed turkeys is valuable from the marketing standpoint and also provides a terminology and a method by means of which turkey growers, breeders, investigators, and marketers may conduct their work and communicate with each other in respect to market quality in turkeys. These grades are uniform regardless of variations in conditions surrounding the production and marketing operations or due to seasonal changes. Briefly, U. S. Grades* are now three, formerly four, given in descending order of quality:

U. S. "AA" formerly Special (now discontinued)
U. S. Grade A formerly Prime
U. S. Grade B formerly Choice
U. S. Grade C formerly Commercial

---

[1]An illustrated leaflet on boning and preparation of a "turkey roll" is available for 6c in stamps from the Turkey World, Mt. Morris, Illinois.

*Described in "Regulations governing the Grading and Inspection of Poultry and Domestic Rabbits and Edible Products Thereof," a printed leaflet of the Poultry Division, Agr. Mktg. Service, U. S. Dept. of Agr., (Wash. 25, D. C.) giving the latest requirements of the various grades; and standards for grading live poultry of all kinds including turkeys, effective May 15, 1953.

Table 6.6—Summary of Specifications for Standards of Quality for Individual Carcasses of Dressed and Ready-to-Cook Turkeys*

(Minimum requirements and maximum defects permitted)

Factor	A Quality		B Quality		C Quality	
Conformation:						
Breastbone	Normal Slight curve, $\frac{1}{4}$" dent		Practically normal Dented, curved, slightly crooked		Abnormal Seriously crooked (If fairly well fleshed)	
Back	Normal (except slight curve)		Moderately crooked		Seriously crooked (If fairly well fleshed)	
Legs and Wings	Normal		Moderately misshapen		Misshapen	
Fleshing:	Well developed moderately long and and broad breast		Fairly well fleshed on breast and legs		Poorly fleshed	
Breastbone	Not prominent		Not prominent		May be prominent in young toms	
Pouchiness	Slight in young toms		Moderate in young toms			
Fat Covering:	Well covered—some fat under skin over entire carcass. Young toms only moderate covering In fryer-roasters it may be somewhat lacking.		Sufficient fat on breast and legs to prevent a distinct appearance of flesh through skin		Lacking in fat covering over all parts of carcass	
Pinfeathers:	Breast and Legs	Elsewhere	Breast and Legs	Elsewhere	Breast and Legs	Elsewhere
Dressed: Pins and hair	Pract. free	Pract. free	Relatively few	Slight scattering	Numerous	Numerous
Ready-to-cook: Non-protruding pins and "hair" feathers	Pract. free	Pract. free	Few scattered	Few scattered	Scattering	Scattering
Protruding pins and hair	Free	Free	Free	Free	Free	Free
Cuts and Tears:[1]	Free	3" long	3" long	6" long	No limit	
Disjointed bones	1		2		No limit	
Broken Bones	None		1 Non-protruding		No limit	
Discolorations:[2]						
Flesh bruises	None	1" circle	1" circle	3" circle	No limit[3]	
Skin bruises	$\frac{3}{4}$" circle	$1\frac{1}{2}$" circle	$1\frac{1}{2}$" circle	3" circle	No limit[3]	
All discolorations	2" circle	3" circle	3" circle	6" circle	No limit[3]	
Freezer Burn	Few small ($\frac{1}{8}$" diameter) pockmarks		Moderate-dried areas not in excess of $\frac{1}{2}$" in diameter		Numerous pockmarks and large dried areas	
Reddening of feather follicles on wings and on neck near head	Slight		Moderate		No limit but must not cause unfitness for food	
Wing tips	May have been removed		May have been removed		May have been removed	

The quality designations specified herein are not applicable to birds possessing any of the following conditions: dirty or bloody head or carcass, dirty feet or vent, fan feathers or neck feathers or garter feathers, or feed in the crop.

[1]Total aggregate length of all cuts and tears including incision for removal of the crop or its contents.

[2]Maximum diameter of aggregate areas of all flesh bruises, skin bruises, and discolorations.

[3]No limit on size and number of areas of discoloration and flesh bruises if such areas do not render parts of carcass unfit for food.

*The U. S. Dept. of Agriculture, Poultry Division, Agr. Mktg. Service. These standards became effective May 15, 1953.

In practice, very few turkeys ever were marketed as AA, even though they may have possessed that quality. These top-quality birds were and are included with Grade A and marketed as such, except in special situations where there is a demand for birds equal to the U. S. Grade AA. Such turkeys are in demand by a large group of consumers which is catered to by many individual producers. The AA grade also is useful in grading turkeys for experimental and selection purposes.

The turkey formerly known as Grade AA or U. S. Special dressed turkey may be described as (1) fully fleshed, the breast broad and full-meated, not pointed across the top; (2) fully covered with fat all over the body; (3) free of noticeable pinfeathers; (4) free of roost dents over 1/8 inch deep; (5) free of noticeable breast calluses, breast blisters, and skin pigmentation (blue-black); (6) possessing no crooked or curved keel, crooked leg, crooked back, seriously crooked feet or toes, or other noticeable deformity; and (7) breast free from all dressing defects, and balance of carcass practically free, that is, must be fully bled, extra well dressed, possess no flesh bruises, no open tears, no sewn skin, no broken or disjointed wing or leg, and may have only very slight skin abrasions, and very slight skin bruises, none of which shall be on the breast. Flesh bruises do not move with the skin, skin bruises do. In addition, AA dressed turkeys must be slack-scalded or dry-picked, dry-packed, show no marks of freezing, and have no flabbiness. They are "commercially perfect" specimens. Eviscerated turkeys of AA quality have the same requirements but they may be medium scalded if they are wrapped air tight before the skin dries.

Grade specifications as described in the 1953 Regulations (mentioned above) and summarized in Table 6.6 are quoted in the following paragraphs. (The paragraph numbers are those of the U. S. Register.)

§ 70.104. United States standards for quality of dressed poultry and ready-to-cook poultry—(a) General. (1) The United States standards for quality contained in this section are applicable to individual carcasses of dressed poultry and ready-to-cook poultry of the kinds and classes set forth in § 70.101.

(2) Carcasses found to be unsound, unwholesome, or unfit for food shall not be included in any of the quality designations specified in this section. If the carcass is dressed poultry, determination of unsoundness or unwholesomeness will be based on external characteristics only.

(3) The quality designations specified in this section may not be made applicable to dressed poultry that is not free from the following conditions: Dirty head; bloody head; dirty carcass; bloody carcass; dirty vent; dirty feet; fan feathers on the wing tips; garter feathers around the hock joints; neck feathers; and, if the crop is not removed, feed in the crop.

(4) The A Quality designation may not be made applicable to any poultry carcass if the poultry was wet picked in such a manner that the skin has been damaged by extended immersion in, or high temperature of, the water which resulted in immediate discoloration or may result in later objectionable discoloration.

(5) The following factors are considered in ascertaining, pursuant to this section, the quality of an individual carcass: (i) Conformation; (ii) fleshing; (iii) fat covering; (iv) the degree of freedom from pinfeathers and vestigial feathers (i.e., hair or down, as the case may be); (v) the degree of freedom from tears, cuts (including, but not being limited to, any cut for the removal of the crop), disjointed bones, and broken bones; (vi) the degree of freedom from discolorations of the skin and of the flesh and of blemishes and bruises of the skin and flesh; and (vii) the degree of freedom from freezer burn.

(6) In interpreting the respective requirements specified in this section for A quality, B quality, and C quality, the intensity, aggregate area involved and locations of (i) discolorations (whether or not caused by dressing operations), (ii) bruises, (iii) pinfeathers, and (iv) freezer burn, as such defects individually, or in combination, detract from the general appearance of the carcass, will be considered in determining the particular quality of an individual carcass.

(b) **Standards of quality—(1) A Quality.** To be of A Quality the carcass:

(i) Is of normal physical conformation except that it may have a slightly curved breastbone or other slight abnormality in the shape of the breastbone which does not interfere with the normal distribution of the flesh. The carcass may also have a very slightly curved back. There may be a dent in the breastbone which does not exceed ⅛ inch in depth except that for turkeys the depth does not exceed ¼ inch.

(ii) Has a well-developed, moderately broad and long breast, well-fleshed throughout its entire length, with the flesh carrying sufficiently well up to the crest of the breastbone so that the breastbone is not prominent; and, with respect to young tom turkeys, there may be a slight thickening and slight pouchiness of the skin on the forepart of the breast. The legs are well covered with flesh.

(iii) Has the breast, back, hips, and pin bones well covered with fat except that chicken broilers or fryers, and young tom turkeys may have only a moderate amount of fat covering these parts, a turkey fryer or roaster may be somewhat lacking in fat covering, and a hen, stewing chicken, or fowl does not have excessive abdominal fat.

(iv) Is practically free from pinfeathers and vestigial feathers, especially on the breast, if the carcass is dressed poultry. If the carcass is ready-to-cook poultry, it is free from protruding pinfeathers, practically free from nonprotruding pinfeathers and vestigial feathers, especially on the breast.

(v) Is free from skin tears and cuts on the breast and legs; however, elsewhere on the carcass there may be tears and cuts (exclusive of the cuts usually made to remove the neck and viscera in the production of eviscerated poultry) the aggregate length of which does not exceed 1½ inches except that, with respect to any turkey carcass or goose carcass, such aggregate length does not exceed 3 inches. There are no sewn tears or cuts. The carcass has no disjointed bones or broken bones except that it may have one dispointed bone in either a leg or wing but only if there is no evidence of a related bruise or blood clot; and, if the carcass is of a chicken broiler or fryer, it may have one nonprotruding broken bone in a wing in addition to such disjointed bone but only if there is no evidence of a related bruise or blood clot. The wing tips may have been removed.

(vi) Is free from bruises and discolorations of the flesh on the breast and legs; however, elsewhere on the carcass there may be bruises and discolorations of the flesh showing not more than a slightly reddened color the aggregate area of which does not exceed he area of a circle ½ inch in diameter, except that, with respect to any turkey or goose carcass, such aggregate area does not exceed the area of a circle 1 inch in diameter. The carcass is free from skin bruises, on the breast and legs, the aggregate area of which exceeds the area of a circle ½ inch in diameter, and from skin bruises, elsewhere on the carcass, the aggregate area of which exceeds the area of a circle ¾ inch in diameter. With respect to any turkey or goose carcass, such aggregate area on the breast and legs does not exceed the area of a circle ¾ inch in diameter; and elsewhere on the carcass such aggregate area does not exceed the area of a circle 1½ inches in diameter. Notwithstanding the foregoing, the total aggregate area, on the breast and legs, of all such flesh bruises, skin bruises, and all other discolorations and blemishes of the skin, is not in excess of the area of a circle 1 inch in diameter; and elsewhere on the carcass such total aggregate area is not in excess of the area of a circle 1½ inches in diameter. Furthermore, with respect to any turkey or goose carcass, such total aggregate area on the breast and legs is not in excess of the area of a circle 2 inches in diameter; and elsewhere on the carcass such total aggregate area is not in excess of the area of a circle 3 inches in diameter. The skin may show only slight reddening in the feather follicles on the neck, near the head, and on the wings because of improper bleeding.

(vii) Shows only slight freezer burn, or evidence thereof (i.e., a few pockmarks, or evidence thereof, none of which exceeds the area of a circle ⅛ inch in diameter).

(2) **B Quality.** To be of B Quality the carcass:

(i) Is of at least practically normal physical conformation, except that it may have a dented, curved and slightly crooked breastbone which does not seriously interfere with the normal distribution of flesh. The carcass may also have a moderately crooked back or misshapen legs or misshapen wings.

(ii) Is sufficiently well-fleshed on the breast and legs so as to prevent a thin appearance and a prominent breastbone; however, a young tom turkey may have a pouchy, thick, and somewhat flabby skin on the forepart of the breast.

(iii) Has a sufficient coverage of fat on the breast and legs to prevent a distinct appearance of the flesh through the skin.

(iv) Has not more than a slight scattering of pinfeathers and vestigial feathers over the entire carcass with only relatively few on the breast, if the carcass is dressed poultry. If the carcass is ready-to-cook poultry, it is free from protruding pinfeathers, but may have not more than a few scattered nonprotruding pinfeathers and vestigial feathers.

(v) Is free from tears and cuts, on the breast and legs, the aggregate length of which exceeds 1½ inches; however, elsewhere on the carcass there may be tears and cuts (exclusive of the cuts usually made to remove the neck and viscera in the production of eviscerated poultry), the aggregate length of which does not exceed 3 inches except that, with respect to any turkey or goose carcass, such aggregate lengths do not exceed 3 inches on the breast and legs and 6 inches elsewhere on the carcass. There are no sewn tears or cuts. The carcass may have not more than a total of 2 disjointed bones in either the legs or wings, or both, but only if there is no evidence of a related bruise or blood clot, and, in addition, 1 broken bone in a leg or wing, but only if it is nonprotruding and does not show an excessive related bruise or blood clot. The wing tips may have been removed.

(vi) Is free from bruises and discolorations, of the flesh on the breast and legs, showing not more than a slightly darkened color and which in the aggregate is in excess of the area of a circle ½ inch in diameter; however, elsewhere on the carcass there may be bruises and discolorations of the flesh the aggregate area of which does not exceed the area of a circle 1½ inches in diameter, except that, with respect to any turkey or goose carcass, such aggregate area on the breast and legs does not exceed the area of a circle 1 inch in diameter, and, elsewhere on the carcass, it does not exceed the area of a circle 3 inches in diameter. The carcass is free from skin bruises, on the breast and legs, the aggregate area of which exceeds the area of a circle ¾ inch in diameter, and from skin bruises, elsewhere on the carcass, the aggregate area of which exceeds the area of a circle 1½ inches in diameter. With respect to any turkey or goose carcass, such aggregate area on the breast and legs does not exceed the area of a circle 1½ inches in dimeter, and, elsewhere on the carcass, such aggregate area does not exceed the area of a circle 3 inches in diameter. Notwithstanding the foregoing, the total aggregate area on the breast and legs of all such flesh bruises, skin bruises, and all other discolorations and blemishes of the skin is not in excess of the area of a circle 1½ inches in diameter; and elsewhere on the carcass such total aggregate area is not in excess of the area of a circle 3 inches in diameter. Furthermore, with respect to any turkey or goose carcass, such total aggregate area on the breast and legs is not in excess of the area of a circle 3 inches in diameter; and elsewhere on the carcass such total aggregate area is not in excess of the area of a circle 6 inches in diameter. The skin may show not more than moderate reddening in the feather follicles on the neck, near the head, and on the wings and thighs because of improper bleeding.

(vii) Shows no more than moderate freezer burn, or evidence thereof, on any part of the carcass and no dried area in excess of the area of a circle ½ inch in diameter.

(3) **C Quality.** A carcass that does not meet the requirements of B Quality maybe of C Quality and such carcass may:

(i) Be of abnormal physical conformation (i.e., possess serious abnormal physical conditions, including, but not being limited to, a crooked breastbone) if it is fairly well fleshed.

(ii) Be poorly fleshed and a young tom turkey may have a thick, coarse skin and extended breast that is pouchy or flabby.

(iii) Be lacking in fat covering, over all parts of the carcass.

(iv) Have numerous pinfeathers and vestigial feathers scattered over the entire carcass, if the carcass is dressed poultry; if ready-to-cook poultry, the carcass is free from protruding pinfeathers but may have a few vestigial feathers and may have nonprotruding pinfeathers that do not seriously detract from the appearance of the carcass.

(v) Have torn skin, disjointed bones, and broken bones, but only if there is no evidence of a related severe bruise or blood clot. There are no sewn tears or cuts. Wing tips may have been removed.

(vi) Have numerous and large discolored areas or blemishes of the skin which may be accompanied by some reddening and darkening of the flesh beneath, if such discolored areas and blemishes do not render any part of the carcass unfit for food.

(vii) Show more than moderate freezer burn or evidence thereof (including, but not being limited to, numerous pockmarks or large dried areas) on any part of the carcass.

Undergrade or below-grade dressed turkeys are birds not qualifying for grade C and include specimens (1) birds with feed in the crop; (2) with feathers anywhere, except filoplumes and those under the headwrap; (3) with dirty feet, dirty vents, and dirty or bloody heads and bodies; (4) birds having an internal temperature above 36° F. unless it is certain they will be properly cooled before packing; (5) birds that are emaciated, show evidence of disease, have more than 25% of the carcass inedible, or have a combination of defects or any special condition such as spoilage that render them unwholesome or unfit for human food. Spoiled turkeys first darken, then turn greenish on the area around vent and over ribs and the crop. Such greenish birds are called greenstruck.

The grading and classifying of dressed turkeys is best learned through attendance at grading schools or personal instruction by competent graders with access to dressed carcasses for grading practice. However, where such instructions cannot be obtained it is suggested that the foregoing specifications be studied along with the following textual material and representative dressed carcasses.

To determine the degree of finish, that is, the quantity and distribution of fat in the skin, is very important and is probably the most difficult phase of dressed-turkey grading. To accurately determine finish, the dressed bird should be thoroughly chilled. To grade *AA in finish* the turkey should have a thick coating of fat so that it does not look thin-skinned anywhere except possibly at the junction of wing and body and on the underside of the first and second wing-segments, not the web which is well fatted in grade AA turkeys. These are the last places to take on finish and will be covered with a noticeably thick layer of fat only in very fat birds which will grade High AA (subgrade 1) in the grading plan to be described in the following paragraphs. In other words, the AA turkey will be uniformly finished so it appears uniformly white, or yellowish white, or yellow, depending on the quantity of yellow in the skin and fat. The *grade A* turkey likewise will not look thin-skinned except on the shoulder joint and the underside of the first two wing segments (not the webs which are fairly well fatted in grade A). However, the drumsticks and inner surfaces of the thighs will appear to have less finish than the breast, and the space between the feather tracts of the breast will be noticeably less thickly finished than the feather tracts. In other words, the A turkey is "well covered" with fat in all sections (except the ventral wing surfaces and shoulder joint) but the difference in degree of finish between the parts mentioned is plainly evident, which is not true of Special. The *grade B* turkey shows decided contrasts in finish. The skin of the drumsticks will carry only a thin covering of fat. The same may be said for the skin of the breast between the feather tracts and of the inner surface of the upper thigh. However, there will be a light coating of fat in these places and the feather tracts of the breast, thighs, back, and neck and will have a moderately heavy coating of fat.

Almost any well-grown turkey that is not emaciated through disease or starvation will have at least light fat deposits in the skin of the densely feathered areas called pterylae, or feather tracts, of the neck, back, thighs, and breast even though there may be none at all on the drumsticks, on the inner surface of the thighs, and on the skin between the densely feathered areas. Birds that have these feather tracts connected even with a light covering of fat in the skin can be graded B; while C birds will have the skin of the drumsticks and of most of the sparsely feathered areas of the body practically devoid of fat deposits which will be skimpy to absent in the feather tracts.

The grade of dressed turkey commonly recognized on the various markets as No. 1 corresponds to the U. S. grade A and includes also all birds worthy of AA. The grade called medium, when such a grade is used, corresponds to the U. S. grade B and the No. 2 turkey is U. S. grade C or Commercial. In markets where no "medium" grade is recognized, birds qualifying for U. S. grade B are classified as No. 1 or reduced to No. 2, depending upon supply on that market and to some extent upon market conditions generally. The general use of a grading system free from the influence of individual market preferences, and from the tone of the market, and uniform on all markets would be most helpful to consumers, distributors and producers alike. The government standards and grades for dressed turkeys are designed to fill this need.

GRADES FOR DRESSED TURKEYS

The 3 or 4 grades of the Federal grading system are fully adequate for market operations. In fact, it would not be practicable commercially to separate turkeys into more than four grades and it would be still less practicable to try to market them in more than four categories. However, to classify turkeys in an adequate manner for showing, judging, the selection of breeding stock and experimental purposes more than these 4 grades are desired and to this end a numerical grading plan for dressed turkeys is offered by the authors in Table 6.7.

Table 6.7—Suggested Numerical Grades for Dressed Turkeys* with Emphasis on Finish.

Numer-ical grade	Relationship to nearest U.S. Grade	Description
1	High Special (AA+)	The fully finished, perfectly formed, and perfectly dressed turkey. It is extra well finished and fleshed, smooth and symmetrical, free from pinfeathers and blemishes, and exceeding in quality the former specifications for AA (page 380). The fat coating appears very smooth and uniform except for the underside of the fleshy and bony portions of the first and second wing segments (not including the web) which may have only a rather thin (No. 6) coating of fat. The drumsticks and inner surfaces of the thighs will have a heavy coating of fat through which the dark color of the musculature does not show.
2	Average Special (AA)	Fully conforming to the former U. S. specifications for AA. The undersides of the fleshy and bony portions of the first two wing-joints may possess only a No. 7 coating of fat and the drumsticks and inner surfaces of the thighs, though fully covered with fat, may appear to be slightly less fully covered than the outer sides of the legs, the breast, sides, and back, which are thickly and uniformly covered.
3	High Prime (A+)	Exceeding in quality the U. S. specifications for A, but not quite reaching the former AA. The fat coating is almost but not quite uniform in appearance over the body. The shoulder joint and the underside of the fleshy and bony portions of the wings may have only a No. 8 coating of fat and the drumsticks and sparsely feathered areas (apteria) of the breast will show slightly less finish than the heavily feathered areas (pterylae) so that a fairly discernible contrast appears. Exceptions to this are the apteria of the sides over the ribs which are always comparatively heavily fatted.
4	Average Prime (A)	Fully conforming to U. S. specifications for A. The Prime or "A" dressed turkey is well covered with fat everywhere except on undersides of the wing segments and on shoulder joints which may have little or no fat in the skin. There will be a fairly distinct contrast in finish between the sparsely feathered portions and the heavily feathered portions of the breast; and the breast apteria and the drumsticks which though well covered appear slightly dark in contrast to the pterylae.

*It should be kept in mind that the numerical subgrades given are not part of the official U. S. Grading System but are based on it.

Table 6.7—(Continued)

Numerical grade	Relationship to nearest U.S. Grade	Description
5	Low Prime (A−)	Barely conforming to U. S. specifications for "A." The dressed Low Prime turkey shows a distinct contrast in finish between the heavily fatted pterylae and the moderately well covered apteria of the breast but there is sufficient fat in the apteria to permit grading above B+. The drumsticks and inner surfaces of the thighs, though actually well covered with fat, appear noticeably dark in contrast to the pterylae.
6	High Choice (B+)	Exceeding in quality the U. S. specifications for B. but not quite reaching A. In the dressed B+ turkey the contrast between the finish in the apteria and the pterylae of the breast is very distinct but the latter are heavily covered with fat as in Prime stock although the apteria are sufficiently lacking in thickness of finish (only fairly well covered) to fail to grade No. 5. The drumsticks and inner surfaces of the thighs are faintly bluish in appearance though they are actually fairly well covered with fat.
7	Average Choice (B)	Fully conforming to U. S. specifications for B. The dressed Choice or "B" turkey shows a very distinct contrast in finish between the lightly covered apteria and the well fatted pterylae. The drumsticks appear somewhat bluish though they actually are fairly well covered with fat.
8	Low Choice (B−)	Barely conforming to U. S. specifications for B. The dressed B− turkey is a bird of very definite contrasts in finish between apteria and pterylae. The apteria have only a very light covering of fat but they are definitely covered. The well-fatted pterylae are not so heavily covered as in Grade No. 7 and may be no better covered than in many "C" specimens, but the apteria are definitely covered with fat even though it may be quite thin. The drumsticks appear decidedly bluish although actually possessing a light covering of fat.
9	High Commercial (C+)	Subgrades for Commercial turkeys are needed only for grading turkey fryers or light roasting turkeys. In the dressed turkeys of Grade C+ the breast apteria will have a very light, barely discernible fat coating. The feather tracts will have a fairly prominent deposit of fat but these deposits are connected only by barely discernible fat deposits in the intervening unfeathered areas. The legs always appear rather dark.

Table 6.7—(Continued)

Numerical grade	Relationship to nearest U.S. Grade	Description
10	Average Commercial (C)	In dressed turkeys of average Grade C, the breast apteria appear to have no fat covering but the feather tracts have a fair but rather narrow deposit of fat. The legs appear dark with no noticeable finish.
11	Low Commercial (C –)	Turkeys of this grade appear to have no finish anywhere except in the feather tracts which may have a thin, narrow, deposit of fat. Legs appear very dark with no finish apparent.

A summary of the U. S. grades and authors' subgrades as applied to live and dressed turkeys and the factors of condition and conformation that determine the qualifications for them are given in Table 6.8.

U. S. CLASSES FOR DRESSED TURKEYS

U. S. classes for dressed turkeys and ready-to-cook turkeys are as follows according to the amended 1950 Standards (AMS-USDA, 1950.)

(1) Fryer. A fryer-roaster is a turkey (usually under 16 weeks of age), of either sex, that is tender-meated with soft, pliable, smooth-textured skin, and flexible breastbone cartilage.

(2) Young hen turkey. A young hen turkey is a young female turkey (usually under 8 months of age) that is tender-meated with soft, pliable, smooth-textured skin, and breastbone cartilage that is somewhat less flexible than in a turkey fryer.

(3) Young tom turkey. A young tom turkey is a young male turkey (usually under 8 months of age) that is tender-meated with soft, pliable, smooth-textured skin and breastbone cartilage that is somewhat less flexible than in a turkey fryer.

(4) Hen turkey. A hen turkey is a fully matured female turkey (usually over 10 months of age) that is less tender

Table 6.8—Requirements for the Grades and Subgrades of Live, Dressed, and Ready-to-Cook Turkeys

U.S. Grade	Subgrade No.	Subgrade Name	Fat grade No.	Feather grade No.	Breast grade No.	Leg grade No.	Deformities permitted[1]	Roost dents permitted	Picking defects permitted[1]	Type of dressing
AA	1	High Special	1	1 or 2	4 or better	4 or better	None	None noticeable	None noticeable	Dry-pick or slack-scald[3]
	2[2]	Special	2	3 or better	5 or better	5 or better	None on breast, very slight ones elsewhere	None over ⅛ inch deep	None on breast, very slight ones elsewhere	Dry-pick or slack-scald[3]
A	3	High Prime	3	3 or better	5 or better	5 or better	None on breast, slight ones elsewhere	None over ¼ inch deep	None on breast slight ones elsewhere	Dry-pick or slack-scald[3]
	4	Prime	4	4 or better	6 or better	5 or better	Very slight ones on breast, slight ones elsewhere	None over ¼ inch deep	Very slight ones on breast, slight ones elsewhere	Dry-pick or slack-scald[3]
	5[2]	Low Prime	5	4 or better	7 or better	6 or better	Same as for Grade 4	None over ¼ inch deep	Same as for Grade 4	Dry-pick or slack-scald[3]
B	6	High Choice	6	5 or better	7 or better	6 or better	Slight ones on breast, moderate ones elsewhere	May be over ¼ inch deep[4]	Slight ones on breast, moderate ones elsewhere	Any method Appearance must be good
	7	Choice	7	6 or better	8 or better	May be grade 7	Moderate ones anywhere	May be over ¼ inch deep[4]	Moderate ones anywhere	Any method Appearance must be reasonably good
	8[2]	Low Choice	8	6 or better	8 or better	May be grade 7	Same as for grade 7	May be over ¼ inch deep[4]	Same as for grade 7	Any method. Appearance must be reasonably good
C[5]	9	Commercial	9, 10 or 11	None specified	9 or better	None specified	Serious (major) ones if bird is of a higher grade otherwise	May be over ¼ inch deep[4]	Serious (major) ones if bird is of a higher grade otherwise	Any method Appearance may be poor

[1]See U.S. Grade specifications for details. Deformities are due to inheritance, injury, or conditions existing during the growth of the turkey. They include disfigurements of skin, flesh, and bones. Defects are due to poor workmanship during preparation for market.

[2]The requirements for these subgrades 2, 5, and 8 are the minimum requirements for the regular U.S. Grades of AA, A and B respectively.

[3]Medium scald may be used on ready-to-cook turkeys if they are wrapped air-tight before skin becomes discolored.

[4]But must not interfere with normal distribution of the flesh.

[5]For description of the subgrades 10 and 11, see preceding outline.

meated than a young hen turkey, has a hardened breast bone, and may have coarse-textured skin and patchy areas of surface fat. All breeder hens are in this class.

(5) Tom turkey. A tom turkey is a mature male turkey (usually over 10 months of age) with coarse skin, toughened flesh, and hardened breastbone. All breeder toms are in this class.

Suggested weight specifications, for dressed and ready-to-cook turkeys are contained in Table 6.9 following:

Table 6.9—Suggested Weight Specifications for Dressed and Ready-to-Cook Turkeys

Class	Weight range per carcass	
	Minimum	Maximum
Dressed fryer-roasters	None	6 pounds
	Over 6 pounds	8 pounds
	Over 8 pounds	10 pounds
Dressed young hen turkeys or young tom turkeys	None	6 pounds
	Over 6 pounds	8 pounds
	Over 8 pounds	10 pounds
	Over 10 pounds	12 pounds
	Over 12 pounds	14 pounds
	Over 14 pounds	16 pounds
	Over 16 pounds	18 pounds
	Over 18 pounds	20 pounds
	Over 20 pounds	22 pounds
	Over 22 pounds	24 pounds
	Over 24 pounds	None
Dressed yearling or old hen turkeys and mature or old tom turkeys	None	10 pounds
	Over 10 pounds	15 pounds
	Over 15 pounds	20 pounds
	Over 20 pounds	None
Ready-to-cook turkeys, all classes	None	4 pounds
	Over 4 pounds	6 pounds
	Over 6 pounds	8 pounds
	Over 8 pounds	10 pounds
	Over 10 pounds	12 pounds
	Over 12 pounds	14 pounds
	Over 14 pounds	16 pounds
	Over 16 pounds	18 pounds
	Over 18 pounds	20 pounds
	Over 20 pounds	None

Turkeys may be classified as to method of chilling as follows:

Fresh chilled—not hard chilled or hard frozen but chilled to 40° F. internal temperature.

Fresh hard chilled—frozen less than 60 days.

Storage—stored over 60 days at a low temperature or showing evidence of deterioration from freezing regardless of time held. Maximum freezing temperatures are −10° F. with circulated air or at −20 F. with still air; desirable freezing temperatures are −20 and −40 respectively. Birds under 8 pounds should freeze solid in 30 to 40 hours, those over 8 pounds in 48 to 60 hours. Frozen dressed poultry should be stored at a constantly maintained maximum of 0° F., preferably lower, −10 to −20.

The effects of ordinary cold storage on turkeys are as follows: (1) loss of body bloom; (2) slight darkening of the flesh; (3) drying out of skin; (4) dry, scaly spots called freezer burn. Modern methods of quick freezing and storage, including the use of air-tight wrappings have markedly improved the quality of storage turkeys and eliminated to a large extent the effects of ordinary cold storage.

There has been a tendency in recent years to store a greater volume of turkey, much of which consists of large toms. These have a year-round wholesale outlet principally to hotels and restaurants; and also may have a retail outlet through quarters, halves, and parts. In spite of increased storage volume only about one bird in five goes into storage, the other four being consumed shortly after slaughter.

Other classifications for dressed and drawn turkeys are made according to:

A. Plucking:

Scalded—immersed for picking in water over 130° up to 180° F. The "medium" scald is done at 138 to 140° F.

Semi-scalded or slack-scalded—Immersed for picking in water 123 to 130° F.

Dry-plucked—Not immersed in water during the picking process

"Wet picked" means, with respect to poultry from which the feathers are removed, that the birds were immersed in or sprayed with water for removal of the feathers.

B. Dressing:

"Dressed poultry" means poultry which has been slaughtered for human food with head, feet, and viscera intact and from which the blood and feathers have been removed.

"Ready-to-cook poultry" means any eviscerated poultry or any cut-up or disjoined portion thereof. It may be government graded and grade-labeled, in form of a black shield with white letters, without being inspected for wholesomeness by a Federal veterinarian. Ready-to-cook poultry also may be inspected for wholesomeness without being graded and may be labeled with a circle containing approximate wording. Ready-to-cook poultry which is government-graded and government-inspected may be labeled with a shield within a circle.

"Giblets" means the liver from which the bile sac has been removed, the heart from which the pericardial sac has been removed, and the gizzard from which the lining and contents have been removed: Provided, that each such organ has been properly trimmed and washed.

C. Packing:

Dry—without contact with ice or water.

Iced—in contact with ice.

MISCELLANEOUS GRADING AND PACKING SUGGESTIONS AND RULES

Peaked breast should be considered the same as "slightly crooked." It causes the bird to be graded below A. This condition consists of a severely depressed rear half of the keel and is usually accompanied by more or less of a curvature of the keel.

Crooked toes and feet are of no importance if slight; if serious they bar the specimen from AA, but not A.

Crooked legs, if severe, bar the specimen from A; if slight, from AA.

Dressed birds with *improper headwraps* may be graded only when the grader is certain this condition will be corrected before the birds are packed. Headwraps should be made of good quality, sixty-pound, water-finish fiber or brown waxed paper.

Scaly legs reduce the specimen one grade.

Internal feather growth, when severe, reduces the specimen to U. S. Grade C. Ingrown feathers often can be removed with a crochet needle without damage to the bird's skin.

Swollen joints, if bad, bar the specimen from A; if slight, from AA.

Pigmented skin, or *blue-back,* is not permitted in AA; only on tail in A, and no further forward than the hips on B. Turkeys showing this condition should be packed in separate containers and properly labeled.

Roached back if slight is not important; if serious it will bar the specimen from AA but not from A.

Decidedly *crooked backs* and decidedly *crooked keels* reduce the specimens possessing them to C.

Breast calluses and blisters, if inconspicuous, are of no importance; if slightly conspicuous they bar the specimen from AA; if serious they cause a reduction in grade to B or C and may even be serious enough to bar the bird from any U. S. grade as when the bone is affected and/or pus or a large quantity of fluid is contained.

Line birds or *liners* are those whose grade is in doubt or borderline. They should always be placed in the lower grade, rather than lower the average of the higher grade.

Individual turkeys may be marked with a tag showing the proper U. S. grade only when they have been graded by a person authorized by the U. S. Department of

Agriculture and when authorized tags and methods of attachment are used. Grade tags must state age of the bird and they must be applied by the grader or by persons directly under his supervision.

Boxes of graded turkeys may be stamped with the proper U. S. grade even though individual birds are not so marked. One end of each box should have the U. S. grade stenciled on it in ¾-inch letters and in addition the net weight, number of birds, and class, should be stamped or stenciled. Private bands and additional descriptive material may also be added.

Packing. All dressed turkeys should be firmly packed on their sides in rump-lock-style in single-layer, clean, strong, odorless boxes of uniform size, color, and markings. All turkeys in each box must be of the same class, grade, skin color, and weight class (exception is made for occasional mixed boxes). Boxes should be lined with good-quality, thirty-pound waxed or parchment paper.

Regrading. Where previously graded turkeys are regraded and show effects of holding, these characteristics shall be considered as factors of condition and not of grade and shall be so noted on the grading certificate.

Grading Memorandum. Licensed dealers should make out a detailed grading certificate for each carlot or less-than-carlot shipment.

Labeling. Purple may be used on special labels to designate U. S. Special, or AA grade; blue for U. S. Prime, or A; and green for U. S. Choice, or B. There is no color or special label used for U. S. Commercial, or C.

AGE AND SEX DETERMINATION IN DRESSED TURKEYS

Young turkeys have fine-grained, soft, tender flesh that is easily indented with the finger. The rear end of the keel bone is more or less flexible and the skin adheres to the flesh and is soft and velvety. Young birds that are well matured will have soft, tender flesh but the end of the keel may be only slightly flexible.

Old turkeys have a rather coarse, sometimes scaly skin under which the flesh is firm to the touch and in very old poorly fed stock may be hard and rubbery. The rear end of the keel bone is fully ossified so that it is completely inflexible. The skin tends to be rough, dry, and somewhat scaly, and in addition decidedly flabby in males. The feather follicles may appear deeper and farther apart. The shanks tend to be rougher and coarser, and except in white-shanked varieties, lighter in color. In white-shanked birds the shank tends to become deep pink with age. Prominent spur and headgear development is characteristic of fully grown males. The bursa of Fabricius, the gland-like, thick-walled appendage of the upper side of the cloaca is reported to be present in young birds (under about 8 months) but absent in older ones. It is difficult to locate unless the bird can be opened up, but careful probing with a blunted nail in the slightly everted vent may disclose its status.

Sex usually can be determined at the vent, if not apparent otherwise. In the dressed bird the "double-barrelled" male organs can be seen in the lower wall of the cloaca if the vent is manipulated so as to partly evert the cloacal wall. The female organs are so inconspicious that they usually are not in evidence. When sex is in doubt the bird should be put in the class to which its body shape best conforms. For further data on sex determination see Chapter **13.**

CHAPTER 7

MARKETING DATA AND PRACTICES

OUTLINE

Of greatest interest to producer, processor, and consumer alike are detailed data on processing losses and edible percentages of the various classes of turkeys. It also is of interest to compare turkey processing losses and meat yields with those of chickens. In an attempt to supply this information a number of tables have been prepared from various publications and are presented along with comments and explanatory remarks.

SHRINKAGES DURING PROCESSING

The data in Tables 7.1 and 7.2 were obtained from standardbred Bronze turkeys that were raised in Montana on limited range with full feeding. Most of the birds were slaughtered in November and December between the ages of 26 and 32 weeks after an overnight fasting period. They were dry-picked, air-chilled, and graded according to fat and fleshing only, no attention being paid (for the purposes of the experiment) to picking defects, blemishes, or pinfeathers.

Table 7-1—Weight Loss of Adult Turkeys During Slaughtering, Dry-Picking, and Overnight Chilling

(1929-1930)

Classification	Birds	Mean live weight at time of slaughter (after fasting)	Weight loss
	No.	Pounds	Percent
Yearling and two-year-old females slaughtered in June (Prime and Choice grades A or B respectively)...........	48	13.20	6.70
Yearlings females slaughtered in Nov.-Dec., 1930 (Prime and Special grades)...........	81	13.35	9.80
Yearling males slaughtered in Nov.-Dec., 1930 (Prime and Special grades)...........	16	28.71	8.82

The weight loss of 271 standardbred Bronze turkeys of both sexes during overnight fasting (not shown in the tables) averaged 3.33 per cent of the live weight before fasting. There were no significant differences due to age or sex.

The weight loss of 65 young standardbred Bronze turkeys of Grade A and AA quality during slaughtering, dry picking, and overnight air-chilling (the three comprising dressing shrinkage) averaged 9.82 per cent of the live weight after fasting. Males (32) averaged 9.91 per cent, females (33) 9.73 per cent; a non-significant difference. Large birds consistently lost smaller percentages than small birds and very well-fattened (U. S. Grade AA) birds lost smaller percentages than thin birds and those having ordinary good to fair (U. S. "A" to "B") finish. There were no significant differences due to sex or to shape of the breastbone but the hens lost slightly less than the toms during dressing and lost more during drawing while total loss from live to drawn was slightly less for the toms.

Table 7.2—Weight Loss of Young Standardbred Bronze Turkeys During Slaughtering, Dry-Picking, Overnight Chilling, and Full-Drawing [1,4]

Classification	Mean live weight before picking[3]	Mean dressed weight after chilling	Percentage wt. loss due to dry picking and chilling (L)[2]	Percentage wt. loss due to full drawing (L)[2]	Percentage wt. loss due to full drawing (D)[2]	Percentage wt. loss due to dry-picking, chilling, and drawing (L)[2]
Special (AA) and Prime (A) grades	lbs.	lbs.	per cent	per cent	per cent	per cent
Toms under 20 lbs.[3] with straight or only slightly deformed breastbones	15.90	14.28	10.19	13.24	14.75	23.43
Toms 20 pounds and over with straight or only slightly deformed breastbones	21.36	19.32	9.55	12.79	14.15	22.36
Toms under 20 pounds with severely crooked breastbones	17.05	15.35	9.97	13.33	14.81	23.32
Toms 20 pounds and over with severely crooked breastbones	20.15	18.15	9.93	12.98	14.41	22.90
All AA and A grade toms, average of the four classes	18.62	16.78	9.91	13.09	14.53	23.00
Hens under 11 pounds[3] with straight or only slightly deformed breastbones	9.82	8.86	9.79	15.14	16.78	24.93
Hens 11 pounds and over with straight or only slightly deformed breastbones	13.09	11.91	8.91	12.57	13.81	21.56
Hens 11 pounds and over with severely crooked breastbones	11.90	10.65	10.50	13.03	14.55	23.53
All AA and A grade hens	11.60	10.47	9.73	13.58	15.05	23.34
All AA and A grade hens and toms—average of both sexes	15.11	13.63	9.82	13.34	14.79	23.17
Miscellaneous						
Choice (B) and (C) grade toms with straight or only slightly deformed breastbones	13.38	12.06	9.87	15.47	17.16	25.34
B and C grade hens with severely crooked breastbones	7.93	7.10	10.41	15.71	17.54	26.12
Very large AA toms with straight or only slightly deformed breastbones	33.28	30.70	7.74	11.07	12.00	18.81
Large AA toms with straight or only slightly deformed breastbones	27.08	24.86	8.19	13.25	14.43	21.44

[1]The full-drawn carcass consisted of the body, wings, legs to the hock joint, cleaned giblets (heart, liver, and gizzard), the edible abdominal fat, and the neck up to its junction with the skull. The preen gland, entrails except giblets, the windpipe, crop, esophagus, beard, shanks, feet, head, and caruncular skin of neck comprised the weight loss due to drawing.

[2](D) indicates percentage based on dressed weight after chilling; (L) indicates percentage based on live weight after overnight fasting.

[3]Live weight after fasting.

[4]From "Dressing shrinkages of Bronze turkeys" by S. J. Marsden, Poultry Science 15:112-117, 1936.

The weight loss due to overnight air-chilling (not shown in the tables) was very small, averaging 0.23 per cent of the fresh-dressed weight of 461 young Bronze turkeys, with no sex differences.

Weight loss due to full-drawing (the cleaned giblets, complete neck, and edible loose abdominal fat being included with the edible portion), averaged 14.53 per cent of the dressed-and-chilled weight of 32 fat young Bronze toms and 15.05 per cent of the dressed-and-chilled weight of the 33 fat young Bronze hens; probably a non-significant difference. Drawing losses tend to be less in broad-

Table 7.3—Dressing and Drawing Percentages of Turkeys of Four Varieties of Turkeys[1]

Variety or strain, age[2], and condition	Sex	Average live weight (fasted)	Dressing shrinkage[3]	Average dressed weight	Drawing shrinkage[4]	Average drawn weight[5]	Total drawing and dressing shrinkage
		Pounds	Per cent of live weight	Pounds	Percent of dressed weight	Pounds	Per cent of live fasted weight
Broad Breasted Bronze: Av. age, 29 wks.	Males	24.00	9.45	21.73	13.00	18.91	21.21
	Females	15.75	8.25	14.45	12.45	12.65	19.68
Av. fat grade No. 5 or A−	Both	19.88	8.85	18.12	12.73	15.81	20.47
Standardbred Bronze: Av. age, 29 wks.	Males	19.80	10.30	17.76	14.06	15.26	22.93
	Females	12.56	9.45	11.37	13.95	9.78	22.13
Av. fat grade No. 4 or "A"	Both	16.18	9.88	14.58	14.00	12.54	22.50
Standardbred White Holland: Av. age 29 wks.	Males	18.50	10.95	16.47	14.16	14.14	23.57
	Females	11.15	11.90	9.82	14.35	8.41	24.57
Av. fat grade No. 3 or A+	Both	14.83	11.43	13.13	14.26	11.26	24.07
Beltsville Small White: Age 28 weeks	Males	15.25	11.15	13.55	12.86	11.81	22.56
	Females	9.05	11.55	8.00	14.25	6.86	24.20
Av. fat grade No. 2 or "AA"	Both	12.15	11.35	10.77	13.56	9.31	23.37

[1]From Harshaw, H. M., et al. 1943. The composition of turkeys of different varieties and strains. Poultry Science 22:126-136 (The U.S.D.A. Research Center, Beltsville, Md.)

[2]Figures are averages for 10 birds of each sex, years 1940 and 1941 combined.

[3]Blood and feather loss.

[4]Loss due to head, shanks, feet, inedible viscera, and oil gland.

[5]Drawn weight includes the cleaned carcass, giblets, and neck. The original data had most of the neck included with the inedible portion. In this table adjustments were made to include the whole neck (without its skin) with the edible.

Dressing and drawing percentages of Standardbred Bronze in this report, compare favorably to those reported earlier (Table 7.2). In both of the large varieties, two types of Bronze, the overall dressing loss was less and the females dressed out better than the males while in the smaller varieties the overall dressing loss was greater and the males had less dressing loss than the females. In loss of weight from live to drawn, the same trend was evident. Those losses were less, significantly so, in the Broad Breasted Bronze than in the standardbred Bronze. The same relationship, though less striking, existed between the relatively broad-breasted Beltsville White and the White Holland, which was less broad breasted.

breasted young hen and tom turkeys (see Table 7.3), more in breeding birds and considerably more in fryer, or light roaster, turkeys.

The percentage differences in dressing and drawing losses in favor of large birds indicate that the larger specimens of each sex, other factors being equal, have slightly but significantly lower percentages of dressing and drawing losses, and consequently slightly higher percentages of edible meat.

Table 7.3 shows weight losses due to dressing and drawing as reported for four varieties of turkeys by the Beltsville Station.

PARTS AND ORGANS OF TURKEYS

The following tables 7.4 and 7.5, show average weights and measurements of certain parts and organs of fat,

Table 7.4—Weights of Parts and Organs of Standardbred Bronze Turkeys[1]

	Average Weight in grams		Percent of live wt.		Percent of dressed wt.	
	Males	Females	Males	Females	Males	Females
Live bird after over-night fasting	10273.0 (22.65)*	5765.0 (12.71)*	100.00	100.00	110.20	109.70
Dressed and chilled carcass	9324.2 (20.56)	5255.0 (11.59)	90.76	91.15	100.00	100.00
Shanks and feet	299.5	144.6	2.92	2.51	3.21	2.75
Head severed at junction with skull	172.8	95.0	1.68	1.65	1.85	1.81
Esophagus, crop, and proventriculus	81.8	42.1	.80	.73	.88	.80
Trachea	16.4	8.9	.16	.15	.18	.17
Lungs	30.3	15.4	.29	.27	.32	.29
Gizzard lining with contents	105.3	58.9	1.03	1.02	1.13	1.12
Edible portion of gizzard	196.3	112.2	1.91	1.95	2.11	2.14
Liver	114.6	59.5	1.12	1.03	1.23	1.13
Heart	36.8	19.1	.36	.33	.39	.36
Spleen	5.9	2.4	.06	.04	.063	.045
Pancreas	12.3	9.0	.12	.16	.13	.17
Kidneys	27.2	17.0	.26	.29	.29	.32
Intestines, ceca, and cloaca with contents	241.9	150.9	2.35	2.62	2.59	2.87
Unclassified material including loose fat	431.9	363.5	4.20	6.31**	4.63	6.92**
Carcass ready for cooking without giblets or loose fat	7551.2 (16.65)	4156.3 (9.16)	73.50	72.09	80.98	79.09

By using the percentages given, the approximate weights of the various parts and organs of turkeys of any size can be calculated.

*Figures in parentheses indicate weight in pounds.

**The greater percentages in females were due to the ovary and oviduct.

[1]From "Weights and Measurements of Parts and Organs of Turkeys," by S. J. Marsden, 1940 Poul. Sci. 19(1):23-28.

Table 7.5—Weights and measurements of Organs of Fat, Young, Standardbred Bronze Turkeys[3]

	Averages		Regression Coefficient*	
	Males	Females	Males	Females
Weights in grams:				
Live bird after overnight fasting	11277.0 (24.862)**	5590.4 (12.325)		
Dressed and chilled carcass	10269.3 (22.640)	5106.3 (11.258)	.95215	.96839
Abdominal fat surrounding gizzard and extending to vent	235.900 (.52)[1]	182.500 (.40)[2]	.01372	.11728
Measurements in inches:				
Pre-esophagus (throat to crop)	14.300	11.025	.00012	.00075
Crop (width between inlet and outlet)	1.925	1.525	.00014	.00006
Post-esophagus (crop to proventriculus)	7.275	5.575	.00014	.00027
Proventriculus	2.500	2.100	.00003	—.000001
Small intestine	107.525	81.600	.00099	.00040
Large intestine, cloaca and anus	8.450	6.750	.00009	.00030
Total length of alimentary canal from tip of beak to anus***	147.625 (12.3 feet)	113.075 (9.4 feet)	.00156	.00197
Cecum (not included in total length of alimentary canal)	18.350	12.950	.00022	.00050
Trachea (windpipe)	17.150	13.875	.00067	.00110

*By multiplying the regression coefficient by the number of grams a bird weighs dressed above or below the average dressed weight given, and adding to or subtracting from (as the case may be) the average lengths given, the approximate lengths of the organs of turkeys of any weight may be estimated.

**Figures in parentheses indicate weight in pounds except where otherwise indicated.

***Includes mouth and throat estimated to be approximately 4¾ and 4½ inches in length for males and females respectively, and the distance between inlet and outlet of the gizzard estimated at 1 inch for the males and ¾ inch for the females.

[1]2.30% of dressed weight.

[2]3.55% of dressed weight.

[3]From S. J. Marsden, 1940. Poul. Sci. 19(1): 23-28.

(U. S. "AA" grade) young roasting turkeys, 10 of each sex.

According to the author's observations (not shown in the tables), the complete neck (including bones, but without skin) from its junction with the skull to junction with the body, averages 4.4% of the dressed weight of fat young standardbred toms averaging 17.3 pounds and 2.8% of fat young hens averaging 9.7 pounds. For both sexes the unweighted average is 3.6% of which about 70% is edible. In large Broad Breasted Bronze toms the neck averages about 4% of the eviscerated weight (including giblets) and 3.5% of New York dressed weight. In corresponding hens the percentages are 3 and 2.5 respectively. The giblets in Broad Breasted Bronze turkeys average about 4.25% of eviscerated weight and 3.75% of the dressed weight of each sex.

Edible Meat Yields

Table 7.6 contains additional anatomical data obtained from 26 turkey hens.

Table 7.6—Weights of Certain Organs and Systems of Standardbred Bronze Turkey Hens—24 to 33 Weeks of Age

(Data of University of Nebraska*)

Name of organ or system	Weight in Grams	Percentage of net body weight
Net body weight**	4125.0	100.00
Feathers	214.3	5.195
Head	67.0	1.624
Entire musculature	2364.7	57.326
Entire ligamentous skeleton (bones, ligaments, and tendons)	388.5	9.418
Skin (excluding skin of shanks and feet)	388.3	9.413
Empty digestive tube	184.7	4.478
Empty crop and esophagus	24.8	.601
Empty intestines	57.0	1.382
Empty proventriculus	7.0	.170
Empty gizzard	95.8	2.322
Liver	92.3	2.238
Heart	21.3	.516
Lungs and trachea	25.6	.621
Pancreas	6.13	1.486
Spleen	1.97	.0478
Thyroid gland	.293	.00710
Thymus gland	2.26	.0548
Suprarenals	.311	.00754
Hypophysis (pituitary gland)	.0189	.000458
Ovary	29.0 (range 10.2 dormant to 881.4 active)	.703
Oviduct	36.04 (range 1.1 dormant to 103.5 active)	.882
Brain	6.31	1.530
Spinal Cord	5.40	1.309
Eyeballs	9.79	2.373

*Latimer, Homer B., and J. A. Rosenbaum, 1926. A quantitative study of the anatomy of the turkey hen. Anat. Rec. 34:15-23.

**Live field weight less contents of the digestive tract. The live field weight of these birds is estimated to have been approximately 4400 grams.

Table 7.7 shows the percentage of breast and leg meat in dressed standardbred turkeys and Table 7.8 shows these percentages separately for four varieties of turkeys including the broadbreasted types, large and small.

Leg meat includes that of the drumstick, thigh and oyster, which is an integral part of the thigh musculature. The ordinary carver, however, obtains only about 16 per cent leg meat because he leaves some meat in the oyster, amounting roughly to 2 per cent of the dressed weight.

Table 7.7—Miscellaneous Data on Raw Breast and Leg Muscles of Fat,[1] Young, Roasting Turkeys of Several Standard Varieties[2]

No. of birds in group	Grouping of birds: Breast type	Dressed weight in pounds (mean)	Percentage[3] breast musculature (mean)	Percentage[3] leg musculature[5] (mean)	Percentage[3] breast and leg musculature (mean)
	Normal males				
4	Moderately broad (Grade 4)	17.04	21.97	16.92	38.89
7	Medium (Grades 5 and 6)	17.22	19.58	18.07	37.65
2	Narrow (Grade 7)	19.79	18.41	19.16	37.57
	Weighted averages for 13 normal males	17.56	20.14	17.88	38.02
	Normal females				
2	Broad (Grade 3)	10.84	20.69	17.45	38.14
4	Moderately broad (Grade 4)	9.00	19.45	18.58	38.02
2	Medium (Grades 5 and 6)	9.24	18.86	17.25	36.11
2	Narrow (Grade 7)	8.02	17.45	18.38	35.83
	Weighted averages for 10 normal females	9.22	19.18	18.05	37.23
	Unweighted averages for normal males and females	13.39	19.65	17.97	37.63
	Abnormally narrow-breasted birds				
5	Males (Grades 9 and 10)	15.48	11.63	20.84	32.47
1	Females (Grade 9)	9.11	14.05	19.29	33.27

[1]Grading U. S. Prime and Special in quantity of external fat carried.
[2]Standardbred Bronze, White Holland, and Black. From S. J. Marsden, Poultry Science, 1940.
[3]Percentage of dressed (undrawn) weight.
[4]One bird had exceptionally plump legs, 21.48 percent of the dressed weight consisting of leg meat.
[5]Total leg meat including that comprising the "oysters."

Depth of body, width of breast, and length of keel are the three measurements which combine to produce a high percentage of breast meat. Selection should be based on all three rather than on breast width only in order to produce a balanced bird.

The effect of size on the percentage of breast and leg meat has not been definitely determined. However, since large birds show smaller percentages of dressing and drawing losses it appears that they should produce a slightly higher percentage of edible meat. Available data indicate that they do, other factors being equal. Conformation, however, has a much greater effect on meat yield than does mere size as is indicated by the data in Table 7.8.

The data in Table 7.8 show variety comparisons in yield of net breast and leg meat as dissected from the dressed carcasses of available market type birds of four

Table 7.8—Yield of Raw Breast and Leg Meat of Four Varieties of Turkeys.

	Average New York dressed weight in pounds			Percentage of New York dressed weight								
				Breast muscle			Leg muscle			Breast and leg muscle		
	Males	Females	Both sexes	Males	Females	Both sexes	Males	Females	Both sexes	Males	Females	Both sexes
Broad Breasted Bronze—average age 29 weeks[1 2]	21.74	14.45	18.07	21.90	22.80	22.35	20.50	19.70	20.10	42.40	42.50	42.45
Standardbred Bronze—average age 28 weeks[1 3]	17.85	11.37	14.61	18.00	18.35	18.18	19.45	19.10	19.28	37.45	37.45	37.45
White Holland—average age 29 weeks[1 3]	16.48	9.83	13.15	18.35	18.45	18.40	18.70	18.10	18.40	37.05	36.55	36.80
Beltsville Small White—age 28 weeks[1]	13.55	8.01	10.78	19.95	18.55	19.25	18.40	17.95	18.18	38.35	36.50	37.43
Unweighted averages—4 varieties	17.41	10.92	14.15	19.55	19.54	19.55	19.26	18.71	18.99	38.81	38.23	38.53
Beltsville Small White male (1). 26 wks. old, avg. quality; breast grade 5, leg grade 4, fat grade 4	12.51			21.74 (25.61% of drawn wt.)			19.58 (23.07% of drawn wt.)			41.32 (48.7% of drawn wt.)		
Beltsville Small White female (1) 30 wks. old, very broadbreasted; breast grade 1, leg grade 4, fat grade 1		9.30			23.23 (26.70% of drawn wt.)			19.89 (22.87% of drawn wt.)			43.12 (49.57% of drawn wt.)	

[1]From H. M. Harshaw, et al., 1943. The composition of turkeys of different varieties and strains, Poultry Science 22(2):126-136. (The U.S.D.A. Research Center, Beltsville, Md.)

[2]Stock of average broad-breasted type; greater yields would be obtained from the better broad-breasted strains.

[3]Typical of standardbred turkeys widely produced in 1940-41.

[4]From unpublished data, Beltsville Station, 1952.

varieties in 1940-41. In combined yield of breast and leg meat, both sexes, the Broad Breasted Bronze exceeded the standardbred Bronze and Beltsville Small White by about 5% and the White Holland by 5.65%. In breast-meat yield the BBB exceeded the Beltsville White by about 3% and the other two varieties by about 4%. In leg meat yield the BBB exceeded the BSW and White Holland by about 2% and the Bronze by about 1%. Sex differences were not striking. BBB hens had about 1% more breast meat but about 1% less leg meat than the corresponding toms, making the totals of leg and breast meat about the same for males and females. Roughly the same situation existed in the Standard Bronze and White Holland, but the differences were less, 0.10 to 0.60%. BSW males, however, out-yielded the BSW females in breast meat, leg meat, and total breast and leg meat by 1.40, 0.45, and 1.85% respectively. In all four varieties combined, the two sexes were equal in percentage of breast meat but in leg meat the males exceeded the females by 0.55%. The limited (1952) data in the second part gives an indication of the effects of selection for greater meat yields. The one Beltsville White male of average conformation in 1952 yielded about 3 per cent (of New York dressed weight) more breast and leg meat combined than did the average Beltsville White male of 1940-41, while the one Beltsville female of exceptionally good meat conformation in 1952 yielded about 7% more breast and leg meat combined than did the average Beltsville hen of 1940-41. On the basis of drawn weight (including giblets and neck) the average good quality Beltsville White turkey well matured for market yields almost 49% in choice cuts of lean meat (breast and legs). Yields equalling or exceeding these figures can be expected from any variety of turkey selected for good meat conformation. For example, the very large Broad Breasted Bronze toms (Table 7.20) yielded 30% of their eviscerated weight in boneless breast meat including attached skin and 27% in thighs and drumsticks combined including bones and skin. Assuming that the breast meat is 8% skin, this gives a lean breast meat yield of 27.60%. Assuming that the whole drumsticks and thighs are 8% bone and 10% skin, this gives a lean leg meat yield of 22.14%. The total of leg and breast muscle therefore is 49.74%, or about 1% more than the combined leg and breast meat of the 1952 Beltsville White tom and the same as that of the 1952 Beltsville White hen.

As ordinarily carved for serving, only about 80 per cent of the thigh meat is removed. The other 20 per cent remains in the "oyster" which consists primarily of the section of thigh musculature that fills the large (forward) concavity of each half of the hip bone. The oyster properly includes also the small quantity of leg musculature that is found in the smaller (rear) concavity of the hip bone and on the side of the tailhead. When the meat usually left in the front and rear sections of the oyster during carving is considered as part of the thigh, the latter possesses considerably more meat than the drumstick. As ordinarily carved, however (leaving some of the thigh meat in the oyster), the thigh supplies approximately the same quality of meat as the drumstick. Of the total leg musculature about 56 per cent is thigh meat, including that of the oyster, while only about 50 per cent is thigh meat as usually carved. The sexes appear not to differ significantly in the relative proportions of thigh and drumstick meat.

Turkeys rank high in the proportion of edible meat to dressed weight. Edible meat consists of giblets and drawn carcass minus the bones. Edible percentages of different types and classes of poultry are shown in Table 7.9.

The skeleton, since it is inedible, represents waste. According to Cline (1933) the bones of 10 to 12 pound drawn (oven dressed, including neck, giblets and loose fat) prime turkey hens comprised about 10 per cent of the dressed, undrawn weight, or about 11.5% of drawn weight (Standardbred Bronze). In Rhode Island Red cockerels averaging 3.88 lbs. drawn (including giblets), the

Table 7.9—Percentage of Raw Meat to Blood and Feather Dressed Weight*

Squab pigeons	73.94%
Fattened capons	67.46
Turkeys (prime roasting)	66.53
Geese	65.07
Fattened hens	64.22
Fattened roasters	63.07
Fattened broilers	60.73
Squab guineas	60.25
Ducks	60.17
Unfattened roasters	56.86
Unfattened broilers	54.27

*From M. A. Jull and W. A. Maw, Sci. Agric. (Canada) 3(10):329-338, 1925. The turkeys were not broad-breasted. The hens, roasters and broilers in the table are chickens.

bones of the drawn carcass weighed 0.57 lb. or 14.7% of drawn weight. In Beltsville White turkey males 14 weeks old averaging 4.56 lbs., drawn with giblets, the bones weighed 0.67 lb. or 14.7% of drawn weight. (Data of McNally and Spicknall, 1949.) At the Beltsville Station (1952 unpublished data) the bones in the drawn carcasses, including neck and giblets, of 26-week-old Beltsville White toms of average good market type and grade A fatness (11.23 lbs. drawn) averaged 10.69% of the New York dressed weight or 12.63% of the eviscerated weight. The drumsticks and thighs complete averaged 11.87% bones and 7.45% skin; the boneless breast complete with skin cut off even with the shoulders averaged 6.93% skin. On the same basis the bones in an eviscerated Beltsville White hen 30 weeks old, of extra good market type (breast grade 1, fat grade 1) comprised 7.23% of the dressed weight or 8.32% of drawn weight. The thigh and drumstick were 7.61% bones and 9.39% skin; the boneless breast was 8.51% skin. In all of the data on bones, only those in the eviscerated carcass including those of the complete neck were considered. Bones of head, shanks, and feet were not counted. For weight of the complete skelton, see Table 7.6.

In roasting turkeys corresponding to U. S. "A" grade, the edible internal fat, most of which lies in the abdomen surrounding the gizzard and extends toward the vent, is of relatively little importance and in U. S. "C" and "B" grades it is absent or nearly so. In very fat (U. S. "AA" grade or fatter) turkeys, however, the loose fat is of considerable quantity. In the data reported this fat averaged about 3.55 per cent of the dressed weight of hens and 2.3 per cent of that of toms. It is noted that fat (U. S. "AA" and "A") turkeys in general lose smaller percentages of their weight during dressing and drawing

Table 7.10—Edible Percentages of Broad Breasted Bronze Turkeys[1]

Size of turkeys, New York dressed weight	Oven-drawn weight[2]	Whole roasted weight	Total Edible Roasted meat[3]
Data from 21 young toms about 29 weeks old			
Pounds	Per cent of dressed weight	Per cent of dressed weight	Per cent of dressed weight
32	90.6	63.1	52.2
30	90.0	63.0	51.3
28	89.6	62.5	50.3
26	88.2	62.3	49.6
24	88.3	61.7	48.3
22	87.7	60.8	47.3
20	87.0	60.5	45.5
18	85.5	59.4	43.3
16	84.4	58.8	41.2
Avgs. 24	87.9	61.3	47.7
Data from 21 young hens about 29 weeks old			
22	87.3	62.7	47.7
20	87.0	62.0	47.0
18	86.6	61.1	46.7
16	86.3	60.0	45.6
14	85.7	58.6	42.9
12	85.0	57.5	43.3
10	84.0	55.0	41.0
8	82.5	51.3	37.5
Avgs. 15	85.6	58.5	44.0

[1]From "Influence of size and sex of turkeys on yields of edible turkey meat," by L. E. Cline, Mimeo. Circular of Norbest Turkey Growers Assn., Salt Lake City 11, Utah (1947); also "What kind of turkey gives buyers the most meat for their money, by L. E. Cline, U. S. Egg and Poultry Mag. 53(6)18-21, 1947.

[2]Includes giblets and neck.

[3]Total edible consisted of all meat including neck muscles, giblets, and skin. The turkeys were freed of all meat that it was practicable to secure.

than do thin ("C" grade) or "B" birds. These differences probably are not unrelated to the extra fat accumulated. If the loose abdominal fat is considered as inedible and discarded, the smaller drawing losses shown by fat birds would be reduced materially.

Some evidence has been presented to show that, regardless of sex or type, the larger the turkey the lower is the dressing shrinkage and the higher is the percentage yield of raw edible meat. In Table 7.10 the hens as a group show greater dressing shrinkages and lower percentage yield of cooked edible meat than do the toms as a group. However, data from other sources show that it is only within male and female groups that the percentage of raw meat yield is positively correlated with body weight, that is, the larger toms yield a higher percentage of raw edible meat than do the smaller ones (the same for hens) but the hens as a group average as high as, or a little higher than the toms in spite of a much smaller average weight. The Norbest data also indicate that not only is the yield of raw edible meat related to body size but that the proportion of **cooked** edible meat follows the same trend. However, data of the Nebraska Station (1946), the Beltsville Station (1948) and Ontario Station (1949) show that the females produce about the same percentage yields of cooked meat as do the males except in Beltsville Whites where the hens yielded better than the toms.

In the two varieties studied at the Beltsville Station (data not shown), the raw leg muscle yield has been shown to be in proportion to plumpness and size of the leg but cooked leg-meat yield showed a reverse tendency, the plumper the leg the lower the percentage yield of cooked muscle. The apparent explanation for this and also the higher cooked yield of total edible for the Beltsville White hen (Tables 7.11 and 7.12) is that the small hens tended to cook evenly and were cooked for a shorter length of time hence lost less weight through drying out. The opposite trend indicated by the Norbest data may possibly

Table 7.11—Yield of various cooked edible portions expressed as per cent of New York dressed weight of Beltsville Small White and Broad Breasted Bronze roasting turkeys.[1]

Variety or Strain	Sex	Number of turkeys	Average New York dressed weight	Average yield muscle excluding neck and giblets.			Average yield skin	Average yield muscle from neck	Average yield giblets	Average yield total edible portion
				Total white and dark	Breast and other white	Leg and other dark				
			pounds	per cent	per cent	per cent	per cent	per cent	per cent	per cent
Beltsville Small White	Male	8	12.9	34.4	20.9	13.5	6.7	3.0	3.2	47.4
	Female	3	7.0	41.1	24.4	16.7	6.5	2.6	2.8	53.0
Broad Breasted Bronze	Male	5	24.3	35.9	22.6	13.3	7.0	2.4	2.4	47.7
	Male	11	25.6	33.8	....	....	7.9	......	......	......
	Female	6	13.3	37.5	23.2	14.3	5.5	2.6	2.5[3]	47.9[3]

[1]See footnotes to Table 7.12.

Table 7.12—Roasting conditions, cooking shrinkage, and yield of muscle and skin of Beltsville Small White and Broad Breasted Bronze turkeys—oven dressed weight basis.[1]

Variety or strain	Sex	Number of turkeys	Average oven dressed weight[2]	Average oven temperature	Average cooking time		Average consumption gas per pound	Average cooking shrinkage	Average yield total muscle[2] and skin	Average yield muscle[2]	Average yield skin
					Total	Per lb.					
			pounds	°F.	hours	min.	cu. ft.	per cent	per cent	per cent	per cent
Beltsville Small White	Male	8	9.7	296	3.8	23.3	2.3	27.9	55.3	46.3	9.0
	Female	3	5.5	321	2.0	21.9	3.0	22.5	60.6	52.4	8.2
Broad Breasted Bronze	Male	16	19.6	262	7.7	24.0	2.1[3]	29.7[4]	53.9	44.2	9.7
	Female	6	10.3	285	4.1	23.6	2.2	30.1	55.7	48.6	7.1

The most striking finding is that on the oven-dressed basis, Beltsville White and Broad Breasted Bronze females yielded a considerably higher percentage of cooked lean meat than did the corresponding males.

[1]From "Yield of cooked edible portion of young roasted turkey, by L. M. Alexander and others." Poultry Science 27(5): 579-587 (1948). U.S.D.A. Research Center, Beltsville, Md.

[2]Excludes neck and giblets

[3]Average of 5 turkeys.

[4]Average of 13 turkeys.

be due to more favorable cooking conditions which prevented drying out of the meat, particularly the leg meat, of the larger birds. It is of interest that average percentage yield of cooked, lean meat of Broad Breasted Bronze toms is the same in the Norbest and Beltsville data but that of the Norbest BBB hens is about 4 percent less than the BBB hens cooked at Beltsville. The reason for the lowered yield of the hens in this report is not apparent.

The turkeys reported on by the Ontario Station all were grade A young birds which were killed, picked, wrapped in heavy waxed paper, stored at -10° F. for around 8 months, then thawed, drawn, trussed but not stuffed, covered with vegetable shortening, and roasted in a large baker's oven employing the revolving spit principle; 325° F. was used for about 2 hrs., after that 300° F. The smaller birds were cooked 3 hrs. 10 min., the medium-sized ones 4 hrs., 20 min., and large ones 4 hrs. 50 min. An internal temperature of 185° F. as registered by the thermometer inserted in the thigh meat was attained. Giblets were roasted in about 2 hours. The (covered) cooked turkeys were kept overnight in a refrigerator at 45° F. then weighed and dissected leaving the skin and its adherent fat with the various parts. Results are given in Table 7.13.

The close ranking of both sexes and various sizes of Broad Breasted Bronze in percentage of cooked edible meat based on drawn weight is noteworthy. The unweighted averages of the 2 classes of males and 4 classes of females are about the same, around 51.5% cooked edible of drawn weight. The standardbred Bronze males show about 5% lower percentage of edible. The capons show a higher percentage of cooked edible than do the turkeys. However, if the fat and skin had not been weighed with the lean meat, the capons probably would have shown a lower percentage of cooked lean meat. The relatively short cooking periods probably account for the relatively high turkey meat yields as compared to the Beltsville and Norbest yields.

Table 7.13—Ontario Data on Percentage Composition of Turkeys*.

	Av. drawn wt. including neck and giblets	Bones	Giblets	White meat[5]	Dark meat[5]	Total cooked wt.	Cooked edible[5] of cooked weight	Cooked edible[5] of live weight	Cooked edible[5] of chilled N. Y. dressed weight	Cooked edible[5] of drawn weight
	Pounds	Per cent[6]	Per cent[6]	Per cent[6]	Per cent[6]	Pounds	Per cent	Per cent	Per cent	Per cent
3 females[1]	8.0	25.95	4.43	38.51	31.10	17.3	74.05	38.05	43.01	52.23
3 females[1]	15.0	23.11	3.46	40.98	32.48	30.7	76.93		45.40	51.88
7 males[1]	22.5	26.88	3.79	37.95	31.31	116.14	73.12	42.85	46.78	53.57
Unweighted averages	15.2	25.93	3.89	39.15	31.63	164.14	74.70	40.45	45.06	52.56

	Av. live weight lbs.	Av. N.Y. dressed weight	Av. drawn wt. including neck and giblets	Bones	Giblets	White meat[5]	Dark meat[5]	Total cooked weight	Cooked edible[5] of cooked weight	Cooked edible[5] of live weight	Cooked edible[5] of N.Y. dressed weight	Cooked edible[5] of drawn weight
	Pounds	Pounds	Pounds	Per cent[6]	Per cent[6]	Per cent[6]	Per cent[6]	Pounds	Per cent	Per cent	Per cent	Per cent
4 females[1]	16.25	14.8[4]	13.3[4]	20.18	4.29	44.20	31.32	34.16	79.67	42.02	45.95	51.33
3 females[1]	19.13	17.7[4]	15.7[4]	20.74	3.40	46.29	29.58	30.31	79.28	41.88	44.91	51.00
2 males[2]	19.31	17.3[4]	16.0[4]	23.20	4.16	39.05	33.60	19.53	76.80	38.77	43.16	46.51
5 males[1]	26.40	24.0[4]	22.1[4]	22.69	3.63	43.00	30.63	70.50	77.26	41.30	45.38	49.28
Unweighted averages	20.52	18.5	16.8	21.70	3.87	43.13	31.28	(Total) 154.50	78.25	40.99	44.85	49.83
10 capons (Barred Rock)[3]	Not given	7.3	6.07[4]	24.86	4.74	33.47	36.92	43.50	75.14	Not given	44.39	53.36

*From "Range to table—what proportion can you eat?" by Earle S. Snyder, Ont. Agr. Col. Guelph. Canadian Polutry Review 73(1); 7-10, Jan., 1949.

[1]Broad Breasted Bronze.
[2]Standardbred Bronze.
[3]Cooked with the turkeys for 2 hrs., 50 min.
[4]Calculated.
[5]Includes skin with adherent fat.
[6]Calculated on total cooked weight.

In an experiment at Beltsville involving the cooking of 40 turkeys[1] it was found that female turkeys when slaughtered near the usual market age of 28 weeks yielded 39% cooked muscle, not including neck and giblets, of New York dressed weight. Very young crossbred turkeys 8-9 weeks old yielded 27%, well-matured toms 39-56 weeks old yielded 34%, hens the same age 32%, and adult hens over 3 years old yielded 31%. In males 9 to 13 months old (beyond the usual market age), the cooked skin with its adhering fat was 12% of the New York dressed weight while in similar hens it was only 6%. To convert the above percentages to a ready-to-cook basis less neck and giblets, increase them by 30% so as to compare them with the following figures at the Beltsville Station in which one Rhode Island Red roasting chicken, a cockerel 36 weeks old when roasted 36.4 minutes per pound of ready-to- cook weight yielded 43.8 per cent cooked muscle and 6.7 per cent cooked skin with adhering fat of ready-to-cook weight (4.86 pounds) excluding neck, giblets, and stuffing; and a Cornish cockerel 44 weeks old weighing 5.58 pounds ready-to-cook excluding neck, giblets and stuffing yielded almost the same percentages 43.7 and 6.4 respectively. Cooking shrinkages were 34.3 and 36.6 per cent respectively of the same ready-to-cook weights. The Nebraska Station (1946) found that 100 pounds of live grade A fowl (chicken hens) delivered 23.85 pounds of cooked edible product (muscles, giblets, and skin); 100 pounds of live grade A BBB toms delivered 39.33 pounds; and 100 pounds of live grade A BBB hens delivered 40.5 pounds. (To convert these percentages to a New York dressed basis increase these by 11 per cent and to an eviscerated basis (including neck and giblets) increase them by 27 per cent). As per cent of New York dressed weight the chicken hens would have yielded 26.47 per cent, the tom turkeys 43.65 per cent, and the hen turkeys 45.00 per cent; as per cent of

[1]From "Yield of cooked edible portion of turkey according to sex, age, fatness, and breast type," by Lucy M. Alexander and others, U.S.D.A. Research Center, Beltsville, Md. Poul. Sci. 30(2); 187-195, March 1951. All turkeys were of standard (non-broad-breasted) varieties or crosses between them.

Table 7.14—Percentage Composition of the Cooked Edible Portion (Lean Meat, Giblets, Skin, and Fat).[1]

	Cooked giblets	Cooked white meat with skin and adhering fat	Cooked dark meat with skin and adhering fat
	Per cent of total cooked edible	Per cent of total cooked edible	Per cent of total cooked edible
13 turkeys 1947 (B.B.B.)	5.13	52.18	42.47
14 turkeys, 1948 (12 B.B.B. 2 standard Bronze)	4.86	55.53	39.60
10 capons, 1948 (Barred Rock)	6.30	44.55	49.14

[1]From "Range to table—what proportion can you eat?" by Earle S. Snyder, Ont. Agr. Col., Guelph. Canadian Poultry Review 73(1):7-10, Jan. 1949.

ready-to-cook weight including giblets, and neck, the percentages would have been 30.29, 50.55, and 51.44 respectively.

As to breast type, USDA data (Poul. Sci. 30:187, 1951) showed that the broader the turkey the greater the percentage yield of cooked muscle (excluding neck and giblets) of New York dressed weight. Taking males and females of all ages (mostly 26 to 43 weeks old) together, those grading 2 and 3 (broad to moderately broad) yielded 36 per cent; those grading barely broad-breasted (breast grade 4) yielded 34.6 per cent; those grading average (grade 5) yielded 34.1 per cent; those slightly under the average (grade 6) yielded 33.9 per cent, and those classed as narrow breasted (grade 7) yielded 33 per cent. The highest yields, 38.5 to 41.9 per cent were obtained from males and females between 26 and 30 weeks old, between breast grades 2 to 5, and between fat grades 3 (high A) and 6 (high B). Extremely fat turkeys (grade 1) yielded only moderately well, averaging 33.8 per cent.

Recent work at the Virginia Station (Davis and Coe) has been reported by Sweet (1953)*. Forty Beltsville

*From "Poultry Processing and Marketing," Jan. 1953, pp. 16, 28-29.

White turkeys were carefully weighed, dressed and checked for edible meat yield.

"In one group, they had 10 toms that averaged 9.3 pounds New York dressed weight. The second lot contained 10 toms that averaged 7.2 pounds. The third lot contained 10 hens that averaged 7.0 pounds. The fourth lot was made up of young turkey hen broilers that averaged 4.05 pounds.

"There seems to be no relation between the New York dressed weight of the birds and the evisceration yield including the giblets. For example, the heavier toms that averaged 9.3 pounds New York dressed weight had an evisceration yield of 78.5 per cent. Toms averaging 7.2 pounds had a yield of 80.4 per cent and the hens averaging 7.0 pounds had a yield of 80.8 per cent whereas the small hens averaging 4.5 pounds had a yield of 78.3 per cent. These yields do not correspond with yields we find with chickens, for in general, heavier roasters will give a higher yield than lighter birds such as fryers.

"These turkeys were all chilled in ice water before drawing. For the cooking part of the test they were placed in partially-covered deep well bake pans without water and roasted at 300 degrees F. until well done. The carcasses of the cooked birds and the giblets were weighed separately. The meat was removed from the bones. The meat and bones were weighed separately for the usual cutup portions.

"There was very little difference in the cooking loss that could be attributed to the size of the bird. For example, the heaviest birds had a cooking loss of 25.5 per cent; the lightest birds had a cooking loss of 26.9 per cent; and the birds intermediate in weight had greater cooking losses.

"In this test heavier toms (Lot 1) and the light hens (Lot 4) gave a higher percentage of meat than either the toms or hens of intermediate weight. It is believed that at least part of the higher meat yields for the smallest hens was due to the less time required for cooking and that the heavier toms had a higher yield because they had less fat to lose during the cooking process."

How Much Turkey to Order[1]

As has been shown, the breed, age, feeding, type, fatness, and cooking influence the yield of cooked meat. However, studies made on turkeys indicate that well-meated turkeys will yield about two 2½-ounce carvings of choice meat per pound of New York dressed weight.

The following data, compiled from a study made on turkeys in a commercial cafeteria, show the approximate serving yield from a 17-pound New York dressed bird:

[1]Extract from, "Serving Many," December, 1945. Published by: U. S. Department of Agriculture, Agr. Mktg. Service, Wash., D. C.

Kind of Meat	Weight
New York dressed weight of turkey	16 pounds 15 ounces
Ready-to-cook weight	14 pounds 11 ounces
Weight of turkey after roasting (hot)*	10 pounds 6½ ounces
White meat, carved	3 pounds
Dark meat, carved	2 pounds
Giblets	7 ounces
Hash meat	1 pound 4 ounces
Neck	3 ounces
Skin	1 pound 1 ounce
Carcass	2 pounds 5 ounces

Yield: 32 servings of sliced meat. weighing 2½ ounces each, plus 8 servings of extended meat made from trimmings (such as hash or loaf). For home serving allowing liberal portions with leftovers, allow one pound of New York dressed turkey per person or ¾ pound of ready-to-cook turkey per person.

PHYSICAL AND CHEMICAL COMPOSITION

Chemical analyses of raw turkey meat from Broad Breasted Bronze reported by Harshaw and others (Poultry Science 22(2):126-136, 1943) are given in Table 7.15. Data on raw turkey meat from crossbred turkeys of five different ages are given in Tables 7.16 and 7.17.

Table 7.15—Average Chemical Composition of Raw Meat from Grade A Broad Breasted Bronze Turkeys of Both Sexes.

	Protein, per cent	Fat, per cent	Ash, per cent	Water, per cent
Breast muscle	24.53	1.31	1.10	73.48
Leg muscle	20.48	4.21	1.02	74.10
Remaining edible (skin, loose fat, muscle other than breast and leg, and giblets)	16.58	30.63	0.71	52.10
Total edible portion (breast muscle, leg muscle, and remaining edible)	20.65	11.73	0.95	66.75

The table does not show variety comparisons of chemical composition but it was found that the percentage of protein and ash in the total edible portion of males and females combined was very uniform for Broad Breasted Bronze, Standardbred Bronze, White Holland and Beltsville Small White. The percentage of fat in the total

*(61.4% of dressed or 70% of ready-to-cook weight.)

Table 7.16—Physical Composition of Raw Small-type Crossbred Turkeys as Affected by Age and Sex[1]

Age in weeks	Live, fasted weights, lbs.		Breast muscle as per cent of dressed weight		Leg muscle as per cent of dressed weight		Distribution of Miscellaneous Tissues as Per cent of Dressed Weight: Giblets		Internal fat		External fat		Skin		Muscle other than breast and leg[2]	
	Males	Females	Males	Females	Males	Females	Males	Females	Males	Females	Males	Females	Males	Females	Males	Females
16	7.66	5.73	16.7	18.6	18.9	19.8	3.9	4.1	0.2	0.2	0.2	0.2	5.2	5.2	5.8	6.9
20	10.04	7.08	17.8	17.7	19.6	19.3	4.1	4.4	0.4	0.4	0.6	0.4	5.6	5.7	6.1	6.3
24	14.03	7.46	20.3	20.9	19.4	18.9	4.0	4.2	0.3	1.0	0.7	1.2	5.2	6.5	6.0	6.7
28	16.50	10.26	20.1	21.8	19.0	18.6	3.5	3.5	1.9	2.4	2.7	2.4	6.8	7.2	6.0	6.9
32	17.48	10.76	19.7	18.7	16.9	16.0	3.1	3.3	2.4	4.7	5.2	7.0	9.1	9.0	5.7	5.7

Table 7.17—Chemical Composition of Turkeys as Affected by Age and Sex[1]

Age in weeks	Per cent dressed weight of fasted live weight		Breast Muscle: Per cent protein		Per cent fat		Per cent ash		Per cent water		Leg Muscle: Per cent protein		Per cent fat		Per cent ash		Per cent water	
	Males	Females	M	F	M	F	M	F	M	F	M	F	M	F	M	F	M	F
16	87.1	87.1	24.2	24.6	0.34	0.37	1.25	1.33	74.3	73.4	20.6	21.1	2.23	2.36	1.07	1.09	75.1	75.3
20	87.1	88.4	24.7	24.3	0.46	0.44	1.25	1.24	73.4	73.9	20.4	20.3	2.56	2.44	1.15	1.19	75.2	75.6
24	88.4	87.6	24.8	24.6	0.24	0.42	1.26	1.22	73.6	73.0	20.5	21.0	2.43	3.34	1.09	1.13	75.4	73.8
28	91.0	89.7	24.6	25.1	0.70	0.75	1.38	1.41	73.1	72.4	20.8	20.7	3.80	3.83	1.18	1.08	73.2	73.2
32	91.0	91.4	24.7	24.5	1.13	0.85	1.31	1.46	72.5	72.3	20.9	20.8	4.79	5.90	1.18	1.14	72.1	71.7
	Per cent drawn weight of live fasted weight		Remaining edible portion[2] (giblets, internal and external fat, skin, and muscle other than breast and leg.)								Total edible portion (Breast and leg muscle plus remaining edible portion)							
16	72.9	72.9	20.9	20.8	8.9	11.2	0.96	0.93	68.0	65.8	21.9	22.2	3.61	4.38	1.10	1.13	72.7	71.8
20	73.3	71.2	20.5	20.7	12.5	11.6	0.96	0.90	65.0	65.4	21.9	21.7	4.94	4.70	1.11	1.11	71.5	71.8
24	74.8	72.3	20.0	17.9	12.5	21.6	0.87	0.94	64.9	58.5	21.9	21.3	4.56	8.20	1.09	1.09	71.7	68.6
28	77.6	75.4	15.9	13.5	33.9	34.1	0.78	0.74	48.4	48.6	20.4	19.7	13.2	13.5	1.10	1.08	64.5	64.1
32	79.9	79.1	13.3	9.9	47.0	57.9	0.64	0.51	39.7	31.8	19.0	16.8	21.0	28.4	1.00	0.93	58.9	53.5

[1]From "The composition of turkeys as affected by age and sex," by H. M. Harshaw and R. R. Rector, U.S.D.A. Research Center, Beltsville, Maryland. Multigraphed release (1940) and Poul. Sci. 19 (6): 404-411 1940.

[2]Does not include the neck which, although included in the drawn percentages given, in Table 7.17 was omitted from the "remaining edible" data. Data on neck composition will be found on page 402.

edible was greatest in the fattest appearing birds and this reduced the percentage of water.

Females generally were fatter than males and had from 4 to 12 per cent more fat in the total edible portion with correspondingly less water. The percentage of fat in the breast muscle was greatest (1.31) in the Broad Breasted Bronze, intermediate (1.22) in White Holland and least in Standardbred Bronze (1.07) and Beltsville Small White (1.06): in leg muscle it was greatest in White Holland (4.69), least in Beltsville Small White (3.92), and intermediate in Broad Breasted Bronze (4.21) and Standardbred Bronze (4.24%). In percentage of fat in the breast muscle, males and females were not significantly different in the four varieties studied but in percentage of fat in the leg muscle the females tended to possess from ½

Table 7.18—Average Composition of Turkeys

Chemical composition U.S.D.A. Circ. 549 (1940)	Refuse[1] %	Water %	Protein %	Fat %	Ash %	Calories per pound
Roasting birds, medium fat:						
Edible portion (skin, giblets, lean meat, and fat)	....	58.3	20.1	20.2	1.0	1,190
Live bird—as purchased	39	35.6	12.3	12.3	.6	730
Dressed bird—as purchased	33	39.1	13.5	13.5	.7	800
Drawn bird—as purchased	19	47.2	16.3	16.4	.8	960
Flesh and skin	....	63.0	22.8	13.0	1.1	940
Flesh only	....	68.6	24.0	6.7	1.1	710
Light meat only	....	69.2	24.5	4.6	1.2	630
Dark meat only	....	68.0	23.2	9.4	1.1	800
Fat young birds—edible part	....	50.7	18.4	29.3	.9	1,530
Thin young birds—edible part	....	69.9	20.6	7.8	1.1	690

[1]Inedible parts including blood, feet, intestines, and bones. From U.S.D.A. Cir. 549.

Mineral content of flesh of chickens and turkeys (in per cent).

Kind of fowl	Sodium (Na)	Potassium (K)	Magnesium (Mg)	Calcium (Ca)	Iron (Fe)
Chickens	.091	.401	.0288	.0119	.0047
Turkeys	.130	.367	.0282	.0383	.0038
	Phosphorus (P)	Sulphur (S)	Chlorine (Cl)	Iodine (I)	Manganese (Mn)
Chickens	.266	.303	.074	.00015	0
Turkeys	.320	.234	.123	.00012	.00005

(From A.N.D. No. 7, U.S.D.A.)

to $1\frac{1}{2}\%$ more than the males. The tendency of the Broad Breasted Bronze, although not so fat appearing, to rate high in percentage of fat in the muscles is of interest since this is believed to contribute considerably to palatability in the cooked meat.

Miscellaneous data on chemical composition of turkeys will be found in Table 7.18, Average Composition of Turkeys.

PALATABILITY OF ROASTED TURKEY MEAT

In extensive experiments at the Beltsville Station the same four varieties of turkeys from the same lots as above were compared as to palatability of the roasted meat. (Marsden et al, 1952). In over-all desirability of breast and leg meat, Beltsville Small White hens took a commanding lead with 130 ratings of "excellent" out of a possible 300. Beltsville White males were next with 103, then Broad Breasted Bronze females with 83, Broad Breasted males with 75, White Holland males with 64, Standardbred Bronze males with 42. These data point out the "eating" superiority of the two newest and most popular varieties of turkeys which it is estimated together comprised about 90% of turkeys raised in 1951 in the United States.

PACKAGING DRESSED TURKEYS

Barrels were formerly quite widely used for packing dressed turkeys but are rapidly being replaced by boxes. To barrel-pack, use strong, lightweight barrels and line with waxed, parchment, or ordinary wrapping paper. Place birds on their sides, backs outward with heads and legs extended and against side of barrel. When full, remove top hoop, cover with clean burlap, and nail hoop back on. The trade shows a distinct preference for boxes over barrels; thus causing many packers and cooperative marketing associations to use boxes exclusively. The number of turkeys per box varies with the size of the box, the method of packing and weight of the turkeys.

Only birds of approximately the same size and grade should be placed in one box. A handy table giving the proper box size has been arranged by Benjamin and Pierce (Table 7.19). Wire-bound boxes of one size are now popular. If nailed boxes are used, 6-penny cement-coated nails should hold sides and bottoms while 5-penny uncoated nails may be used on the tops. Turkey boxes should be made of wood that does not possess a noticeable odor. All boxes should be lined with parchment or waxed paper and stamped with appropriate data regarding the contents.

In the one-sized box, very large turkeys lie on their sides, very small ones on their backs, and intermediate sizes at various degrees of slope between back and side positions. The rump-lock pack is used in all cases.

Table 7.19—Dimensions of Boxes for Packing Turkeys*

(Single layer, breast up and side pack)

Weight of Birds (pounds)	Inside Measurements (in inches)			Thickness of Lumber (in inches)	
	Length	Width	Depth	Ends	Top, Bottom and Sides
BOXES FOR 10 TO 12 TURKEYS					
Under 9	30	18	7	7/8	1/2
9-12	30	19	7½	7/8	1/2
12-15	32	20	8	7/8	1/2
15 and over	32	21	8½	7/8	1/2
BOXES FOR 6 TO 8 TURKEYS					
Under 9	25	18	6½	3/4	7/16
9-12	26	19	7	3/4	7/16
12-15	27	20	7½	7/8	1/2
15 and over	28	21	8	7/8	1/2

*From "Marketing Poultry Products," 1937. John Wiley and Sons, Inc., New York, N. Y.; by permission.

EVISCERATION AND FREEZING

Ready-to-Cook, formerly designated as *eviscerated,* turkeys must have head, shanks, crop, windpipe, gullet, all entrails, lungs, reproductive organs, oil gland, pinfeathers, and hair feathers removed; must be washed, and, with or without giblets, be ready to cook without need of further processing. The removal of kidneys is optional.

Although the kidney, according to government rules, may be left in the carcasses of ready-to-cook turkeys, there is considerable opposition to this practice and customers frequently object. Kidney removal is recommended for those wishing to market ready-to-cook poultry of highest quality.

Ready-to-cook turkeys may be federally inspected when not federally graded and vice versa. The terms *drawn, full-dressed, eviscerated* and *pan-ready* mean the same as ready-to-cook.

Packaged, completely ready-to-cook turkeys are now on the market. They should stimulate consumption because of their attractiveness and convenience to the housewife. Preparation of ready-to-cook turkeys is a series of operations consisting of dressing, grading, chilling, pinning, singeing, evisceration, cleaning, wrapping, packaging, quick-freezing, and sometimes tendon pulling. First, the birds are dressed, usually by slack-scalding and mechanical or hand picking, and then graded. Only birds with good finish, heavy fleshing, and freedom from pinfeathers and other defects and deformities are selected. The selected birds, hanging on mechanical conveyors, are singed by a gas flame and all remaining pinfeathers removed by hand. In some dressing plants the tendons are then pulled. The birds are placed in individual rustless steel trays and an opening is made in the abdomen between vent and keel and the viscera withdrawn and laid beside the carcass. An automatic conveyor carries the opened birds on trays to a government inspector.* Unfit birds are condemned. After inspection the trayed carcasses move on to operators who remove the heads and feet.

*Not all eviscerated turkeys are government inspected. Inspection service is optional with the processor, who must bear the cost thereof.

On the next table the neck is removed, cut off at its base, and the viscera carefully removed under running water, great care being taken not to break the intestines or gall bladder. The conveyor then carries the edible carcass and giblets to high-pressure washing tanks where they are thoroughly washed, then drained of surplus moisture. The giblets and neck are wrapped in waxed paper and returned to the body cavity, the wings are locked, the legs and tail tied together and the carcass wrapped, in a bag of plastic or other air-tight, waterproof material sometimes with a stockinette covering for protection of the wrapping, less commonly in stockinette only.

Wrapping is a most important operation and is done by hand. If wrapped in stockinette only, the carcass is thoroughly dried; in plastic or cellophane a little moisture is left on the bird to reduce freezer burn and promote a tight fit of the wrapping material. The wrapping frequently is vacuumized which is desirable; so also is the stockinette over-wrap, especially over a type of wrapper that becomes brittle when frozen. The wrapped carcass is then weighed, labelled, and packed for freezing, either in an individual, telescoped, waxed, fiber box or in a container usually holding six birds. Quick freezing is done by several processes, chief of which are the air-blast at 40°F., the sub-zero brine spray, and the direct contact between refrigerated surfaces under pressure at sub-zero temperatures. When direct contact or air-blasts are used, an airtight waterproof wrapping material is put on the birds before freezing to prevent freezer burn. If the birds are retailed frozen, which is desirable, the airtight waterproof wrapper is an advantage but if they are retailed thawed, which is not desirable, the wrapper should not be airtight. If the birds are wrapped, boxed, and frozen at about zero without air circulation, non airtight wrappers such as waterproof cellophane are used and birds thus frozen may be retailed in ordinary showcases at a temperature of 40 to 50° F. Quick freezing is much superior.

PROCESSING AND COLD STORAGE

At the U. S. Western Utilization Research Branch (1950) young hen turkeys were eviscerated while warm and split; one half immediately wrapped in MSAT cellophane with an over-wrap of butcher paper, and quick-frozen in a forced-draft tunnel. The other halves were held unwrapped at 36° F. for 4, 12, or 30 hours, then wrapped as above and frozen. After 8 months of storage at -10° F, no rancidity was detected, and after 13 months at -10° F, tasting tests showed no detectable differences between the turkey halves held unwrapped for 4 to 30 hours and those wrapped and frozen immediately. Chemical tests showed low rancidity for all turkeys. After 18 months at -10° F, differences still were not significant. There was found to be no advantage in spraying the turkey halves just before freezing with an antioxidant nordihydroguairetic acid (NDGA) in propylene glycol or with the latter alone. In other tests, turkeys eviscerated warm were superior, either fresh or after 6 to 12 months of storage in air-tight wrappings, to those eviscerated after holding New York dressed for 26 hours at 36° F only in peroxide values (measure of fat deterioration). So far as cooked off-flavor and off-odor were concerned, even after 12 months of storage, there were no large or significant differences between the warm and chilled eviscerated turkeys. Similar results were obtained with well-wrapped turkeys in comparing storage temperatures of -10, 0, and 10° F,—there were no significant differences except in peroxide values but all had deteriorated as compared to fresh-killed turkeys. In contrast to this, turkeys stored at -30° F for 8 months in air-tight vacuumized bags were equal in quality to freshly dressed controls. Cooked turkey meat and fat were found to be much less stable than were cooked chicken fat and meat under ordinary cold-storage conditions, but turkey fat rendered at 65° C (149° F.) in vacuum was relatively stable.

The California Station (1951) reported that mixed tocopherols (vitamin E), about half alpha the other half beta and gamma forms, added to the all-mash growing diet of well-matured (8-month-old) turkey hens at 0.1% for 35 days or 1.0% for 21 days just prior to slaughter reduced rancidity markedly in cooked carcasses that had been in cold storage for 5 or 9 months.

TURKEY STEAKS

Turkeys may be processed into steaks by cutting the frozen eviscerated carcass crosswise into steaks 5/8-inch thick of which about 20 can be cut from a 22-pound bird. The first steaks from the front half of the bird will be all, or nearly all, white meat, while the last 5 will be all, or nearly all, dark meat. However, since wings and drumsticks do not lend themselves well, due to tendons and bones, to steak making, it may be more efficient to market them whole. A 20-lb. eviscerated tom turkey (without giblets) will yield as follows: Wings, 2 lbs.; drumsticks, 2.5 lbs.; light-meat steaks, 6 lbs.; dark-meat steaks, 2.5 lbs.; carcass (soup bones), 4.5 lbs.; and skin and fat, 2.5 lbs. Bone-free steaks may be made by cutting the meat from the bones and forming it into 4-oz. steaks by a steak machine. At the Western Utilization Research Lab. Branch (1950) crosscut steaks stored in sealed MSAT cellophane bags at different temperatures were cooked and tested. Storage for 3 months at 0° or + 10° F. showed slightly but significantly greater off-flavor than did the controls stored at -30° F. but it required 6 months at -10° F. to produce significant off-flavor which in most cases was described as one of fat rancidity. Dark meat deteriorated more rapidly than light meat.

Deboned knitted steaks made with the steak machine showed highly significant differences in favor of the -30° F. controls over those held at 0° F. and + 10° F. temperatures during 3 months of storage and longer. Those stored at -30° F. remained pink and fresh looking, while those stored at 0° and + 10° became dull and yellow-gray.

The meat from eviscerated turkeys stored whole, then cut just before cooking, in all cases was superior to meat cut, packaged, and stored as steaks. About the some protection was accorded meat in storage by aluminum foil, polyethylene, and cellophane, while parchment paper gave decidedly less protection. Moisture loss of deboned steaks stored for 3 months at 0° F. averaged 4.2% of the original weight when wrapped in waxed paper and only 0.85% when wrapped in cellophane.

Deboned knitted steaks in which either 5% of rendered turkey fat or 10% of neck tissue fat had been incorporated were no more juicy nor flavorful than non-treated steaks when cooked fresh. After 3½ to 8 months of storage at 0° F. the flavor deteriorated slightly to significantly in both light- and dark-meat steaks with rendered fat added and in the light-meat steaks with tissue fat added.

Knitted steaks frozen at -30° F. were thawed at room temperature, refrozen at -30° F., and stored at 0°, then subjected to this treatment daily for 5 successive days. Part of the steaks were thus treated before 4 months storage and part were treated after 4 months storage, then thawed and cooked. The flavor of cooked light-meat steaks was not affected by this treatment but that of the treated dark-meat steaks deteriorated significantly with both types of thawing and freezing.

BONED TURKEY

"Turkey logs" are long, cylindrical packages of boned turkey meat wrapped in aluminum foil. The roll is roasted intact in its wrapper which results in superior flavor, retention of nearly all nutrients and flavor, and a reduced cooking shrinkage (about 18% in contrast to the usual 30 to 40%). The foil-enclosed roll is best cooked in a special roaster which prevents break-up of the roll and permits slicing.

Boned "turkey roll" may be prepared by boning the entire bird arranging the light and dark meats as desired in the skin and rolling it tightly starting at one side into a loaf-shaped mass with ends of skin tucked in, and tying

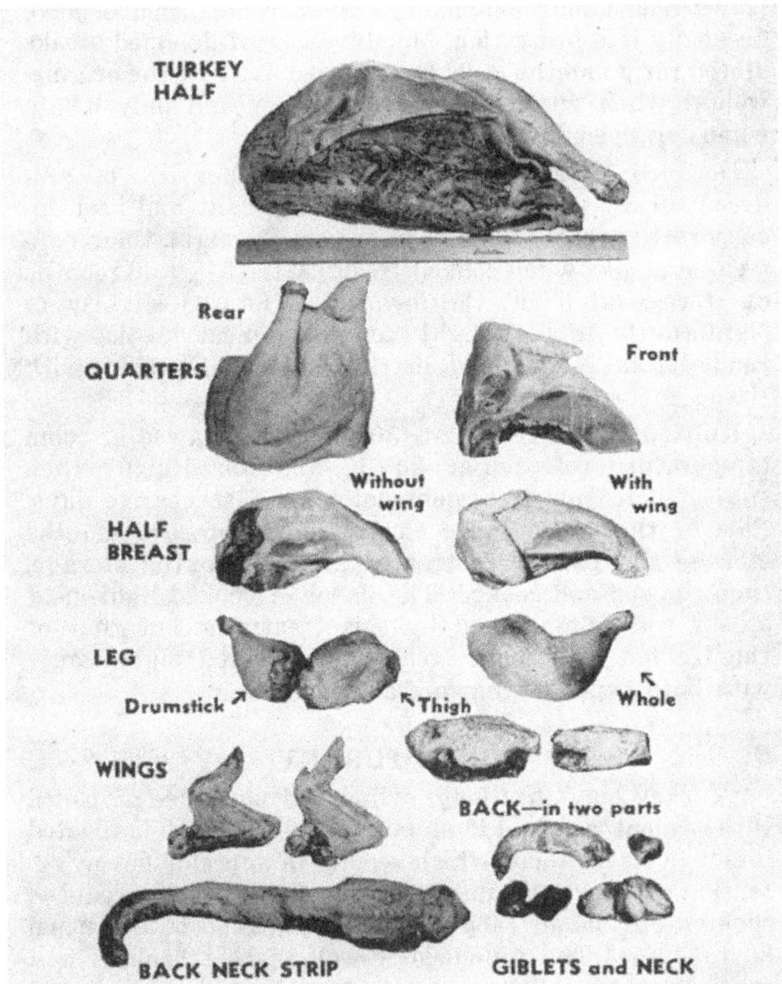

Fig. 7.1. Suggested cuts for cut-up turkey. (Photo from U. S. Dept. of Agr.)

with clean butcher's cord. The roll may be made of all light or all dark meat or both. Dressing may be placed on the skin and meat before making up the roll.

TURKEY PARTS

Large turkeys are well adapted to merchandizing in parts. They may be sawed into halves or quarters while in frozen eviscerated form with a mechanical band-saw, with a hand meat saw, or with an ordinary carpenter's saw. These parts make excellent small roasts, which are ideal for locker storage when properly wrapped to exclude air and retard loss of moisture. The turkey may be further reduced to smaller parts suitable for roasting, braising, and other types of cooking. The yield of these parts along with cost figures is shown in Table 7.20. The cuts are shown in Fig. 7.1.

Table 7.20—Estimated Costs of Turkey Parts Based on Various Wholesale Prices of 25 pound Eviscerated Turkeys*

Turkey parts	25 lbs. drawn weight	.49	.53	.57	.61	.65	.69
	Per cent	Cost Per Pound of Eviscerated Turkey					
Boneless breast	30	.76	.81	.87	.90	.94	1.10
Thighs	16	.62	.68	.72	.78	.83	.89
Drumsticks	11	.52	.55	.59	.63	.67	.71
1st Section Wing (Wingstick)	5	.48	.51	.55	.59	.63	.67
2nd Section Wing (flat wing)	4	.37	.41	.45	.49	.53	.56
Cutlets	5	.44	.48	.52	.56	.60	.65
Back strip	9	.17	.19	.23	.27	.31	.34
Neck	4	.21	.23	.27	.31	.35	.39
Heart and gizzard	2	.27	.29	.31	.34	.38	.41
Liver	1	.59	.63	.67	.71	.75	.79
Soup bones	11	.06	.08	.10	.12	.15	.17
MOISTURE LOSS	2						
	100%						

*Turkeys under 20 pounds are not recommended for cutting up to sell as parts. If done, however, adjust parts costs by adding 5%.

Printed with permission of Poultry and Egg National Board, Chicago, Illinois.

The Ontario Station (Circular 138, 1952) reported the following yield of parts of 35 Broad Breasted Bronze toms 24 weeks old.

Table 7.21—Weight of Parts of Dressed Turkeys

Parts	Average Weight lbs. - ozs.	% of Chilled Dressed Weight
Live weight	21 - 15	
Chilled dressed	19 - 11	
Feet	- 12.9	4.11
Head	- 10.6	3.36
Neck	- 12.9	4.10
Neck skin	- 3.8	1.19
Wings	1 - 12.3	8.98
Wing tips	- 3.6	1.14
Thighs	2 - 8.8	12.91
Drumsticks	2 - 3.5	11.26
Total breast	5 - 11.9	29.20
Breast meat	4 - 14.5	24.94
Breastbone	- 12.9	4.10
Front of back	1 - 1.5	5.57
Rear of back	1 - 1.8	5.65
Tailhead	- 4.6	1.46
Heart	- 1.4	.44
Liver	- 3.5	1.10
Gizzard	- 5.7	1.82
Total giblets	- 10.7	3.39
Abdominal fat	- 2.9	.93

TURKEY BROILERS

The marketing of immature turkeys in sizeable quantities first got under way in 1948 and by 1951 had become nation-wide and of economically significant proportions, with possibly four million produced. The term *turkey broiler* was first used to identify these immature turkeys and the name persists although the U. S. Agricultural Marketing Service adopted the term *fryer-roaster* and commercial concerns adopted names such as *light roaster, spring turkey,* and *junior turkey.* Roasting is the commonest method of cooking these young turkeys but they can be broiled, fried, or braised, especially the smaller sizes. Turkeys weighing four to about 11 pounds alive (or three to eight pounds ready-to-cook) are in demand as fryers or light roasters, the six- to eight-pound ready-to-cook class being very popular. Demand is best from January through September, the smaller sizes early in the year, the larger sizes later. By October the demand turns more to regular roasting turkeys from 8 pounds to 14 and on up, ready-to-cook, although the market-mature small-type hen weighing 5½ to 7½ pounds ready-to-cook enjoys considerable popularity during the fall holidays.

The Beltsville Small White variety is by far the most popular as a "broiler" for which it is well adapted because of its inconspicuous pinfeathers, reasonable poult cost, general availability of poults, well-fleshed carcass, and ability to lay on an appreciable amount of surface fat at the ages, usually marketed, 14 to 18 weeks. Feed conversion of young small-type turkeys is about equal to that of the larger varieties. At 12 weeks of age under favorable conditions about 2.70 pounds of feed; and at 14 weeks about 2.86 pounds are required for each pound of live turkey produced either by the Broad Breasted Bronze or the Beltsville Small White, according to data on large flocks reported by the Grange Company*. At 16 weeks the BBB required 3.06 pounds and the Beltsville Small White 3.16 pounds; at 18 weeks the figures were 3.27 and 3.53 and at 20 weeks, 3.51 and 3.88 pounds respectively. The high feed efficiency along with relative freedom from disease are some advantages of turkey broiler production. These high feed efficiencies are not always obtained however. At 14 weeks a feed efficiency of 3.00 to 3.20 is more likely to be obtained with either large- or small-type turkeys under average practical conditions. In dressing, drawing, and edible percentages, 14-to-16-week old turkeys compare favorably with chicken fryers and roasters. Their chief advantage over the chicken is the higher (3 to 4 per cent) raw breast-meat yield which advantage is partially lost by about a 2-per-

*From "Comparative mash consumption, grain consumption, and feed efficiency of Broad Breasted Bronze and Beltsville Small White Turkeys" by H. J. Almquist, The Grange Company, Modesto, California. Special bulletin, 1951.

Table 7.22—Dressing, Drawing, and Edible Percentages of Chickens and Turkeys

Class	Live fasted weight	Dressed chilled weight	Drawn or ready-to-cook weight[4]	Weight of total raw edible meat[1]	Per cent dressed of live weight	Per cent drawn of dressed weight	Per cent drawn of live weight	Per cent total raw edible of live weight	Per cent total raw edible[1] of dressed weight	Per cent total raw edible of drawn weight	Weight of giblets	Per cent giblets of drawn weight
	lbs.	lbs.	lbs.	lbs.							lbs.	
Rhode Island Red chickens[2]												
Males and females	2.50	2.13	1.66	1.39	85.20	77.93	66.40	55.60	65.26	83.73	0.13	7.83
Males and females	3.50	3.04	2.40	2.03	86.86	78.95	68.57	58.00	66.78	84.58	0.15	6.25
Males and females	4.50	3.94	3.14	2.68	87.56	79.70	69.78	59.56	68.02	85.35	0.18	5.73
Males and females	5.50	4.85	3.88	3.31	88.18	80.00	70.55	60.18	68.25	85.31	0.19	4.90
Males and females	6.50	5.76	4.61	3.95	88.62	80.00	70.92	60.77	68.58	85.68	0.21	4.56
Males and females	7.50	6.66	5.35	4.58	88.80	80.33	71.33	61.07	68.77	85.61	0.23	4.30
Average for R.I. Red chickens	5.00	4.40	3.51	2.99	88.00	79.77	70.20	59.80	67.95	85.19	0.18	5.13
Beltsville White turkeys												
Males 12, 14, and 16 weeks old	6.45	5.77	4.56	3.89	89.52	79.02	70.75	60.31	67.37	85.31	0.23	5.04
Females 12, 14, and 16 weeks old	4.83	4.29	3.38	2.87	88.76	78.55	69.99	59.42	66.95	84.91	0.18	5.33
Unweighted averages, young males and females[3]	5.63	5.02	3.97	3.38	89.14	78.79	70.37	60.04	67.34	85.14	0.21	5.29
Males (2) average conformation (breast grade average 5.5, fat grade average 3.5) 26 weeks old [6]	14.81	13.27	11.23	9.82	89.65	84.64	75.88	66.26	73.92	87.34	0.43	3.83
Females (1) very broadbreasted (breast grade 1, fat grade 1) 30 weeks old[6]	10.20	9.30	8.09	7.41	91.18	86.99	79.31	72.65	79.68	91.59	0.30	3.71
Broad Breasted Bronze turkeys[3]												
Males 12, 14, and 16 weeks old	8.71	7.78	6.08	5.15	89.37	77.23	70.13	59.13	66.16	84.70	0.32	5.26
Females 12, 14, and 16 weeks old	7.06	6.33	4.97	4.29	89.67	78.65	70.72	60.76	67.76	86.32	0.26	5.23
Unweighted averages males and females	7.89	7.06	5.53	4.72	89.52	77.94	70.43	59.82	66.83	85.35	0.29	5.24

[1]Total raw edible was determined in all cases by subtracting from the drawn carcass weight (including giblets and neck) the weight of the bones which was obtained by cooking the carcass, stripping meat from the bones and weighing the cooked bones (which weigh about the same as they do raw).

[2]From "Meat yield from live, dressed, and eviscerated Rhode Island Red Males," by E. H. McNally and N. H. Spicknall, Poul. Sci. 28(4): 562-567 (1949).

[3]From unpublished data, Beltsville Station (1950).

[4]Includes giblets and neck.

[5]From Harshaw et al., Poul. Sci. 22(2): 126-136 (1943).

[6]From unpublished data, Beltsville Station (1952).

cent-less leg-meat yield. In total edible percentages the two species are equal. The Beacon Milling Company (Beacon Turkey Flashes, 1952) reported an average feed efficiency of 3.16 pounds of feed per pound of gain in five commercial flocks (four BSW and one White Holland) marketed at an average age of 14 weeks. The efficiencies ranged from 2.85 to 3.55 pounds.

Large and medium-sized varieties and strains of turkeys sometimes are marketed successfully as fryers or light roasters but the smaller types are more popular and better adapted. Jersey Buffs make good fryers as they are free from conspicuous pinfeathers, have good fleshing qualities and the poults could on competition be produced at very reasonable prices. Some excellent feed conversion figures have been reported for Jersey Buffs (N. J. State Turkey Experiment Station, Millville). The Jersey Buff is slightly larger than the Beltsville White (see weight table). Dark-plumaged varieties with dark undercolor of feathers are at a disadvantage, because of dark pinfeathers, and so are large varieties because of higher poult cost and less finish at fryer or light roaster age. Large appearing legs and feet also detract from their appearance. (See Table 7.22 for meat-yield comparisons.)

In the production of fryers or light roaster turkeys, standard methods of brooding and feeding are generally used. Often the poults are brooded to 8 weeks in a brooder house with or without porches, then transferred to range until marketed; or they may be raised to market age in brooder houses with or without porches, allowing 2 square feet per bird of floor and porch space combined to 14 weeks and 2½ to 3 square feet if kept to 16 to 18 weeks. A low-fiber starting feed, preferably with an antibiotic included, and with 26 to 29% protein is recommended. It should be fed to 8 weeks, then supplemented by whole oats or other grain feed free choice with the mash until the turkeys are marketed. Many growers believe oats and other grains improve the finish and tend to cheapen the diet. In lieu of diluting the starting mash with oats and/or other grains as above, the starting diet may be continued until the turkey "broilers" are slaughtered. Another alternative is to reduce the protein content at 8 or 10 weeks to 22 to 24% (Almquist, H. J., 1952). It doubtless could be reduced again at 12 weeks if desired, to around 20%. There seem to be no appreciable differences in the dressed appearance or of feed efficiency of these young turkeys slaughtered at 14 to 16 weeks when fed according to the three different feeding plans discussed above. Attention should be paid to the fish-products content of turkey broiler mashes. Fish meal or fish solubles or both should be limited to 5% and fish oil or fish-liver oils should be limited to about 0.15% to avoid possible off-flavor in the cooked flesh. Gravel or other insoluble grit should be made available to the poults whenever whole or cracked grains are fed.

TREATMENT TO IMPROVE MARKET QUALITY

The treatment of turkey fryers with an estrogen (female hormone) to improve finish is not practiced to any appreciable extent to date. The implantation of one 15-milligram pellet of diethylstilbestrol (or simply stilbestrol) under the skin at rear of the head of tom turkeys 3 or 4 weeks prior to marketing will improve the finish and the conformation somewhat, making the toms look more like hens when dressed. It also improves the appearance of the hens slightly. Growth and feed efficiency usually are affected little and palatability when cooked is not improved. The extra labor and special management problems associated with the treated birds, and unfavorable public reaction to hormone treatments probably more than offset the advantages gained. Toms when treated at 8, 10, or 12 weeks of age show effects within 4 or 5 days. Strutting becomes almost continuous, the mating instinct is stimulated, and the vent enlarges markedly. However, if there are no hens in with the treated toms, the latter do not as a rule tread each other to a serious degree. It is necessary however to remove all hens whether treated or not, to avoid treading.

The North Dakota Station (Bimo. Bull. 1953) obtained the following results with Beltsville Small White turkeys in an experiment extending from 10 to 14 weeks of age.

Table 7.23—Feed Consumption and Gain with BSW Turkeys

	Lot 1 18% grower[2]	Lot 2 22% grower plus grain[3] free-choice	Lot 3 22% grower only	Lot 4 Stilbestrol[4] 22% grower plus grain[3] free-choice
Average gain per bird (lbs.)	2.01	2.35	2.19	2.31
Feed efficiency[1]	3.75	3.42	3.58	3.99
Per cent grain consumed	0	32	0	20
Per cent mash consumed	100	68	100	80
Pounds of feed consumed per bird	7.5	8.0	7.8	9.2
Protein intake level (per-cent)	18.0	18.9	22.0	20.1
Feed cost per lb. gain, cents	13.5	13.6	16.1	16.6

[1]Feed efficiency is determined by dividing the pounds of feed by the pounds of gain in weight.
[2]Obtained by adding ground wheat and proso (equal parts) to the 22% protein growing mash to make an 18 per cent protein mash.
[3]Wheat and proso each in a separate feeder.
[4]One 10-milligram pellet injected at 10 weeks.

These results support growers' observations that grain feeding (when grain is relatively cheap) reduces feed costs without impairing feed efficiency, or growth. Observations of fleshing and finish indicated no differences between the groups. It also was shown that mixing the ground grain with the mash (which in this case resulted in a premature reduction of protein level) was slightly inferior to free-choice feeding of the mash and grains. The turkeys ate the proso before they did the wheat and consumed 56% proso to 44% wheat during the four-week period. Implanting the stilbestrol was of no value.

The Clemson (S. C.) Station (1953) reported no improvement in weight, feed conversion, beast width, or dressing percentage when 10-week-old BSW toms were implanted with 30 milligrams of diethylstilbestrol and marketed at 14 and 16 weeks.

The Washington Station (Stadelman, 1952) reported that the improvement in finish was slight in Broad Breasted Bronze fryers treated from 7 to 11 weeks with 0.2% thiouracil in the all-mash diet and with 12, 24 or 36 milligrams of stilbestrol implanted. Growth was slightly, probably not significantly, retarded by the combined treatment; was not affected by the stilbestrol; and was decidedly retarded by the thiouracil. Feed efficiency was slightly but not significantly improved by the stilbestrol alone, and was decreased significantly by the combined treatment and by the thiouracil alone. In Broad Breasted Bronze turkeys treated the same way from 13 to 17 weeks and from 17 to 21 weeks, a marked improvement in finish, and little effect on growth rate and feed efficiency resulted from the combined treatment while thiouracil alone decreased the weight gain markedly and the feed efficiency slightly. The stilbestrol alone had little effect on weight gains, reduced feed efficiency slightly, and improved the finish to an appreciable degree. Torn backs were more common with stilbestrol only than when thiouracil also was fed. It was concluded that it is possible to finish Broad Breasted Bronze turkeys (as light roasters, over 82% grade "A") at either 17 or 21 weeks of age by the use for 4 weeks prior to killing, of thiouracil in the diet at 0.2% combined with implantation of 24 milligrams of stilbestrol.

The Beltsville Station (M. Swickard, et al, 1952) reported that the estrogen treatment (by implantation of one 15-milligram pellet at 8, 10 or 12 weeks, three weeks before slaughter) had no significant effect when compared to the control birds, on the tenderness, moistness, or turkey flavor of the cooked (braised) flesh but did tend to accentuate any off-flavor latent in the turkeys such as that engendered by fish products in the feed. In these BSW turkeys the treatments failed to produce significant improvements in growth, finish, and conformation.

The Grange Co. (Poul. Sci., 1952) reported the following from Beltsville White turkeys treated at 12 weeks of age.

Table 7.24—Gain from Hormone Treatment

Experimental groups[1]	Diethylstilbestrol per bird	Grain per bird 12 to 16 weeks	Feed Conversion 12 to 16 weeks	Increased income from treatment per bird[2]	Average dressing shrinkage[3]
	milligrams	pounds	pounds[4]	cents	per cent
Toms........	0	2.07	5.28	0.0	13.0
Toms........	15	2.95	4.74	18.1	12.2
Toms........	30	2.42	5.51	0.1	11.2
Hens..........	0	1.72	5.89	0.0	14.8
Hens..........	15	2.11	5.01	10.3	13.0
Hens..........	30	1.78	5.66	-1.1	12.3

[1]About 300 birds per group.
[2]Increase in market value, less feed and treatment cost as compared to controls. All birds sold at 40c per lb.
[3]Decrease from live to N. Y. dressed weights.
[4]Pounds of feed per pound of gain 12 to 16 weeks.

Finish (fat in the skin) was slightly improved by the stilbestrol treatment after 2, 3 and 4 weeks but the authors did not consider the treatment worthwhile from the standpoint of finish alone. However, better growth and feed conversion and lower dressing shrinkage in the treated turkeys combined to make the treatment economically advantageous when 15 milligrams of stilbestrol were used. The side effects of the hormone made strict control necessary to avoid carcass damage and mortality due to fighting and attempts at mating which occurred indiscriminately in both sexes. These favorable effects of early stilbestrol treatment are at variance with those reported by the other investigators. Need for further tests is thus evident. Optimum dosage and age at implantation however might have accounted for the favorable results since the other investigators used more or less than the optimum 15 mg. used by the Grange Company and implanted the pellets at different ages.

Hormone treatment to hasten the finishing process has been tried on older turkeys more or less successfully. The synthetic female hormone diethylstilbestrol alone has given only mediocre results, likewise the synthetic thyroid inhibiting hormone thiouracil. The Washington Station has shown (1951-52) that a combination of the two drugs gives good results in the fattening of pre-roaster and roaster-age BBB turkeys (page 436) and that neither drug affects the keeping quality of turkeys held in storage. Best results were shown by a group of turkeys selected at 17 weeks of age and given a combined treatment of 0.2% thiouracil in the pelleted all-mash feed and the implantation of pellets containing 24 to 36 milligrams of stilbestrol under the skin at the back of the head. About 80% of the turkeys of both sexes graded U. S., A Grade at 21 weeks of age. Similar good results could be expected with turkeys treated 4 weeks starting later perhaps at 20 weeks. A limit of 30 milligrams of diethylstilbestrol per bird and removal of thiouracil from the feed at least 72 hours before slaughter are necessary in order to comply with U. S. Food and Drug regulations. The combined hormone treatment was considered feasible for toms 17 weeks of age but not for hens.

A combination treatment of 16-week-old Broad Breasted Bronze toms with an implantation of 36 milligrams of stilbestrol (three 12-milligram pellets) and 0.2% thiouracil (a synthetic thyroid inhibiting hormone) in the all-mash feed containing 16% protein for 2 weeks (to 18 weeks of age) and 14.5% protein 2 weeks more (to 20 weeks of age) resulted (Alberta Station, 1951) in 76% Grade A birds averaging 16 pounds New York dressed at 20 weeks. Similar toms given only the 36-milligram pellet implantation were nearly as good, 71% A grade and a 15.4-pound average dressed weight. The thiouracil alone resulted in 41% "A", and a 15-pound average while the untreated controls were only 9% grade "A" and averaged 15 pounds.

Turkey Capons

Surgically caponized turkeys have not been placed on the market to any appreciable extent. They have no advantages over uncaponized young toms and are relatively expensive to produce. "Chemical caponization" by female hormone (stilbestrol) pellets or gel implanted under the skin at the back of the head of toms is discussed on pages 434-438.

Feathers

Turkey feathers have considerable market value but it is seldom worthwhile for any except plant operators to attempt to market them. When marketed, they either are picked without soiling or else they are washed and dried, the stiff feathers being separated from the "cotton" feathers. Chicken feathers contain approximately 73.15% protein, 21.6% water, 1.01% fat and 1.11% ash. Turkey feathers doubtless are of similar composition.

Work at the U. S. Western Regional Laboratory at Albany, California, indicates that processed chicken feathers make a good supplemental feed containing about 62% protein of which 9 to 12% is cystine, an amino acid needed for feather growth. The process of preparing feathers for use as feed is too costly at present, in relation to the value of the resulting protein concentrate. At the same laboratory (1950) it was found that wet-calcined turkey feathers produced a good fertilizer and that bristles for paint brushes could be made from feathers according to alcohol-keratin studies.

Feathers make an effective fertilizer when applied at the rate of two tons of the damp product per acre, allowing several months for incorporation into the soil.

A new use for 100,000,000 pounds of chicken feathers and 30,000,000 pounds from turkeys that are now wasted every year has been found by researchers of the U. S. Rubber Company at Passaic, N. J. The feathers were processed into a fabric which is described as warmer, softer, and lighter than wool, possessing a brilliant luster, can be dyed any color, and can be laundered with no more shrinkage than cotton. The new fabric would be suitable for suits, dresses, and sweaters. It is not yet in production but is one of the post-war developments that may possibly become a source of additional revenue to poultry producers.

Feather Yields. At the Beltsville Station (1944) a group of 30 Beltsville Small White young hens averaging 9 lbs. fasted live weight yielded 3.10 per cent blood and 6.93 per cent feathers; young toms gave similar percentage yields; a group of adult males of several colored varieties averaging 23.8 lbs. fasted live weight yielded 4.18% blood and 4.58% feathers. Feather yield of 6 month-old large type White Holland turkeys dry-roughed was reported by the Beacon Experimental Farm (1950) as 75 pounds of marketable body feathers and 68 pounds of marketable wing and tail feathers from 245 birds of both sexes. Sale value was $52.00.

FORECASTING

Forecasting the market involves information, thought, and study. Some of the necessary facts that one should have before reaching a conclusion as to when to market are (1) the approximate size and quality of the season's crop, (2) the cold storage holdings, not only of turkeys and other poultry but also other competing meats, (3) the purchasing power of consumers, (4) the cost of feeding and caring for the birds if held for a later market. Other factors of interest are the turkey poult hatchings as reported by the hatcheries, the carry-over of breeding stock and the season's average mortality.

Not all of these facts are available each season. The wisest procedure for the average turkey grower is to keep in touch with (1) the Agricultural College or Experiment Station of his own state, asking for their Outlook Reports and any news on the market situation; (2) the Agricultural Economics Division of the Agricultural Marketing Service, Washington 25, D. C., which prepares national outlook reports of cold storage holdings, breeders tested by months, seasonal reports of settings and hatchings of turkey eggs in hatcheries, a monthly report of "The Poultry and Egg Situation," crop reports and other reports of interest and value to the turkey raiser, (3) some good turkey journal that follows the market situation closely, and (4) the poultry columns of the farm papers and the turkey columns of the poultry journals. Addresses of these agencies are listed in the Appendix. The National Turkey Federation, Mt. Morris, Illinois publishes timely marketing and production information.

Irrespective of the conclusions that may be reached as to the state of the market, it must be kept in mind that young turkeys should be marketed when "ripe" since the feed cost is proportionately much higher thereafter. However, storage facilities sometimes may be utilized in an economical manner to hold dressed birds in anticipation of an improved market.

COST OF PRODUCTION

Table 7.25—Estimated Cost of Growing Turkeys

(From the Univ. of Nebraska)

Factors	Low Estimate	High Estimate
Poult cost per turkey raised (25% above actual cost when 80% of poults are raised)	$ .835	$1.25
Feed cost for an 18 lb. live-weight average		
(a) 40 lbs. of mash @ 3 to 4½c per lb.	1.20	1.80
(b) 40 lbs. of grain @ 2 to 2½c per lb.	.80	1.00
Fuel, litter, gravel, trucking, power	.02	.08
Interest and depreciation on equipment (25% charge)	.10	.25
Land rental including preparation and seeding	.01	.05
Interest on invested and borrowed capital	.02	.06
Insurance	.07	.10
Labor, paid and unpaid, and supervision	.20	.30
Estimated costs per bird	3.255	4.89
Estimated costs per lb. live wt.	.18	.27

Average of these estimates is $4.07 per bird or 22.5c per lb.; turkeys raised on range, in 1945.

Cost of production records in 1945 on 17 Illinois flocks (16 Broad Breasted Bronze and 1 Beltsville White) revealed that the average flock contained 3,704 turkeys at market age; the average cost, including labor, per pound of live BBB turkey produced was 26.4c; the average selling price was 33.9c; the mortality 25.6%; the average live weight when marketed, 18.9 lbs.; and the feed consumed per pound of live turkey produced was 5.8 lbs., of which 50% was mash.

Similar cost studies in 1946 in Illinois revealed that with 15 flocks (14 BBB and 1 Beltsville White) the average size of the flocks when finished for market was 3,857 birds; total cost, including labor, per pound of live turkey produced was 29.1c; average selling price was 33.3c; mortality was 27.3%; live weight when marketed was 18.5 lbs.; and feed consumed per pound of live turkey produced was 5.9 lbs. of which 48% was mash.

Table 7.26—Cost of Raising Turkeys in Illinois, 1947

Cost Item	Totals for the average flock—dollars	Totals per bird—dollars	Percentage of total cost
Poults started	3,794.37	1.02	15.02
Feed	17,566.09	4.73	69.66
Labor	2,623.82	.71	10.46
Transportation (tractor, truck, horse)	230.89	.06	.88
Use of buildings and equipment	421.68	.11	1.62
Miscellaneous	591.45	.16	2.36[2]
Total cost	25,228.30	6.79	100.00

[1]From "Turkey Production Costs in Illinois, 1947," by R. H. Wilcox and H. M. Scott. University of Illinois mimeographed circular.
[2]Included in this charge are 0.45% (of total cost) for fuel (soft coal and fuel oil), 0.42% for electricity, and 0.18% for litter.

In Illinois in 1947, the cost study involved 15 flocks with an average of 3,717 birds raised to market age, all except one Broad Breasted Bronze. The results are shown in Table 7.26. Mortality was 25.6; average weight when sold was 19.0 lbs. Net cost per pound of live turkey produced was 34.9c, the average selling price per pound was 41.8c. Average feed costs per ton were: starting mash $103.83, growing mash $102.82, protein concentrate $103.10, oats per bushel $1.06, corn per bushel $2.08.

A survey in North Carolina indicated that the over-all cost, including labor, of producing Broad Breasted Bronze roasting turkeys in 1946 was 34c per pound (live basis). Feed was 73% of the total.

Cost of production of market turkeys in Virginia in 1948 varied in 50 flocks from 28.1 to 43.1c per pound of live turkey produced. Costs were broken down as follows:

	% of total cost
Feed (including medicated feeds)	72.7
Poults	15.5
Labor	5.2
Use of buildings and equipment	1.8
Interest on operating capital	1.2
Miscellaneous (litter, electricity, transportation)	3.0
Range charge	0.6
Total ($5.48 per bird, or 33.0c per pound live weight)	100.0

Table 7.27—Breed Comparisons—Production Costs

	B. B. Bronze	Black	Beltsville White
Number of flocks	27	16	7
Number poults started per flock	1765	1511	2045
Mortality, per cent	15.3	14.8	17.1
Age marketed, weeks	25.7	26.4	23.1
Weight at sale, live, lbs.	18.3	16.3	11.2
Feed eaten per lb. sold, lbs.	4.9	5.3	4.9
Hours labor per 100 poults started	54.1	47.7	33.1
Receipts per lb. sold, cents	48.5	49.9	49.2
Total production cost per lb., cents	32.7	33.1	36.8
Net profit per lb. sold, cents	15.8	16.8	12.4
Return to labor per hour, dollars	5.02	5.38	3.99

From "Market turkeys—an economic survey," by James S. Plaxico, Va. Agrl. Exp. Sta., Blacksburg. Bull. 440 (1950).

These comparisons indicate that Beltsville White turkeys were produced at a cost of 4c per pound (live weight) greater than that of Broad Breasted Bronze. This extra cost was due to poult and overhead charges which were higher per pound of meat due to the smaller bird unit. In this particular instance no premium was obtained for the small-type turkeys which rendered them less profitable than the larger varieties.

The cost formula was as follows:
5.0 x cost of feed per pound = cost of feed.
0.07 x cost per poult = poult cost.
0.03 x cost of labor per hour = labor cost.
Total of above 3 items x 1.07 = estimated cost of production per pound.

Charges for use of buildings and equipment, taxes, and insurance were 10c per bird sold. Value of all houses and equipment used was $151.68 per 100 poults started. Interest on operating capital was 6c per bird sold.

Table 7.28—Cost of Raising Turkeys in Western Oregon, 1949*

Item	Cost per turkey raised	
	Bought as day old	Bought as started or custom brooded
Number of flocks	79	8
Number of birds raised per flock	2655	3233
Mortality, per cent	14.7	(not reported)
Labor	$ .63	$ .37
Feed	4.79	4.23
Poults	.98	1.90
Equipment and buildings	.33	.15
Miscellaneous	.28	.21
Total cost per bird	7.01	6.86
Less manure credits	.06	.06
Net cost per bird	6.95	6.80
Total returns per bird	6.69	6.68
Net profit per bird	−.26	−.12

*From "Cost of raising market turkeys in western Oregon, 1949," by Edgar A. Hyer. Oregon State College Circ. of Information 493, April, 1951.

Net profit per bird, all flocks $-.25 (loss)
Cost per pound of live turkey produced, all flocks 37.7¢
Total returns per pound of gain, all flocks 37.3¢
Average age when marketed: Hens 27.8 weeks, toms 29 weeks.
Averaged dressed weight when marketed: hens 13.9 lbs. (live, about 15.3 lbs.); toms 24.2 lbs. (live, about 26.6 lbs.). Feed was 68% of total cost and amounted to 107.6 lbs. per bird raised; 40.2 lb. of grain and 67.4 lbs. of mash and pellets. Grain cost $3.25 per cwt., and mash and pellets averaged $4.80 per cwt. Labor requirement for the 79 flocks bought as day-old averaged per bird raised as follows:

Brooding .20 man hour
On range .31 man hour
Indirect labor .10 man hour
Total .61 man hour

Average cost of labor per hour was $1.03.

The National Turkey Federation (1951) as the result of a nation-wide survey estimated that the average total cost per pound of live Broad Breasted Bronze roasting turkeys produced in the United States in 1950 was 34.55c of which 31.41c was for out-of-pocket costs which included feed, poults, litter, medicine, fuel and power. For Beltsville White roasting turkeys these figures were 35c and 32.45c respectively. The difference between the total and out-of-pocket cost includes charges for labor, interest, depreciation, insurance, taxes, and other costs not listed as out-of-pocket. For 1951 total costs were estimated as follows: Broad Breasted Bronze, 38.70c; Beltsville White, 39.97c.

Costs did not vary widely among the various large producing areas of the United States. The out-of-pocket costs for Bronze varied, in the areas surveyed, from 29.84 to 32.36c; for Beltsvilles, from 31.52 to 33.84c. The lower figures were obtained from certain mid-western areas, the higher figures from the east and west coastal areas in which costs were very similar to each other, in 1950.

Hedging

Provision for producers to "hedge" their crop has been made available on the Chicago Mercantile Exchange* since 1949. The trading unit is 10,000 pounds of dressed (N. Y. dressed or eviscerated) toms or hens, containing at least 80% of Grade A (U.S.D.A. graded officially). As set forth in the leaflet of the Exchange:—

"Hedging is buying or selling on an organized central market, when prices are favorable, for the purpose of eliminating or materially reducing the chance of loss to the producer, the wholesale buyer, or manufacturer, due to price fluctuation. It involves the relationship of the cash and futures market to each other. It is an operation that requires careful thought. Be sure to consult your broker before starting a hedging operation.

---

*110 No. Franklin St., Chicago 6, Illinois.

How Does A Producer Hedge? A producer of turkeys knows to within a reasonable fraction what his cost of growing will be before his turkeys are ready for market. Knowing this figure, he watches the futures market while his turkeys are growing. When the price on that market is satisfactory to him he sells at that time, for delivery after his turkeys are grown. Having thus hedged, and protected himself against a decline in market prices, he can make use of his regular distribution channels to market his turkeys.

Why Can He Use Regular Distribution Channels? When a producer sells futures contracts he is not selling his actual production. He is selling a contract to deliver. Details of this contract are specifically set out by Exchange rules. All contracts for delivery in a given month are exactly the same. Therefore these contracts can be sold or bought anytime prior to actual delivery time.

What Does the Producer Do When His Turkeys Are Ready For Market? He has two choices of action. (a) If he decides that he wants to use his regular distribution channels, he may do so. To do this, he must "Buy in" his hedge so that he will not have to make delivery against his sales contract on the Exchange. Then he sells his turkeys on the cash market. (b) If he finds it would be more profitable to deliver against his sales contracts on the Exchange than to market through regular distribution channels, he may do so.

Does a Producer Usually Deliver on a Hedge? No, although each transaction contemplates delivery. However, the cash and future markets may be in such position in the distribution season that it is advantageous for him to buy back his hedge and sell his turkeys on the cash market.

Delivery. When a producer who has hedged his turkeys decides to make delivery, he may do so on any business day of the delivery month. If he fails to make delivery and has not bought in his hedge, he will be in default. The Exchange rules provide that in case of default by the seller, a "settling price" will be established based on the fair market value of the commodity. A penalty which may be up to 5c per lb. but not less than ¼c per lb. must be paid by the seller who is in default.

Where Is Delivery Required? Each transaction is based on delivery in Chicago, so 'par delivery,'—that is, delivery with no penalties or other conditions—will be at the contract point, Chicago."

MARKETING SERVICES AND MARGINS

From the time the farmer in Texas or Minnesota decides to market his turkeys until some consumer in New York City or Chicago sits down to a dinner of roast turkey from one of those birds, many services are performed by many different persons and agencies. Whether the turkeys are marketed cooperatively or through other channels, these essential services must be performed and the cost thereof borne by the consumer. Many of these costs such as (1) transportation, (2) dressing, (3) packaging and (4) financing can be figured in advance quite accurately. Others are unpredictable with accuracy, such as (1) the hazard of anticipating demand of the various sizes and grades, (2) advertising, which may not be necessary if demand is keen and yet essential if the market is sluggish, (3) cost of storage, which may be for a short or relatively long period, and (4) deterioration in quality of a perishable product.

Obviously, after careful study of the contributing factors, all of the necessary expenses of marketing must be taken into consideration and deducted from the price level it is assumed the consumer will be willing to pay, before a price basis to offer the producer can be determined. If cooperative pools and associations take over some of the hazards and the association executives handle the marketing of the crop with shrewd judgment on sound business principles, the member producers may then receive some of the income that now goes to the commission man, jobbers and others. Sound judgment and executive ability based on practical experience must be paid well whether employed by the producer through his cooperative association or by private business houses engaged in poultry marketing.

A study of farm to retail marketing of turkeys from Washington, Oregon, and Utah* shows that wholesalers were the chief, 70 to 90%, receivers of western turkeys in carlots and they distributed them to retail or jobbing outlets. Brokers received most of the remaining 10 to 30% and wholesalers probably distributed them. Wholesalers operate in large volume and on a narrow margin of profit. They are in constant touch with price making markets therefore price charges seem to originate at this stage. Retailing in 1948-49 accounted for more than 40% of the total spread between producers and consumers for dressed young turkey hens and more than 50% for dressed young toms. Following are some of the retailing margins per pound for Grade A N. Y. dressed young turkeys from Western States. The larger, early-

Table 7.29—Retailing Margins of Turkeys in Three Markets

Date	Hens			Toms		
	New York	Chicago	San Francisco	New York	Chicago	San Francisco
	Cents	Cents	Cents	Cents	Cents	Cents
1940						
Sept. 15	16.4	17.5	11.5	15.4	....	0.1
Sept. 27	12.1	8.4	5.9	19.6	....	1.2
Oct. 15	13.3	9.6	7.6	13.8	14.5	0.2
Nov. 2	10.5	5.8	6.0	12.8	10.7	....
Nov. 15	10.1	6.6	7.1	10.0	7.6	7.0
Nov. 23	6.7	2.7	....	9.7	4.2	....
Dec. 15	7.8	9.6	6.5	5.7	12.0	7.3
Dec. 21	5.4	11.6	....	12.3	10.1	....
1949						
Jan. 15	6.7	18.6	....	11.0	10.5	....
Average	8.6	7.9	8.7	11.1	9.0	9.4

*From "Farm-to-retail margins for marketing western turkeys," by John O. Gerald, U. S. Bur. Agrl. Econ., Washington, D. C. Agric. Information Bull. No. 5, December, 1949.

season margins reflect the relatively small volumes of retail business done at that time. The New York and Chicago figures include jobbers' margins; San Francisco figures do not.

Retailing margins per pound on eviscerated young Grade A western turkeys sold in New York prior to Christmas 1948 were about 12c for hens and 13.5c for toms. In San Francisco for the Thanksgiving and Christmas markets they averaged about 6.5c for both hens and toms.

Wholesaling margins made up about 14% of the total spread for N. Y. dressed hens sold in New York and for both hens and toms in San Francisco, 5% for hens sold in Chicago, and 1.4% for toms sold in New York. In New York, wholesaling margins between September 15, 1948, and January 15, 1949, averaged 2.64c per pound on New York dressed hens in New York, 0.73c in Chicago, and 2.57c in San Francisco. Comparative figures for New York dressed young toms were 0.24c, -0.57c, and 2.67c, respectively. Speculative activities may account for the low and negative margins on toms in New York and Chicago. Wholesaling margins on eviscerated young hens per pound, just prior to Thanksgiving and Christmas, averaged 2.54c in New York and 4.47c in San Francisco. On eviscerated young toms they averaged 4.19c in New York and 4.87 in San Francisco.

Transportation and icing charges from the West to New York were about 22% of all marketing charges, to Chicago they were about 20% and to San Francisco only 6%.

Margins of processors (not including assembling birds at the plants) were uniform 3.8c per pound (for killing, picking, cooling, grading, boxing and short-time storage) in Utah where custom dressing was the rule, but in Oregon and Washington where processors assembled and bought the turkeys, the margins varied. In Oregon from September 15, 1948, to January 15, 1949, the average processors' margins, including assembling at the plants, were 5.3c for hens and 5.4c for toms; in Washington they were 7.9c and 8.3c, respectively. These assembling and processing charges were about 22% of all marketing charges for birds sold in New York, 26% for those sold in Chicago, and 29% for those in San Francisco. Processing charges for evisceration by processors in Oregon amounted to about 4.5c per lb., dressed weight, in addition to the dressing charges.

Following are the processing charges broken down into defined and non-defined costs, in cents per pound (New York dressed) for specified markets:

Table 7.30—Processing Charges (cents per lb.)

Function	Hens			Toms		
	New York	Chicago	San Francisco	New York	Chicago	San Francisco
Assembling	.30	.30	.30	.30	.30	.30
Killing and dressing	1.25	1.25	1.25	1.00	1.00	1.00
Cooling	.25	.25	.25	.25	.25	.25
Grading and packing	1.25	1.25	1.25	1.25	1.25	1.25
Other (non-defined charges)	1.35	.95	2.05	1.70	1.40	2.70
Totals	4.40	4.00	5.10	4.50	4.20	5.50

Average selling prices in cents per pound for Grade A, New York dressed turkeys from Oregon, Utah, and Washington in specified markets, between September 15, 1948, and January 31, 1949, are given in Table 7.31.

MARKETING BY PRODUCERS

Many turkey growers are successful in marketing their own products to wholesale or to retail consumers thus taking over all or part of the processing, transportation, wholesaling and retailing functions normally performed by others. Nearness to market and the interest and ability to perform these functions are prerequisite to success.

Table 7.31—Selling Price and the Consumer's Dollar

(cents)

Agency	Hens			Toms		
	New York	Chicago	San Francisco	New York	Chicago	San Francisco
Producer, at the farm........	58.5	59.3	59.5	49.8	49.7	48.6
Processor........	62.9	63.3	64.6	54.3	53.9	54.1
Wholesaler......	70.0	67.3	68.2	59.0	56.6	57.8
Retailer...........	78.5	75.2	76.9	70.1	65.6	67.2

Percentage distribution of the consumer's dollar for grade A, New York dressed turkeys from Oregon, Utah, and Washington, specified markets, for September, 1948, to January, 1949, are given below:

(Per Cent)

Agency	Hens			Toms		
	New York	Chicago	San Francisco	New York	Chicago	San Francisco
Producer..........	74.5	78.9	77.5	71.1	75.8	72.3
Processor........	5.8	5.3	6.6	6.4	6.6	8.2
Transportation	5.5	4.3	1.3	6.4	5.0	1.5
Wholesaler......	3.4	1.0	3.3	.3	−.9	4.0
Retailer...........	10.8	10.5	11.3	15.8	13.7	14.0

The gift package business on a wholesale or retail basis is one of sizeable proportions.

Novel ways of representing a more attractive article to the consumer constantly present themselves to the alert turkey grower. An attractive sticker or brand containing the brand name may be placed on the breast or leg of the dressed bird or sealed in the web of the wing. Wrapping in transparent air-tight material increases the attractiveness especially if tied with colored cellophane ribbon and keeps dirt and germs from the carcass. Individual New York dressed birds, either chilled or frozen, may be sent by parcel post or express direct to the consumer in cold weather. Hard-frozen drawn turkeys also may be shipped in well-insulated containers but if shipping should require more than about 24 hours the shipping carton should contain dry ice. The amount of dry ice should range from ½ to 1 pound per turkey, depending on the time in transit. The turkey should be wrapped tightly and sealed in air-tight wrapping material, waxed paper, or parchment, then placed among paper shavings. The box should be insulated and sealed tightly.

Fig. 7.2. An attractive package. (Photo from Omaha Cold Storage Co.)

TURKEY SPECIALTIES

Canned Turkey

Canned turkey meat frequently can be sold at a good price for use in salads and sandwiches. Some turkey growers have found it a profitable practice to kill and can the breeding stock in June as soon as hatching eggs are no longer needed. The cans may be priced according to the flavor and tenderness of the meat from the various sections. Canned white meat should be especially in demand for parties and luncheons.

Smoked Turkey

Smoked turkey is on the market in a limited way. Some is highly spiced which results in a product quite different in flavor from that obtained by a standard

Table 7.32—Weight Changes in Cured and Smoked Turkeys

Cause of weight change	Basis for computation	Average for both sexes	
		Standard bred and Small-type	Broad Breasted Bronze
		Percent	Percent
Overnight fasting or long haul	Loss; of live field weight	3.0	3.0
Dressing (bleeding and picking)	Loss; of fasted live weight	10.0	9.0
Drawing (removal of entrails, head, legs, preen gland, gizzard contents, crop, and windpipe)	Loss; of dressed weight	15.0	13.0
Special drawing prior to curing (removal of neck, giblets, and loose fat)	Loss; of drawn weight Loss; of dressed weight	13.0 11.0	13.0 11.0
Curing in brine	Gain; of pre-curing (specially drawn) weight	5.0	5.0
Smoking 20 hrs. at 110° F.	Loss; of pre-curing weight	7.0	7.0
Aging 28 da. at 68° F.(unwrapped)[1]	Loss; of pre-curing weight	20 to 25	20 to 25
Relationship of ready-to-cure weight to	Drawn weight Dressed weight Live, fasted weight Live, field weight	87.0 73.9 66.5 64.6	87.0 75.7 68.9 66.8
Relationship of weight of carcasses freshly smoked at 110° F. for 20 hrs. to	Drawn weight Dressed weight Live, fasted weight Live, field weight	80.9 68.8 61.9 60.1	80.9 70.4 64.0 62.1

[1]This loss can and should be prevented by as near air-tight wrapping as possible.

brine cured and "cool" smoking process which is described in a leaflet (A.H.D. 28) obtainable free from the U.S.D.A. Research Center, Beltsville, Maryland. Turkeys processed in this manner must be cooked before eating. Smoke-cooked turkeys ready for eating without further cooking may be produced by a method described in Texas Station Bulletin 163 (1948) and also in A.H.D. 28. Smoke-cooked turkeys weigh about half the birds' original live weight. The spice flavoring of smoked turkeys was described by Highlands and Burns in the U. S. Egg and Poultry Magazine, August 1941.

In Table 7.32 will be found the weight changes that may be expected when turkeys are cured and smoked in a standard manner (not smoke-cooked).

Brine Cure

Birds are drawn and chilled. Add 2 lbs. special poultry smoking salt to each gallon of boiling water. Stir to dissolve, but do not boil. Strain through cloth and cool to 40° F. Place drawn, cleaned birds in stoneware crocks and cover completely with brine. Store at 40° F (approx.).

Chickens and turkeys under 12 lbs.—1 week. Larger birds stay in solution 1 day per each 3 lbs. weight above 12 lbs. Remove, rinse, wipe dry and hang in cool room. Wrap in waxed paper, preferably use within 2 weeks.

Preparatory to cooking, soak in cold water to remove some of the salt. Roast on rack and don't baste with fat.

Barbecued Turkey

Popularizing the turkey as splendid meat for barbecuing offers possibilities for increasing the consumption of turkeys. In the January 19, 1935 issue of the *Saturday Evening Post* Hermann B. Deutsch gave an excellent description of barbecuing turkey. It is quoted (by permission) in the Appendix. Additional information on barbecuing is available through the National Turkey Federation and Poultry and Egg National Board.

Fig. 7.3. BSW tom roasted to serve; this bird weighed 7 lbs. (live) when slaughtered at 14 weeks. (Photo from U. S. Dept. of Agr.)

Cooking Broiler Turkeys

The Poultry and Egg National Board* which cooperates with the National Turkey Federation** in the "Eat More Turkey" program, has issued cooking instructions available in pamphlet form on request. With the popularization of the turkey broiler (spring or fryer, see Fig. 7.3) in the period 1949-51 the Board issued cooking instructions based on their laboratory (kitchen-tested) trial as follows:

> "How to broil a small turkey. Choose a young turkey about 4 lbs. Have turkey split in half lengthwise for broiling. Snap drumstick, hip and wing joints to keep bird flat during broiling.
>
> **Flatten halves**—Skewer leg to body. Fold wing tip back under wing. Skewer wing flat against cut edge of backbone.
>
> Season each half turkey with about ¼ teaspoon salt and a sprinkling of pepper.

*185 No. Wabash Ave., Chicago 1, Ill.
**Mount Morris, Ill.—See Section 11 of Appendix for the N.T.F. instructions on cooking turkeys.

Place turkey in broiler pan (not on rack). Brush with melted fat and then place skin side down.

Broil slowly—Place in broiler 7 to 10 inches under heat source. Regulate heat or pan position so that turkey just begins to brown lightly in 15 minutes. Cook slowly.

Turn and brush with fat. Baste frequently during broiling to brown and cook evenly. Total cooking time, about 60 minutes for 4 lbs. ready-to-cook turkey.

Test for doneness—The turkey is done when meat on the thickest part of the drumstick cuts easily and there is no pink color visible.

Serve on warm platter with pan drippings poured over turkey.

If giblets are served with broiled turkey, coat liver and precooked heart and gizzard with fat, season and broil just long enough to brown."

Advertising

Surely no farm product gets more favorable free publicity than the turkey at Thanksgiving and Christmas time. It only remains for the growers, handlers and retailers to pool their efforts to stimulate turkey consumption at other times. This may be done through distribution of cooking instructions, such as those given above, with the name of the producer or the retailer. Chain stores through their national association have done much good along this line and it behooves the producers to solicit their support and good will. The big packing companies through their printed advertising and radio programs have also aided greatly in stimulating consumer demand for turkeys. Turkey consumption has increased 277% from 1930 to 1950. Per capita consumption reached 6 pounds per person in 1950.

Price

As the price of turkeys more nearly approaches that of chickens and other meats with which they compete

Table 7.33—Comparative Turkey and Chicken Prices
(Chickens 100%)

	Farm price cents per pound		Turkey price as percentage of chicken price
	Turkey price	Chicken price*	
1940	15.2	13.9	109
1941	19.9	16.3	122
1942	27.5	19.8	139
1943	32.7	25.3	129
1944	33.9	24.9	136
1945	33.7	27.0	125
1946	36.3	29.0	125
1947	36.5	28.2	129
1948	46.8	32.3	145
1949	35.2	26.6	132
1950	32.8	24.9	132
1951	37.4	27.1	138
1952	33.6	26.2	128
1953	33.6	25.3	133
1954	28.8	21.0	137

*All farm chickens including broilers.

for a place on the family table the number which can be sold will naturally increase. If prices fall, in relation to other food, more turkeys can be sold. The American public is highly desirous of eating turkey and has in the past limited its consumption chiefly because of the price discrepancy. A study of the farm price (see Table 7.33) of chickens and turkeys received during the past 20 years shows very clearly that turkeys remain above the price level of the highest grade of chickens (prime roasters). MacDonald and Cray (1952)* have related all of the predictive factors of value with the price of turkeys. They found U. S. chicken price most closely related to turkey price (95.8%). Other factors related to and affecting the price of turkeys (ranked in order of importance) were found to be:—wage income of industrial workers 85%, U. S. beef price 84.8%, U. S. hog price 79.4%, U. S. corn price 72.6%, size of U. S. population

*U. S. Egg and Poultry Mag. 58(2):16, 31-33.

59.8%, U. S. cold storage holdings of turkey 39.4%, value of U. S. turkey production 36.7% and the general U. S. price level 21.8%. In short, when one tries to "guess" the oncoming market price and situation he should tie to chicken, beef and hog prices and the wage income of industrial workers as the most reliable predictive factors.

Financing

Turkey raising in commercial numbers involves considerable financing which may be carried by the grower himself, the Federal Agencies through Production Credit and Farm Security, feed companies, processing plants, cooperative organizations, banks, finance companies, hatcheries, and landowners or other private individuals. To obtain financing a grower must furnish a detailed statement of assets and liabilities, along with an estimate of costs of the project, which should include cost of poults, feed, transportation, fuel, litter, vaccination, disinfectants, medicines, insurance, and miscellaneous items, the total of which may vary from about $1.25 to raise a small turkey fryer to $8.00 per bird raised, depending on locality, management, and type of turkey raised. He must sign delivery receipts or notes for all goods delivered. In addition to these costs, there are those relating to land, buildings, equipment, and labor. Feed companies usually limit financing to their own products, supervise closely the growing and marketing, and require the grower to furnish not less than 1/3 of the cash cost of raising the birds which might consist of the money for the poults or the grain.

A common type of partnership agreement is that in which the farmer furnishes land, labor, houses, equipment, and transportation, and the financing agency furnishes poults—all at no charge. The cost of raising the turkeys (feed, litter, fuel, vaccine, disinfectants, insurance, and miscellaneous) is then advanced by the financing agency and deducted from the proceeds of the sale of the turkeys. The profit remaining after this deduction is then divided equally between the two parties.

Another popular type of partnership is based on the farmer's furnishing land, labor, housing, equipment, and transportation, and the financing agency's furnishing poults, feed, insurance, and other rearing necessities. When the turkeys are sold, the grower receives five per cent of the total proceeds of the sale and the costs of rearing are deducted from the remaining 95 per cent of the proceeds. The profits resulting from this procedure are then divided 60 per cent to the grower and 40 per cent to the financing agency. If a loss occurs, the financing agency absorbs it but the grower still receives his 5 per cent of gross sales.

In any case, a detailed contract in writing should be drawn up and signed. A chattel mortgage note is usually signed by the grower in favor of the financing agency. Good equipment is commonly demanded of the grower and insurance against losses due to bad weather are usually required. The terms of any contract should be thoroughly studied and understood by both parties for their mutual protection.

Landowners sometimes give their hired men a place to live with garden, milk, eggs, meat, and a monthly salary plus 10 per cent of the net profits. Contract growing is another form of financing in which the financier furnishes everything and pays the grower a stated amount for each turkey raised, the amount increasing if more than 75 per cent of the turkeys are raised.

CHAPTER 8

TURKEY SHOWS AND SHOWING

OUTLINE

Purpose and Value of Shows

Poultry shows provide for competition between the breeders and producers of poultry for the exhibition of breeding stock or potential breeding stock, and for the comparison of results of individual efforts to breed and raise domestic birds. Shows have educational and advertising value. Turkeys are exhibited in almost every poultry show along with chickens and other poultry. They also are exhibited in separate shows—live, dressed, or a combination.

The educational possibilities of shows should not be overlooked. Instructional programs, dinners, panel discussions and round-table discussions add to their value and interest. Among the special exhibits that may be included in addition to the regular show features are crossbred birds, freak specimens, pets, commercial displays (including dressed and cooked specimens if possible), bulletins, and books. Special classes of turkeys for junior exhibitors particularly 4-H Club members and Future Farmers are a good feature.

Organization, Terminology,* and Management of Shows

General Aspects. Since poultry shows present a fine opportunity for constructive building of the turkey breeding industry, the show management should ultilize every chance to make it attractive, to advertise the show so that people will come, to organize and label the coops so that people will understand, and to group the varieties and classes so that the judge can do efficient work and the visitors can readily compare the competing birds. Adequate and well-arranged coops with proper lighting; coops each labeled with the entry number, variety, class, exhibitor's name and address, band number, and weight of the bird; colored stickers indicating the winnings; all these help to make a show successful. Each single entry should have a coop and each pen should have a large pen-size coop. It is not advisable, unless absolutely necessary, to double-deck the coops.

Financing. The cost of putting on a show consists largely of hall rent (frequently donated), coop rent, labor, transportation, judging expenses, premiums, and printing the premium list. Help in defraying these expenses can often be obtained from interested parties or agencies. Credit for advance expenditures sometimes is obtainable through an association. Income may be obtained from entry fees, gate receipts, advertising in catalogs and booths at the show, and donations. The entry fees can be made to contribute largely to the expense of judging, cooping, ribbons, and cash prizes. The gate receipts and advertising fees sometimes will pay the remainder of the expenses. However, a high admission or gate charge may deter many persons whose presence would be to the advantage of the exhibition. Many shows are free; still others use a nominal gate fee of 10 to 25 cents, groups of school children under supervision being admitted free. Cash and donations, such as feed for the birds, and transportation facilities, may be accepted. A show usually

*See Appendix for more complete list of definitions.

has a manager or superintendent, a secretary and assistants to care for the birds and help the judges. Vocational agriculture high school seniors make good assistants and the exhibitors are usually willing to help, due to their natural interest in the show.

The Standard Live Show. The principal, and often the only, feature of many turkey shows is the exhibition section where birds compete chiefly from the standpoint of standard quality, that is, the ideal shape, color, and size for each variety as recognized by the American Poultry Association and described in the "Standard of Perfection" also are provided for non-standard varieties and their showing should be encouraged.

The birds in the various classes of each variety compete among themselves for the regular numbered prizes and the winners of the classes in each variety compete against each other for the Champion and Reserve Champion awards. Sometimes awards for "Best Male" and "Best Female" are given. For example, all the young Bronze toms compete against each other for regular ribbons and after these awards are made, the first-prize young tom of the Bronze competes against the first prize young tom of each of the other varieties for the award of "Champion young tom." Sometimes second-prize birds are considered for Reserve Champion if they happen to be better than any first-prize bird in the other competing varieties or classes.

In the showroom, 3 or 4 females (4 is the number stipulated in the Standard) and a male constitute a pen. There are young pens and old pens, depending on the age of the birds composing them, birds under 12 months of age constituting a young pen and birds 12 months and older an old pen. A trio is a male and two females of the same age grouping; a pair is a male and a female of the same age grouping. There may be young and old trios, and young and old pairs.

Turkeys, however, seldom are exhibited as pens, trios, or pairs. Some shows with limited coop space substitute breeding trios (1 tom and 2 hens) for the pens. Turkeys are most frequently exhibited as singles classified as to age and sex. In Standard live shows the following classes of turkeys usually are provided far each variety: (1) young toms, (2) yearling toms, (3) old (or adult) toms, (4) young hens, (5) yearling hens, (6) old (or adult) hens. Young toms and young hens are birds under 12 months of age; yearling toms and yearling hens are birds 12 months old but under 24 months; old (adult) toms and old (adult) hens are birds 2 years old or older. Usually in small shows all the turkeys of one variety 12 months or older are grouped together as "old hens" and "old toms," respectively. In some shows the above young classes are subdivided into heavy and light classes in the large varieties in which there are many entries. Exhibition classes are supposed to be judged on type (or shape), symmetry, carriage, color, weight, vigor, and condition weighted according to the Standard and Scorecard (p. 469). Condition in the showroom refers to health, cleanliness, sleekness, completeness of plumage, and general appearance. The classes provided, the prizes, entry fees, names of judges, and other pertinent information should be given in an announcement published well in advance of the show.

The Meat Show. The live meat show often is substituted for the Standard live show and may provide for the following classes: (1) young toms, 2 or more weight classes, (2) old toms, all weights, (3) young hens, 2 or more weight classes, and (4) old hens all weights. Live market classes are judged on meat type, fleshing, finish, health, and vigor. Little or no attention usually is paid to the so-called standard qualities, but health and apparent vigor usually receive recognition. A special scorecard such as the one shown on p. 477 may be provided or the Standard scorecard used. The prevailing tendency now is for judges to place the turkeys mainly on body type with

Fig. 8.1. Exhibit of N. Y. dressed turkeys on sun porch enables keeping the birds chilled (30° to 40° F.) while on exhibit in winter months. (Photo by Robt. L. Hogue.)

emphasis on broad-breastedness. All varieties may compete against each other in the live meat show but usually there are separate classes for the different varieties.

The weight classifications suggested in Table 8.1 may be applied to all varieties and ages of turkeys. The weight classification used whether A, B or C, would depend mainly on size of the show. Classification A is suitable for large shows. Classifications B and C are suitable for the small show. Classification C is a simple breakdown into two weight classes which works well, since it results in the large, broadbreasted varieties competing as "heavies" and the other varieties as "lights."

Table 8.1—Showroom Classification in Pounds for Live, Dressed, and Drawn Turkeys

Classification	Live		Blood-and-Feather, or New York, dressed		Full-dressed (drawn, or oven-ready, or eviscerated	
	Toms	Hens	Toms	Hens	Toms	Hens
A						
Very large	36 and over	22 and over	31 and over	19 and over	27 and over	16 and over
Large	28, under 36	17, under 22	24, under 31	15, under 19	21, under 27	13, under 16
Medium large	21, under 28	13, under 17	18, under 24	12, under 15	15, under 21	10, under 13
Medium	16, under 21	10, under 13	14, under 18	9, under 12	12, under 15	7½, under10
Small	under 16	under 10	under 14	under 9	under 12	under 7½
B						
Large	26 and over	16 and over	23 and over	14½ and over	20 and over	13 and over
Medium	18, under 26	11, under 16	16, under 23	10, under 14½	14, under 20	8½, under 13
Small	under 18	under 11	under 16	under 10	under 14	under 8½
C						
Large, or heavy	26 and over	14 and over	23 and over	13 and over	20 and over	11 and over
Small, or light	under 26	under 14	under 23	under 13	under 20	under 11

Sometimes there are classes for Broad Breasted Bronze or Whites and another set of classes for other than broad-breasted varieties.

The Dressed Turkey Show. Dressed shows usually provide no classes for the different varieties because of difficulty in determining variety when feathers have been removed, instead, a number of weight classes are set up within the one or two divisions of the dressed show—I. Dressed and II Drawn.

Dressed turkey classes usually include the ones enumerated above for the live meat show. Each young class in most shows is subdivided into two or more weight classes. The old-turkey classes sometimes are subdivided according to weight but, more frequently, due to small participation, are set up to include all weights. Suggestions for weight classes are given in table 8.1. "New York" or blood-and-feather-dressed or simply "dressed" should be neatly head-wrapped and clean. Full dressed and trussed, or "drawn," ready-to-cook turkeys should have head, shanks, and inedible viscera removed; giblets cleaned, neck removed at base, and giblets and neck placed in the body cavity or crop space. They may or may not be wrapped in transparent covering. The show management usually publishes pertinent rules relating to shipping, wrapping, freezing et cetera. Usually New York Dressed specimens are exhibited unfrozen while drawn specimens frequently are exhibited frozen. Frozen turkeys are difficult to judge but less likely to spoil. For efficient judging, dressed or drawn turkeys should be thawed at least on the surface. All wrappings that are not transparent should be removed.

Old turkey classes could be divided in the manner shown in table 8.1 by adding 3 lbs. to the above weight requirements of hens and 6 lbs. to those of toms.

Exhibits of drawn birds must be kept cold after they are arranged according to class in refrigerated showcases at a temperature of not over 40° to 45° F., preferably about 35°, or they may be kept hard-frozen. When refrigeration is not available it is practically impossible to preserve drawn birds and difficult to properly preserve dressed birds, although the latter is possible in cold weather by maintaining the temperature of the showroom for drawn birds at about 50° F. during the day and lower at night. Dressed birds will keep for several days if well chilled after slaughter. They may be displayed on tables. If they must be kept in the room with the live birds (at about 60° F.) they are not safe for more than two days unless they can be placed under refrigeration each night at a temperature of 35° to 40°. After such treatment during the show period, they should be consumed immediately. It is not advisable to try to exhibit drawn birds without adequate refrigeration (unless weather is very cold). Refrigerated showcases are best although bakery showcases may be chilled sufficiently with dry ice.

Setting Up and Managing the Live Show. Running the show is largely the duty of the secretary and superintendent. The secretary generally handles the following items:

(1) Solicits exhibitors
(2) Solicits donations for show expenses and prizes
(3) Puts out the catalog (a booklet or merely a handbill) listing the prizes, judges, classes to be shown, place and date of show, deadline for entries, rules of the show, fees, explanations, and advertising (which usually pays for the catalog).
(4) Advertises the show through newspapers, handbills, and signs
(5) Secures the judges
(6) Aids amateur exhibitors to select and prepare their birds
(7) Secures showroom, coopage, etc.
(8) Classifies and coops the birds
(9) Makes out list of awards as soon as the judges place the birds

Consult the Appendix, Section 9 for simple rules and plans for turkey shows.

Judging Live Turkeys. Poultry is now judged mainly by comparing the birds with each other. All points are considered and final selection made to determine which are the best birds in each class of each variety. The judge if competent is able to determine which bird is nearest the ideal in color, type, or shape, condition, and size. To provide a guide, a turkey scorecard has been perfected by the American Poultry Association which shows the value of color and type in all sections of the bird's body.

For example, in placing an exhibition class of Bronze young toms, the judge makes the placings by comparison of the birds, one with the other, but he keeps in mind that, among other things, 8% of the exhibition value of a turkey is placed on weight, that is, standard weight. Consequently, it is very difficult in close competition, for a judge to place an undersized bird very high no matter how good it may be otherwise. On the other hand, an extremely large bird should receive no additional consideration merely because of its weight. It should be compared on other points with those competing birds that are up to standard weight. The judge recognizes the fact that according to the official scale of points, 85% of the show points of turkeys (all varieties) are type points including 10 points for symmetry, 10 points for condition and vigor, and 8 points for weight. Although color is important it is not so important as type (which includes size) when one sums up the points allotted in the Standard. If the birds

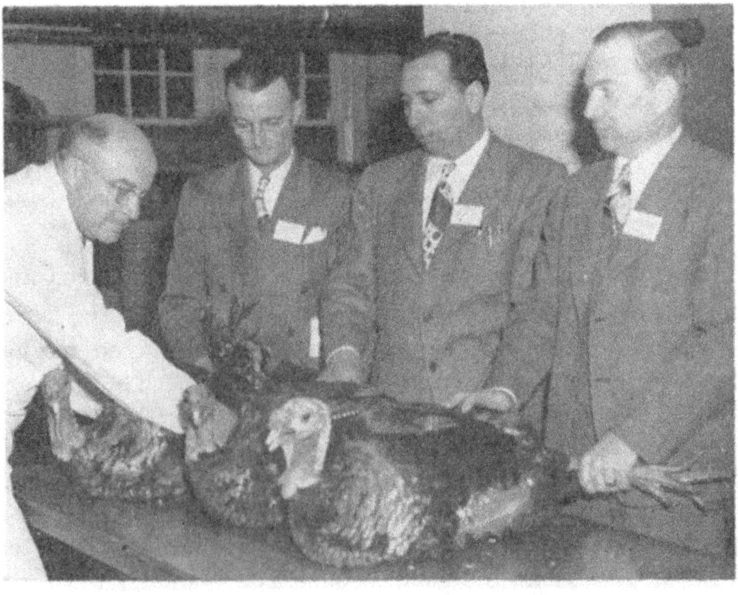

Fig. 8.2. Judging live turkeys promotes good sportsmanship and has educational value. (Photo by Robt. L. Hogue.)

were not weighed when cooped the judge should have them weighed. Ringside judging as commonly practiced is interesting and instructive to the exhibitors and the spectators. In ringside judging, the competing specimens in each class may be held and displayed by their owners or other interested persons. A feed sack loosely filled with straw may be placed on the table to avoid injury to the bird's breast while it is resting on the judge's table. Body type, symmetry, fleshing and balance can be judged most accurately if the bird is held suspended by the legs, by an assistant.

Table 8.2—Standard Scorecard for Exhibition Judging of All Varieties of Turkeys (From 1953 Standard)

Section	Shape Points	Color Points	Total Points
Symmetry and carriage	6	0	6
Condition	10	0	10
Weight	8	0	8
Head	3	1	4
Eyes	1	1	2
Throat wattle	1	0	1
Neck	3	0	3
Back and spring of ribs	10	3	13
Tail	3	3	6
Wings ahd shoulders	6	2	8
Breast and keel	25	3	28
Thigh and fluff	6	1	7
Legs and toes	3	1	4
Totals	85	15	100

The exhibition scorecard and the illustrations and descriptions in the Standard of Perfection together constitute the basis for the judging of exhibition classes. They embody features both beautiful and useful. The ideal standard turkey is not only a beautiful bird but an excellent market bird as well. The standard scale of points was changed in 1946 and 1953 to describe more adequately the modern meat-type turkey. The scorecard is given in Table 8.2.

The accompanying tentative scorecards, explanatory remarks, and record forms are presented as an attempt to set up grades and standards for the exhibiting and judging of turkeys with what is believed to be proper emphasis on market quality.

Judging Dressed Turkeys

Dressed classes, both blood-and-feather-dressed, and full-dressed, are judged on type, fleshing, finish, and dressing. The following scorecards for dressed market classes were adopted by the National Turkey Federation in 1952.

SCORECARD

Blood-and-Feather Dressed

Entry No.________________ Class__________

Sex and Age________________ Variety__________

Exhibitor's Name __________________________

Address__________________________________

		Cuts	Entry Score	Perfect Score
FLESHING AND FINISH—55 points				
Thickness and distribution of flesh				35
Breast	— 18 points			
Thighs (upper and lower) and wings	— 12 "			
Shoulders and oysters (back)	— 5 "			
Quality and distribution of fat				20
In feather tracts*	— 6 points			
Between feather tracts*	— 6 "			
Abdomen**	— 5 "			
Back	— 3 "			
DRESSING AND CONDITION—25 points				
Bleeding				5
Bloom and color of skin				5
Pinfeathers, "ink", stubs, and hair				5
Picking defects and injuries				5
Cleanliness, empty crop, and headwrap				5
TYPE OR CONFORMATION—20 points				
General symmetry				5
Shape, length, and flexibility of keel				10
Neck and legs				3
Back				2
TOTALS				100

*Cut toms for excessive spongy skin on breast.
**Cut for excessive fat; also hens with large anal opening.

SCORECARD

Ready-to-Cook or Drawn

Entry No.______________________ Class______________

Sex and Age______________________ Variety______________

Exhibitor's Name ______________________________

Address______________________________________

	Cuts	Entry Score	Perfect Score
FLESHING AND FINISH—55 points			
Thickness and distribution of flesh			35
Breast — 18 points			
Thighs (upper and lower) and wings — 12 "			
Shoulders and oysters (back) — 5 "			
Quality and distribution of fat (cut for excessive fat)			20
In feather tracts* — 6 points			
Between feather tracts* — 6 "			
Abdomen** — 5 "			
Back — 3 "			
DRESSING AND CONDITION—25 points			
Bleeding			5
Bloom and color of skin			5
Pinfeathers, "ink", stubs, and hair			5
Dressing and other injuries			5
Cleanliness and fresh appearance			5
TYPE OR CONFORMATION—20 points			
General symmetry			5
Shape, length, and flexibility of keel			10
Neatness of trussing			3
Back			2
TOTALS			100

*Cut toms for excessive spongy skin on breast.
**Cut for excessive fat; also hens with large anal opening.

EXPLANATION OF CUTS USED IN JUDGING DRESSED AND DRAWN TURKEYS*

A. Fleshing and Finish—55 points — Points to Cut Score

1. Amount and distribution of flesh—35 points
   a. Breast to be fully fleshed—18 points.
      Cut ½ to 12 points for deficiency in fleshing.
   b. Legs and wings fully fleshed, drumsticks well rounded —12 points.
      Cut ¼ to 8 points for deficiency in fleshing.
   c. Shoulders and oysters well filled and smooth—5 points.
      Cut ¼ to 3 points for lack of fullness.
      Cut ¼ to 1 point for roughness over back.
2. Finish—quality and distribution of fat—20 points.
      Note. Color of fat and skin optional, white or yellow.
   a. Feather tracts well covered with fat—6 points.
      Cut ¼ to 4 points for lack of fat.
   b. Even layer of fat in skin between feather tracts and on thighs and drumsticks—6 points.
      Cut ¼ to 4 points for lack of fat in these sections.
   c. Back well covered with fat but not excessively bunchy —3 points.
      Cut ¼ to 2 points for lack of fat.
      Cut ¼ to 1½ points for excessive bunchiness.
   d. Abdomen, fat but not excessively so; vent of normal size—5 points.
      Cut ¼ to 3 points for overfatness or distended vent.

B. Dressing and Condition—25 points.

1. Bleeding—5 points. (See disqualifications.)
      Cut 1 to 2 points for slightly poor bleeding in neck.
      Cut 2 to 3 points for slight evidence of poor bleeding in other sections.
2. Bloom and color of skin—5 points.
   a. Skin smooth in texture, even in color, and free of gall spots (greenish areas) and flakiness.
      Cut ¼ to 1½ points for flakiness or roughness of skin.
      Cut ¼ to 1 point for uneven skin color.
      Cut 1 to 2½ points for gall spots.
      Cut ½ to 3 points for lack of freshness.
3. Pinfeathers, "ink," stubs, and hair—5 points.
   a. Carcass to be practically free of pinfeathers, stubs, hair, and feather pigment.
      Cut ½ to 3 points for scattered pinfeathers on breast.
      Cut ¼ to 3 points for scattered pinfeathers elsewhere.
      Cut ½ to 1½ points for feather pigment on breast.
      Cut ¼ to 1 point for feather pigment in other sections.
4. Picking defects—5 points. (See disqualifications.)
   a. For scuffs, abrasions, and slight bruises on breast—cut ½ to 2 points.
      For scuffs, abrasions, and bruises in other sections—cut ¼ to 2 points.
      For removed crops—cut ½ to 1 point.

*According to foregoing scorecards.

b. For tears neatly sewn, and skin and flesh bruises in sections other than breast—cut ½ to 1½ points.
c. For full-dressed carcesses—cut ¼ to 2 points for excessive cutting of skin and lack of neatness in removing neck and crop.
Cut ¼ to 2 points for excessive cutting of skin and lack of neatness in removing entrails.
Cut ½ point for not removing lungs.
Cut ¼ to 1½ points for giblets badly cut.

5. Cleanliness, headwraps, and empty crops—5 points. (See disqualifications.)
   a. Cut ¼ to 1½ points for slightly bloody or dirty head.
   b. Cut ½ to 2 points for slightly soiled or dirty carcass.
   c. Cut ½ to 1½ points for slightly soiled vent or vent filled with fecal matter.
   d. Cut ¼ to 2 points for absence of, or improper or stained headwraps.
   e. Cut ¼ to 1 point for poorly cleaned body cavities in full-dressed birds.

C. Type or Conformation—20 points. (See disqualifications.)
(Due allowance to be made for sex and age.)

1. General symmetry—5 points.
   a. Cut ¼ to 3 points for general lack of symmetry or poor proportions.
2. Shape, length, and flexibility of keel—10 points.
   a. Keel to be straight or slightly arched, of good relative length, free of dents, knots, and curves, parallel to back, and with rear tip flexible in young birds.
   Cut ¼ to 2 points for decidedly arched keels.
   Cut ½ to 4 points for each deep dent or knob.
   Cut ½ to 4 points for curved or slightly crooked keels.
   Cut ½ to 4 points for decidedly short keel.
   Cut ½ to 2 points for rigid keel in young birds.
   Cut ½ to 4 points for pouchiness and flabby breast skin in young males.
   Cut ½ to 4 points for keels noticeably higher in front than in rear.
3. Neck and legs—3 points.
   a. Neck and legs to be medium in length.
   Cut ¼ to 3 points for excessively long neck.
   Cut ¼ to 3 points for excessively long legs.
   Cut ¼ to 1½ points for slightly crooked toes and other slight deformities.
4. Trussing—3 points.
   Cut ¼ to 3 points for defective or no trussing.
5. Back—2 points.
   a. Back fairly broad and straight, free from deformities.
   Cut ¼ to 1½ points for lack of breadth or for very slight deformities.
   Cut ¼ to 1½ points for arched or roached back.
   Cut ¼ to 1½ points for roughness over back.

SERIOUS DEFECTS

Carcass slightly lacking in fleshing.
Carcass slightly lacking in finish.
Carcass not fully bled.
Pinfeathers or feather pigment on breast.
Slight bruise or bruises.
Noticeable tears or bruises in skin on any part of carcass.
Slight scuffs or abrasions on breast.
Presence of noticeable amount of blood on head.
Noticeable amount of dirt on carcass or fecal matter in vent.
Decided dent or knob on keel.
Curved or very slightly crooked keel.
Slight pouchiness in young males.

DISQUALIFICATIONS

Positive evidence of spoilage.
Crop containing feed.
Carcass not grading U. S. Grade A or better.*
Specifications for U. S. Grade A or better:

(a) Must be dry-picked or semi-scalded; may be sub-scalded if drawn, and wrapped air tight.
(b) Keel bones straight; slight curves and dents permitted if not over ¼ inch deep.
(c) No serious physical deformities.
(d) No broken wings (except tips), or legs.
(e) No torn skin, except slight tears on back.
(f) No skin or flesh bruises, abrasions, discolorations, dirt, or blood; except slight in extent.
(g) Well fleshed especially on breast, entire carcass well covered with fat.
(h) Well bled, practically free of pinfeathers and "ink," especially over the breast.
(i) Young males, no more than slightly pouchy.

Judging Live Turkeys for Market Quality

For judging live turkeys on the basis of market quality, an age schedule and scorecard are offered by the authors. To place emphasis on quick maturity is a good plan but one that is very difficult to carry out. For one thing, sworn statements of age (by weeks) are difficult to obtain and requiring them may cut down the number of entries. If obtained, it will take a great deal of experience and good judgment on the part of the judge to make proper allowances for the variety of ages that are certain to appear in each class. Age by weeks is almost necessary if accuracy is to be obtained. Age by months is likely to be too indefinite to permit fine distinctions to be made. If

*See Chapter 7.

it is assumed that young turkey toms of medium and large-sized strains should be of U. S. Grade A quality at 28 weeks and young hens U. S. Grade A at 26 weeks, a definite basis for judging birds of these ages is acquired. But when ages vary from 20 weeks to 32 weeks or even to a greater extent, the judge is faced with a very great task. In addition to being a feather and type judge he has to be a judge of live market quality. However, the difficulties are probably not insurmountable and if the judge will undertake it, a start can be made, the judge attempting to correlate the birds' ages with the state of the fleshing, fat, and feathering.

Following is a tentative scheme by the authors for judging maturity in turkeys raised by standard methods. For small-type turkeys use the hen schedule for both hens and toms, since both sexes should mature at the same rate. By using high-energy feeds and/or an antibiotic, maturity may be stepped up enough to warrant advancing both schedules of ages by two weeks.

Tabl 8.3—Judging Turkeys for Rate of Market Maturity

Age of birds of large and medium sized varieties	Age of Beltsville White turkeys[2]	Young toms	Young hens
		U.S. Grade including feathering[1]	U.S. Grade including feathering[1]
18 weeks	14	C or 10	C+ or 9
20 weeks	16	C+ or 9[3]	B− or 8
22 weeks	18	B− or 8	B or 7
24 weeks	20	B or 7	B+ or 6
26 weeks	22	B+ or 6	A or 4
28 weeks	24	A or 4	A+ or 3
30 weeks	26	A+ or 3	AA or 2
32 weeks	28	AA or 2	AA or 2

[1]See Chapter 6 for description of grades and methods of judging on basis of market grades.
[2]Also other small-type turkeys bred to the same weight standard.
[3]The numbers refer to the numerical grades described in Chapter 7.

The relationship between grade and age is clearly shown in Table 8.4. When the age of the dressed birds is unknown the table is of special value.

Table 8.4—Scorecard for Judging Live Market Turkeys

Show__________ Date__________ Band No.__________

Variety*__________ Class†__________ Weight__________

Section No. and Name	Point value in scoring		Explanation
I Standard quality	Vigor and condition	5	This much weight is given to the general appearance of the bird. If crossbred turkeys are scored, the color is not considered. When exhibited in market classes as purebreds, specimens should creditably represent the variety and be free from serious disqualifications. Standard disqualifications, unless so serious as to impair the appearance or actions of the bird are not to disqualify. Bird should have normal gait and stand high at front of back
	Variety color and size	5	
	Symmetry and breed type	5	
	Balance and action	15	
	Total	30	
II Finish (fat)		15	A specimen should grade fully U. S. "AA" to rate 15.
III Feathering		15	A specimen should be completely free of pinfeathers to rate 15. Short pinfeathers are especially undesirable.
IV Fleshing, especially breast and legs		30	A specimen should have a broad, fully fleshed breast, the width carrying out well towards the rear, and be fully fleshed in all other sections of the body. Drumsticks should be plump and moderately short.
V Skeletal development and proportions		10	As measured by sliding-jaw calipers, the length of the keel bone should equal, or exceed, the length of the shank. The depth depends upon size and should approximate the figures given in table 6.2, page 347. The shank is measured by flexing it at a right angle to the tibia, bending back the outer toe, and measuring the distance from back of hock to the walking surface of the base of the toe, pressing calipers firmly against the parts. The keel is measured by holding the calipers against the rear end of the bone and the sliding jaw against the front end. The depth is measured from front of keel to middle of back. The keel should be parallel or nearly so to the back and should be free from curves dents, and knobs. It should be straight except for the rear portion which may curve slightly toward the body.
Total		100	

*Any standard or non-standard variety or crossbreds permitted. †Sex and age.

In making these appraisals as to U. S. grade and subgrade the judge must determine (1) whether or not the bird can be picked clean enough for the proper grade, (2) whether or not it carries enough flesh for that grade and (3) whether it carries enough fat for that grade. As each bird is examined, the judge may if he desires, record the scores section by section according to the score card and make the placings according to the record. If no attempt is made to require age certificates, the scorecard is still applicable to the judging of market turkeys as then each section still has the same relative importance in turkeys intended for market even though the age is not known.

Body weight is given but small value since size is assumed to be a comparatively unimportant point in judging for market quality. A value is placed on weight when appraising the bird for section I. For Section V, calipers are desirable and with a good instrument the measuring takes but a few seconds. The relative length of these parts may be estimated with the hands if calipers are not available. The keel-back relationship is obtained by handling, running one hand over the keel simultaneously with the other hand over the back. The bird is held on its side for this appraisal.

Disqualifications:

Severely crooked breastbone.

Seriously deformed leg, beak, wing, or back.

General inferiority—not fit to represent the breed and variety (off-colored birds should be exhibited as crossbreds or not at all).

Deficiency of fleshing and finish sufficient to cause specimen to be graded U. S. Commercial.

Judging Trends. Prior to about 1940 the prevailing tendency among turkey judges was to emphasize plumage color far in excess of the 32 points out of 100 that the Standard exhibition scorecard then allowed, and to pay little or no attention to type, the points for which amounted to 46. Weight, or size, which rated 12 points, and condition and vigor 10 points, were given appropriate recognition. Under this system the live turkey show was a handsome display of well-groomed, full-feathered birds, the plumage pattern of which followed the Standard description and often were marvels of perfection and beauty. White birds frequently had their plumage washed and those of all colors were in full plumage and had their legs and feet cleaned, manicured, and polished. However with few exceptions these exquisitely colored birds were poorly fleshed even when fat and so, when the importance of meat quality began to be realized, the plan changed and judges began to place turkeys in the live shows almost entirely on conformation, principally the breast, with the same disregard for color and show condition points that had been the lot of the type points in the prior era. As a result the live turkey shows deteriorated in appearance but led the way with a trend toward better meat type. The new Standard Scale of Points provides a (reasonable) 15-point allowance for color, 10 points for "condition" (including vigor) and gives proper weight to all the important features of a dressed turkey good to look at and good to eat. Its use would promote a middle-of-the road approach to the judging and breeding of turkeys.

The accompanying tentative scorecards, explanatory remarks, and judging hints are presented as an attempt to set up grades and standards for the exhibiting and judging of turkeys with what is believed to be proper emphasis on market quality.

Day-old Poult Shows

Day-old poults make an interesting exhibit either as a separate show or in connection with day-old chick shows. The following scorecard indicates the manner in which poults were judged at the Midwest Egg, Chick, and Poultry Show, Iowa State College.

Scorecard for Judging Day-old Chicks and Poults

	1953	1954
Sturdiness and activity	25	40
Perfection of development[1]	25	30
Weight before feeding[2]	10	10
Uniformity of Size	10	10
Trueness of color[3]	15	0
Uniformity of color	15	10
Total	100	100

[1]Score down for poorly healed navels.

[2]Desired weight 21.2 ounces (600 grams) per entry of poults. (Authors suggest that 600 grams be specified for large varieties, 500 grams for small varieties.) Allowance is made for loss of weight of entries shipped long distances.

[3]Description of down patterns of day-old poults is given under each variety in Chapter 10.

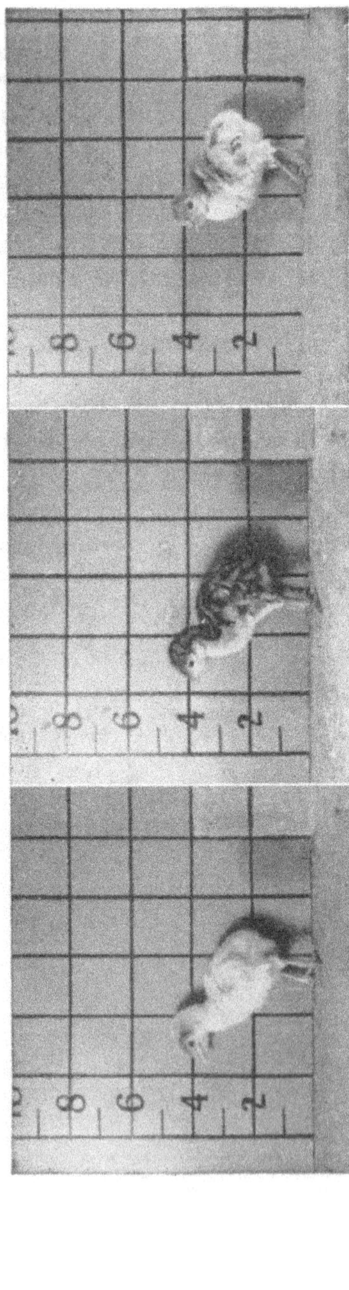

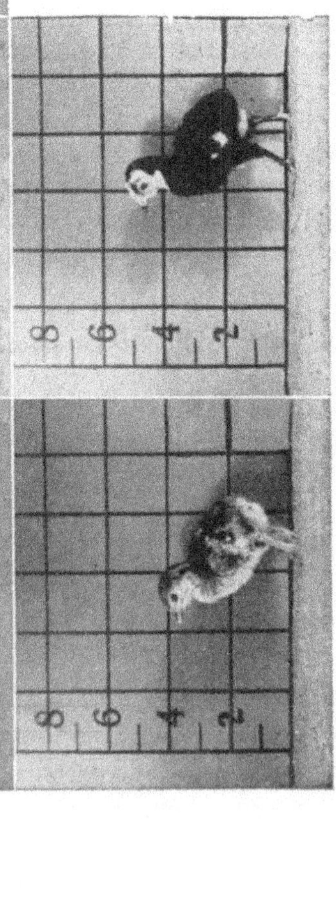

Fig. 8.3. Poults of the more popular varieties: left to right, White Holland, Bronze, Bourbon Red, Narragansett, Black. (Photo from Larrowe Research Farm.)

Fig. 8.4. Small-type Bronze tom, ready for market and weighing 16 lbs. at 24 weeks. Note upstanding carriage, deep breast, and strong but refined head. (Photo from U. S. Dept. of Agr.)

Disqualifications

Any entry will be disqualified if any of the following conditions apply:

1. Entries showing evidence of having been fed.
2. Entries more than two (2) poults short of ten (10). In entries having less than ten (10) poults, the shortage be considered as poults dead and ten points deducted for each poult missing.
3. Entries which have been sexed.

Selecting Baby Chicks and Poults for Exhibition

STURDINESS AND ACTIVITY. Select only vigorous, active and healthy chicks or poults.

PERFECTION OF DEVELOPMENT. The navel should be perfectly healed. No pasty vents should be present.

The down should be fluffy, dry and show no signs of stickiness, or other deformities.

UNIFORMITY OF SIZE. All chicks or poults should be as near the same weight as possible.

APPARENT BREED TYPE. The American Standard of Perfection lists the standard type and color for most breeds of chickens; for poult color see Chapter 10.

UNIFORMITY OF COLOR. It is desirable to examine about twice the number of chicks required for an entry and then select those most uniform in color.

TRUENESS OF COLOR. All chicks or poults should be representative of their breed. Compare the shank color with the average of the entry.

CHAPTER 9

CLASSIFICATION, ORIGIN, AND HISTORY

OUTLINE

All living creatures are divided into two great groups: (1) the plant kingdom and (2) the animal kingdom. Each kingdom is divided into a number of large subdivisions called *phyla*. The animal kingdom is divided by most zoologists into eleven phyla. The highest ranking phylum of the animal kingdom is the *Chordata* which is divided into two *subphyla*, the Prochordata and the Vertebrata (vertebrates). The vertebrates are divided into seven *classes:*

Class 1—Cyclostomes, i.e., lampreys and hag-fishes (Cyclostomata)

Class 2—Elasmobranchs, i.e., sharks (Elasmobranchia)

Class 3—Fishes (Pisces)

Class 4—Amphibians, i.e., frogs, toads, and salamanders (Amphibia)

Class 5—Reptiles, i.e., snakes, turtles (Reptilia)

Class 6—Birds (Aves)

Class 7—Mammals (Mammalia)

The class Aves (birds), is divided into two subclasses: (1) the Archaeornithes, consisting of two genera of prehistoric reptile-like birds called Archaeopteryx and Archaeornis; and (2) the Neornithes, all other birds both living and extinct. The subclass Neornithes contains 32 *orders* of which 3 are known only as fossils. The order Struthioniformes contains the ostriches; the order Anseriformes the ducks, geese, and swans; the order Columbiformes the pigeons; and the order Galliformes the chickens, guineas, peafowl, and turkeys. The birds of the order Galliformes are fowl-like (gallinaceous) birds with feet adapted to perching. The order Galliformes is subdivided into a number of *families* of which the family Meleagrididae contains the turkeys; the family Phasianidae the chickens, peafowl, and pheasants; and the family Numididae the guinea fowl.

The family Meleagrididae (turkeys) is represented by only two living genera. One genus called *Agriocharis* is represented only by one species called *ocellata,* the beautiful Yucatan or ocellated turkey. The other genus called *Meleagris* is represented by only one species, *gallopavo,* the North American turkey, of which there are five wild varieties (or subspecies, or races). A sixth, *Meleagris gallopavo onusta,* has been described by Moore, but has not as yet received recognition as a distinct race.

The Ocellated Turkey

The ocellated turkey, *Agriocharis ocellata* (Figs. 9.1 and 9.2) is confined to the Yucatan peninsula of Mexico and the adjoining territory and is an inhabitant of the tropical forests. It is smaller than the common wild turkeys—the maximum weight being reported as 15 to 18 pounds. It lacks the breast tuft (beard), is brighter-colored, has two fleshy appendages on the head instead of one, and has eyed (ocellated) main tail feathers. Apparently it has been domesticated little if any. Crossing between it and the domestic turkey has been reported, but not authentically. The coloring rivals that of the peacock but the color pattern resembles that of the Bronze very closely although the colors are much different. Its flesh is considered a delicacy.

Following is a brief description of the male. The main tail feathers are indistinctly but evenly pencilled, black on light grey, except at the ends and near the skin. There is a broad, rounded tip of brilliant, coppery red with green reflections below which is a wide, bluntly diamond-shaped subterminal band of black extending completely

Fig. 9.1. The ocellated turkey male. (Photo from U. S. Biological Survey.)

across the feather. The center of this black band is largely occupied by the eye-shaped ocellus, or eye, the long axis of which lies crosswise of the feather. The ocellus consists of bright, iridescent, purplish-blue bronzing. Next to the skin, on the portion of the tail feather normally hidden from view, the color is a medium shade of brownish gray. The greater tail coverts are similar to the main tail feathers in shape and coloring except that the ocellus is much larger, extends nearly across the feather, and is partially divided at the shaft of the feather on the side next to the body, each half being convex proximally (on the side nearer the body). The lower section of the black bar follows the outline of the ocellus. In the lesser coverts, the tips become nearly square and the same color scheme is found except that the ocellus extends completely across the feather, thus dividing the black, subterminal band completely in two. In the back feathers nearest the tail the tips are square, as are all the other contour feathers, and the ocellus becomes a fairly wide bar of purplish-blue bronzing extending completely across the feather and bordered front and rear with a black band. The color of the tips remains as in the tail feathers but the pencilling is only faintly in evidence. As the shoulders are approached the pencilling practically disappears, the tips of the (back) feathers show a predominance of the greenish reflections over the coppery red, and the bar of bronzing becomes narrower and greenish blue rather than purplish blue. The neck feathers are iridescent greenish black with faint red reflections narrowly tipped with an iridescent green edging. The breast and body feathers are black, tipped with a brilliant coppery-red edging which shows faint greenish reflections. The wing bows are brilliant bluish green with yellowish reflections and the wing bar (consisting of the secondary coverts) is brilliant coppery red with greenish reflections. The primaries are barred black and white almost exactly like those of the domesticated Bronze. The secondaries are barred and mottled (black and white) and possess a little of the greenish iridescence much as in the Bronze. The undercolor throughout is a medium shade of brownish gray. The unfeathered head and upper neck are described as blue, studded with orange warts.

The female is colored almost exactly like the male although in brilliancy she is not quite his equal.

As with domesticated turkeys, the usual number of main tail feathers of the ocellated turkey is 18. On each wing there are 10 primaries and usually 16 secondaries including the smaller secondaries which are graduated as to length and number about six. The average of the measurements of 3 adult male skins in the National Museum at Washington, D. C., are as follows: folded wing 15.6 inches (from wing bow to end of the longest wing feather) ; tail (length of the two middle tail feathers from end of longest feather to tailhead) 14.0 inches; metatarsus (back of hock to base of large toe) 6.2 inches; length of middle toe (base of toe to base of toenail) 2.9 inches; spurs 1.1 inches; beak (upper mandible) from the tip to the flesh at the base 1.0 inches. The measurements of an adult female are as follows: 13.6 inches, 11.4 inches, 4.6 inches, 2.6 inches, 0 (no spurs), .8 inch, respectively. The eggs are reported as similar to those of Meleagris, but smaller.

The National Zoological Park in Washington, D. C. has some ocellated turkeys and a few offspring have been obtained. The California Station (1954) has successfully crossed modern day Oscellated turkeys with Bronze; the female offspring being much nearer the weight of the males than with popular varieties.

Turkey-like Birds

A number of extinct species of true turkeys have been identified from fossils, but among living turkeys there are only two genera and two species. Other turkey-like birds have been called turkeys by various writers and these accounts have led to popular misunderstanding. An example of this is the so-called "brush turkey" of Australia which is not a turkey at all but belongs to an entirely different family, the *Megapodiidae.* Another example is the reported discovery of a new "turkey" called *Pauxi unicornis.* This bird is not a turkey but a guan of the family *Cracidae.* It is found in South America.

The North American Wild Turkey

The North American wild turkey and all varieties of domesticated turkeys are called *Meleagris gallopavo* in scientific terminology, the first name, *Meleagris,* being that of the genus, (always capitalized) and the second name, *gallopavo,* (not capitalized), that of the species. To distinguish the varieties, two more names are added, making four in all: The third name, not capitalized, is the variety name, and the fourth, capitalized, is that of the scientist who first described it. There are five wild varieties (races) of Meleagris now recognized, plus a sixth proposed by Moore, *Meleagris gallopavo onusta.* This newly described race is found in the western slopes of the Sierra Madre mountains in Mexico. These birds resemble *merriami,* but have no cinnamon subterminal bar on the main tail feathers and no black bar on the median pair. The tips of the upper tail coverts and main tail feathers are pure white. The margins of the primaries and secondaries have less cinnamon than *Merriami.*

These races are listed in the following outline taken from the 1931 check-list of the American Ornithologists Union, the recognized authority on the classification of American birds.

Family: Meleagrididae (Turkeys)

Genus: *Meleagris*

Species: *gallopavo*

Varieties:

1. *Meleagris gallopavo gallopavo* Linnaeus. The Mexican turkey. Range, central Mexico.

2. *Meleagris gallopavo merriami* Nelson. Merriam's turkey. Range, mountains of southern Colorado, New Mexico, Arizona, western Texas, and nothern Sonora and Chihuahua, Mexico.

3. *Meleagris gallopavo intermedia* Sennett. The Rio Grande turkey. Range, middle northern Texas to Coahuila, Nueva Leon, and Tamaulipas, Mexico.

4. *Meleagris gallopavo osceola Scott.* The Florida turkey. Range, Florida at least as far north as Gainesville.

5. *Meleagris gallopavo silvestris* Vieillot. The eastern turkey. Present range, southeaastern United States to eastern Oklahoma, eastern Texas, central Missouri, central Kentucky, West Viriginia, central Pennsylvania, and northern Florida; formerly to Kansas, Nebraska, southeastern South Dakota, southern Wisconsin, southern Minnesota, southern Michigan, southern Ontario, central New York and New England to southern Maine.

In summary the zoological classification of turkeys is shown in the following outline:

Kingdom: Animal
  Phylum: Chordata
    Subphylum: Vertebrata
      Class: Aves (birds)
        Subclass: Neornithes
          Order: Galliformes (fowl-like birds)
            Suborder: Galli
              Super family: Phasianidae
                Family: Meleagrididae (turkeys)
                  Genus: (a) *Agriocharis*
                    Species: *ocellata* (the ocellated turkey)
                      Subspecies: None

                  Genus: (b) *Meleagris*
                    Species: *gallopavo* (the North American turkey)

                    Subspecies: Five in number, full names as follows:

                    (1) *Meleagris gallopavo gallopavo* Linnaeus —the Mexican turkey

                    (2) *Meleagris gallopavo merriami* Nelson—Merriam's turkey

                    (3) *Meleagris gallopavo intermedia* Sennett —Rio Grande turkey

                    (4) *Meleagris gallopavo osceola* Scott—the Florida turkey

                    (5) *Meleagris gallopavo silvestris* Vieillot—the eastern turkey

Range: The majority of evidence indicates that the North American turkey, taken it as a whole, including its five subspecies, formerly ranged over all of Mexico except the extreme southern and western parts, and over the United States south and east of a line extending north through central Arizona; diagonally northeast across Colorado and Nebraska to northern Iowa including a small portion of southeastern South Dakota, to southern Wisconsin, southern Michigan, southern Ontario (Canada), central New York, southern Vermont, southern New Hampshire, and southern Maine. Some evidence indicates that the range may have extended to central Minnesota, northern Wisconsin, the Lake Superior region, northern New York and northern Vermont and New Hampshire. For color description and habits of wild turkeys see Chapter 10.

Classification of the various wild varieties (races, or subspecific forms) of *Meleagris gallopavo* is not settled but the classification given in "The Birds of North and Middle America" (Bulletin 50, U. S. National Museum, 1946, Part X, by Ridgway and Friedmann) is the basis of the outline that follows.

In this outline recognition is given to two more subspecies, *Meleagris gallopavo onusta* Moore and *Meleagris gallopavo Mexicana* Gould, making seven in all instead of five.

Key To The Forms of Meleagris Gallapavo
(Linnaeus)

a. Tail tipped with deep rusty, its coverts and the feathers of lower rump tipped with rich dark chestnut.

b. Primaries broadly barred with white, white bars nearly or quite as broad as the dusky interspaces and extending to shafts of quills . . . .

Meleagris gallopavo silvestris Vieillot, the eastern turkey. Present range, southeastern U. S. to eastern Texas, eastern Oklahoma, southern Missouri, central Kentucky, West Virginia, central Pennsylvania, and south to include northern Florida; formerly ranged to eastern Kansas, eastern Nebraska, southeastern South Dakota, southern Minnesota, southern Wisconsin, southern Michigan, southern Ontario, central New York, New England to include southern Maine, and south to include northern Florida.

bb. Primaries narrowly barred with white, white bars very much narrower than the dusky interspaces and not extending to shafts of quills . . . *Meleagris gallopavo osceola* Scott, the Florida turkey. Range, Florida at least as far north as Gainesville.

aa. Tail and tail coverts and feathers of lower rump tipped with light cinnamon-brown, buffy, or white.

b. Tail, upper tail coverts, etc., tipped with light cinnamon brown, cinnamon, or cinnamon-buff; rump almost wholly solid glossy black (feathers tipped with gray in female and young . . . *Meleagris gallopavo intermedia* Scott, the Rio Grande turkey. Range, central Texas, south to include the Mexican States of Coahuila, Nueva Leon, and Tamaulipas.

bb. Tail, upper tail coverts, etc., tipped with white or pale buffy.

c. Lower back and rump bluish black without reddish and greenish-golden metallic reflections.

d. Upper body plumage purplish bronzy.

e. Narrow bars on basal three-fourths of undersurface of rectrices (main tail feathers) more grayish than rufescent (reddish) . . . *Meleagris gallopavo onusta* Moore, Moore's turkey. Range, western slopes of the Sierra Madre mountains in the Mexican States of Durango and Chihuahua, north to southern Sonora. It is a mountain bird, resident at 8500 to 4000 feet, occasionally to 2500 feet.

ee. Narrow bars on basal three-fourths of undersurface of rectrices more rufescent than grayish . . . *Meleagris gallopavo merriami* Nelson, Merriam's turkey. Range, mountains of southern Colorado, south through New Mexico, southwestern Texas, into northern Chihuahua, Mexico, west through Northern Sonora, Mexico, and north to include most of Arizona. It is a mountain bird.

dd. Upper body plumage highly glossed with greenish and reddish-golden reflections, less purplish bronzy . . . *Meleagris gallopavo gallopavo* Linnaeus, the south Mexican turkey. Range, southern Mexico from about the latitude of Mexico City south to include the Mexican States of Veracruz and Oaxaca.

cc. Lower back and rump with reddish and greenish-golden metallic reflections, (not bluish black as in *intermedia, merriami,* and *gallopavo*) . . . *Meleagris gallopavo mexicana* Gould, Gould's turkey. Range north central Mexico east of the Sierra Madre mountains including southeastern Chihuahua, eastern Durango, and south to northern Jalisco. This variety, m.g. mexicana (or *m.g. gallopavo* if *m.g. Mexicana* is not given recognition as a separate variety but is considered synonymous with the former) undoubtedly is the parent stock of the domesticated turkey. The 1931 check-list of the American Ornithologists Union recognized each of the four United States varieties (*silvestris, osceola, intermedia,* and *merriami*) but classified all Mexican wild turkeys as one variety, *Meleagris gallopavo gallopavo* Linnaeus instead of as three varieties as listed here. (*onusta, gallopavo,* and *Mexicana*).

History of Domesticated Varieties

Turkeys, Canada (Gray) Geese, and Muscovy ducks (which originated in South America) constitute the total contribution of the western hemisphere to the list of species of poultry. The complete list includes chickens, turkeys, guineas, peafowl, ostriches, ducks, geese, swans, and pigeons. America, however, while contributing only three species, has taken the lead in the development of new breeds and varieties and in the improvement of poultry, especially turkeys.

The turkey existed both wild and domesticated (by the Indians) in America at the time of its discovery in 1492. Many excavations in the Tennessee Valley by Webb (1938) contained turkey bones. The evidence is clear that wild turkeys served as a source of food for the Indians as early as 1000 A. D. and perhaps earlier. According to Robinson, turkeys were first brought to Spain from America in 1498. Historical evidence has been presented by Wright indicating that it was 1519 or 1520 before turkeys were shipped to Europe. Later importations followed, probably all from the turkeys domesticated by the natives of Mexico. These domesticated turkeys undoubtedly belonged to the Mexican subspecies, *Meleagris gallopavo gallopavo*. From Spain the turkeys spread over Europe, being introduced into England between 1524 and 1541, according to D. E. Hale.

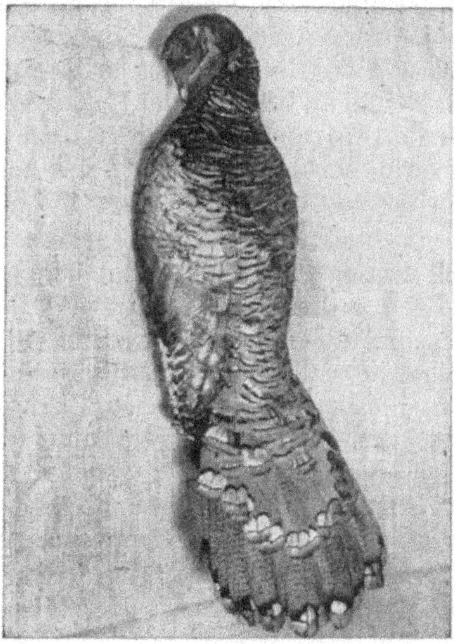

Fig. 9.2. Ocellated turkey tom; dorsal view showing details of color pattern. (Photo from U. S. Dept. of Agr.)

The discovery of a frieze with 8 medallions depicting American turkeys painted below a 13th century mural in St. Peter's Cathedral at Schlesing opens up the possibility of live turkeys having been sent from "Markland" or some other part of North America to northwest Europe by the Norsemen prior to 1280 A. D. Authorities decided that the medallions were not a later addition (16th century) but were painted around 1280 A. D. This finding opens up a field for speculation regarding the development of turkeys in Europe, which at best is only speculation, for there is no other evidence that indicates the presence of turkeys in Europe prior to 1498.

Several European varieties were developed, among them being the Norfolk Black, Cambridgeshire Bronze, White Austrian, Buff, Blue, and Ronquieres (a small vari-colored Belgian variety).

Seven domesticated varieties are recognized (1953) by the American Poultry Association. These are the Bronze, Narragansett, White Holland, Beltsville Small White, Bourbon Red, Black, and Slate. The Buff was admitted to the Standard in 1874 and dropped in 1915. In addition there are several non-standard varieties and the wild turkeys (described in Chapter 10).

Apparently the first domesticated turkeys to be grown in the United States came not from the wild turkeys indigenous to the United States but from Europe. They were of the bronze and black color varieties. For a long time, it appears, no serious attempts were made to cross the domesticated European strains of turkeys with the native wild turkeys, then early in the nineteenth century turkey breeders began to cross the domesticated stock with the wild. In the region of Narragansett Bay, Rhode Island, according to Robinson, there was developed a local "breed" of turkeys by crossing the domesticated with the wild. This local breed was the forerunner of the Narragansett and the Bronze varieties. A little later (1830-40) there was developed, in the vicinity of Point Judith, Rhode Island, another local race by further crossing of the domesticated Narragansett Bay turkeys with wild turkeys and these birds were called for the first time by the name Bronze. This race was preserved and improved by selection and further crossing with the wild birds. About 1860 the Rev. R. H. Avery of Wampsville,

N. Y., exhibited what was then a superior strain of Bronze turkeys larger than the common domesticated or wild turkeys. From this strain, according to Robinson, the modern Bronze turkey was originated (Fig. 9.3). The Bronze was recognized and described in the first American (then called Lockwood) Standard of Perfection, which was issued in 1871.

Fig. 9.3. Turkeys of 100 years ago, as illustrated in Prairie Farmer of 1851. (Photo from Prairie Farmer.)

The Narragansett, it is believed, originated (as noted above) in the Narragansett Bay region by breeders who preserved the Narragansett color factor when its presence was discovered among the Bronze. While the original Narragansett Bay race of turkeys appears to be the ancestor of both the Bronze and Narragansett, the latter variety was not so named until after the Bronze had become quite well known. The Narragansett was admitted to the Standard in 1874.

The tendency for Bronze turkeys to throw "sports" of a partridge and of black, white, and gray color pattern frequently almost devoid of bronzing provides an interesting field for the development of new varieties. Several such varieties described in Chapter 10 have been developed. Inbreeding the Bronze may cause gray, black, and white *sports** and also sports with partridge markings to appear.

Little is known of the origin of the White Holland variety which was admitted to the Standard in 1878. Robinson is of the opinion that it was developed in the United States from white sports of the Bronze variety. Although white turkeys had been bred in Holland and other parts of Europe, in the eighteenth century, it appears that these did not contribute to the development of the modern White Holland variety of the United States.

The Bourbon Red is believed to have originated in Bourbon County, Kentucky. Robinson believes that it comes either from an older variety called the Tuscawara Red developed in Pennsylvania from the reddish and brownish turkeys often found in mixed flocks. It was first described in the 1910 Standard. It is interesting to note that Bourbon Red turkeys can be produced by crossing the Bronze and Buff, and, in reverse, buff turkeys can be produced by crossing Bourbon Red and Black. Like other varieties, genetic factors are responsible for the varietal patterns. The originators merely preserved and improved the original stock.

The Black turkey presumably was developed by selection from wild turkey stock. Apparently it was the first variety to be developed with a color pattern different from the Bronze. It has been and still is popular in Europe. It was admitted to the Standard in 1874.

---

*A sport or mutation is a sudden deviation from type, that proves to be transmitted to the offspring.

No information appears to be available on the origin of Slate turkeys. This variety was admitted to the Standard in 1874. Either of two different genetic factors causing dilution of the bronze and black color patterns is responsible for most slate-colored turkeys.

A discussion of the genetic basis of color pattern inheritance will be found in Chapter 11.

CHAPTER 10

VARIETIES

OUTLINE

According to the American Standard of Perfection, turkeys constitute a single class of fowls. Within the class there is properly only one breed within which seven standard varieties are recognized by the American Poultry Association in the 1953 Standard of Perfection. The varieties, popularly called "breeds," are the Bronze, Narragansett, White Holland, Beltsville Small White, Bourbon Red, Black, and Slate. The Blue*, Buff and several

*The so-called blue turkey is acceptable as Slate.

other varieties are not included in the Standard, neither is the wild turkey. Excellent illustrations of some of the standard varieties and detailed descriptions of the ideal color and shape of all of the standard varieties are to be found in the American Standard of Perfection (1953) published by the American Poultry Association which maintains offices in Oklahoma City.* It is the recognized reference work for all questions relating to showing, judging, and breeding for standard quality, of poultry, including chickens, ducks, geese, and turkeys.

The extremely popular Broad Breasted Bronze is not recognized as a separate variety in the Standard of Perfection, it being the sense of the Standard makers that the official description of type, or shape, in turkeys is fully applicable to the Broad Breasted Bronze and that the color pattern of the latter is identical, or should be, with the Standard description for the Bronze variety. The National Turkey Improvement Plan lists several varieties of Bronze turkeys, (1) the Broad Breasted Bronze, (2) the small-type B. B. Bronze (prescribing a minimum breast width) and (3) the standardbred Bronze which differs, in practice, from the Broad Breasted Bronze only by its relative lack of heavy fleshing qualities.

Within each variety there are many *strains*. The term strain is properly applied to the individuals of a family composing a reasonably distinct line of breeding. To constitute a true strain, a flock of birds must be of common ancestry, within the same breed and variety, and have certain distinguishing characteristics. The term strain is often loosely applied to the product of any breeder who has been in the business of breeding the same variety of livestock or poultry for several years. Such a breeder's stock if it has been carefully bred and selected toward a certain ideal for several generations without too-frequent introduction of blood from other breeders, can justifiably be considered as a strain. A cer-

*Dr. L. H. Ritzhaupt, Pres., and Edna Cline, Secy., Box 968, Oklahoma City, Okla.

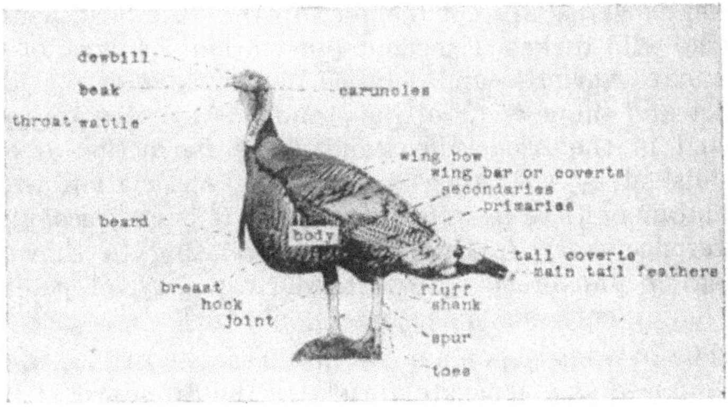

Figure 10.1. Nomenclature of the turkey.

tain (not necessarily high) percentage of inbreeding is implied when the term is properly used, for if a flock of turkeys be truly outbred it cannot possibly possess sufficient distinction to be designated as a distinct strain.

A male turkey is commonly called a tom or gobbler. It is less commonly called a cock or cockerel depending upon age. Female turkeys are called young hens or old hens and less commonly pullets or adult hens, respectively. The terms young tom and young hen; yearling tom and yearling hen; old tom and old hen have come into common usage and acceptance for birds in or approaching their first, second, and subsequent breeding seasons, respectively. In practical usage a bird goes into the next higher age class sometime during the summer. For example, a young hen remains a young hen (or pullet) throughout her first breeding season but sometime before the show or fall marketing season opens, she becomes a yearling.

The reader is referred to the Appendix for definitions of descriptive terms, particularly those of anatomical nature.

Shape of Male and Female Turkeys

The head of all varieties should appear strong, rather wide, fairly deep, and rugged but not coarse. The headgear is of moderate size and carunculated. The beak is strong-looking, medium long, and well curved. The dewbill of the male is of good size and very extensible; of the female, much smaller and not very extensible. The eyes are slightly oval, set well out, and alert in appearance. Neck is of moderate length. The male should have a beard; the female may have one but absence of it is preferred. The good-sized, strong-appearing wings blend smoothly into the remainder of the plumage and have smoothly overlapping feathers of good width. The back is broad, well-fleshed, and when the bird is standing normally, slopes decidedly in a slightly curved line from front to rear at a 35° angle above the horizontal.

The full-fashioned tail when relaxed is carried low so that it continues the downward curve of the back. When strutting, the tail feathers (along with most of the contour feathers of the neck, breast, back, wings, and body) are raised and spread. The breast is a most important section and the standard description of it tallies well with accepted standards for market quality. Medium length along with good breadth, fullness, and straightness or a very slight rocker effect are desirable features. The body as a whole is blocky. The fluff feathers are rather short to conform with the neat and smooth appearance of the remainder of the plumage. The upper and lower thighs are of medium length and well fleshed to conform with desirable market standards. The shanks and toes are strong-looking, smooth, straight, and of moderate length with good width between them; four toes on each foot.

DESCRIPTION OF STANDARD VARIETIES

In addition to the seven standard varieties of turkeys in the United States there are the wild turkey and a number of other non-standard varieties. The following brief color descriptions of specimens of the various varieties are an attempt by the authors to describe in common terminology prize-winning specimens as seen in the show-

Fig. 10.2. An exceptionally fine Slate tom. (Photo from U. S. Dept. of Agr.)

room. For details of requirements of color and type in grown turkeys consult the Standard of Perfection.

The Bronze

The standardbred Bronze, now almost completely superseded by the Broad Breasted Bronze in the United States, is colored as follows. *Male*: Plumage on the neck and upper breast is black, the exposed tips overlaid with iridescent, red-green bronzing. That of the lower breast, wing bows, wing fronts, secondary coverts, and front half of the back is similar to that of the neck, but there is an edging of velvety black added to the red-green bronzing. The feathers of the rear half of the back have black undercolor and black surface color which is bordered with a broad, brilliant, copper-colored band of bronzing which in turn is edged with a narrow band of black. The upper thigh feathers are like the foregoing but have in addition a terminal edging of white. The tail coverts and main tail feathers are black in the lower one-third to one-half which normally is not exposed. The exposed portion is black, evenly penciled with narrow strips of brown until near the end where there is a broad band of the brilliant, copper-colored bronzing which is bordered front and rear with black. The tips have a broad edging of white. The primaries, primary coverts, and pinion feathers are evenly marked with dull black and white bars of uniform width. The secondaries are also barred, dull black and white, but those nearest the body have indistinct barring with reduced white bars and enlarged black portions which are suffused with red-green bronzing. The

Fig. 10.3. A Broad Breasted Bronze tom of near ideal type. Note good balance, vigorous head and moderately short shanks. (U.S.D.A. photo.)

Fig. 10.4. "Double-rainbow" tails of the exhibition strains of Bronze turkeys present a most attractive picture. (Photo from Turkey World.)

lower thighs (drumsticks) are black with grayish edgings, and the fluff is dull black.

The plumage of the *female* is similar to that of the male, except that an edging of white is added to the black bars on the back, wing bows, secondary coverts, and breast. This white edging is narrow in front and gradually widens toward the rear of the body. White tipping is controlled by sex hormones since the normal black breast feathers of males grow in with white tips if the birds are injected with the female sex hormone oestrin.

Both sexes have the same color pattern in the large wing feathers and in the main tail feathers.

In both sexes the beak is brownish yellow at the tip, shading to brownish black at the base, eyes dark brown, the beard black, and the shanks almost black in young birds and blackish pink in older birds. The skin normally is white but may be yellowish white to decidedly yellow, depending largely on the feed eaten.

The head, snood, throat wattle, and unfeathered upper position of the neck of the male is red, changeable to white overlaid with bright blue, the blue most intense on the face. The female does not normally show the blue color to a noticeable degree, the head, snood, throat wattle, and upper neck being red, changeable to pale flesh color almost white.

Creaminess, yellow or brown (instead of pure white) in the edging of the main tail feathers and coverts is a serious defect in the standardbred Bronze. Lack of brilliant, copper-colored bronzing and a tendency for it to be greenish or lacking in substance also are serious color defects.

In addition to the general disqualifications, Bronze turkeys are disqualified for the following: White feathers anywhere; solid-colored feathers (one or more) in main tail feathers or coverts or in primaries or secondaries, not counting the axial feather in young birds; lack of barring more than half the length of the primaries; white or gray markings (not counting the terminal edgings of white) that can be seen on the main tail feathers beyond the coverts; lack of the characteristic white edging on the female breast feathers.

Color Description of Day-Old Bronze Poults*

Beak: Lower half flesh color; upper half medium brown, lighter at tip.

Head: Light brown, streaked and mottled with dark brown; top of head with three parallel, irregular, dark streaks, the center one broadest.

Eyes: Iris lead color; pupil darker, almost black.

Throat: Light yellow, or gray tinged with yellow.

Upper (dorsal) parts: Light brown everywhere marked with very dark brown; back and neck with a broad, irregular dark streak down the center, and two narrower streaks, one on either side; wings with two dark streaks on center portion and a dark spot near

*See Chapter 8 for poult score card.

the tip; undercolor medium to dark gray, corresponding to shade of surface color.

Under (ventral) parts: Neck to tail inclusive, yellowish white to light yellow on surface, light to medium gray underneath.

Shanks: Flesh color overlaid for the most part with brownish-slate pigment.

Feet and toes: Upper portion same color as shanks; bottom pale flesh color.

Off-colored poults occasionally appear. The commonest types are (1) birds that have in general the same markings, but the ground color is yellowish white instead of light brown (as in the Crimson Dawn) and (2) birds that have normal ground color but the markings are lighter-colored than normal. The markings usually are fairly typical in shape and size in both types. Off-colored poults frequently develop into off-colored adults.

The White Holland

The White Holland ranks third in popularity in the United States. The plumage of the White Holland is pure white, free from cream and other off-color; shanks and toes are pale pink; the eyes are dark brown; the beak is light pinkish horn in color; and the beard is black. The skin is white but may be yellowish white depending largely upon the nature of the feed eaten. Both sexes are the same color in all respects. Disqualifications for White Hollands are the general disqualifications which are listed and described on page 545. Gray or black specking in moderation is of little significance genetically but affected feathers should be plucked before birds are shown. Headgear coloration is like that of the Bronze and is the same in all varieties of turkeys including the wild.

Color Description of Day-Old White Holland Poults

Beak: Pale flesh color, brownish through middle of the upper half.

Fig. 10.5. White Holland yearling hen showing desirable type. (Photo from Mrs. Homer Price.)

Head: Medium brownish buff on sides and top except for an irregular pale dusky yellow streak in center of rear portion extending into upper neck.

Eyes: Iris, lead color; pupil darker, almost black.

Throat: Yellowish white.

Upper parts: Neck to tail, pale dusky yellow lightly suffused with light brownish buff; tendency to possess two wide, irregular, indistinct light brownish buff streaks, one on either side of the back and neck; wings lightly suffused with light brownish buff.

Under parts: Yellowish white.

Shanks: Pale flesh color.

Feet and toes: Pale flesh color throughout.

Occasionally poults occur that are pure lemon yellow in color without the buffy dorsal markings. Such poults carry the dominant black or the recessive black-winged bronze factor rather than the bronze from which most white turkeys are derived. There is no objection to such poults except from the standpoint of uniformity.

The above color description for White Holland poults applies equally well to poults of all other white varieties including the Beltsville and the large, broad-breasted white strains.

Fig. 10.6. Jersey Buff tom. (Photo from N. J. Exp. Station.)

Beltsville Small White

The U. S. Department of Agriculture starting in 1934 developed through crossbreeding and selection a small-type, white-plumaged turkey. The standardbred Bronze (2 strains), Broad Breasted Bronze (1 strain), Charlevoix (Canadian small-type) Bronze (1 strain), White Holland (4 strains), Black (1 strain), Narragansett (1 strain), wild (4 strains), and White Austrian (1 strain), a small-type white turkey imported from Scotland were used in this project. The Beltsville Small White was first introduced as a new variety in 1941 and was admitted to the Standard in 1951.

Table 10.1—Weight of Beltsville Small White Turkeys*

	At 14 weeks for marketing as fryer-roasters (broilers)	Live field weight	Dressed undrawn weight	Ready-to-cook weight including neck and giblets
		Pounds	Pounds	Pounds
Toms......	Normal range in weight	6 to 9	5 to 8	4 to 7
	Average weight	7½	6½	5½
Hens........	Normal range in weight	4½ to 6½	4 to 6	3½ to 5
	Average weight	5½	5	4¼
	At 24 weeks for marketing as well-matured roasters			
Toms......	Normal range in weight	13 to 17	11½ to 15	9¾ to 12¾
	Average weight	15	13¼	11¼
Hens........	Normal range in weight	7½ to 11	6½ to 10	5½ to 8½
	Average weight	9¼	8¼	7
	At first-year breeding age, 8 months, or 35 weeks			
Toms......	Normal range in weight	18 to 24	16 to 22	13¾ to 19
	Average weight	21	19	16½
Hens........	Normal range in weight	9½ to 13½	8½ to 12½	7¼ to 10¾
	Average weight	11½	10½	9

The 8-month weights given here are somewhat higher than those adopted as "standard weight" for young Beltsville White turkeys by the American Poultry Association. These two sets of weights may be reconciled by assuming that the A. P. A. standard weights are attained at about 30 weeks of age instead of at about 35 weeks as with the larger varieties.

*Based on typical Beltsville Small White stock fed antibiotics and/or high-energy diets.

The Beltsville Small White is distinguishable from the White Holland only by its smaller size—color and shape specifications are identical for the two varieties. Therefore the weight standard should be rigidly adhered to in selecting and breeding this variety. To accurately gauge the true size of turkeys, either in judging or selecting as breeders, age and environment as well as weight should be taken into account. To facilitate this procedure in the case of the Beltsville White and some other varieties, data on average weights at various ages will be found in Table 10.1.

Breeding standards for the Beltsville Small White at the Research Center are as follows; conformation—compact body with smooth, uniformly broad breast, legs to match in size and plumpness, and keel of moderate length, free from knobbiness. Shanks should be clean-cut, refined, and of moderate length and the bird should present a sprightly, upstanding appearance with neat but vigorous appearing head without coarseness or excess

Fig. 10.7. Beltsville Small White tom. (Photo from U. S. Dept. of Agr.)

headgear. Body depth (as measured from front point of keel vertically to the back, over the feathers) should be maintained at a ratio of approximately one-fifth greater than length of shank (as measured from rear of hock to base of outside toe next to sole pad when shank is held at right angle to drumstick) in well-matured specimens 24 weeks or older. The keel should be of good length, straight or slightly convex, parallel to back or nearly so, approximately equal in length or slightly greater than that of the shank in fully matured specimens 34 weeks or older, equal to or slightly shorter than the shank in younger specimens, and should extend well back between the thighs when bird is suspended by the shanks. Legs should be placed well forward to give the bird good balance. When walking, gait should be easy and free from any tendency to waddle or "paddle." Breast width at widest part (measurement taken at a point 1¾″ above the keel) should be 3½ to 4″ at 22 to 24 weeks and 4 to 4½″ at 35 weeks for both sexes.

Fig. 10.8. Beltsville Small White turkey hen. (Photo from U. S. Dept. of Agr.)

Early maturity:

Both males and females to be finished (U. S. Grade A) condition with respect to fat, fleshing, and feathering at the age of 24 weeks under average conditions or at 22 weeks when antibiotics and/or high energy diets are fed throughout the growing period.

Performance over a 5-month period from first egg:

Egg production—60 per cent.

Egg characteristics—strong, smooth shell; normal shape, medium size; 32 oz. per doz. or 75.6 gm. ea. These egg weights apply to stock brought into production at 34 to 36 weeks of age. If production does not occur until the hens are about 40 weeks or older, average egg weight will be about 34 oz. per doz. (80.3 grams each). If forced into production earlier than about 34 weeks, egg weight will be correspondingly smaller. The eggs of hens starting to lay at 24 weeks will average about 25 oz. per doz. (59 grams each) after egg size has been established (after about a month's production). Shell color tends to be light with inconspicuous spotting in stock artificially lighted starting at 30 weeks or so; it is darker with more conspicuous spotting in stock not artificially lighted.

Fertility—95 per cent or greater.

Hatchability—85 per cent or more of fertile eggs.

Viability—85 per cent or more, at 24 weeks, of poults started.

The ideal early winter weights (adopted as standard) are given on page 544. At market age (22-24 weeks) males and females attain weights that make them fully acceptable on the retail market. In feeding efficiency they are competitive with the larger varieties on account of their shorter growing period. At broiler, or light roaster, age (14 to 16 weeks) both males and females have attained desirable weights, plump conformation and acceptable finish. The lower poult cost, good finish and feed conversion, broad-breasted conformation, and absence of noticeable pinfeathers all continue to make this variety the favorite as a broiler turkey.

The general disqualifications apply and in addition Beltsville White males are disqualified for more than 3 pounds overweight, females over 2. This variety is now widely bred and produced and it was estimated that about 10% of the turkeys raised in 1950 were Beltsville Whites, increasing to about 30 per cent in 1954.

The Bourbon Red

The Bourbon Red *male* is a rich, dark, brownish red, narrowly edged with black in all sections except the large wing and tail feathers and the neck. The primaries and secondaries are solid white, and the main tail feathers are white except for an indistinct bar of red crossing each feather near the end. The neck feathers are brownish red without the black edging. The undercolor is red, fading to salmon color at the base of the feathers.

The color of the *female* is similar to that of the male, except there are no black edgings and there is a very narrow edging of white instead of black on the breast.

In both sexes the eyes are a rich medium brown and the beak and beard are the same as in the Bronze; the shanks and feet are pink with reddish cast in grown birds and dark reddish horn in young birds.

More than ⅓ of any off-color showing in the large feathers of wing or tail is a special disqualification.

The rich, reddish-brown color is difficult to obtain. Black ticking or flecking is a rather common fault. A faded red, approaching buff, is very undesirable. The skin usually is white to yellowish white as in the other varieties. However some breeders believe that there is a strong inherited tendency in some strains to deposit yellow pigment in the fat and skin.

Color Description of Day-Old Bourbon Red Poults

Beak: Pinkish brown.

Head: Light brown to medium reddish brown, usually

with a darker spot on back of head. Down over eye, brown to tan; under eye, creamy.

Eyes: Leaden color with center or pupil almost black.

Throat: Creamy yellow to yellowish white.

Upper (dorsal) parts: Back and neck light yellowish brown with a tendency to possess a small, medium to dark brown stripe near either edge and a broad stripe down the center. Tail section creamy yellow. Wings dark brown at front, light brown in the middle, and creamy at the tips.

Under (ventral) parts: Throat, breast, thighs, and fluff light yellow with a tendency to be grayish under the wings.

Shanks, feet, and toes: Dark flesh color. Shanks and bottoms of feet should be free from pronounced yellow.

The Narragansett

The Narragansett resembles the Bronze in color pattern, but has neither the iridescent red-green sheen nor the copper-colored bronzing which are characteristic of the latter variety. The Narragansett colors are metallic black, light silvery gray, and white.

The color of the *male* is as follows: The neck has black undercolor and the surface color is light gray with a narrow edging of black which is very narrow near the throat and becomes broader toward the back. The colors of the wing front, wing bow, and secondary coverts are like those of the neck but there is little or no black undercolor and the gray portion is much wider so that the general appearance is gray with black edging. The gray of the secondary coverts forms a broad silvery bar across the folded wings. The breast has black undercolor and light gray surface color which has a black edging narrow near the front and broader toward the rear. The gray also becomes darker in the rear portion. The back is rich metallic black, free from bronzing. The saddle feathers are black, with broad, silvery-gray tips. The body feathers, including those of the thigh proper, are black ending in a broad silvery-gray band tipped with black. The fluff and lower thighs (drumsticks) are black with light edgings. The large wing feathers, the primary coverts, and the pinion feathers are evenly barred with black and white (as in the Bronze), the barring of the upper secondaries becoming indistinct as the back is approached. The main tail feathers and greater and lesser tail coverts are evenly penciled, a medium shade of tan on black, and have broad, black tips broadly edged with white (Fig. 10.16).

Fig. 10.9. A Narragansett hen showing typical markings. Wild, Bronze and ocellated turkeys have the same general pattern although the colors are different. (Photo from U. S. Dept. of Agr.)

The plumage of the *female* is similar to that of the male except that an edging of white is added to the black tips of the feathers on the back and breast. The light edging is narrow toward the front of the bird and broader toward the rear. The tan of the main tail feathers and of the greater and lesser tail coverts usually is of a lighter shade than in the male. Due to the white edgings on back and breast the female appears lighter-colored than the male.

Where black is called for, it should be clear black free from greenish cast, bronzing, and purple lustre. The offspring of a Narragansett mating sometimes possess bronzing, but such birds should not be kept for breeders.

The eyes are medium dark brown, rimmed with gray; the beak is bluish gray; beard is black; shanks and feet in grown birds are medium pinkish brown, in young birds somewhat darker. The skin is white or yellowish white.

In the showroom, Narragansetts are subject to the general disqualifications and in addition are disqualified for the same off-colors in wings and tail that disqualify the Bronze and for complete lack of black bands on the largest tail coverts.

Color Description of Day-Old Narragansett Poults

Beak: Lower half flesh color; upper half flesh color at tip, brown mottling at base.

Head: Yellowish gray, streaked and mottled with dark brown; top of head usually has 3 parallel dark streaks, the center one broadest.

Eyes: Iris lead color, pupil darker, almost black.

Throat: Light yellow to nearly white.

Upper parts, neck to tail incl.: Light grayish brown everywhere marked with very dark brown. Back and neck have a broad, irregular, dark streak down the center and two narrow streaks, one on either side. Each wing has two dark streaks on center portion and a dark spot near the tip. The undercolor is grayish corresponding in shade to surface color.

Shanks and toes: Same as the Bronze.

Under parts, neck to tail inclusive: Yellowish white to nearly white on surface; light gray undercolor.

The chief differences between Narragansett and Bronze poults are: (1) breast tends to be more nearly white than it is in the Bronze; (2) the grayish undercolor of neck and back comes nearer to the surface in the Narragansett than it does in the Bronze; and (3) the light ground color on the upper parts is grayish in the Narragansett, giving the appearance of a slightly light-colored Bronze poult. Poults of the two varieties cannot be accurately segregated either when just hatched or for several (5 or 6) weeks thereafter.

The Black

The Black, known in England as the Black Norfolk turkey, is black with greenish lustre in all sections of the plumage of *males* and *females*. The undercolor is black. The shanks and toes are pink in mature birds and almost black in young birds; the beak is slaty black; the eyes are very dark brown, almost black, and the beard is black. The skin is white or yellowish white as in the other varieties. Any variation from the solid black color should be carefully avoided in breeding Black turkeys. Solidly off-colored feathers disqualify. Purple lustre, brownish tinge,

Fig. 10.10. Pedigreed Black turkey poults a few days old. (Photo from U. S. Dept. of Agr.)

and slight white tipping, while undesirable, do not disqualify.

Color Description of Day-Old Black Poults

Beak: Pale flesh color with varying amounts of black, always some black near the tip.

Head: Yellowish white and black. Either color may predominate but there is always a band of the light color across the top of the head and at base of beak.

Eyes: Iris, lead color; pupil darker, almost black.

Throat: Yellowish white.

Upper parts: Black except tips of wings which are yellowish white.

Under parts: Black and yellowish white. Breast is black; abdomen varies from almost all yellowish white to black with only a trace of the lighter color. (The forward half of the under parts is called the breast, the rear half the abdomen.)

Shanks: Black, with or without a small amount of pale flesh color.

Feet and toes: Upper surface black and pale flesh color, the latter predominating; bottom pale flesh color.

The Slate

The Slate, sometimes called Blue Slate, is described as staty blue in all sections of both sexes. Solidly off-colored feathers of any other color such as white, buff, black, or red are a standard disqualification. This variety breeds true to color in its pure (homozygous) form, although white and rusty brown markings are normal and are a color defect. These off-colored markings follow the plumage pattern of the Bronze. Light slate birds are generally preferred because they breed true and more nearly meet standard requirements. The dark slate specimens, being heterozygous, do not breed true and hence should be eliminated from Slate flocks. Black specking is a defect occurring mainly in males and is almost impossible to eliminate.

The shanks and toes are pink, the beak is horn-colored (gray), and the eyes are medium dark brown. This variety, like the others, is subject to the general disqualifications. See page 618 for the genetic aspects of slate plumage. A description of Blue turkeys which are acceptable as Slates will be found on page 535.

Color Description of Day-Old Slate Poults

Beak: Pale flesh color.

Head: Light yellowish brown to rich tan color with faint, light gray markings corresponding in shape and size to those of the Bronze.

Eyes: Leaden color with pupil almost black.

Throat: Pale yellowish white to light rich yellow.

Upper (dorsal) parts, neck to tail inclusive: Yellowish white with surface tinge of tan and faded, light gray markings corresponding in shape and size to those of the Bronze. Some specimens show only faint traces of these markings. Undercolor is pale gray or yellowish, corresponding to surface color.

Under (ventral) parts, neck to tail inclusive: Yellowish white.

Shanks, feet, and toes: Pale flesh color.

The above color description for Slate poults can not be applied to Blue poults which are clear pale gray with white where the Black poult has it; beaks, legs, and feet are like the Black but gray rather than black.

DESCRIPTION OF NON-STANDARD VARIETIES

Broad Breasted Bronze*

This variety was developed in the Pacific Coast States. It is distinguished by extremely heavy muscular de-

*Of English origin, being imported to Canada from England in 1927 by Jesse Throessel, Aldergrove, British Columbia.

Fig. 10.11. An exceptionally good meat type BBB tom with poor balance. (Photo from U. S. Dept. of Agr.)

velopment all over the body, especially on breast and legs. The body is exceptionally blocky in conformation, which is enhanced by the possession of relatively short legs and neck. The percentage of leg and breast meat (combined) in the average dressed 28-week-old turkey Broad Breasted Bronze carcass appears to be around 43% in contrast to about 37% for ordinary Bronze, and other non-broadbreasted varieties and strains. Evidence shows that laying ability and hatchability tend to be low and that the Broad Breasted Bronze matures at about the same age as do most standardbred turkeys. As to ability to utilize feed more efficiently, the evidence available indicates that the Broad Breasted Bronze is somewhat more efficient than turkeys of the Bronze or other standardbred varieties. Its heavy fleshing qualities are transmitted genetically, the mode of inheritance not being sex-linked and due to more than one pair of genes which act independently of size or weight. Since 1945, the heavy fleshing of the Broad Breasted Bronze has been introduced into the Beltsville Small White, the Broad Breasted White Holland, and certain strains of some other varieties so that this characteristic no longer is confined to the Broad Breasted Bronze.

Relatively large size also characterizes the Broad Breasted Bronze, the usual weight at market time—28 weeks of age—being approximately 27 pounds alive (24 lbs. dressed) for toms and 16½ lbs. alive (15 lbs. dressed) for hens. The heavy fleshing and blocky conformation produce an exceptionally attractive carcass which has met with great favor on the markets. The adoption of the Broad Breasted Bronze was phenominal in its speed and completeness, practically replacing the standardbred Bronze as a commercial turkey within a period of five or six years following 1939. The fact that, from the standpoint of lean meat carried, the birds can be butchered profitably as early as 10 weeks is a point stressed by the breeders. Due to the larger size and distinctly different conformation, large, broad-breasted turkeys require somewhat different methods of management such as providing no roosts or very low roosts, preferably of poles 2 to 4 inches in diameter.

In poult color and type the Broad Breasted Bronze cannot accurately be told from Standardbred Bronze but in general tends to be slightly darker, to have less distinct markings, and to be more plump. In 1940 the American

Poultry Association requested its judges to use the Broad Breasted Bronze Club standard for judging. This standard, prepared by a committee of breeders* is reproduced herewith:

Scale of Points

Breast	30
Symmetry, including Carriage	20
Back and Spring of Rib	15
Condition and Vigor	10
Legs and Thighs	10
Breed Character[1]	5
Head and Neck	5
Wings and Shoulders	5
	100

Disqualifications

Any evidence of crossbreeding
Any physical deformities
Any evidence of disease

[1]Plumage color 3, refined ruggedness in bones and headgear, 2.

Standard Description

Breast—30 points

The breast should be wide, with extremely heavy flesh, and carry its width well back. The front of the breast, over the wishbone, should be heavily fleshed. The flesh should extend well over the ribs, smoothly and symmetrically, with the face of the breast practically flat. Any tendency for the flesh to cut away sharply at the junction of the ribs should be discouraged. The front of the keel bone should be carried well forward to avoid a long slope to the shoulders. The keel bone itself should be straight except for a slight upward curvature at the rear, with no tendency toward a knob on the front end. The rear end of the breast should be blunt, not narrow out to a long point. The breast should be parallel to the back. The keel bone should be of sufficient length to intersect the flesh of the thighs when the bird is suspended. The tendency toward an "hour glass" breast (extreme width in front and swiftly tapering) should be severely dis-

*Consisting of R. D. Mitchell, Oak Harbor, Washington; H. P. Griffin, Sonoma, California; A. L. Hamilton, Chehalis, Washington; M. M. Lyons, Portland, Oregon; Joseph Kupetz, Goshen, Oregon; and Ward Cockeram, Oakland, Oregon.

couraged. Too short a breast should also be heavily cut, for while width is perhaps more to be desired than length, it is possible to obtain the two in one bird. The breastbone should be free from roost dents and the flesh should be firm and smoothly placed. Knobs and roughness on the front of the keel of toms or hens are undesirable.

Symmetry or "Balance"—20 points

Body and entire general appearance of the bird should be balanced and symmetrical. In order to determine this the bird should be observed while walking on the ground or in a typing cage where natural action will disclose faults in symmetry. While it is permissible for the back to be practically horizontal, any tendency to tilt forward or towards stilted walk should be severely cut. Any tendency to bowed and crooked legs or enlarged hock joints can be cut here as well as under "Legs and Thighs." The general outline of the body should be rectangular and massive. The breast should be practically parallel to the back and the back moderately short. The common tendency for the body to fall away rapidly in depth toward the rear of the carcass should be severely cut. In old hens the tendency to greater abdominal depth toward the rear is permissible. The depth should be sufficient to make the bird symmetrical; slightly in excess of the width.

Back and Spring of Rib—15 points

Back should be broad and relatively flat. The width should carry well back to the tailhead and be free from tapering. The upper back should be flat between the shoulders, with ample width. A bird lacking in width of back and spring of rib cannot carry the type of breast required in this breed. A narrow, sloping back denotes lack of room in the body cavity for either feed consumption or the proper operation of the digestive tract. A back that tapers toward the rear indicates that the entire body tendency is toward a light rear end and lack of balance or symmetry. Heart girth is important in that any deficiency here not only denotes lack of constitutional vigor, but destroys the rectangular shape sought in body development. Lack of spring of rib means there is lack of body capacity and less room in the body cavity for lungs and other essential organs.

Condition and Vigor—10 points

Condition and vigor denote the health and well-being of the bird. They are indicated by alert appearance, the silken lustre of the feathers, the width of feather and its quality whether soft or brittle, expression of the face,

Fig. 10.12. Broad Breasted Bronze breeding toms. (Photo from R. D. Mitchell.)

color and condition of the caruncular development, ease and readiness of locomotion, and the general attitude of the bird. Native stamina and vigor are essential in the propagation of the strain. A bird lacking in vigor quite often is deficient in the ability to put on flesh or to fatten properly, and quick and early development of both these qualities is essential in this breed. Early maturity as evidenced by close feathering and freedom from pinfeathers is one of the outstanding qualities of this type of bird.

Legs and Thighs—10 points

The upper and lower thighs should be thick-fleshed, the latter tapering sharply to clean-cut hocks. Shanks should be of moderate length in proportion to the balance of the bird. Leg-bones should be strong but show no coarseness; the feet fairly large to hold the heavy body, but not big and coarse; the toes straight; legs attached well forward on the body to balance the bird and straight with no tendency to bow; hocks clean-cut with no puffiness; action free and easy.

Breed Character—5 points

Color and other breed characters are in general conformity with existing standards of the Bronze variety, except that copperish bronzing in back, tail, and upper thighs is not required as it is in the Standard Bronze. Cleancut ruggedness in headgear and bony structure are specified.

Head and Neck—5 points

The shape of the head denotes feeding qualities and, in the tom, masculinity. The head should be short, broad, and massive with strong, curved beak. The eyes should be alert and bright, denoting health, vigor, and stamina. The neck should be of medium length, strong and in proportion to the balance of the bird. Both extra long, unduly short, or slender necks should be cut.

Wings and Shoulders—5 points

Wings should be proportional in size to balance of the body and be strongly and firmly carried, as droopy wings denote lack of vigor and stamina. Shoulders should be square, wide, and flat. While the shoulders carry little meat, it is impossible to get a rectangular body and width of breast without square, wide shoulders.

The Wild Turkey

The eastern wild turkey, (*Meleagris gallopavo silvestris*) which appears to be the subspecies most commonly raised in captivity in the United States (see Fig. 1.3) is described by Barrows as follows: "Adult male: Head with its bristly bare skin, fleshy appendage, wattles, etc., red, blue, and white, as in the domestic turkey; a large tuft of coarse black bristles hanging from the center of the upper breast; general plumage dark brown, with rich metallic lustre, showing burnished bronze, copper, blued steel, or gold, according to the angle at which the light strikes it; most of the feathers of the under parts, and especially those of the wing coverts, lower back and rump, tipped with velvet black; upper tail coverts tipped with chestnut; primaries and secondaries slate-colored, barred with white, the white bars broadest and most conspicuous on the secondaries; tail brown, narrowly barred (penciled) with black, with a broad subterminal black band and tipped conspicuously with bright reddish brown or chestnut. Adult female similar but much smaller, duller . . . . no trace of the beard on the chest. Length of male 48 to 50 inches; wing (length of the folded wing front to end of the longest feather) 18½ inches; weight 16 to 40 pounds . . . . eggs average 2.55 by 1.79 inches."

As a general statement it may be said that the color patterns of all 5 to 7 subspecies (races or varieties) of the North American wild turkey are similar to that of the exhibition Bronze without the heavy, coppery-red bronzing and pure white edgings of the latter. The four subspecies found wild in the United States differ prominently from the Bronze in having cream color to deep rust or iodine color in place of the pure white edgings. The Mexican wild turkeys except *Meleagris gallopavo onusta* (Moore) resemble the Bronze more closely than the other subspecies, the main points of difference being the lack of the copper bronzing on the back and lack of bronzing and distinct penciling in the main tail feathers. None of the wild turkey races carry the large amount of copperish bronzing that is found on the back and tail of the exhibition Bronze but they all have some of it. However, all of the wild turkeys are what is called "blackbacked" by the fancier, that is, carrying practically no copper bronzing on the back feathers. The Broad Breasted Bronze with its lack of copper bronzing on back and tail and its lack of pure white edgings tends to revert back to color of its Mexican wild turkey ancestors.

The eastern wild turkey (*Meleagris gallopavo silvestris*) has dark chestnut or iodine color in the tips of all feathers where white tips are specified in the Bronze, and the white bars on primaries and secondaries are as wide as or wider than the dark bars. The Florida wild turkey (*Meleagris gallopavo osceola*) is similar to *silvestris* but is distinguishable from it by the narrow white bars of the large wing feathers which are much narrower than the dark bars and do not cross the shaft of the feather. The Rio Grande wild turkey (*Meleagris gallopavo intermedia*) resembles the eastern turkey but its main tail feathers and coverts are tipped with light chestnut or cinnamon color instead of iodine and the lower back feathers present a bluish-black appearance. Merriam's wild turkey (*Meleagris gallopavo merriami*) also resembles the eastern turkey but the tips of the tail and tail coverts are creamy to buffy white, not cinnamon or iodine.

The three Mexican wild turkey varieties along with Merriam's turkey differ from the other three varieties by their white to almost-white tipping instead of cinnamon to iodine. The central and south Mexican sub-species (*M.g. mexicana* and *M.g. gallopavo*) are distinguished by lack of distinct penciling in the main tail feathers which were described as "mottled and maculated" by an early author. However as judged by skins in the National Museum, the tail feathers of these Mexican wild turkeys are in fact indistinctly penciled. In contrast, all four United States varieties of wild turkeys have very distinctly penciled main tail feathers and so has Moore's *Meleagris gallopavo onusta,* also a Mexican sub-species and the latest (1938) to be described. It has no cinnamon subterminal bar on any of the rectrices and no black bar on the median pair as does *merriami*; the tips of the upper tail coverts and main tail feathers are pure white and the margins of the remiges (quill feathers of the wing) have less cinnamon than *merrami*. Onusta, mexicana, merriami, and silvestris all are about the same size and considerably larger than *osceola* and *intermedia*. The south Mexican turkey (*M.g. gallopavo*) is a little smaller than the first four.

In appearance all wild turkeys differ rather strikingly from the domesticated birds in the slimness of the head and the greater proportionate length of neck and legs. In addition, the bones appear to be smaller and more refined.

In size the wild turkey unquestionably averages smaller than most domesticated strains even when reared in captivity and full-fed. It also is slower to attain sexual maturity and market condition, and does not lay so well as most domestic strains. Genetically the wild strains tend to transmit these qualities, consequently the introduction of wild blood tends to delay maturity, reduce the size, and lower the egg production, at least of the better standardbred strains. McIlhenny, who it appears had first-hand information regarding the American wild turkey, makes a number of interesting comments as follows: "By the first of November the young gobbler weighs 7 to 9 pounds, the hens 4 to 7 pounds; in December and January the former weighs 12 pounds and the latter 9 pounds. . . . The tuft or beard does not appear in the southern climate until late October or November. . . . The gobbler's beard grows rapidly until the end of the third year, then slowly until 11 or 12 inches long, when it seems to stop.

McIlhenny was of the opinion that the males do not breed the first year, but that the females do. He mentions the "breast sponge" which is a reservoir of energy consisting of a mass of fibrous fatty material located over the crop. This is observed readily in domestic turkeys. In a young Bronze tom weighing 18 lbs. dressed, the breast sponge weighed 2 pounds in early March. It constitutes the "flabbiness" found in mature males. McIlhenny remarks that wild males eat but little during the breeding season resulting in rapid loss of weight. This is in accord with the behavior of domesticated turkeys even though full-fed. However, observations of wild turkeys raised in captivity indicate that the young males are fully capable of breeding their first year. At the Beltsville Station, young wild toms produced semen* when they were 38 to 40 weeks of age. In the wild state the young toms though probably capable of breeding are prevented from doing so by the older males.

McIlhenny observes that the eggs are laid 8 to 15 to a setting, are a little smaller than those of the domestic turkey, are quite pointed at the end, and are "spotted and blotched all over with reddish brown and sometimes lilac." Bent reported the normal setting to be 8 to 15 eggs with a maximum of 18 to 20. Barrows states that the number laid is 10 to 18, which are "light buff, thickly speckled or sprinkled with brown." Blakey reports that average clutches range from 9 to 15 eggs each, with 15 most frequent.

STANDARD FOR THE NORTHEASTERN WILD TURKEY[1]

By C. L. SIBLEY, *Secretary*
American Pheasant Society, Wallingford, Conn.

Shape of Male and Female

Head: Small, slim, snaky, somewhat carunculated (the less the better) with tubular leader on top of the head at the base of beak, the size of which is subject to expansion or contraction, according to the will of the bird.

Beak: Strong, curved, well set in head.

Eyes: Round to oval, full, prominent.

Throat-Wattle: Very small, almost lacking in best specimens, and with few or no caruncles.

Neck: Long, slim, held erect.

---

*When manipulated according to the artificial insemination technique described in Chapter 13.

[1]Published by permission of Modern Game Breeding, Doylestown, Pa.

Beard[2]*:* Long, bristly, prominent in old males; may or may not be found in females.

Wings: Large, powerful, smoothly folded, carried well up on sides of body.

Back: Rounded, sloping from juncture of the neck in a convex curve to the tail.

Tail: Long, feathers broad, carried low (except when displaying) in a continued graceful curve with top line of back; tail coverts broad and abundant, extending well over base of tail.

Breast: Somewhat rounded, more so in adult males.

Body and fluff: Body long (about 48 inches for adult tom), not too broad; fluff short.

Legs and toes: Lower thighs long and muscular; shanks slim, long, strong; toes straight, long, strong, and well spread.

Carriage: Upright, nervous, alert; backline of body and tail carried at about 55 degrees to the horizontal.

Color of Male and Female

NOTE: Serious defects are white feathers in any part of plumage; flight or secondary feathers showing clear black or brown in one or more; absence of white or gray bars more than one-half the length of the flight feathers; white tips on rump or main tail feathers; absence of buff or brown edging on breast of female; shanks other than red or pink in adults (young birds, especially hens, may show a grayish or brownish tinge on shanks for the first year).

Color of Male

Head: Blue, changing to reddish purple, the bluer the better.

Beak: Light brown at tip, darker at base.

Eyes: Dark brown.

Throat-Wattle: Blue, with few bluish purple caruncles, the fewer the better.

Neck: Rich brilliant iridescent bronze.

Beard: Black.

Wings: Fronts and bows, rich copperish bronze, ending in a narrow band of black; coverts bright gleaming coppery bronze, forming a broad bronzy band across the wings when folded, each feather terminating in a black band, forming when wing is folded, a distinct black mark separating them from the secondaries; flight feathers each crossed throughout its entire length with distinct parallel black and white bars of about equal width, running straight across the feather; secondaries alternately crossed with black and white bars, the white and black becoming less distinct on the shorter top secondaries, the black bars becoming wider and taking on a rich bronze cast.

Back: From neck to middle of back, rich coppery bronze, each feather terminating in a narrow black band; from middle of back to tail coverts, black, each feather having a broad,

---

[2]Mosby reports that toms with beards over 6 inches long are over 3 years old, beards 4 to 6 inches long indicate 2-year-old toms, while beards under 4 inches long indicate a young tom. Hens with beards over 2 inches long are over 2 years old.

brilliant coppery band extending across it near the end, feathers ending in a distinct black band, gradually narrowing as the tail coverts are approached.

Tail: Main tail and greater coverts dull black, each feather marked tranversely more or less distinctly with parallel lines of brown, each feather having a wide band of iridescent coppery bronze across it near the end, a distinct black band bordering the bronze and terminating in a broad edging of deep chocolate brown; the more distinct and brilliant the colors, the better. Young birds may not show a dark brown terminal edging to main tail feathers and tail coverts as do older birds.

Breast: Rich, iridescent bronze, feathers of lower breast terminating in a black band across the end of the feather.

Body and fluff: Body black, each feather with a wide, iridescent, coppery bronze band extending across it near the end, a distinct band of black bordering the bronze and terminating in a narrow edging of chocolate brown; fluff, dull black.

Legs and toes: Lower thighs similar to breast but less brilliant; legs and toes in adults red or pinkish red; in young birds pink, sometimes overcast with gray or brown.

Color of Female

Beak, eyes, head, legs, and toes: Same as in male, except legs sometimes paler in color and sometimes overcast with grayish or horn color.

Plumage: Similar to male but less brilliant. Also a narrow edging of buff to chocolate on feathers of back, wing bows, wing coverts, breast, and body. This edging, narrow in front, gradually widens as it approches the rear of the bird, where it merges into the deep brown terminal edging of tail coverts and main tail feathers.

THE NEBRASKAN

A pied color mutation apparently produced by a non-sex-linked recessive gene appeared in 1947 in a closed flock of Broad Breasted Bronze. Thirteen poults were hatched the first year, and these formed the basis of a new true-breeding variety named the Nebraskan by the originator R. H. Jandebeur of North Platte, Nebraska. Some back-crossing to the BBB was employed to broaden the genetic base for the new variety which has all the characteristics of the parent strain except color of shanks, feet, eyes, and plumage. Plumage color of both sexes is described as white, including all feather shafts, with black surface specking, more or less in all sections, the main tail and main wing feathers and their coverts being nearly white. Flecking is most abundant on back.

Fig. 10.13. Nebraskan tom (photo from John M. Lukert, East Moriches, N. Y.).

Undercolor is white in all sections except the back where it may be slate-colored. Color selection should be toward the lighter-colored specimens which show a fairly uniform, lightly stippled pattern with white undercolor everywhere and without conspicuous dark colored areas anywhere. These light-colored specimens will produce no dark-colored pinfeathers. Shanks, feet, and back are white without dark spots or tendency toward foreign color. Eyes are bluish pearl. The day-old poult is creamy white with a brown spot on the head.

In weight and type the Nebraskan closely resembles the typical Broad Breasted Bronze strain from which it was derived. The originator suggests that at 8 months (35 weeks) of age the females should weigh 17 to 20 pounds, the males 28 to 35 pounds with emphasis on heavy bone, strong legs, good balance, broad (but not excessively broad) breast, good body depth, and uniform heavy fleshing in all sections.

THE NITTANY TURKEY*

This variety was developed at Pennsylvania State University through selection from the native wild turkey. Selections were made with the idea of developing a domesticated variety similar to the wild turkey in size and color pattern but with a more docile temperament, better egg-producing ability, and better ability to take on finish. Nittany females were found to be ready for market at 24 weeks of age. At 28 weeks, the normal market age for standardbred stock, Nittany turkeys produced a pound of turkey meat on the same quantity of feed as did the Bronze.

The color of the **male** is as follows:—Head: red, changeable to bluish white. Beak; light horn at tip, dark at base. Eyes; dark brown. Throat wattle; red, changeable to bluish white. Neck; iridescent bronze with a black edging added as the breast and shoulders are approached. Wing fronts, wing bows, and secondary coverts (wing bar); iridescent bronze with a black edging. Primaries, primary coverts, and pinion feathers; each feather throughout its entire length crossed alternately with parallel black and light gray bars of equal width across the feathers. Secondaries; outer secondaries barred like primaries, barring becoming less distinct on inner secondaries which have a copperish-bronze surface color. Back; from base of neck to thighs iridescent bronze edged on the outer side with a velvety black band which is wider toward the thighs; from the thighs to tail coverts each feather crossed with a copperish-bronze band edged on the outer side with a velvety black band that gets narrower as the tail coverts are approached. Tail; each covert and main tail feather marked transversely with brown and black penciling and tipped with a solid black band which in turn is edged completely with chestnut brown. Breast; iridescent bronze, each feather edged with velvety black, the edging becoming wider toward the fluff. Body and fluff; body dull black; upper thighs black with a copperish-bronze band on surface bordered with band of black and terminating in an edging of chestnut brown; fluff black, tipped with chestnut brown. Undercolor; dull black in all sections. Legs and toes; lower thighs (drumsticks) dull black lightly edged with chestnut brown; shanks and toes in mature specimens pinkish red, in young specimens dark, almost black. Beard; black. Skin; white or yellowish white depending on the nature of the feed eaten.

The color of the **female** is similar to that of the male, except that there is an edging of chestnut brown on the black tips of the feathers of back, wing bows, secondary coverts, and breast. The edging should be narrow in front, gradually widening as it approaches the rear of the specimen.

Proposed standard weights for Nittany turkeys are as follows: Adult tom, 20 lbs.; yearling tom, 17 lbs.; young tom, 15 lbs.; adult hen, 12 lbs.; yearling hen, 11 lbs.; and young hen, 10 lbs. These weights show the smallest sex differences of any variety of turkeys and indicate that the Nittany is the smallest domesticated variety of turkey in America today. It appears to be adaptable to the retail trade since both hens and toms at the usual market age are within the weight requirements, but has never been produced in commercial quantities.

*Data furnished by P. H. Margolf and H. H. Kauffman of Pennsylvania State University.

THE ROYAL PALM[1]

The Royal Palm (called Desert Palm in California) (Fig. 10.14) was developed from a single male sport or mutation occurring in a mixed flock consisting of Bronze, Black, Narragansett, and Wild turkeys. This single Royal Palm sport was mated to unrelated, off-colored Bronze hens and from this mating off-colored Bronze sons and Royal Palm daughters were obtained, indicating sex-linkage and recessiveness of the genetic factor or gene responsible for the new variety. These daughters when mated back to their sire produced only Royal Palms and since then the variety has bred true. When the Royal Palm is crossed with pure Bronze, all of the first-generation crossbreds are Bronze-colored in appearance.

Fig. 10.14. Royal Palm turkey tom. (Photo from Enoch Carson.)

The Royal Palm color pattern may be the result of the action of a modifying gene, or factor, on the Bronze pattern which is plainly in evidence as a background. The gene which produced the Royal Palm undoubtedly has appeared in Europe since a variety similar to the Royal Palm but called the Crollwitz turkey was exhibited at the Sixth World's Poultry Congress at Berlin in 1936. For discussion of the "Palm" and "Red Palm" color patterns, see Chapter 11.

The color of the Royal Palm **male** is as follows: The neck plumage is silvery white except at the point of union with the back plumage where a small iridescent black tip appears. This narrow black tip is bow-shaped and increases in size down the back so that at the rear it is about three-fourths inch wide, giving the main central portion of the back the appearance of being solid black on the surface. Starting with the lesser tail coverts adjoining the back plumage, there appears a narrow white edging on the

[1]Data furnished by the originator, Enoch Carson, Lake Worth, Fla.

black tip which at this point has become a black bar. This white edging increases in width as the main tail coverts are approached. The single row of main tail coverts, however, is clear silvery white. The main tail feathers are silvery white except for an iridescent black bar which is about 1½ inches from the end of the feather. The breast is white with a small black tip which increases in width to a fairly wide bar as the thigh is approached. The upper thighs are white; the lower thighs are white with shafted gray underneath shading to white at base, with tips black.

The wing fronts and bows are white, tipped very lightly with black toward the front, the width increasing somewhat as the wing bar is approached. The secondary coverts forming the wing bar are white with conspicuous black tips forming a dotted line across the wing. The primaries are solid, dark slate with a white shaft. The secondaries nearest the back are white with a black spot near the end. The outer secondaries are similar and in addition the center portion of the outer web is dark slate. The undercolor is light gray except on the back where it is a medium shade of slate. The legs are light reddish color. The beak is horn color. The eyes are dark brown.

The **female** is colored like the male except that her neck, breast, upper back, and wing fronts are clear, silvery white. In other sections black markings are present as described for the male but are not so prominent.

The color of the Royal Palm **poult** is clear light yellow, or in a few cases light yellow with traces of black on the back.

The proposed standard weights for Royal Palms indicate greater sex differences than in the other varieties of turkeys but in general the Royal Palm may be considered as a rather small variety. In market type as well as in ornamental value this new variety appears to rank high. Its relatively small size makes it suitable to the retail trade.

The Royal Palm probably is the most strikingly attractive true-breeding variety of turkeys so far produced. It is regrettable that such a beautiful and practical color pattern is not in vogue among the commonly grown varieties of turkeys. The white and black pattern should produce relatively few dark pinfeathers combined with easy picking qualities.

THE BLUE TURKEY

This little-known variety, like the Buff, is very rare but can be produced readily by crossbreeding. To produce Blues, the Slate should be crossed with Black, and the black-splashed, brown-tinted, blue-gray, crossbred offspring bred together. One-sixteenth of the offspring of this cross will be homozygous, true-breeding Blues which when day-old are a clear pale gray with white where the pure black poult has it—on the head, wing tips, and abdomen. When grown, these Blue males are blue gray with a tendency to possess more or less brownish-gray shafting and tipping. The Blue females are light bluish gray with a tendency to possess a little white specking in wings and tail. Possibly through selection, the homozygous Blues obtained from the Slate-Black cross could be improved in color to produce clear blue plumage. Homozygous blue turkeys meet the Standard requirements for Slates, hence the Standard description for the latter could be used by breeders wishing to develop Blues. Blues, since they carry the factor for black color, have very dark (almost black) eyes.

THE JERSEY BUFF TURKEY

A buff variety of turkeys once was recognized by the American Poultry Association but was dropped from the Standard in 1915. It is now being bred at the New Jersey Station (Millville) and has been named the Jersey Buff. Emphasis is being placed on good reproductive and market qualities and medium-small size. In 1951 and 1952 on the basis of National Turkey Improvement Plan participation, Jersey Buffs were fourth in popularity in the U. S.

The Buff can be produced by crossing the Black and the Bourbon Red, then mating the rusty-black crossbreds together (no sex-linkage involved). One-sixteenth of the resulting F_2 offspring will be pure, true-breeding Buff and two-sixteenths will be heterozygous Buff. The two kinds (homozygous and heterozygous cannot be segregated accurately at hatching time or even when adult, therefore the problem of producing buffs that will breed true can be solved only by a breeding test by which the homozygous Buffs could be recognized and retained for breeding. If a buff turkey is mated to a Bourbon Red and produces 100% buff offspring, he or she is pure for buff color. Average weights of Jersey Buff turkeys are given (See Fig. 10.6).

The color and special disqualifications for the Jersey Buff turkey as described by the New Jersey Experiment Station are as follows:

Special Disqualifications

More than one-third any other color than white showing in either primaries, secondaries or main tail feathers.

Color of Male

Head: Red, changeable to bluish white.

Beak: Light horn at tip, darker at base.

Eyes: Very dark brown, shading almost to black.

Throat and Wattles: Red, changeable to bluish white.

Neck: Deep reddish buff.

Beard: Black

Wings: Fronts, bows and coverts, rich reddish buff; each feather having a very narrow edging of black; primaries and secondaries, pure white with some buff shading toward tips.

Back: Rich reddish buff; each feather from point of shoulder to base of main tail, having a very narrow edging of black.

Tail: Pure white, with a dimly outlined bar of soft buff crossing each main tail feather near the end; coverts, deep reddish buff.

Breast: Rich reddish buff, feathers having a very narrow edging of lustrous black.

Body and Fluff: Body, deep reddish buff; each feather edged with a fine line of black; fluff, shading from buff to light creamy white at base of each feather.

Legs and Toes: Lower thighs, dark reddish buff; shanks and toes, mottled slate with some reddish horn.

Undercolor in all sections: Creamy white.

Color of Female

Similar to that of the male except there is no black edging in any section. Most buff hens have narrow white edgings on the breast and body feathers and this usually has to be accepted as a necessary evil, although selection should be toward the elimination of the white edgings.

The Buff poult is a medium to dark shade of rich, reddish tan with white where the Black poult has it—on head, wings, and abdomen. There is a slight tendency to possess dark brown markings on the head and back but the clear, reddish-buff color is preferred.

Fig. 10.15. A Crimson Dawn adult tom. (Photo from U. S. Dept. of Agr.)

THE CRIMSON DAWN[1]

(Black-winged Bronze)

The shape of the Crimson Dawn (Fig. 10.15) turkey is similar to that described for all turkeys in the Standard of Perfection except that the breast is described as exceptionally long and broad, the breadth being most apparent near the front of the keel and surrounding the furcula, or wishbone. The breast should narrow somewhat toward the rear of the keel so that it does not interfere with the locomotion or balance of the specimen. The feathers, especially those of the tail coverts and back, are described as broad, any tendency toward narrowness being considered a serious defect and extreme narrowness a disqualification. The Crimson Dawn color pattern, technically known as black-winged bronze, is the result of the action of a genetic factor recessive to the bronze and black patterns and part of a triple-allel series of color genes.

[1]Data furnished by the originator, Tom Stodghill, R. R. 4, Terrell, Texas.

The color of the Crimson Dawn *male* is described as follows: Head; red, changeable to bluish white. Beak; light horn at the base; pale, almost white, near the tip. Eyes; dark brown. Throat wattle; red, changeable to bluish white. Neck; black, overlaid with lustrous bronzing of a dawn shade of crimson gold[1] except near the shoulders where each feather is brilliant black finely penciled with white or buff, the latter preferred. The penciling is single to quintuple, the more markings the better, and follows the contour of the feather. Wing fronts; penciled, similar to lower neck feathers. Wing bows and secondary coverts; black, overlaid with crimson-gold bronzing, and lightly edged with velvety black. Primaries, primary coverts, and pinion feathers; solid black. Secondaries; lower webs black, over-laid with crimson-gold bronzing; upper webs black finely stippled with buff, tips pure white forming a line across the wing. Back; forward portion between the shoulders penciled similar to lower neck; center and rear portion black ending with a bar of rich crimson-gold bronzing edged in turn with a narrow strip of black. As the tail coverts are approached there is an edging of pure white added to the black tips. Tail coverts; body of the feathers brilliant tan or tan irregularly penciled with black, the former preferred; tip possessing a band of rich crimson-gold bronzing one-half inch wide, bordered front and rear by a narrow black band and terminated by a broad band of pure white. Main tail feathers; similar to tail coverts, the rich crimson-gold band becoming somewhat wider and the white edging approximately three-fourth inch in width. Breast; forward half black overlaid with lustrous crimson-gold bronzing; rear half similar to forward half but with a narrow edging of velvety black which becomes wider as fluff is approached. Some white penciling may appear on the breast, but this is not desirable. Body and fluff; body dull black; upper thighs black with a bar of crimson-gold bronzing bordered on the outside with a narrow band of black and terminated by an edging of pure white; lower thighs (drumsticks) penciled white on dark gray; fluff, dull black. Undercolor; light gray becoming almost white near the skin. Shafts of all feathers white in unexposed portions, black in surface plumage. Shanks and toes; dark pinkish gray in young birds, deep pink in adults. Toenails; light horn color, almost white, the whiter the better. Beard: black. Skin; white or yellowish white.

The color of the *female* is similar to that of the male except, (1) the breast is distinctly penciled, white on black, similar to lower neck, and (2) the penciling of the lower neck and shoulders in general is more distinct.

The color of the Crimson Dawn *poult* is similar to the Bronze except that the ground color is yellowish white instead of light brown, producing the appearance of a pale colored Bronze poult.

Special disqualifications: Solid white feather or feathers anywhere; one or more primaries or secondaries (not counting the axial feather in young birds), barred for more than one-half the length of the feather; white or gray markings that can be seen in the tail beyond the coverts; one or more solid-colored main tail feathers; complete or almost complete lack of the crimson-gold bar

---

[1]The "dawn shade of crimson-gold" resembles the red-green bronzing of the Bronze but the greenish iridescence is partly suppressed in favor of rose pink.

in the center and rear feathers of the back; complete or almost complete lack of the white penciling in breast of female; failure of the black or crimson-gold band to extend completely across each main tail feather or greater tail covert; main tail feathers and greater tail coverts conspicuously narrow in shape.

THE SILVER BRONZE

This new color variation appeared as a mutation in the Broad Breasted Bronze flock of George R. Roberts, Lancaster, California in 1948. In color it resembles, but is not identical with, the Palm (not Royal Palm) turkey described by Asmundson and appears to be produced by an autosomal (non-sex-linked) recessive gene. In size and type the Silver Bronze is identical with the typical BBB strain from which it originated. In color the male is white on neck and forward half of back; black on rear

Fig. 10.16. Narragansett tom showing fine exhibition color. (Photo from A. D. Hudson, Tangent, Oregon.)

half of back; tail coverts are black with white tips; main-tail feathers are white with small amount of grey and have a black band near the ends. Primary and secondary wing feathers are white with a little gray and have black tips; wing coverts are black, wing bows, breast, and body are dark gray with a little white mottling. The female has the same color scheme as that of the male, but has dark gray where the male has black, medium gray where the male has dark gray, and in general shows more gray throughout, lacking the male's nearly white tail, wing feathers, forward half of the back, and neck feathers. Undercolor of both sexes is white in all sections.

Day-old Silver Bronze are partially auto-sexing, the males tending to be a light gray with dark gray stripes, the female plain light gray.

LARGE BROAD-BREASTED WHITE STRAINS

Great interest recently (1950-1955) has developed in white-plumaged turkeys intended to approach or equal the Broad Breasted Bronze in performance, conformation and size. Several such strains have been developed, among them the following. The Empire White by E. Y. and Don Smith, Genoa, N. Y.; the Broadwhite by Jos. L. Thompson, Pottsville, Arkansas; the Lancaster White by Geo. R. Roberts, Lancaster, California; and the Michigan White by Clarence Orcutt, Bellaire, Michigan. The first two strains were developed with the aid of crossing the BBB with White Hollands, the last two were derived from white sports that appeared in Broad Breasted Bronze flocks. Other large white strains are being developed by individuals and by experiment stations, among them North Dakota and Cornell, at which the Empire White was first developed. A medium-sized broad breasted white turkey is being bred at the Maryland Station.

Some breeders of White Hollands have improved their stock to the point where it has characteristics similar to those of the special strains mentioned above. It appears highly improbable that separate variety status will ever be attained by any of the special strains of large white turkeys since they have no characteristics that would serve to distinguish them readily and definitely from each other and from other large or medium-sized white turkeys which normally are identifiable as White Hollands (The Beltsville Small White attained variety status by virtue of its decidedly smaller size.)

NEW SMALL-TYPE STRAINS

In addition to the Beltsville Small White, Royal Palm, and Jersey Buff described previously, other small-type strains have been developed. Among the breeders who have developed small-type Broad Breasted Bronze are H. P. Griffin, Sonoma, California. G. W. Andrews (Pozo Turkey Ranch) Santa Margarita, California, Victor Ryckebosch, Lancaster, California, and Steenbergen Turkey Farm, Ironton, Ohio. The small strains of Broad Breasted Bronze are only slightly heavier than the typical Beltsville White. West and Son, Stacy, Minnesota have produced a small-type Narragansett with broad-breasted characteristics. In addition, both West and Andrews have small-type broad-breasted strains with plumage patterns resembling the Narragansett but of light coloring, especially the undercolor. Other small strains probably exist that have not come to the attention of the authors.

STANDARD WEIGHT

Adherence to a weight, or size, standard is necessary in breeding and judging poultry if uniformity of breed and variety characteristics is to be obtained and retained. The original standard weight requirements as nearly as can be determined are based on the weights that were expected of poultry including turkeys when exhibited in winter shows (usually held in December and January). This meant that the prescribed standard weights were winter weights, which implied that young birds were well matured and old birds had regained lost weight. With turkeys, these winter weights usually are considerably higher than the weights at which they are generally marketed and, under certain conditions of climate or management are close to the maximum first-year weight attained. If the climate and/or management are such as to promote early egg production, these weights (of both males and females) will be reduced. If breeding and egg production do not start before March and if feeding is liberal, turkeys continue to put on weight through winter and early spring. In Montana it was noted that with no artificial lighting, average date of first egg was about March 26 and maximum first-year weight of males and females was attained the first week in April. When lighted starting the first week in February, maximum weight and average date of first egg were attained around March tenth (breeding-stock weight, pgs. 544 and 691). In Maryland when lighting is started January 15, maximum weight usually is attained in mid-February; when started December 1, maximum body weight is reached the first week in January.

These and other data indicate that maximum weight of both males and females usually is attained at about the average date of first egg or shortly thereafter. Growth records indicate that turkeys of large and medium size attain the prescribed standard weights about 8 months or 35 weeks, provided feeding is liberal and egg production does not start prior to that age. If production starts prior to that age, the standard weight probably will not be attained that year. If on the other hand production is delayed until spring, "standard" weight is likely to be exceeded considerably. Once production starts, body weight of both sexes (if toms are managed in the same manner as the hens) decreases and usually continues to decrease until mid-summer or early fall, when, if feeding is liberal, it will be regained. Usually by mid-November the maximum weight attained the previous season will be regained. Then during the fall and winter a further increase in weight usually occurs so that turkeys approaching their second and third breeding season are heavier than they were the previous seasons as is indicated by the greater standard weights specified for yearling and adult turkeys. The observation that males whether mated or isolated from the females provided they are treated the same way, show eight changes comparable to those of the females is of considerable interest.

Since the Standard of Perfection does not specify either the age or the date at which standard weight should have been attained, the individual breeder and judge must make his own interpretation. For turkeys a practical application of the weight standard, based on average gains made by well-fed stock, would be to use either the 35-week, or roughly 8-month weights, or the weights attained between January 15 and January 31 for comparison with the standard requirements. For example a young standardbred Bronze tom that attains a weight of 25 lbs. at the age of 35 weeks (8 months) or any time during the last half of January should be considered a standard-size bird. Likewise a Beltsville White tom that weighs 21 lbs. at 34 to 36 weeks, is of the proper weight. Such a bird would have weighed about 15 lbs. at 24 weeks and 18 lbs. at 28 weeks. The standard weights of 19 and 11 pounds (respectively) are reached by the Beltsville White young toms and hens at about 30 weeks (7 months).

Weight records of well fed Broad Breasted Bronze stock indicate that, on the average, young toms gain 2½ lbs. each 4 weeks and hens 1 lb. each 4 weeks (small-type turkeys 1½ lbs. and ⅝ lb., respectively), between 28 and 36 weeks of age and also slightly less on the average, between 28 weeks and the time egg production starts if it does not start till after the age of 36 weeks. Thus a 28-week-old BBB young hen weighing 16½ lbs. between November 15 and November 30 would weigh 18½ lbs. between January 15 and January 31 (at the age of about 35 weeks) and at the earlier date and age mentioned would also be considered as a standard-sized turkey. In the same manner a small-type young hen weighing 10½ lbs. at 28 weeks would weigh about 11½ lbs. late in January (at about 35 weeks of age.)

Table 10.2 gives the weights prescribed for the standard varieties and proposed for non-standard varieties. If, as suggested, the 8-month or 35-week weights are considered to be standard weights, a definite basis for selection and comparison is established. It is suggested by the authors that the term "breeding weight" be used to avoid the looseness of understanding that now occurs in connection with the term "standard weight," and that it represent the weight at about 35 weeks (8 months) for all varieties except Beltsville Small Whites which reach standard weights at 30 weeks or approximately 7 months. Other small varieties could be treated similarly.

Table 10.2—Standard Weights[1] in Pounds

	Young hen	Yearling hen	Adult hen	Young tom	Yearling tom	Adult tom
Broad Breasted Bronze, Broad Breasted White, Silver Bronze and Nebraskan[2]	18	20	22	32	38	41
Beltsville Small White[3]	11	12	13	19	22	23
White Holland[3]	14	17	18	25	30	33
Jersey Buff[4]	12	13	14	21	24	25
Bourbon Red, Narragansett, Black and Slate[3]	14	16	18	23	30	33
Bronze, standardbred[3]	16	18	20	25	33	36
Royal Palm[4]	11	13	15	22	26	30

[1]Interpreted to mean maximum weights attained as young, yearling, and adult turkeys or when turkeys are approximately 8, 20 and 32 months old respectively.

[2]Non-standard varieties. Weights given here are derived from weights of typical specimens and proposed as "standard" by Marsden and Martin.

[3]These are the varieties recognized by the American Poultry Association and described in the Standard of Perfection (1953).

[4]Non-standard varieties. Weights given here are those proposed by the originators.

General Disqualifications of Turkeys When Shown in Exhibition Classes*

In judging exhibition classes, specimens of all varieties of turkeys are disqualified if they possess one or more of the following faults:

1. General unfitness—not qualified to represent the variety.
2. Evidence of unfair altering.
3. Permanently deformed beak, tail, back, or wings; including roach back and wry tail.
4. Clipped wing, slipped wing, or split wing.
5. One or more permanently twisted main tail or large wing feathers.
6. Weight more than 6 pounds below the standard for all except Beltsville Whites. According to the Standard, young birds are not disqualified for underweight until after November 30th of each year. In practice neither young nor old birds are disqualified for underweight before December 1st. Beltsville Whites are not disqualified for underweight but are disqualified for overweight, more than 3 pounds in toms, more than 2 pounds in hens, at any date shown.
7. Feather or stub of feather on shanks or feet; or on hocks if disconnected from normal feathering.
8. Decidedly crooked breastbone.
9. Pronounced bowlegs or knock-knees.
10. Enlarged or deformed shank or hock.
11. More or less than four toes on one or both feet.
12. Black in quills of primaries or secondaries or foreign color, except slight gray ticking, anywhere in the plumage of white varieties.
13. Main tail feathers completely missing.
14. Presence of crest.

---

*This listing is adapted from the 1953 Standard of Perfection.

15. Shank, shanks, foot, feet, or toes of color foreign to the variety.
16. Definite evidence of infectious disease or presence of a tumor affecting health or market value.
17. Blind in both eyes.
18. Penulous crop.

Serious General Defects of Turkeys When Shown in Exhibition Classes

1. Breastbone knobbed or dented.
2. One or more missing main tail, primary, or secondary feathers (except disqualification 13, above.)
3. Tail not grown out fully.
4. Tendency toward bowlegs or knock-knees.
5. Crooked toes.

Variety and Strain Differences

Significant, consistent differences between varieties in reproductive ability, ability to utilize feed, and palatability of the flesh may or may not exist. It more often happens that significant differences exist in different strains. Certain strains of a variety may mature earlier than others. Some may have superior reproductive ability or produce more desirable market carcasses. Selection of a breeding source according to strain characteristics is important after the variety has been chosen on the basis of its general traits.

Most strains of white turkeys have proportionately more feathers and seem to be somewhat slower in maturing all their feathers than are the colored varieties, and in addition their feathers remain "set" in the skin with more persistence, making a "good stick" difficult to attain during dry picking. These two characteristics make them harder to pluck. However the semi-, sub- and full-scalds if properly done, practically eliminate this difference, and any remaining disadvantages are compensated for by the fact that pinfeathers remaining after picking are not conspicuous and that skin discoloration due to feather pigment does not occur. Pinfeathers also are inconspicuous in the Buff, Nebraskan, Royal Palm, Slate, and Bourbon Red, but so-called "blue-back," when it occurs may be noticeable. The broad-breasted strains of turkeys produce, on the average, a market carcass superior in appearance and possessing a somewhat higher percentage of breast and leg meat. The relatively small trade in live turkeys does not favor white or partly white turkeys.

Strain and varietal differences in palatability of the flesh were studied at the Beltsville Station (Marsden *et al*, 1952) ; see Table 10.3. Broad Breasted Bronze (BBB), standardbred Bronze (SBB), Beltsville Small White (BSW), and White Holland (WH) were used. From the average scores obtained from tasting the roasted meat it appears that in tenderness of breast meat, BBB and BSW tied for first and were noticeably more tender than the other two strains. In tenderness of dark meat, strain differences did not appear significant but the BSW ranked first in juiciness of white meat, differences did not appear significant but again the BSW ranked first. In juiciness of dark meat, BBB was scored slightly lower than the other three which were about the same. This difference probably was due to the fact that in the BBB the legs frequently were overcooked and dried out while the thick breast meat was being cooked to tenderness. In flavor of light and dark meat the differences did not appear significant but the BBB and BSW were slightly superior to the other two. In general conclusions regarding the desirability of the cooked white meat as food the BBB and BSW were considered as slightly but significantly better than the other two varieties. In general conclusions on the dark meat the BSW was the outstanding favorite while the other three were about the same.

In final analysis of the tasting tests, slight but demonstrable differences in favor of Broad Breasted Bronze

Table 10.3—Average Scores of the Judges in Tasting Tests of Cooked Turkey Meat

Palatability factor	Variety or strain	Score*	Rank	Palatability factor	Variety or strain	Score*	Rank
Tenderness of breast meat	B.B.B.	3.90	(2)	Juiciness of breast meat	B.B.B.	2.95	(4)
	S.B.B.	3.44	(4)		S.B.B.	3.02	(2)
	S.B.W.H.	3.45	(3)		S.B.W.H.	3.00	(3)
	B.S.W.	3.91	(1)		B.S.W.	3.13	(1)
Flavor of breast meat	B.B.B.	2.86	(1)	General conclusions on breast meat	B.B.B.	4.22	(1)
	S.B.B.	2.75	(4)		S.B.B.	3.84	(3)
	S.B.W.H.	2.76	(3)		S.B.W.H.	3.79	(4)
	B.S.W.	2.85	(2)		B.S.W.	4.18	(2)
Tenderness of upper thigh	B.B.B.	4.32	(2)	Juiciness of upper thighs	B.B.B.	3.06	(4)
	S.B.B.	4.33	(3)		S.B.B.	3.35	(3)
	S.B.W.H.	4.28	(4)		S.B.W.H.	3.40	(2)
	B.S.W.	4.46	(1)		B.S.W.	3.45	(1)
Flavor of upper thigh	B.B.B.	2.77	(2)	General conclusions on upper thigh meat	B.B.B.	4.03	(2)
	S.B.B.	2.68	(3)†		S.B.B.	4.00	(3)†
	S.B.W.H.	2.68	(3)†		S.B.W.H.	4.00	(3)†
	B.S.W.	2.82	(1)		B.S.W.	4.54	(1)

*Note: Perfect score for flavor of light and dark meat is 3:00; all others 5:00.
†Tied for third and fourth.

and Beltsville Small White over the other two strains were indicated by the judges. White Hollands and standardbred Bronze had close to the same scores in all tests, and ranked third and fourth in most of them. The outstanding fault of these two varieties was the low scores for tenderness of white meat.

The Utah Station (1951) reporting on the first year's work of a long-time project found that there were significant differences between well-known "utility" strains of Broad Breasted Bronze turkeys in respect to growth rate 6 to 30 weeks of age, mortality 0-30 weeks, susceptibility to hexamatiasis, tendency to lose weight during the breeding season, number of good eggs laid during the breeding season, age at first egg, per cent fertility of eggs, and per cent hatch of eggs. Significant differences were not found as to shank length, keel length, breast width, and average egg weight.

Selection of Variety, Strain and Source

The grower who raises turkeys for market and does not intend to keep breeding stock is interested primarily in obtaining poults that will live well and grow into a desirable type of market bird. Poults or hatching eggs should originate in flocks blood-tested under Federal-State or reliable private auspices and found free from indications of pullorum-typhoid disease. Those tested under the National Turkey Improvement Plan (NTIP) are rated as U. S. Pullorum-Typhoid Clean or U. S. P-T Passed. Poults so designated must be hatched in a hatchery that is participating in the NTIP. Pullorum ratings obtained through private testing cannot carry the "U. S." label or the National Plan terminology but they can be fully reliable. A high pullorum rating from any source is not a guarantee of high livability but it is a reliable indication of freedom from destructive pullorum disease. Nearly all turkey breeding birds in the United States are blood-tested and it is estimated that about 73% of all the breeding stock kept for the 1952 season was tested under the NTIP, information concerning which can be obtained by writing to the U. S. Department of Agriculture, Beltsville, Maryland. The plan is a two-pronged voluntary program involving (1) blood testing stock for pullorum-typhoid disease, along with incubation of the eggs under supervised sanitary conditions; and (2) selection of breeding stock for Standard and meat production qualities. Under this plan poults labeled "U. S. Approved Pullorum-Typhoid Clean" or "U. S. Approved Pullorum-Typhoid Passed" are suitable for the commercial grower. Pullorum-Typhoid-Clean poults may be sold only by hatcheries dealing with the "clean" classification. Poults with the Pullorum-Typhoid rating only are available in some States, and these too are acceptable. They sometimes are of high quality, depending on the breeder who produces them. Poults rated "U. S. Certified Pullorum-Typhoid Clean" or "U. S. Certified Pullorum-Typhoid Passed" are sometimes available. These are poults out of U. S. Approved hens and sired by U. S. Record of Performance (ROP) males or from performance-tested stock and usually bring a somewhat higher

*The four breeding classes are Approved, Certified, Performance-Tested, and Record of Performance (R. O. P.).

price. Limited numbers of pedigreed hatching eggs, poults, and breeding stock are obtainable from R. O. P. breeders.

As to variety, the commercial or farm turkey grower is not now offered a very wide choice but he does have the choice of a large number of strains within the three most common varieties produced in the U. S.: Broad Breasted Bronze (BBB), Beltsville Small White (BSW), and White Holland (WH). There is a small production (but a fairly wide distribution) of the new Jersey Buff (J.B.) and Nebraskan varieties but only a very small and scattered production of the other varieties. If the grower wants to raise large turkeys he has a choice of BBB, large-type or Broad Breasted W.H., and Nebraskan. If he wants a medium-sized turkey, he can select the average W.H., or the now relatively rare standardbred Bronze, the Black, Narragansett, Slate, or Bourbon Red. Small-type turkeys are practical for use as turkey "broilers" (fryers or light roasters). They also are suitable for a fair share of the holiday trade which can utilize both toms and hens of the small varieties if they are well selected for good market type and are properly processed. Of the small-type varieties the Beltsville White predominates to an extent rivalling the predominance of the BBB in the large-turkey field. The Jersey Buff, a little larger than the BSW, is next in popularity among the small-type varieties and there also is a small scattered production of small-type BBB, small-type Narragansett, small-type white strains (most of them a little larger than BSW), Royal Palm, and Small Grey. Wild turkeys are produced on a few game farms but these birds are not well suited to commercial production since they are poor layers, difficult to manage, slow to mature, and have long necks and legs.

The prospective turkey grower should investigate the market demands and facilities in the area in which he is located or intends to locate and to plan to grow a turkey that will be most acceptable to that market. As aids in the selection of a strain there are (1) the turkey shows in which the breeders can display their wares; (2) the turkey meetings, local to national in scope; (3) the poultry magazines, with their articles and advertising and (4) personal visits and discussions.

There are good opportunities in the turkey breeding field for (1) independent specialized breeders, (2) commercial breeders or reproducers. For the former, personal liking and strong belief in the future of the chosen variety and strain would be of primary importance and no expense or care should be spared in the selection of foundation breeding stock. For the commercial breeder, first consideration would be to select a variety and strain the products of which will find ready sale. If he produces or intends to produce eggs for a hatchery or a hatching egg dealer he must cooperate with the management thereof in the selection of the breeding stock; if he sets up his own breeder hatchery he should select varieties and strains that are already in good demand in the area or have good possibilities for sales promotion.

At present (1955) there is great interest in large, broadbreasted white turkeys of which several different strains recently have been produced. There is some reason to think that these large white and nearly white turkeys will in time largely replace the Broad Breasted Bronze, which, in a very few years, starting about 1939, almost completely replaced the ordinary non-broad-breasted Bronze which at that time (1939) comprised at least 65% of all turkeys raised in the U. S. For convenience these old-style Bronze frequently were and are called, without much justification, standardbred or Standard Bronze. However there were at that time, prior to the advent of the BBB, a few flocks of well-meated but not exhibition-colored Bronze called Baby Beef Bronze by certain breeders. There also were many flocks of genuine Standard Bronze of the beautiful exhibition

Table 10.4—Variety Distribution of Breeding Hens and Toms for the 1952 Season in the U. S.[1]

State and Geographical Div.	Broad Breasted Bronze	Standard Bronze	White Holland	Beltsville Small White	Jersey Buff	Bourbon Red	Black	Narra-gansett	Nebras-kan	Crosses	Others	Totals
Maine	2,045	....	952	10,882	626	....	....	....	98	....	....	14,603
New Hampshire	1,457	....	4,274	120	148	....	....	....	....	....	....	5,999
Vermont	2,675	3,700	310	100	....	....	....	....	39	....	....	6,824
Massachusetts	7,986	625	12,651	878	580	....	....	....	142	16	....	22,878
Rhode Island	734	....	831	9	209	....	....	40	....	....	....	1,823
Connecticut	4,156	....	3,651	725	285	136	....	540	....	....	Slate 4 [2]Roy.Pm. 128	9,625
New York	11,017	....	20,032	721	358	....	....	....	56	....	....	32,184
New Jersey	5,318	....	4,174	1,814	1,562	....	....	....	202	....	....	13,070
Pennsylvania	26,898	....	10,789	5,147	79	....	....	....	809	....	Nittany 1,025	44,757
North Atlantic	62,286	4,325	57,664	20,396	3,847	136	....	580	1,346	16	1,167	151,763
Ohio	59,253	....	18,198	18,594	....	....	....	....	201	913	....	96,952
Indiana	21,179	....	....	4,930	....	....	....	....	....	....	....	26,109
Illinois	31,203	....	1,012	3,136	....	35	....	....	190		Med. Small White 57	35,633
Michigan	70,980	....	4,639	6,875	182	....	....	....	....	....	....	82,676
Wisconsin	6,544	....	200	150	....	....	....	....	705	....	....	7,599
East North Central	189,159	....	24,049	33,685	182	35	....	....	1,096	913	57	249,176
Minnesota	158,847	204	3,495	43,354	372	....	60	1,881	5,182		[3]S.T.B.B.B. 4,022	217,417
Iowa	74,800	....	....	....	....	....	....	....	....	....	....	74,800
Missouri	143,061	....	4,908	21,589	293	....	....	....	889	....	....	170,740
North Dakota	3,977	....	....	3,885	325	....	....	....	....	544	....	8,731
South Dakota	4,269	....	....	....	....	....	....	....	1,926	....	....	6,195
Nebraska	21,369	....	84	....	....	....	....	....	1,519	....	....	22,972
Kansas	21,822	....	961	4,200	....	....	....	....	506	....	....	27,489
West North Central	428,145	204	9,448	73,028	990	....	60	1,881	10,022	544	4,022	528,344
Delaware	2,577	....	....	6,450	....	....	....	....	....	....	....	9,027
Maryland	26,108	....	1,794	6,096	....	318	....	....	....	....	....	34,316
Virginia	86,657	....	2,499	237,995	364	....	....	....	1,199	....	....	328,714
West Virginia	....	2,711	207	13,399	....	....	....	18	....	....	....	16,335
North Carolina	17,989	2,451	....	5,775	420	....	....	23	....	....	....	26,658
South Carolina	53,638	....	....	26,868	....	....	....	....	1,819	....	....	82,325
Georgia	3,152	....	....	1,426	....	....	....	....	....	....	....	4,578
Florida	1,230	58	44	3,486	....	....	....	....	....	....	[3]S.T.B.B.B. 68 [2]Roy. Pm. 7	4,893
South Atlantic	191,351	5,220	4,544	301,495	784	318	....	41	3,018	....	75	506,846

Table 10.4—Continued (Variety Distr.)

State and Geographical Div.	Broad Breasted Bronze	Standard Bronze	White Holland	Beltsville Small White	Jersey Buff	Bourbon Red	Black	Narragansett	Nebraskan	Crosses	Others	Totals
Kentucky	26,350			4,052								30,402
Tennessee	1,270											1,270
Alabama	8,401		459	2,224					35			11,119
Mississippi	3,030		124									3,154
Arkansas	9,689		2,718									12,407
Louisiana	21											21
Oklahoma	25,648		706	8,050		57			18		Slate 27	34,506
Texas	168,910		157	17,251		417	222					186,957
South Central	243,319		4,164	31,577		474	222		53		27	279,836
Montana												
Idaho	2,460			560	57							3,077
Wyoming												
Colorado	389		30	958								1,377
New Mexico												
Arizona	1,810			297								2,107
Utah	38,354			2,388								40,742
Nevada												
Washington	64,634		4,217	19,239					566			88,656
Oregon	251,008		2,343	18,514	23	30		12			[4]1,732	273,662
California	829,658		3,403	86,088	19,148			62	850		[2]Roy. Pm. 39	939,248
Western	1,188,313		9,993	128,044	19,228	30		74	1,416		2,705	1,348,869
United States 1952	2,302,573	9,749	109,862	588,225	25,031	993	282	2,576	16,951	1,473	7,119	3,064,834
Percentage of U.S. Total												
U. S. 1947, Pct.	88.8	3.6	4.6	1.9	.1	.3	.3	.2	0		.2	1,890,879
U. S. 1948, Pct.	72.5	14.8	5.2	6.2	.2	.3	.2	.4	0		.2	1,095,938
U. S. 1949, Pct.	77.0	15.0	3.8	3.6	.2	.1	.1	.1	0		.1	1,849,907
U. S. 1950, Pct.	75.6	13.5	3.6	6.5	.4	.1	.1	.1	0		.1	2,132,644
U. S. 1951, Pct.	76.6	1.5	4.4	16.3	.6	.1	.03	.1	.2	.1	.1	2,264,376
U. S. 1952, Pct.	75.12	.32	3.58	19.19	.82	.03	.01	.08	.55	.05	.23	3,064,834
U. S. 1953, Pct.	76.0	.2	4.0	17.4	.2	.003	.1	.1	.9	.1	1.0	2,596,694

[1]From National Turkey Improvement Plan, U. S. Dept. of Agriculture, Beltsville, Md.
[2]Royal Palm
[3]Small-type Broad Breasted Bronze.
[4]Mostly small-type BBB with a few Black, Jersey Buff, Bourbon Red, Narragansett, and Mutation BBB.

bronze pattern. Then there were just Bronze turkeys, birds without exhibition color or good meat qualities. Probably at least 60 per cent of the turkeys produced in the nineteen-thirties were in the latter category. The Broad Breasted Bronze changed all of this but still the predominating plumage color remained bronze. White turkeys were not generally popular except in the northeastern States (New England, N. Y., and Pennsylvania) until the appearance of the Beltsville Small White which grew in popularity from no commercial production at all in 1946 until in 1951 this variety comprised an estimated (by the U. S. Division of Agricultural Economics) 20 per cent of all turkeys produced in the U. S. In 1954 the estimate was 30%. Starting about 1939 the Standard varieties seem to have decreased in popularity. According to a nation-wide survey made by the Turkey World among its subscribers, 63.9% of the 1938 turkey crop consisted of Bronze turkeys, 13.5% White Hollands, 9.7% Narragansetts, 9.1% Bourbon Red, and 3.8% Black. A survey made about the same time in Connecticut indicated that only about 43% of the turkeys raised in that state were Bronze, while 29% were Bourbon Red, 17% Narragansett, 8% White Holland, and 3% other varieties and crossbreds. In 1929, 91% of Connecticut's turkeys had been Bronze, 3% Bourbon Red, 1% Narragansett, 3% White Holland, and 2% miscellaneous.

VARIETY DISTRIBUTION

The only information on turkey variety distribution in the United States available at present (1955) is that from the records of breeding turkeys tested for pullorum disease under the National Turkey Improvement Plan. Table 10.4 shows as nearly as can be ascertained the numbers of turkey breeding birds, males and females combined, tested and retained for breeding, arranged according to variety, State, and geographical division for the 1951-52 breeding season.

Since in 1952 about 73% of the USDA estimate of about 4,200,000 breeding turkeys (the DAE breeder hen estimate of 3,836,000 hens plus 10% for toms) in the U. S. were under NTIP supervision, the plan figures probably may be considered as generally representative of variety distribution in the U. S. as a whole. However there probably are actually more flocks of the less common varieties than are shown in the Plan figures since small-scale specialty breeders and fanciers are less likely to participate than are commercial breeders who usually have the common varieties. In estimating the number of turkeys produced from the number of breeders kept it will be noted that although Beltsville Whites made up about 19% of the breeding birds kept they comprised about 24% of the turkeys raised (by numbers not tonnage). The reason for this is that the Beltsville White breeder hen, due to higher reproductive performance on the average produces more poults than does the average breeder hen of the large and medium sized varieties.

SELECTED REFERENCES

American Ornithologists Union, 1931. Check list of North American Birds, published by Am. Ornith. Union, Lancaster, Pa.

American Poultry Assn., 1953. Standard of Perfection. Pub. at 16 So. Park St., Sapulpa, Okla.

Barrows, W. B., 1912. Michigan Bird Life. Mich. State Coll., Bul. (791 pages, illus.)

Bent, Arthur C., 1931. Life histories of North American gallinaceous birds. Bull. 162 of U. S. Natl. Museum, Washington, D. C.

Jull, M. A., 1947. Raising turkeys, ducks, geese, game birds. Pub. by Doubleday, Page & Co., Garden City, N. Y. (245 pages, illus.)

McIlhenny, Ed. A., 1914. The wild turkey and its hunting. Pub. by Doubleday, Page & Co., Garden City, N. Y. (245 pages, illus.)

Mosby, Henry S., and Charles O. Handley, 1943. The wild Turkey in Virginia. Pub. by Div. of Game and Inland Fisheries, Richmond, Va. (260 pages, illus.)

Myrick, Herbert and others, 1897. Turkeys and how to grow them. Pub. by Orange-Judd Co., New York, N. Y.

National Science Museum, 1946. Birds of North and Middle America. Bull. 50 (Washington, D. C.)

Wright, A. H., 1914-15. Early records of the wild turkey. The Auk, 31:334, 463; 32:61, 207, 348.

CHAPTER 11

BREEDING PRINCIPLES AND PRACTICES

OUTLINE

Breeding includes not only reproduction but the maintenance of desirable qualities and general improvement. Continued improvement can result only when a constructive organized breeding program is followed. The guiding principle in turkey breeding is the rule that nearly all important characteristics are to some degree heritable and passed from parent to offspring. The essentials for success are careful observations, accurate record keeping, careful selection on family and individual basis, systematic mating, good feeding, and careful management. Feeding and management, while not breeding problems, are inseparably linked to them. Parent stock must be well-nourished and properly managed if the breeding value is to be accurately appraised. Proper nourishment and care, in order to allow full expression of hereditary characteristics, are equally necessary for young stock. Only by providing favorable environment is accurate selection made possible.

Age of Breeding Stock

Well-matured young hens and toms generally are more satisfactory for the production of hatchable eggs and vigorous poults than are older birds. The Purdue Station found that poults from young turkey hens were as large at 16 to 24 weeks as were those from older hens. The fertility and hatchability of eggs produced by matings of young birds are usually better than those produced by yearling or older hens and in addition, young hens lay more eggs. In a comparison of young and yearling BBB hens the Utah Station (Stephenson et al., 1951) found that the yearling hens were two pounds heavier, produced 11½% fewer poults and laid 25% fewer eggs which were 8% larger.

The commercial breeder seldom would find it profitable to retain either hens or toms for re-use. The pedigree breeder, however, often can utilize to good advantage the best performing hens for a second or even third year's breeding. He might even find it desirable to retain the most successful breeding pens intact. Toms are less likely than hens to be profitable beyond the first season and rarely are retained.

Young breeding turkeys should be reasonably well matured if they are expected to produce heavily and efficiently over a long season. Young hens and toms of the large- and medium-sized strains should be 33 to 37 weeks (average 8 months) old when first exposed to artificial light; those of small varieties should be 31 to 35 weeks (average 7½ months). If exposed only to natural light, the ages mentioned should be attained about two weeks ahead of the average expected date of first egg which varies according to latitude and amount of housing provided. In the latitude of Maryland (approx. 39° latitude) for young, well-matured turkey hens on open roosts, or housed but with free access to outside yards, average date of first egg is about March 8. For best and most economical results then, poults of the large and medium sized varieties intended for use as breeders under these conditions, should be hatched around June 21; those of the small varieties should be hatched around July 5. In the latitude of central Montana (approx. 47°), for example, average date of first egg under the above-specified conditions is about March 29; except for western U. S.* Prospective breeders, therefore, could be hatched about three weeks later than in Maryland. However adverse fall and winter growing conditions might make it impracticable to grow to maturity poults hatched later than about July 1. Turkeys younger than the ages mentioned lack the necessary maturity to make good breeders; while those much older are expensive to raise due to the unnecessarily long maintenance period before production starts. In addition, the toms of the heavy, broadbreasted strains in particular get too heavy and clumsy to make good breeders if they are more than 9 to 10 months old when mating starts.

SEX RATIO IN BREEDING FLOCKS

A well-matured vigorous young male turkey will serve up to 20 or even more females, the number depending upon his age, type, and size. For single-male matings the following suggestions are made : — large broad breasted types 10 to 15 hens; medium-sized standardbred turkeys, 16 to 18 hens; and small-type strains, 19 to 20 hens. There is no objection to using fewer hens but the larger numbers are desirable at the start so as to allow for mortality and culling early in the season for such reasons as low egg production, development of leg defects, infertility, poor egg-shell quality, and low hatchability. For flock matings of large broad breasted turkeys a common plan is to provide 1 tom for 8 to 10 hens, then

*At about 45° latitude in western Oregon, date of first egg is the same as in Maryland, about March 8th.

divide the toms into 2 groups and alternate them once or twice weekly. In this manner, all the toms are used but they are never in the flock in a higher proportion than one tom to 16 to 20 hens. Some operators use all the toms until the first flush of mating is over, then start alternating. Others put in more toms than they intend to use then cull them out after the intense initial mating activity is over. Excellent results frequently are obtained with heavy turkeys by setting up the matings at the rate of 1 to 12 and allowing all the toms in the flock at one time throughout the season. This plan works well in commercial flocks not of extreme broad breasted type. For flock matings of standardbred turkeys a ratio of 1 to 13 or 14; for small-type turkeys 1 to 15 or 16 is suggested. It seldom is necessary to alternate the males in such matings but in cases where there is undue interference, alternating the toms in the manner suggested for the large broad-breasted types may be beneficial.

Yearling and adult toms will serve fewer hens, on the average about two-thirds as many as the young males. Artificial insemination (described in Chapter 13) may be employed to obtain offspring from valuable males which are too old or crippled to mate naturally but should not be used in ordinary breeding operations because of the possibility of developing strains unable to breed naturally. It also may be used initially to augment or take the place of natural mating in breeding flocks.

THE SELECTION OF BREEDING STOCK

Breeding stock should be selected for market qualities, standard qualities, and qualities contributing to efficient production and reproduction. Market and standard qualities are determined by physical inspection of prospective breeding birds (Section 1); the qualities which contribute toward efficient production and reproduction can be ascertained only with the aid of records (Section 2).

Section 1. Physical Selection for Market and Standard Qualities. (a) Market type, size, and vigor. Preferably between the ages of 22 and 30 weeks, the breeding stock should be selected on the following points:

(1) Fullness of breast, leg and back fleshing—determined by handling. Breast should be smooth and broad, its width carried out well to the rear end of the breastbone (grade 4 or better). A slight tapering of breast width toward the rear of the keel is normal and necessary for locomotion and proper leg placement. Breast can be too narrow at the rear or two broad and square. Drumsticks should be plump, well meated toward the hocks, and of sufficient size to produce a well-balanced carcass. (Grade 4 or better). The back should be uniformly wide, straight, and smooth. (See grading plans for breast and leg fleshing in Chapter 6).

A method of judging the quantity of meat carried on the breast has been described by Kansas State College. A piece of No. 12 fuse wire about 9 inches long is placed under the feathers and painstakingly molded to the shape of the bird's breast at its widest part about one inch back from the front of the keel. The wire is carefully removed, laid on a piece of paper and traced with a pencil. On this paper a vertical line 3 inches long is extended from the peak of the resulting figure and midway between the edges of it. Another line is then extended across the figure to intersect its two edges and the free end of the 3-inch line to which it is perpendicular. By tracing the outline of the resulting figure (which is a cross-section of the bird's breast) with a mechanic's planimeter, the area in square inches can be obtained. A home-made planimeter can also be used. This procedure probably has rather limited possibilities because it is not very accurate with live birds, and too time-consuming to use in selections involving large numbers of birds.

An instrument to measure body depth and half the breast width at one-fifth the body depth was designed at the Dominion Experimental Farm, Ottawa, Ontario, Canada by S. Bird. This instrument was tested and found to be accurate and practical to use for taking measurements of breast width and body depth of prospective turkey breeding stock (El-Ibiary and Jull, 1948).

(2) Desirable skeletal development and proportions including reasonable shortness of neck and legs, moderate depth and good breadth of body at the ribs and over the hips, and good length of breastbone (keel). The development should be such that when bird is suspended by the legs there is no conspicuous gap between legs and rear point of keel, and keel and back are parallel. This is part of the "balance" factor.

At 24 weeks or later the keel of females should equal or slightly exceed the shank in length; in males, at 30 weeks. To measure the right shank, hold it at a right angle to the drumstick, then bend back the outer toe so that it is at a right angle to the shank (which is about as far as it will go), hold it firmly in this position and measure the distance between (1) its walking surface at a point very close to the sole pad and (2) the outer, or rear, surface of the hock joint, pressing the calipers firmly against the parts mentioned. This method of measuring is the one upon which the shank measurements in Tables 11.1, 11.2, and 11.3 are based. Measurements to surface of sole pad will be slightly greater and less accurate than those made to the outer toe next to the sole pad, as recommended. The keel is measured from front to rear points, using firm pressure of calipers (See Figs. 11.1 and 11.2).

Shanks can be too short. Asmundson reported a recessive non-sex-linked, sublethal genetic factor in Bronze turkeys called "S" which produced short leg and wing bones and very low hatchability and livability. Mature males pure for the S factor had shanks 1½ inches under the average for the variety while mature hens had shanks about 1 inch shorter than normal. It is suggested that the approximate lengths given in Table 11.1 be used as a guide after making proper adjustments as indicated for the broad-breasted and small-type strains. A caliper suitable for this purpose as well as for taking other measurements is shown in Fig. 11.1.

Depth is measured from front point of keel vertically to the back, pressing the calipers firmly onto the feathers.

Because the keel continues to lengthen and the body to deepen as the turkey matures, measurements of these parts must be closely checked against age in order to use them to good advantage. In Table 11.3 will be found useful data on age changes in weight, depth and keel length.

An important point in selection on skeletal development is that the keel be parallel or nearly so, to the back; not turned in toward the back to a conspicuous degree. However, the rear tip of the keel, especially in toms, should turn in slightly toward the back.

Carriage or balance is associated with skeletal development, and is a highly important point to observe, especially in broad-breasted turkeys. There is no means of measuring this trait. Good balance means upright carriage, the back sloping at about a 35° angle above horizontal. A very undesirable type of poor carriage is a condition where the body is tipped forward and the breast is carried low. Observing the birds while walking is important in the selection for "balance" and in detecting leg abnormalities. When the bird is suspended by the legs the rear of the keel should intersect the front line of the legs leaving no conspicuous gap between.

The *back* should be relatively broad and there should be a good spring of rib. Measured over the feathers of the back, at the widest part, should be about 6 inches in 20- to 22-lb. standardbred toms and 4½ to 4¾ inches in well matured 12- to 14-lb. standardbred hens about 30 weeks old. In large, broad-breasted, 24-week-old toms weighing 20 to 22 pounds, back width over the feathers should be about 7 inches—for comparative hens 12 to 14 pounds, 5½ inches. In Beltsville Small White of broad breasted conformation, comparative figures at 24 weeks

Table 11.1—Desirable Skeletal Dimensions of Live Turkeys*

(Body measurements 24 Weeks after hatching)

Males, standard varieties**				Comparable Females, standard varieties			
Weight (lbs.)	Approx. Depth (inches)	Approx. Shank (inches)	Approx. Keel (inches)	Weight (lbs.)	Approx. Depth (inches)	Approx. Shank (inches)	Approx. Keel (inches)
9	6.9	6.4	5.6	5½	5.4	4.9	4.7
10	7.2	6.6	5.8	6	5.6	5.0	4.8
11	7.4	6.8	6.0	7	5.9	5.2	5.1
12	7.6	7.0	6.2	7½	6.0	5.4	5.2
13	7.8	7.2	6.4	8	6.1	5.5	5.3
14	8.0	7.4	6.5	8½	6.3	5.6	5.4
15	8.1	7.5	6.6	9½	6.5	5.8	5.6
16	8.3	7.7	6.7	10	6.6	5.9	5.7
17	8.3	7.9	6.9	10½	6.7	6.0	5.8
18	8.6	8.0	7.0	11	6.8	6.1	5.9
19	8.8	8.1	7.1	12	7.0	6.3	6.0
20	8.9	8.3	7.2	12½	7.1	6.4	6.1
21	9.0	8.4	7.3	13	7.2	6.5	6.2
22	9.2	8.5	7.5	14	7.4	6.6	6.4
23	9.3	8.7	7.6	14½	7.5	6.7	6.5
24	9.4	8.8	7.7	15	7.6	6.8	6.5
25	9.6	8.9	7.8	15½	7.6	6.9	6.6
26	9.7	9.0	7.9	16½	7.7	7.0	6.7

(Body measurements 28 Weeks after hatching)

Males, standard varieties**				Comparable Females, standard varieties			
Weight (lbs.)	Approx. Depth (inches)	Approx. Shank (inches)	Approx. Keel (inches)	Weight (lbs.)	Approx. Depth (inches)	Approx. Shank (inches)	Approx. Keel (inches)
11	7.4	6.3	6.0	7	5.7	4.9	5.0
12	7.6	6.5	6.2	8	5.9	5.1	5.2
13	7.8	6.7	6.4	8½	6.0	5.2	5.3
14	8.0	6.8	6.5	9	6.1	5.3	5.4
15	8.2	7.0	6.7	9½	6.3	5.5	5.6
16	8.4	7.1	6.8	10½	6.5	5.6	5.7
17	8.5	7.3	7.0	11	6.6	5.7	5.8
18	8.7	7.4	7.1	11½	6.7	5.8	5.9
19	8.9	7.5	7.2	12	6.8	5.9	6.0
20	9.0	7.7	7.3	12½	6.9	6.0	6.1
21	9.2	7.8	7.5	13½	7.0	6.1	6.2
22	9.3	7.9	7.6	14	7.1	6.2	6.3
23	9.4	8.1	7.7	14½	7.3	6.3	6.4
24	9.6	8.2	7.8	15½	7.4	6.4	6.5
25	9.7	8.3	7.9	16	7.5	6.5	6.6
26	9.8	8.4	8.0	16½	7.6	6.6	6.7
27	10.0	8.5	8.1	17	7.7	6.7	6.8
28	10.1	8.6	8.2	17½	7.7	6.7	6.8
29	10.2	8.7	8.3	18½	7.8	6.8	6.9
30	10.3	8.8	8.4	19	7.9	6.9	7.0
31	10.4	8.9	8.5	20	7.9	7.0	7.0
32	10.5	9.0	8.6	21	7.9	7.0	7.1

*All of the data in this composite table are taken from Mimeographed Circular 49 (1939) by Jaap, Milby, and Thompson of the Oklahoma A. and M. College, Stillwater.

**To convert these figures for use in selecting large-type broad breasted turkeys, take the actual weight of the tom and use the figures given in the tables for a tom 3 pounds lighter; for hens subtract 2 pounds. For small type broad breasted turkeys, subtract 2 and 1 pounds respectively. However, shank length in large and small-type broad-breasted strains is ⅛ less on toms and in large-type hens and 1/10 less in small-type hens after the 1- to 3-pound subtractions in weight are made. For example, take a Beltsville White tom weighing 15 pounds at 24 weeks. Subtract 2 pounds, leaving 13 pounds for which the shank length is 7.2 inches. Subtracting one-eighth or 0.9 inch, leaves 6.3 inches, the desired length. A 9-lb. Beltsville White hen less one pound is 8 pounds for which the shank length is 5.5 inches. Subtracting one-tenth or .6 inch, leaves 4.9 inches, the desired shank length.

Table 11.2—Desirable Skeletal Dimensions of Live Turkeys (cont.)

(Body measurements **32 Weeks** after hatching)

Males, standard varieties**				Comparable Females, standard varieties			
Weight (lbs.)	Approx. Depth (inches)	Approx. Shank (inches)	Approx. Keel (inches)	Weight (lbs.)	Approx. Depth (inches)	Approx. Shank (inches)	Approx. Keel (inches)
14	8.0	6.5	6.7	9	6.1	5.1	5.4
15	8.2	6.6	6.8	9½	6.2	5.2	5.4
16	8.4	6.8	7.0	9½	6.2	5.2	5.4
17	8.5	6.9	7.1	10	6.3	5.3	5.5
18	8.7	7.1	7.2	11	6.5	5.5	5.7
19	8.8	7.2	7.4	11½	6.6	5.6	5.8
20	9.0	7.3	7.5	12½	6.8	5.7	5.9
21	9.1	7.4	7.6	13	6.9	5.8	6.0
22	9.3	7.5	7.7	13½	7.0	5.9	6.1
23	9.4	7.7	7.9	14½	7.1	6.0	6.2
24	9.6	7.8	8.0	15	7.2	6.1	6.3
25	9.7	7.9	8.1	16	7.4	6.2	6.4
26	9.8	8.0	8.2	17	7.5	6.3	6.5
27	9.9	8.1	8.3	17	7.5	6.3	6.5
28	10.1	8.2	8.4	17½	7.6	6.4	6.6
29	10.2	8.3	8.5	18½	7.7	6.5	6.7
30	10.3	8.4	8.6	19½	7.9	6.6	6.8
32	10.5	8.5	8.8	20	7.9	6.7	6.9
34	10.7	8.7	8.9	21	8.1	6.8	7.0
36	10.9	8.9	9.1	23	8.3	7.0	7.3

(Body measurements **36 Weeks** after hatching, including approximate first year maximum weights)

Males, standard varieties**					Comparable Females, standard varieties				
Wt. (lbs.)	Approx. Depth (inches)	Approx. Shank (inches)	Approx. Keel (inches)	First Year Wt. (lbs.)	Wt. (lbs.)	Approx. Depth (inches)	Approx. Shank (inches)	Approx. Keel (inches)	First Year Wt. (lbs.)
16	8.2	6.6	7.0	20½	10	6.2	5.2	5.4	11
17	8.4	6.7	7.1	21½	10	6.2	5.2	5.4	11
18	8.6	6.8	7.3	22½	10½	6.3	5.3	5.5	11½
19	8.7	6.9	7.4	23½	11½	6.5	5.4	5.6	12
20	8.9	7.1	7.5	25½	12½	6.6	5.6	5.8	13½
21	9.0	7.2	7.7	26½	12½	6.6	5.6	5.8	13½
22	9.2	7.3	7.8	27½	13½	6.8	5.7	5.9	14
23	9.3	7.4	7.9	29	14	6.9	5.8	6.0	15
24	9.4	7.5	8.0	30	14½	6.9	5.9	6.1	16
25	9.6	7.6	8.1	31½	15½	7.1	6.0	6.2	16½
26	9.7	7.7	8.2	32½	15½	7.1	6.0	6.2	16½
27	9.8	7.8	8.3	34	16½	7.3	6.1	6.3	17½
28	9.9	7.9	8.4	35	17	7.4	6.2	6.4	18½
29	10.1	8.0	8.5	36½	18	7.5	6.3	6.5	19
30	10.2	8.1	8.6	38	18	7.5	6.3	6.5	19
31	10.3	8.2	8.7	39½	19	7.6	6.4	6.6	20
32	10.4	8.3	8.8	40	19½	7.7	6.5	6.7	21
33	10.5	8.4	8.9	42½	20½	7.9	6.6	6.8	22
35	10.7	8.5	9.1	44	21½	7.9	6.7	6.9	23
37	10.9	8.7	9.2	47	22½	8.1	6.8	7.0	24
39	11.1	8.8	9.4	48½	23½	8.2	6.9	7.1	25

**See footnote on preceding page.

for 15-pound toms are 6 inches, for 9-pound hens, 4½ inches.

The *keel* should be straight as seen by parting the feathers and should be parallel or nearly so to the back

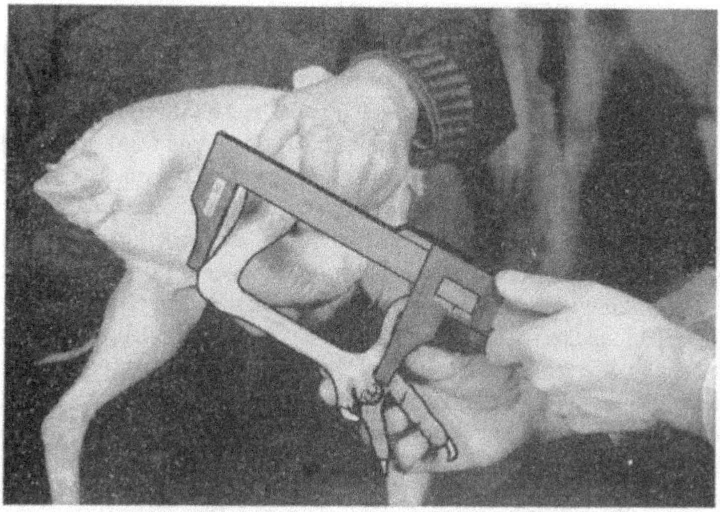

Fig. 11.1. Measuring length of shank. The calipers measure the distance from rear of hock joint to the lower surface of the outer tow close to the sole pad. (Photo by U. S. Dept. of Agr.)

as determined by placing one hand on the back and the other on the keel. Roost dents are undesirable but slight ones may be tolerated; crooked or slightly crooked keels should not be; likewise keels with rear portion noticeably turned in toward the body or turned out away from it.

A convenient table of measurements (Table 11.2) has been developed by the Oklahoma Experiment Station, which used the methods of measurement that are described herein. This is offered as a guide to aid in the selection of breeding stock. They show the proper relationships of depth, length of keel, and length of shank for birds of various sizes and ages of both sexes. To utilize these data efficiently the turkeys must be measured with calipers at the correct ages and proper adjustments made for the broad-breasted and small-type strains as indicated in the footnote of Table 11.1.

(3) Sound legs. The modern broad-breasted turkey not only carries a relatively heavy weight but the heavy breast fleshing tends to put extra strain on the legs. The

result is a tendency toward leg trouble of various kinds. The breeder can help the situation by careful selection for strong, straight legs with no tendency toward deformed hocks. The pedigree breeder should note all leg deformities in the records and weed out all families that produce offspring with deformed legs, particularly of the type known as perosis, the tendency toward which, according to observations at the Beltsville Station, is inherited. Severe leg deformities are easily detected but the slight deformities are hard to spot. The only way they can be detected is to "walk" the bird, getting behind (or in front) of it so the gait can be observed. As suggested elsewhere, birds selected as breeders should be forced to walk or run for a few feet (perhaps 15 or 20) before they pass the final test. Look for and cull all birds that (1) throw one or both legs out to the side (paddle or toe-out) in a conspicuous manner as they walk or run (2) appear bow-legged, (3) limp severely or (4) show crookedness. Nearly all extremely broad-breasted birds "wabble" (body showing a sidewise movement) as they walk or run but if their legs are sound and straight, their feet will be placed nearly straight ahead with each step, not thrown outward to a conspicuous degree. Males are more prone to leg deformities than are females but both sexes undoubtedly share alike in the genetic trans-

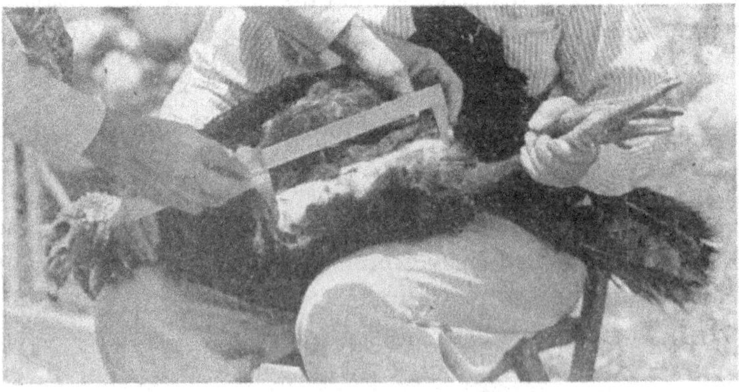

Fig. 11.2. Measuring length of breastbone. (Photo from U. S. Dept. of Agr.)

mission of the tendency toward or susceptibility to leg deformities.

If turkeys are selected at a relatively early age, say 22 to 28 weeks, they should be re-checked at about 32 to 36 weeks since leg deformities often appear during that interval.

(4) Finish and state of feathering are extremely important items in the judging of turkeys for rate of maturity and for market quality, but unless the age is known and unless judging is done at an acceptable age these items are not of great practical value in selecting breeding stock. However, when selection is made at the proper age, 22 to 28 weeks, the finish and the state of feathering become important selection points.

A quick estimate of market maturity can be made by parting the feathers about half way down the side of the breast near front tip of keel (method described and illustrated in Chapter 6). If the skin appears creamy and thick and there are few or no short pinfeathers the bird is ready for market. The general condition of the plumage being smooth, full, and complete in the well-matured market bird is in itself a rough indication of market maturity. Birds that are obviously slow maturing should not be selected as breeders. Small-type turkeys should be mature for market at 22 to 24 weeks of age, large- and medium-sized strains at 26 to 28 weeks.

(b) Standard quality. This term includes variety color, breed type, weight (or size) symmetry, carriage, health, and vigor as specified for turkeys in the American Standard of Perfection. Action, which is not described in the Standard, is important and should be considered as part of this section. For the details of standard quality the reader is referred to Chapter 10 and the American Standard of Perfection.

If the turkey breeder wishes to breed strictly to the color standard, then the standard description should be adhered to. If the breeder does not intend to specialize in exhibition stock he may give only nominal attention to that part of the standard description dealing with color and minor points of conformation. Since color is the only means of separating one variety from another, and since it contributes much to flock uniformity and pride of ownership, at least reasonable adherence to the color standard is highly desirable. Disqualified birds and those noticeably off-colored should not be used as breeders even in flocks not being bred especially for color.

The size (or weight) desired may be standard (medium), above standard (large), or below standard (small). Age at time of selection must be considered because both young and adult stock continue to put on weight during fall and winter. From the standpoint of meat production the weight attained at 24 to 28 weeks of age is most important, and the individual breeder may set his own standard. Weight at turkey "broiler" age is important in strains designed to be used as fryer-roasters.

Health and vigor are indicated by the head, the actions, and the fleshing. The bird should show no signs of disease. The head should be refined yet strong (rugged-refined) with a full, round, bright eye and good depth. Extreme coarseness should be avoided, also its opposite, an over-refined appearance denoting disease or lack of vigor. Likewise crow-head should be avoided. The carriage should be normal for turkeys, the back sloping at 35° from shoulders to tail (as emphasized under balance). The bird should be active, have a normal gait, and be free from deformities, such as crooked legs or back, pendulous crop and roached back.

Section 2. Record of Performance Selection for qualities measured through trapnest, hatching, and growing records. Before birds are selected on basis of appearance and action (as outlined in Section 1) the trapnest, hatchability, growth, and weight records, if available, should be studied. The purpose is to consider as possible breeders only birds whose dams have shown satisfactory breeding performance, who themselves have desirable physical characteristics, and whose sibs (brothers and sisters) also have produced a high family average in physical characteristics. This is the record-of-performance and progeny test as applied to turkeys. The performance of the sire is not considered except as it may be so poor as to justify discarding all the pen progeny. This not infrequently happens and when it does a new pen is made up from extra progeny of the better performing pens. Reasons for discarding single pens or even lines consisting of two or more pens are; poor breeding results, low average grades for the progeny, or the appearance of highly objectionable traits.

The desirable characteristics that may be ascertained with the aid of a complete set of records, are as follows:

(a) Hatchability record of dam:—A minimum of 80% of fertile eggs set during the normal hatching season should hatch strong poults. Persistence of good hatchability through a minimum hatching season of about three months is important.

(b) Fertility record of dam:—A minimum of 90% fertility of eggs set during the normal hatching season. Persistence of good fertility throughout the normal hatching season is important. Proper care of hatching eggs is an essential factor to consider when measuring persistence of fertility and hatchability because unfavorable holding or handling conditions reduce both hatchability and apparent fertility, especially in warm weather. When low fertility is general and obviously the fault of the tom, it should not be charged against the hen. However, if some hens are fertile and some not, those that are not should be counted out. Selection on basis of percentage hatch of total eggs set automatically includes fertility and may be used in preference to separating the two factors of fertility and hatch of fertile eggs. Records, however, should show fertility and hatchability of fertile eggs separately and in combination. Poor management, particularly that involving lighting of the breeding stock, may cause a shortened breeding season and generally poor breeding results. Management therefore should be the best possible and also be uniform for all pens so that the record-of-performance selection may be effective.

(c) Egg production of dam:—Young hen, a minimum of 60% production from the time laying starts in late winter or spring to June 1st, or 55 eggs during any period of 13 consecutive weeks between January 1st and June 30th; yearling hens should do 75% as well.

(d) Sexual maturity of dam:—The offspring of hens that start to lay significantly later than average should be discarded unless it is due to accident or management differences.

(e) Egg characteristics of dam:—

(1) Size:—Egg size is immaterial if selection can be based on hatchability records of the individual hens. Without such records it would be advisable to eliminate the extremely large and extremely small eggs, aiming for the average, or slightly below average, size for the strain.

(f) Livability of sibs:—Birds whose brothers and sisters (sibs) show a non-accidental mortality significantly higher than the average for the flock should not be considered as possible breeders.

Table 11.3—Approximate Changes in Body Weight After the Shank Ceases Growth in Length*

Males						Females					
	Weeks of age						Weeks of age				
Shank Length inches	24 (lbs.)	28 (lbs.)	32 (lbs.)	36 (lbs.)	First Year (lbs.)	Shank Length inches	24 (lbs.)	28 (lbs.)	32 (lbs.)	36 (lbs.)	First Year (lbs.)
6.4	9	11½	13½	15	18½	5.2	7	8½	9½	10	11
6.6	10	12½	15	16	20½	5.4	7½	9	10½	11½	12
6.8	11	14	16	18	22½	5.6	8½	10½	11½	12½	13½
7.0	12	15	17½	19½	24½	5.8	9½	11½	13	14	15
7.2	13	16½	19	21	26½	6.0	10½	12½	14½	15½	16½
7.4	14	18	21	23	29	6.2	11½	14	16	17	18½
7.6	15	19	22½	25	31½	6.4	12½	15½	17½	19	20
7.8	17	21	24	27	34	6.6	14	16½	19½	20½	22
8.0	18	22½	26	29	36½	6.8	15	18½	21	22½	24
						7.0	16½	20	23	24½	26½
8.2	19½	24	28	31	39½						
8.4	21	26	30	33	42½						
8.6	22½	28	32½	36	45½						
8.8	24	30	35	39	48½						

*Based on standard varieties, at the Oklahoma Station.

Table 11.3—(Continued)*

Approximate Age Changes in Depth

Males						Females					
	Weeks of age				Adult body depth inches		Weeks of age				Adult body depth inches
Shank Length inches	24 (lbs.)	28 (lbs.)	32 (lbs.)	36 (lbs.)		Shank Length inches	24 (lbs.)	28 (lbs.)	32 (lbs.)	36 (lbs.)	
6.4	6.9	7.4	7.8	7.9	8.7	5.2	5.9	6.0	6.2	6.2	6.3
6.6	7.2	7.7	8.2	8.2	9.1	5.4	6.0	6.2	6.4	6.5	6.5
6.8	7.4	8.0	8.5	8.6	9.4	5.6	6.3	6.5	6.6	6.6	6.7
7.0	7.6	8.2	8.7	8.8	9.6	5.8	6.5	6.7	6.9	6.9	7.0
7.2	7.8	8.4	8.9	9.0	9.9	6.0	6.7	6.9	7.1	7.1	7.2
7.4	8.0	8.7	9.1	9.3	10.1	6.2	6.9	7.1	7.4	7.4	7.4
7.6	8.2	8.9	9.4	9.6	10.4	6.4	7.1	7.4	7.6	7.6	7.6
7.8	8.5	9.2	9.6	9.8	10.8	6.6	7.4	7.6	7.9	7.9	7.9
8.0	8.6	9.3	9.7	10.1	10.9	6.8	7.6	7.8	8.1	8.1	8.1
8.2	8.8	9.6	9.8	10.3	11.2	7.0	7.7	7.9	8.3	8.3	8.3
8.4	9.0	9.8	10.1	10.5	11.4						
8.6	9.3	10.1	10.3	10.8	11.8						
8.8	9.4	10.3	10.7	11.1	11.9						

Approximate Age Changes in Keel Length

Males						Females					
	Weeks of age				Adult keel length inches		Weeks of age				Adult keel length inches
Shank Length inches	24 (lbs.)	28 (lbs.)	32 (lbs.)	36 (lbs.)		Shank Length inches	24 (lbs.)	28 (lbs.)	32 (lbs.)	36 (lbs.)	
6.4	5.6	6.0	6.5	6.7	7.4	5.2	5.1	5.3	5.4	5.4	5.5
6.6	5.8	6.3	6.8	7.0	7.7	5.4	5.2	5.4	5.6	5.6	5.6
6.8	6.0	6.5	7.0	7.3	7.9	5.6	5.4	5.6	5.8	5.8	5.9
7.0	6.2	6.7	7.2	7.7	8.2	5.8	5.6	5.9	6.0	6.0	6.1
7.2	6.4	6.9	7.4	7.9	8.3	6.0	5.8	6.1	6.2	6.2	6.3
7.4	6.5	7.1	7.6	8.1	8.6	6.2	6.0	6.3	6.4	6.4	6.5
7.6	6.7	7.3	7.8	8.2	8.8	6.4	6.1	6.5	6.6	6.6	6.6
7.8	6.9	7.5	8.0	8.3	9.1	6.6	6.4	6.7	6.8	6.8	6.9
8.0	7.0	7.6	8.2	8.5	9.2	6.8	6.5	6.9	7.0	7.0	7.1
8.2	7.1	7.8	8.4	8.7	9.4	7.0	6.7	7.0	7.2	7.3	7.3
8.4	7.3	8.0	8.6	8.9	9.6						
8.6	7.6	8.2	8.8	9.2	10.0						
8.8	7.7	8.4	9.0	9.4	10.2						

*Based on standard varieties; skeletal changes probably are similar in the broad-breasted and small-type strains.

(g) Livability of dam:—All surviving offspring of hens that die from non-accidental causes or become sick during their first breeding season should be marketed for meat. Breeding only from yearlings and older hens accomplishes the same result but this is impracticable in turkeys. It is usually impracticable to consider livability of the sire. Sometimes the widespread occurrence of infectious disease makes it impracticable to consider livability of the dam.

(h) Freedom from abnormalities:—Birds that themselves have at any time possessed any serious defect such as pendulous crop, spraddle legs, crooked legs, or deformed back should be discarded along with all members of families producing abnormalities.

(i) Rate of market maturity:—Between the ages of 22 and 28 weeks select as breeders the toms and hens that have the fewest pinfeathers and the fullest covering of fat. Standardbred and large-type young toms should reach U. S. grade "A" at 28 weeks, young hens at 26 weeks; small-type young turkeys of both sexes should reach market maturity at about 24 weeks. Age identification and inspection at the ages mentioned are necessary if this type of information is to be obtained and if it is, the inspection to determine fat and feathering grades should be made in connection with the determination of market qualities. Turkeys fed modern diets with an antibiotic and/or high energy content throughout the growing season should reach market maturity two weeks earlier than indicated above.

SELECTION PROCEDURE

The selector should have both hands free, therefore an assistant (the holder) should hold the turkey suspended by its legs (carefully with heavy broad-breasts) the shanks held apart about the normal distance, the hocks even (on the same horizontal plane). If the holder stands at a higher level than the selector or, better, the latter stands in a hole, the work will be done more easily and more accurately. The holder can help by checking the *condition of the shanks and feet* for infection, bumblefoot, injury, crooked toes or other abnormality. The selector will be able to spot many discards at a glance without handling, but most birds will be handled and for this a routine procedure should be developed. With one hand on the bird's back, the other on the breast, the selector determines the relationship of back to keel (they should be approximately parallel to each other) and with one hand checks for rough, crooked or arched (roached), or injured *back*. With the other hand he goes over the *breast* checking for knobbiness, curved or crooked keel, width of breast, fleshing over the entire length of keel, breast blisters or calluses, pendulous crop, and off-color. He will note, by looking and feeling, whether, in the suspended bird, the rear end of the *keel*

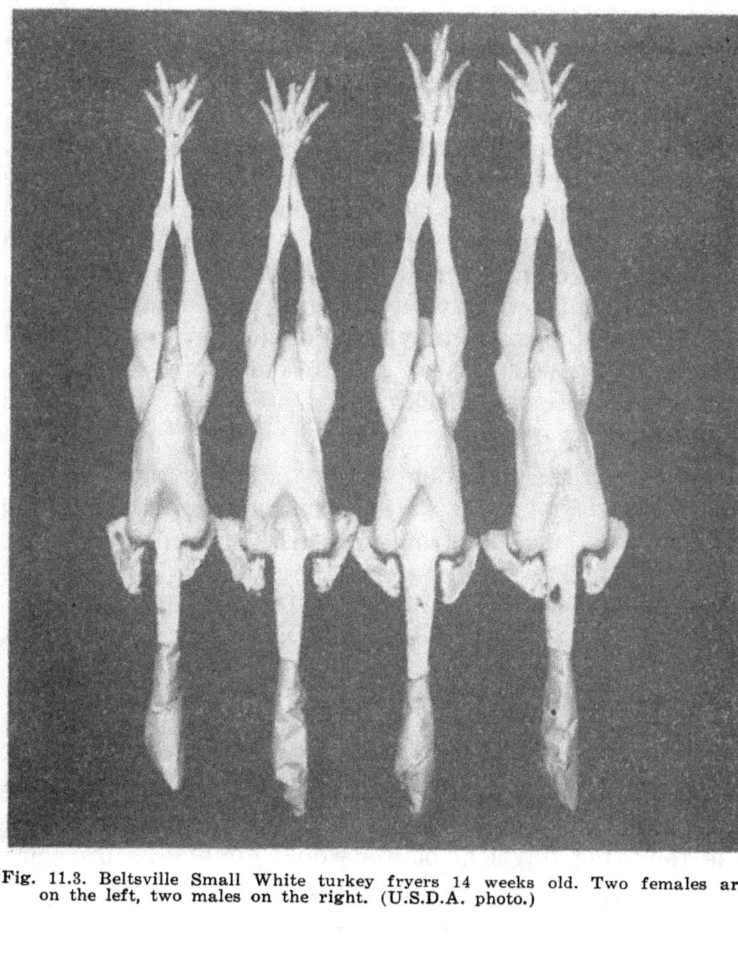

Fig. 11.3. Beltsville Small White turkey fryers 14 weeks old. Two females are on the left, two males on the right. (U.S.D.A. photo.)

extends to the legs or up between them so there is no conspicuous gap (part of the balance factor). He will feel or at least observe the *drumsticks* to see if they are plump and of proper size for the weight and type of turkey and will observe the *hocks* for any swelling or abnormality and the *shanks* for straightness, proper color, refinement, and proper (medium-short) length, not real short.

If *market maturity* is to be determined, he will part the feathers and pull a few half way down the side of the breast near the front to determine if there is a good finish (fat in the skin) between the feather tracts and freedom from short pinfeathers. For the latter he may also examine the shoulder joint at the base of the wing, and the drumsticks. If the *wings* are not already spread, the selector spreads them (one in each hand) and checks for split wing, twisted quill feathers and off-color. He then grasps the bird's neck and turns the *head* so as to check for blindness, crow head, coarseness, eye color, beak deformity, injury, and signs of disease such as sinusitis and pox. Finally, the holder swings the bird around so the selector can see the *back, tail, wing bows, and neck feathers,* checking for off-color, and twisted main-tail feathers.

Following the handling, birds that meet the requirements *should be observed as they walk or run* into the holding pens (see diagram of selection equipment, Fig. 6.9). Birds that do not pass the walking test can be hooked out or shunted into the cull pen by an assistant.

CONDENSED SELECTION OUTLINE

For the Farm Breeder

At 22, or 24 to 28 or 30 weeks or later, but before marketing any birds, the choicest specimens selected on the following physical points would qualify as breeders.

1. Broad, smooth, well-fleshed breast free from

knob or callus and plump, well-fleshed drumsticks of appropriate size.

2. Back smooth, straight, uniformly broad and free from injury or deformity.
3. Breastbone straight, parallel to back, and extending back to or between the legs when bird is suspended.
4. Reasonably short legs and neck, relatively good length of keel, and moderate depth of body.
5. Upright carriage, or balance.
6. Straight, strong legs, as demonstrated by a walking or running test.
7. The proper size for the variety or strain at the age examined.
8. Proper color for the variety and freedom from defects, deformities, and signs of disease or weakness.
9. Early market maturity. If the age of the birds can be identified and if selection is made between 22 and about 28 weeks, it is easily possible to pick the comparatively quick-maturing specimens. If age cannot be identified, or if birds are selected even after the slowest-growing individuals have matured, this point (No. 9) cannot be given much attention. However, obviously slow maturing birds usually can be recognized and should be culled.

For The Pedigree Breeder

From the pedigree and rearing records available, all offspring of dams that performed unsatisfactorily in the breeding pen or proved to be undesirable for some other reason should be marketed. Young birds which themselves made unsatisfactory records should also be marketed.

This "record selection" is based on the following points:

1. Good hatchability of dam's eggs; minimum 80 per cent of fertile eggs.
2. High fertility of dam's eggs, minimum 90 per cent.
3. Good shell quality and shape of dam's eggs.
4. Good egg production of dam, 60% production for at least 13 weeks.
5. Normal sexual maturity of dam (starting to lay at proper time).
6. Livability of dam during first season (accidents not counted).
7. High livability of sibs (family livability).
8. High average market quality of sibs (family quality).
9. Freedom from abnormalities up to the time of selection. Families producing such defects as pendulous crop and perosis should be discarded.

Fig. 11.4. A typical "gangly" young tom and hen. Note long legs and neck in relation to body depth.

10. Fast rate of growth (early market maturity). At a set age (usually between 22 and 26 weeks) all birds should be graded and measured, according to Nos. 1, 4 and 9 in physical selection column above and family averages calculated. Only members of the high-performing families should be considered as possible breeding birds.

The birds retained as breeding-pen candidates as a result of this "record selection" are then further selected as outlined under the nine points listed under physical selection, paying attention not only to measurements, early maturity, and market quality, but to the points not included in the family averages. The best individuals from the best families are thus selected.

BREEDING PLANS FOR THE FARM FLOCK

One or more matings provided with more or less range are the rule in commercial farm flocks. These matings vary in size from a few hens up to several hundred, even 1,000 or more. About 500 hens makes a good practical commercial breeding flock. When a mating regularly contains about 100 hens, random-selection breeding within that mating may be practiced safely for an indefinite number of generations. There probably is no advantage in maintaining more than one breeding unit if the whole flock consists of 100 or more hens. With smaller flocks than this, more than one breeding pen, preferably four, not less than three, is desirable, however, and for this, pen pedigreeing, which is done by marking the eggs from each pen, hatching those from each pen separately, and toe-marking or tatooing, the poults, or raising them in separate units, is necessary. When pen pedigreeing is thus used, female offspring from one breeding flock can be mated to males from another each year in systematic rotation. This procedure avoids close inbreeding.

For commercial and farm flock matings the use of well-matured young hens and toms each season is advised. This permits disposing of all breeders each season as soon as the demand for hatching eggs subsides. There is seldom any profit in holding breeders and marketing them in the fall, because of the feed cost, labor cost, and normal summer mortality charge that must be borne by the surviving breeders. Furthermore, marketing the breeders in early summer lessens the possibility of the spreading of disease, an important point to consider.

In some cases, especially if moderate or close confinement of breeding stock is to be practiced, one-male matings or two-male matings, the males alternated weekly or semi-weekly, may pay well in spite of the extra equipment and labor required. Thirteen to twenty* hens and one young tom make a good unit. For yearling and old toms (rarely used), 10 to 12 hens are sufficient. When thirteen to twenty hens per young tom are used in one-male matings, or 26 to 40 in two-male alternated matings, the resulting reduction of overhead costs is an important item. Pen pedigreeing as noted above is necessary in order to obtain system in the breeding work and greater opportunity for improvement. The first few eggs laid by the hens in each mating may be set for preliminary fertility tests which can be made after 24 hours of incubation or even in the fresh eggs by breaking and examining them (Chapter 14). These tests make it possible to replace early in the season males giving low fertility or otherwise unsatisfactory performance. In a large breeding project, the one-male matings may be organized into breeding blocks or groups each consisting of two or more matings. The males within each block may then be rotated by removing each one to the next pen twice a week throughout the season, always in the same (clockwise) direction. This system results in regular rotation of the males without fighting or other disturbance. Each breeding block would constitute a separate breeding group and the offspring would be marked with the block number instead of the male or pen number. Pedigrees on the male

*The smaller numbers for large-type, the larger numbers for small-type turkeys.

side are impossible in such cases which is a disadvantage of rotating males. This system may be easily practiced by calling the pens A, B, C, D, and E. On each Monday and Thursday the tom in A is shifted to B, the one in B to C, etc., the one in E being moved to A. By this method every hen is almost sure to be mated by one of the 5 toms that heads her pen within 2½ weeks, which may help fertility where preferential mating is a problem.

New blood should be introduced only when necessary to effect desired improvement. Many breeders introduce new blood frequently as a matter of routine, sometimes to the detriment of the flock, both in quality of the stock and in diseases and parasites that may be brought in. Whenever a breeder has stock that is giving satisfaction and is free of infection he should set up a self-perpetuating breeding plan to avoid the expense and dangers connected with frequent introductions of "new blood." As mentioned, a single large mating of 100 or more hens is self-perpetuating for many generations if a new group of young breeders is selected at random each season and mated in the single large group. Any danger from too close inbreeding with such a plan is less than the danger connected with frequent new introductions. With breeding flocks smaller than 100 hens the danger from inbreeding with such a one-pen plan is somewhat greater and for such flocks the multiple-pen, single or double-male plan just described is preferable.

For the small, one or two male farm breeding flock, fairly frequent introductions of new blood, preferably through pullorum-free hatching eggs or poults, seem advisable but such introductions need not be made more often than once every 2 to 4 years if the following plan is used: Retain the original (presumably outcrossed) mating a second year, then select a new mating at random from the second-year offspring and retain it for a second year, introducing new blood the fourth year. If it is not deemed desirable to retain matings for a second year's use, introductions of new blood could be cut to every second year by making one random-selection mating between introductions. A third alternative, if the breeder wished to maintain a small flock by line-breeding, would be to use the family-group breeding plan described later.

If three pens are maintained, for example, A, B, and C (using toe punch markings or numbered wing bands for identification of the poults) then it is a simple matter to do progressive breeding work and maintain a closed flock by using what will be called the Systematic Intercrossing Plan of breeding. This consists of selecting each season the best young hens and the best young toms out of the offspring of each of the three pens which should be of dissimilar blood lines at the start. The selected young hens would then be placed in the pens from which they originated, that is, back to pens A, B, and C, respectively. The selected young toms would then be moved over one pen in the same direction each season. That is, the toms from A would go to B, those from B to C., and those from C to A. The next year the same performance would be repeated, moving the toms always in the same direction and same order of rotation. This procedure could be continued for at least five or six years if the matings are used for only one season, or 10 to 12 years if the matings are all used for two successive seasons each (and consequently a rotation of males made only once every two years).

With four or more pens, a program based on the Systematic Intercrossing Plan would last many years as it would be safe to make one such move or rotation each season for many years without danger of close inbreeding. It cannot be stated definitely how long this plan could be used, but to multiply the number of pens by four provides a fair estimate. A four-pen project could thus be maintained for 16 years if the matings are used only once. By keeping each mating intact for a second season's use, and selecting breeding stock only from the second season's offspring, breeding only from adult stock can be accomplished and the plan can be followed twice as long (32 years) without close inbreeding. More than four pens are preferred if a long-time project is being considered.

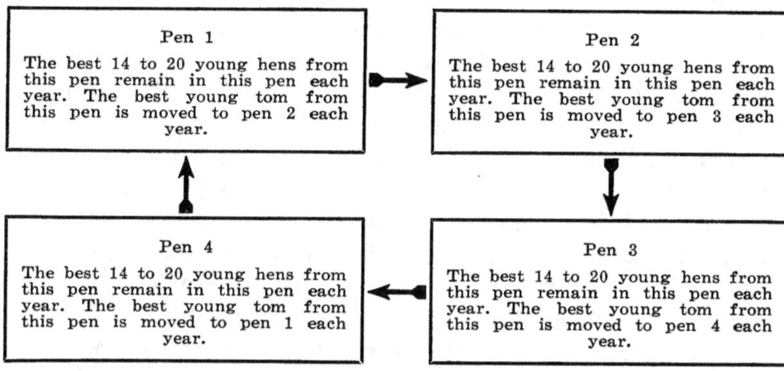

Fig. 11.5. This diagram illustrates the systematic intercrossing plan. By using this plan, it is not necessary to introduce new blood. A breeder can continue to breed his own stock for many years without enough inbreeding to be of any significance. Large breeding flocks may be intercrossed in exactly the same manner.

This plan is recommended for three reasons: (1) it avoids close inbreeding, (2) it is simple, and (3) if each mating is retained for two seasons, selecting the next generation only every other year (from the yearling matings), it permits, as far as selections for future breeders is concerned, selection from tested breeders—birds that show livability and longevity. On the other hand, if the matings are discontinued after the end of the first breeding season (using only young stock as breeders) and all breeders disposed of at the end of the normal breeding season, advantages in disease and parasite control and savings in feed and labor costs are gained. The breeder must decide which he prefers, to breed from tested yearlings, or to dispose of his breeding stock each spring at the close of the hatching season and breed from young birds each year. The latter probably would be chosen in most cases.

Pedigreeing (pen or individual) is essential except when the various breeding pens or flocks and their offspring can be raised in entirely separate locations, such as different farms. It also is essential that each mating be preserved intact or legbanded or otherwise marked (tattooed) if it is to be used again another season. Toe-punching is recommended as being the cheapest and most permanent method of marking. The toms could not be rotated from pen to pen during the season with this plan and great care would have to be taken to see that no mismating occurs. If it does, the resulting eggs should be discarded for a period of five weeks so far as the breeding project is concerned, selling such eggs as hatching eggs or marking for market all poults from such mismatings. However, according to the Oregon Station, if artificial insemination can be used two times, 11 to 14 days apart, with the semen of the right tom or toms, only the 11 to 14 days loss of time would be required to obtain 95% accuracy of replacement by the replacing tom. (Harper and Parker, 1952). The same procedure should be effective if for any reason a change of toms is desired.

BREEDING PLANS FOR THE PEDIGREED FLOCK

Three possible breeding systems are suggested for the professional turkey breeder who wishes to develop and improve his own strain without the frequent introduction of new blood, and to be able to furnish purchasers with comparatively unrelated stock year after year.

1. The Systematic Intercrossing Plan

This is fundamentally the same simple intercrossing plan suggested and described in the preceding paragraphs. However, for the pedigree breeder it involves trapnesting, one-male matings, individual pedigreeing, rigid selection, and careful record keeping. By systematic intercrossing a distinctive strain of turkeys may be developed. It must first be decided how many breeding birds can be maintained and how many young birds can be raised. An average of 10 to 15 (25 if possible) poults should be raised from each breeding hen whose performance is satisfactory up to the end of the breeding season. A good breeding project consists of 7 to 10 one-male matings. A good practical mating consists of 15 to 20 hens (the number depending on variety and other factors) and one tom. It must be decided whether young birds will be selected as breeders each year, using each mating for one season only, designated as Plan A; or whether each mating will be used for two seasons, Plan B. If Plan A is used, the turnover is quicker (a generation each year) and the progress could be more rapid, hence it would be preferred in most cases. If Plan B is used, the turnover takes twice as long with the result that the introduction of new blood can be postponed a proportionately greater length of time. This is an advantage if the original stock is satisfactory but may be a disadvantage if the breeder is a long way from his ideal and wants to make more rapid progress toward it. With five or more pens, preferably seven or more, at the start, this intercrossing system could be used indefinitely. A small percentage of inbreeding occurs but it is built up very slowly and would never become high enough to cause concern.

For the start, if seven pens are to be maintained, seven unrelated or distantly related lines of breeding stock secured preferably through the purchase of hatching eggs, should be obtained. Many breeders of turkeys can furnish two or more distinct lines. The seven breeding pens, each consisting of 10 to 20 young hens from one breeding line and the best-obtainable young tom from another would be mated, using these unrelated or distantly related lines in such a way that no two matings are similar, that is, that they do not consist of toms and hens from the same combination of blood lines. At least two full or half brothers of each young tom selected as the next season's breeder should be retained. Sisters of each of the females in the breeding pen should also be retained as reserve breeders wherever possible. These reserve birds of both sexes can be used when needed as substitutes for breeding birds that need replacing on account of mortality, injury, poor performance, or other causes. They may be kept in general flock matings, the eggs from which can be used for sales purposes or to produce market turkeys, none of the offspring ever being used in the regular breeding pens. Each season one of these flock matings may consist, for example, of the reserve hens from pens 1 to 4 mated to the reserve toms from pens 5 to 7; the other would consist of the reserve hens from pens 5 to 7 mated to the reserve toms from pens 1 to 4. This arrangement would avoid inbreeding which might conceivably, though it is not at all likely to, be harmful in the flock matings. These flock matings of reserve breeders are not necessary, especially if large matings of 17 to 20 hens are used in the regular pens. However, they are likely to be needed for sales or market purposes. Reserve toms must be retained for substitution as breeders when necessary for a full or half brother, so that the breeding program will not be interrupted. If a hen is changed from the flock mating to a breeding pen or is exposed to mismating, she must be allowed to remain in the right mating for about five weeks before her eggs are pedigreed; and even then there may be an occasional poult from the former tom. (This loss of time may be reduced by artificial insemination.) The hens in the breeding pen should be trapnested during the breeding season to June 1 or July 1 (longer if desired) and all breeding birds should be weighed individually at a certain date each year. The date of weighing should be that at which maximum body weight is attained. This is usually at about the average date of first egg. In the northern States, without artificial lights, this usually occurs in late March or early April, and in the middle States, early March. The eggs should be marked with the hen's legband or saddle number and that of the sire or pen, on the small end of the egg, as the egg is removed and the hen released. Adding the date laid also is desirable. All the eggs from one hen in any one hatch are placed in the same pedigree basket or compartment on the 23rd or 24th day of incubation, so that the offspring may be pedigreed individually (see Chap. 13). Ten to 20 eggs from each hen should be weighed and inspected for shell quality and shape during the third month of production. Any abnormalities noted during the setting operations should be recorded so that such characters may be considered along with the hen's egg record.

The system of breeding after the first season's outcross matings are made is as follows (consult diagram): Each year for Plan A, or every other year for Plan B, the best young female offspring from each mating are selected for the next season's breeding pens, and along with them, as many reserve pullets as may be desired. They should be full sisters of the females selected for the breeding pens. Selection is based upon superior family performance in respect to reproduction, market quality of offspring, and freedom from hereditary defects. The individual selected, of course, would have to meet a high physical requirement. Outstanding hens are often retained in the pen for a second or even a third season. For example, if there were 16 hens in the original pen and 7 of them performed in a superior manner, the breeder would select the two best daughters from each of these hens. This would make only 14 young hens so 2 more pullets from the offspring of the outstanding dams would be added to make up the new pen of 16 young hens, provided 16 is the number decided upon for the regular breeding pens each season. Five of the breeding hens would thereby be represented by 2 daughters and 2 of them by 3 daughters. One dam in addition would be represented by a son which would go into the next pen. Or, the breeder might use one or more of the best of the old hens which would cut down on the number of pullets used. Another use for R.O.P. qualified old hens would be to produce R.O.P. eggs and poults for sale. The selected breeding females are then placed in the pen from which they originated. The best young tom from each pen would be selected along with two, or preferably three, of his full or half brothers. Selection of the young tom is based on (1) performance of his dam (2) livability and other characteristics of his sibs (sisters and brothers), and (3) his own physical characteristics. He should be selected from one of the outstanding hens in the pen. His full sisters are therefore valuable as breeders or as sales stock. They should be used in the breeding pens unless one has a profitable sales outlet for choice breeding pullets. The selected males then are placed with the selected pullets but they are moved over one pen, all in the same direction. The male from pen 1 would go in with the pullets from pen 2, and so on, completing the circle, with the male from pen 7 going into pen 1. Each generation this performance is repeated, moving the males always in the same order of rotation each successive year. If Plan B (breeding from yearlings) is used, this performance is repeated only every other year, because the same matings when once established are used two successive seasons. Each season the offspring from each pen will be sufficiently unrelated so that purchasers of the stock may safely mate them with the offspring from almost any other pen. The necessity for accurate pedigreeing cannot be stressed too strongly. Birds whose pedigree is for any reason in doubt must not be chosen for the pedigree pens.

Those families from each breeding pen coming nearest to the ideal should be retained as breeders. In the case of pens below the flock average in performance, only the families well above the pen average should be selected. However, whenever certain proof of weakness is shown, it is wise to discard the entire line. In actual practice the systematic shifting of males and females as outlined above probably would not occur regularly. Sometimes the offspring of one or more pens must be dropped, wholly or partially substituting stock from the pens that have a surplus of good offspring. In such cases relationships should be such that closely related individuals never are mated and, in small projects, that no two pens consist of the same combination of bloodlines.

A desirable modification of this intercrossing system in operations consisting of 14 or more pens is to organize the pens into seven or more breeding groups consisting of two or more adjacent pens each. The annual rotation is then made by groups rather than by single pens, thus allowing the greatest use of the top ranking families; and easier elimination of the low ranking families and pens.

The use of two or even three full brothers rather than just one, to head breeding pens may sometimes be desirable. The full brothers may be alternated semi-weekly or weekly or they may be used simultaneously if they get along together. Larger numbers of hens may then be used in the pens so there may accrue better chance to obtain high fertility. There is no objection to such a practice if the breeder finds it desirable. However, retention of the single-male mating appears desirable, using full brothers only if and when the original male must be replaced.

It should be noted that the heart of the system described above is selection based on (1) individual breeding-hen performance and (2) the quality of her offspring (hen-family). Thus the performance of the sire is more or less ignored and so is the performance of the sisters (if any) or the individual breeding hens. Of course certain pens sometimes perform so poorly on the average that they must be discarded—all or in part—in this way poor sire performance is taken into account. While **performance** of the sire is necessarily ignored in the breeding plan as explained above, his **selection** is of vital importance and the breeder should spare no effort to pick the best possible males to head the breeding pens.

II. The Family-Group Breeding Plan

(A form of linebreeding)

Breeding experiments conducted by the California Station and by the senior author at the U. S. Range Livestock Station in Montana have shown that turkeys will stand considerable inbreeding, much more than generally believed. However, inbreeding is a practice to be avoided by farm or commercial turkey raisers. It should be used only in connection with well-organized pedigree breeding work. The following paragraphs contain a description of a plan for the breeding of turkeys within family bloodlines. The plan involves introduction of unrelated blood only at the start.

First, the breeder must decide on the number of breeding families or lines to be maintained, and also which of two alternatives to use: (1) mating young hens and yearling toms; or (2) yearling hens and young toms. The number of breeding hens to be maintained may be reduced by choosing plan 1 (yearling toms on young hens) as will be shown. However, most breeders would have need for general flock matings to supply hatching eggs and market stock and would therefore prefer plan 2. Furthermore, matings of yearling hens and young toms would probably be more satisfactory in the long run than matings between young hens and yearling toms. The description in the following paragraph is based on the assumption that plan 2 (young toms on yearling hens) is selected and one-male matings used.

At the start, the foundation pens of females are established by selecting, for example, five breeding pens of young hens of the same, or preferably, different ancestry. Each pen of hens is mated to an unrelated or distantly related tom. The toms should not be related to each other, thus permitting the different lines to be crossed later as the program progresses. These matings, which are assumed to start with the 1956 season,* would be called the P_1, or parent generation. From the 1956 offspring of each pen, called the F_1 generation, are selected (in the manner described for the selection of breeding stock) the best pullets, at least 20 from each pen and

*If the program is started later merely move up the dates: 1957—P_1 (original parent stock), 1958—F_1 (first filial generation), 1959—second generation.

preferably more, depending upon how many are needed for the flock matings. These F_1 young hens are kept during their first laying season (1957) in the flock matings mated with young males from the other lines to produce only market stock or hatching eggs to sell. The regular 1957 pens which produce the next generation, would consist of the original P_1 hens (grown to yearlings in 1957) mated preferably to outside unrelated young toms or to F_1 toms from the other families. Two full brothers of each male used in the regular breeding pens must be retained, and if it is desired to lessen the percentage of inbreeding, or if there is a good sales outlet for choice breeding pullets, the full sisters of these breeding-pen males may be sold. They would be the choicest young hens that the breeder can spare for sales purposes. The 1958 matings would consist of the hold-over F_1 hens (grown to yearlings by 1958) mated to the second-generation young toms from the same line to produce the third generation. In 1958 and succeeding years the same plan would be used, always mating the holdover yearling hens with the current year's young males and always keeping within the same family. However, the families could and probably should be crossed every 4 to 6 generations to reduce the percentage of inbreeding. If this crossing is done no oftener than this, the breeder will be able to sell stock from several comparatively unrelated lines. Such a family breeding plan, if carefully carried out with rigid selection for vitality and reproductive ability, should be effective for an indefinite number of generations without introducing new blood except that obtained by occasionally crossing the various lines.

Plan 1 could be used with the same chance for success by using young hens for breeding stock instead of yearling hens and keeping over the toms instead of the hens, thus reducing the number of breeders kept. The matings each season after the first would then consist of young hens and the hold-over yearling toms instead of yearling hens and young toms. The beginning would be made in the same way except that in the fall of 1956, instead of retaining the original pullets for a second year's breeding, the original toms along with two full brothers of each would be retained. The 1957 breeding pens, instead of consisting of the original hens mated to outside toms or young toms from another line, would consist of the original toms (grown to yearlings by 1957) mated to outside pullets or pullets from another line to produce the next generation. In 1958 pens instead of consisting of the yearling F_1 females mated to the second-generation young males, would consist of the second-generation pullets and F_1 yearling toms. In this way only the toms, along with not less than two full brothers (preferably 3) of each for reserve breeders, need be kept over. Half brothers could be used, though full brothers are preferred. Each season the matings would consist of the best hold-over yearling toms mated to the "cream of the crop" of young hens.

As with all breeding programs, emphasis must be placed on rigid selection and accurate pedigree work with the elimination from the breeding flocks of all birds whose pedigrees are in any way doubtful. The words "pedigree uncertain" should be placed opposite the record of any birds whose identity cannot be full vouched for by the operator in charge.

III. A Double-Mating Plan for Linebreeding

Two breeding pens constitute a unit in this plan. The two original pens should be unrelated to each other and to other units of the breeding flock. Each season the best young tom from one pen is mated to the best young hens from the other pen, eliminating the full sisters of the tom selected as a breeder. These sisters can be sold as high-class breeders or used in extra matings to produce poults for sales or market purposes. Each season the procedure is the same, simply mating the best young tom from one pen to the best young hens from the other pen. This procedure builds up a fair percentage of inbreeding which often is an advantage genetically. To prevent deterioration from this inbreeding the various two-pen lines may be crossed every 3 or 4 generations. A minimum of four two-pen lines should be maintained.

SIZE OF PEDIGREE BREEDING OPERATIONS

How many single-male breeding pens are needed to accomplish significant genetic improvement is a question frequently asked. The Beltsville Small White turkey was developed by the use of an average of seven one-male breeding pens containing about twenty hens each over a period of six years. Further crossing and breeding for improvement of body type, in egg production, and in non-broodiness had been carried on successfully up to 1952 for ten more years with an average of six pens of one male and twenty hens each. In each case only one male was used in each pen each season except where, in a very few cases, replacement by a full or half-brother was necessary. Nordskog and Wyatt (1952) after an analysis of chicken breeding operations concluded that "size of breeding operations in terms of number of breeding pens has only a slight influence on possible greater gains from selection unless the number (of pens) is very small . . . Compared with selection intensity as a factor influencing selection pressure, size of operation is only a secondary force . . . Perhaps 8 or 10 pedigree breeding pens . . . would adequately take care of the inbreeding problem." So far as turkey breeding is concerned the experiences and observations of the authors indicate that a relatively small operation is more likely to succeed than a large one. There probably is no advantage genetically in maintaining more than fifteen to twenty turkey breeding pens in one project unless more are needed or desired for the commercial production of pedigreed eggs, poults, or breeding stock. Even then better dividends in genetic progress might result if it were decided to limit the main breeding operation to ten to twenty well-handled breeding pens from which the young stock necessary for good selection would be raised. Extra R.O.P. toms and other products could be produced for commercial purposes from special trapnested matings which would not play a direct part in the breeding program.

Outbreeding and Crossbreeding

Outbreeding is a general term used to denote the breeding of unrelated individuals of the same species. It includes the practices of outcrossing and crossbreeding, the latter being its wider form (see breeding chart).

Outcrossing as the term is commonly used in poultry breeding consists of the mating of comparatively unrelated or distantly related birds of the same variety. In turkey breeding to practice outcrossing involves the regular and frequent introduction of new blood of the same variety from strains of bloodlines of distinctly different breeding. Outcrossing is the plan of breeding commonly used by farmers and commercial breeders and, if not overdone, is the safest and most satisfactory plan for them to use. It may be combined with other forms of breeding as in the systematic intercrossing plan described in this chapter.

Crossbreeding of turkeys is the crossing of two different varieties. It results in crossbred offspring which are generally produced with the idea of obtaining superior market birds or stock to utilize in developing a new variety. The crossbreds are popularly supposed to be more vigorous than either parent. The existence of hybrid vigor (or heterosis) has been demonstrated by tests with chickens. It may manifest itself by improvement in egg production, fertility, hatchability, viability, growth rate, or resistance to disease. Some matings result in the appearance of hybrid vigor to a greater or less extent, while other matings produce no significant improvement. The results appear to depend upon how widely the mated individuals differ in their genetic constitution. As a rule, the more distant the relationship of the birds mated, the greater is the difference in their genetic makeup and con-

Fig. 11.6. White Austrian adult tom imported by the U. S. Dept. of Agr. to cross on White Holland and Black hens to produce small-type turkeys. Note blocky type, long breast, and short legs. Keel length 6.9 in., shank, 6 in., maximum weight 21.8 lbs. (Photo U. S. Dept. of Agr.)

sequently the greater the hybrid vigor resulting from the mating.

The favorable results due to crossbreeding, when they occur at all, show up chiefly in the first generation. If the resulting crossbred offspring of a mating are bred together the hybrid vigor tends to disappear and as a rule no advantage is gained. It is evident then that in order to obtain hybrid vigor, two separate varieties, two unrelated strains of the same variety, two separate and distinct sets of crossbreds, or a set of crossbreds and a third (entirely different) variety, must be used. The latter plan, three-way crossing, has been shown to be of value in hog and chicken breeding.

There is a small amount of scientific evidence available on the subject of the practical value of crossing the different varieties of turkeys. The Nebraska Station reported on the mating of Bronze with Bourbon Red and White Holland and the Kentucky Station reported on the Bronze-White Holland cross. No advantages over the purebred matings were found. The California Station reported on the results of crossbreeding experiments: "There is no indication that better results will be obtained by crossbreeding than by outbreeding within the Bronze variety at least. The results of crossbreeding Bronze and Bourbon Reds show considerable variability but no improvement in growth or meat quality." The growth of the crossbreds was influenced by the rate of growth of the parent strains. In later studies some slight evidence of heterosis was noted in crosses between Black, Bourbon Red, and Lilac and between Black and Crimson Dawn and White Holland. However, no practical advantages were demonstrated. These results together with experiences and observations of the authors, lead to the conclusion that, for the commercial production of turkeys, the use of well-bred strains of the same variety for new blood, which may be referred to as outcrossing, will produce as good results as crossbreeding. Since there are many disadvantages of crossbreeding it is not advised except for experimental purposes.

Topcrossing

Topcrossing is the practice of continuously mating the females of an outbred flock to males of another strain or breeding, which is usually inbred or linebred. Males to be added are usually drawn from the same source, while the females are of the original group or descendants of that group. Topcrossing is of value only when the source from which the males are obtained is superior to the average of the females to which they are mated. It is a good plan to use for the improvement of turkey flocks when the owners of them do not wish to conduct breeding work.

Inbreeding and Linebreeding

Inbreeding is the mating of comparatively closely related individuals. The percentage of inbreeding obtained depends upon the number of generations during which it has been carried on and the closeness of relationship of the individuals mated. Inbreeding occurs, usually to a mild degree, in most well-known strains of turkeys. The same may be said of most of the famous strains of other domestic animals. The closest forms of inbreeding consist of brother-sister matings and the mating of parent to offspring (see Fig. 11.7). Linebreeding is a form of inbreeding consisting of the mating of related individuals with the object in view of perpetuating the blood of some particularly desirable individual. It of necessity involves mild inbreeding, as the mated individuals are related to the much-admired foundation animal. Its greatest use has been in animal breeding. The ability of the breeder to choose as the foundation animal from which he linebreeds, one that possesses vitality, market type, and fine performance measures to a large degree the success of any linebreeding program. The tendency of this foundation animal, usually a great male, to transmit his desirable qualities is in the final analysis the true measure of his worthiness as a focal point for a linebreeding program. If he should prove a disappointment when judged by his sons and daughters as they approach maturity, he should be discarded and some other more worthy animal selected as the point of departure of the linebreeding program. Lush states that "linebreeding is accomplished by using for parents, animals which are closely related to the admired ancestor but are little if at all related to each other through any other ancestor." The most common form of linebreeding involves the mating of an outstanding male to his progeny for two or more generations which results in the preservation of his blood. Changes, occasional or regular, are made in the female line in order to avoid close inbreeding. Linebreeding is a type of breeding commonly practiced for the

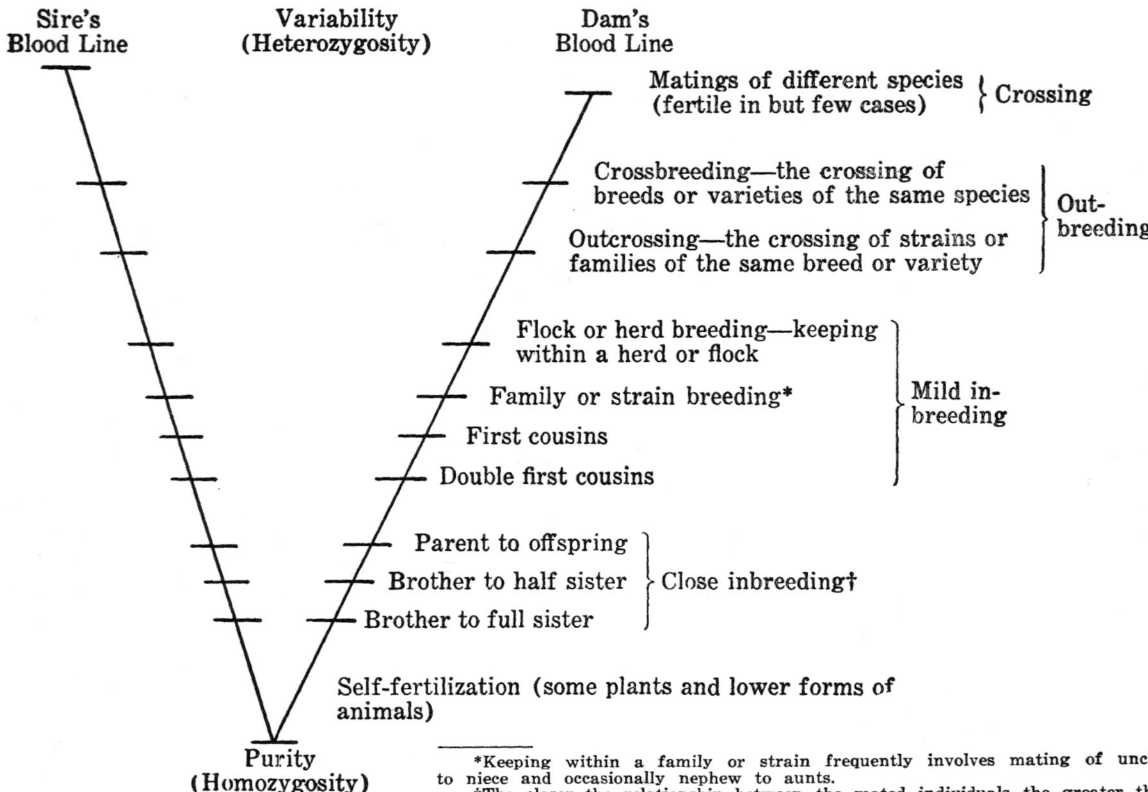

*Keeping within a family or strain frequently involves mating of uncle to niece and occasionally nephew to aunts.

†The closer the relationship between the mated individuals the greater the degree of intensification of those characters (good and bad) which the mated individuals have in common.

purpose of preserving strain characters without resorting to close inbreeding. Inbreeding in any form has the effect of "fixing" characters. Therefore, if the characteristics of the foundation stock are good, linebreeding is a desirable practice because linebred (mildly inbred) individuals are more likely to "stamp" their characteristics on their offspring, than are outbred individuals. Linebreeding and other forms of inbreeding are therefore desirable but they are practices to be undertaken with caution and accompanied by rigid selection and painstaking pedigree work.

Inbreeding and linebreeding are tools for the skilled breeder. That close inbreeding may have practical possibilities is indicated by results with chickens. In certain experiments, after five consecutive generations of brother-sister matings, a turning point was reached and marked improvement started.

Inbreeding has been used by the corn breeder to improve crops and uncover weaknesses of the parental lines. After many of these undesirable characters are eliminated from the strain by careful family selection in the inbred line it is then crossed with other inbred lines. The resulting hybrid corn outyields either parental line but subsequent generations are of little value for the production of seed corn. Perhaps the day will come when the advanced turkey breeder will make use of inbreeding, for it determines quickly the weaknesses of his strain which may by selection be removed from the strain. The practical breeder will do well to remember that inbreeding is merely a means to an end and not the end in itself; that end, in many cases, being the incross hybrid; now quite common in chickens. While the practice of inbreeding may intensify and purify the strain more quickly for the characters sought, it may also lead to a reduction in vitality and hatchability, and the loss of other desirable traits, unless constant selection is practiced in the opposite direction. At present it seems that inbreeding in turkeys should be left to the advanced breeder and research worker; and then used only when trapnesting and pedigreeing are done, in order that the parents of undesirable offspring as well as the undesirables themselves may be culled from the bloodlines. At the California Station the low-producing strain of Standard Bronze turkeys withstood inbreeding better than the heavy layers. In livability the inbred strains of Bronze was decidedly inferior to the non-inbred strain. It responded to a high-protein diet less dramatically, had shorter legs, larger Bursa of Fabricus, but keel length was the same for both strains according to Asmundson and Lerner, (1940).

A recent development in poultry breeding for the creation of superior strains of poultry is the building up of inbred strains for the purpose of producing hybrids by crossing the inbred strains. However, it is much more difficult to build up inbred lines of chickens or turkeys than it is of corn. In addition to the fact that individual turkeys are more costly than the individual in corn, fewer individuals can be used and selfing (self-fertilization) can not be practiced. Furthermore, the chromosome count is considerably less in corn than in turkeys, making homozygosity (purity) more remote in the inbred turkey. Whether the crossing of inbred strains of turkeys will yield anything superior to our present stocks remains for the future to determine. The authors have

Table 11.4—Fertility, Hatchability and Egg Production for 773 Inbred and Outbred Bronze Turkeys

(U. S. Dept. of Agr. data, Marsden and Knox)

Degree of inbreeding	Coefficient* of inbreeding	Fertility	Hatchability of fertile eggs	Production of eggs to June 1	Average egg weight
	Percent	Percent	Percent	Number	Grams
Outbred	0 - 6.3	87.8	67.6	47.7	83.6
Mild	12.5-21.8	75.4	51.7	42.2	82.7
Close	25 -41.1	82.8	52.8	39.0	82.1
Intensive	50 -67.2	69.3	34.9	41.4	81.5

*The offspring of sib matings is 25% inbred, the second generation is 37.5% and the third 50% inbred. The first generation of self-pollinated corn is 50% inbred.

observed a small amount of experimental work in which the crossing of inbred strains gave only mediocre results. However, the California Station has reported considerable stimulation to growth rate, although it has not yet recommended the practice as a means of securing superior stock.

Although it is well nigh impossible to catalog each character in turkeys, the scientific breeder may accumulate and purify the good characteristics of the turkey by the process of inbreeding. The inbreeding process fixes various complements of good and bad traits of the parents in the different inbred progeny. In many of the progeny there may be such an accumulation of bad traits that they must be discarded, and the parents considered undesirable breeders. Other families may accumulate many different desirable traits and when these are crossed with others, superior hybrid individuals may be produced.

When two varieties or two strains are crossed and additional vigor as expressed in rate of growth, high livability or hatchability, the resulting increase over and above the best of the two strains crossed is referred to as *heterosis* or hybrid vigor. Since these two strains are said to "nick," perhaps the breeding for actual ability to nick is a program that should be considered quite as much as breeding for the physical traits and egg production. It is known that "nickability" is inherited in many plants and animals, which suggests selection for the ability to transmit favorable combining values. If such a breeding program is followed and reciprocal matings are made (using the males of each of the respective lines on the females of the opposite lines) one would in reality be practicing what the geneticist knows as **recurrent reciprocal selection.** This phrase is common in the corn breeding programs and is now finding its way into our poultry terminology. The recurrent selection to a common tester flock or strain will perhaps be the most feasible program for the turkey raiser.

Some results of inbreeding in standardbred Bronze

turkeys at the U. S. Range Livestock Experiment Station (1930-1935) are shown in table 11.4. In this case even a relatively small (mild) percentage of inbreeding (12.5 to 22%) reduced fertility, hatchability and egg production noticeably. This amount of inbreeding was obtained by mating half-brother-sisters for one to 3 generations. Increasing the inbreeding to 25-41.1% did not further decrease the reproductive qualities, but intensive inbreeding (50 to 67%) reduced further the fertility and hatchability but not the egg production. However, the Colorado Station (1950) reported that "comparisons among progeny from 3 strains of (Broad Breasted Bronze) turkeys showed that moderate inbreeding (37.5%) has thus far not resulted in any significant influence on reproductive performance." The West Virginia Station (1952) reported that full- and half-brother-sister matings of Broad Breasted Bronze showed, after three generations, significant lowering of hatchability of fertile eggs but no significant effects on fertility or egg production.

Overdominance is a relatively new concept in breeding parlance. It is illustrated by the interaction of one or more pairs (allels) of genes in which the heterozygote or hybrid is superior, for the traits concerned, to either parent. When such interaction has been found it is possible to select the parent stock for superior combineability; which, of course, necessitates continuous crossing to produce the overdominant character.

THE BASIS OF HEREDITY

It is beyond the scope of this work to supply a detailed explanation of the mechanism of heredity. Only a brief discussion of this important subject will be attempted.

The *chromosomes* are minute bodies found in the nucleus of every cell of all plants and animals. The chromosomes contain the *genes* or determiners of characters or characteristics. The genes are the bearers of heredity that are thought to be arranged within the chromosomes in linear order. When body cells multiply, as in growth, each chromosome and with it each gene, divides into two equal parts, one going to the nucleus of each of the new cells that are formed in this ordinary cell division or *mitosis*. Thus each new body cell has the same chromosome count and the same genes, as the cell from which it was formed.

In all body or *somatic* cells each chromosome is a member of a pair (except in the case of the unpaired sex chromosome explained later) and the number of chromosomes in the paired condition is called the *diploid* or *2n* number which is that usually referred to when speaking of chromosome counts. This diploid number of chromosomes occurs in all cells except the sperm and the unfertilized egg by means of which the genes are transmitted from parent to offspring. Sperm and egg each contain only *one* member of each pair of chromosomes, and consequently only half as many chromosomes called the *haploid* or *n* number. The reduction in the number is effected through *reduction-division*, a specialized process involved in the formation of the reproductive cells (sperm and egg) from the testes and ovaries, which have the diploid number. During reduction-division the chromosome pairs come together in synapsis, then separate, one member of each pair going into one of the two cells or *gametes* (sperms or eggs) that are formed. When sperm and egg unite in fertilization, the diploid or *2n* number is restored. Turkey sperm are similar morphologically to those of other birds. There are four parts listed in order from front to rear: Acrosome ("nose"), head, body (midpiece), and tail. Abnormalities have been observed in all four sections and it is believed by some that the occurrence of abnormalities is negatively correlated with fertility.

The chromosomes, as stated, are paired and consequently even in number except in the case of the sex chromosomes. In one sex the number is always even, and in the other sex it is usually considered to be uneven because the sex chromosome may be unpaired or if it is paired, the other member of the pair carries no genes for known characters and therefore may be considered as not functioning. As with other birds, turkey females have only one gene-bearing sex chromosome while the males have two. Sex is believed to be determined by chance, individuals receiving a functioning sex chromosome from the dam as well as the sire being males, while those failing to receive a functioning sex chromosome through the egg (from the hen) being females with only one sex chromosome, that one always being received from the sire. The sex chromosomes are sometimes designated as the *XX* pair in male birds, and the *XY* pair in female birds, the *Y* referring to the non-existent, or perhaps existent but non-functioning, sex chromosome.

Each known gene on a chromosome (autosome) has its *allele* (or allelomorph) which is a gene found in the same relative position on the other member of the chromosome pair. These two paired genes, alleles to each other, are determiners either for the same or for opposite characters. When both genes are determiners for the same character, the individual is said to be *homozygous* or pure for that particular characteristic; when the genes are determiners for opposite characters the individual is *heterozygous*. For example, in a purebred Black turkey the gene for black color is found at the same relative point or *locus* on both chromosomes of the pair that carries those particular genes. If a turkey is the result of a cross between purebred Bronze and purebred Black it will be heterozygous for black because it will have a gene for black color, (inherited from its Black parent) and a gene for bronze color (inherited from its Bronze parent) opposite each other in the chromosome pair. Thus if a bird be homozygous for a character, his sperm (or eggs* if a female) will all be the same as far as the gene for that character is concerned. If a bird be heterozygous, however, half his sperm (or eggs* if a female) will carry the

*The term egg in this discussion refers to the single female reproductive cell before fertilization is effected.

gene for one of the unlike characters while the other half will carry the gene for the other character. A crossbred Black-Bronze male turkey therefore produces in equal numbers on the average, two kinds of sperm as far as genes for black and bronze are concerned. Likewise the crossbred female produces two kinds of eggs.

In the case of the genes carried on the unpaired sex chromosome of the female turkey, half of her eggs contain none of them because they contain no functioning sex chromosome, or at most a Y chromosome carrying no genes. Each egg of this group therefore, is capable upon fertilzation of developing only into a female because the only sex chromosome the resulting individual can possess will be the one supplied by the sperm.

The two paired genes occurring at the same relative point on both members of the chromosome pair are, as stated above, alleles of each other and may carry opposite characters such as black and bronze. When they represent opposite characters, one of them dominates the other to some degree (which is sometimes complete) insofar as the expression of the character is concerned. Dominance is lacking in some cases, however, resulting in a blending effect. The gene that dominates is called dominant while the one that is dominated so that the character determined by it usually recedes from view is called recessive. Dominant genes always are represented by a capital letter and recessives by a small letter. For example, black color in turkeys (B) and bronze (b) are alleles along with a third allele b^1 or black-winged bronze, the three constituting a triple allele series with dominance in the order: B (black) b(bronze), and b^1 (black-winged bronze). Black color is almost but not completely dominant over the recessive black-winged bronze or bronze color, and bronze is completely dominant over black-winged bronze. In a triple allele series only two of the three alleles ever appear in the same bird. Whenever the gene B occurs along with b or b^1 the bird will be almost completely black. Any completely recessive gene, to show its effects must occur in double dosage, that is, on both members of the chromosome pair as bb or b^1b^1 (except when sex linkage is involved) which results in a Bronze bird insofar as the genes for color located at these similar points or loci (plural of locus) on the chromosome pair are concerned. A purebred or homozygous Black turkey will carry the gene B in double dosage, BB. A purebred Bronze will be bb; a purebred black-winged bronze will be b^1b^1. A crossbred Black-Bronze will be Bb. If these crossbreds (Bb) are mated together (one-fourth of the offspring will on the average be pure Bronze (bb), one-half hybrid or heterozygous Black (Bb), and one-fourth pure Black (BB). This is simple Mendelian inheritance, and the resulting ratio is 3 Black: 1 Bronze.

The term overdominance is now playing a large part in both plant and animal breeding. It may be said to be the extra vigor or "kick" which the offspring receives over and above the more favorable dominant of its parent. Saying it another way, overdominance is that amount of the dominance which is accounted for by the superiority of the hybrid over the better of the two parents.

It is known that there is more than one point or locus on the same or other chromosomes that carry genes for color. For example, at other loci the genes for Bourbon Red and the other colors occurring in turkeys are found. When a gene prevents the expression of genes other than its allel it is said to be *epistatic* to those genes. It may be completely epistatic as in the case of the white of any ordinary white turkey such as the White Holland or Beltsville White over any other color factors or incompletely epistatic, which is sometimes referred to as factor interaction.

The important characteristics of turkeys are all believed to be inherited, one or more genes being responsible for the transmission of each from parent to offspring.

Each species has a characteristic number of chromosomes but the exact number has been determined for only a few species. There is no general agreement among investigators as to the number of chromosomes of the turkey or of any other species of poultry, even the chicken with which the most investigational work has been done. For turkeys (and ducks) Werner counted 76 in the cells of males and 77 in those of females. For turkeys Shiwago (1929) reported 46; Scaccini (1939) 50 to 55; and Yamashina (1946) 82 for males and 81 for females. Sokolov and co-workers expressed the belief that the chromosome complement of turkeys is similar to that of chickens, peafowl, and pheasants, but stated that an accurate count could not be made.

With the exception of color inheritance and a very few others the genetic behavior of the various characters of turkeys has not been determined in detail. We are therefore able to apply to turkey breeding only the fundamental principles of genetics which though worked out with other species apply equally to turkeys.

With this brief discussion of inheritance it is hoped that the reader may be able to understand the discussion of the inheritance of color that follows.

The Inheritance of Plumage Color in Turkeys

The first published reports on color inheritance in turkeys, so far as can be found, were made by W. R. B. Robertson in the Anatomical Record in 1920.* In 1943, Robertson's work was reviewed, checked, and published in the *Journal of Heredity* (34:246-256) under the title, "The Inheritance of Plumage Color in the Turkey," by W. R. B. Robertson, B. B. Bohren and D. C. Warren. Following this, V. S. Asmundson of the University of California reviewed all previous publications, conducted many experiments, and published in 1945, a comprehensive report entitled, "A Triple-allele Series and Plumage Color in Turkeys" (*Genetics* 30:305-322). In 1950 As-

---

*Anatomical Record: Vol. 17(1920) 336-337; 20 (1921) 213-214; 23 (1922) 98; 31 (1925) 556; 44 (1929) 289 and Mo. Exp. Sta. Bull. 236 (1926.)

mundson published "Sex-linkage in the Turkey," in *The Journal of Heredity* (44:205-207). For detailed information on color inheritance in turkeys the reader is referred to these publications which are reviewed in the following pages.

Domesticated turkeys usually are divided on the basis of plumage color into varieties, seven of which are now (1953) recognized by the American Poultry Association and are described in the Standard of Perfection. An eighth, the Buff, once was recognized and described but was dropped in 1915. The seven Standard varieties are: Bronze, White Holland, Black, Bourbon Red, Narragansett, Slate, and Beltsville Small White. Each of the colored varieties and the two white varieties considered as one so far as color inheritance is concerned has a color pattern or combination of genes (genetic determiners). Since 1929, a number of new genes producing different colors of plumage have been described and a number of others have been observed but not described. There is the recessive gray, or slate (sl sl) of Asmundson (1940) which produces a lighter-colored Slate than the older, better known dominant Slate,*DD* (pure) or *Dd* (heterozygous). The Black-winged Bronze (Asmundson 1939) is determined by a recessive gene designated as b^l, and is part of a triple allele series with *B* and *b*. Hutt and Mueller (1942) reported a sub-lethal partial albino pattern determined by a completely recessive sex-linked gene designated as *al* which was investigated further by Asmundson (1950) who reported a new triple allele series consisting of *N*, non-Narragansett (Bronze); *n*, Narragansett; and n^a, albino, with dominance in that order. The 75% of the albino, $-n^a$, females that die in incubation and the females that do hatch are completely blind, hence almost impossible to raise. The day-old albino poults are whitish with the bronze pattern faintly showing. Adult albinos are white with subdued dusky pigmentation corresponding to the black markings of the Bronze. The eyes are almost devoid of pigment but what little is present is confined to the iris which is a pale blue-gray while the pupil being free of pigment appears red in color as in all true albino animals. Still another type of albino turkey was found in a flock of White Holland turkeys at the Ohio Experiment Station at Wooster in 1953. It has pure white plumage and unpigmented eyes; it is blind, very nervous, and its gene is non-sex-linked. Ordinary white

Fig. 11.8. Black tom of ideal type. (Photo from Turkey World.)

turkeys are not true albinos since they have normally pigmented (medium brown) eyes denoting their Bronze origin. The Narragansett has gray-rimmed brown eyes; the Black has very dark brown (almost black) eyes, the factor for which normally is linked with that for black plumage.

The bronze color pattern is basically the wild pattern and the color differences between the seven reported wild turkey color patterns (and the various strains of domesticated Bronze including the Broad Breasted Bronze) are believed to be determined by multiple modifying genes. The gene responsible for the basic bronze pattern, however, is designated as *b*, a gene recessive to black color *B* but dominant over b^1, the black-winged bronze factor. These three genes constitute a triple-allele series meaning that at most only two of the three genes can be carried in the paired chromosomes of any one bird and that any cross involving birds of the 3 different colors could result in the recovery of only 2 of the 3 color patterns that form the triple-allele series.

The gene producing brownish-red (Bourbon Red) color in the turkey is called *r* and is recessive to non-red, *R*. *R* however is only imperfectly dominant over *r*. The Bourbon Red is derived from the Bronze and the genes responsible for it are designated as *bb rr* while Bronze is *bb RR,* the Black variety is *BBRR,* and the Black-winged Bronze is b^1 b^1 *RR*. Bourbon Red x (crossed with) Bronze results in reddish bronze (bronze-red) offspring, *bb Rr*. The Bourbon Red x Black results in "rusty black" offspring *Bb Rr,* which when bred together produce several colors of offspring including pure black, black-bronze, rusty black, black-red, pure bronze, bronze-red, pure buff, heterozygous buff, and Bourbon Red as shown later.

When Bourbon Red, *bbrr,* and Black-winged Bronze, b^1b^1RR, are crossed, the first generation (F_1) hybrids are bronze-red bb^1 *Rr* (colored just like those from the Bronze-Bourbon cross). However, when crossed together those hybrids produce a variety of colors including bronze,

bronze-red, black-winged bronze, black-winged bronze-red, Bourbon Red, and a new true-breeding color, white-downed red (b^lb^lrr). These latter when day-old have pure yellowish-white down but the adult feathers come in a rather uniform light brownish red, or buff, with but little white in wings or tail but some black in the primaries. The eyes are rich medium brown as in the Bourbon Red and the shanks have little or no dark pigment. This type of buff color should, theoretically at least, make a better buff turkey than that produced from the

Fig. 11.9. Bourbon Red tom with especially fine wings and tail. (Photo from Turkey World.)

Bourbon Red and Black because it is a solid buff color and breeds true; there being no possibility of the heterozygosity which can not be removed from the Black-Bourbon Red type of Buff except by special breeding tests with the Bourbon Red.

The black-winged bronze pattern (named Crimson Dawn by Tom Stodghill of Terrell, Texas, probably around 1930) is colored like the Bronze except that the primary wing feathers are black, the secondaries white-tipped black with upper surfaces buff-stippled, the lower neck and wing front black penciled with buff, and the beak and toenails whitish. The Black-winged Bronze turkey is a handsome true-breeding variety much more attractive than any Bronze turkey except the old exhibition Bronze. Day-old poults have the bronze pattern but the ground color is yellowish white instead of light brown resulting in a very pale-colored bronze poult.

Crosses of white turkeys with Bronze, Black, and Black-winged Bronze show that white color in turkeys is not a color but simply absence of color due to a gene called *c* or *w*, which is completely recessive to its allele *C* or *W*, which permits color to be expressed. The symbol c is used by Asmundson and by Robertson, Bohren, and Warren. The symbol *w* is used by some other authors, obviously to designate the same gene. White poults that have buff in the down were shown to carry the bronze factor pure (*bb*) or with b^1 ($b\ b^1$) and so, except for the *cc* gene that prevents the color from showing, react the same genetically as the pure Bronze or the Black-winged Bronze-Bronze cross respectively. Poults with clear yellow down without any buff carry either the black factor, pure (*BB*) or with the bronze (*Bb*), or with the black-winged bronze (Bb^1), or the black-winged bronze factor pure (b^1b^1). When these clear yellow poults matured, those found to have bluish-gray eyes were shown to be the ones carrying the black gene either pure (*BB*) or heterozygous (*Bb* or Bb^1) while those pure yellow poults that developed normal brown eyes were derived from the

Black-winged Bronze (b^1b^1). At the Beltsville Station there have been observed a few black turkeys heterozygous for white (Ww) that have blue-gray eyes; and a few white turkeys with the very dark brown eyes characteristic of black-plumaged turkeys and of buff turkeys produced by crossing Bourbon Red with Black. These findings would indicate some crossing-over, or breaking of linkage, among the genes involved in the plumage color-eye color relationship.

In the turkey, the gray gene *D* acts on the recessive bronze gene (*b*) to produce the Slate turkey, and on the dominant black gene (B) to produce the Blue turkey. It also acts on the recessive Narragansett factor (*n*) to produce still another type of gray or slate plumage about which little is known; and to make the picture still more complicated, a typical slate color may also be produced by the action of an entirely different *recessive* gene for gray color, Asmundson's *sl* on the Bronze. Whether or not the recessive gene *sl* could be combined with black to produce a blue has not been reported.

Purebred (homozygous) Blue turkey *toms* (the product of the dominant gene *D* with black *B*) are a medium shade of bluish gray with blackish tipping on the back, neck, wing, and tail feathers. Homozygous (pure) Blue hens are a beautiful clear, even shade of light bluish gray without black tipping. At hatching, both sexes of blues are an even, pale, bluish gray with white on the head, wing tips, and ventral (lower) surfaces. Since the Blue males vary considerably in the amount of the dark tipping carried, there would seem to be a possibility of improving them through selection so that Standard requirements for a clear bluish slate plumage could be met.

The purebred (homozygous) Slate turkey tom and hen (the result of a combination of the gene *D* with Bronze, b) both are a fairly even, rather light shade of slaty blue but have rather inconspicuous brownish and white markings wherever the Bronze has them. Thus the Slate tom has white to yellowish or brownish white in the large wing feathers, and on tips of the thigh feathers, the main tail feathers, and the tail coverts. The slate hen has the same pattern and in addition has white tips on back, wing-bows, secondary coverts and breast, and has reddish brown in the body of the main tail feathers and tail coverts. It probably would be impossible to eliminate these markings since they are the inevitable expression of the genetic factors involved. Day-old Slate poults of both sexes have clear yellowish-white underparts and yellowish-white upper parts with a buff tinge and pale gray markings following the Bronze pattern and located where the Bronze poult has its dark brown markings, thus indicating its Bronze origin. The gene D acting on Bourbon Red produces a reddish turkey with slate-tipped feathers.

When Bourbon Reds (rrSlSl) were mated to recessive slates (RRslsl) and the hybrid offspring crossed, there is obtained, among others, the Lilac turkey (bbrrslsl) a true-breeding reddish-slate bird. Slates produced by a combination of the recessive slate (sl) with black-winged bronze (b^l) are a uniform light slate (nearly devoid of brownish red) called slate-gray (b^lb^lslsl) which is true breeding, as are those slates (BBslsl) derived from black (B) with recessive slate (sl). It has been observed that Slate turkeys produced by the dominant gene *D* tend to carry a factor producing one defective eye which is reduced in size and sightless or nearly so. Possibly Slates produced by Asmundson's recessive slate factor *sl* would not carry this factor hence might be a more desirable type to breed.

The Palm, darker than the Light Palm and Royal Palm is a mutation derived from Bronze flocks but the gene responsible for it is not an allele with the bronze, black, and black-winged bronze factors. In color the Palm resembles Bronze but has less white barring in the secondaries and lacks the bronze sheen. It may be the same as the "Silver Bronze" (see Chap. 10). The day-old poult has grayish down with a purple tinge. The palm gene (p) appears to be autosomal recessive. Its allel is non-palm, *P*, over which black-winged bronze, black, and bronze are epistatic. The typical Palm is homozygous as to the *R* gene as are Bronze, Black, Black-winged Bronze and Narragansett. The Red Palm (pprr) obtained in the second generation from crossing Bourbon Red with Palm is white with pale brownish red where the Palm has black. The white-and-black Desert Palm, or Royal Palm (see Chapter 10) is the lightest in color of the three Palm patterns. The light Palm is intermediate being gray and white, superficially resembling the Narragansett. The day-old Light Palm poult is yellowish white with inconspicuous light brown to medium brown dorsal markings distributed in the manner of those of the Bronze poult. There is some indication that the Desert or Royal Palm pattern is an unfixable pattern (Asmundson, 1951). However, it has been described as true by Carson.

The Brown turkey also is a color mutation first noted in a Bronze flock. Its pattern follows that of the Bronze, the black of the latter being replaced largely by brown while retaining the bronze surface sheen. It is an interesting though not particularly attractive color pattern. It appears to be due to a sex-linked recessive factor named *e*, which is completely suppressed by its dominant allele *E* or non-brown. The brown turkey is homozygous for *R*, the non-red factor, which means that it carries no factor for brownish red as do the Bourbon Red and Buff. It is a true-breeding variety. The day-old Brown poult has light brown ground color as in the Bronze but the dorsal markings which follow the Bronze pattern are medium brown instead of very dark brown, almost black, as in the Bronze. Auto-sexed poults can be produced by mating Brown males with Bronze females, all the resulting male poults being typically Bronze in color while all the female poults are brown.

With respect to linkage it appears that the triple-allele series (B, b, b^l) is on a different chromosome from, and consequently not linked to *R, r; P, p;* or *D, d* but may be linked with *Slsl*. Apparently *R*, *r* and *Sl*, *sl* are not linked. Other linkage relationships have not been worked out. A light brown turkey can be produced by combining the Narragansett factor *n* with the brown gene *e*.

The four mutant genes; *B*, b^1, *p*, and *n* (Narragansett) are primarily pattern genes and reduce the intensity of red pigmentation in the plumage. The mutant genes of the triple-allele series *B* and b^1, produce more uniformly slate-colored birds than the "wild" gene *b* (Bronze), and the b^1 (Black-winged Bronze) gene produces a more uniformly buff-colored turkey than does *B* (Black).

To the number of known color genes should now be added that producing the pied "Nebraskan" which appeared in a broad Breasted Bronze flock and is evidently a simple recessive. It has a background of white fairly well spotted with black, apparently without a definite pattern (see description).

A new mutant in a BBB flock named "Faded Bronze" was reported from Pennsylvania (Keeler et al, 1949) but was not investigated thoroughly. The evidence available suggests it was due to a non-sex-linked recessive factor or factors. The day-old poults were small, had reddish brown where the Bronze has dark brown, had down noticeably shorter than normal, especially on the head, showed low livability and appeared to be partially blind. The adult plumage was bronze in pattern but the normal black color was faded out to a reddish brown resembling the Brown turkey in this respect. Feathers tended to be rough and easily broken. Beak, toes and toenails were pinkish white and the iris color was lighter than normal for Bronze and had a definite reddish cast.

Bourbon Red color is due to an incompletely recessive gene *r*. Its imperfectly dominant allel is not another color but an extension factor *R* which permits the presence of overlying dark pigment, thus obscuring the underlying red. Whenever this imperfectly dominant extension factor *R* occurs in double dosage (*RR*) a bird does not show any red color unless modified by pattern factors; but when it occurs only in single dosage as *Rr*, some red may show. The gene *r*, while incompletely recessive to *R*, is however much more potent when it occurs in double dosage (*rr*) than is the gene for bronze (*b*), so much so that when a bird possesses both *rr* and *bb*, the bronze fails to show, *r* being completely epistatic over *b*. The *rr* combination also covers up the black factor *B* and the Narragansett factor *n*. These last two, however, modify the *rr* somewhat, resulting in light colored reds (*r* imperfectly epistatic to *B* and *n* resulting in dilution of the red.) When both *BB* (or *Bb*) and *rr* occur the result is light-colored red or buff. When pure black (*BBRR*) is mated to pure Bourbon Red (*bbrr*), the offspring are hybrids carrying the *Rr* and *Bb* combinations. These birds are rusty black in which there is a slight trace of white barring in the wing feathers and of barring with brown (but not the brown of the factor *e*) to a slight extent in the remainder of the plumage.

Buff color may be produced by first crossing Black with Bourbon Red. The rusty-black offspring may then be crossed among themselves as illustrated later or they may be mated back to the Bourbon Red, resulting in four types of plumage: Rusty blacks (BbRr), Bourbon Reds (bbrr), bronze-reds (bbRr), and heterozygous buff (Bbrr). When these heterozygous buff turkeys are mated together, one-fourth of the offspring should be Bourbon Reds, one-half heterozygous buffs (Bbrr) and one-fourth homozygous buffs (BBrr). The homozygous buffs which breed true give rise to one type of Buff turkey of which the Jersey Buff is representative.

The dominant factor for slate color (D) is inherited independently of the 4 factors for bronze, black, Narragansett, and Bourbon Red color and is a factor diluting the pigment of the four colors and their hybrids. Its recessive allelomorph is non-slate (d) which permits the various patterns to be expressed to their full intensity. In combination with black, the slate dilution factor D produces a handsome bluish-slate color; in combination with bronze it produces a slate color, tending to show the bronze pattern; and in combination with Bourbon Red a bird similar to the Bourbon Red except that the contour and covert feathers are tipped with slate. However if the recessive slate factor of Asmundson *sl* is used, such a slate bird crossed with non-slate individuals would result in all non-slate offspring. If two of the resulting heterozygous birds were subsequently crossed, however, one slate individual would be obtained to each three non-slate and these Slates would breed true.

The pigment factor W which dominates completely over its allel *w*, the factor for white (some geneticists use *C* and *c*) causes certain cells to deposit pigment in plumage, shanks, feet, and beak. It is distributed independently of the factors for black, bronze, slate, Narragansett, and Bourbon Red. It may possibly be an oxidase factor without which no color (except white, which is lack of color) can be expressed. Its recessive allel *w* when present without the dominant *W* prevents almost completely the appearance of any pigment in the plumage, resulting in a white bird. This factor usually does not affect eye color, however, so that white turkeys tend to have the eye color determined by the pattern factor carried but suppressed by the white *ww* factor. Since most white turkeys originated in Bronze flocks they usually have the medium dark brown eye characteristics of the Bronze. An exception perhaps, to this rule, is the possible effect of *w* on the black eye color that goes with the black plumage factor *B*. (See previous remarks on eye color.)

Another sex-linked recessive factor for color is that of the Narragansett pattern *n* which forms a triple-allele series with *N* and n^a (previously discussed). Males that show Narragansett color must have a double dose, *nn*, because they have two functioning sex chromosomes. Its dominant allelomorph is *N*, non-Narragansett or Bronze. Narragansett females have *nY*, only one *n* because they (as well as female birds of all species thus far studied) have only one functioning sex chromosome, the other, if it exists, being the non-functioning Y chromosome. The recessive factor *n* when present without its dominant allel *N* functions as a modifier of bronze, black, and red plumage, with by far the greatest visible effect on the bronze (producing Narragansett) and least on the red.

To produce Narragansett males, there must be at least one Narragansett parent, the dam, the sire may be some other color but must carry the recessive Narragansett factor. Narragansett females, however, may be produced when neither parent is Narragansett if the male parent carries the recessive *n* factor. For example, a white tom carrying but not showing the *n* factor will produce some Narragansett female offspring when mated to Bronze hens. The sex-linked Narragansett pattern theoretically may be used to tell sex in day-old poults. Narragansett toms mated to Bronze hens give male poults of the bronze pattern and female poults of the Narragansett pattern, the poult resembling the parent of the opposite sex (criss-cross inheritance). However, due to the difficulty of separating Bronze and Narragansett day-old poults, this method has no practical value. By using some color pattern other than Bronze, to cross with Narragansett, auto-sexed poults possibly could be produced as can be done by using the Brown turkey male on Bronze hens.

The genetic formulae for birds pure for color and color patterns of some of the different varieties of turkeys are given. The *Y* in the formula for the females designates the non-functioning sex chromosome.

	Males					Females				
Bronze†	WW	dd	NN	bb	RR	WW	dd	NY	bb	RR
Bourbon Red	WW	dd	NN	bb	rr	WW	dd	NY	bb	rr
Narra-gansett	WW	dd	nn	bb	RR	WW	dd	nY	bb	RR
Black	WW	dd	NN	BB	RR	WW	dd	NY	BB	RR
Slate*	WW	DD	NN	bb	RR	WW	DD	NY	bb	RR
Blue*	WW	DD	NN	BB	RR	WW	DD	NY	BB	RR
Buff	WW	dd	NN	BB	rr	WW	dd	NY	BB	rr
White	ww (with any combination of th› others)					ww (with any combination of the others)				

*Either of these colors is acceptable as "Slate" in the Standard.

†The Palm has the same formula as the Bronze, except that it is recessive for palm pattern (pp), whereas the Bronze is dominant (PP); black-winged bronze has the same formula as Bronze except it has b^1b^1, where Bronze has bb.

Y represents the non-functioning sex chromosome of the female.

Calculating Mendelian Expectancies

The checkerboard method is a simple way to compute the genetic constitution of the offspring from a cross. First, the genetic formula of each specimen crossed is determined. The different kinds of gametes that will be produced by each parent are listed on two adjoining edges of the checkerboard. Each kind of gamete listed on the left edge is then combined in the proper space within the checkerboard with each kind of gamete on the upper edge. In this manner all possible combinations of genes involved will be disclosed in such a way that the expected ratio can be determined. The following examples serve to illustrate the process.

The simplest type of Mendelian inheritance involves only two genes allelic to each other. Since the sex chromosomes are not involved, the sex of the parents makes no difference. The first example shows the result of a cross between a purebred Bronze (*bb*) tom and a purebred Black (*BB*) hen. Since both parents are homozygous there is only one kind of sperm and one kind of egg*

*Sperm and eggs are called gametes, which means marrying cells.

produced, at least as far as the bronze and the black color factors are concerned. Therefore there is only one kind of hybrid offspring which has the genetic formula *Bb*, as illustrated herewith:

		Male gametes (of one kind only, derived from Bronze, *bb*)
		b
Female gametes (of one kind only, derived from Black, *BB*)	*B*	*Bb*

These (*Bb*) hybrids are black but always have a few bronze and partly bronze feathers by which they can readily be identified when well feathered out, but not at hatching time. If they are crossed *inter se* (among themselves) a segregation of factors occurs as follows:

		Male gametes (two kinds, derived from *Bb*)	
		B	*b*
Female gametes (2 kinds, derived from *Bb*)	*B*	*BB*	*Bb*
	b	*Bb*	*bb*

There are obtained then:—3 offspring that are predominately black (1 pure black and 2 black with a few bronze and partly bronze feathers, like the parents) and one that is pure bronze. This is called the 3:1 or phenotypic* monohybrid ratio, which can be broken down into

*Phenotype refers to combinations of visible characteristics while genotype refers to combinations of the genes themselves. Genotype is the genetic constitution of an individual; phenotype is the physical expression or appearance.

a 1:2:1 genotypic monohybrid ratio by separating the homozygous blacks from the heterozygous blacks.

When Black and Bourbon Red are crossed, the picture is a little more complicated as it involves two pairs of genes. When a purebred Black male (*BBRR*) and a purebred Bourbon Red female (*bbrr*) are crossed, all F_1 (first cross) offspring are alike genetically (*BbRr*). This is always the case when both individuals crossed are pure or homozygous for the factors involved. The genetic formula of these F_1 hybrids is *BbRr* and they were named rusty black by Robertson. Their origin is explained as follows:

		Male gamete (one kind, derived from Black, *BBRR*)
		BR
Female gametes (one kind derived from Bourbon Red, *bbrr*)	*br*	*BbRr*

When these rusty blacks are crossed *inter se* the following segregation of the color factors takes place:

		Male gametes (four kinds, derived from rusty black, *BbRr*)			
		BR	*Br*	*bR*	*br*
Female gametes (four kinds derived from rusty black, *BbRr*)	*BR*	*BBRR* pure black	*BBRr* black red	*BbRR* black with a few bronze feathers	*BbRr* rusty black
	Br	*BBRr* black-red	*BBrr* homozygous buff	*BbRr* rusty black	*Bbrr* heterozygous buff
	bR	*BbRR* black with a few bronze feathers	*BbRr* rusty black	*bbRR* pure bronze	*bbRr* bronze red
	br	*BbRr* rusty black	*Bbrr* heterozygous buff	*bbRr* bronze-red	*bbrr* pure Bourbon Red

There are obtained then:—9 offspring that are predominantly black (1 pure black, 4 rusty blacks, 2 black-reds, and 2 black with a few bronze feathers); 3 that are predominantly bronze (1 pure bronze and 2 bronze-red); 3 that are buff (1 homozygous or pure and 2 heterozygous or impure); and 1 Bourbon Red. This is known as the 9:3:3:1 or dihybrid (phenotypic) ratio. It will be noted that from a cross using no bronze or buff stock at any time, pure bronze and pure buff, as well as hybrid buff are obtained. There are nine different combinations of genes or genotypes.

If 3 different pairs of genes are involved, the ratio (phenotypic) is trihybrid or 27:9:9:9:3:3:3:1, with a total of 27 different combinations of genes of which only 8 (phenotypes) could be separated according to plumage color if dominance were complete and no other genetic phenomena entered the picture.

Sex-linkage is illustrated by the following examples. If a purebred Narragansett male, which carries the factor for Narragansett color (n) on both sex chromosomes is mated to a Bronze hen, which possesses only one functioning sex-chromosome (in each cell) and carries the non-Narragansett factor N, the resulting male offspring are all bronze in color while the females are all Narragansett.

		Male gametes (only one kind, derived from Narragansett, $bbnn$)
		bn
Female gametes, always of 2 kinds as far as the sex chromosomes are concerned, one with a sex chromosome and the other without it. In this case the sex chromosome carries the N (non-Narragansett) factor. The gametes are derived from Bronze, ($bbNY$)	bN	$bbNn$ Males, bronze color, possessing two sex chromosomes
	bY*	$bbnY$ Females, Narragansett color, possessing one sex chromosome which is always derived from the male parent

*Y represents the non-functioning sex chromosome.

The female offspring are pure Narragansett while the male offspring are hybrid bronze carrying the Narragansett factor.

If a purebred Bronze male is mated to a purebred Narragansett female (the reciprocal of the cross used in the preceding example), all the offspring are bronze in color as shown herewith.

		Male gametes (only one kind derived from Bronze, *bbNN*) *bN*
Female gametes (of two kinds, derived from Narragansett, *bbnY*)	*bn*	*bbNn* Males, hybrid bronze
	bY	*bbNY* Females, pure bronze

In the female offspring the Narragansett factor has been completely eliminated. The male offspring, although bronze in color, still carry the recessive Narragansett factor.

Plumage Color—Mosaics

A mosaic hybrid is one which exhibits certain alternative characters (alleles) of both parents in juxtaposition unblended. Mosaics are rare in poultry but bilateral mosaics differing in size and in skin color of right and left sides and the color of right and left shanks have been reported in chickens. Mosaics are considered by some to be due to somatic gene mutations although other explanations such as "nuclear shifting" are tenable.

Asmundson (1937) described a female Bronze-Bourbon Red mosaic as follows "The anterior and posterior extremities were typically Bronze in color while there was a complete band of Bourbon Red plumage around the middle of the bird. There were, however, in the wing, areas of Bronze plumage completely surrounded by Bourbon Red feathers. Several mosaic feathers were observed." When mated to a Bourbon Red male, all the progeny obtained were Bourbon Red in color.

The senior author obtained a female bronze-white mosaic, which is shown in Fig. 11.10. One shank was dark-colored as in the Bronze, the other was white. The right half of breast and the right lower thigh were Bronze; their left counterparts were Bronze and white. Genetically this hen behaved as a white. Her dam was a Narragansett derived from a cross between a Bronze hen and a White Holland tom which carried the Narragansett factor. Her sire was black with a few Bronze feathers, the result of a cross between a white tom and a black hen.

George Crandall of Haddonfield, N. J. reported (1947) a mosaic tom buff in color except for a 6″ to 8″ band of white plumage across the back starting just to the rear of the shoulder joints. This male was not tested genetically but probably would have behaved as a Buff.

Hairy-type plumage has been found to be due to a single pair of autosomal recessive genes by Smyth (1954). At the Massachusetts Station this type of plumage was found in Bronze turkeys. In the poults the down appeared wiry, similar to coarse hairs laying close to the skin. In

Fig. 11.10. Bronze and white mosaic feathering. (Photo from U. S. Dept. of Agr.)

adults the feathers have a coarse ragged appearance (hair-like) and the plumage looks and feels damp-like. Cannibalism is hard to control and the experimental birds of this strain could not withstand heavy rains.

Heritability

Unfortunately for the breeder, not all characters are inherited in as simple a manner as comb type in chickens or feather pattern in turkeys. Many factors (pairs of genes) influence such characters as rate of growth, egg production, fertility and hatchability. These characters are influenced to a great degree by environment whereas the environmental conditions have practically no effect on feather color and but little affect on egg size and meat type. The latter characters and many others affected but little by nutrition, climate, housing and management are said to be highly heritable. Heritability is the degree (or per cent) of the influence of heredity on the transmission of a given character. It is measured in per cent and traits of 40 to 60% heritability are said to be highly heritable. Fifty per cent (50%) heritable means about half of the gain that can be accomplished may be made through breeding while environment will determine or circumscribe the other half. In turkeys, body weight at 26 weeks of age is 52% heritable in Broad Breasted Bronze toms, whereas in Beltsville Small White males at that age it is but 20%, according to Kosin (1951), whose data are given in Table 11.5. Thus, with the same degree of selection with both breeds one could make much more rapid progress toward a given "size" goal with Broad Breasted Bronze males than with B.S. White males; the same selection pressure being applied to both breeds.

Selection pressure is the force which the breeder places upon the population with which he is dealing in choosing the parents of the next generation. As a simple illustration, if a turkey breeder raised 1,000 turkeys to six months of age and retained 400 breeding hens from among the 500 he raised, he would exert but little selection pressure against the female population; merely cull out 100 for market (20% in this case) of the least desirable females. If, on the other hand, he actually placed in his breeding pens with these 400 hens 25 toms, he would have the opportunity to exert terrific pressure against the male population of the same flock; since he actually held and used as breeders only 25 birds (5%) of 500 available in November when he started his selection program. Selection pressure is of little value unless the breeder is wise in his choice of the individuals saved to carry on his strain.

The rapid growth in popularity of the Broad Breasted Bronze turkey and the rapidity with which the broad breast type was fixed within several strains proves that this trait is highly heritable, and, second, that selection pressure was wisely used. The breeders of the Northwest (U.S.A. and Canada) including Jesse Throessel of British Columbia, who originally imported birds of this strain which he had selected in England for meat type, took full advantage of these two genetic tools. In short, they coupled the science of genetics with the art of selection and gave us our modern day broad breasted turkey.

Characters of turkeys which may be spoken of as medium heritability are the weight of both Bronze and Beltsville White females at eight weeks as well as at 26 weeks. The Washington Station (1951) has shown that breast width at 26 weeks is 23 per cent heritable in Bronze males and 13 per cent in females. The depth of the breast meat is 29 per cent heritable in Bronze males and 40 per cent in females. The length of the keel is 22 per cent heritable in the males and 18 per cent in the Bronze females; whereas shank length is 25 and 28 per cent respectively. All of these data are for Broad Breasted Bronze at 26 weeks. They also found in the Bronze that hatchabilty was 12 per cent heritable for eggs stored for 8 to 14 days. For all Bronze eggs set fertility proved to be 19 per cent heritable and egg weight 32 per cent (see Table 11.5).

Pedigree breeders who are carrying on family breeding and sib testing programs should concentrate on traits of low heritability in their pedigree selection. However, they should also use mass selection or physical selection (judging) for the traits of high heritability which they desire to incorporate in their strains.

The Inheritance of Body Weight, Fleshing and Skeletal Developments

The California Station (1945) reported that width of breast in turkeys (24 weeks of age when observed) is determined by the combined effect of more than one pair of genes which results in a blending type of inheritance without any dominance or sex linkage in evidence. At least some of these genes act independently of genes responsible for body size or weight, thus making it possible to obtain small turkeys carrying broad-breasted characteristics. When broad-breasted turkeys were crossed, either way, with non-broad-breasted turkeys the resulting offspring averaged midway between the two parent strains in broad-breastedness with considerable variation among the progeny.

In another report from the California Station (1948) the results with 24-week-old birds indicated that heavy body weight tended to be dominant over light body weight in this type of cross (BBB x standard Bronze) and that there is no sex-linkage in weight (or size) inheritance. Results also indicated that long shank length is dominant over short shank and that there is no sex-linkage involved. In length of keel the parental strains did not differ materially, hence the evidence is inconclusive but the F_1 and F_2 data suggest that long keel is dominant over short keel and that sex linkage is not involved.

Breast width was shown to be non-dominant and non-sex-linked, with more than one pair of genes involved; the width of breast of first crosses between the broad and narrow strains being intermediate between those of the parental strains. Females had broader breasts than males but males were larger and had longer shanks which, however, were shorter than those of the females in proportion to the body weight.

Full brother-sister families of Broad Breasted Bronze dams showed significant variations in weight, length of keel, length of shank, and breadth of breast, indicating that hen-family selection should be effective in changing these characters.

In general, the larger the bird the longer were the keel and shank and the broader the breast. However, when the effect of weight was eliminated there was found to be no correlation between length of shank and width of breast or between length of shank and length of keel but the length of keel and width of breast were negatively correlated, that is, the longer the keel the narrower was the breast. While width of breast was found to be positively correlated with body weight, there was enough independent variation to indicate that good length and width of breast may be successfully combined with small size. Applied to selection and breeding, these results would seem to indicate that if turkeys were to be selected mainly on shortness of shank, their weight likely would be decreased; that width of breast and moderate length of keel could be selected for without affecting the weight; that some selection for moderate length of keel along with that for broad breast would appear desirable; and that constant selection of both male and female breeders for the desired physical characteristics is necessary to keep up standards for body type. The fact that all these body type factors are heritable indicates that selective breeding should be quite successful in improving and maintaining good meat conformation.

The findings (above) that larger size is dominant over smaller size and long shanks over short shanks in the type of crosses (two types of Bronze) used in the California experiment were not found to be true in crosses between Broad Breasted Bronze and Beltsville Small White at the Beltsville Station (Knox and Marsden, 1944). These crosses resulted in intermediate-sized offspring which were nearer the smaller parents in weight and shank length, thus indicating a blending type of inheritance with a tendency for small size to be dominant over large size and short shanks over long shanks. The explanation for this difference awaits further research. Breast width (as in the California experiment) was found to be non-dominant, non-sex-linked, and apparently due to more than one pair of genes.

The Maryland Station (1948) reported on the inheritance of body conformation in turkeys as obtained from 28-week-old crosses between Broad Breasted Bronze and Beltsville Small White turkeys at the University of Maryland. The following were significant positive correlations in males: Breast width x body depth, breast width x keel length, breast width x body weight, body depth x shank length, body depth x keel length, body depth x body weight, shank length x keel length, shank length x body weight, and keel length x body weight. In females all of these pairs of characters also had significant positive correlations except breast width x body depth (slightly negative but insignificant) and breast width x keel length (slightly positive but insignificant). All the above include the effects of body weight. Breast width and shank length showed no significant relationship in males or females separately but when both sexes were considered together, there was a significant negative correlation. When the variations in body weights were eliminated it was found that the correlations between shank length and keel length were, in males and females, still significantly positive, but between breast width and shank length it was significantly negative in both males and females and that breast width and keel length were still significantly positively correlated in males but not in females in which the correlation was significant and negative.

Where comparable these results agree fairly well with those of Asmundson, the chief points of disagreement being that the latter found no correlation between shank length and keel length and none between breast width and shank length but found a negative correlation between breast width and keel length in both sexes. Possibly differences in breed, age of birds measured, or method of measuring may explain the discrepancies between the two reports.

Of greatest interest in this report are the findings that the positive correlations between body depth with shank length, with keel length, and with body weight are highly significant. This would indicate where birds are selected for moderate length of keel and shank, body depth should take care of itself and conversely if keel and/or legs are unduly shortened there is danger of losing depth of body, which is necessary to a well-balanced turkey and a high percentage of edible meat. A combination of five measurements should be very effective in breeding for desirable market type; body weight, breast width, body depth, keel length, and shank length.

The relationship of breast width and body depth is not clear since when the sexes are combined the correlation is negative and significant, but when considered separately the males show a barely significant positive correlation and the females an insignificant negative correlation. If the thesis that the combined data for both sexes are more reliable is accepted, it is indicated that there is definite danger of losing body depth if selection for wide breast is made without considering good length of keel and moderate length of shank also.

Heritability values for body weight and measurements in two types of turkeys were reported by the Washington Station (1952). Several heritability estimates on body weight at certain ages are given in Table 11.5.

Table 11.5—Heritability Values in Turkeys

Data obtained at the Washington Agricultural Experiment Station (1947-1950)

Characteristic	Heritabilities			
	BBB Males	BBB Females	BSW Males	BSW Females
Body Weight— 4 weeks	49.12%	38.31%	25.80%	18.83%
8 weeks	55.95	27.55	33.13	16.21
14 weeks	71.53	39.27	18.69	16.07
26 weeks	52.33	29.29	20.03	24.12
Breast width* 26 weeks	23.44%	13.27%		
Breast depth 26 weeks	29.49	40.10		
Keel Lengths 26 weeks	21.52	17.68		
Shank Length 26 weeks	25.01	27.65		
Hatchability:				
Eggs stored 0-7 days	2.62%			
Eggs stored 8-14 days	12.01			
Fertility	19.00			
Specific Gravity	51.88			
Egg weight	31.91			

*As measured by the instrument of Bird (1948).

There was no advantage found in basing heritability estimations on weight gains rather than on absolute body weights as above.

The heritability estimates of all measurements were made from data taken when the turkeys were 26 weeks of age.

The low heritability values for breast width in BBB may be due to a high degree of homozygosity for the genes contributing to breast width (the result of long continued accurate selection). On the average the heritability values are high enough to suggest that selection should be effective in controlling these relationships in turkeys.

According to Lush (1947), "Paradoxical though it may seem, family selection is most helpful and most needed for characteristics in which the members of the family resemble each other the *least*. When family members resemble each other closely and families are therefore conspicuously distinct from each other, either the characteristic is so highly heritable that individual selection is not needed, or each family is affected so strongly by its own special environment that the phenotypic average of the family is but slightly correlated with its average breeding value."

The Inheritance of Reproductive Factors

Egg production in turkeys is inherited in the same manner as it is in chickens, through autosomal and sex-linked genes and the breeding methods used in improving egg production in chickens are equally effective with turkeys.

The Oregon Station (1949) suggests that quickness of response to artificial lighting is a desirable selection point in breeding for increased egg production. It was found that well-matured BBB hens that responded to artificial lighting in 19 to 24 days produced the most eggs and the most poults for the season. The relative absence of pauses in production (both broody and non-broody types) is necessary to high production; so is persistency through a reasonably long season of 5 months and a high intensity of production or eggs per clutch, that is, eggs laid on successive days. Poor producers lay many of their eggs in one-egg clutches, high producers lay mostly in clutches of 3 eggs and more with only one day between.

Breeding males should be selected with due regard to their dam's egg records. However, egg production being only one of the many qualities desirable in market turkeys, it is seldom used as a basis for selection to the detriment of fertility, hatchability, and market quality. These qualities also are heritable and should be used as the basis for selection according to family performance insofar as possible.

The effect of broodiness on egg production, fertility, and total poults produced is shown in Table 11.6 obtained from six pens totaling 59 Broad Breasted Bronze young hens at the North Dakota Station (1945). The hens were artificially lighted starting January 3.

Table 11.6—The Effect of Broodiness on Reproduction

Broodiness Classification	No. of hens	Av. No. eggs to June 1	Per cent fertile	Av. No. of poults produced
Hens non-broody	15	70.5	82.4	42.9
Hens broody once	16	56.1	91.6	38.4
Hens broody twice	18	52.9	91.9	35.4
Hens broody three times	9	37.7	92.7	25.8
Hens broody four times	1	34.0	90.0	18.0

Broodiness in this case, while it apparently increased fertility, decreased significantly the number of poults produced per hen in addition to requiring extra labor and equipment to control it.

The South Dakota Station (1946) found from analyses of trapnest records of BBB turkeys, that broody hens, "pause" hens, and those showing both pause and broody periods did not hatch any better, were no more fertile, and did not lay at a higher rate than those which had no broody or pause periods.

At the Beltsville Station it has been observed that, while some non-broody Beltsville White hens are less fertile than broody ones, a good share of them are fully as fertile. In pedigree breeding these fertile, non-broody hens can be identified, and they and their progeny selected for breeding to eliminate genetic broodiness and at the same time retain high reproductive ability. At the same Station this procedure has been followed with Beltsville Whites, the percentage of non-broody hens being increased over a period of several years from about 20 per cent to 75 per cent without loss of fertility or hatchability and with a large increase in egg production.

Body weight at time of first egg of White Holland and BBB hens was found to be significantly and negatively correlated with the percentage hatch of all eggs set (Funk, 1950) the larger hens thus tending to produce fewer poults from a given number of eggs set (fertility and hatchability both involved). Egg production however was found not to be related to size of the hen, neither was viability of poults. It was also shown that large White Holland hens tended to lay large eggs; in the BBB this relationship existed but was not considered significant. In both varieties studied there existed a negative correlation not quite significant statistically between egg weight based on hen averages and percentage hatch of all eggs set, hens laying the smaller eggs (but not the *smallest*) tending to hatch better than those laying larger eggs. This also was found to be the case at the Beltsville Station (1950) with both Beltsville Whites and Broad Breasted Bronze. Highest hatchability of fertile eggs was obtained from hens of both varieties that laid eggs in the weight class of 75 to 79.9 grams (31.8 to 33.8 oz. per doz.). In the same report it was shown that hatchability of a hen's eggs was not affected by the male to which she was mated but apparently was conditioned solely by genetic factors carried by the dam assuming environmental factors to be favorable and equal.

The N. D. Station (1946) reported on the hatchability of 3000 BBB eggs and the Kentucky Station (1936) found the following hatchabilities of fertile eggs among 7000 eggs of standardbred Bronze (Table 11.7):

Table 11.7—Hatchability of Bronze Turkey Eggs

Egg weight range in grams	Percentage Hatchability of fertile eggs	
	Kentucky	North Dakota
70– 74.5	61	79
75– 79.5	75	81
80– 84.5	71	83
85– 89.5	68	81
90– 94.5	69	82
95– 99.5	65	75
100–104.5	59	64

Those two reports, based on individual egg weight (not hen averages as reported above by the Beltsville Station) indicate that eggs somewhat smaller than average tend to hatch better than those real small and those of average size and larger. The 75 to 79.5-gram group it will be noted, had the highest hatchability in the Kentucky experiment and rated among the highest in the North Dakota data.

In a previous report from Beltsville (Byerly and Marsden, 1938) based on 4341 eggs from small-sized crossbred turkeys, highest hatchability, 80.4 to 82.6% of fertile, was obtained from eggs in the weight range of 73 to 87 grams, those over 87 and under 73 grams hatching about 74 and 71% respectively. On a hen basis results were similar; eggs from females whose eggs averaged 73 to 87 grams showed slightly higher hatchability than eggs from females with averages of less than 73 or more than 87 grams. In this experiment, average hatchability was relatively high, about 80%, and average egg size relatively small, 77.5 grams—almost exactly in the mid-range of the 75- to 79.5-gram groups showing the highest hatchability in other experiments.

That fertility, or factors contributing to it, is inherited seems probable. The Oregon Station (1950) reported evidence to this effect through the artificial insemination at 2-week intervals of 10 families of 2 to 6 sisters each. One family of 6 sisters showed an early-season fertility of 60.9% while 3 families totaling 13 half-sisters all out of the same male (different from the sire of the low-fertility family) had an average fertility of 97.3%. It was demonstrated in this experiment that fertility persisted through relatively long pauses, the hens laying fertile eggs after a broody or other pause period without being inseminated again. The N. C. Station (1951) reported that in standardbred Bronze turkeys the heritability of fertility was 30%; selection on basis of fertility was 30%. Selection on basis of hatch of total eggs set, which would include both fertility and hatchability, was recommended.

The Oregon Station (1950) reported that BBB hens with breast widths of 3.5 to 4.0 inches at a point 1¾ inches above the keel, measured at 30 weeks of age, laid eggs that were 87.5% fertile while hens of breast widths of 4.25 to 4.75 inches showed 80.0% fertility. Hatchability of fertile eggs however was 66.3% in the broader hens and 62.6% in the narrower hens. In overall reproductive ability—percent hatch of total eggs set—there was no significant difference between the broader and narrower hens. It should be noted however that the broader hens in this experiment could by no means be classified as of extreme broad-breasted type. In the same experiment it was found that hens weighing on the average 19 lbs. and over at about 30 weeks of age laid eggs that were 78.3% fertile and hatched 60.4% of fertile, while the hens weighing 15 to 18.85 lbs. at 30 weeks showed an average fertility of 83.4% and an average hatch of fertile eggs of 65.8%. These differences were not statistically significant but they indicate a trend toward better reproductive ability in hens of moderate size and body type as compared with those of larger size and wider breast measurements.

The *heritability,* or *heritability value* (H^2) was defined by Lush (1949) as the fraction of the observed (phenotype) variance which was caused by differences between the genes or genotypes of the individual. Such estimated values are likely to vary considerably in turkeys according to variety, age, sex, environmental conditions and method of computation. Heritability of individual differences in hatchability of BBB and BSW turkeys (combined data) was reported by the South Dakota Station (1946) to be 26%, and the correlation between egg production and hatchability small but positive, indicating that the two characteristics are not incompatible. The Beltsville Station (Shaklee, Knox, and Marsden, 1952) reported the heritability of egg production in Beltsville Small White (BSW) turkeys to be about 40%. For example, if average egg production of all breeder hens this year is 80 eggs to June 1 and we select for next season hens whose dams' average egg production is 90 eggs we could expect an increase of 40% of the difference which would be 4 eggs, or 84 eggs for next season.

CROSSING WITH OTHER SPECIES OF POULTRY

Except for a brief mention by a Russian (Ogorodny) of a live hybrid obtained from a mating of a rooster with a turkey hen, no authentic report of the successful crossing of turkeys with other species of poultry has come to the attention of the authors. Several newspaper accounts of a turkey-chicken hybrid (turken) have been traced to their source by the authors and found to be without basis of fact. Various workers have reported hatching peacock-chicken hybrids, the latter sometimes being designated as "turkens." However, none of these species crosses proved to be of practical value.

The so-called "Turken" (or "Churkey" in England) is not the result of a cross between turkeys and chickens. It is a breed of chickens called the "Naked-neck" that originated in Europe (Transylvania) and which possesses a dominant genetic factor producing lack of feathering on the neck. These "turkens," in red or white plumage patterns, are bred by a few fanciers in the United States.

The Kansas Station reported an unsuccessful attempt to cross chickens and turkeys by artificially impregnating* White Leghorn and Rhode Island Red hens with Bronze turkey semen and by impregnating Bronze turkey hens with the semen of Rhode Island Red, White Leghorn, and Rhode Island Red-White Leghorn crossbred cockerels. One fertile egg was obtained from the turkey male-chicken hen cross. The embryo, however, died very early in incubation. Several fertile eggs were obtained from the rooster-turkey hen cross and one of the embryos lived through 22 days of incubation. No hybrids were hatched. The 22-day hybrid embryo was described as follows: "—the embryo was fully covered with down and had

*Artificial insemination was used in order that a large number of females could be mated to the male of the other species. When penned together a tom and chicken hens or a rooster and turkey hens rarely mate.

Fig. 11.11. Fully developed turkey-chicken hybrid (left) compared with a 27 day turkey embryo (right). The extra toes can be seen on the hybrid's left leg. Instead of having one rear toe the hybrid has a group of 3 toes. (Photo from U. S. Dept. of Agr.)

most of the characteristics of a normal (Bronze) turkey except that the down was generally lighter in color, especially about the head and neck. . . . The entire yolk stalk had been drawn into the body as well as some of the yolk. This does not occur in the normal turkey embryo until about the 25th day of incubation."

The U. S. Department of Agriculture reported an unsuccessful attempt to cross turkeys and chickens. Rhode Island Red hens were artificially inseminated with mixed semen from several Bronze turkey males. Bronze and White Holland turkey hens were artificially inseminated with semen of Rhode Island Red, White Wyandotte, and crossbred males. In the crosses of chicken males and turkey hens, apparent fertility on the average was 20% but only two embryos lived past the third day of incubation. Recent findings of Olsen and Marsden (1953) indicate that most of the apparent fertility in these eggs probably was not fertility at all but consisted of developmental forms resembling early dead germs; which were of a parthenogenetic nature. One of these lived to 6 days and the other lived until nearly ready to hatch. It died sometime between the 23rd and 28th day of incubation. This hybrid, a photograph of which is shown in Fig. 11.11, was intermediate in conformation between the chicken and turkey and had two extra toes on the left foot. It more nearly resembled its chicken sire, i.e., short neck, short, broad body, rose comb, no snood, yellow shanks and feet, and reddish-brown dorsal plumage. In the crosses of turkey males and chicken hens the fertility on the average was only three-fourths of one per cent and none of the embryos lived beyond the third day of incubation.

SUGGESTED READING

To obtain an understanding of the fundamentals underlying poultry breeding it is suggested that the turkey breeder read one or more of the following:

"Practical Poultry Breeding" by D. C. Warren (1953), pub. by The MacMillan Co., New York, N. Y.

"Poultry Breeding," by M. A. Jull (1940), pub. by John Wiley and Sons, Inc., New York, N. Y.

"The Principles of Heredity," by L. H. Snyder (4th Ed. 1951), pub. by D. C. Heath Co., Chicago, Ill.

"Animal Breeding Plans," by Jay L. Lush (3rd Ed. 1945), pub. by Iowa State College Press, Inc., Ames, Iowa.

"Family Merit and Individual Merit as Bases for Selection, II," by Jay L. Lush, The Amer. Naturalist, 81(5): 362-379, Sept., 1947.

"Simplified Poultry Breeding," by F. A. Hays and G. L. Klein (1943), pub. by Poultry Tribune, Mt. Morris, Ill.

"Yearbook, U. S. Dept. of Agr."
Better Plants and Animals I, 1936.
Better Plants and Animals II, 1937.

"Genetics—The Breeder's Tool," by I. L. Kosin. Turkey World 26(9):18, 81-87, Sept. 1951.

"Genetics of the Fowl," by F. B. Hutt (1949), pub. by McGraw-Hill Book Co., New York, N. Y.

"The New You and Heredity," by A. Scheinfeld (1950), pub. by J. B. Lippincott Co., Philadelphia, Pa.

"Principles of Commercial Poultry Breeding," by I. M. Lerner (1951) Calif. Exp. Sta., Berkeley, Calif., Manual No. I, pp. 1-47.

CHAPTER 12

MANAGEMENT OF BREEDING STOCK

OUTLINE

FERTILIZATION AND DEVELOPMENT OF THE TURKEY EMBRYO

A fertile egg is one in which the egg cell or *ovum* of the female and the spermatozoon of the male have made a living union. Each normal unfertilized egg yolk or ovum has on its surface a germ spot (germinal disc, or blastodisc) in which is located the chromosomes of the dam and from which the embryo starts its development. The lesser density of the blastodisc and of the white yolk underlying it, causes the yolk to turn so that the disc remains on the upper surface of the yolk as long as the latter remains undisturbed. The yolks are produced in the hen's ovary which is attached to the back just to rear of the lungs. When the yolk completes its development it is released directly into the oviduct or into the body cavity (ovulation). In the latter case it usually is immediately picked up by the expanded forward end or funnel of the oviduct. If the hen has been previously mated either normally by the male or artificially by insemination, there will be spermatozoa in the funnel. If one of them succeeds in entering the ovum at the blastodisc and uniting with the female pronucleus the egg becomes fertile. At mating the semen is deposited in the outer morning-glory-shaped orifice of the oviduct which the hen extrudes from the vent if she is receptive to mating. Through the action of cilia, waving, hair-like structures in the oviduct, the spermatozoa and possibly other components of the semen are conveyed the length of the oviduct to its forward end or funnel where fertilization takes place. Several spermatozoa may enter the ovum but only one takes part in the actual fertilization. The remaining spermatozoa migrate to the margin, undergo cleavage, and then disintegrate.

The result of this union or fertilization is the blastoderm which immediately starts to grow while the egg is being formed in the hen's body. By the time the egg is laid, about 24 to 26 hours after ovulation, considerable growth has taken place, resulting in a blastoderm that is circular, smooth-edged, dull white, about one-eighth inch in diameter, and composed of thousands of cells arranged in a thin layer. This development in the unincubated egg cannot be discerned by candling but if the unincubated egg is broken and the fertile disc examined it will be seen to consist of a regularly outlined, wide, circular band or collar of dull white embryonic tissue, the periphereal *area opaca,* which encloses a center space, the semi-

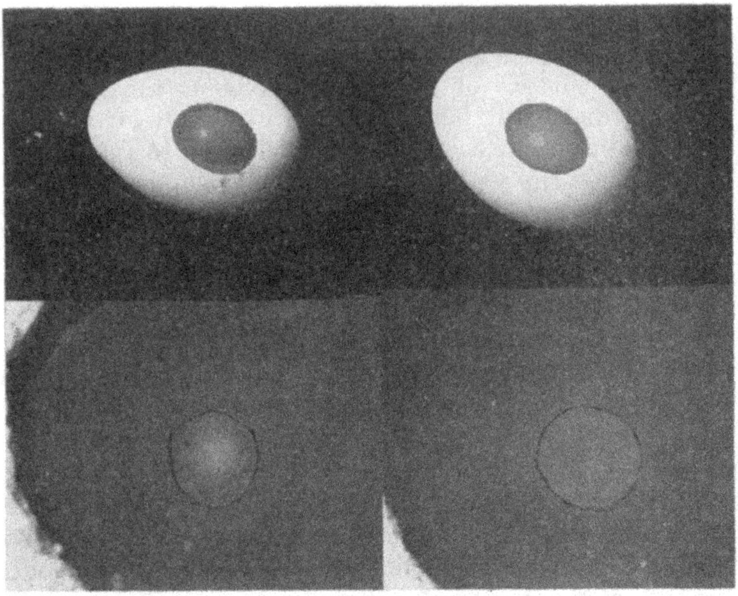

Fig. 12.1. Germinal discs of fresh turkey eggs (on the left infertile, on right fertile). The top row shows sides of shell broken away for photographing. Lower row shows enlarged discs (the small spots near the center of upper circles). This is the site of the embryo; all else is food and protection. (Photo from U.S.D.A.)

translucent *area pellucida*. The nucleus of Pander, a mass of white yolk material, usually appears in the center portion of the area pellucida, although it actually is under it, in the form of sizeable rounded dull white spot which appears to fill, partially to almost completely, the center area as shown in Fig. 12.1. Fertile eggs with embryos dead or dying (moribund), *early dead germs*, have a good-sized area of china-white dead tissue inside of or cutting into the outer collar of the area opaca. In an egg that has not been fertilized, the germinal disc starts to disintegrate while the egg is being formed in the hen's oviduct so that after laying (oviposition) the infertile disc consists of a conspicuous irregular-shaped, china-white vacuolated spot about 1/16 inch in diameter which may or may not be bounded by an inconspicuous thin whitish ring (the periblastic ring), which has a diameter of about 1/8 inch—about the same as that of the blastoderm of the fertile egg just after oviposition. Typical discs of infertile and live-germ eggs are shown in the figures. If the unincubated eggs are broken and carefully examined in line with the above discussion, an accurate estimate of fertility may be obtained. A large hand magnifying glass is useful in examining the discs. In the infertile egg, look for the lacunae (vacuoles or holes) which are characteristic and fairly conspicuous under a magnifying glass.

If after laying, the fertile egg is cooled to below 80° F. (Funk) development ceases and the blastoderm becomes dormant (Edwards reports 68° F.). If the holding temperature is maintained within favorable limits, somewhere between 50 and 60° F., the blastoderm may remain alive for several weeks. However, if after laying, the fertile egg is subjected to long-continued holding temperatures much below 50 or much above 60° F., the ability of the blastoderm to remain alive and strong is impaired and hatchability decreases rapidly as the holding period is extended. The moving air temperature at which reasonably normal development takes place is within the short range of about 98 to 100.9, and for optimum results it must be tween 99 and 100° F. (See Chapter 14 for egg holding and incubation practices.)

When eggs are placed under conditions favorable to incubation, the fertile ones start to develop immediately so that after a relatively few hours of incubation the germinal disc, frequently called blastoderm, has increased considerably in size. After 2 or 3 days of incubation there is an influx of water from the albumen into the yolk which gives the yolk the appearance of being mixed with water, which actually is the case. Such yolks show a tendency to settle to the bottom of the egg and to appear crescent-shaped when the large end of the egg is held up to the candler. As will be pointed out later, there sometimes occurs a parthenogenetic growth of extra-embryonic membranes in infertile eggs that also causes the yolk to appear as described above.

Measurements of the blastoderms of fertile Beltsville Small White eggs at the Beltsville Station indicate averages as follows for normally developing embryos.

Table 12.1—Blastoderm Development

Hours of incubation	Diameter of disc	
	Millimeters	Inches
12	4.5	0.18
18	7.0	0.28
24	8.5	0.33
30	9.5	0.37
36	12.5	0.49
42	17.2	0.68
48	26.0	1.02

DETERMINATION OF FERTILITY AND PARTHOGENETIC DEVELOPMENT

When eggs are broken out and examined after 24 to 48 hours of incubation there is seen to be considerable variation in the diameter of normal blastoderms. Those blastoderms that are considerably under the diameters indicated in the tables above are likely to contain weak embryos.

If turkey eggs are candled carefully between 24 and 48 hours of incubation, a fairly accurate estimate of fertility may be obtained without having to sacrifice the eggs. The operator should wear blue-green sunglasses in a well blacked-out room and use a powerful candler gently rotating the eggs so as to see the dark, apparently floating blastoderm, which is indicative of fertility. The undeveloped germinal discs of infertile eggs are not visible while those of weak embryos appear smaller and less conspicuous than those of the normal strong embryos. The chalazae cords attached to the yolk (to hold it in the center of the egg) sometimes are mistaken for the blastoderm. However, they usually reflect the light slightly which the blastoderm does not do. Sometimes blastoderms and chalazae appear together.

In the authors' experience, even with greatest care, a few infertile eggs will be misjudged as fertile and vice versa; but the two will tend to cancel each other, thus giving a fairly accurate over-all picture of fertility. Because of this apparently unavoidable inaccuracy, it is well, if poults are valuable, to return all eggs for further incubation, to 7 to 10 days, when the eggs containing normal embryos can be spotted with almost 100% accuracy by candling. Greatest accuracy in determining fertility by candling probably is obtained after 30 to 32 hours of incubation. However, if incubation is to be interrupted for any reason such as transportation to a new location or to delay date of hatch, the 24-hour embryo probably is less liable to damage. The 48-hour embryos are much larger than the 24 and 32 hour embryos and are somewhat more difficult to identify by candling. Since the parthenogenetic development to be described later does not start until after about 72 hours of incubation, this phenomenon is not a complicating factor in determining fertility, when eggs are candled after 24 to 48 hours of incubation.

When turkey eggs are incubated for 7 to 10 days, then carefully candled, rotating them gently back and forth, they may be divided into three categories as follows: (1) Clear (infertile) eggs, the yolk fairly well centered in the egg and seen as a rounded floating shadow with no tissue masses in evidence; when broken, the germ spot will appear approximately as described previously for the unincubated infertile egg. (2) Eggs with normal live embryos identified by the decidedly reddish appearance of the egg contents. As the egg is held up to the candler large end upwards into the light, there will be seen the spider-webbed network of blood vessels which appears to occupy about two-thirds of the interior. The crescent-shaped yolk settles toward the small (lower) end and appears quite dark. The embryo itself usually can be seen in the upper portion near the air cell although sometimes it is concealed underneath the air cell. The embryo when visible appears as a dark, loosely anchored, quivering lump which sways freely about if the egg is moved gently back and forth. (3) "Dead germ" eggs in which dark tissue masses can be seen more or less distinct, along with (nearly always) a yolk that is crescent-shaped and settled toward the small (lower) end as the large end is held up to the candler. These eggs contain a lump, spot, or blood-ring, or occasionally an embryo with a network of blood vessels but much below normal size, or a combination of two or more of these.

However, not all of the eggs in category 3 are true dead germ eggs; part of them are almost certain to be non-typical infertiles which can not be separated from the dead germs by candling, except with a considerable element of error. At present it appears that the most accurate method of telling by candling which of these dead germ eggs are really fertilized eggs in which the embryo has died is to so classify (as true "dead germs") only those eggs in which there can be seen a blood ring or part of one, a dark-colored embryo or part of one; a definite network of blood vessels; one or more blood islands; or a combination of two or more of these along with the sunken crescent-shaped yolk. All other "dead germs" that is, those with only faintly discernible dark-colored spots with or without the more or less sunken crescent-shaped yolks probably should be considered as infertile since most of them appear to be unfertilized eggs containing foreign infections (the products of chemical changes not due to embryonic growth) a blood deposit, a meat spot, or the parthenogenetic development to be described later. Any egg in category 3 which is classified as a dead germ should show the more or less settled crescent-shaped yolk without which the egg should be classified as infertile regardless of any "spots" or apparent tissue masses that may be seen.

If the eggs candled out as "dead germs" after 7 to 10 days of incubation (category 3 above) are broken and examined there may occasionally be found infertile eggs that contain brownish white, easily broken-up foreign infection on the surface of the yolk resulting in a blemished appearance of the otherwise normal-appearing infertile yolk. In most of these eggs the typical infertile germinal disc, often outside of the area of infection, can be seen. Another type rarely found is an egg, often a typical infertile, containing a sizeable meat spot or mass of blood originating from the hen's ovary or oviduct. The great majority of the eggs in category 3, however, when examined after breaking will show signs of embryonic development but not all of these eggs, as indicated above, necessarily contain tissues resulting from true embryonic growth in a fertilized egg; some of these "dead-germ" eggs appear to be the result of a parthenogenetic development in an infertile egg which apparently takes place without union with the male element. True embryonic development can be told with fair accuracy from the pseudo development in broken-out eggs by careful examinaiton. Dead germs in eggs broken out after 7 to 10 days of incubation will have the following characteristics; (1) in germs of a size indicating 24 to 48 hours of development, the blastoderm is more or less circular and has relatively smooth edges; (2) in dead germs 3 days or older, a nearly complete blood ring with the question-mark-shaped complete embryo is visible approximately in the center of the blood ring. In contrast, eggs with the parthenogenetic development usually have irregular-shaped germinal discs irregularly outlined and may superficially resemble eggs containing foreign infections. The infection is brownish and easily broken up but the pseudo development consists of a tough whitish tissue mat similar to that of true embryonic growth. On infrequent occasions there will be a well-defined blood ring or sizeable blood island and very rarely an embryo or part of one. The latter type of parthenogenetic development is almost impossible to distinguish from true embryonic growth but its rarity makes it of little practical significance.

The parthenogenetic development appears to start slowly, presumably after 72 hours or so of incubation. At the Beltsville Station (Olsen and Marsden, 1953, 1954) 16 per cent of several hundred Beltsville White turkey eggs known to be infertile contained these pseudodevelopmental forms which normally would be considered as dead germs. About 11 per cent of the eggs from young virgin hens of two strains of Broad Breasted Bronze and one of Light Palm turkeys contained these parthenogenetic forms while about 30 per cent of the eggs from virgin hens of a selected strain of Beltsville Whites contained them. From the latter strain one fatherless male poult was hatched in 1954 and four other fatherless male embryos developed almost to the point of hatching. The one hatched poult died soon after hatching. It was a male and appeared to be normal except for its weight, about 30 grams compared to 50 grams for an average normal Beltsville White poult. Further studies at Beltsville showed that parthenogenesis occurs not only in infertile eggs from virgin hens but also in the infertile eggs laid by mated hens in which the incidence appears to be around twice as great as it is in the eggs from unmated hens.

Of these parthogenetic forms, possibly 99 per cent show tissue mats (but no embryos) of a size that would indicate 1, 2, 3, or 4 days of growth in a normal embryo but a few, perhaps 1 per cent, show germinal disc enlargement with an embryo or part of one. Usually the development does not exceed that of normal 5- or 6-day embryos but, as shown above, development has proceeded much further in a very few cases, even up to the production of a full-term embryo that hatched in a normal manner.

The distribution and extent of parthogenetic development in turkey eggs from mated flocks in general is not known but, in about 2400 Beltsville Small White eggs at the Beltsville Station laid in March and April, 1953, 37% of the 182 (7.6%) candled out at 7 days were infertiles 30% were true dead germs, and 33% contained partheno genetic development. (The eggs were broken and examined in making these tests.) Since the latter were infertile it will be seen that actual infertility in these Beltsville White eggs was almost twice that which would be obtained by the usual candling procedure at 7 days. In eggs laid by hens that had not been inseminated for a considerable period of time, perhaps two months, or had never been inseminated, the percentage of forms ran lower, 16% of total eggs (all of them infertile) as noted above. Parthenogenesis in eggs from mated hens (even though the fertility may be high) appears high enough to be of considerable practical importance in determining fertility. The situation awaits clarification but even in its present state more doubt than normal is placed on our ability to determine fertility which emphasizes the value of selection based on hatch of total eggs set rather than on hatch of fertile eggs. The occurrence of these forms may mean there is no peak of embryonic mortality on the fourth day of incubation as has been believed and that part, probably about half, of the so-called early embryonic mortality as determined by candling, or even by breaking and examining in the usual manner after 7 days of incubation, is true infertility. It seems to mean that the hatcheryman's method of considering all reasonably clear-appearing eggs candled out after 22 to 25 days of incubation as infertile is more accurate than heretofore believed. This may be because most of the slightly cloudy eggs which usually are considered as infertile contain parthenogenetic development and actually are infertiles; not early dead germs as heretofore believed by some investigators including the present authors.

If parthenogenesis proves to be widespread in turkey eggs and to occur at the rate indicated by the preliminary investigations (as indicated above) most of the existing figures on fertility and on hatchability based on fertile eggs will be in need of revision, likewise much of the discussion and figures on fertility contained in this work.

THE INHERITANCE OF FERTILITY

High fertility is characteristic of some strains of turkeys and of certain families within strains, and there are great individual differences among toms and hens. Upward of 90% fertility is common during the normal breeding season in the standardbred Bronze, White Holland, and other standard varieties and tendencies toward similar high fertility are fairly well established in the new Beltsville White and Jersey Buff varieties. The Broad Breasted Bronze in general tends to show relatively low fertility, 85% is considered good. However, some strains of Broad Breasted Bronze are highly fertile, and also certain families within other strains of the large broad-breasted varieties. The Washington Station (Abplanalp and Kosin, 1953) reported evidence indicating that fertility is inherited to a fairly high degree, hence can be improved by selection based on hen-performance data. As an inherited character it should be considered in breeding operations. The use of percentage hatch of total eggs set, as a selection criterion accomplishes this. Differences in fertility due to management and/or nutrition should be separated and considered separately from those which may be of genetic origin. It also is necessary in rating the hen on fertility, to consider the performance of the male to which she is mated. In some cases, low fertility is obviously the fault of the male, therefore may not be charged to the hen.

The North Carolina Station estimated the heritability of fertility in turkey hens (selected standardbred Bronze) to be 30% at the 95% confidence level with the effect of the male eliminated, thus demonstrating important differences in hen performance as regards fertility.

FACTORS AFFECTING FERTILITY

Management may affect fertility, as will be discussed later. Yearling and older toms frequently are sterile or nearly so and frequently become too heavy and clumsy to mate well. For these reasons, among others, toms seldom are retained for breeding beyond their first season. Yearling and older hens on the contrary, seem to be as fertile as the young hens when mated to young toms. Size of egg in itself probably is not related to fertility. Inbreeding has little effect on fertility until it reaches a high degree, 50% or more coefficient of inbreeding.

A premature molt of breeding toms is at least partly responsible for the commonly observed seasonal decline in fertility according to the work at the Beltsville Station (Olsen and Marsden, 1952). That overlighting of males hastens the premature molt in turn is indicated. The synthetic drug thiouracil, fed continuously at 0.1 or 0.05% was found to delay the molt and to prevent the usual seasonal loss of weight of turkey breeding males but when given continuously to mated flocks it also adversely affected fertility and hatchability. Feeding the males 0.1% thiouracil in the total diet on alternate weeks when they were not with hens (which received no thiouracil) appeared in one experiment to have possible practical application, as a means of prolonging the breeding season of toms. However these results were not verified in a second test.

In a study of fertility in White Holland breeders, the Pennsylvania Station (1951) reported that "preferential mating did not contribute toward infertility, rather the source of infertility was low effectiveness of males in mating accompanied by a reduction in sex drive of the female following uncompleted matings that was comparable to that following completed matings and (b) low sex drive in some females. Effectiveness of various males in naturally inseminating the hens ranged from 15 to 97% of the attempted copulations. In a group of 12 females representing a wide range in sex drive, the average interval between matings following a completed copulation was 9 days (range 3 to 17 days). The average interval for the same hens following incomplete copulations was 7 days (range 2 to 12 days). Limited observations of 100 other females supported those data.

Mating frequency, controlled mainly by the hen, and mating efficiency, mainly a function of the tom, were the most important factors involved in determining fertility in Jersey Buff breeders, according to the Massachusetts Station (Poul. Sci., 1953). During a 23-week period, January 29 to June 8, 1952, hens averaging 95 to 100 per cent fertility mated 22 times and 78 per cent of the matings were completed. Hens with 85 to 94.9 per cent fertility mated 20 times with 63 per cent of the matings completed. Hens with 70 to 84.9 per cent fertility mated 13 times with 59 per cent completed. Hens 8 to 69.9 per cent fertile mated 11 times with 48 per cent completed. The best tom studied made 208 matings, completed 88 per cent of them and produced 98 per cent fertility. The next best tom made 287 matings, completed 75 per cent of them and produced 97 per cent fertility. The poorest tom made 195 matings, completed 48 per cent and produced 71 per cent fertility. The average per cent of matings completed was 68 and average fertility was 86 per cent. Of total matings, 27 per cent occurred during the first flush of mating prior to egg laying. Broodiness in this stock did not affect the total number of matings or the fertility of the eggs. Number of eggs laid did not affect frequency of matings or fertility, neither did the afternoon room air temperatures.

Fertility may be adversely affected by the weather if the turkeys are not protected against extremes. Shade and well-ventilated roosting and nesting quarters are necessary in hot weather; likewise protection from protracted rains and from sleet and snow. Extremely cold weather may cut down mating activity somewhat but if the males are sexually mature they usually mate quite well in spite of bad weather, provided they are given protection from it.

A rapid increase in testicle size and proliferation of the sperm forming tubules results within two weeks after toms are exposed to artificial light in the winter. Hence toms should be lighted two or three weeks ahead of the hens. Sexually active toms should be placed with hens even before the latter become sexually active. This may lessen early-season infertility which may result from failure of the hens to be successfully mated in the period before they start to lay. The sex drive or tendency to mate on the average is less pronounced among the hens after they start laying and a few hens apparently do not mate at all after the first few weeks of the breeding season. It also seems quite probable that the hens may share in the blame for lowered fertility and hatchability toward the end of a long breeding season due to the drain on their vitality and reduced sex drive.

Protection from the cold appears to be desirable where winter production of hatching eggs is sought in the northern States. Large matings of several hundred to a thousand hens are believed by some breeders to be more conducive to high fertility than medium-sized matings of 50 to 200 hens. It should be kept in mind that turkeys cannot mate successfully on steeply sloping ground; fairly level ground is an advantage.

Diseases and parasites affect fertility adversely. Lowered fertility and hatchability have been observed as a result of pox or sinusitis infection.

The following table summarizes the suggestions on fertility:

Table 12.2—Causes of Low Fertility

Cause	Remedy
Extremes of temperature and other weather hazards.	When winter production of hatching eggs is sought in cold climates, provide well-built houses and prelight the toms for 3 weeks before they are placed with the hens in unheated but protected quarters. Preheating the males at 65° F., during the prelighting period of 2 or 3 weeks may be beneficial, according to the Washington Station. Plenty of shade and well ventilated nesting and roosting facilities are desirable if good fertility is to be secured in hot weather. Provide protection from rain, sleet, and snow.
Debility due to parasites and disease	Make examinations for worms, lice, and other parasites. If present, take proper steps to control them. Prevent diseases and parasites through sanitation, good management, and appropriate vaccinations.
Hereditary tendencies of males and females	Replace the offending individuals if they can be located. Do not breed from them or select their offspring as breeders.
Fighting or interfering of males in large matings	Provide a good-sized range lot or roomy confinement quarters, and insofar as practicable, distribute feeders, waterers, and nests about the area. Alternating males may help. Always put the males together in the flock well in advance of the season. Remove quarrelsome and ineffective males. In restricted quarters solid "baffle" boards may help.
Fighting, interference, and distraction through fences	Separate all matings from each other and from all other turkeys by a fence solid three feet up from the ground or by an empty yard, about 10 feet wide or wider.
Too many or too few males	In large matings select heavy breed males at the rate of 8 or 10 hens to one male, divide in two groups, and alternate them weekly or semi-weekly. In this manner the hen-tom ratio of the mated flock is 16 to 20 to 1. Select small-type males at the ratio of 14 to 1 and leave them all in the flock, alternating usually not being necessary.
Molting of males, causing premature decline of fertility	When males enter a general molt their fertilizing ability is impaired. To delay molting, use proper light intensity, keep feeding and management uniform and adequate.
Injury to females	Use canvas saddles on the hens or if this is not possible or advisable, trim to smooth bluntness the toenails and, if necessary, spurs of toms. Cut nearly to the "quick," and file smooth. File once or twice during the season, if necessary. The electric debeaker can be used to cut off toenails even with the flesh.

Table 12.2—(Continued)

Cause	Remedy
Nutritional deficiency	Feed the breeders a complete diet including fresh green feed or substitutes for it. Antibiotics are not indicated in breeder diets. They may reduce hatchability.
Sexual immaturity of breeding birds	Use well-matured males that are 8½ to 9 months old when first expected to mate; females 8 to 8½ months. Toms should be prelighted and possibly preheated also, 3 weeks ahead of the hens in cold climates from November through February to bring them into sexual maturity in time to fertilize the first eggs laid. In moderate or mild climates preheating of the males would not be necessary but prelighting is desirable even in stock not given any artificial light.
Overfatness and lack of exercise	There is no evidence to show that this is an important cause of low fertility or low hatchability so far as hens are concerned. However, over-matured toms may become too heavy, fat, and clumsy for best breeding results. Full feeding of a complete diet containing not over 30% corn is recommended.
Preferential mating	There is no evidence to indicate that this exists or is an important factor in fertility.

ONSET AND DURATION OF FERTILITY

Fertility persists after the removal of males for a considerable length of time. The senior author (1934) hatched poults from standardbred Bronze eggs laid 41 days after removal of the toms, and observed apparent fertility in eggs laid 59 days after removal. The fertile eggs obtained between 42 and 59 days were incapable of hatching as all the germs died early in the incubation period. Recent work indicates that a large share of these apparent dead germs probably were infertile eggs containing parthenogenetic development.

The Kansas Station (1937) reported no reduction in fertility the first two weeks after the males were removed. Starting the third week, fertility gradually became lower but was 40% on the 34th day after removal of males.

At the Massachusetts Station (1951) the average duration of fertility of eggs laid by Jersey Buff turkeys was 47.5 days following an insemination. For the first 30 days following the insemination, 96.1 per cent of 466 eggs laid were fertile; of eggs laid 30 to 40 days afterward, 80.8 per cent were fertile, and after 40 days fertility dropped rapidly to a very low level. In these some birds' hatchability held up well for 30 days after the insemination. At the California Station (1949) with Broad Breasted Bronze, insemination with .025 c.c. of semen maintained over 90% fertility for 3 weeks; during the fourth week it declined slowly, dropped rapidly thereafter, and disappeared by the end of the eighth week. When .0125 cc. of semen was used, fertility through 2 weeks was excellent but dropped abruptly thereafter.

The time elapsing between mating and the occurrence of a high percentage of fertility has been observed. In two experiments at the Beltsville Station it required an average of 7 days to obtain normal (over 90%) fertility in small-type crossbred turkeys when a tom was placed with 20 previously unmated hens in laying condition. In a few cases fertile eggs were laid 24 hours after males were introduced. When toms are exchanged it may require up to 2 to 3 weeks for the new male to mate with all the hens in the flock. Eight weeks should be allowed to practically completely eliminate fertility from the tom that was removed; although after five weeks only an occasional fertile egg will be produced from the former male, after he has been replaced by a new male. This time lapse may be cut down to 11 to 14 days by artificial insemination, according to the Oregon Station, two inseminations 11 to 14 days apart, being employed.

The California Station (Lorenz, 1950) found 11.4% fertility in 51 Broad Breasted Bronze hens on the first day following artificial insemination with .05 or .025 cc. of undiluted semen from crossbred toms. On the second day after insemination fertility was 87.2%, which level was held for 3 weeks. During the 22nd to 26th days fertility dropped to 84.1% after which it dropped rapidly, being 50% at 38 days, and "essentially zero" at 57 days. It was stated, however, that "fertile eggs have been observed rarely as late as 72 days after insemination." There is a possibility that these "fertile eggs" may have been the parthenogenetic forms reported by the Beltsville Station (1953).

The Pennsylvania Station (1951) obtained the following results from a single insemination with 0.05 ml. (or cc.) of undiluted pooled semen using several White Holland toms and 19 hens all 36 weeks old when artificially lighted on Nov. 15, 1949. Insemination was done on Jan. 4, 1950. Eggs were set bi-weekly.

Table 12.3—Results of Single Inseminations (Penn. Exp. Sta.)

Days after insemination	No. of eggs set	Per cent fertile	Per cent hatched of fertile eggs
1–7	102	87.2	80.9
8–14	86	97.7	78.6
15–21	69	100.0	78.2
22–28	65	90.8	74.6
29–35	74	89.2	84.8
36–42	86	80.2	86.9
43–49	93	64.5	80.0
50–56	79	20.2	56.2
57–63	80	2.5	0.0
64–70	62	0.0	0.0

The last poult to be hatched was from an egg laid 53 days after insemination. The last fertile egg was laid 59 days after insemination but the embryo died. Eight of the hens were broody once during the 10-week period of the test and 11 were non-broody. There were no significant differences in degree or duration of hatchability and fertility between the broody and non-broody hens. Each of the broody hens remained fertile after her broody period as a result of insemination.

TESTING FOR FERTILITY

The usual method of making these tests is to incubate for 7 to 10 days the first eggs laid, then candle them. The appearance of infertile, dead germ, live-germ, infected eggs and those with pseudo-development have been described. A better and quicker method of testing for fertility when all available poults are needed is after 24 to 30 hours of incubation. The operator should (as previously noted) wear dark sunglasses in a completely blacked-out room and use a well-made candler with a 75-watt bulb. A dark, floating germ spot appearing to be $\frac{1}{8}$ to $\frac{1}{2}$ inch in diameter with an average of about $\frac{3}{8}$ inch indicates fertility and absence of it infertility, as shown previously.

The quickest test of all and probably the most accurate (if the eggs can be sacrificed) is to break and examine the freshly laid eggs. As described previously, the germ spot of infertile eggs is an irregularly outlined, small but conspicuous, china-white spot about 1/16 inch in diameter, which may or may not be bounded by an inconspicuous, irregularly outlined whitish ring with a diameter of about $\frac{1}{8}$ inch. In fertile eggs, the disc is circular, smooth-edged, dull white, and also about $\frac{1}{8}$ inch across, consisting of a regularly outlined wide outer ring enclosing a center space more or less filled with dull white yolk material often in the form of a spot. Fertile eggs with embryos dead or likely to die have a good-sized spot of china-white inside the outer ring. Sometimes 24-day candling, common in hatchery practice, may be preferred. If carefully done, all eggs that do not contain a conspicuous blood ring or a sizeable embryo, may be considered infertile (commonly called **clears**). Those 24-day eggs that appear only cloudy before the candler probably contain parthenogenetic development; hence are infertiles, not dead germs.

FACTORS AFFECTING HATCHABILITY

Hatchability as the term generally is used refers to the percentage of fertile eggs that hatch but it sometimes refers to the percentage hatch of total eggs set. Due to the difficulty in determining true fertility, it is better to base hatchability percentages on total eggs set. Low hatchability may be due to a variety of causes and may or may not be associated with low fertility. Hatchability in turkeys is inherited as noted previously. Two lethal factors were found in turkeys at the California Station. Both were recessive (and not sex-linked), causing death of the embryos which had deformed legs, beak, and head. Abnormally short shanks also were found to be a recessive semi-lethal character, causing the death of a large percentage of embryos. High hatchability is a most important point to consider when purchasing breeding stock. In good flocks from 80% to 90% of the fertile eggs should hatch under satisfactory incubation conditions.

Size of egg apparently has some effect on hatchability. As shown previously, data from the Beltsville Station (1950) indicated that regardless of variety, whether Broad Breasted Bronze or Beltsville White, the hatchability of hens with first-year egg weight averages of between 75 and 80 grams was higher than that of hens with either lower or higher averages. Data from the North Dakota Station (1946) on the hatchability of fertile eggs of 129 young Broad Breasted Bronze turkey hens indicated a similar trend. Based on average egg weight of the individual hens, the following results were obtained.

Table 12.4—Hatchability and Egg Weight (N. D. Exp. Sta.)

Average egg-weight Class	Number of hens	Average hatchability of fertile eggs
Less than 75 gms. (31.8 oz. per doz.)	1	64.6
75 to 79.9 gms. (31.8 to 33.8 oz. per doz.)	5	81.7
80 to 84.9 gms. (33.9 to 35.9 oz. per doz.)	26	82.5
85 to 89.9 gms. (36.0 to 38.0 oz. per doz.)	40	77.2
90 to 94.9 gms. (38.1 to 40.2 oz. per doz.)	24	77.8
95 to 99.9 gms. (40.2 to 42.3 oz. per doz.)	7	65.9

Specific gravity, as determined by the floating character of turkey eggs, is not related to hatchability but is related to shell thickness. Hatchability also was found to be unrelated to degree and color of spotting and to shell texture, whether smooth-shelled or possessing large calcareous deposits. However, chalk-white porous-shelled eggs showed lowered hatchability.

Length of holding reduces hatchability markedly, if holding temperature is above 60° F. and/or the relative humidity low (80% relative humidity with 55° F. temperature recommended). Eggs should be held in wire baskets or well ventilated on trays until their internal temperature reaches the desired 55° F. For holding and shipping, aluminum-lined cases are considered much better than unlined cans but eggs should be brought to the right temperature before casing.

The following table summarizes information available for quickly determining the cause of low hatchability.

Table 12.5—Causes of Low Hatchability

Cause	Remedy
Hereditary tendencies	Obtain breeding stock possessing inherent high hatchability; discard breeding stock that does not possess it. As far as it is possible to do so, make high hatchability a requirement for the selection of breeding stock.
Size of eggs	Where pedigree records are not available, selection of hatching eggs of young hens for medium-small size 75 to 85 grams (31.75 to 36 oz. per doz.), the lower half of the range for small-type the higher half for large-type stock.
Inbreeding	Do not inbreed closely unless individual pedigreeing is done so that high hatchability of parents can be made a requirement for every bird selected as a breeder.
Fighting, interference, and distraction of males	Adjust environmental conditions as suggested for the improvement of fertility.
Debility due to diseases and parasites	See suggestions for the improvement of fertility.

Table 12.5—(Continued)

Cause	Remedy
Nutritional deficiency	See suggestions for the improvement of fertility. Check vitamins A, D, E, and riboflavin.
Premature molt of males resulting in a decline of hatchability	See comments on this subject in the preceding table and discussion of lighting on page 666.
Injury of females	See comments on this subject in the preceding table.
Overfatness and lack of exercise	See comments on this subject in the preceding table.
Immaturity of breeding stock	See comments on this subject in the preceding table.
Improper care of hatching eggs.	Gather 4 times (or more) daily; do not store on large end or in a temperature over 60° or under 48°F. (preferably about 55°). Set every week but if kept longer than 10 days turn once daily. Temperatures between 29° and 47° will not injure hatching eggs unless maintained for more than about 12 hours at a time. Storage on their sides is recommended.
Conditions during incubation	Unfavorable incubation conditions will cause lowered hatchability (see chapter on incubation).

CARE OF BREEDING MALES

It has been suggested that prospective breeding toms should have roosts during the growing period in order to develop their leg muscles and promote better balance and ability to mate well naturally (MacIlraith, Calif. Turkey Growers Assoc., 1953). It also was recommended that breeding males be pre-selected at 16-20 weeks then placed on a non-fattening diet such as a relatively high-fiber growing mash plus oats free choice. The authors suggest that prospective breeding birds be grown to market age on the usual feeding program, then given a standard growing mash plus grain both free choice, the grain to consist of barley alone, or any two or three other grains, not including corn, either each free choice or in a mixture fed free choice with the mash.

Special feeding of active breeding males is not needed under ordinary circumstances. The toms appear to do well on the feed given the hens, even though it be an all-mash laying diet. Reserve toms when kept in a group by themselves, are usually fed a complete breeding or growing diet either all-mash or mash and grain (not corn) all kept before them at all times. When allowed free choice, males will eat more of the scratch grain than the hens will as their requirement for protein is not so high as that of laying hens. At the Beltsville Station, breeding males do very well on the recommended all-mash breeding diet.

Turkey males of strains medium or small in size seldom damage the backs of the hens unless they are mated to comparatively undersized hens or are themselves old or clumsy. Toms of large strains are more likely to give trouble in this respect so the hens usually are saddled. In lieu of this the eight toenails of each breeding male should be carefully trimmed and smoothed. The spurs of the male, if sharp, should be sawed and filed, or otherwise blunted. The toenails may be cut completely off, up to the flesh, and spurs of old males may be blunted, with an electric debeaker. It is best not to clip the wing of a breeding male but where necessary, clipping the inner half dozen primaries of each wing is recommended. While it is not desirable for future breeders, some flocks have the last wing segment cut off or the wing tendons severed (tenotomy or wing notching). Available field reports indicate that these procedures do not impair the breeding ability of toms to an appreciable degree even in large, broad-breasted turkeys.

If tailhead picking of the toms by the hens becomes a problem, the latter should be debeaked moderately or the tailheads of the toms may be painted with pine tar or with pine tar to which bitter aloes has been added. Another good mixture for this and for prevention of other types of flesh picking (suggested by C. G. Lewis) consists of equal parts of tree tanglefoot and cup grease to which a little lamp-black has been added. Toms which are badly picked in the tailhead are useless as breeders and should be removed.

CARE OF BREEDING HENS

The saddling of the breeder hens of the larger strains now is common practice. It is being practiced increasingly with the smaller strains. Saddles of various styles are on the market. Some have an egg-shaped hole at the proper location for each wing, while others have narrow criss-cross strips that go over the wings and around the neck. In one of the most satisfactory types the wing holes are bounded on the forward edge by a free-sliding tape thus allowing ease in installing and removing and preventing choking (see Fig. 12.2). The use of tested commercial saddles of heavyweight material is recommended as regular practice to prevent "blueback" and loss of production in flocks where heavy toms are used. However, proper trimming of toenails of young toms and in addition the spurs of old males usually affords sufficient protection, even with the heavy strains. Saddles with non-expandable wing-holes often cause the loss of a few hens from choking.

Injured hens can be spotted by their lameness or stiff gait. The torn skin should be washed, disinfected with a

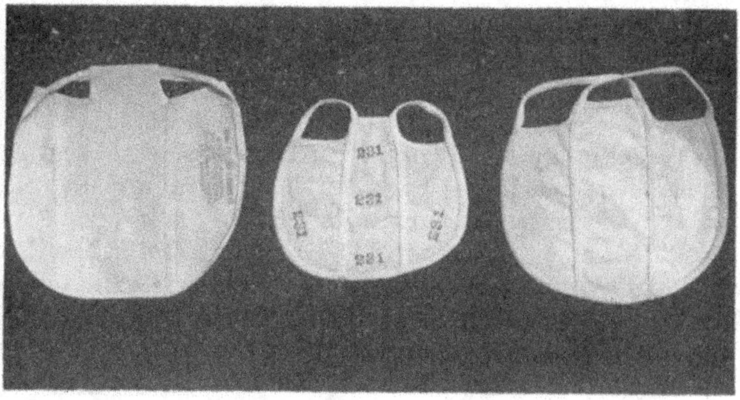

Fig. 12.2. Saddles may be marked with breeding hen's band number.

3% compound cresol or other disinfectant solution, trimmed, and carefully sewed up. The hens usually recover quickly if treated soon after injury. It is a good practice to examine the breeding hens soon after mating starts, to see if torn backs are common. It is still better practice, especially with large strains of turkeys, to saddle the hens or treat the males' feet before mating starts.

Clipping the outer half of the primaries (flight feathers) of one wing of each breeding hen is good practice to follow in many situations; in fact it is almost a necessity in many situations to keep them in the proper quarters. There appears to be no objection to clipping the hens' wings and it is better to clip them than to have mismating occur. Satisfactory results may be secured by clipping the inner seven primaries of both wings which reduces ability to fly but does not handicap the bird too much.

MANAGEMENT OF BROODY HENS

Prompt discouragement of broodiness is necessary if good egg production is desired. The management of broody hens in trapnested breeding pens is discussed in Chapter 13. In commercial and farm flocks the management of broodies should follow the same pattern. The usual procedure is to provide quarters without nests or litter but containing the regular laying feed and water. A wire or slat floor is preferred though bare or lightly littered yards often are successfully used. Some growers keep a few toms in the broody pens to assist in breaking up the hens and insure early fertility. Broodiness is indicated by persistent nesting, hissing, and walking on "tiptoes." However, this is not a reliable nor practical method of identifying the sitters. A good plan for handling broodiness is to provide four equal broody pens. Each evening at sunset all hens found on the nests are driven off. The broodies will return, but the later layers tend to go to the roosts. The returning hens are assumed to be broody and are driven into a broody pen and kept there one day, then driving them into the second pen and so on.

Another plan that possibly works better is to inspect the nests at daybreak while the flock as a whole is confined to a roosting corral. All hens found on the nests at daybreak are considered as broody—the late layers will have left the nests. The broody hens then are driven into one of a series of four broody pens and put through the four-day routine with feed and water before them as suggested above, returning them to the flock after four days. Daily inspection of the nests to identify broody hens may not always be necessary but it undoubtedly results in greater efficiency and less hen-time lost through broodiness. Moving the breeding flock weekly and the use of tip-up semi-trapnests are practices that tend to discourage broodiness according to field reports.

Breaking up broodiness in turkey hens by giving a subcutaneous injection of 200 milligrams of diethylstilbestrol in corn oil per hen was reported as successful by Blakely *et al* (1951). The hens returned to production in a period of time comparable to those given the broody coop treatment. Other methods of using stilbestrol (e.g. in gel form with "pump gun") to break up broody hens have been developed. On the basis of present knowledge the gel should be tried on a small scale on a few birds at the onset of the season; at daybreak only. If the results are satisfactory then flock treatment on all birds found in the nests at daybreak may be used.

ALTERNATING MALES

When flock matings are used, alternating the males is sometimes practiced to remedy low fertility. Providing a tom for every 8 to 10 hens; putting half the males with the hens for 3½ to 7 days, then alternating with the other half of the toms is widely practiced with heavy broad breasted turkeys in some sections. A recent survey in Washington indicated that such alternation in purebred Broad Breasted Bronze flocks was a profitable practice. However this conclusion was not supported at the Oregon Station (Harper, 1951), which reported no benefit from alternating Broad Breasted Bronze males as compared to "stationary" matings. Best results with natural matings were obtained in non-alternated matings with a 1 to 5 (tom to hen) ratio at the start, up to include the first 3 weeks of production, followed by a 1-10 ratio. Average fertility was 84.2 compared to 76.0% with a 1 to 10 ratio throughout and 87.7% with artificial insemination. Where alternating is practiced a sorting gate to eliminate handling the toms is desirable. This slatted gate allows hens but not the toms to pass through to the roosts. The toms are then driven to the bachelor quarters for the rest period and the alternate group driven to the hen quarters. Semi-weekly alternation seems to be favored over weekly when 2 groups of toms are used.

ARTIFICIAL LIGHTING

Artificial lighting may be used to aid in obtaining out-of-season eggs. During fall and winter but not before November 1, lights starting at about 4:00 in the morning so as to provide a 13 to 14-hour day during the shortest days are satisfactory and practical. About a month before egg production is desired the lights may be turned on at the desired hour or they may be turned on gradually by starting a half-hour before natural daylight and moving the time back 15-30 minutes each day until about 4:00 a.m., is reached. The latter method is more conservative. The lights should come on at the same hour until natural daylight replaces the artificial light, not reduced gradually or turned off prior to this time. Usually lights are turned off on June 21, or sooner if the breeders are marketed or eggs no longer needed. Regularity is an essential feature. With electric lighting, a reliable alarm clock or a time switch is all the special equipment needed. Reflectors or a reflecting ceiling may be used but usually are omitted in turkey lighting. Lights should be adjusted so that the light shines evenly over the roosting area and provides an average of 1.7 to 2.5 foot-candles of intensity at roosting bird height, the roosts being on the same level and preferably 6½ feet below the bulbs. A light meter should be used to take readings at various positions in the roosting area. Lights also may be made to shine on the feeders and on the waterers though this is not necessary and may complicate matters in some situations. The ideal arrangement probably would be to have the light intensity the same over the roosting, feeding, and watering areas with feed and water always available. Next best probably is to have the roosting area properly and evenly lighted and confine the birds to it until natural daylight arrives. In either situation it is desirable to confine the turkeys to the lighted area during darkness since they sometimes stray away from the lights into the darkness when permitted to do so. As soon as natural daylight arrives, full access to it should be allowed since incandescent light alone without natural daylight supplementation may not be satisfactory for turkeys.

Distribution of the lights to provide a uniform intensity is most important. If the intensity varies greatly at roosting bird height, it is possible for some birds to be overstimulated while others are understimulated. The closer the lights are to the turkeys the smaller and more numerous should be the light bulbs. Where the distance from bulbs to roosts is a uniform 6½ feet as in the example following, 60-watt bulbs 10 feet apart provide good light distribution. However if the distance is 4½ to 5½ feet it would be better to use 25-watt bulbs and space them closer, possibly 6 or 7 feet apart. If the distance is 8 to 12 feet, larger bulbs of 100 to 200 watts may be used. In any case a sensitive light meter should be used whenever possible to gage the intensity accurately.

As indicated above, the light intensity at roosting bird height and also at the level of the feeders and waterers may vary from 1.7 to 2.5 foot-candles. If the light distribution is even, 2.5 foot candles is satisfactory and will promote a desirable rate of lay without overstimulating the toms. If light distribution is poor however, an average intensity of 1.7 to 2.0 foot candles is safer for males.

The wattage required depends primarily on whether or not reflectors or reflecting ceiling are used and how far lights are from the roosting turkeys. Without reflectors, or reflecting ceiling, 60-watt light bulbs 6½ feet above the roosts should provide 2.5 foot candles. Without reflectors or reflecting ceiling, lights an average of 5 feet above the roosting birds and the floor, two 25 watt lamps will light a 14 x 14 breeding pen or a roosting area the same size. Such rule-of-thumb directions as the above for rigging up lighting systems are at best unreliable—it is far preferable to obtain light-meter readings so as to be sure that the proper intensity is obtained because while hens respond, well and without harm, to a wide range of intensity, the toms are very sensitive and are easily overstimulated by too much intensity resulting in a shortened season of usefulness as breeders.

All lighting equipment must be well protected or suspended freely as the turkeys may fly against it. Gasoline and gas mantle lamps and other sources of light may be used provided they do not constitute a fire hazard. Kerosene lanterns are not satisfactory. A fairly good poultry house and water warmers or flowing water should be used in cold climates. Waterproof lighting fixtures should be used especially if they are exposed to rain.

At the Beltsville Station (1952-54) effective breeding seasons of five to seven and sometimes eight months have been obtained from well-matured Beltsville White breeding toms properly lighted starting about December 1. An effective season of about four months is normally obtained from similar stock not lighted artificially but housed and allowed access to outside yards. Shortened seasons of only two to three months were obtained from toms exposed to too much artificial light intensity. There also are indications that it may pay to judiciously prelight toms for three or four weeks prior to their use on *unlighted* hens. Such prelighting insures against early-season infertility and may help to extend the toms' breeding season.

The Kansas Station showed that while the use of morning lights (4:30 to daylight) on turkeys made winter egg production possible, the stimulation causing it was due only to the increased light, not to increased feeding periods. While it is not necessary to have feed and water before the birds during the early morning hours when lights are on, most growers feel that it is desirable. If this is done, light intensity over feeders and waterers (as noted above) should be the same as that over the roosts. White light and red light are effective while blue light is not. Good housing alone is of no value in stimulating early egg production as compared to no housing at all. However, in the experience of the authors, housing is of value in connection with lights; without lights, housing actually delays the onset of production if the birds are confined. Lighting toms 3 to 4 weeks ahead of the hens is standard practice in the northern areas when turkeys are started under lights between November 1 and February 1. The toms, after 3 to 4 weeks of lighting, same as that provided for the hens, should be placed with the hens when the latter are first lighted.

Lighting turkeys in summer and early fall has recently been tried by breeders in attempts to produce out-of-season poults for the fryer and light roaster trade. In many instances growers have found it difficult to profitably produce summer and fall turkey hatching eggs. The early hatched hens which mature in the long summer days, cannot be induced to attain physical maturity before they lay. They usually are under the normal body weight, have a short producing season and a low rate of lay of small eggs 2 oz. or so per dozen. Methods of overcoming these troubles await development. There is a probability that restricting the hours of light of the developing birds (both male and female) to a moderate amount, say about 8 hours daily, might hold back sexual maturity and at the same time allow body growth to proceed normally so as to attain average weights for the strain at about 30 weeks of age at which time they might respond satisfactorily to about 14 hours of natural or natural plus artificial light. This light restriction could be accomplished easily by confining the birds in a darkened building, allowing them only about 8 hours of natural daylight starting at about 3 months of age on up to about 7 or 8 months, then gradually lengthening the light day, by 30-minute increments, until about 14 hours of light are allowed. Such a program should result in conditions similar to those successfully imposed on normal-season birds that are given artificial lights in late fall or winter.

Artificial lighting properly employed does not shorten the breeding season of the toms as noted above; on the contrary it often extends it. For example, if turkeys are lighted properly so mating starts December 15th, the toms should give 5 to 7 months of continuous good breeding performance. However, if and when the fertility declines prematurely, it may help to replace the original toms by an unbred, unlighted, 8-to-9-month-old group of toms when fertility of the first group of toms starts to declined. In this way it should be possible to prolong the period of high fertility by another two or three months.

Turkey toms not artificially lighted should provide high fertility for an average of about four months of egg production, a somewhat shorter period than is obtained from properly lighted stock. Hot weather in itself may adversely affect fertility; there is little evidence on this subject. This might mean a shortened breeding season in areas where high temperatures occur early in the season. In such cases providing protection from extreme heat might be worthwhile.

HEATING OF BREEDING STOCK

The Washington Station (Kosin *et al,* 1952) reported increased fertilizing capacity over a 5-month season from prelighting Broad Breasted Bronze toms 13 hours daily for 3 weeks and preheating them (at 65 degrees F.) for five weeks. From January 22 to June 25th, when three of these preheated toms were continued under similar conditions (to the end of cold weather with about 25 hens) the average fertility was 90% and the hatch of fertile eggs was 69%, while in another pen where the three toms were preheated and prelighted but placed with about 25 hens in lighted but unheated quarters, the fertility averaged 81% and the hatch of fertile eggs averaged 72%. In contrast to those results a similar mating of unheated but prelighted toms with unheated but lighted hens, produced fertility of 61% and an average hatch of fertile eggs of 62%. A fourth mating of non-preheated but prelighted toms with heated and lighted hens produced an average fertility of 76% and an average hatch of fertile eggs of 71%. The effect of the 65° F. heated environment on the hens however was adverse in that broodiness was greatly increased, resulting in loss of egg production. As a result of this and other experiments, the Washington Station suggests that in cold climates turkey males be placed in quarters heated to 65° F. for 5 weeks prior to placing with artificially lighted (but not heated) hens. The lights should be placed on the toms only for the last 3 weeks of the 5-week conditioning period and should be of moderate intensity to provide a 13-hour day as suggested elsewhere. Heating of breeder-hen quarters, on the basis of the knowledge available, is not indicated due to increase in broodiness, apparently caused by the uniform and mild temperature of the quarters heated to 65° F.

MATING ACTIVITY AND ONSET OF EGG PRODUCTION

It requires about 30 days on the average, to bring well-matured turkey hens into full production by use of artificial lights. Average date of first egg will vary from 30 to 35 days after lighting starts, the length of time depending on several factors including time of year, strain and individual differences, intensity and hours of artificial lighting, and sexual development of the hens at the start of the experiment.

In one series of experiments with White Holland Turkeys, the Pennsylvania Station found the average date of first egg was 20 days after morning lights were turned on (4 a.m.) January 18th, and peak of egg production was reached on February 28th. Broodiness appeared about February 28th. The first hen to lay laid her first egg 14 days after lighting period started and the last hen 30 days. Peak of mating activity was reached within the week just prior to onset of egg production which was four weeks prior to the attainment of peak of egg production. Frequency of mating, on basis of one mating per day per hen, attained a peak of 60% just prior to onset of egg production then dropped rapidly off to an average of about 20% throughout the remainder of the season studied, to March 18th. In another experiment starting December 10th, the same patterns of egg production and mating prevailed but the frequency of matings leveled off to an average of about 30% instead of 20%. Increased mating activity followed broody periods in individual hens. The number of matings between onset of lighting and first egg varied from 1 to 12 in the various hens with an average of 5. Fertility during the first 14 days of egg production varied from 55 to 100% and was not related to the frequency of pre-laying matings. Of 7 toms tested, 5 showed about equal mating ability, making 30 to 35 matings each during the period between onset of lighting and onset of egg production. One male made 23 matings and another only 7. Sexual maturity of White Holland males and females was reached at an average of about 7 months (30½ weeks) of age provided they received enough light stimulation. Early-hatched (February and March) toms attained sexual maturity in early fall, then as days became shorter they regressed toward an inactive sexual status.

Confining turkeys to ordinary poultry houses without artificial lighting appears to delay the onset of egg production in the spring. At the Beltsville Station (1952) in a group of young well matured Beltsville White hens housed but given access to outside yards, the average date of first egg was March 6th (normal for that locality). In another group of hens of the same breeding, age, and treatment, except that they were confined to the same house as the other group, the average date of first egg was April 4th. The California Station reported similar results with confined, unlighted Bronze breeding stock. At Miles City, Montana average date of first egg of Standardbred Bronze young hens not lighted artificially and housed, but given access to outside yards, was March 29 in 1930 (an average winter) and March 24 in 1931 (an exceptionally mild winter).

EGG PRODUCTION

It has been observed that turkeys tend to lay later in the day than do chickens which lay about six per cent more of their eggs in the morning than in the afternoon. The California Station reported that from 1934 to 1938 (at Davis, California), standardbred Bronze turkeys not artificially lighted laid 37 to 43% of their eggs before noon (P.S.T.). The average was 40% and no seasonal trends were noted except that fewer eggs (29%) were laid in the mornings after August 31st. Over the entire season, for the five years recorded, an average of 19% of the eggs were laid up to 10 a.m., 21% from 10 a.m. to noon, 38% from noon to 3 p.m., and 22% after 3 p.m Time of day laid did not influence fertility or hatchability.

The Washington Station (1951) reported the results of a detailed study of egg production in two varieties of turkeys. Beltsville Small White (BSW) and Broad Breasted Bronze (BBB) young turkeys confined to houses and given 2.5 foot-candles of light from 7 a.m. to 6 p.m.

at bird height starting January 10, 1949, were trapnested at hourly intervals between 8 a.m. and 5 p.m., and finally at 8 p.m. (after dark) for four months (Feb.-May, inclusive) and all eggs weighed. In the egg laying clutches (eggs laid on successive days) of both varieties, first egg usually was the heaviest and the following ones tended to decrease gradually in size except the last one which usually was noticeably up in size over its predecessor in the clutch but seldom equalled the first one. In BSW the egg size of the clutch tended to be larger as clutch length increased but in BBB the reverse tendency was seen. In BSW, the average weight of the eggs in the 2-egg clutches was 73.77 grams while that of the 7-egg clutches was 75.98 grams. In BBB the 2-egg clutches averaged 85.30 grams, while the eggs of 7-egg clutches averaged 83.27 grams. Average of all BSW eggs was about 74.50 (31.5 oz. per doz.) and that of BBB, 84.80 grams (35.7 oz. per doz.).

The time required to produce an egg decreased as the length of the clutch increased. In BBB the second egg of the 2-egg clutches took 27.71 hours; the average of the last 7 eggs of the 8-egg clutches was 25.21 hours. In BSW the figures were 27.55 and 25.13 hours respectively. In both breeds the time required to produce the second and last eggs of the 4- to 8-egg clutches invariably was greater than that required by the intervening eggs, the last egg requiring considerably more time, by 1 to 3 hours, than that required by the preceding egg. The interval between the third and fourth egg in the 5- to 8-egg clutches averaged the shortest, being slightly under 24 hours (23.60 and 23.69 hours in the 7-egg clutches of the BSW and BBB varieties respectively.

The average hour of lay for both varieties was 1:45 p.m. for the four month period, but during February the average for BBB was 1:57 p.m.; for the BSW, 12:56 p.m., almost an hour earlier. For the other 3 months of the period the two varieties did not differ in this respect. Taking the average of the four months, both varieties, about 8% of the eggs were laid between 8 and 10 a.m.; 21% between 10 and 12; 27% between 12 and 2; 22% between 2 and 4; 9.5% between 4 and 5; and 12½% between 5 and 8 p.m.

Length of pause between clutches affected the hour of lay of the first egg of the following clutch. In BSW following a 2-day pause, the first egg of the next clutch was laid at 11:17 a.m.; after a 3-day pause, 11:58 a.m.; and after a 4-day pause, 12:20 p.m. No further retardation was noted following pauses longer than 4 days. In February the first eggs of all clutches in BSW were laid relatively late in the day and then progressively earlier in the day during March, April and May. BBB showed a trend in the opposite direction.

Average clutch length in BSW increased as the laying season progressed, but that of BBB decreased and the decrease was due to the higher percentage of short clutches. As a result, the BBB hens reached their maximum egg production in March followed by a decline, while the BSW hens increased their egg production in April and May. In May, for example, 51% of the BBB clutches consisted of 1 egg while only 25% of the BSW clutches were of 1 egg. Average length of clutch for the season was 2.13 eggs for BBB and 2.53 for BSW. However, the clutch patterns of individual hens showed that the good layers of the two varieties were strikingly similar, likewise the indifferent layers, the differences between breed averages apparently being due to a higher percentage of good layers among the BSW hens. The good layers in many cases showed no long periods of broodiness; instead they changed their normal pattern of long clutches to short ones, the changes apparently being caused by delayed rupture of the yolk follicles in the ovary resulting in a one-day pause.

The data as a whole show a high degree of agreement between the egg laying patterns of chickens and turkeys and between the two varieties of turkeys.

From egg production data obtained with Standardbred Bronze by Marsden (1936) at Miles City, Montana, the following observations were made.

(1) When exposed to electric lights from 4:00 a.m., until daylight beginning the first week in February, the following egg production to June 1 was obtained, broodiness being discouraged promptly:

Young hens (first laying season)	50 eggs
Yearling hens (second laying season)	36 eggs
Old hens (third laying season)	29 eggs
Old hens (fourth laying season)	24 eggs
Old hens (fifth laying season)	14 eggs

(2) Hens in their first laying season not exposed to artificial lighting laid an average of 39 eggs to June 1, and 56 eggs during the entire season.

(3) Turkey hens when fed and managed to obtain maximum egg production did not differ significantly in their egg-laying habits from chickens. Some birds showed much more broodiness than others, just as chickens do, and the same variations in intensity and persistency of production were readily observable.

(4) A study of trapnest records of more than three hundred young standardbred Bronze hens artifically lighted (starting first week in February), and managed for maximum egg production, shows that from the start of production to May 31, inclusive, 48% of the hens laid without a pause in production of five days or longer. The pauses in the rest of the flock were due to broodiness in the great majority of cases. Of the 52% that showed one or more pauses in production, the number of litters* varied from one to four in number (up to May 31). The number of eggs laid in a litter varied from 5 to 63 eggs. One-pause hens averaged 27.5 eggs in their first litter; two-pause hens, 20.0 eggs; and three-pause hens, 10.8 eggs. The second litter of the two-pause hens averaged

*__Clutch__ refers to the eggs laid on successive days; __litter__ refers to the eggs laid between pauses (usually due to broodiness) of 5 days or more; __pause__ is a period of 5 or more days during which no eggs are laid; __cycle__ refers to seasonal egg production such as the spring cycle.

15.5 eggs; of the three-pause hens, 13.0 eggs. Three-pause hens averaged 9.3 eggs in their third litter. The fourth and succeeding litters were not measured as trapnesting ceased June 1st. Pauses in production showed great variability, the first pause (when more than one was observed) being shorter than succeeding pauses. The first pause averaged 17.5 days in a group of hens pausing once before June 1st; 14.0 days in a group of hens pausing twice; and 8.9 days in a group pausing three times. The second pause averaged 15.3 days in the group of two-pause hens and 10.6 days in the group of three-pause hens. Data on third and succeeding pauses were not obtained. One hen laid 82 eggs from start of laying to June 1. Records of more than 65 eggs were frequent.

(5) Trapnest records of 171 standardbred Bronze turkey hens in their first laying year, fed and managed for egg production and not lighted, showed that on the average the clutch (eggs laid on successive days) was 2.39 eggs. Of the 2,465 clutches recorded, 37.7% were one-egg clutches, 26.0% were two-egg clutches, 17.2% were three-egg clutches, 9.3% were four-egg clutches, 4.5% were five-egg clutches, and 3.0% were six-egg clutches. Clutches of seven to fourteen eggs were of infrequent occurrence. One clutch of nineteen eggs was observed. On rare occasions hens were observed to lay two normal eggs in one day.

That year-long egg production can be obtained from turkeys with the aid of artificial lighting and modern methods of feeding has been demonstrated. The Purina Mills Experimental Farm obtained the following egg production from a flock of 65 standardbred Bronze turkey hens (hatched in May, 1936) during the 1936-37 season:

Jan. %	Feb. %	Mar. %	Apr. %	May %	June %	July %	Aug. %	Sept. %	Oct. %	Nov. %	Dec. %
26.6	30.7	30.1	43.6	51.1	46.0	28.9	21.7	13.9	25.8	27.0	29.0

Eggs were hatched every two weeks from these hens and the average percentage hatch of all eggs set was 65.7% for the year. The lowest individual hatchability was 43.1% and the highest 83.6%. Morning lights were used in such a way as to supply 12 to 15 hours of light each day, regardless of season. Average production for the year was 123 eggs. During the calendar year 1939, a flock of 110 Bronze and White Holland hens averaged 160 eggs each and, along with 11 toms, ate an average of 183.1 lbs. of feed per bird. Artificial lights were used to provide about a 13-hour day. In 1940, one pen of 29 standardbred Bronze hens (with 2 toms) showed an average annual fertility of 88.0% and a hatchability of fertile eggs of 82.9%.

In 1940-41, some interesting data on year-round production, fertility and hatchability were reported by the Purina Experimental Farm as follows. Standardbred Bronze turkeys were lighted 13 hours daily whenever natural light did not provide that much, and carefully managed to obtain best results. The highest yearly record was 251 eggs and several records between 200 and 215 eggs were made.

Table 12.6—Production, Fertility and Hatchability of Bronze Turkeys (Purina Farm Data)

Month	Production	Fertility	Hatchability of fertile eggs
August, 1940	38.7%	81.1%	81.6%
September	18.9	81.2	69.8
October	20.4	76.0	89.5
November	30.6	76.3	79.3
December	42.1	88.1	78.9
January, 1941	63.0	88.2	77.0
February	61.2	89.6	79.6
March	65.4	92.3	76.1
April	77.0	82.7	83.4
May	69.1	83.2	78.8
June	54.7	82.6	75.0
July	42.6	73.5	65.2
Average for year	47.4% or 173 eggs per hen	82.95%	77.1%

In 1941 and 1942 at Beltsville, Md., young turkey hens given morning lights (from 4 a.m.) starting January 15 produced, to June 1, as follows: Standardbred Bronze, 76.1 eggs; standardbred White Holland 63.2; Beltsville Small White, 62.1, and Broad Breasted Bronze, 58.5. The hatch (of all eggs set) was 71.0, 73.5, 73.7, and 50.0, respectively. These results show the high productivity of "old-style" turkeys which was not equalled even by the Beltsville White in this test.

The average egg production for the entire laying season from a flock of unlighted standardbred Bronze turkey females kept through five years at the California Station was 77 eggs, 50 eggs, 44 eggs, 45 eggs and 28 eggs for each of the five years, respectively. The Oklahoma Station (1941) obtained an average of 86 eggs per hen from December 1 to July 1 and an average production of 125 eggs per hen with the maximum production of 205 eggs from one hen in a full calendar year. The birds were exposed to morning lights from December 1 to June 1 from 4:30 a.m., until daylight. The average date of hatch of the hens was April 18th and the average age at first egg was 278 days. The first hen to lay was 237 days old and the last laid her first egg at 324 days.

The Utah Station (1951) reported an average of 43 eggs per hen from young Broad Breasted Bronze hens from January 20 to May 3 when lighted 14 hours per day with morning lights starting January 10. Yearling hens laid 32 eggs. In per cent hatch of total eggs set the young hens were 43 and the old hens 38 with an average egg weight of 87 and 94 grams respectively.

The North Dakota Station (1945) reported the following performance from 60 Broad Breasted Bronze hens divided into six one-male matings from January 1 through May (150 days) lighted 14 hours per day (morning and evening) starting January 3:

Production per hen	60.4 eggs
Fertility	88.2 %
Hatch of total eggs	65.5 %
Hatch of fertile eggs	74.3 %
Average No. of poults per hen	36.4 poults

Two of the six matings had access to outside yards throughout the season, the other four were confined to unheated uninsulated houses providing 11 sq. feet of floor space per bird. The hens averaged 17.04 lbs. each on January 3 when lighting started, and 16.96 lbs., on June 1; the toms lost only 0.8 lb. during the same period. Temperature of the outside air ranged from 18° below zero to 38° above zero during February when fertility averaged over 90%. The lighting in 1945 was considered to be somewhat inadequate. In 1946 with management and the test period the same as in 1945, but with adequate lighting, confined Broad Breasted Bronze breeders at the North Dakota Station averaged 78 eggs for the season while hens allowed access to range lots averaged 65 eggs. Fertility in the two groups averaged about the same, around 90%, and hatchability of fertile eggs the same, around 76%. With lighting started January 3, first eggs were laid on the average, January 24; 41% production was attained by Jan. 31; 72% by mid-February until mid-March when it dropped to 66%, increasing again until 72% was again attained in late April after which it dropped gradually to 55% in mid-June.

At the Oregon Station (1947) well-matured Broad Breasted Bronze young hens (160) were started under 14-hour lighting on January 14, housed in two 20 x 30′ open-front shelters, with access to grass range. Toms had been prelighted 3 weeks. One pen of 80 hens averagin 19 lbs. at date of first egg was headed by 8 toms averaging 35 lbs.; another pen of 80 hens had 8 toms averaging 30.7 lbs. each. From January 14 to May 17 (124 days or about 18 weeks) average production for the 160 trapnested hens started was 50 eggs per hen (7½% of the hens died during the period). Of the eggs laid, 94.5% were normal and sound, with 5.5% cracked, broken, double-yolked, odd-shaped, or dwarf. In settings 10 days apart, fertility averaged 81.8%, hatchability of fertile eggs 65.2%, and hatchability of total eggs set, 53.4%. Fertility was the same in the two pens headed by large and small toms respectively. Of total eggs, 8.6% were laid outside the trapnests. An average of 24.3 poults per hen started was obtained. Fertility started at 81%, rose to 87% by March 18 then decreased gradually to 75% on May 17. Hatch of fertile eggs started at 78% and with minor fluctuations decreased gradually to 59% on May 17. In the laying period of these birds, 103 days, egg production varied from 2 to 79 eggs with 9% of the hens laying over 70 eggs. In fertility, 47% of the hens showed over 90%, while 10% showed under 50%. One hen was completely infertile. In hatch of fertile eggs, 15% hatched over 80% and 20% hatched under 50%.

The Nebraska Station reported a first-year average of 95 eggs per hen from young unlighted Standardbred Bronze stock.

In 1950-51 at Beltsville, Md., 95 young Beltsville Small White young hens laid an average of 100 eggs to June 1 with morning lights from 3 a.m., Dec. 1 to Jan. 2 and from 4 a.m. afterward, to about June 1. Average date of first egg was December 26. Fifty three per cent production was reached on December 30 and 73% on January 2. In contrast, 32 unlighted young Beltsville White hens averaged to lay 55 eggs to June 1. Average egg production for the full year of the 95 (surviving) lighted hens was 142 eggs. This represented an increase through selective breeding of 55% over the 1944 annual production of 91 eggs (Shaklee, *et al,* 1952).

On a monthly basis these same hens laid as follows:

Table 12.7—Egg Production of Lighted and Unlighted BSW Hens (USDA Data)

Month	Number eggs per hen		Percent production	
	Lighted hens (95)	Unlighted hens (32)	Lighted hens (95)	Unlighted hens (32)
December	3.0	0	9.61	0
January	19.8	0	63.84	0
February	18.1	0.4	64.71	0.13
March	20.8	12.7	67.06	40.87
April	19.5	21.6	64.97	71.93
May	19.4	20.4	62.45	65.74
June	16.5	17.5	55.03	58.43
July	13.3	15.2	43.00	49.00
August	7.3	11.8	23.39	37.97
September	3.6	8.7	12.07	29.07
October	0.6	7.5	1.97	24.03
November	0.1	0.5	0.03	0.15
Totals for the year	142.0	116.3	—	—

EGG SHELL COLOR

Turkeys lay spotted eggs except in rare instances. The spotting and a coat of ground color of light yellowish brown (ecru) are believed to be laid on shortly before egg laying. If an egg be obtained at the instant of laying and rubbed with a cloth it is possible to wipe off the ground color and the spotting.

The ground color most often is light yellowish brown but varies from an off-white (rarely pure white) to a medium yellowish brown, sometimes with a purplish cast. Spotting most often is what may be described as moderately thick but varies from no spots at all (rare) to very thick. The color of the spots usually is a medium to dark brown but varies from very pale brown to very dark brown. There is sometimes a purplish cast. Size of spots varies from a very fine stippling to a decidedly coarse spotting. The average is a combination of medium and small spots. The Kansas Station has shown that turkeys on a diet low in vitamin A produced eggs lighter in color and with much less spotting than normal.

At the Beltsville Station it has been observed the eggs of Beltsville White hens that are lighted starting in December average considerably lighter in color and are less heavily spotted than eggs of hens of the same stock that are allowed to come into production naturally without the aid of artificial lighting and they also average slightly smaller according to observations taken in April and May. Shell color tends to lighten as the season progresses; by midsummer the eggs are nearly white with very inconspicuous spotting.

Each turkey hen has a tendency to lay eggs of reasonably uniform appearance and size, but the eggs from each hen do not show sufficient individuality to permit identification for pedigree purposes. However, there is sufficient individuality to be useful in detecting mistakes in pedigree sorting and in correcting some mistakes in trapnesting. Such corrections while they may be counted in recording egg production should never be relied upon to establish pedigrees.

Strain differences in egg-shell color are quite noticeable, some strains laying rather pale-colored eggs while others tend to lay eggs that are noticeably darker. The general shade of color is influenced by the shade and density of the spotting or by the ground-color or both. Color of shell or nature of the spotting does not appear to be related to hatchability.

EGG SIZE AND SHELL TEXTURE

The egg is usually but not always laid small end first. In shape, turkey eggs are much the same as chicken eggs but have more of a tendency to possess a distinct point, as shown in Fig. 14.4 (p. 791). There is seldom any doubt as to which is the small end of a turkey egg, whereas there often is that doubt in chicken eggs. Egg shape and size are inherited and can be influenced by selection. Shell texture usually is quite uniform if feeding is adequate but there is some tendency among individual hens for the shell texture to vary from thin to strong and from smooth to rough or lumpy. The California Station (1940) found that eggs with calcareous lumps hatched as well as those with smooth shells and concluded that the tendency to lay such eggs is inherited. Lumpy shells are unsightly, however, and are often associated with thinness of shell, hence should be guarded against in breeding operations. Newcastle disease in a light form may temporarily cause thin shells, blind checks, and abnormal shapes. Shell quality tends to deteriorate toward the end of a long season, more thin shelled eggs being produced during midsummer and fall.

The variation in egg weight among individual hens of the same strain is quite pronounced and in general heredity seems to be the most important factor controlling egg size. However age at which egg production starts also influences egg size. At the California Station (1946), Bronze hens (standardbred Bronze × Broad Breasted Bronze crosses) that were first exposed to artificial lights at 7 months of age (in November and December) laid December and January eggs averaging 30½ to 31½ oz. per dozen (72 to 74 grams each). By April the weight had increased to 34.7 oz. per doz. (82 grams each). Hens of the same breeding and hatched at the same time that came into production in late February averaged 35.3 oz. per doz. (83.3 grams each) for their April eggs while hens not lighted at all and starting to lay in March averaged 36.0 oz. per doz. (85 grams) for their April eggs. (See egg-weight conversion table in Appendix.)

At the North Dakota Station in 1946, with 14-hour lighting started January 3 on Broad Breasted Bronze hens 8 months old at the time, egg size of first eggs laid averaged 79 grams (33.4 oz. per doz.) increasing rapidly to 84 grams (35.6 oz. per doz.) in early February, then more gradually to 89 grams (37.7 oz. per doz.) in mid-April, and 90.4 grams (38.3 oz. per doz.) in mid-June.

At the Oregon Station in 1947 egg weight of BBB hens started at 86.4 grams (36.6 oz. per doz.) in early February rose gradually to 91.3 grams (38.7 oz. per doz.) on

April 17, and declined gradually to 90.2 grams (37.7 oz. per doz.) on May 17. Average for the 124-day period was 89.1 grams (37.7 oz. per doz.). These hens were about 8 months old when started under 14 hours of light on January 14th.

Double-yolked eggs and pewees are laid infrequently. The former contain two (rarely 3) yolks, and the latter a very small one or none at all. Such eggs rarely, if ever, hatch.

The average weights of fair samples of April and May eggs of 531 April to June hatched standardbred Bronze hens artificially lighted starting about Feb. 5, with average date of first egg about March 8, are shown in Table 12.8. (Marsden, 1934 unpublished data, U. S. Range Livestock Experiment Station, Miles City, Montana.)

Table 12.8—Weight of Standardbred Bronze Turkey Eggs

Classification	No. of hens	Per cent of Total	Average weight	
			in grams	in ozs.[1]
Hens laying very small eggs—below 70 grams	4	1	67.00	2.36
Hens laying small eggs—70-79 grams	127	24	76.17	2.69
Hens laying medium sized eggs—80-89 grams	298	56	84.01	2.96
Hens laying large eggs—90-99 grams	98	18	92.36	3.26
Hens laying very large eggs—100 grams and over	4	1	101.50	3.58
All hens	531	100	83.68	2.95

[1]28.35 grams equal one ounce. See conversion table in Appendix.

The lowest individual-hen average was 65 grams, the highest 103 grams. Average weight of April and May eggs of 18 young hens of a completely unrelated strain of standardbred Bronze averaged 83.9 grams. (Marsden, 1942.) From these data it may be estimated that the spring eggs of the majority of young standardbred

Bronze hens are within the limits of 80 and 90 grams and that the average weight of all eggs is nearly 84 grams—close to 3 ounces each. There are marked differences in various strains, however. The authors have noted that certain strains of White Holland turkeys lay eggs averaging 88 grams midway during the first year's production. The minimum weight for turkey hatching eggs (not including small-type turkeys) has been set at 30 oz. per doz. (2.5 oz. or 70.9 gms. each) by one western cooperative association, and by the National Turkey Improvement Plan. (See Appendix for Plan details.) For hatching eggs shipped interstate the small-type minimum is 28 oz. per dozen.

Large-type turkeys tend to lay larger eggs than small-type turkeys. For the latter (midway in the hen's first season) 76 grams (32 oz. per doz.) may be considered as average with a desirable range of 70 to 82 grams, a minimum of 66 grams (2⅓ oz.; 28 oz. per doz.) and maximum of 85 grams (3 oz.; 36 oz. per doz.). For Broad Breasted Bronze these figures would be respectively, in grams 87.5 (37 oz. per doz.) 75 to 95, 70, and 100. (See conversion table in Appendix.)

The California Turkey Growers Association (1951) found that 14% of the eggs from their flocks (mostly Broad Breasted Bronze) laid Jan. 15 to May 30 weighed 30 to 34 oz. per dozen and were classified as small; 52% weighed 34-38 oz. per dozen and were classified as medium; and 34% weighed 38-42 oz. per dozen and were classified as large. The average was 36.8 oz. per doz. or 87 grams each. Lower incubator humidity, and four hours earlier setting than for average eggs were recommended for hatching the eggs in the "large" class.

Age influences egg weight markedly. Yearling hens lay eggs that are about 7.0% heavier than the eggs they laid their first season. During their third season their eggs are about 2.1% heavier than they were the second season. Stated conversely, first-year weight is about

93.5% of the second-year average, and second-year average is about 98% of the third-year weight.

Because of the tendency for egg characteristics to be inherited, it is wise to select for the desirable characters during breeding operations.

Table 12.9—Dimensions of Typical Turkey Eggs

Size classification	Weight when fresh	Length in inches	Diameter in inches
Very large	106 grams (3.74 oz.)	2.81	2.03
Large	97 grams (3.42 oz.)	2.75	2.00
Medium large	92 grams (3.25 oz.)	2.66	1.96
Medium	85 grams (3.00 oz.)	2.56	1.92
Medium small	80 grams (2.82 oz.)	2.52	1.88
Small	70 grams (2.47 oz.)	2.47	1.78
Very small	64 grams (2.26 oz.)	2.31	1.75

If the shape varies noticeably from normal the measurements in Table 84 will vary accordingly. The measurements for the 70-gram egg correspond closely to Barrows' (1912) figures for average egg size of wild turkeys. Blakey (1937) reports average weights and measurements of 500 random pen-run eggs of wild turkeys raised in semi-confinement as follows: 59.74 mm. (2.35 in.) in length; 45.69 mm. (1.80 in.) in width: and a weight of 69.19 grams (2.37 oz.). Fifty native wild turkey eggs produced on open range with natural feed averaged 58.2 mm. (2.29 in.) in length, 46.1 mm. (1.82 in.) in width, and 66.55 grams (2.28 oz.) in weight. These figures indicate that eggs of wild turkeys tend to be shorter and thicker than those of domesticated turkeys.

Composition of Turkey Eggs

Turkey eggs laid by well-fed stock are of excellent flavor when fresh, fully comparable to those laid by chickens. They are free from all objectionable characteristics except for the shell membrane which is very strong and tough, much more so than that of hens' eggs. Turkey egg albumen is not any tougher when cooked but the yolk tends to be more firm when raw and when cooked (except when hard-boiled). However, the thin, colorless vitelline membrane which surrounds the yolk of the turkey egg is not so strong as it is in chicken eggs, therefore the yolk of uncooked eggs tends to break more easily even though the contents are less fluid. The albumen of turkey eggs can be beaten as easily and the whole turkey egg is fully adaptable in cookery. Two turkey eggs of average size may be used in place of three chicken eggs.

The following data of Asmundson (1943) were obtained by examining and averaging 12 normal fresh turkey eggs:

Average weight of eggs, 88 gms. (3.1 oz.).

Average weight of yolks, 27.7 gms.

Average color of yolks—light orange (color index 13.7; data of Nestler).

Average weight of shell, 7.32 gms.; membranes 1.31 gms.

Average weight of whole albumen, 51.7 gms.

Average proportion thick albumen of total albumen, 52.50%; determined by sieve method from eggs used for yolk-color determinations (at Beltsville Research Center).

Average per cent yolk of whole egg, 31.43%.

Average shell and membranes of whole egg, 9.81%.

Average proportion albumen of whole egg, 58.76%.

The California Station reported the following comparative data for eggs of several species of birds as follows:

Table 12.10—Size of Eggs of Various Species (Cal. Exp. Sta.)

	Turkey	Chicken	Ring-Necked Pheasant	Quail (Lophortyx California)
Number of eggs	196	384	23	5
Length, mm.	65.65	57.29	42.77	35.00
Diameter, mm.	47.47	41.64	33.65	24.19
Diam.-length index, %	72.3	72.7	78.7	69.1
Egg wt., grams	81.76	55.81	26.66	10.34
Yolk weight, grams	25.99	17.21	8.71	3.17
Albumen weight, grams	47.21	32.97	15.15	6.18
Shell membrane, wt., gms.	1.32	0.35	0.22	0.27
Shell weight, gms.	7.25	5.27	2.58	0.72
Shell thickness, mm.	0.35	0.31	0.26	0.21
Spec. grav. of eggs	1.075	1.082	1.085	1.043
Spec. grav. of egg shell	2.174	2.235	2.325	1.843
Per cent yolk of whole egg	31.79	30.84	32.67	30.66
Per cent albumen of whole egg	57.74	59.06	56.83	59.77
Per cent shell mem. of whole egg	1.61	0.63	0.83	2.61
Per cent shell of whole egg	8.87	9.46	9.68	6.96

The weight required to break the egg shell when pressure is applied along its long axis is 1 to 2 kilograms greater (in chicken eggs) than when pressure is applied the short way. Breaking strength (along the long axis) is as follows for several species of poultry (from *The Avian Egg,* Romanoff and Romanoff, 1949, by permission).

Table 12.11—Breaking Strength of Egg Shells of Various Species (Romanoff)

Species	Weight of eggs (grams)	Breaking Strength	
		kgm.	lbs.
Ostrich	1400	55.0	121.3
Peafowl	95	10.0	22.0
Turkey	85	6.0	13.2
Chicken	60	4.1	9.0
Ring-necked Pheasant	30	3.5	7.7
Virginia Quail	9	1.3	2.9

The following information on the chemical composition of turkey eggs was reported by Hepburn and Miraglia (1937).

Table 12.12—Chemical Composition of Turkey Eggs

	Whole egg less shell and shell membranes	White (or albumen)	Yolk
Total solids, per cent	28.64	13.78	51.74
Moisture, per cent	71.36	86.22	48.26
Ether extract (fat) per cent	12.29	.05	33.35
Crude protein, per cent	12.97	11.38	15.24
Ash, per cent	.66	.51	1.29
Dextrose, per cent	.3629	.4255	.2703
Calories per 100 grams:			
Fat	114.3	0.5	310.2
Protein	53.2	46.7	62.5
Carbohydrates	1.5	1.7	1.1
Total calories (per 100 gms.)	169.0	48.9	373.8

WEIGHT AND CONFORMATION CHANGES IN BREEDING STOCK

The following table of average weights of standard-bred Bronze breeders obtained in a period of several years from the records of the U. S. Range Livestock Experiment Station, Miles City, Mont., (1936) shows the changes in weight that may be expected under average conditions of management. These birds were full-fed throughout the period except from mid-June to mid-September, when they were fed a restricted whole-grain maintenance diet and given access to growing green feed. The feed given (ad libitum) from mid-September to mid-November consisted of a mixture of whole grains. Access to growing green feed was provided whenever it was available. During winter and spring, all-mash diets were fed and very little feed was obtained from the yards. The birds were exposed to artificial morning lighting from about February 6 through April.

Table 12.13—Body Weight of Standardbred Bronze Breeding Stock (USDA Data)

Approximate dates of weighing	Average Weight of toms	Average Weight of hens
	Pounds	Pounds
Dec. 12-17 (Av. age approx. 32 wks.)	23.05	13.00
Jan. 9 to 14	24.47	13.47
Feb. 6 to 11	26.63	14.55
March 6 to 11	29.08	15.68
April 3 to 8	28.39	14.68
May 1 to 6	26.65	13.65
May 29 to June 3	24.34	12.83
June 26 to July 1	22.69	11.91
July 24 to 29	22.31	11.58
Aug. 21 to 26	23.83	11.90
Sept. 18 to 23	23.94	12.40
Oct. 16 to 21	28.97	14.90
Nov. 18	29.90	15.50

Some facts brought out by these data are the seasonal decline in weight of both hens and toms and the rapid gain when put on an *ad libitum* fattening diet after being held on a restricted diet during the summer. At the Purdue Experiment Station (1934), Bourbon Red turkeys showed weight changes similar to those reported above when conditions of feeding and management were similar. It was also observed (at the Miles City and Purdue Stations) that nothing was gained by feeding mash during the summer and fall when the birds had access to green pasture and grain unless egg production was desired.

Changes in Body Weight and Conformation of Broad Breasted Bronze Turkeys During the Breeding Season

At the Oregon Station (1950) typical Broad Breasted Bronze stock 8 months old, toms pre-lighted 3 weeks, were placed under lights Jan. 14th and given a 14-hour day. One group of 80 average hens was mated to 8 of the smallest toms available. In each case 4 toms were placed with the 80 hens at one time and were alternated every 4 days with the other group of 4 toms. Standard mash-grain diets were used and each group was housed in a

20 x 30′ open-front house with a concrete floor. Weights and measurements were taken at 2-week intervals to May 20. Egg production started the first week in February and on a hen-housed basis, the average was 53 eggs per hen. Average weight of the hens was as follows:

January 14	17.18 lbs.
January 28	18.08 lbs.
February 11	18.98 lbs.
February 25	18.60 lbs.
March 11	18.15 lbs.
March 25	17.64 lbs.
April 8	17.68 lbs.
April 22	17.30 lbs.
May 6	17.18 lbs.
May 20	17.14 lbs.

These figures show a gain of 1.8 lb. per hen between Jan. 14 and Feb. 11, after which there was a gradual decline. Of the 1.8 lbs. of gain during the first 4 weeks, it was shown that about 1 lb. was due to the expansion of the ovary and oviduct.

Feed consumption of the 21% protein breeder mash was .339 lb. per bird in the mated pens (80 hens and 4 toms); .215 lb. of grain also was consumed, a total of .545 lb. of feed per bird per day. The toms held by themselves ate, from Dec. 20 to Jan. 14, an average of .39 lb. mash and .54 lb. of grain per day, a total of .93 lb. of feed. From Jan. 15 to May 20, the toms ate an average of .51 lb. of mash and .68 lb. of grain (both fed free-choice) or a total of 1.19 lb. of feed per day.

In width of breast 1¾″ above the keel bone, the hens averaged 4.29 in. Jan. 14, 4.15 in. Mar. 25, and 4.19 in. May 20. Keel length averaged 6.36 in. Jan. 14 and 6.28 in. May 20. Shank length (from rear of hock joint to base of outer toe) 5.86 in. Jan. 14 and 6.02 in. May 20.

Tom weights were as follows (in pounds):

Table 12.14—Body Weight of BBB Toms (Ore. Exp. Sta.)

	Jan. 14	Jan. 28	Feb. 11	Feb. 25	Mar. 11	Mar. 25	Apr. 8	Apr. 22	May 6	May 20	Pounds gain or loss during season
8 largest toms	34.59	34.61	34.95	34.39	34.43	33.26	31.93	31.83	31.04	31.45	−3.14
8 smallest toms	30.22	30.40	30.76	30.10	30.12	29.11	28.81	29.20	28.56	29.02	−1.19
4 reserve toms*	31.81	32.90	33.33	33.35	32.98	32.68	32.10	31.95	31.67	32.68	−0.87
Total of 20 toms	32.21	32.65	33.01	32.66	32.76	31.68	30.95	30.99	30.42	31.05	−1.16

*Not mated.

The toms gained an average of only .8 lb. during the first four weeks, then declined gradually, losing an average of 1.16 lbs. from Jan. 14 to May 20. This relatively small weight loss may be due to the 4-day alternation in which the toms get to rest half the time.

Average width of breast of the toms, measured at the widest point 1¾ in. above the keel bone was as follows: 4.19 in. Jan. 14, 4.21 in. Feb. 11, 4.38 in. Mar. 11, 4.13 in. Apr. 8, 4.03 in. May 6 and 4.18 in. May 20. Breast width seemed to be correlated to weight changes, losses in weight resulting in less breast width and vice versa. Maximum breast width was attained on Mar. 11, 8 weeks after lighting started.

Keel length of toms averaged 8.54 in. Jan. 14 and 8.61 in. May 20; shank length 7.50 in. Jan. 14 and 7.88 in. May 20.

THE ECONOMICS OF HATCHING EGG PRODUCTION

From a survey in Washington State, Berryman and Frasier reported that the average cost of producing Broad Breasted Bronze hatching eggs during the 1942 breeding season from 23,000 hens was 16.46c per egg of which 33% was feed; 27% labor; 31% depreciation, death loss, and interest on breeders; 4% use of buildings and equipment; and 5% operations and miscellaneous expenses. Average egg production was 41, producing 22 salable poults; feed consumption per bird in the mated flocks was 72.7 lbs. for the 139-day season. The largest flocks, 400 or more hens, with 10 or less hens per tom and toms rotated, produced 65 poults per 100 eggs set as against 46 to 50 for non-rotated flocks and flocks rotated but with more than 10 hens per tom. Costs per salable poult were 33c for large flocks (over 400 hens), 38c for flocks of 200-300 hens, and 48c for flocks of less than 200 hens; the average was 38½c. Labor requirements per breeding bird varied from 1.9 man-hours in large flocks to 3.9 man-hours in small flocks.

At the Beltsville Station (1944), it was found that feed consumption of mated pens during the breeding season, and during the summer if fed *ad libitum,* may be calculated accurately on the basis of .030 pound of feed per day per pound of live turkey at maximum seasonal weight, irrespective of variety, size, or age of stock. In mated flocks, during the breeding season Jan. 15 to June 1, 19 hens to 1 male, Broad Breasted Bronze turkeys consumed 0.54 lbs. of feed per bird per day; standard-bred Bronze 0.51, White Holland 0.44, and Beltsville White 0.37 lbs. During the winter, prior to production, the consumption is slightly higher.

At the North Dakota Station (1946) during the wintering period of 46 days, Nov. 8, 1945 to Jan. 2, 1946, 59 Broad Breasted Bronze breeding hens and 11 toms hatched May 1, 1945 were housed separately without artificial lighting and were fed grain and mash free choice. The hens, weighing 15.0 lbs. at the start, gained 3½ lbs. in weight and ate .66 lb. of feed per day, 17.3% of which was mash and 82.7% grain. These hens, 27 weeks old at the start, gained weight at the rate of 10.6 lbs. of feed per pound of gain. The toms, the same age, averaging 23.4 lbs. at the start, gained 7.9 lbs., ate 1.11 lbs. of feed per day, 30.1% of which was mash and 69.9% grain, and gained a pound of weight on 7.9 lbs. of feed. During the laying season of 149 days (about 5 months or 21 weeks), January 3 to May 31, the birds were artificially lighted; 44 hens and 4 toms were kept together in a flock, and were given free access to laying mash and fed grain at the rate of 4 to 4½ oz. per bird per day. The hens averaged to lay 81 eggs each which produced 51 poults. The mated flock ate 9.44 oz. (0.59 lb.) of feed per bird per day, of which 60% was mash. Six toms housed separately and given free access to grain and mash, ate 13.12 oz. (0.82 lb.), of which 37½ per cent was mash, per bird per day. Feed consumption of the mixed flock, restricted to about ¼ lb. of grain per bird per day, was 52.8 lbs. of mash and 37.8 lbs. of grain per bird for the 149-day season (0.61 lb. per day).

At the Oregon Station (1947) feed consumption of Broad Breasted Bronze breeders averaged 8.9 oz. (0.555 lb.) of feed per bird per day in mated pens during the lay-

Table 12.15—Summary Factors—Oregon Turkey Hatching Egg Study, 1949

(From Oregon Station Circular 488)

Item	All flocks	Average number of hens in flock			
		100–399	400–699	700–999	1000 or more
Hens per flock (average number)	674	264	522	860	2113
Toms per flock (average number)	82	32	64	106	254
No. hens per tom	8.2	8.4	8.1	8.1	8.3
Total pounds of feed per bird	128.5	128.2	128.2	128.8	128.6
Pounds of feed per bird day	.689	.701	.666	.671	.736
Per cent mash of total feed	52	54	59	45	50
Per cent protein in total feed	15.1	15.1	15.5	14.7	15.1
Total hours of man labor per bird	2.38	3.67	2.72	2.20	1.68
Hours of labor per day for 100 birds	1.28	2.01	1.41	1.15	.96
Date hens were hatched	May 13	May 10	May 18	May 18	June 3
Breeder hens separated, date	Nov. 16	Nov. 15	Nov. 7	Nov. 8	Dec. 4
Age of hens when separated (weeks)	27	27	27	27	26
Breeder hens sold	May 20	May 17	May 18	May 18	May 27
Breeder flock season for hens (days)	186	184	193	192	175
Hens first lighted	Dec. 25	Dec. 24	Dec. 23	Dec. 18	Jan. 3
Toms first lighted	Dec. 13	Dec. 15	Dec. 10	Dec. 7	Dec. 23
Lights turned off	Mar. 29	Mar. 27	Mar. 28	Mar. 22	Apr. 8
Total lighting season for hens (days)	95	94	96	95	96
Average light day (nat. plus art.) (hours)	14.4	14.2	14.2	14.8	14.5
Mating started	Dec. 24	Dec. 23	Dec. 21	Dec. 20	Dec. 31
Started to lay	Jan. 19	Jan. 13	Jan. 15	Jan. 14	Feb. 1
Length of laying season (days)	122	125	124	125	116
Per cent lay	43.9	47.2	46.2	42.8	40.9
Total eggs laid per hen	53.6	59.0	57.3	53.5	47.4
Cull eggs per hen at the farm	3.7	3.6	3.8	3.5	3.8
Salable hatching eggs per hen	49.9	55.6	53.5	50.0	43.6
Rejects at hatchery per hen	1.2	1.1	1.1	1.1	1.5
Total eggs set per hen	48.7	54.5	52.4	48.9	42.1
Fertile eggs per hen	34.7	39.5	37.7	35.6	28.7
Fertility (per cent)	71.3	72.4	71.9	72.9	68.2
Death loss of hens (per cent)	7	5	4	9	8
Death loss of toms (per cent)	14	15	13	16	14
Average depreciation per bird	$5.20	$5.06	$4.67	$5.13	$5.94
Cost of production per hatching egg:	Cents	Cents	Cents	Cents	Cents
Labor costs:					
Operator's labor, direct	2.53	4.29	3.64	1.84	.81
Operator's labor, indirect	1.17	2.66	1.26	.77	.82
Unpaid family labor	.67	1.06	.85	.74	.16
Hired labor	1.49	.30	.89	1.63	2.75
Total labor cost	5.86	8.31	6.64	4.98	4.54
Total feed cost	13.86	12.95	12.87	13.45	16.19
Total miscellaneous cost	1.34	1.34	1.31	1.30	1.43
Total depreciation	12.40	10.74	10.37	12.16	16.43
Total interest	.65	.57	.59	.66	.79
Total gross cost	34.11	33.91	31.78	32.55	39.38
Less total credit	1.23	1.07	1.18	1.26	1.33
Total net cost per egg	32.88	32.84	30.60	31.29	38.05
Total net cost per breeder bird	$14.64	$16.30	$14.59	$13.94	$14.79

ing period, while "resting" (alternated) toms ate 8.16 oz. (.51 lb.) of mash and 10.88 oz. (.68 lb.) of grain per day, both fed free choice, a total of 1.19 pounds.

A survey conducted by Mumford of the Oregon Station (1950) involving 51,844 turkey hens in 284 flocks, practically all Broad Breasted Bronze, for the 1949 season yielded much valuable information concerning the production of turkey hatching eggs. Some of the results are shown in Table 12.15.

On an individual flock basis it was found that turkey growers who obtained fewer than 40 salable hatching eggs per hen produced them at an average cost of 40c per egg whereas the growers who obtained 70 or more eggs per hen produced them for 24c each. Profit per egg depended largely on how many fertile eggs were obtained per hen. It took 35 fertile eggs to break even. The returns to the operator per hour for this labor depended upon whether the hens laid fewer than 30 or as many as 50 fertile eggs each. The average return for all 77 flocks was $1.46 per hour to the operator for his labor and management. The four best operators earned $4.16 per hour, the 16 poorest lost $0.73 per hour. On basis of hatching eggs sold, the average profit (77 flocks) was 0.76 cent per eggs The four best operators made a profit of 9.24c per egg and the 16 poorest lost 4.04c per egg sold.

In Illinois, according to surveys conducted in 1945 1946 and 1947 (Ill. Farm Econ. Apr. 1949) average production of marketable eggs per Broad Breasted Bronze hen was 32 and average hatchability of all eggs set was 54.6%. The breakdown of costs of producing hatching eggs was as follows:

Cost Items	Per cent
Feed	52
Man labor	11
Mortality	13
Buildings and equipment	5
Breeding flock depreciation	6
Other costs	13

Following are data from Illinois farm surveys (Mimeo. Ext. Leaflets U. of Ill., 1945, 1946 and 1947):

Table 12.16—Factors in Hatching Egg Production (Ill. Data)

	1945	1946	1947
Cost of producing a hatching egg, dollars	0.155	0.211	0.280
Cost of producing a poult, dollars	0.377	0.446	0.580
Net cost of keeping a breeder hen, dollars	6.62	5.99	7.37
Eggs produced per hen during the hatching season, number	41	30	26
Feed cost per hatching egg, dollars	0.097	0.115	0.128
Hatchability of all eggs set, per cent	54.9	54.8	54.0
Mortality in the breeder flock, per cent	9.6	14.4	13.8
Fixed capital in the plant per breeder hen, dollars	2.20	1.52	1.58
Hours of labor per breeder hen, number	172	107	109
Length of time breeders were kept on farms, days	152	126	132
Breeder hens per flock, average number	976	1,266	603
Average profit flock, after charging interest on capital, dollars	6,298	4,584	1,501

A survey in Virginia for the season 1948-49 (Va. Bull. 443, 1950) indicated that the total cost of producing a Broad Breasted Bronze hatching egg varied from 15.3 to 31.3c with an average of 22.3c; total cost per Beltsville Small White hatching egg varied from 13.4 to 22.2c with an average of 16.2c.

The survey involved four northern Virginia counties, three in the Shenandoah Valley with one adjoining, and thirty flocks, 22 Broad Breasted Bronze and 8 Beltsville Small White. The average number of breeding hens started per flock was 402. Feed was the largest single item of cost. From time of selection in late fall of 1948 to marketing in late spring of 1949, 11 Bronze flocks required 148 lbs. of feed per hen purchased; average of two White flocks 102 lbs. per hen purchased. Mash or pellets comprised 54.2%, grain 43.7%, and milk products 2.1% of the diet of 11 Bronze flocks; and 61.3%, 34.2% and 4.5%, respectively of that of 2 Beltsville White flocks. Per 100 marketable eggs produced, the Bronze required 325 lbs. of feed, 6.44 lbs. of grit, and 7.06 lbs. of shell; the

Whites 165 lbs. of feed, 13.03 lbs. of grit, and 13.02 lbs. of shell.

Flock depreciation included mortality and lowered price received per pound. It was the second ranking item in cost of production of eggs. Ninety per cent of the hens originally purchased were sold. The Bronze hens averaged 14.2 lbs. at the time of selection (at market age) and 16.4 lbs. when sold at the end of the breeding season. The White hens averaged 8.3 lbs. when selected and 9.4 lbs. when marketed. The value per pound of Bronze hens when selected was 54.5c; when sold it was 41.8c. The White hens were worth 67.4c per pound when selected and 34.2c when sold. Net flock depreciation for Bronze was $148.80 per 100 hens purchased compared to $318.16 for White.

All labor was charged at 50c per hour; 90% of which was family and 10% hired. In all 30 flocks labor breakdown was as follows:

Table 12.17—Labor Distribution (Va. Exp. Sta. Data)

Type of job	Hours of labor per		
	100 hens bought	100 poults produced	Percent
Chores	64.8	2.19	38.1
Gathering eggs	62.8	2.13	37.1
Marketing eggs	14.4	.49	8.5
Selecting and testing breeders	7.0	.24	4.2
Preparing range	6.8	.23	4.0
Hauling supplies	6.2	.21	3.7
Moving range equipment	6.0	.20	3.5
Marketing hens	1.6	.05	0.9
Totals	169.6 hrs.	5.74 hrs.	100%

The pro-rata share of the equipment value chargeable to the turkey breeding flock averaged $284.64 per farm which includes depreciation, interest, etc. Per 100 hens purchased it was $66.06. The cost of the use of this equipment was $98.31 per farm or $22.82 per 100 hens purchased.

Costs and returns are as follows:

Table 12.18—Costs and Returns in Hatching Egg Production

(Va. Exp. Sta. Data)

Item	Average of	
	22 Bronze flocks	8 Beltsville White flocks
Number hens bought	402	402
Eggs produced per hen bought	44.1	61.4
Percent hatch	60.5	60.8
Poults produced per hen bought	26.7	37.4
Pounds feed per hen bought	134.3	101.3
Pounds feed per bird bought	122.0	92.3
Hours labor per 100 hens bought	160.9	193.4
Weeks birds kept in breeding flock	25.8	28.5
Percent mortality of hens	9.2	11.2

	Average of 22 Bronze flocks			Average of 8 White flocks		
	Per 100 hens bought	Per 100 eggs produced	Percent	Per 100 hens bought	Per 100 eggs produced	Percent
Costs:						
Feed	$671.50	$15.22	68.3	$ 486.84	$7.92	49.0
Flock depreciation	148.80	3.37	15.1	318.16	5.18	32.1
Labor	80.65	1.83	8.2	96.70	1.57	9.7
Equipment use	25.27	.57	2.6	16.07	.26	1.6
Interest on value of flock	20.45	.47	2.1	12.37	.20	1.2
Use of motive power	11.57	.26	1.2	28.12	.46	2.9
Miscellaneous	20.58	.47	2.1	27.45	.45	2.8
Use of range	4.43	.10	0.4	6.62	.11	0.7
Total costs	$983.25	$22.29	100.0	$ 992.33	$16.15	100.0
Receipts:						
Egg sales	$1265.41	$28.68	—	$1566.33	$25.50	—
Net gain	282.16	6.39	—	574.00	9.35	—
Return to labor	362.81	8.22	—	670.70	10.92	—

Per cent hatch ranged from 58 to 68 in White flocks and 44 to 84 in Bronze flocks (which were kept a shorter time by about 3 weeks on the average.)

Artificial lighting was used by 15 of the 28 producers involved, 10 starting in December, 1 in January, and 4 in February. In feed required per hen (purchased) per month for the period, the Bronze flock varied from 17 to 31 pounds, the White flocks 13 to 20 pounds, the latter on the average being much more efficient. Failure of the White flocks to show better hatchability than that of the Bronze is contrary to the usual experience of breeders but in spite of this, profits per breeder hen were nearly double for the Whites over the Bronze.

YARDING FOR RANGED BREEDING STOCK

Well-drained land should be selected and any puddles that form around the waterers or elsewhere should be eliminated. For one-male matings, rather long narrow yards, two for each pen or shelter, serve the purpose well and are adaptable to plans for yard rotation and separation of breeding pens from each other by an unoccupied space. In situations where grass grows well, the yards may be about 14 by 80 to 100 feet or of different shape but of equivalent area. They are ample, when alternated back and forth, for a tom and 12 to 20 hens. Each bird should be allowed 85 to 100 sq. feet of yard area in each of the alternate yards. Sufficient exercise and green feed will be obtained, and the size is small enough so that laying in the regular nests will be encouraged and the eggs, if laid in the yard, can be easily found. If trapnesting is done, the yard must be small enough so that the birds keep the green crop fairly well grazed. For flock matings where several toms are kept in the flock at a time it is desirable to use alternate yards, each of which provides about 200 square feet per bird in order that chances for interference of toms be minimized. Tall vegetation and irregular land contours are a decided disadvantage in one-male matings, especially where trapnesting is done but they may be desirable as an aid to obtaining good fertility in large matings. More trouble will be experienced in finding the eggs, however, so many breeders prefer to use level open fields for the breeding stock. A good method of yarding for a breeding flock of any size, including the largest now in vogue is to set up four yards of a size to allow 200 square feet per bird. The breeding flock and its equipment (roosts, feeders, waterers, nests) are then moved into a different lot every four to seven days. This system helps to control broodiness, promotes good sanitation, and makes efficient use of the green feed. Quadruple sets of equipment can be set up to avoid moving it, moving only the breeding birds.

Six-foot, tightly stretched fencing is best for confining breeders, especially in pedigree projects. It provides good protection and prevents mixups. However, five-foot fencing usually can be made to do, especially for flock matings. If gates have a board or iron pipe at the top, a three-or four-foot strip of lightweight fencing should be put on top as an "antifly," fixed to swing with the gate. As a rule, however, gates made of regular five or six-foot fencing with two 1″ by 2″ strips nailed together with the wire between (or a single 2″ x 2″ or 2″ x 3″) to form one of the ends (the other end wired to the gate post) will serve the purpose and are quite inexpensive. They can be used where gates are indicated in the accompanying diagram; (Fig. 12.3). "Antiflies" of three or four-foot lightweight fencing sometimes are needed on all buildings the tops of which come within about seven feet of the ground. However, with one wing of breeder hens clipped even those of the light varieties can do little flying. The flight feathers of one wing or, as preferred by some, the inner seven primaries of both wings of the hens should be clipped but the toms should remain unclipped. Toms of the heavily fleshed modern strains canont fly over five- or six-foot fences even if their wing feathers remain unclipped. Some unclipped toms and the clipped hens can fly up on solid-top gates and on buildings lower than about 7 feet, hence the need of antiflies there.

Double yards for breeders serve several purposes and are almost indispensable on well-equipped pedigree breeding plants. They help to preserve soil cleanliness, they allow a vacant yard between all breeding groups, and they tend to prevent mismating because any bird that gets out of its pen will usually be found in a vacant yard and be returned to the proper pen without loss of egg pedigree. If alternated weekly, the green feed is well utilized and broodiness may be discouraged to some extent.

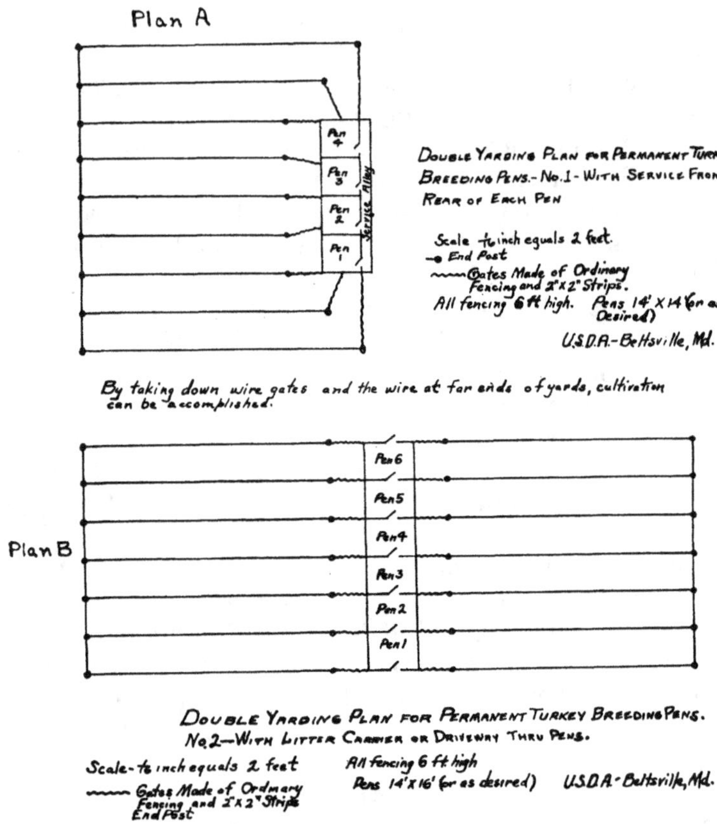

Fig. 12.3. Yard arrangement for breeding pens. (This scale "1/16-inch equals 2 feet" applies on the blueprint; not to this reproduction.)

Figure 12.3 shows two different plans, designated as A and B, for one-male matings during the breeding season. Either plan may be adapted easily to the use of larger flock matings. Plan B necessitates the use of a shelter that can be serviced from the ends or else so constructed that a wagon, truck, or litter carrier may pass freely from one end to the other. The shelter may be divided into breeding pens, each 12 to 14 feet wide with large hinged double doors between the pens, each door equipped with a counterbalance and pulley for automatic closing. All the doors on one side can be made to swing one way and those on the other side to swing the other way, thus facilitating passage through the pens.

Alternating the yards with either of the double yarding plans (A or B) is easily accomplished. There are two turkey doors (large enough for a man to pass in and out—18″ wide by 60 to 70″ high) in each breeding pen, each one opening into an alternate yard. The birds can be forced to use whichever yard the attendant desires by opening the proper door and closing the other one. It probably matters little what system of yard rotation is used, be it weekly, fortnightly, or monthly. Monthly and fortnightly rotations have been used successfully at the Beltsville Station. As indicated previously, weekly rotation of yards may help to discourage broodiness but this has not been proved. When the birds occupy the same shelter continuously and nest there, the yards only being changed, it is doubtful if any such benefit accrues. However, if the nesting facilities are alternated, there is more chance that the weekly change may help to discourage broodiness.

For the sake of soil sanitation in permanent yards it is best that the breeding pens should be used only six or seven months of the year. The other months they should be allowed to rest or be pastured by stock, preferably other than swine or sheep. Since sheep and swine harbor the swine erysipelas organism that sometimes affects turkeys, it is advisable not to pasture them on land to be used by turkeys, or to range turkeys on such pasture if erysipelas has ever appeared.

CONFINEMENT FOR BREEDING STOCK

While keeping turkey breeding birds confined to pole barns or shelters and sunporches is less desirable than permitting them access to range lots, confinement under protected sanitary conditions probably is better than giving the breeders access to contaminated range or range that exposes them to losses from predatory animals. Good results can be obtained from confined breeders provided feeding and management are fully adequate to meet the needs of the birds and ample floor space is allowed. Possibly the cheapest and most satisfactory type of confinement equipment for breeding stock would consist of a shelter with a sunyard attached, the yard surrounded with a tightly stretched 5- or 6-foot fence. Cobblestones and mold-free sawdust each has been reported satisfactory as ground covering for well-drained outside yards at a depth of one foot. Coarse gravel probably would be satisfactory if it were not allowed to become contaminated by continuous heavy use and is placed on a very well-drained site. Usually breeder yards can be rested several months of the year so that a build-up of disease organisms and parasites does not occur. Inside the shelter, about 10 square feet of floor space per large-type bird and 8 for small-type are needed for roosts, nests, feeders, waterers, and broody coops. The outside yard should provide about the same, or more, floor space per bird. The inside quarters may be littered with straw, peanut hulls, sawdust, cane litter, hay, or almost any available roughage which may be built up, starting with 3 or 4 inches and adding to it as required till the end of the season, then re-

Fig. 12.4. Nebraska roosting shed with sanitary yard attached. Note slatted floor (Photo from University of Nebraska.)

moved. Litter of all types (especially sawdust) must be free of molds. The shelter need not be elaborate or expensive but some shelter (see Fig. 12.4) is needed even in the dry, warm southwest where it provides needed shade. In other areas the type of shelter will depend somewhat upon the climate and whether or not winter eggs are to be produced. In the cold climate of southeastern Montana, a shed with single-board battened side walls and windows that could be closed in bad weather was satisfactory for breeders lighted starting early in February.

Pole barns or sheds (see Fig. 12.5) have become popular for rearing turkeys and should be well adapted to the housing of breeding stock with or without outside yards. They may be used for both rearing and holding breeding stock if the old stock is removed at least 2 weeks prior to brooding and the litter changed.

Complete confinement of turkey breeding stock in buildings with fully satisfactory results is somewhat difficult

Fig. 12.5. Pole shed for rearing (from 6 weeks) or breeding stock; in Buckinghamshire, England. (Photo from Hoppers Turkey Farm.)

of accomplishment. If such a project is undertaken, the following special suggestions are made. First, provide sufficient window or preferably, wire-covered open wall space, so that nearly all available natural daylight may be utilized. Second, provide ample floor space, about 12 square feet for small-type and about 15 per large-type turkeys in the breeding pens. Half this much space has frequently been recommened. If this practice is followed, both hens and toms should be debeaked. Third, in large units use one of three alternatives as follows: (a) Select males at 8 to 1 ratio and alternate them semi-weekly using half of them in the flock at a time; (b) supplement the former with artificial insemination if necessary; or (c) use artificial insemination only. In lieu of large breeding units, two-tom groups of about 24 large-type or 30 small-type hens, toms alternated twice weekly should be practical. If alternation is not practiced, one-tom units of 16 large-type or 20-small type hens, the tom in continuous use, also should work well. In the latter case, if desired, toms may be rotated from pen to pen at weekly intervals within breeding groups of 5 or more pens each. This latter system is intended as a means of neutralizing the ill effects of one or more ineffective tom within each breeding group. However, if toms are not rotated, those that are ineffective can be replaced after identification by means of pen pedigreeing of the eggs.

The reason for utilizing natural daylight as much as possible is that the breeding value of turkey males exposed only to incandescent electric illumination (by ordinary light bulbs) may possibly be somewhat impaired. To avoid this, use artificial lighting only in the morning, or morning and evening, so that about 14 hours of light per day are provided, counting the artificial light and the natural daylight. As with other types of management, proper light intensity is important, likewise ample ventilation, cleanliness, and diets that are nutritionally complete.

If porches are used, floor space allotment of 16 and 20 square feet for small and large type turkeys respectively seems advisable, about half of it devoted to the porches which may be floored with 2″ slats, 1¼″ apart or with a plastic-or rubber-covered heavy gauge welded wire. Inside quarters should have litter.

Batteries for breeders. Batteries for laying turkeys have been tried experimentally at the Oregon Station with good results for BBB. In 4 months egg production equalled that of breeders in floor pens or outside range lots but hatch of fertile eggs averaged about 10% less. The latter however was offset by 6 more eggs per hen and 17% higher fertility from the batteries so, in poults per hen, the battery birds equalled the range birds.

The principles of management and feeding of battery-housed turkeys are similar to those discussed above under complete confinement. Artificial insemination probably would be the most practical means of fertilizing hens kept in batteries although stud mating might have possibilities. The Oregon Station inseminated every 3 weeks with .03 cc. of semen and obtained 80 to 90% fertility in Broad Breasted Bronze and above 90% in Beltsville White hens. There would seem to be little practical use for batteries in commercial hatching-egg production; but as a means of obtaining completely accurate egg and nesting records on individual hens with a minimum of labor, they might have possibilities. For large-type-hens the Oregon Station (1954) suggests individual cages 18 x 28 inches; for small-type hens 16 × 24 in. A nutritionally complete diet is a prime essential for battery-housed breeders.

While it is generally regarded as more difficult to obtain good results from confined breeders than from ranged or yarded birds, two experiment stations, New Hampshire (1938) and North Dakota (1946), have reported more poults per hen produced per season from breeding turkeys confined to buildings as compared to similar stock given access to range lots. However, these flocks were small and had an allotment of floor space of 11 square feet per bird (BBB) in North Dakota and 13 in New Hampshire (standard Bronze). In both cases the confined stock laid about 20% more eggs, showed an average of 5% lower fertility, and an average of 2.5% higher hatchability of fertile eggs. The confined breeders in New Hampshire ate 1.8 lb of feed per egg produced while the range birds ate 1.9 lbs. during the 24-week period starting December 17th.

SHELTER AND EQUIPMENT

The Purdue Experiment Station found that turkey breeding stock that was housed produced no better results than unhoused stock. Artificial lights were not used in either case. However, experience has shown that some shelter for breeding flocks pays dividends in almost all situations. Even in mild climates, prolonged rainy weather may throw unsheltered breeding flocks out of production and cause considerable mortality. In mild climates the protection of a roof with a good-sized overhang is sufficient. In cold climates it seems advisable to provide a single-walled shed with windows, a reasonably tight roof, and a dry floor, arranged so as to provide protection against storms. This type of shelter should give sufficient protection for ordinary turkey flocks in almost any climate even when winter egg production is desired. Artificial lights and the addition of heating devices to keep the water supply open at all times complete the list of equipment necessary to obtain hatchable winter eggs in cold climates. Yarding equipment, soil sanitation, and feeding have been discussed elsewhere. Protection from predatory animals should be provided by tight fences, watchdogs and outdoor lights. Good material for predator proofing is 1½″ hexagonal mesh 16-gauge galvanized wire or one by two inch 14-gauge welded wire. If small portable shelters are used, they may be built on skids according to the standard colony-house plan altered somewhat to fit the needs of the turkeys.

Nests: Individual nests may be 10″ wide by 20″ long for small-type turkeys, or 10″ by 24″ for large-type turkeys but unless the tip-up front or trapnest front is installed, individual nests are not recommended. Community nests built to accommodate a number of hens are generally more satisfactory. The bottoms covered with 2 or 3 inches of sand with a relatively small amount of straw on top help to prevent breakage. Other suitable nesting materials are 5 or 6 inches of plain straw, a mixture of equal parts of dry soil and sawdust, and sugarcane litter. Planer shavings, sawdust, or rice hulls, into which the eggs will settle, are preferred by some operators. Nest units 2 to 3 feet wide by 6 to 8 feet long with an 8-in. baseboard will accommodate 15 to 20 hens. When the outside yards are comparatively small and devoid of tall vegetation, the nests should be placed inside the shelters. In large matings provided with large yards, some nests may be placed inside and some out, or all outside. In any case the operator must be on the lookout for stolen nests. Nesting houses which are small buildings, about 8 by 12 feet, placed in desirable locations and provided with clean litter (usually no nest partitions) are favored by many growers; likewise individual semi-trapnests with tip-up fronts, one for each 3 or 4 hens.

The semi-trapnest or tip-up front may consist of a 4 x 4″ block 12″ long with two 1 x 3″ strips nailed to each of two adjoining sides of the block, the strips 5¼″ apart for large turkeys, 4½″ for small-type. The pivots consist of a 20-penny spike with head cut off driven into each end of the 4 x 4″, in the lower inside corner as far from the slats as possible. The pivots are held to the nest framing by large staples or nails bent over. Several galvanized washers are placed on the pivot pin between the block and the side of the nest. The device should be adjusted so that when the hen on the outside pushes against the slats, they swing up, allowing her to enter the nest, while the other two slats swing down behind her to rest against the board at the bottom of the nest framework, preventing

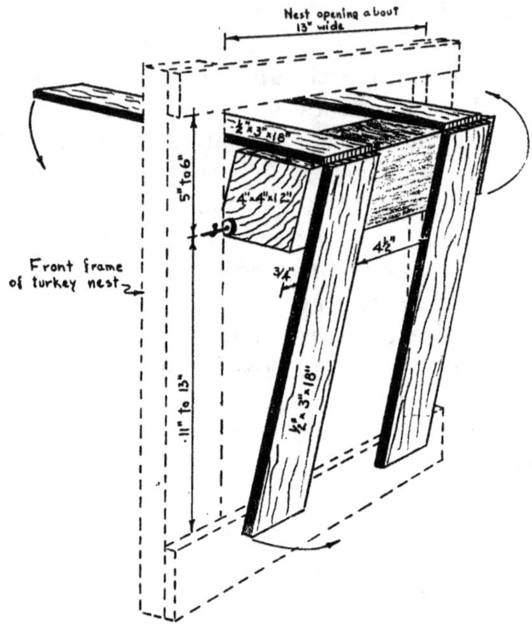

Fig. 12.6. Tip-up semi-trapnest front. (Photo from Va. Exp. Sta.)

other hens from entering. When the hen on the nest wants to leave, she pushes against the slats which swing up, letting her out and leaving the nest set for another hen to enter. Sawdust or some other similar litter is best since it is not dragged out to clog the tip-ups (see Fig. 12.6).

Roosts and Droppings Pits: Twelve lineal inches of roost space for small-type breeders, 15 for medium, and 18 for large, per bird, are needed. Stepladder-type roosts are desirable especially for small and medium sized varieties. The lowest roost should be about 24 inches above the floor or the ground; the succeeding ones each 4 to 6 inches higher. They should be about 2 feet apart center to center and the top roost 18 inches out from the back wall. The roost supports may be fastened at an angle of about 15 degrees and the roosts nailed directly to them without notching. This arrangement, when roosts of square-cornered sawed material are used, provides the desired tilt to the roosting surface. Two-by-threes or two-by-fours, unbeveled, wide side up, are commonly used, but poles 2″ to 3½″ thick make the best roosts; two-by-fours, narrow side up, with top corners beveled, are also good.

Low roosts with the upper surfaces horizontal and 14 to 18 inches above the floor and all on the same level are well adapted to use by large-type turkeys and make it easy to obtain uniform distribution of artificial light. No roosts at all in breeding stock shelters are preferred by some, especially for large, broad-breasted turkeys. In such cases the floor litter should be quite thick and kept clean and dry by frequent additions on the built-up plan. However, future breeding toms should be allowed to roost early according to some observers.

In permanent buildings, space under the higher roosts, when roosts are used, should be "wired in" to form a droppings pit to keep turkeys out of their droppings, to reduce the amount of cleaning necessary, and to eliminate one type of undesirable nesting place. The top of the pit is formed by heavy galvanized wire nailed to the undersides of the roosts themselves. For this, several types of wire may be used among them 1½″ hexagonal mesh 16-gauge woven, and 1 x 2 or 1½ x 1½″ 14-gauge welded, the latter preferred. Two-inch mesh wire of the same types and of the same (or better) heavier gauges may be used but eggs laid on the wire will fall through. The sides of the pit should consist in part at least of removable wooden panels covered with 2″ hexagonal mesh 16-gauge woven wire. The panels are removed when the under-roost droppings are removed, usually once at the end of the breeding season.

Floors: The floor of the shelter should be dry. Well-packed dirt floors are satisfactory. Board or concrete floors provide better protection and simplify sanitation. Wire floors or slatted floors as described elsewhere may be utilized if desired in permanent plants in locations where winters are mild and the birds have access to outside yards of soil or gravel. Such floors save labor, eliminate the cost of litter, and provide more sanitary conditions. However, littered floors are generally preferred for breeding stock.

Water Supply: A clean source of drinking water is necessary. Natural sources of any kind are dangerous, as parasites and diseases may be found there. Consequently, locations containing streams, ponds, or lakes should be avoided. Pure water should be hauled or piped to the turkeys' quarters. Bubblers with drains are the best waterers. Drip fountains or troughs with floats often are practical. Ordinary water pails, tubs, or troughs will serve the purpose. They can be set on wire platforms and in a frame or on a hook to prevent tipping. Electric water heaters may be used during cold weather in water pails. Bubblers, if properly installed and adjusted are frost-proof to sub-zero temperatures and are easy to care for. A method of heating drinking water by running electric heating cable inside the water pipes is described by the California Station in *California Agriculture* (Jan., 1954). Cistern water is not suitable for turkeys if pigeons or other birds roost upon the building from which the water drains, as these wild birds may carry certain filth-borne diseases that are transmissible to turkeys. Watering equipment should be kept clean by daily washing and rinsing. Occasional disinfection of waterers may be indicated if disease is present.

FEEDING PROSPECTIVE BREEDING STOCK

Liberal feeding of prospective breeding birds from the time they are first selected is a desirable practice. Since all breeding birds, both males and females, normally lose weight during the laying season is appears desirable that they should be brought to a prime condition in order to withstand, for as long as possible, this drain on their vitality. With these facts in mind, the following suggestions are made for the feeding of selected breeders as soon as they are segregated from the market birds. In situations where winter weather is mild and the birds are able to obtain liberal amounts of direct sunshine and green feed from the yards, a simplified high-fiber mash such as growing mash No. 26 (Chap. 3) is recommended, keeping it and scratch grain before the birds at all times along with water and grit. In this and other breeder maintenance diets, the mash and grain together (the total feed intake) should not contain over 25 to 30 per cent corn. For confined birds or stock held over in cold weather, a complete growing mash is advised, feeding it and grain free choice. An all-mash diet suitable for breeder maintenance is breeder diet No. 13 (Chap. 3) with all of the dried milk (or its substitutes) and three per cent of the limestone, replaced by oats. Such a diet would not be particularly fattening and would have plenty of vitamins and minerals. When egg production starts the dried milk or its substitutes and the limestone should be restored.

A good grain mixture to feed with the growing mash suggested as a breeder maintenance mash is one composed of equal parts of yellow corn (or one of the grain sorghums), barley, wheat, and oats. Barley alone is excellent.

FEEDING OF PRODUCING BREEDING STOCK

The feeding of a breeder mash or pellet containing around 22% protein free choice along with enough (measured) grain so that a 50-50 mash to grain consumption is obtained probably may be considered as the "standard method" of the 1940-50 period. It is an effective method but has the disadvantage of being difficult to administer. The Minnesota Station (1951) in one year's trial obtained best overall results by keeping before the turkeys a 50-50 mixture of 22 per cent protein breeder mash and grain (which was half whole corn and half whole oats). Other methods tried (with poorer results) were (1) all-mash, (2) free choice of corn, oats, and a 28% protein breeder mash, and (3) the 50-50 grain-mash mixture with extra vitamin supplements added. The experiment ran January 10 to April 25 and the birds were lighted for a 14 hour day. The variety was Broad Breasted Bronze and feed consumption per bird per day in the mated pens (2 toms and 25 hens) varied from 10.2 oz. (.638 lb.) in the all-mash pen through 10.7 oz (.667 lb.) in the 50-50 grain-mash-mixture pen to 11.6 oz (.723 lb.) in the pen fed a 28% protein mash with corn and oats, all free choice.

The Beltsville Station has used all-mash breeder diet for many years with uniformly good results and with the least possible amount of labor in feeding. For feed formulas and other data on feeding breeders, see Chapters 3, 4, and 5. Feeder space should be furnished at the rate of 12 lineal feet per 100 birds.

The feeding of antibiotics in turkey breeder diets apparently is not indicated. The Montana and Ontario Stations (1952) reported reduced hatchability from feeding aureomycin and penicillin respectively, in turkey breeder diets.

SUMMER CARE AND FEEDING OF HOLDOVER BREEDING STOCK

Because of feed, labor, and mortality charges for sun mer maintenance and fall fattening, it is seldom profitab to keep breeding birds through the summer for fall marketing. Furthermore, the hold-over birds are a disease hazard. If, however, it is desirable to keep some or all of the breeding stock for fall sales, for further breeding or to obtain livability data beyond the normal breeding season the best method is to concentrate them in two or more large groups, males separated from females, and give them access to a large clean range lot containing growing green feed. The hens of all except the large strains may be given ¼ lb. (⅓ lb. for larger strains) of a grain mixture per hen per day and the toms ½ lb. (⅔ lb. for larger types) each day. A mixture of 40 pounds wheat, 40 pounds corn and 20 pounds oats is satisfactory. It should be fed in open troughs once daily with enough trough space so all the birds may eat at the same time. At the U. S. Range Livestock Station (Mont.) it was found that both hens and toms did better on this treatment than they did when fed mash and grain *ad libitum*. Moulting was completed earlier and regaining the lost weight proceeded at a more rapid rate. For fattening, the grain mixture may be offered ad libitum starting four weeks before they are to be marketed. If they are to be held for breeding the following season, the maintenance diet can be continued until the pasture is bare, or until about 10 weeks before eggs are desired. At this time growing mash and scratch grain should be fed, both ad libitum until time to feed the laying diet. As with young breeders, the percentage of corn in the maintenance diet should be held down to 30% or less.

Under favorable conditions adult turkey hens will maintain themselves through the summer on free range without any feed at all but the losses from predatory animals, theft, and straying are likely to more than offset the value of the feed saved. Free range without the constant attention of a guard is very hazardous for turkeys of any age at any time.

The summer range lots should be used only half the year or else alternate yards or portable equipment should be provided in order to keep the soil from becoming contaminated.

Adult males are especially subject to heat prostration which usually is fatal. To prevent it, ample shade should be provided in the summer yards and plenty of ventilation in the summer roosting quarters. The hens should receive similar attention. During very hot weather, large, well-ventilated nests should be used and on excessively hot days the birds shut out of the nests. Piling up in the same nest box must be avoided. Unless summer eggs are valuable the nest boxes should be removed or closed before very hot summer weather arrives.

FATTENING AND MARKETING BREEDERS

Breeding birds should be marketed as soon as their eggs are no longer in demand or when rate of lay and/or percentage hatch of eggs set fall to an unprofitable level. It does not as a rule pay to attempt to fatten breeders at the end of their production season either by normal feeding methods or by estrogen treatment. The North Dakota Station (1950) found that the injection of two 15-mgm. pellets of diethylstilbestrol per BSW hen, 27 days before marketing did not influence egg production, fertility or hatchability but did cause a weight increase averaging ½ pound per hen along with somewhat better market grades. However, unless a premium could be obtained for estrogen treated breeders, the small weight increase obtained would not pay for the treatment. Holding breeding stock through the summer even on a maintenance diet, for the fall marketing season is not regarded as a profitable operation under most conditions. Although the breeder hens, if held until the cool fall weather, can be fattened to make very attractive dressed birds, they seldom bring as high prices as do the young birds. Given free access to mixed grains plus alfalfa hay or whatever green feed is available from a range lot, breeders, both hens and toms, will fatten to Grade A condition in four weeks at the rate of about 5½ lbs. of grain per pound of gain. The gain in weight should amount to approximately 20% of their starting weight in September or October.

VACCINATING BREEDERS

The Washington Station (1952) suggests that turkey breeding stock be vaccinated with live-virus Newcastle vaccine at the time breeders are selected. It may be done along with pox vaccination (see Chapter 17). Installing turkey bits, or moderate debeaking of breeders is often desirable. Severe debeaking (nearly to nostril) is not recommended for breeder or other turkeys. Revaccination is not necessary. Vaccination against Newcastle disease should be regarded as an individual problem and veterinary counsel should be sought in determining whether or not it should be done and how to vaccinate.

DISEASE PREVENTION

Sanitation is the foundation upon which success in turkey raising is built. In the U. S. in 1950 the loss of breeding turkeys during the year (of turkeys on hand January 1, mainly breeding stock) was about 9%, which indicates that disease among breeding stock is fairly well controlled. Fortunately turkeys over about 7 months of age normally develop considerable resistance to such parasites as the blackhead organism, cecal worms, and crop worms so it often is possible to keep mature breeding stock in range lots under conditions that would not be tolerated by growing stock. In the management of breeding stock the chief factors in the prevention of diseases are as follows:

(1) New blood should be secured through hatching eggs or day-old poults from pullorum-clean stock in preference to older birds.

(2) If breeding birds are acquired, they should be quarantined for 3 weeks in easily cleaned quarters and dusted with sodium fluoride or some other good louse killer twice, three weeks apart. If the presence of disease is demonstrated during quarantine, the new stock should not be retained. Treatment of new stock for worms with a good vermifuge is advisable. In this way, the presence of worms (if any) will be revealed. If worms are present, every effort should be made to eliminate them before the new stock is allowed to mix with that already on hand.

(3) In breeder houses, built-up litter may be used for

the entire season. The application of 1 lb. of hydrate of lime per 100 sq. ft. every month helps to keep it sweet. Wire floors and pits for droppings improve the sanitation and lessen the labor.

(4) Portable yards and equipment should be moved to clean ground each season. If permanent quarters are used, double yards alternated every 1 to 4 weeks should be afforded the breeders. After the breeding season, these yards should remain idle for 3 to 5 months each year. If year-round use of the breeding quarters is necessary, cobblestoned, gravel floor, or slatted sunporches or sunyards should be used for the 3 to 5-month yard resting period so that the regular yards will not be pastured continuously. Porches can be used throughout the year; likewise thick gravel and cobblestoned yards, and pole sheds.

(5) The turkeys should be kept away from all natural sources of drinking water no matter how clean the water may appear. Rain water should drain away promptly from all yards.

(6) Water and watering equipment must be clean. Wash waterers daily with brush or cloth, and rinse. Damp spots around it may be avoided by use of wire platforms, dry wells or drains.

(7) All feed should be fed in water-proof, contamination-proof feeders or hoppers.

(8) In so far as possible, visitors should be excluded from the shelters and enclosures.

(9) Chickens and turkeys should be separated. The same attendant should not care for both.

(10) The growing stock should be kept entirely separate from the breeders.

(11) The breeders should be protected from inclement weather and not overcrowded.

(12) The breeders should be fed a complete, well-balanced diet to aid in promoting high fertility and hatchability and resistance to disease.

(13) Whenever the breeding program permits, all

breeding birds should be marketed in the early summer as soon as the demand for hatching eggs ceases. This helps to break the chain of infection of diseases and parasites. Birds held over should be kept through summer and fall in yards other than the regular breeding yards and away from the area devoted to young stock.

(14) Turkey yards should be kept free from brush, vines, and trash of all kinds which may harbor predatory animals, disease organisms, parasites, or insects that are hosts of parasites.

(15) Breeding stock never should be retained from a growing flock in which a communicable disease has been diagnosed. If this happens in extremely valuable stock from which breeders must be chosen, every effort should be made to eradicate the disease through sanitation, rigid selection, and special tests when such are available. For example, a paratyphoid-infected flock, if of ordinary quality, would be completely eliminated as a source of breeding stock. However, if it were extremely valuable, special blood tests could be made to eliminate the carriers, using as breeders only the non-reactors; and transferring them to remote quarters, preferably another farm.

ARTIFICIAL BREEDING*

Artificial breeding is important to animal industry because it brings into better control a most vital step in animal production. By its use, the fertilizing capacity of valuable males may be increased, and females that would not otherwise produce offspring can often be fertilized, and certain crosses may be made that would otherwise be impossible or impracticable.

Although artificial breeding of certain mammals has been practiced for many years, and although chickens were artificially inseminated in an experimental way as early as 1913 by Ivanov, the artificial breeding of birds has but recently been made practical. The chief obstacle

*Much of this material on artificial breeding originally was supplied by the late William H. Burrows formerly of the U. S. Dept. of Agriculture.

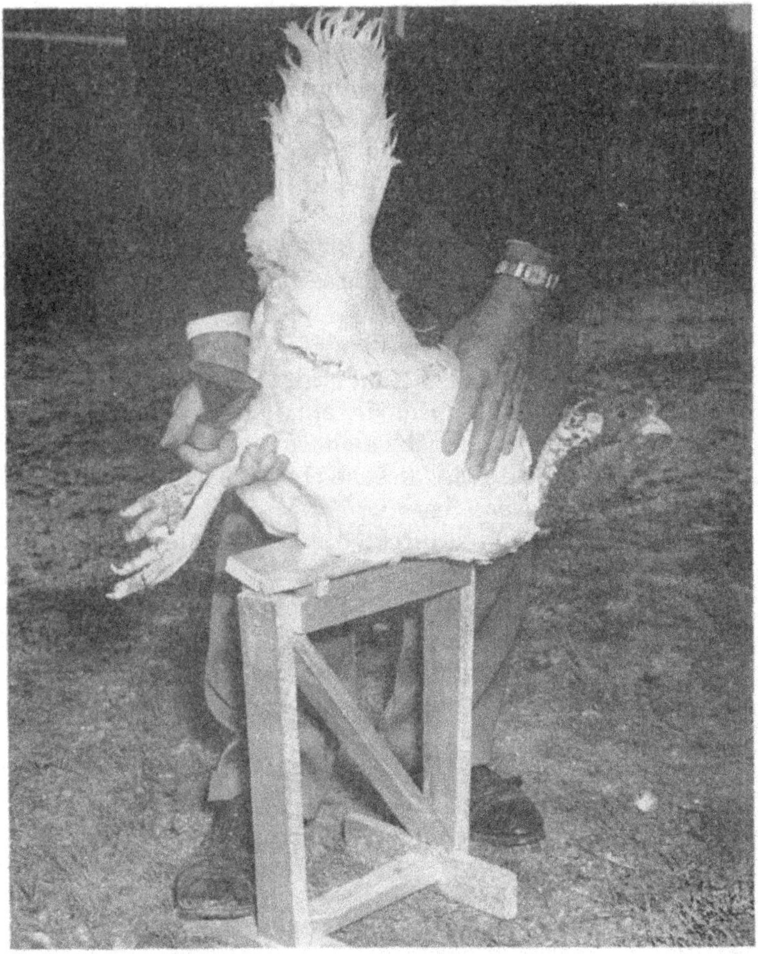

Fig. 12.7. Tom being placed on "milking stool" preparatory to obtaining semen for artificial insemination, as described on pages 724-730. The operator straddles the stool (and the tom) and holds the stool firm by standing on the forward cross-piece. (Photo from U. S. Dept. of Agr.)

in the work with birds, for some time has been the obtaining of semen.

Ivanov obtained semen by killing cocks and squeezing it from the vasa deferentia. Payne (1914), Craft, McElroy, and Penquite (1926), and Jull and Quinn (1931), mated cocks to hens and recovered the fluid from the cloaca of the hen. Amantea (1922) and Dunn (1927) inserted a glass dish between the cock and the hen during copulation and obtained small amounts of semen this way. Ishsikawa (1930) and Adamstone and Card (1934) used an artificial cloaca on the hen.

Workers at the U.S.D.A. Research Center (Burrows, Quinn, and Marsden) developed a practical method of artificial breeding of chickens and turkeys consisting of stimulating the male to ejaculate semen and exposing the hen's oviduct orifice to receive it.

By the use of these latter methods the artificial breeding of chickens has been successful both in experimental work and commercial practice. Turkeys have been successfully inseminated in experimental work and can be just as successfully inseminated for commercial purposes. However, artificial breeding is not recommended as a substitute for natural mating in the breeding of turkeys for economic qualities. In commercial reproduction it may have a place.

Basis of the Method

The method of obtaining semen as described by Burrows and Quinn (1935) is essentially the same for the male chicken and the turkey tom. The apparatus necessary for the work consists of a small glass funnel (about 2 inches wide across the top) sealed with paraffin at the base of the V, a one-cubic-centimeter tuberculin syringe without the needle and a pyrex glass test tube. It is important that the test tube be of pyrex glass if the semen is to remain long in it. It should be of such a size and (short) length that the semen can be withdrawn easily with the syringe. If the semen is not used immediately and the weather is cold, the pyrex test tube may be inserted through a hole in the cork into a thermos bottle containing water at 106° F. In any event the semen should be used within 30 minutes of collection and the holding container (test tube or small beaker) should be kept covered to prevent drying out of the semen. In lieu of the collecting funnel, a glass whiskey jigger (or wine glass) with a rounded bottom may be used.

The method is based upon reflexes. Many who have attempted to practice artificial insemination have experienced difficulty in obtaining results when attempting to follow written instructions. To overcome this difficulty, one must understand that to obtain semen from a bird entails the stimulation of a reflex action, and that a strong reflex always inhibits a weaker one. The pressing down upon one's foot to maintain balance is a reflex action, and this reflex is stronger than the one set up by the pain caused by the sharp stone. Therefore the balance reflex can inhibit the protective reflex and the foot is pressed down upon the sharp object to maintain body balance and is thus injured. If the pain reflex were the stronger, one would fall rather than injure the foot.

To obtain semen from the male, the ejaculatory reflex is stimulated by a gentle massage of the soft part of the abdomen with the right hand while gently stimulating the sides of the vent with the thumb and fingers of the left hand (details given later). However, adverse reflexes may be set up which will entirely inhibit the response to such stimulation. The strongest adverse reflex the operator has to contend with is that of self-preservation caused by fear. Any bird when first picked up is startled. This is not fear, but rather a state of confusion which does not seem to inhibit the ejaculatory reflex. However, upon being held for any length for time confusion grows into apprehension and then into fear, after which it is very difficult and usually impossible to stimulate the desired reflex. The assistant or the operator may hold the bird in a tight, rigid grip and so set up adverse reflexes that stimulation of the desired reflex is impossible.

These are the two conditions to avoid if success is to be obtained. The second is easily avoided, but the first is not so simple. To excite the proper reflex without delay is difficult for the inexperienced. The simplest way to overcome the difficulty is to try a number of birds until one is found that is easily stimulated. After the operator has trained his hands on a bird that is easily stimulated, he will then find no trouble in manipulating those that are more inclined to be fearful. After the desired reflex has been stimulated a few times and the bird learns that he is not going to be harmed, the ejaculatory reflex can more and more easily be brought about. The use of toms that are tame and friendly is a long step in the right direction when learning the technique of obtaining semen.

ANATOMY OF THE MALE SEX ORGANS

Contrary to popular belief, the turkey has a copulatory organ of appreciable dimensions. This organ is located on the ventral (or lower) wall of the cloaca and is normally distinguished only as flaccid folds of the tissue which are concealed within the cloaca. Normal erection takes place with great rapidity and persists for only a few seconds, making observation very difficult and a thorough examination impossible.

However, with the development of a method of manual stimulation, it has been possible to maintain a condition of erection extending over several minutes, and with a clamping-off process, to study the organ in some detail.

Figure 12.8 shows the copulatory organs of the rooster (A) and of the turkey tom (B). It will be noted that these organs are very similar except in shape and size. The testes in market-age turkeys are small, averaging 2.3 grams for both organs in seven 28-week-old standardbred Bronze toms averaging 19 lbs. each alive. Length averaged 0.75 (¾) inch long, 0.40 (13/32) inch broad, and 0.3125 (5/16) in. thick. Testes size increases greatly with the onset of sexual maturity. Five small-type sexually mature toms weighing an average of about 19 lbs. alive had testes averaging 23 grams for both. Length was

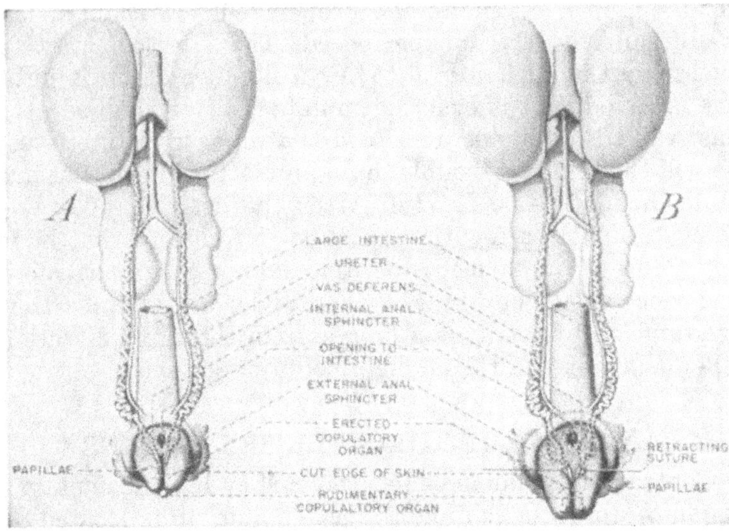

Fig. 12.8. Diagrammatic drawing of the genital organs of the male chickens (A) and of the male turkey (B). (Photo from U. S. Dept. of Agr.)

about 1.5 inches and width about 0.75 inch. In relative size the two organs may be nearly the same size or they may vary up to 50% difference in size, the left one larger.

It is assumed that under normal conditions, the semen, coming down from the testes, is stored in the posterior positions of the vasa deferentia which are shown in the figure—as convoluted ducts. From these vasa deferentia, the bulbous ducts, shown by dotted lines, are filled when the bird is sexually stimulated. At copulation the internal sphincter of the vent contracts, forcing the semen out of the bulbous ducts through the papilla, from which it follows the groove to the tip of the organ.

Obtaining Semen

The first step in obtaining semen from the bird is to hold the tom in the proper manner. The thighs are held loosely in each of the two hands of the holder, operator 1, with as much weight as possible supported without any particular strain on the legs of the bird. To avoid undue strain on both tom and holder the latter may sit down, or as the California Station suggests, the tom may be placed in a canvas sling mounted on a framework at a convenient height. Another satisfactory method of holding the tom is between the operator's legs (Fig. 12.7); the bird's feet off the floor, his breast resting on a specially constructed stool, head to rear and preferably against a wall. It is possible in this way for one man to collect the semen unaided. However the operation is more efficient if a helper collects the semen and aids in pushing aside the feathers and removing fecal material which often is released prior to the ejaculation of semen. The assistant also should hold down the bird's legs as shown.

The protrusion of the copulatory organ is brought about by massaging the soft part of the abdomen with the free fingers and thumb of the right hand; the container being held in the crook of the index finger or by an assistant who also should hold the bird's legs down. At the same time an upward stroking motion is started at the vent, thumb on one side, fingers on the other. At the end of the stroke the tail is pushed back over the bird's back with "heel" of the left hand. Through repeated strokings the bird is further stimulated and the thumb and fingers are in position to "milk" the semen from the bulbous ducts the instant the organ is protruded (Fig. 12.9). Considerable force and a strong, deep grip are necessary. Every attempt however, should be made to stimulate the bird to ejaculate—without a voluntary ejaculation the semen yield will be small. This is important—if the organ is not extruded well, poor results will be had.

During the first few attempts to obtain semen from a bird, be it the first time the bird has been used or the first time after a long rest, the defecation reflex is very likely to be stimulated. However, after a few stimulations, a conditioned reflex (or habit) is formed and the bird can then be stimulated to ejaculation without the complicating factor of defecation. It is obvious that it is unwise to attempt to collect semen for artificial insemination until the conditioned reflex has been set up and the operator can collect samples of semen comparatively free from either feces or urine. A few white flecks of urates (urine of the bird) apparently do not affect the semen so as to make it unfit for fertilizing. Toms should be removed from the hens two, or preferably three, days prior to milking. The Texas Station (Draper, 1953) suggests allowing the toms no green feed during this time and removing feed and water at noon of the day before they are milked. This procedure is intended to reduce fecal contamination of the semen.

It is good practice to withdraw the semen from the collecting funnel with a 1 cc. tuberculin syringe and store it in the test tube as soon as the collection from one bird is made. The semen has a tendency to dry quite rapidly and there is always the danger of contaminating the sample during the next operation, thus losing two or more samples instead of one. Although work at the Nebraska Station indicates that turkey semen can be diluted successfully, best results have been obtained by using the semen undiluted and as soon as possible after it has been collected, not even attempting to transfer it from one farm to another. If it can not be used immediately, the holding temperature probably should be between 98 and 106° F. If storage is attempted, a temperature of 50 to 60° F. might be best, as was determined for undiluted rooster semen for 3 to 4 hours at the Maryland Station (1950).

The turkey tom produces semen which is creamy whitish or yellowish white, thick, and of a somewhat sticky nature. The volume of semen obtained per ejaculation varies considerably. An average of 0.25 cc. from Broad Breasted Bronze toms was reported by the California Station for the normal season, with less than this at the beginning and towards the end. The maximum reported at one handling was 0.5 cc. The North Dakota Station (1946) collected small volumes of semen, .05 to .133 c.c. by the abdominal massage method, from 6 unmated Broad Breasted Bronze toms Dec. 11 to 18, 1944, when they were 7 months old and before lighting was turned on, on Dec. 18 to give a 14-hour day. On Jan. 1, 2 weeks later, the volume had increased to .265 c.c. per ejaculation. On Jan. 15, an average of .295 c.c. was obtained, and on Feb. 12, .34 c.c. From Jan. 15 to June 6 the average volume per collection at 2-week intervals, the males being ejaculated 48 hours before each regular collection, was .33 c.c., the average density 8.4 millions of sperm per cubic millimeter of semen, the average number of sperm per ejaculate (of .33 c.c.) was 2.8 billions, and the average motility score of the sperm 3.9. Motility of sperm was relatively low, 2.8 to 3.3, until the collection of Jan. 29 when it was 4.8. From then on it varied from 4.8 down to 3.7, except for 3.0 at one collection (on Apr. 24). Density of sperm was low, 1.2 to 1.43 millions per cu. m.m., prior to the artificial lighting. After 2 weeks of lighting it had increased to 4.53 millions and tended to increase up to the end of the experiment (June 6), when it was about 10 millions. The 6 males varied but little in volume of semen, sperm density, sperm motility, or body weight, which averaged 30.3 lbs. on Jan. 3 and 30.4 lbs. on June 2. The smallest volume of semen per ejaculate between Jan. 15 and June 6 was .14 c.c., and the greatest .50 c.c. The author remarked that the volume might have been greater had not the males been ejaculated 48 hours prior to each recorded collection.

At the Beltsville Station, with Beltsville Small White turkeys, an average of 0.26 c.c. per ejaculation was obtained over a period of 42 weeks beginning Dec. 19, 1951 when artificial lighting (40 watts for 14 x 14 pen) was started. The largest amount obtained at one manipulation was 0.75 c.c. Semen production and body weights of these toms at 2-week intervals are shown in Table 12.19.

Table 12.19—Average Semen Production and Body Weights of 8 Unmated Beltsville Small White Toms at 2-week Intervals

Date	Average semen production in cc.	Average body weight in lbs.
Dec. 19, 1951	.11	17.60
Jan. 2, 1952	.18	19.60
Jan. 16, 1952	.19	20.80
Jan. 31, 1952	.27	20.60
Feb. 13, 1952	.24	20.60
Feb. 27, 1952	.22	20.40
Mar. 14, 1952	.28	19.90
Mar. 26, 1952	.36	19.50
Apr. 9, 1952	.37	19.30
Apr. 23, 1952	.37	18.90
May 7, 1952	.31	18.30
May 21, 1952	.31	18.30
June 4, 1952	.31	18.80
June 18, 1952	.34	18.60
July 2, 1952	.35	19.10
July 16, 1952	.22	19.60
July 30, 1952	.23	19.20
Aug. 13, 1952	.26	19.40
Aug. 27, 1952	.28	19.90
Sept. 10, 1952	.16	20.00
Sept. 24, 1952	.07	20.40
Averages, 42 weeks	.26	19.50
Averages, 22 weeks, Jan. 31 to July 2	.31	19.40

The Washington Station (1951) studied the morphology of turkey spermatozoa and found them similar in gross morphology to those of roosters. The normal turkey spermatozoon is long and very slender. First is the V-shaped tip or *acrosome* which is 1.4 to 2.1 microns* long and attached to the head which is 9 to 12 microns long with a width of about 1.5 microns at the rear, becoming slightly narrower at the front. Next is the body or midpiece about 1.5 microns wide and 3.5 to 5.5 microns long, bordered fore and aft by the proximal and distal centrioles (dark-staining narrow bands). The body or midpiece is believed to contain in its center an *axial filament,* a

*A micron is one-millionth of 39.37 inches (one meter).

thread-like structure apparently connecting the two centrioles. The tail is joined to the distal centriole where it is about 1.5 microns wide tapering gradually to a thin filament which is moderately curved. Total length of tail is usually over 50 microns.

Some abnormalities of sperm are hooked, swollen or missing acrosome; vacuolated, giant, or double head; body bare to the axial filament, giant, bulbous, or bent; tail beaded, coiled, absent, or consisting only of a filament. General abnormalities are immature sperm, giant sperm, and reverse staining where the head takes the acid stains, appearing light instead of dark while the body takes the basic (alkali) stains, appearing dark instead of light. Presumably the presence of abnormal sperm in a semen sample would indicate reduced fertilizing ability.

Average age at sexual maturity in Broad Breasted Bronze turkeys, as evidenced by semen production appeared to be about 7 months (30½ weeks) under average normal conditions in California. This suggests testing the toms for semen production at 7½ to 8 months, saving only good producers. Some toms reach sexual maturity much earlier than this, as early as about 17 weeks in small-type turkeys according to observations by the authors. The duration of viable semen production may vary according to management or hereditary tendencies and may extend from about 8 weeks to 8 months. Overstimulation by artificial lighting may cause it to be short (Olsen and Marsden, 1952). Under favorable conditions, toms may be expected to remain in good breeding condition for at least 4 to 5 months. If a breeding season longer than this is contemplated, a younger, unlighted set of toms should be held in reserve. It may even be practical to keep reserve toms on a schedule restricting the light to 8 or 9 hours daily through the winter, gradually increasing the light over a 30-day period, to that provided by normal daylight. Since artificial insemination functions only when the toms are producing viable semen, the care of the toms is just as important then as in natural matings. Furthermore, accidents, weather, and/or mismanagement may shorten the toms' effective breeding season, therefore a reserve group of unlighted or light-restricted younger toms is good insurance in any breeding program that utilizes artificial lighting to advance the breeding season or to produce out-of-season hatching eggs.

INSEMINATION

To inseminate the hen, the oviduct tip is forced out and the semen (usually 0.025 to 0.05 c.c.) is injected with a one-cubic-centimeter turberculin syringe without any hypodermic needle. Some operators fasten a ¾ inch piece of tapered, dull-pointed glass tubing to the syringe tip by a piece of thin rubber tubing. This extends the length of the syringe tip by about an inch and is believed by some to make insemination easier. When the funnel-shaped end of the oviduct appears, the syringe tip or the glass tip of the extension is inserted gently and as the oviduct returns to its normal internal position, the syringe follows it, the operator pressing the plunger to release the desired amount of semen when the oviduct is withdrawn and the pressure against the syringe is relaxed.

The optimum dosage of semen originally was believed to be about 0.05 cc. but the California Station (1950) found .025 cc. to be fully equal to 0.05 and .0125 cc. to be equal to the larger dosages for 2 weeks but not longer. The Pennsylvania Station (1952) obtained about the same results (with bi-weekly inseminations) when .01, .0125, .025, and .05 cc. (or ml.) were used. With semen at a premium, as it usually is, the lowest dosages given at two-week intervals might be more practical than the larger dosages at three-week intervals.

Insemination in the afternoon rather than the morning so as to find fewer hens with hard-shelled eggs in their oviducts has been suggested, but the North Dakota Station (1946), found that morning insemination (7-8 a.m.) gave practically the same results as afternoon insemination (4 to 5 p.m.) and that artificial insemination (0.05 cc. at 2-week intervals) gave the same fertility (about 90 per cent) in Broad Breasted Bronze as did natural matings. Failures in insemination often are due to the presence in the oviduct of an egg about to be laid. Afternoon insemination controls this situation quite well but it is relatively easy to tell by inserting the forefinger into the vent which hens are about to lay because the hard-shelled egg will be felt close to the vent. Such hens can be put aside until they lay, then inseminated.

The use of Ringer solution as a cleansing agent for insemination equipment is suggested. However as a diluent for turkey semen it appears to produce only mediocre results. According to Bonnier and Trulsson (1939), a solution consisting of 30% rooster semen and 70% Ringer solution gave excellent results in chicken insemination. Ringer solution may be made up as follows. In 10,000 cc. of triple-distilled water dissolve the following pure chemicals:

NaCl (common salt)	68.00 grams
KCl (potassium chloride)	17.33 grams
Ca Cl_2 (calcium chloride)	6.42 grams
Mg SO_4 (magnesium sulphate)	2.50 grams
NaH CO_4 (sodium acid carbonate)	24.50 grams

Dissolve each chemical separately in the distilled water and combine. Keep solution cool.

Successful dilution of turkey semen was reported by the Nebraska Station (1954). A solution called Earle's Bicarbonate Buffer was made up as follows:

Ca Cl_2	.020 gm.
Mg $SO_4 \cdot 7H_2O$	.020 gm.
Na H_2 $PO_4 \cdot H_2O$	.014 gm.
K Cl	.040 gm.
NaCl	.680 gm.
Fructose	.100 gm.
Dihydrostreptomycin	.060 gm.
$NaHCO_3$	.220 gm.

Add glass-distilled H_2O to make 100 cc. CO_2 is bubbled through HCO_3-solution before adding it to the dissolved salts to adjust the pH to 7.4. The resulting diluent apparently should be held 72 hours before being used. A suggested method of use is to mix 1 cc. of semen with 9 cc. of the diluent, then inseminate each hen with 0.05 cc. of the diluted semen. With this system, 200 hens could be inseminated with 1 cc. of semen which is about 5½ times the number that could be inseminated with 1 cc. of undiluted semen at .025 cc. per hen.

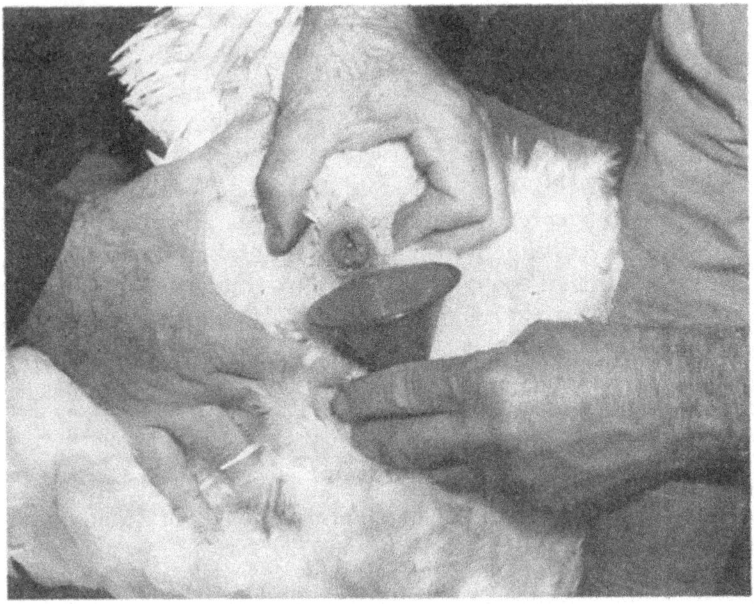

Fig. 12.9. Position of operator's two hands (above) and left hand of assistant with funnel to collect the semen. (Photo from U. S. Dept. of Agr.)

To hold the hen, one operator, either sitting or stooping, places her between and on his thighs and knees, her head toward his body. With thumb and fingers of his right hand on either side of the vent, he pushes her tail and tailhead back toward her head at the same time applying pressure on her abdomen from beneath to force the oviduct out and hold it there until another operator makes the insemination as described above. When the oviduct is extruded sufficiently, the morning-glory-shaped orifice will be seen on the bird's left side of the vent. The oviduct usually can be everted with little difficulty if the hen is in full laying condition but even if she is not in full lay it still can be forced out but to do so requires the use of considerable force. If, however, the oviduct is totally inactive it cannot be forced out.

Insemination can be done quickly, efficiently, and at a reasonable cost by a well-trained crew. According to various estimates, a trained four-man crew (a "milker"-inseminator, a holder, and two helpers) can inseminate 100 to 125 hens per hour. With eight men, one group can "milk" the toms to obtain semen while another inseminates, resulting in a turnout of 250 to 300 hens per hour. With twelve men (a milker and two inseminators, each with a holder and two helpers), 400 to 500 hens per hour can be treated. Costs of insemination vary considerably, from 5 cents to 15 cents per hen per insemination, with possibly an average of 10 cents. This cost is reasonable enough to justify an all-insemination breeding program without natural mating, the resulting reduction in the number of toms needed helping to pay for the cost of insemination. Insemination services have been established in some breeding areas.

Good fertility has been obtained in experimental insemination of small-type turkey hens (Burrows and Marsden, 1938). Many of the hens showed 100 per cent fertility for two to four weeks after once becoming fertile. Out of 948 eggs in one experiment, the fertility was 80 per cent, fairly satisfactory but not so high as that obtained in natural matings. In large-type turkeys, artificial insemination properly done usually is as effective as or better than that obtained by natural mating.

In one experiment at Beltsville the fertility of eggs laid by a flock of ten turkey hens mated to an old tom was increased from 7.5 to 88.4 when artificial insemination was substituted for natural mating. The old male in this instance was unable to properly fertilize by natural means more than one of the ten hens in the pen, while by artificial means he produced semen sufficient to fertilize all ten. Such as increase in fertility would be of great value with a progeny-tested old tom.

The Oregon Station (1947) compared during two seasons four natural matings of 11 Broad Breasted Bronze hens and one tom with two groups of 8 Broad Breasted Bronze hens, each hen artificially inseminated with .05 cc. of semen biweekly. Artificial lighting was started Jan. 3rd to give 14 hours of light per day; stock was 7½ to 8 months old at the start. Fertility in the naturally mated pens started at 81.3% in late January, increased rapidly to 90.6% in early February, and maintained this level until May when it dropped to 85.2%, then to 79.6% in late May and 69.1% in early June. In the artificially inseminated pens, fertility started out at 94.3%, and maintained a level of about 90% through to early May when it dropped to 83.9% then to 80.4% in early June. In this case fertility was generally improved by artificial insemination.

The California Station (1951) as a result of studies on artificial insemination suggests that in an exclusively artificial breeding program, the toms be stripped three times per week which, based on an average yield of 0.2 ml. (or c.c.) of semen per bird per ejaculation, would result in each tom maintaining fertility in 72 hens by artificial insemination only, inseminating them once every three weeks with 1/40 (0.025) cc. of undiluted semen. However, more toms than this should be kept, a figure of one tom to 40 hens being suggested. This is only about 25 percent of the number required for natural mating. Males should be handled in advance of the breeding season (when about 7½ months old) and selected, among other things, on basis of good semen production. Hens to be inseminated can be divided into nine groups, the oldest hens in the first groups, and each day's semen collection used to inseminate one group of hens. Some breeders prefer a two-week insemination schedule as a safety precaution, using only six groups of hens instead of nine, and possibly a few more toms if necessary.

The Pennsylvania Station (1951), using White Holland turkeys lighted starting November 15th and confined to buildings, found that artificial insemination (.05 cc. of pooled undiluted semen bi-weekly from other toms) combined with natural matings (2 toms with 16 hens) produced over an 18-week season starting Nov. 15, an average of 14.4 more poults (55.2) per hen than did natural mating (41.8 poults) and 8.5 more poults than artificial insemination alone (46.7 poults). These differences were due mainly to fertility which averaged 81.3% for the natural matings, 91.7% for the artificial only, and 95.7% for the combination. Egg production and hatchability were not significantly affected by the method of mating. The differences in fertility between the combination method and the straight insemination method were not significant hence there would seem to be no advantage in using the combination method as compared to the artificial insemination alone. When the naturally mated hens, after 18 weeks of production, were inseminated, fertility increased 12% for the next 4 weeks.

Supplementary insemination sometimes is helpful where fertility is not satisfactory. A single round of insemination often will bring up fertility noticeably as indicated by the Pennsylvania results. To do this, isolate the males in the flock for two or three days, then strip the males and inseminate all hens, returning the males to the flock the following day. This system works if the males are producing viable semen. If for some reason they are not, then other males which are in breeding condition must be used. The rapid fertility test can be used on each hatch of eggs to determine the need for artificial aid in maintaining high fertility. To do this the California Station suggests using a powerful candling lamp with light blue filter. After 24 hours or a little more of incubation, lay the eggs on their sides for 2 or 3 hours, then candle without turning them over. Fertile eggs have a small dark spot on the surface of the yolk.

The California Turkey Growers Association (MacIlraith, 1952) in tests with 28 flocks containing 33,000 Broad Breasted Bronze breeding hens of similar breeding reported that supplementary artificial insemination in addition to natural mating increased fertility about 15% over that obtained by natural mating alone. The toms supplying the semen were also of the same breeding. Supplementary insemination plans used were: (1) Inseminating the entire flock twice during the 18-week season (2) inseminating only the broody hens after they were broken of their broodiness, and (3) inseminating the entire flock once during the 18-week season. All three were successful; the choice depended upon the situation in each flock.

REFERENCES

Asmundson, V. S., 1945. Inheritance of breast width in turkeys. Poul. Sci. 24:150-154.

Johnson, A. S., 1953. The artificial insemination of turkeys, Canad. Poultry Review 77(6): 31.

Kosin, I. L., 1951. A study of the morphology of moribund turkey blastoderms. Reprint from 9th World's Poultry Congress.

Misner, E. G., 1946. Costs and returns for the turkey enterprize. Cornell. Bull. 827, Ithaca, N. Y.

Mumford, D. C., 1952. Costs and net returns in producing turkey hatching eggs. Oregon Sta. Bull. 524.

Newlon, W. E., and V. S. Asmundson, 1948. Selecting turkeys for breeding improvement. Calif. Ext. Circ. 143. Davis.

Plaxico, Jas. S., 1950. Turkey breeding—an economic survey. Va. Bull. 443, Blacksburg.

Romanoff, Alexis L., and Anastasia J. Romanoff, 1949. The Avian Egg. Pub. by John Wiley & Sons, N. Y., 872 pp. illus.

CHAPTER 13

TRAPNESTING, PEDIGREEING, SEXING AND CAPONIZING

OUTLINE

Pen pedigreeing is the process of identifying, as a group, the poults hatched from eggs laid by the hens in different breeding groups. These groups may contain one or more males. By this method the pen from which the poults were hatched may be determined. In the case of one-male matings the sire may also be identified unless the males have been rotated. Pen pedigreeing does not require much extra labor and is of considerable value in breeding projects where more than one breeding unit is maintained. By means of the resulting records one may carry on more intelligent selection and mating of breeding stock. Where one-male matings are used without rotation of males, pen pedigreeing is especially valuable because then a definite breeding plan may be followed and

Table 13.1—Toe-Punch Pedigree Charts

(This chart may be used as a permanent record by inserting pen numbers in the blank spaces).

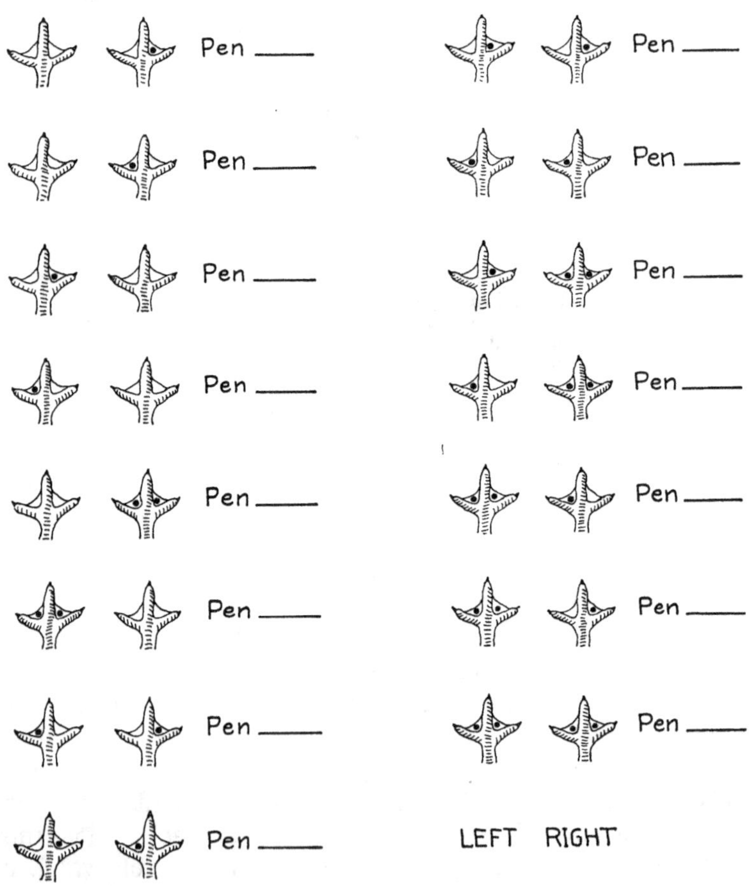

the breeding value of the males may be determined through progeny testing.

In pen pedigreeing, the eggs laid by the hens in each mating are marked (preferably on the small end) with the mating or pen number. They may be marked according to the male number if the matings contain only one male that is left in the pen throughout the season. All the eggs may be incubated together in the incubator until the 22nd to 24th day. At that time they should be sorted according to pen, placing the eggs from each mating in a separate tray or a compartment formed by wire partitions in the tray. After hatching is over, the poults are banded or toe-marked. Toe-marking may be done by punching holes in the webs between the toes, by slitting the webs, or possibly by clipping off the tips (including the toenails) of one or two toes. Fifteen different combinations of punchmarks or slits are possible, in addition to the poults that may be left without marks; many more if 6, or all 8, toes are utilized for clipping. However, it is best not to use lack of toemarks as identification because toe-punch marks and slits occasionally grow over, and clipped toenails may be regenerated if they are not clipped to include the whole nail. In toe-punching, if the circular piece of flesh cut out by the punch is completely removed, the punch mark will not grow shut. A record should be kept showing the relation of the different toe-mark combinations to the pen or male numbers.

INDIVIDUAL PEDIGREEING

To the specialized breeder, individual pedigrees are of great value and to the research worker they are indispensable. Through pedigrees the ancestry may be traced and evaluated, progeny testing may be carried on, and the growth rate and other characteristics be recorded.

The steps in the process of pedigreeing are taken up in chronological order, beginning with the mating.

The Matings

One-male matings are essential and they must be kept intact throughout the season. If a hen is found in the wrong pen with another male her eggs must not be used for pedigreeing for four* weeks after she is put back with the right male. If a male is found in the wrong pen, all the eggs laid by the hens in that pen must be withheld from pedigreeing for four* weeks after the removal of the visiting male. Obviously then, it is most important that all possible precautions be taken to keep each mating intact throughout the entire breeding season. Double yards so arranged that there is a vacant yard or space surrounding each mating are very desirable; so are six-foot fences and "antiflies" on buildings and solid-top gates. If these precautions are taken and in addition the flight feathers of one wing (or the inner 7 flight feathers of both wings) of each hen are clipped, there will be positively no mixing of the matings in the outside yards. On the inside, precautions must also be taken. The best method of construction is for each breeding pen to open into an alleyway and the pen door to open only one way—inwards—with spring or counterbalance to hold it shut. If the breeding pens open into each other which is undesirable, then precautions must be taken to see that the doors are self-closing and will stay closed. Doors, each equipped with a counterbalance to close them and hold them closed, work quite well. The bottoms of the doors should be high enough above the litter to insure no interference in closing. The doors and partitions between pens must be solid for at least 3 feet up from the floor to prevent fighting and there must not be any appreciable crack between the door and door framing, otherwise the toms may fight through this and force the door open.

Each bird should be banded on the right leg, the band being placed upside down so the number can be read

*For ordinary breeding operations; for extreme accuracy 8 weeks would be required or else two inseminations from the proper male two weeks apart should be done.

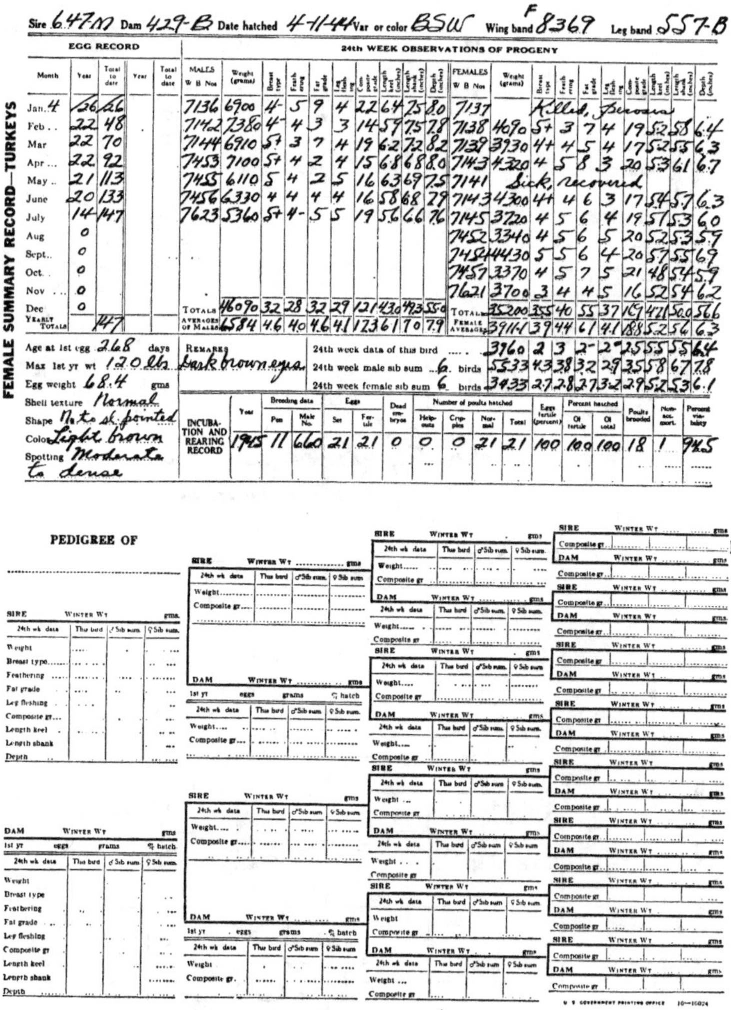

FEMALE SUMMARY RECORD—TURKEYS

Sire 647-M Dam 429-B Date hatched 4-11-44 Var or color BSW Wing band F 8369 Leg band 557-B

EGG RECORD					24th WEEK OBSERVATIONS OF PROGENY																			
Month	Year	Total to date	Year	Total to date	MALES W B Nos	Weight (grams)	Breast type	Feathering	Fat grade	Leg fleshing	Composite grade	Length keel (inches)	Length shank (inches)	Depth (inches)	FEMALES W B Nos	Weight (grams)	Breast type	Feathering	Fat grade	Leg fleshing	Composite grade	Length keel (inches)	Length shank (inches)	Depth (inches)
Jan. 4	26	26			7136	6900	4-	5	9	4	22	6.4	7.5	8.0	7137	Killed, Kerosene								
Feb.	22	48			7142	7380	4-	4	3	3	14	5.9	7.5	7.8	7138	4090	5+	3	7	4	19	5.2	5.8	6.4
Mar	22	70			7144	6910	5+	3	7	4	19	6.2	7.2	8.2	7139	3930	4+	4	5	4	17	5.2	5.5	6.3
Apr	22	92			7453	7100	5+	4	2	4	15	6.8	6.8	8.0	7143	4320	4	5	8	3	20	5.3	6.1	6.7
May	21	113			7455	6110	5	4	2	5	16	6.3	6.9	7.5	7141	Sick, recovered								
June	20	133			7456	6330	4	4	4	4	16	5.8	6.8	7.9	7143	4300	4+	4	6	3	17	5.4	5.7	6.3
July	14	147			7623	5360	5+	4-	5	5	19	5.6	6.6	7.6	7145	3720	4	5	6	4	19	5.1	5.3	6.0
Aug	0														7452	3340	4	5	6	5	20	5.2	5.3	5.9
Sept.	0														7454	4430	5	5	6	4	20	5.7	5.5	6.9
Oct.	0														7457	3370	4	5	7	5	21	4.8	5.4	5.9
Nov	0														7621	3700	3	4	4	5	16	5.2	5.4	6.2
Dec	0				TOTALS	46090	32	28	32	29	121	43.0	49.3	55.0	TOTAL	35200	35.5	40	55	37	169	47.1	50.0	56.6
YEARLY TOTALS		147			AVERAGES OF MALES	6584	4.6	4.0	4.6	4.1	17.3	6.1	7.0	7.9	FEMALE AVERAGES	3911.1	3.9	4.4	6.1	4.1	18.8	5.2	5.6	6.3

Age at 1st egg 268 days
Max 1st yr wt 12.0 lbs
Egg weight 68.4 gms
Shell texture Normal
Shape Nat. to sl. pointed
Color Light brown
Spotting Moderate to dense

REMARKS Dark brown eyes

	Weight	Breast type	Feathering	Fat grade	Leg fleshing	Composite grade	Length keel	Length shank	Depth
24th week data of this bird	3960	2	3	2-	2	25	5.5	5.5	6.4
24th week male sib sum 6 birds	5533	4.3	3.8	3.2	2.9	35	5.8	6.7	7.8
24th week female sib sum 6 birds	3433	2.7	2.8	2.7	3.2	29	5.2	5.3	6.1

INCUBATION AND REARING RECORD

Year	Breeding data: Pen	Breeding data: Male No.	Eggs: Set	Eggs: Fertile	Dead embryos	Number of poults hatched: Helpouts	Cripples	Normal	Total	Eggs fertile (percent)	Percent hatched: Of fertile	Of total	Poults brooded	Non-acc. mort.	Percent viability
1945	11	660	21	21	0	0	0	21	21	100	100	100	18	1	94.5

PEDIGREE OF

SIRE Winter Wt gms · DAM Winter Wt gms · 1st yr eggs grams % hatch · 24th wk data: This bird, ♂ Sib sum, ♀ Sib sum · Weight · Breast type · Feathering · Fat grade · Leg fleshing · Composite gr · Length keel · Length shank · Depth

U. S. GOVERNMENT PRINTING OFFICE 10—16024

Fig. 13.1. Female Summary Sheet. (From U. S. Dept. of Agr.)
This record summarizes all the data obtained on each breeding hen's performance in the breeding of turkeys for economic qualities. It is the key to the selection of breeding stock. A similar sheet is made for each breeding male to include performance data for all hens in the pen.

UNITED STATES DEPARTMENT OF AGRICULTURE
BUREAU OF ANIMAL INDUSTRY
A. H. Form 307
HATCHING RECORD
Breed Br. X Wh. Hol.

YEAR 1936 MATING NO. 5 SIRE NO. 362 HEN NO. 1574

DATE SET	EGGS SET	INFERTILE	DEAD GERMS I	DEAD GERMS II	DIED IN SHELL	CRIPPLES	GOOD CHICKS	DATE HATCHED
Mar. 9	12	0	0	1	2	1(H)*	8	Apr. 6
" 24	6	0	0	0	0	1(H)*	5	Apr. 21
Apr. 8	7	0	0	0	0	0	7	May 6
Apr. 21	5	0	0	0	0	0	5	May 19
June 1	9	0	0	0	2	0	7	June 29
June 9	5	0	0	0	0	0	5	July 7
Totals	44	0	0	1	4	2	37	
						Hatchability	88.6 %	
TOTALS								

Egg data—April:
Av. weight 85 gm.
Normal shape
Normal shell texture
Pale to light brown color
No spots to moderately spotted

U. S. GOVERNMENT PRINTING OFFICE 1937 * Helpout

UNITED STATES DEPARTMENT OF AGRICULTURE
BUREAU OF ANIMAL INDUSTRY
WING AND LEG BAND INDEX
A. H. Form 310

Hatched Apr. 13, 1938.

Leg No.	Wing No.	Dam	Sire	Date of Disposal	Reason	Leg No.	Wing No.	Dam	Sire	Date of Disposal	Reason
6	6137	1860	430				6178	1872	430		
	6138	1860	430				6179	1874	430		
	6139	1860	430				6180	1874	430		
	6140	1860	430				6181	1876	430		
	6141	1861	430				6182	1876	430		
	6142	1861	430				6183	1876	430		
	6143	1861	430				6184	1879	430		
	6144	1861	430				6185	1879	430		
	6145	1862	430				6186	1881	430		
	6146	1862	430				6187	1881	430		
	6147	1862	430				6188	1886	433		
	6148	1863	430				6189	1886	433		
	6149	1863	430				6190	1887	433		
	6150	1863	430				6191	1887	433		
	6151	1864	430				6192	1888	433		

Fig. 13.2. Individual hen and wingband records. (From U.S.D.A.)

easily when the hen is lifted from the trapnest. Even if saddle numbers rather than legband numbers are used in trapnesting, each hen should have a permanent serial number stamped on a sealed metal legband. Marked saddles avoid the necessity of handling the hens when trapping. White saddles should be stamped with figures about 1 inch high in opaque black waterproof rubber-stamp ink near the edge at the lower rear on both sides and in the center-front, middle, rear—reading from the rear.

Trapnesting

Turkey trapnests, a photograph of which is shown in Fig. 13.4, may be built in units of from two to eight, as desired. Each nest occupies 14½ inches of linear space except the two end nests which occupy 16 inches including the siding of half-inch boards. A battery of eight nests would be 119 inches long, 45 inches high and 26 inches wide. Detailed plans for trapnest batteries may be obtained by writing to the U.S.D.A. Research Center, Beltsville, Maryland. Resilient false bottoms are desirable.

Either a home-made or a commercial turkey trap-nest front may be used. If it is of different dimensions from the one illustrated appropriate changes should be made in the dimensions of the battery. The front illustrated measures 13½ inches by 17 inches over all and consists of a wooden framework of 1 x 1 inch material with a thin board at the bottom. The opening between the top framing and the bottom board is 10½ inches wide and 11 inches high, and contains a folding two-piece galvanized metal trap-nest device without guide wings. The battery of trapnests usually sits directly on the floor of the pen with its back against a wall or against the back of another battery. (The guide wings shown in Fig. 13.3 may be eliminated.)

Trapnesting procedure can be adjusted to suit varying conditions. Possibly the most accurate and efficient method is as follows. Have the breeding hens fitted with well-marked saddles, the numbers on which can be read

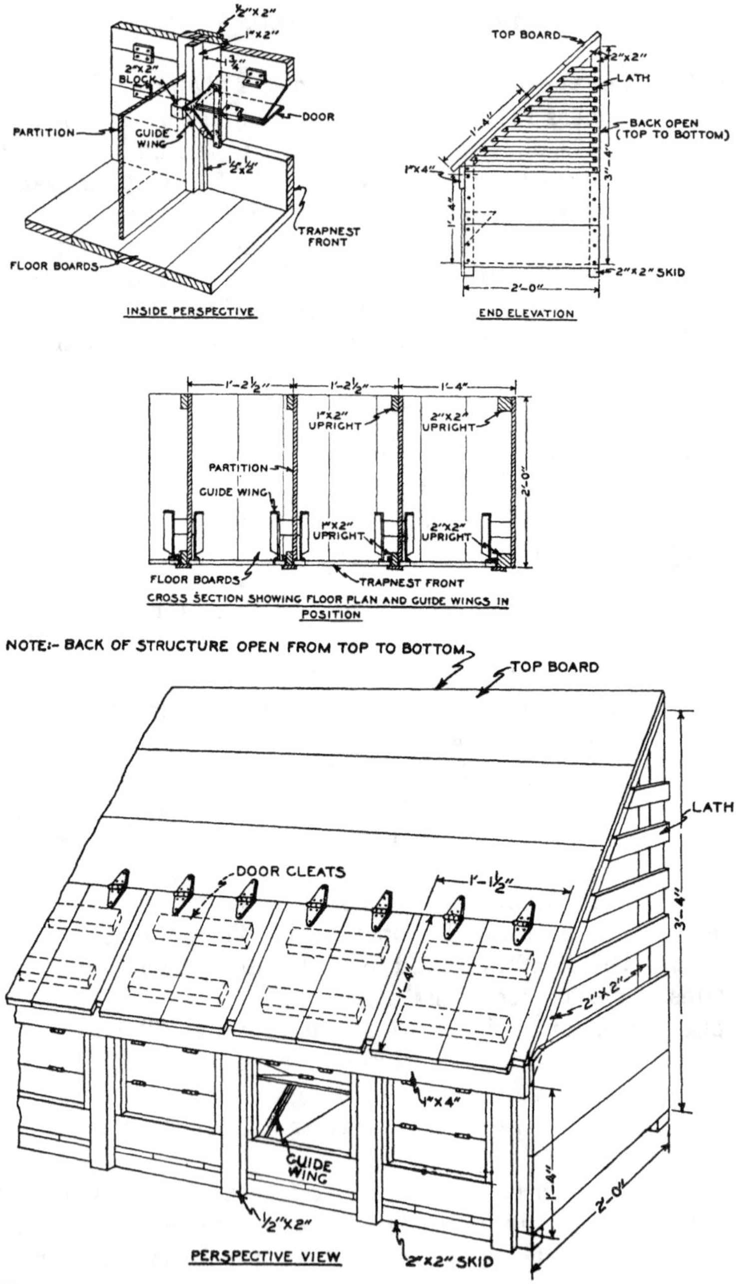

Fig. 13.3. Plans for turkey trapnest used by U. S. Dept. of Agr. (Bill of materials available from U.S.D.A.)

easily and accurately through a $1\frac{1}{4}$-inch-wide opening in the center of the hinged trapnest cover. (This opening is not shown in Fig. 13.4.) First, remove the egg through the trapnest front opening, resetting the trap in the process. Then read the hen's saddle number and write it on the *small end* of the egg along with the pen or male number and date laid. Finally, open the hinged lid which allows the hen to jump out of the nest. The lids can remain open until the trapnesting is completed in the pen. If the hens are not equipped with marked saddles it is necessary to handle them by one wing so as to read the legband numbers. If the hens are inclined to be flighty the above procedure is almost necessary to hold down egg breakage to a minimum. However if they are quiet and gentle it may save time and labor to first open the top door, read the hen's numbers, and lift her out of the nest, obtaining the egg last rather than first. When the eggs are all obtained and marked, the operator should take them at once to the egg record card (which is securely fastened in a slotted frame in the pen). Each egg is then credited to the hen which laid, in the proper date column on the trapnest sheet. As the eggs are recorded they are placed in a receptacle in the pen for picking up later. To indicate an egg laid, the last digit of each hen's number is used, which helps to avoid crediting a hen with eggs laid by another. Marking the egg record at the time of trapping is important. It helps to detect errors in reading bands for if the number on the egg is not on the egg-record card or if the hen has already laid that day, the error will often be discovered in time to mark the egg doubtful or as is often the case the hen's number can be re-checked. On rare occasions a hen may lay two normal eggs in one day but as a rule two eggs from the same hen the same day means that an error has been made. In such cases both eggs should be marked doubtful and not used for pedigreeing.

Floor eggs may become a serious problem. They may be laid in the yard, in the pen, or from the roost and can be pedigreed only for sire. They are, therefore, of littl value in the breeding project but may be sold or utilize for the production of market birds. To prevent laying outside the trapnests is one of the trapnester's greates problems. Usually turkey hens, if they start to lay be

Fig. 13.4. Hen entering trapnest to lay, the door closes (as at left) when the hen enters. After laying, she is removed through the top door (shown open).

fore the outside vegetation has grown very much, take readily to the nests and will continue to use them as long as no tall vegetation or other hiding places become available. It is on account of this that the yards for trapnested matings must be rather small and contain no tall vegetation. The trapnests should be in place and a few of the traps fastened open some time before production starts to acquaint the hens with them.

When production starts, the traps are set for operation and the trapnester makes frequent visits to get the hens properly started. Some hens will try to make nests in the litter, in the yard, or even on the wire under the roosts. When a hen engaged in this undesirable practice is discovered she is caught and examined for laying condition. This is done by gently inserting the forefinger (with closely clipped finger nail) into the vent and carefully exploring the interior in an upward and inward direction. It is very easy to determine whether the hen will lay that day as the vent will be pliable and moist, and the hard shell of the egg can be felt, less than half a finger's length from the vent. Sometimes an egg can be felt but feels soft and appears to be located farther inside the bird. In this case the hen probably will lay the following day. If it appears that an egg will be laid the same day the hen is then taken to the trapnest and guided into it where she is left until she lays. All inviting corners should be eliminated, and each stolen nest destroyed after removing the eggs. The hen should be excluded from the site of the stolen nest in some convenient way so that she is not likely to use it again. The yard must be scrutinized frequently for eggs, especially if the record card shows some hens not laying in the trapnests and are not broody.

Two hens may get into the same nest, due to too few nests, poorly functioning traps or nest openings that are too wide. If both hens have laid, each hen may be credited with an egg on the record card but the eggs must be marked doubtful and not pedigreed except for sire. If only one egg is found in a nest with two hens, the vent test and the egg record card will often decide the issue; sometimes it cannot be decided and the egg is considered a "floor egg." The egg must not be marked as a pedigreed egg when there is the least doubt as to its source. Occasionally the hens in a pen will lay so many floor eggs that it becomes necessary to examine them early each morning for a few days to determine, by the vent test, which will lay and then make them walk into the nests to remain there until they lay. Persistent floor layers should be removed as soon as they can be identified.

The vent test is almost indispensable in determining when to put the hens into the broody coops. Turkey hens do not indicate their broodiness in a readily descernible manner at first and if they are put in the broody coops while still laying, the pedigree and egg record will be lost if there is more than one hen in the broody coop. Consequently, when a hen appears to be broody she should be examined. If her vent is constricted and no egg is felt she has ceased production and should be put in a broody coop with a floor of wire or slats for four days with mash and water before her. The broody coop should be placed in good light to encourage return to production. A steel exhibition coop with a 3′ x 3′ wire floor section for a base makes a good broody coop. The regular feed and water receptacles often can be placed next to the broody coop so that the hens therein may have a continuous supply of feed and drink without extra labor.

Trapnesting should be done three or four times daily in regular practice although sometimes more frequent trappings are necessary at the start if floor eggs are being laid. Under most management plans the first trapping should be started about 9 a.m., the second at about 12:30, the third about 2:30, and the fourth about 5:00 or a little later. Trapping may be reduced to three times daily (9:00, 1:30 and 5:00) to reduce the amount of labor required; but four times a day tends to give better results. A supplementary evening trapping to release layers is helpful but as a rule it tends to be impractical. One trapnest for every two hens is needed, otherwise the rounds must be made more often than three or four times daily.

During the first morning trapping, the following duties are regularly performed:

(1) All hens in the nests that have not laid are removed and examined.

(2) Broody hens, if any, are identified (by their actions, the vent tests, and lack of production on the egg-record sheet) and placed in the broody coop.

(3) Hens in the broody coop are released when they have been confined four full days.

During the mid-day trapping (or trappings) as a rule, only hens that have laid are removed, the others being allowed to remain until evening.

During the last trapping all hens not having laid are removed and examined. For each hen found in the nest but not laying, a dot is placed on the egg record under the proper date to show that she nested but did not lay. If the following morning the same hen is found in the nest and has not laid, she is suspected of broodiness. If the vent test indicates an inactive oviduct and she already has two or three dots to her credit she is adjudged broody and placed in the broody coop. The turkey hen does not cluck when broody as does the chicken hen. Her gentle call is heard only when she has poults with her. Broodiness is not indicated in any certain manner, hence is hard to identify. Walking on tiptoe, hissing and persistent nesting without laying are indications of broodiness. If floor eggs are being laid, all hens having a hard-shelled egg in the oviduct close to the vent are returned to the nest to be removed the following morning or during a late evening supplementary trapping when this can be done conveniently. In pens where eggs are not being laid outside the nest, less care need be taken in returning to the nests hens that have eggs ready to lay. However, floor eggs are such a serious problem that it is usually best to return such birds to the nest at the time of examination. Trapnests save many eggs from breakage and becoming soiled. It is estimated that one operator can trapnest, care for, and handle the eggs from 400 hens.

Holding and Sorting Pedigreed Eggs

Egg-holding racks are not essential but they are convenient. The 2-strip tracks are 3¼ inches on center, 10 tracks per row, 14 eggs per track. Each of the hens from which pedigreed eggs are being saved has a place in numerical order, in the egg racks. These racks should be located in an egg room that can be maintained at a temperature below 65°, preferably between 50° and 60° F. After gathering, the eggs are placed on the racks, in rows according to hen number. The freshest eggs are placed on the near end of the row, rolling the older eggs back in the operation. Consequently each egg is turned whenever a fresh one is added, and the operator is able to tell at a glance just how many eggs are available from each hen. If holding racks are not used (and often there is not space in the egg room for them), the pedigreed eggs may be placed in egg cases or on trays according to pen number. They may be sorted according to hen number when set or, if number set per hen is not needed at the time, sorting may be deferred until basketing after 22 to 24 days of incubation. The number set may be determined when the hatching data are all in. If sorted at the time of setting, the eggs, if not already sorted according to pen number, are so sorted and then sorted according to the individual hens. A large egg-sorting rack or table with 1⅝ in. holes, 2⅛ in apart sidewise and 3 in. apart front to rear, on centers, permits arranging the eggs on their sides end to end for easy checking. The number of holes per row should be sufficient to handle the maximum set from one hen at one time.

Selecting and Setting the Eggs

The date set, hen number, pen number, and number of eggs set are recorded on an incubation-record sheet, or individual hen record (top, Fig. 13.2) at the time of setting. The eggs are first checked for both male or pen and hen number, soundness of shell, and general appearance. Soundness of shell may be determined by gently tapping each two eggs together end to end. A hollow sound indicates a cracked egg. As a rule, cracked eggs will not hatch, but in the case of extremely valuable eggs the crack may be covered with a little collodion or a small piece of scotch tape. With this treatment it is possible to hatch many of the cracked eggs if they are carefully handled. Errors in marking, in trapnesting, or in sorting can often be detected by checking each hen's eggs for uniformity of appearance. Any egg or eggs that differ markedly from the other eggs in the group should be looked upon with suspicion. If the difference is great, they should be discarded or at least not pedigreed. The same is true of eggs that are marked incorrectly. These precautions contribute toward accuracy in pedigreeing without which no pedigree system is worthwhile.

Some time during the laying season after each hen has laid for a few weeks, the egg shape, weight and shell texture should be determined and recorded as an aid to selection of breeding stock. The occurrence of abnormal eggs should also be noted on the hen's record. Attention to these points will prevent undesirable egg characters from becoming established in a strain. A fair sample of ten or more typical eggs should be used to obtain shell-quality and egg-weight data.

The tendency to lay thin-shelled, off-sized, off-shaped, or rough-shelled eggs should be eliminated by marketing all offspring from hens that lay eggs defective in these respects. This tendency, if it occurs, should be noted during the course of the breeding season to be sure that the hens and their offspring are eliminated. A hen that starts out laying eggs of poor quality should immediately be replaced in the breeding pen, or at least her eggs shoulc not be set for pedigree purposes.

Testing

Testing after 7 to 9 days of incubation and agaiı on the 22nd to 24th day is useful in pedigree work The first test serves to eliminate the infertile and earl; dead-germ eggs. The second eliminates the few embryo: that die during the middle of the incubation period anc leaves for pedigreeing only those eggs with strong em bryos. The results of these tests should be recorded (fo: each hen) on the incubation or individual-hen recorc sheets. Some breeders may consider that more than on test, on the 22nd to 24th day, is impractical in force draft incubators, since the eggs are packed (trayed] tightly on end in each tray and the tray turned. Th 7th-day test would involve retraying which, however usually is justified because of the more prompt and accu rate fertility data secured. After testing (on the 7tl to 9th day, all the eggs originally grouped together iı each incubator tray (when the eggs were set) should b returned to the incubator in the original groups, or i: desired, two or more groups may be consolidated anc sorted together at the time of basketing. If the egg: when set were taken off the holding or sorting racks iı numerical order and the original grouping maintained then at sacking or basketing time all the eggs from on hen will be found in the same group and can be put in th same basket or basket compartment. When the eggs ar placed in the pedigree sacks or baskets no further turn ing is necessary.

Basketing

Between the twenty-second and twenty-fourth day o incubation (preferably the twenty-third day) the pedi greed eggs are removed from the incubator and sortec numerically according to the hen number on special sort ing boards, preferably at a room temperature of abou 75° to 80° so that the eggs will not become chilled. Eacl hen's eggs are placed flatwise in a row on the sorting board and then checked. If two persons are sorting (and this is an advantage), one may start with the lowest hen numbers and the other with the highest hen numbers until all eggs in the group are on the sorting boards. Then each operator should check his own work to be sure that all of the eggs in one row were laid by a certain hen. Here again the uniformity of the eggs from each hen can and should be checked for the sake of accuracy, eliminating any doubtful eggs. After each operator has checked the eggs he has sorted, he should check the other operator's work. This double check is most important. One person working alone should carefully recheck his own work. When the eggs are all sorted and checked, each hen's eggs in a separate group, a slip of paper about 1 inch square is placed with each single egg (a case where there is only one from a hen). On this identification slip is marked the hen's number and the male's number or pen. When the eggs are sorted the slip is placed face down in the compartment with the egg. It is to be used to identify the eggs, and consequently the poults hatched from them, in case the markings on the eggs become obliterated during hatching. When the group consists of two or more eggs, the chances of not finding at least one egg shell with the marketings intact are so small that an identification slip is not needed. Marking the egg near the *small end* facilitates identification as this end of the egg shell is seldom broken during hatching, whereas the large end almost always is. Normally the poults pip the egg near the large end, cutting it in the region of the normal position of the air cell. When identification slips are prepared and checked to their respective egg groups, the basketing itself is done, always by the most reliable person available, preferably the one in charge of breeding operations.

Rectangular pedigree baskets made of 1/4 to 1/3 inch mesh hardware cloth generally are used. They may be purchased or made. In either case they should fit the incubator trays so that there will be a minimum of waste space. A basket 3½ in. high by 5¼ in. wide by 9 in. long will hold 5 turkey eggs of average size, 6 eggs a little under average, or 4 extra large eggs (see Fig. 13.5). If provided with removable partitions made of ¼ to ⅓ in. mesh hardware cloth, it is a satisfactory size to use for general pedigree work where only one size of basket is employed. Baskets as long as the incubator trays, with sliding cover and easily adjustable partitions are very satisfactory. One manufacturer makes these long baskets, 3¾″ high, 5½″ wide and 25⅜″ long, to fit the hatching trays without waste of space. Baskets higher than about 3¾ in. are a disadvantage because poults are more likely to trample each other. About 9 square inches of floor space in the baskets are needed for each egg, allowing a little more space for the eggs larger than average and a little less space for the eggs smaller than average. Baskets, or compartments of various sizes in the baskets, are needed to fit the number of eggs from each hen generally available at basketing time. Under most conditions one or two sizes of baskets are all that are needed. If two

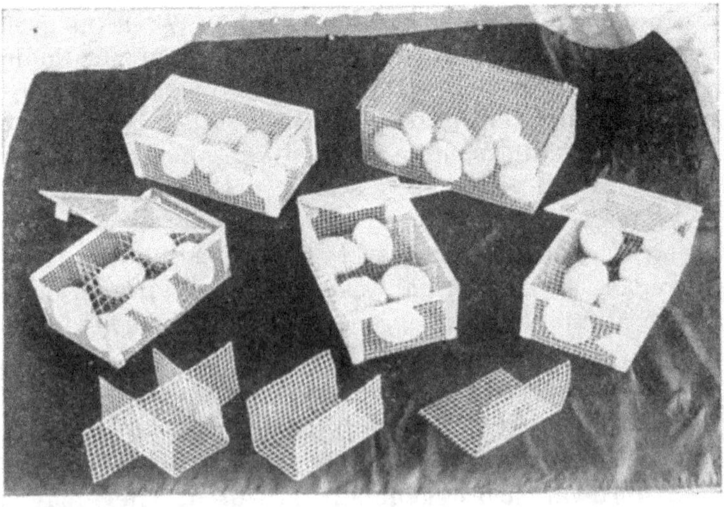

Fig. 13.5. Pedigree baskets and removable wire partitions to fit them. (Photo from U. S. Dept. of Agr.)

sizes are used, about half of them may be of fairly large size (6″ x 10″ x approximately 3½″ high) to hold 6 to 8 eggs, and the other half of a smaller size (5¼″ x 9″ x approx. 3½″ high) to hold 5 or 6 eggs. Removable "L" and "U" shaped partitions (see Fig. 13.5), to place in both sizes of baskets are provided so that all groups of the various sizes can be accommodated without great loss of space. These partitions may be placed in the positions needed to accommodate the various groups of eggs. When there are groups of nine or more eggs, two of the baskets may be used. If home-made baskets or machine-made baskets without fasteners are used, the lids can be fastened down with a piece of No. 24 copper wire about 6 in. long. These wires are put through the basket and the lid and given just a half turn. This holds them securely and permits them to be opened quickly. The copper wire may be used over again many times. Large baskets without reinforced edges should be wired near each end, small baskets only at the middle. Care must be taken that the lids fit tightly so the poults cannot get their heads through and hang themselves. Machine-made baskets with smooth edges and handy friction fasteners can be purchased and are the most satisfactory type to use. They should be ordered in the right height, length, and breadth to fit the trays. Partitions for them can be made or purchased. As mentioned above, when space between trays permits, the pedigree baskets should be approximately 3½″ high, allowing ⅜″ or more of clearance above the lids of the baskets.

Construction of home-made baskets is not difficult once a pattern is made out of stiff cardboard, sheet metal, or other suitable material. The material for the box itself is cut out, then carefully bent to shape and fastened securely with No. 18 galvanized wire. The covers are made to fit the tops exactly and are hinged with loops of the No. 18 wire or with small lock-washers pinched together to form a ring. It is desirable to fasten folded strips of lightweight galvanized sheet metal over the edges and around the cover to make the boxes easier to handle.

Pedigree baskets shaped like strawberry baskets and about the same size can be neatly stored in stacks by "nesting" when not in use. Light weight sheet metal should be used and holes ($\frac{1}{2}$ to $\frac{3}{4}''$) punched in the metal before cutting the sheets to the pattern; to allow air circulation. The basket is placed over the eggs of one hen; with what is commonly thought of as the bottom of the basket turned up.

Sometimes pedigree trays are used that have, instead of separate baskets, merely separate compartments with a single cover for the whole tray. These are not recommended unless the cover is the sliding type. Sometimes the spaces bounded by the pedigree baskets and the sides and regular partitions of the trays are used in lieu of filling the tray with baskets. This is not desirable, since there may be room for the poults to crawl up over the baskets or tray partitions and get into the wrong compartment. When these spaces are used, identification slips should be included at basketing time and each space must be carefully covered as the tray is taken from the incubator. Then each egg shell from which a poult was hatched must be checked against the number of poults found in the group.

Pedigree baskets should be cleaned and sterilized after use. The quickest and most satisfactory way to clean them is to brush out loose material (shells, droppings, etc.) with a wire brush then pile the baskets in a cabinet and treat them with live steam for one hour. If live steam is not available, a satisfactory method of caring for the baskets is to stack them in piles and expose them to fine spray from a hose for an hour or two. This loosens the dirt so that they can be cleaned quickly with a brush or by pressure spray from a hose. Following the cleaning the baskets may be dipped in a 3% cresol or stock dip solution or any good non-corrosive disinfectant. Disinfecting probably is not necessary if washing is thorough and the baskets are dried well after washing; assuming there has been no infection in the incubator.

Circumstances may occur under which one does not wish to go to the expense and trouble of making wire pedigree baskets. In such a case, pedigree sacks may be made of mosquito netting, crinoline or the material in common use in laundries for bags to keep identity of the several lots. Sacks should be about 8 in. by 12 or 14 in., sewed on both sides and one end. The other end is folded over (corners first) and pinned with a safety pin after the eggs are placed in it on the 22nd to 24th day of incubation. Such sacks can only be used once or twice as they soon become soiled. They cannot be washed as the stiffness is thus lost and the poults may work their way out. Never more than 5 eggs should be placed in one sack. Sacks made from heavy mosquito netting used by the Army may be laundered.

Wingbanding

Banding is done after all the poults are hatched and thoroughly dry. This is usually early for small-type or late for large-type—in the 28th day of incubation. Bands are available in various styles. The No. 3½ turkey wing-bands of extra-heavy aluminum appear to be one of the best types now available for turkeys. They are large and strong and the lettering is large enough to be read easily. They are ordered stamped with a series of numbers starting with a letter prefix and containing not over four figures; for example, A-1 to A-9999. When one series is exhausted, another may be started using a different prefix. Chick-size bands are too small for turkeys but may be used in emergencies, replacing them later by turkey-size bands.

Some breeders prefer to stamp, or have stamped, the bands to show the dam and hatch number and possibly pen and poult number (from each dam) for direct reading without referring to records. This system is workable but extra labor in stamping bands is required or else wastage of bands, due to failure of some hens to produce poults as estimated, is certain to occur. There are other disadvantages, so bands stamped in straight numerical sequences are recommended.

There are several types of pedigree wingbands on the market with a pointed end that can be pushed easily through the web of the wing, without the use of a knife. However with turkey poults sufficiently accurate placement cannot be made by punching the hole with the band. The band may be sealed or clinched on the wing of day-old poults, making the puncture with a knife, as described in the wing-band method. The clinch-type band is preferred to the sealed type, unless the latter is required by regulations governing official pedigree work. Bands may be put on the leg of the poult at first and changed to the wing at three or four weeks of age or put directly into the wing at hatching time. The second method is preferred, but it requires more skill.

Circular-shaped wingbands should be 9/16″ for small-type or 10/16″ in diameter inside for large-type and made of aluminum or aluminum alloy ¼ in. wide. Rectangular-shaped wing-bands should be 21/32 in. for small-type to 11/16 in. long inside for large-type and ¼ in. wide with figures 3/16 in. high. These specifications produce bands that are just large enough for turkeys' wings, provided they are placed *exactly 3/32 in.* in from the edge of the web. At the same time there is a minimum amount of slipping over the poults' wings when bands are first put on at the day-old stage. However, due to the difficulty in inserting wingbands in day-old poults at exactly the right point it is more practicable to use a rectangular band about 1″ long or the circular No. 3½ band ¾″ in diameter. In either case squeeze it together near the middle when first inserted, then open it up fully when birds are about 8 weeks old. The circular band is flattened by applying pressure on the clinched or sealed joint and the portion opposite the joint.

Legband method: The bands (which must be of the clinch type) are first sorted numerically (if they are not already arranged that way), made round (using a penholder to shape them), clinched, and strung on a wire in numerical order. The toes of one foot of the poult are held together with the fingers and the circular bands slipped over them. The band is then flattened, using more pressure on one side than on the other, so that it will not slip off over the foot but at the same time will allow for about ten days' growth of the shank. The thumb is placed at the point where the band is clinched, the finger is placed on the opposite side, and the band flattened as directed above. This is somewhat difficult to do and much care is necessary at the start until the operator learns how. Each band should be tested to see that it will not come off. At about ten days of age, the poults are all handled and the bands loosened (with a dull knife blade) to allow for about ten more days' growth. Then, at about three weeks of age, the bands are transferred to the wings in the following manner: They are unclinched, smoothed out with small pliers and made circular again, but not reclinched until placed in the wing. With a sharp thin knife blade or scalpel a slit a little longer than the width of the band is made 3/16 inch back from the edge of the web of the wing about half way between the first and third joints, care being taken to avoid cutting the large wing vein. The open band, already made semicircular, is slipped into the slit, clinched and made round or rounded. This completes the pedigreeing process. No antiseptic treatment is needed. A little bleeding may occur but of thousands of poults banded by the authors not one has ever bled seriously or developed infection.

Wingband method: The bands in groups of one hundred are first sorted and strung numerically on a piece of wire or string (unless they are already arranged numerically.) The poult may be held by one operator while another person stretches the poult's wing so that the web is moderately taut, or better, one person should do the

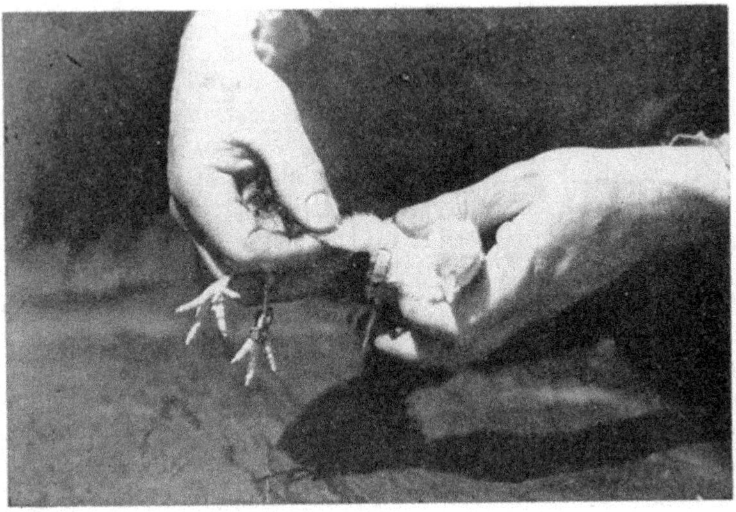

Fig. 13.6. Pedgiree band placed in wing (right) and on leg (left). (Photo from U. S. Dept. of Agr.)

banding, holding the wing taut as shown. A slit, a little longer than the band is wide, should be made in the web about half way between the first and third joints, *exactly 3/32-inch* from the edge. In learning the operator should measure the distance or use a 3/32″ width gauge. The best instrument to use is a small botanical scalpel with a sharp thin blade just the width desired for the length of the cut. A very thin-bladed pocket or pen knife kept sharp will do. When the cut is made, the open band is slipped into it, clinched and, unless of the rectangular shape, is made circular with the aid of small pointed pliers. If of the smaller size suggested (on page 758) it is left this way, but if it is of the larger (¾ inch) size it should be squeezed together until the indented portions touch, then opened up at 8 weeks. If the slit is too far back from the edge of the web there will not be room for normal growth of the wing tissue within the band when the poult grows larger. To be sure about this, the poults should all be examined at about sixteen weeks of age. If, at that age, any of the bands are so far back so that the flesh nearly fills the band, the band should be unclinched and replaced in a new slit made about 3/16 to 1/4 inch from the edge of the web at a different point in the same wing or in the other wing. If the bands are placed too close to the edge of the web they will be lost. In the brooding quarters care should be taken that there are no wires, nails, or other objects that may catch the bands. Some metal watering devices have wire fasteners over which the bands will sometimes slip and be torn out of the wings. Floor wire of ½-inch hardware cloth will cause loss of bands which slip down into the mesh when the poults sit down; the bands are caught there when the poult arises.

No matter what the method of banding is to be, great care must be exercised at banding time to obtain accuracy in the pedigrees. The pedigree record book or sheet should be ready (see Fig. 13.2), preferably with the band numbers already recorded. This record may contain columns for weight and other data useful to the pedigree breeder and investigator. A carbon copy should be made when banding is done, and the carbon copy habitually filed in a location separate from the original to prevent total loss of all pedigree records in case of fire or accident.

When the tray is removed from the incubator at banding time, any loose poults and the shells from which they hatched should be removed. Those in open-top compartments should be confined by covering with sections of heavyweight hardware cloth or else each group should be put in a basket along with the corresponding shells or unhatched eggs and checked against the number of egg shells from which poults have hatched. If there is any mistake or if there are any poults loose in the tray without identification, the poults in all groups involved must be recorded as of doubtful pedigree. It is on account of this opportunity for error and consequent loss of pedigree that baskets for the eggs should be used for all pedigree hatching. These loose groups of poults, if any, should be banded first unless they are put in baskets as suggested.

While banding, only one basket is opened at a time. The person doing this (operator No. 1) counts the number of poults and takes that number of bands from the group, checking the number of the bands against the band numbers on the pedigree sheet. Bands and poults are then turned over to the person (operator No. 2) or persons doing the banding while operator No. 1 checks whenever possible each shell from which a poult has hatched to be sure that they all have the same hen and male numbers, and then writes them opposite the proper band numbers on the pedigree sheet. The number of unhatched eggs (containing dead-in-shell poults) from each hen should be marked on the incubation record. This is not essential but does provide a final check for the recording of eggs set and those removed after testing. If this is not done, the number of poults dead in the shell is determined by subtracting, from the number of eggs set, the total of eggs tested out and the number of poults hatched.

Occasionally pedigrees will be lost when the poult loses its band. Duplicate banding, one band on each wing, will prevent this. However, except for especially valuable birds or for critical experimental work, this precaution is not necessary. When a band is lost the bird becomes a non-pedigreed individual, unless it happens that only one band has been lost out of a group of poults, the band numbers of which are known. Even in this case it is safer not to use the bird for pedigree breeding.

Rebanding is often necessary and for this purpose blank bands are useful. The desired numbers can be stamped on them, using a set of dies having 1/8 to 3/16″ figures. It is on account of the not-infrequent necessity for rebanding that the clinch-band types are preferable to the sealed type.

TATTOOING

Tattooing as a means of identification is practicable but its chief use is in identifying turkeys as to ownership rather than as a means of pedigreeing. However, it has some value for the latter purpose but only after poults are partly grown—during the latter part or at the end of the brooding season. If at that time the proper identification is tattooed into the wing-web, a satisfactory and permanent marking is established. The best time to tattoo is at the age of about 8 weeks or when brooding is over. Care must be taken to punch completely through the wing-web and to use enough of the proper kind of tattoo ink. Some states have a tattoo-identification system for protection of poultry against thievery. This fact must be considered if a grower wishes to tattoo his turkeys, and inquiry should be made of the State Police or State Department of Agriculture.

CAPONIZING

Turkeys seldom are caponized surgically since there is no market demand for turkey capons and it is doubtful if they have any practical advantages over uncaponized toms that are marketed at the usual age and in the usual manner. That turkey caponizing has been a known practice for many years and yet turkey capons have not become popular is indicated by the following,* written in 1675: "Turkeys are excellent meat, especially a turkie-capon beyond that, for which eight shillings was given, their eggs are very wholesome." The so-called "chemical" capon is discussed under estrogen treatment (page 434).

Advantages of Capons

The advantages claimed for capons are (1) that the meat is superior to that of hens and of toms the same age; (2) that gains are more economical; (3) that fighting is eliminated; (4) that they do not become flabby over the crop as quickly or as seriously as uncaponized toms, and (5) that the fine quality of flesh is retained for a longer period. However, the last two advantages are slight, as it is rarely economical to hold turkeys past the age at which they first reach U. S. Grade A quality, which is seldom more than 28 weeks, and uncaponized toms

*Josselyn's Voyages—An Account of Two Voyages to New England, by John Josselyn, Ghent.

have lost none of their quality and have not become flabby at that age. There is considerable doubt that the first two claims can be substantiated. The third claim is probably valid but the prevention of fighting without other advantages would not justify the cost of caponizing. Capons do not grow faster than toms but continue to grow over a longer period of time. Turkey capons weighing 60 pounds when 18 months old have been reported.

The North Dakota Station (1941), after experimenting with turkey capons, found that surgical caponizing had no advantages and was not recommended. For a general discussion of chicken and turkey capons and detailed directions for caponizing, U. S. Farmers' Bulletin 849, Kansas Experimental Station Bulletin 274 and publications of the Colorado and North Dakota Stations are suggested.

The Operation

The following comment concerning turkey caponizing is based on information supplied by the Colorado Extension Service.

Birds weighing 2 to 3 lbs. are best for caponizing. At this age sex cannot be accurately determined except by examination of the sex organs. An experienced turkey raiser can pick the large young toms in a group of brids all the same age with a fair degree of accuracy by the head, actions, size, and shape of the hocks. Young males are larger; have broader, flatter hocks; have broader, coarser heads; and to some extent, exhibit characteristic male tendencies. The starving and operating technique is the same as for caponizing chickens and can be done as successfully. The bird's left side should be opened first, making the cut between the last two ribs about half way down on the bird's side just below the "buttons" on the ribs. A stiff adjustable spreader should be used to hold the ribs apart. If a light weight spreader is used the cut may be made back of the last rib. If a female, the ovary will be seen, a flat granular-appearing mass, just back of the lungs and attached to the forward lobe of the kidney slightly to the left of the center of the back. The ovary can be removed, if desired, thus producing a "poulard." If a male, the left testicle will be seen, a small organ shaped like a grain of wheat. It may be light colored, or blackish, or part light and part blackish. The other testicle is a short distance to the right. Until considerable skill has been attained it is well to remove the left testicle from the left side and make another opening on the right side from which to remove the right organ. One should not attempt to caponize turkeys until he has developed skill in caponizing chickens. It is possible to remove both testicles from the left side and as skill is developed, this should be done as it possesses definite advantages. It is important to obtain a good set of instruments,* including a sharp knife or scalpel, an adjustable spreader for holding the ribs and musculature apart, a hooked probe, and forceps for removing the testes. A stand or table for holding the bird in position, and strong light, preferably daylight, passing over the operator's shoulder, are necessary. The feathers are plucked, the skin is drawn back over the hip, a cut about 1 inch long is made, and the spreaders are inserted. The air-sac membranes are then torn to obtain a view of the interior, and the removers (forceps) inserted and placed over the organ, being careful to include all of it and along with it the surrounding tissues but excluding the large blood vessels. The instrument is withdrawn with a rotary motion. Disinfection is not necessary but cleanliness should be maintained and it is well to use a 2 or 3 per cent solution of compound cresol for cleaning the instruments and the area around the wound.

Before the operation, the birds should be fasted overnight (12 to 18 hours) without feed or water up to the time of the operation. After the operation they should be allowed access to mash and water and confined to a small yard or building but not crowded in the house. No roosts should be provided. Wind-puffs, which frequently appear soon after the operation, can be punctured with a scissors or knife every day or two for a week or two until the wind-puffs have subsided. The capons can be reared in the usual manner, preferably not with young uncaponized toms.

*Electric caponizing instruments are more costly and no more accurate or dependable than the other types.

Appearance

Since head furnishings, sex behavior, spurs, and voice are true secondary sexual characters dependent upon male hormones for their expression, capons, which do not produce sex hormones, do not develop these typical male characteristics to any great extent. They do strut but not so frequently as do toms and no more frequently than some hens. Capons are typically male in plumage,* beard and size, while ovariectomized females remain the normal size of the female.

The results of spaying the turkey hen** were reported by van Oordt as follows: Head furnishings and body size show no change, the spurs remain small and do not enlarge as in the ovariectomized chicken hen. Since female plumage is believed to be a true secondary sexual character dependent upon the female sex hormones, the plumage of poulards assumes a neutral greenish metallic color

*Plumage and beard in the male turkey are independent of the male sex hormones, hence they remain typically male after castration.

**The spayed chicken or turkey female is called a poulard.

resembling feathers of males. The left oviduct remains small and thin. The regeneration of the rudimentary right ovary into a testes-like organ as observed in chickens did not take place. The status of the beard in the female is doubtful. Beards are common in white turkey females but uncommon in females of the colored varieties. Spurs and beard in female turkeys probably are independent, or neutral, unaffected by female sex hormones.

SEX DETERMINATION

I. The Vent Method. Sex may now be determined with a high degree of accuracy any time from hatching on, by a method described by Hammond and Marsden (1936). This method consists of an examination of the portion of the sex organs that can be seen in the vent. It is easy to learn but some practice is required to attain 90 per cent (or higher) accuracy. A study of the following description and illustrations, and practice with a few day-old poults that can be killed for checking, will suffice for the average person to master the procedure.

The technique is simple. Strong, non-glaring light is necessary. In the case of day-old poults the bird should be grasped in one hand, its breast toward the palm of the hand, feet out and held between first and second fingers, the breast supported mainly by the first and second fingers, and the back supported by the last two fingers. The neck is bent forward and held between the last two fingers. The thumb and the second finger of the other hand are placed on the right and left sides of the vent, then stretching it very slightly, the vent is pulled back over the tailhead as far as possible without hurting the bird, pressing down lightly on the umbilicus (located beneath the vent) with the thumb of the first hand. This operation if properly done exposes the sex organs without necessitating the squeezing of the abdomen to remove droppings. Gentleness is necessary to avoid injury to the bird and irritation of the anal region. If the vent becomes inflamed, the bird cannot be sexed accurately.

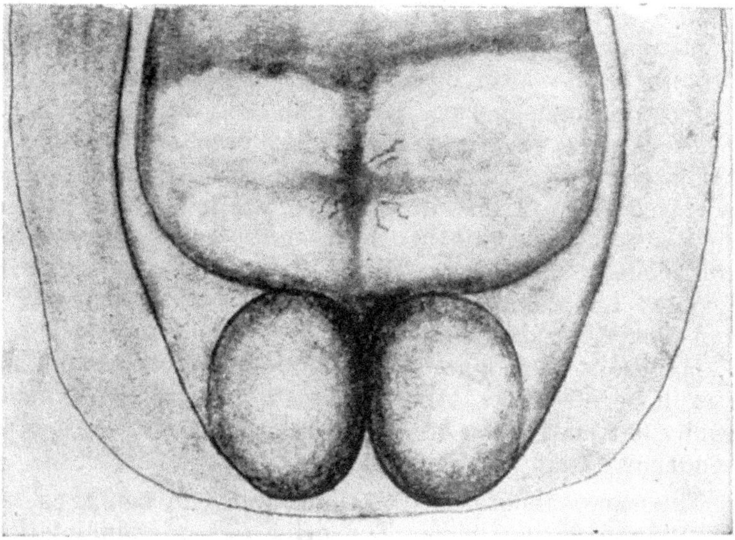

Fig. 13.7. Diagram of male sex organs of day-old poult. Note the two hemispherical papillae at the bottom of the diagram. (Photo from U. S. Dept. of Agr.)

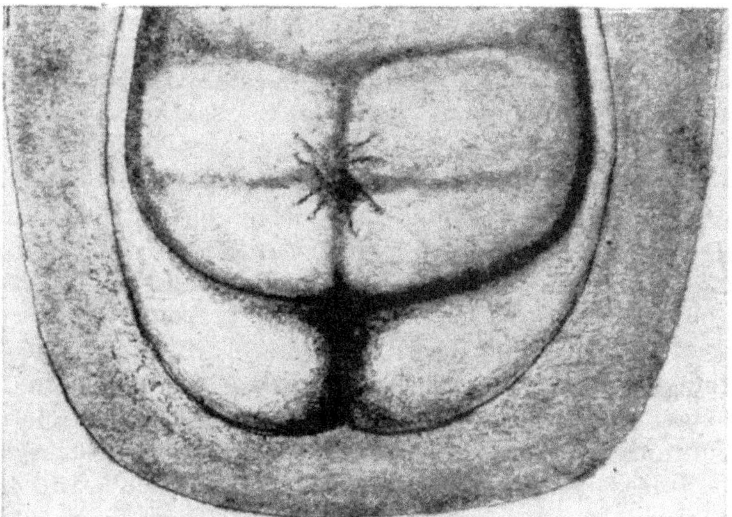

Fig. 13.8. Diagram of female organs of day-old poult. Note flattened eminences as contrasted with papillae of male. (Photo from U. S. Dept. of Agr.)

As will be seen in the illustration, all males and the great majority of the females have two easily discernible prominences. Those of the male, however, are always decidedly hemispherical and composed of firm tissue, whereas those of the female are elongated and composed of flabby tissue. In the great majority of females, the organs are not much in evidence, while in most males they are quite prominent. However, some males have inconspicuous organs that may be identified only by their firmness and hemispherical shape; and some females have conspicuous organs that may be identified only by their flabbiness and elongated shape. If carefully manipulated it will be noted that the female organs are much more easily flattened, whereas the male organs have more of a tendency to retain their shape.

The same general instructions apply to sexing older poults and mature birds. The principles of manipulation are essentially the same from hatching up to maturity. After the age of about 16 weeks, however, the organs are harder to expose so that a combination of the manipulations used in this method and those described for artificial breeding is necessary. However, once the organs are exposed there is no difficulty in telling the sex at any age. Any color pigment found on the sex organs should be disregarded.

II. The color of the adult breast feathers may be used to aid in determining sex after about 12 weeks of age in the Bronze. This method can be used but is less effective with poults of the Wild, Narragansett and Slate (but not Blue) varieties. Fast-growing poults will have some of their adult breast feathers partly grown out at the age of 12 weeks; slow-growing poults require a longer time. The adult feathers of the females will have white tips (or brownish tips in the case of wild turkeys) but those of males will have black tips. The adult feathers must not be confused with the baby feathers, all of which have light-colored shafts and tips; also feathers in the centers of the breast should be examined, not those of the margins of the breast region which are not necessarily different in males and females. In Bourbon Reds the males have black-edged adult breast feathers, while the females lack the dark edgings, having narrow white ones instead. Buff toms have black-edged adult breast feathers like the Bourbon Reds; Buff females have plain buff breast feathers without white tips.

III. General body shape, carriage, and behavior are of value in sex determination though the differences are hard to describe. Males appear to be longer of leg and neck and have characteristic male shape and behavior which can be recognized only with practice and familiarity with turkeys. Strutting, while characteristic of males, is not always confined to them, for many mature females and a few young females strut.

IV. The shape and size of the hocks and shanks are an aid in determining sex. The hocks of the males are broad and flattened, those of the females narrower and more rounded.

V. The beard is not a reliable indicator of sex because frequently hens, even young hens, especially of the white varieties, have beards, though usually they are thinner and shorter than the beard of the male.

VI. The characteristics of the head are of some value in determining sex. In general the head of the male is larger and coarser, has fewer feathers on the crown and on the face, and possesses a plump, prominent snood (nasothelium) in contrast to the inconspicuous, thin snood of the female.

In some cases a combination of characteristics must be used in determining sex, as a small percentage of birds are not typical in one or more respects.

VII. In a dead bird sex can be definitely determined by an examination of the internal sex organs. With small poults, the bird must be opened carefully to avoid destruction or misplacement of the organs. For routine examinations of small poults where great speed is not required the following procedure is satisfactory. After killing the poult by cracking the skull against a hard object and dislocating the neck, a sharp pair of scissors should be used to cut into the abdomen at the rear of the keel. The esophagus where it passes over the lungs should be clipped. All the viscera should be removed to expose the backbone, clipping with the scissors the adhering connective tissue. If the poult is a male, the two testes will be seen adhering to the back just to the rear of the lungs. The testes in small poults are comparatively small organs shaped like slender grains of wheat usually whitish in color although occasionally they will be black or partly black. In the female, the one ovary will be seen adhering to the backbone just to the bird's left of center and just back of the lungs. In small poults the ovary is a flattened, whitish organ of granular texture, somewhat irregular or pear-shaped. As the birds grow the sex organs change somewhat in appearance, but no trouble should be encountered in determining sex if the organs are not removed or destroyed during the preparation for examination.

VIII. In dressed roasting turkeys, sex can be determined accurately by examination of the sex organs in the vent. To examine, grasp the lower edge of the vent and slightly evert the cloacal wall. If the bird is a male the sex organs will be plainly visible although they are flabby and colorless due to absence of blood; if a female, the sex organs will usually be so inconspicuous that they appear to be absent. When the bird is drawn the observation may be checked for accuracy by the internal sex organs.

Body type and other secondary sexual characteristics also are useful in determining sex, especially when it must be done quickly as for instance in the commercial grading of dressed turkeys. The distinguishing external characteristics as seen in live and dressed turkeys are listed in Table 13.2.

In typical birds no difficulty in determining sex is encountered by experienced handlers—only a glance is necessary. However, there are considerable numbers of

Table 13.2—External Sex Characters

Characteristic	Appearance in Males	Appearance in Females
General body shape	Angular, with neck and legs apparently longer than in females.	Rounded, with neck and legs apparently shorter than in males.
Breast shape	Relatively angular, with high keel.	Rounded, with low keel.
Shanks and feet	Comparatively larger and coarser.	Comparatively smaller and more refined.
Hocks	Broad, coarse, and angular; comparatively broad and flat in front.	Narrow, refined, and rounded; comparatively narrow and slightly rounded in front.
Head	Headgear large; very few feathers on face and top of head.	Headgear small; many small feathers on face and top of head.
Dewbill or snood	Plump, large, and very distensible.	Thin, small, and not very distensible.
Beard	Usually good sized, consisting of many bristles.	Usually not present; when present it is likely to be small and to consist of comparatively few bristles.
Spurs	In young males spurs are small and inconspicuous; in old males they are quite large and prominent.	In females of all ages spurs are almost lacking though infrequently adult hens have a well-developed spur on one or both legs.

non-typical birds marketed; with these a careful examination of the various external features will usually suffice to determine the sex. As a last resort an examination of the sex organs at the vent can be made.

The *sex ratio* usually approaches 50:50 in newly hatched poults. The California Station reported 49.2% males and 50.8% females, with several thousand poults.

PHYSIOLOGY AND BEHAVIOR

At hatching time turkeys of both sexes have a body temperature of approximately 103° F. Within a week it rises to about 106°, and by the sixth week it reaches 106½°. At other times the average is approximately 106½° F. except for the females during broodiness, when it is approximately 1° lower, or 105½°. Except at the day-old stage, variations of as much as 1½° either way from the average given are within the normal range*.

Turkeys do not scratch to the extent that chickens do. They will work lightly on the surface among leaves and other loose material but do not dig deep holes, as do chickens except when dusting.

---

*Data collected by the senior author; rectal temperatures taken by inserting a thermometer well into the vent of standardbred Bronze turkeys; 8 a.m. to 5 p.m.

Responsiveness to gentle treatment is a characteristic of turkeys. They become very tractable and tame if handled carefully.

Day-old males occasionally strut and this essentially male behavior continues throughout the life of the bird as long as he remains healthy; a sick tom never struts. During the first few weeks of life, male strutting is only intermittent but as maturity is approached it occurs much more frequently.

In strutting the turkey inflates his crop, lowers his wings so that the tips of the primaries touch the ground, elevates all the body feathers so that they stand at right angles to the body, and bends back the neck until head and neck form a hairpin turn and the back of the head touches the erected back feathers. At intervals subdued hissing sounds similar to the human "tsk-tsk" are made and the bird gyrates for a few short steps, his primaries scraping the ground all the while.

Strutting, though a male characteristic, is not confined to males since hens, especially those over a year old, occasionally strut quite persistently. In one case a hen was observed by the senior author to tread other hens in a very realistic manner. These and other manifestations of male behavior constitute virilism, a term used to indicate the occurrence of male behavior in females.

It is commonly assumed that virilism in humans and perhaps generally in higher animals is due to disfunction of the adrenal (suprarenal) glands, one of which is located on each kidney. Virilism in the female chicken has been reported by McGowan to be associated with a tumor of the suprarenal cortical tissue. There is some evidence that under certain conditions the suprarenals secrete a substance acting in a manner similar to that of the male hormone, which hormone is the normal basis for behavior typical of the male.

Tumors of the normal left ovary of chickens are sometimes associated with the appearance of virilism. In this case there is probably interference with the normal hormone secretion of the left ovary. Reports dealing with virilism in turkeys are scarce but turkey hens not infrequently show it. It probably is due to the same causes as it is in chickens.

Virilism amounting to outwardly complete sex reversal from female to male was induced by Herrick (Poul. Sci., 1951). In this Kansas Experiment, two Broad Breasted Bronze hen turkeys six months old were implanted with two pellets of testosterone (male hormone) but adsorption was very slow so intramuscular injections of 8 mg. of testosterone were given twice weekly. Fighting soon became so intense that one hen had to be removed. The first indications of effect of the hormone were a slight reddening of the skin of the head and gobbling like a male. In a few days the bird was strutting, the skin of the neck became thickened, and caruncles appeared. The snood became enlarged and the "beard spot" on the skin of the breast became enlarged and thickened but only a few "hairs" grew. In appearance and actions this treated female was indistinguishable from a normal tom turkey. The ovary remained small and inactive.

Cases where males develop female characteristics are extremely rare, consequently but little is known concerning this type of sex reversal in poultry. However, implanting stilbestrol in the skin of well grown male turkeys (then about 16 weeks old) causes a partial sex reversal while in younger male turkeys (under about 16 weeks old) the same material causes precocious male behavior and premature development of the seminal vesicles and external sex organs as noted in the discussion of stilbestrol treatment of turkey broilers on page 434. In immature female turkeys of any age stilbestrol implantation causes premature enlargement of the oviduct and external sex organs accompanied by a desire to mate. The older males when implanted with stilbestrol show loss of male head characteristics, increased fat deposition, less strutting, faster feather growth, and their bodies tend to resemble those of hens. In addition they show a tendency to assume the female position in mating.

Sexual maturity, as evidenced by the growth and opening of the oviduct, was studied at the California Station (Asmundson et al, Poul. Sci. 25:346, 1946). It was found that a grayish membrane develops across the lower end of the oviduct as the latter approaches the active stage and that this membrane either ruptures when 1 to 2 millimeters in diameter or grows to 5 or 10 mm. across then balloons out and ruptures when the cloaca is everted as in mating. This rupture of the protective membrane, or opening of the oviduct, occurred 2 to 3 weeks ahead of first egg which indicates that (as may be observed in the field) mating can and does take place 2 to 3 weeks in advance of the first egg.

The weight of the oviduct of turkey hens varies tremendously as does the ovary (p. 403), the size depending on the degree of sexual activity involved. Fully active oviducts of standardbred Bronze turkey hens averaged 106.1 cm. in length (Asmundson, J. Exp. Biol. 83: 287, 1939).

Measurements of the various sections averaged as follows in centimeters: Infundibulum (funnel) 3.7; tube of infundibulum 12.3; magnum (albumen secreting section) 45.9; isthmus 16.4; uterus 13.7; and vagina 14.1.

Food passed through the alimentary canal of laying BBB turkey hens in 3 hours 13 min.; in non-layers 4 hrs. 16 min.; in young turkey hens 2 hrs. 27 min.; in young turkeys, both sexes, on a normal diet, 2 hrs. 57 min.; with penicillin added 3 hrs. 15 min. (Hilleman *et al.* Poul. Sci. 32: 332, 1953).

The crop and gizzard digestion of turkey hens was studied by Russian workers (World's Poultry Sci. Jour. Jan.-Mar. 1952—Abstr.). Turkeys have a neuro-muscular mechanism for regulating entrance of food into the gizzard which contracts regularly 2 to 3 times per minute.

Normal pulse rate in turkeys is about 150, and respiration rate about 42, per minute.

CHAPTER 14

INCUBATION

OUTLINE

The Incubation Process

Although a few poults are still hatched by hens the turkey industry relies on artificial incubation in the U. S. For the few readers who may hatch turkey eggs with hens practical rules and suggestions are given at the end of this chapter.

The incubation period of turkey eggs is considered to be 28 full days. With fresh eggs of average size incubated at normal temperature 28 full days are required for completion of the hatch. However, a number of factors may lengthen or shorten the period, for example, egg size, or weight. Fresh Beltsville Small White eggs averaging 32 oz. per dozen will be through hatching in 27 days 12 hours. Eggs from average and medium-sized turkeys in the weight range of 34 to 38 oz. per dozen tend to hatch in 28 days while large eggs over 38 oz. per dozen may require 28 days plus 4 to 6 hours. Incubation temperatures lower than normal tend to lengthen the incubation period while those higher than normal tend to shorten it—by a few hours in each case. Eggs held for a longer-than-normal time are reported to require a slightly longer time than 28 days to complete their hatching. Wild turkey eggs although small are reported to require full 28 days of incubation under natural conditions. The fact that the eggs usually are incubated on the ground hence likely to be cold on the bottoms may account for these relatively small turkey eggs requiring the full 28-day period. There also are species, variety and strain differences affecting length of the incubation period. For comparison the normal incubation period for average-sized chickens is 21 days; for pigeons 17 to 19 days; for quail eggs, 22½ days; for pheasants 23 to 28 days depending on the variety; for peafowl, 28 to 30 days; for guineas, 28 days; for domesticated ducks, 28 days, except the Muscovy which requires 33 to 35 days; for tame geese, 30 to 34 days. The Ostrich egg is said to be incubated 42 days. In general, but with a few exceptions, the larger the egg the longer is the incubation period required. The day following the day on which eggs are set should be counted as the first day of incubation in order to properly count the full days required. If the eggs are placed in an incubator that is up to temperature they will warm up in a few hours so that the hatch should be completed by the same time of day, the same day in the week, four weeks later. It usually takes about 40 hours from the time the first poult hatches until the hatch is entirely over. If the hatch is spread out over a longer period than this it indicates faulty incubation or that some of the eggs might have been partially incubated under hens before they were placed in the incubator. Embryo mortality in turkey eggs reaches its peak during the 25th to 28th days of incubation. The turkey embryo starts development in the body of the hen before the egg is laid and continues to develop after laying of the egg during extremely warm weather or when kept warm by the same or other hens. Unless the eggs are to be set at once, the germ development should be arrested fairly soon after the egg is laid since it is nature's provision that if development is checked at this point, the embryo will remain vigorous and alive for a considerable time. Because of these facts the eggs which are to be saved for hatching should be gathered 3 or 4 times daily and at the end of the day placed in an open tray or wire egg pail in a cool place. If desired, the eggs may be transferred to an egg case after they are cool to make sure all embryo development ceases. The eggs should be cooled below 60° F., preferably to about 55°.

Care of Hatching Eggs

Temperature of the eggs being held for hatching is very important. Early embryo deaths may be due to poor care of the hatching eggs prior to setting. Hatching eggs should be held in a room below 60° F., temperatures between 50 and 60° being preferred. The germ is injured if held at 29 to 44° F. for more than about fifty

Fig. 14.1. The modern method—nursery tray of electric incubator. (Photo from James S. Kellerman.)

hours. The room in which hatching eggs are held should not be stuffy, nor is excessive ventilation desirable. A well ventilated cellar, milk room, spring house, or root cellar affords a satisfactory place for holding hatching eggs. Relative humidity should be not lower than 60% nor higher than 89%. Where it can be controlled, about 80% may be best. A convenient way to save turkey eggs for hatching is to place them small end down in an egg case in molded 5 x 5 egg trays or duck-egg fillers. If the eggs are to be shipped they are then ready for shipment when the case is full; the capacity being 200 turkey eggs. Extra high cases holding 250 jumbo size chicken eggs may also be used. The case should be tilted against the wall and reversed end for end once each day or cases may be set on large racks that may be tilted back and forth. It is not necessary to turn eggs that will be set before they are 10 days old. Turkey eggs deteriorate rapidly in hatching power after they are two weeks old if the holding temperature is too high or too low. However, they can be saved as long as three to four weeks, but with somewhat reduced hatchability, if the holding temperature is correct. An egg cooler 10′ x 28″ x 32″ deep will hold 300 lbs. of ice and will effectively cool turkey eggs. Electric-powered thermostatically-controlled air cooling units are available. They are reliable, cheap to operate, and very accurate. Such a unit should be installed in an insulated egg room. A humidifying device seldom is needed.

Hatchability is usually measured in per cent; either the ratio of poults hatched to fertile eggs set or the ratio of poults to total eggs set. *Fertility* is the ratio of eggs containing a germ to total eggs set. Hatcherymen are not able to distinguish between infertile eggs and those in which the germ dies prior to transfer to the hatching trays on the 23rd or 24th day of incubation because the accurate determination of fertility of eggs after this much incubation is practically impossible. The common hatchery practice of considering as *dead germs* only those eggs showing a considerable amount of embryonic growth at the 23rd or 24th day, candling is as nearly accurate as any practical method can be. The remainder, those eggs that either are clear or have a slight to moderate amount of cloudiness or an addled appearance without definite signs of embryonic growth which the hatcheryman calls infertile, actually are mostly genuine infertile eggs or infertile eggs in which parthenogenetic development has taken place. (See page 644 for discussion of parthenogenesis and details concerning the appearance of fertile, dead germ, and infertile eggs.)

At the Kansas Station 1,049 turkey eggs were involved in experiments to check hatchability. No noticeable deterioration occurred until the eggs were two weeks old if held in temperatures that did not fall below 55° F. Eggs held between 60° and 75° F. gave the following hatchability; not over 6 days old, 72%; 7 to 13 days, 73%; 14 to 20, 45%; 21 to 27, 14%; 28 to 34, 6%. Eggs held between 55° and 60° F. did not deteriorate so rapidly, the hatchability running 89%, 90%, 85%, 84%, and 86% for each of the respective periods (of fertile eggs); there were only 7 eggs held 28 to 34 days. From these results it appears that temperatures above 60° F. injure the holding quality of hatching eggs. Other tests showed that holding temperatures of 36° F. lowered hatchability about 5% in eggs held for a week and about 13% in eggs held 2 weeks. Turkey eggs freeze at 28° F. A later experiment at the Kansas Station showed that when turkey eggs were held at 54° F. for three weeks before setting, there was little loss of hatchability but after three weeks the decline was quite marked.

The California Station (Poul. Sci. 1948) reported no advantage in setting Broad Breasted Bronze eggs as soon as laid over those held for 8 days at 50 to 55° F. Eggs chilled in open trays at 29 or 30° F. for 6 hours then held at 50 or 55° for 6¾ days averaged 78.9% hatch of fertile compared to 75.2% for those held 7 days at 50 or 55°. Eggs chilled 3, 6, 14, and 48 hours at 29 or 30° then held 1 to 3 days at 50 or 55° hatched 81.6 to 88.9% of fertile compared to 76.1% for eggs held 3 days at 50 to 55°. Eggs chilled in an open tray reached chill-room temperature (29-30°) in about 4 hours while those in paper-lined wooden cases required 14 to 45 hours and those in unlined wooden cases required 11 to 30 hours. Eggs near the center of a half case cooled much more slowly than those in the outer cells.

The California Station (Poul. Sci. 1947) reported that Broad Breasted Bronze eggs held 7 days at 55° F. hatched about 72% of fertile; those held 14 days, 62%; 21 days, 50%; 28 days, 44%; and 35 days, 10%. Apparent fertility held fairly uniform, 85 to 93%, for 28 days then dropped to 50% in the 35-day eggs. Eggs were turned once daily.

The Maryland Station (Poul. Sci. 1948) reported hatchability of fertile eggs from Black, Bronze, and Beltsville White crossbred turkeys held at 50 to 55° F. for 6, 12, 18, and 24 days to be 84.8, 73.3, 62.2, and 54.3% respectively. Eggs held 1, 2, 3, and 4 hours at -1° F. hatched 77.3, 85.7, 79.2 and 76.7% respectively compared to 75% for the 50-55° controls. All eggs were first held 1 to 6 days at 50-55° before being subjected to temperature experiments and were held 3 hours at that temperature after they were removed from the experimental temperature before setting. They were packed for the temperature experiments in wooden egg cases, covers, standard filters, and cup flats. Eggs held 2 days at 20° F. hatched 71.4% compared to 87.5 for the controls while those held 4, 5, 6, and 7 days at 20° F. all gave zero hatchability. Eggs held 9 to 12 hours at -10 to -13° F. hatched 21.4 to 10% of fertiles. The North Dakota Station (Bimo. Bull. 1953) found that hatchability of all eggs set from Beltsville White and large white turkeys (January through May) was 5 per cent less in eggs held 14 days than in eggs held 7 days. Eggs held 21 days hatched 22 per cent less than those held 14 days. Holding temperature was 45 to 50° F., which may have been below the optimum.

The variations in experimental results indicate probable strain differences in ability to withstand holding periods of longer than one week. The results of the various experiments might be intepreted to suggest, that under favorable holding conditions turkey eggs may generally be held for up to 7 or 8 days without loss of hatchability and that eggs held 7 days longer, to 14 days, usually but not necessarily always, show a decline in hatchability possibly on the order of 6% of all eggs set. Both California and Maryland experiments suggest that arresting the development of the germ quickly by short exposures to subfreezing temperatures actually increases the hatchability slightly. Almost all the experimental results point to significantly lower hatchability in eggs held 21 days with still greater losses after 28 and 35 days of holding.

Dirty eggs are unsalable and may transmit disease. To prevent soiled eggs, use well-designed and well-placed nests with plenty of clean litter (rice hulls are very good), gather frequently, and keep turkeys out of mud. Cleaning dirty eggs is not desirable but usually is preferred to setting them badly soiled. If slightly soiled, brush with steel wool or sandpaper. If too dirty to dry-clean, wash by soaking in warm water (105° F.) gently sponging off the dirt under water, rinsing in clean water, and allowing to dry. Dipping clean or cleaned eggs for 5 minutes in sodium orthophenylphenate, 1 per cent solution, or lye, 2%, or quaternary ammonium disinfectant 0.1%, may assist in the control of egg-shell-carried diseases without affecting the hatchability. Dirty eggs washed as above, but not dipped in disinfectant, in one experiment by the senior author hatched 12% less than clean eggs. California experiments (Calif. Turkey News 1949) indicated no advantage in adding detergent or disinfectant to the wash water for which warm water (105°) was much superior to cold water. Dipping washed eggs in quaternary ammonium disinfectant was of no value in decreasing the number of rots or increasing the percentage of hatch. Clean washed eggs hatched as well as clean unwashed eggs, hence washing itself apparently had no ill effect—these came from the dirt which reduced hatchability of fertile eggs by 13 to 23% after they had been washed or dry-cleaned. In another small experiment clean eggs hatched 85% of total set while dirty eggs (untreated) hatched 60%, a loss of 25% due to dirt.

Cracked eggs also are unsalable, often because infected, and hatch poorly or not at all. Severely cracked eggs are practically unhatchable, but slightly cracked ones may sometimes hatch if the crack is covered with collodion or a proprietary preparation such as "New Skin." Scotch tape may be used. Finger nail polish or adhesive tape also might be effective. The same precautions used to prevent dirty eggs also may be used to prevent cracked eggs.

Table 14.1—Summary of Recommendations for the Care of Hatching Eggs

Gathering	Three or four times daily, preferably four.
Temperature during holding:	Not over 60° nor under 50° F. The ideal is 55° F.
Position during holding:	On sides (preferred) or small end down, never the large end down.
Turning during holding:	Once each day when more than 10 days holding is contemplated.
Length of holding:	In general, eggs should be set within two weeks and preferably within a week. However, if holding conditions are satisfactory, fair hatchability may be expected when the holding period extends for three or four weeks.

Exploding Eggs

Exploding and oozing eggs have been studied at the Kentucky Station and by Taggart (1951) who reports that the organisms which cause eggs to ooze or explode come from the ground and may be found in the topsoil almost anywhere in the cultivated sections of the United States. They may contaminate the outside of the egg shell and become active in warm, moist surroundings but dormant in cold or dry situations. Thus eggs laid on warm wet ground and contaminated with the soil under warm, moist conditions may pick up the infection. During incubation the organisms multiply and cause the eggs to ooze or to explode. If an egg explodes in the incubator, other eggs may be infected and the trouble is perpetuated. Infected incubators should be thoroughly cleaned and disinfected. Experiments have failed to disclose any method of washing or disinfecting eggs that is effective. Taggart's recommendations are to produce clean eggs but if dirty eggs are obtained and must be set, he suggests dry-cleaning them by scraping (with sandpaper buffer) immediately after gathering. Washing by any known method seems to aggravate the trouble. Infection by *B. coli* frequently found in adhering droppings may lead to navel trouble.

Shipping Hatching Eggs

The most practical and generally satisfactory method of packing eggs for shipping is to use regular egg cases equipped with the 5 x 5 molded pulp cup or prong trays turkey size. The next best to use is probably 5 x 5 duck-egg fillers and excelsior pads or better, 5 x 5 cup flats. Two pads or cup flats are placed under the bottom filler, one between the intermediate fillers and one or two on top to make a tight pack. Some shippers fill up the fillers with rice hulls, oat hulls, or sawdust after the eggs have been placed in them or line the cases with storage envelopes to retard evaporation. There is some evidence to show that, for ordinary shipping, use of aluminum egg-case liners or moisture pads in fiber cases may be justified on the basis of 4 to 5% dead germs after 7 days' incubation as compared to 10% dead germs from eggs held the same length of time (6 to 9 days) in ordinary fiber cases (Calif. Turkey News, 1950).

If non-regulation egg packages are used, it is best to pack the eggs on their sides so that no matter which side of the package is uppermost, the eggs will be on their side. Shipping eggs with their small ends uppermost is injurious to hatchability.

Pre-incubated turkey hatching eggs may be successfully shipped or held over for 4 days or even longer after they have been incubated 24 to 30 hours provided they are promptly and thoroughly chilled at about 55° F., after removal from the incubator and held at temperatures under 60° F. In this way infertile eggs need not be shipped or held—they can be tested out by the use of a very good candler. All doubtful eggs should be placed with the fertile eggs but those obviously clear, showing no development at all after 24 to 30 hours of incubation, may be removed and if desired, utilized as feed. (See page 644 for information on testing after 18 to 30 hours of incubation.)

Factors in Incubation

There are four principal factors to consider in the artificial incubation of turkey eggs, namely temperature, humidity, ventilation and turning. The first three factors are closely interrelated. A fifth factor, the addition of oxygen should be considered by those operating incubators at high altitudes. "Thin" air is short of oxygen, hence its addition by means of the oxygen tank usually increases hatchability. The Wyoming Station (1947) pioneered in research with oxygen.

Considerable work has been done dealing with the proper temperature at which to run the incubator. Since the measurement of temperature depends entirely upon the accuracy of the incubator thermometer it is extremely important that the thermometer be tested at the beginning of each hatching season. This may easily be done by placing the bulb of the thermometer in a thermos bottle or jug nearly full of warm water beside a doctor's clinical thermometer or any thermometer known to be accurate. Since the mercury in the clinical thermometer does not go down without shaking, the thermometer bulbs should first be placed in water between 95° and

100° F. The bulbs of the incubator and clinical thermometers should close together while stirring the water. If they check below 100° then warm water should be added to the solution, stirring constantly and checking again between 102° and 105°. If the thermometer is accurate in the range a few degrees above and below 100° it may be used. The first incubator thermometer found to be accurate may be used in place of the clinical thermometer to check the others. If it is inaccurate it should be discarded and a new thermometer purchased and checked. Care must be used when testing, since most incubator and clinical thermometers are quickly broken by temperatures over 110° F.

Temperature Recommendations

The temperature recommendations of the manufacturer should be followed, especially if the incubator is of the forced-draft type. Exact temperature is important as 0.2 to 0.3° more heat may result in smaller poults. The manufacturer has run experiments to determine the best temperature for his particular machine. Makers of agitated air and forced draft machines recommend a temperature from 99° to 100° F. for the first 22 to 24 days of incubation. For the last 4 to 6 days, 97° to 98° is commonly recommended, but this can be practiced only when separate hatchers are used. The manufacturer should be consulted about temperature and humidity recommendations before each hatching season to benefit from recent research. In the forced-draft type of incubator the air is of the same temperature in all parts of the machine. Consequently, the incubator is run at a lower temperature than is recommended for the small hot-air or hot-water incubator and the sectional type of machine. In the sectional and small-size machines the air is not forced by means of a fan or paddle and depends upon gravity for circulation, which accounts for the term "still-air" machine. The source of heat is in the top and the heat is forced down against the natural tendency of warm air to raise hence there is a temperature difference of from 5 to 8 degrees between the top and bottom of the turkey egg. For this reason such incubators are run at a higher temperature. Research at the Kentucky Station suggests that machines of this type should be run at a temperature of 100½° the first week, 101½° the second week, 102½° the third, and 103° F. the fourth week, when the bulb of the thermometer is level with the top of the turkey eggs (the bottom of the bulb 1⅞" above the tray). Results at the North Dakota Station indicate that if "still-air" incubators are to be kept at one temperature throughout the hatch, the best results with turkey eggs can be secured at 102° F. In either case care should be taken at the end of the hatch to see that the temperature does not go above 103° as a temperature of 103½° to 104° the last few days has been found to be harmful, lowering the hatch.

Fertile eggs produce heat increasingly as they develop. Some of this heat is radiated to the nearby eggs; whereas infertile eggs produce no radiant heat. Lining the walls of the incubator with shiny aluminum foil reflects the radiant heat back to the eggs, instead of permitting its absorption by the walls to produce cool areas.

Sagging egg trays must be avoided because of the difference between different levels in the natural draft machine. The actual embryo temperature as compared with incubator temperature was determined at the Kentucky Station. The graphs in Fig. 108 show that the embryo gradually increases in temperature as the hatch progresses, irrespective of the type of incubation.

While most thermostats are dependable and accurate, it should be kept in mind that their sensitivity depends on barometric pressure. When the pressure drops, the thermostat will cut off below the point for which it was set; while high barometric pressure allows an excess flow of heat.

Moisture Loss from Eggs

Although turkey eggs remain in the incubator a week longer than chicken eggs, the turkey eggs lose about

3½% less moisture than chicken eggs during incubation, the poult weighing about 66% of the original egg weight. If the air in the machine does not contain enough moisture, the evaporation will be more rapid and the hatch will be reduced. Turkey eggs are very sensitive to lack of

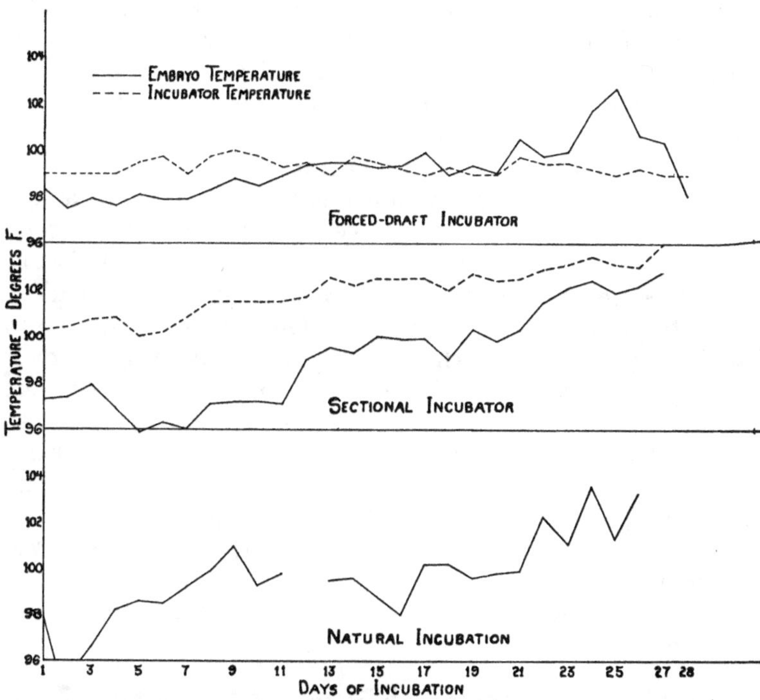

Fig. 14.2. Mean embryo and incubator temperatures. Ky. Exp. Station data. The body temperature of day-old poults hatched in a forced-draft incubator is approximately 103° F.

moisture. Even more care than is taken with chicken eggs should be practiced to make sure there is sufficient humidity. Cold water should not be added to natural-draft incubators as it will not evaporate to any extent until warmed to the temperature of the machine. When filling the moisture pans or when sprinkling the eggs the water should be warmed to the temperature of the incubator (comfortably warm to the hand, approximately

100° F.). If one is to regulate the amount of moisture in the incubator with any degree of accuracy there must first of all be some method of measurement. The amount of moisture in the air is measured in terms of percentage relative humidity. Relative humidity is the relationship between the amount of moisture in the air and the amount of moisture that the air could hold at that temperature. The moisture-holding capacity of air increases rapidly with a rise in temperature. This means if the air in a room at 70° F. contains a sufficient amount of moisture (60%) and this air is warmed up to 100° F. as it goes in the incubator, no additional moisture being added inside the machine, the relative humidity will have been cut to 23%. Consequently, it is necessary that moisture be added to the air inside the incubator. This is usually done by means of moisture pans. These pans should be in such a position that the air is constantly passing over them and evaporating the water from the surface of the pan. In some incubators which have hatched chicken eggs well it may be necessary to add additional moisture to get good turkey hatches. To increase the evaporating surface if this is necessary burlap wicks may be laid in the water trays or extra trays added or both. Wet sponges or cloths are sometimes used. Modern forced-draft incubators usually are equipped with a humidifying device.

Fig. 14.3. Incubator hatched poults may be shipped safely the same as baby chicks (15 per compartment, 60 per box of 100-chick size).

Measuring Humidity

Relative humidity may be measured by means of a hygrometer consisting of wet and dry bulbs. The dry bulb measures the temperature of the incubator. The wet bulb is constantly being cooled by the evaporation of distilled water (clean rainwater will do) from the wick surrounding the bulb. If the air in the incubator contains only a small amount of moisture there will be considerable evaporation which will lower the temperature of the wet-bulb thermometer. The greater the difference in the temperature reading of the yet and dry bulbs the lower will be the relative humidity. This is readily seen in the following table:

Table 14.2—Relative Humidity Table (in per cent)

Temperature of Incubator (Dry Bulb)	Number degrees wet bulb is below dry bulb						
	4°	6°	8°	10°	12°	14°	16°
98° F.	86	79	73	67	61	55	50
99° F.	86	79	73	67	61	55	51
100° F.	86	80	73	68	62	56	51

NOTE: A common recommendation for the forced-draft machine run at 99° F. is a wet bulb temperature of 87° to 88° F., thus giving a relative humidity of 61 per cent, up to the 24th day. The wet bulb is generally raised to 90° F. (9° below dry bulb) on the twenty-fourth day, giving a relative humidity of 70 per cent for the hatch. See Appendix for a more complete table of relative humidity values.

A wet-bulb thermometer is accurate only when the air is circulating rapidly over the wick, hence the wet-bulb type of hygrometer cannot be used successfully except in forced-draft incubators. Wet-bulb wicks should be rinsed daily and washed weekly in soapy water. There are on the market satisfactory metal hygrometers calibrated to read directly in per cent relative humidity which do not require any more rapid circulation of air than is found in the sectional type incubator. If this instrument is used in the sectional type machine it should be left on top of the eggs after they have been turned, and then read and removed immediately before the next turning. The hygrometer is usually set for a machine run at 100° F., but may be set for a higher or lower temperature if requested when ordering. It is only necessary to check the humidity in the incubator once a day, whereas temperature should be checked at least four times daily.

Calculating Moisture Loss

The rate of evaporation of moisture from the turkey eggs affords a valuable means of determining whether the incubator contains sufficient moisture. The size of the air cell indicates the rate of evaporation (see Fig. 14.4). During the first 6 days eggs should lose approximately 2½% to 3% of their weight at the time of setting; at the end of the 12th day the loss in weight should be approximately 5%; at the end of the 18th day 7½%, and on the 24th day when the eggs are turned for the last time, approximately 10%. It is at once apparent that the moisture loss approximates 0.4% per day in the incubator. A simple way to check the rate of loss is to place from 33 to 35 turkey eggs in a tray or other container on an accurate spring scale. A tray should be placed on the scale and the arrow set at zero. About 30 eggs should then he placed on the tray, among the last few, eggs of such size as will bring the arrow to 6¼ lbs. (100 ozs.). With a soft lead pencil a line should be drawn around these eggs so they may be easily selected from among the other eggs on the tray on the days when they are to be weighed. These same eggs should be taken out and placed on the scale on the 6th day and should weigh not less than 6 pounds, 1 ounce (97 ozs.), which allows for 3% evaporation. At the end of the 12th day they should weigh at least 5 lbs. 15 ozs. (95 ozs.); at the end of the 18th day 5 lbs. 12½ ozs. (92½ ozs.), and on the 24th day 5 lbs. 10 ozs. (90 ozs.). All of the eggs originally marked and weighed must be left in the incubator until the 24th day even though they are infertile or contain dead germs. If the eggs weigh distinctly less than the amount indicated it shows that evaporation has been too great. This indi-

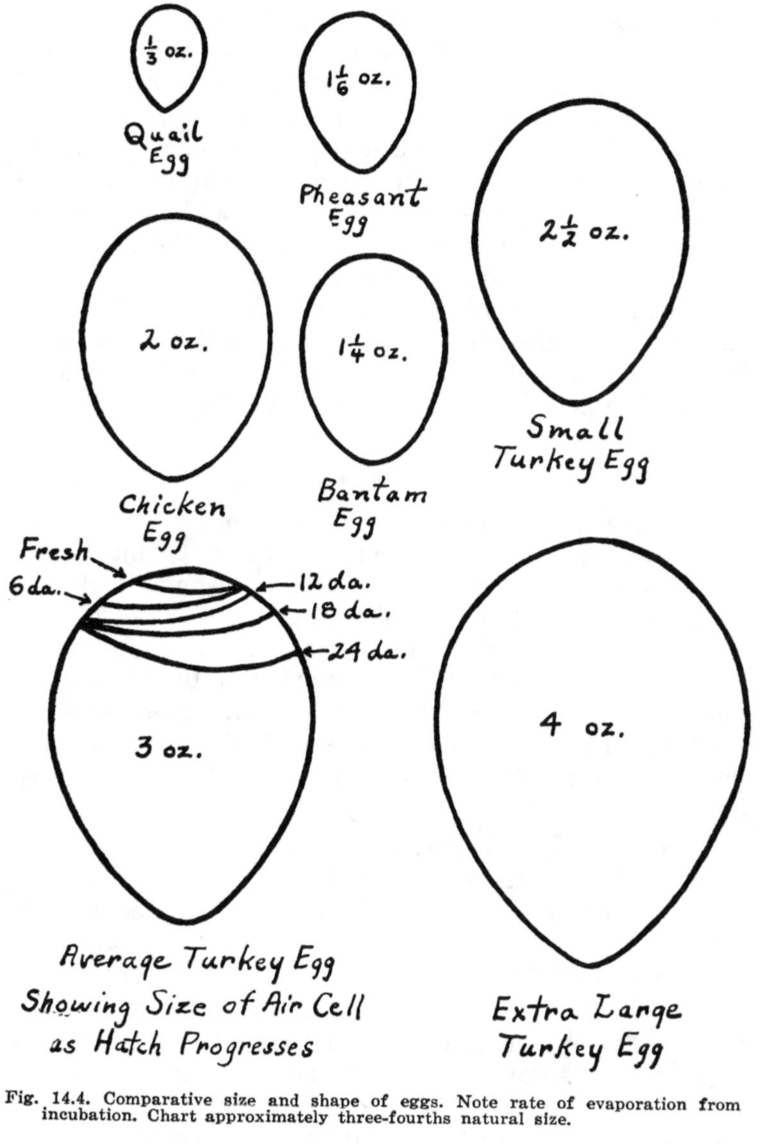

Fig. 14.4. Comparative size and shape of eggs. Note rate of evaporation from incubation. Chart approximately three-fourths natural size.

cates that the incubator is not receiving sufficient moisture and the humidity should be increased by increasing the evaporation surface. This may be done by adding additional moisture pans, being careful not to place them over any of the ventilator holes. If the entire floor of the machine is covered with moisture pans and there is no room to add pans then the logical solution would be to increase the humidity in the air of the room in which the incubator is being operated. This may be done by darkening the room, cutting down on the ventilation in the room and in the incubator if it is greater than necessary and sprinkling the floor of the room two or three times a day. Burlap wicks may also be placed in the water pans. If there is an automatic humidifying device it can be adjusted to provide the right amount of moisture.

Humidity

Moisture loss of Broad Breasted Bronze eggs during the first 24 days of incubation in forced-draft incubators was reported by the South Dakota Station (Poul. Sci. 1944) as follows; at 37% relative humidity, 14.7%; at 46% relative humidity, 13.8%, and at 57% humidity, 11.3%. Hatchability of fertile eggs was 15, 46, and 57 per cent respectively for the three humidities. At 60% relatively humidity as recommended, the moisture loss during the first 24 days should be about 10% as noted above.

A relative humidity close to 60% through the first 24 days of incubation and about 70% the last four days has been found to produce best results. The necessity of increasing the humidity at hatching time is apparent when one considers that the wet poult would dry off too quickly (and chill while drying) in an incubator containing insufficient moisture. Occasionally after pipping, the egg dries out so quickly that the poult is unable to hatch, indicating the need of additional moisture. As the poult dries off, its skin drops toward the wet-bulb temperature of the incubator. Evaporation from the surface of its body absorbs from the poult heat that is required to change water into vapor. If drying off takes place slowly in a moist atmosphere the poult is not chilled or injured.

Ventilation

In practical terms, forced-draft turkey incubators run at 99.5° F. should have a wet-bulb temperature of 87 to 88° for the first 24 days of incubation and a 90° wet-bulb temperature for the last 4 days. If the hatching temperature (25th to 28th days) is reduced to 97°, the wet bulb should read 88° to indicate a relative humidity of 70°. These high humidity requirements are especially important for hatching small eggs such as those of small type turkeys.

Practically all incubators which are now on the market provide adequate ventilation. Any incubator that is well enough ventilated to have given good hatches with chicken eggs, should afford sufficient ventilation for turkey eggs. The chief object of ventilation is to remove the carbon dioxide that is given off by the developing embryos and provide a fresh supply of oxygen. To keep the carbon dioxide content below the harmful point of 1½% requires a hundred times as much fresh air the last day of incubation as the first day. All incubators are well enough ventilated to meet the requirements up until the eggs pip. However, the test of adequate ventilation lies in a sufficiently rapid exchange of air from the time the first egg pips until the hatch is over. The amount of carbon dioxide given off by developing turkey embryos increases rapidly as the hatch progresses and the carbon dioxide content of the air would pass the toleration point (1½%) at the conclusion of the hatch if adequate ventilation were not provided. Since there is no practical means of measuring the carbon dioxide content of the air within the incubator the best practice is to provide ample ventilation in the room and follow incubator directions for ventilating. Fresh air may be conducted directly to the region of the incubator air intakes. In any event, ample ventilation should be provided in the incubator room. Without it good hatches will not be obtained.

Turning

It is customary to turn turkey eggs frequently from the time they are set, through the 22nd to 24th day of incubation. However, work at the Kentucky and Beltsville Stations suggests that turning is not essential after the 18th day. Frequent turning the first two weeks permits the developing embryo to orient itself properly within the egg. Automatic turners should be checked daily to be sure they are working. If the eggs are not turned frequently in the early incubation period some of the embryos may start development in the wrong position and may become malformed. These malformations are frequently referred to as malpositions. There are several causes of malpositions, some of which are hereditary and others caused by faulty incubation conditions. Experiments have shown that it pays to turn turkey eggs more than twice daily. Turning every 3 hours day and night is considered the practical optimum with automatic turning equipment. Without it, five times a day is a very satisfactory number of turnings as it means that the eggs do not rest in the same relative position on any two successive nights. These five turnings should be distributed evenly throughout the day from the first convenient time in the morning to the last convenient time in the evening, preferably covering a 16-hour period so that the eggs do not remain unturned more than about 8 hours. When turning the eggs in the sectional type of incubator it is advisable to rotate the turning crank over, back, and then over again, permitting the turning device to pass across the tray, then back and across again. This insures turning all of the eggs. With the small incubator not equipped with a turning device a few eggs should be removed from the center of the tray, and the others rolled toward the center. The eggs that have been removed should be put back on the tray in the outer corners. Gentleness is necessary in hand-turning to avoid breakage and damage to the developing embryo. Forced-draft incubators are equipped with turning devices which are constructed so as to turn the trays and insure a sufficient rotation of the embryos.

Malpositions

The normal hatching position of the embryos of all birds including turkeys is lengthwise of the egg, head towards large end, and beak under the right wing with its tip turned towards the air cell. Any position differing from the above is termed a malposition. As long ago as 1751 a French scientist, Reamur, observed one of the more common malpositions. These abnormalities have been studied in a series of researches starting in 1925. Most of these have dealt with chicken eggs but the California Station published the results of a study of malpositions in turkey and game-bird embryos that failed to hatch. One important finding was that chicken and turkey embryos do not differ significantly in the nature of their malpositions or in the percentage of dead embryos in each malposition. Turkey embryos, like those of chickens, change position in the egg during the last 3 or 4 days before hatching.

The four general types of malpositions found most frequently in dead-in-shell turkeys were (1) head buried between thighs, (2) head in small instead of large end of the egg, (3) head turned left, over or under the wing, and (4) head turned right, over wing instead of under it. Embryos in positions (1) and (3) seldom pip and apparently rarely hatch. Embryos in positions (2) and (4) are handicapped and have less chance of hatch than do those in normal position. Position 2 was found to be more prevalent in eggs incubated for the first 24 days flat on their sides. Other malpositions found in small numbers are (5) feet over head, resting on the head, and (6) embryo rotated out of position so that beak is turned away from air cell. Embryos in these malpositions also appear to be greatly handicapped in hatching.

The malposition in which the head is turned to left and is under or over the left wing is regarded by some as the most important type of malposition because it almost always prevents hatching and may be present in up to 5% of all chicken embryos alive after 18 days' incubation. In turkey eggs Asmundson found this malposition in 6% of turkey embryos that failed to hatch. Practically none of these embryos pipped the shell. The evidence indicated that size of egg and position during incubation whether large end up or horizontal, did not affect the incidence of this malposition in turkey eggs, but that hereditary factors may be involved. Hereditary factors may also be involved in other malpositions. Selection for high hatchability would tend to eliminate factors responsible for malpositions.

Incubation at High Elevations

Incubation at elevations, over about 3,500 feet above sea level, presents some difficulties due to the "thin" air which does not provide enough oxygen for normal embryo development. The Wyoming and Utah Stations studied this problem and found that by introducing extra oxygen into the incubator the hatchability could be brought up to normal. Normal air at sea level contains about 20% oxygen while at 7,200 feet elevation, for example, it contains only 16% oxygen. By bringing the oxygen in the air in the incubator up to 20% or 25% normal hatchability was obtained. The introduction of oxygen in proper amounts is somewhat difficult and requires careful technique. The carbon dioxide content of the air also is important. The Wyoming Station recently announced the development through selection of a strain of Broad Breasted Bronze turkeys the eggs of which will hatch normally when incubated at high elevations without adding supplementary oxygen.

Eggs from breeding turkeys maintained at high elevations hatch normally when transported to low elevations for incubation and conversely eggs produced at low elevations hatch poorly when incubated at high elevations. Information on adding supplementary oxygen may be obtained from the Wyoming, Colorado, or Utah Stations and from some incubator manufacturers.

Development of the Turkey Embryo

The development of the poult in the shell was studied at the University of Maryland, and reported by Phillips and Williams (1944). A condensed account of the growth process is given herewith:

At the end of the first 24 hours of incubation the primitive streak is barely visible, by 36 hours it is plainly seen with neural folds at head end. At 48 hours there are 7 pairs of somites (body segments arranged in a double row) and brain and eyes have started to form. At 60 hours the embryo is still flat, there are 12 pairs of somites, no blood is present, and the eyes have become prominent. At 72 hours (3 days) there are 22 pairs of somites, the amnion (water-sac) covers the head and heart regions, blood appears, the heart is S-shaped and is pulsating, the upper half of the body is twisted and raised, and ear-pits appear. At 4 days there are 38 pairs of somites, the amnion nearly surrounds the body, nasal pits appear, heart becomes divided, the allantois (primitive breathing apparatus) first appears, blood vessels extend outside the embryo in contact with half of the shell membrane, and the embryo presents a typical question-mark shape with a prominent head and heart. At 5 days there are 45 pairs of somites, wing and leg buds are seen as definite appendages, a liver bulge appears near the heart, the mandibular arches that form the jaws are prominent, and the embryo's tail and head are decidedly curved in toward each other. At 6 days the somites as such are no longer in evidence, the eyes are dark with pigmentation, the allantois is as big as the embryo, the mouth pit is noticeable, and head and tail nearly touch. At 7 days the eyes are solidly pigmented, leg appendages are larger than the wings, and the little ridge-like forerunners of the toes can be seen. At 8 days there may be seen a beak-like structure, greatly bulged eyes, ear holes, leg parts, a distinct neck, blood vessels lining half the shell membrane surface, and an allantois* 3 times as big as the embryo. At 9 days the upper jaw is complete with its "egg tooth" at the top (a temporary structure to aid the poult in breaking out of the egg), the third eyelid appears, and blood vessels line the entire shell membrane of the egg. At 10 days the wings are fully formed, there are feather papillae (or goose-pimples) on the back and wings, a cloacal bulge appears, toes are distinct but webbed, and the drumsticks look typical. At 11 days all toes are separated and feather papillae extend to thighs and tail. At 12 days feather papillae extend to breast, wing, head, and drumsticks, and claws start to form. At 13 days feather papillae cover the body, short plumage appears on back, tail, and thighs, snood (or dewbill) appears, claws are distinct, beak is hardened, and pads appear on soles of the toes. At 14 days the preen gland can be seen and elongated plumage appears on back, thigh, and tail regions and short plumage on neck, breast, drumstick, and wings. At 15 days elongated plumage covers the body except ventral (lower) surfaces of head and neck, tiny leg scales appear, and claw tips begin to harden (keratinize). At 16 days plumage covers the body, the anus (vent) is open, leg scales are prominent and pigmented in black varieties, spurs are present, claws are almost completely hardened, and the eyes are about complete. At 17 days the plumage has grown so long that no bare spaces appear anywhere on the body. At 18 days nearly all the egg albumen (or white) has disappeared. At 19 days the ear openings are hidden by the plumage, nostrils are open, and leg scales are raised and prominent. At 20 days the snood is hidden by the head plumage and plumage in general has white deposits scattered over it. At 21 days the egg albumen has dis-

*The membranous sac-like primitive breathing apparatus containing blood vessels. It is attached to the abdomen of the embryo and appears to line the shell membrane surface.

appeared. At 22 days the nostrils appear as in a newly hatched poult. At 23 days the embryo for all practical purposes is complete and the yolk sac is firm and begins to enter the body. At 24 days 1/4 to 1/2 of the yolk sac is inside the body. At 25 days the beak enters the air cell and 3/4 of the yolk sac is inside the body. At 26 days the poult fills the entire egg, the yolk sac is all inside the body with an umbilical opening 1/2″ in diameter, and pipping begins.

Taking Off the Hatch

The humidity should be raised and turning discontinued on the 22nd to 24th day of incubation, preferably the 23rd day. It is best not to open the door of the incubator while the hatch is coming off, at least not more than once during the hatching period. It is sometimes necessary to remove the poults that are hatched when about half of the hatch appears to have been completed, placing them in chick or poult boxes. If there is a nursery tray in the machine the strip that covers the opening that permits the poults to drop into the nursery should be in place at the beginning of the hatch as the first few poults may drop to the nursery and chill while they are still wet. When the hatch apears to be about half completed the door should be opened quickly and the strips removed so as to let the newly hatched poults drop into the nursery tray. Keeping the incubator darkened at hatching time tends to keep the poults more contended and quiet. The nursery tray should have a rough surface such as hardware cloth or muslin tray cloths, since newly hatched poults will slip on a slick surface such as newspaper and numerous "spraddles" may result.

The young poults should be fed as soon as they are removed from the incubator to the brooder. Poults should never be placed in the brooder house unless feed and water are available as they may start eating the litter. If poults are kept overnight in baby chick boxes, from 15 to 18 may be placed in each compartment that accommodates 25 baby chicks. If the poults are shipped, the

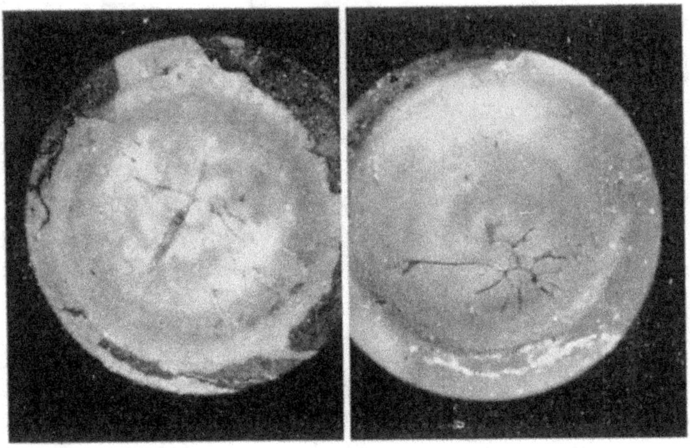

Fig. 14.5. Early development of the turkey embryo. On the left, the 4-cell stage: on the right, the 16 cell. This development takes place about four hours after fertilization, while the egg is still in the body of the hen. (Microphotograph by M. W. Olsen, U.S.D.A. Research Center.)

box should contain this number as any less than 15 in a section may result in chilling and more than 18 in overcrowding. Judgment must be used regarding the number of poults as well as the number of holes in the box, both being adjusted to weather conditions. Poult boxes have compartments about 10″ x 12″ x 8″ for 25 poults. Special shipping boxes accommodating 100 poults are on the market. In size they are about 24″ x 22½″ x 9″ high including the ventilation strip and with pads, weigh about 3.25 lbs. Loaded with 100 Broad Breasted Bronze poults, the total weight is about 15.75 pounds, with 100 small type Whites, about 14.75 pounds. New boxes should be used for each hatch or shipment. If however, they have not left the premises they may be used again if thoroughly "dry cleaned" and then painted inside with a solution of sodium silicate (waterglass) and water, equal parts.

Fumigation

The best disinfectant especially for the control of pullorum disease is formadlehyde. By the cheesecloth method, when no live poults are present, sufficient cheesecloth is used to soak up 13.5 liq. oz. of formalin per 1000 cu. ft. of incubator air space and is hung under the fans for 3 hours with doors tightly closed. When poults are to be disinfected, use half as much formalin. By the potassium permanganate ($KMnO_4$) method, no poults present, 24 liq. oz. (1.5 pts. or 710 cc.) of formalin are added to 24 avdp. oz. of $KMnO_4$ per 1,000 cu. ft. of incubator space and allowed to remain in the tightly closed incubator for 3 hours. To fumigate live poults, use 12 liq. oz. (355 cc.) of formalin and 6 avdp. oz. (175 gms.) of $KMnO_4$ per 1000 cu. ft. Fumigation of poults is not often necessary but routine incubator fumigation should be performed every 4 weeks in the hatching machines, between hatches, more often if disease is present. Incubators should be cleaned well before fumigation is attempted. Incubator air exhausts should be closed the first 10 minutes then slightly opened for the rest of the fumigation. Humidity always must be kept high.

Practical Suggestions

1. Turkey eggs should be gathered at least three times, better four times daily, and stored in a cool place between 50° and 60° (preferably 55°) until set.

2. If turkey eggs are to be held longer than 10 days they should be turned once daily during the holding period.

3. Egg cases, equipped with duck-egg fillers, hold 200 turkey eggs and are quite satisfactory for holding, turning, and shipping turkey eggs. Prong-type molded fiber 5 x 5 egg trays also make excellent fillers for holding and shipping eggs.

4. It is best to set turkey eggs within one week after laying, although they hatch fairly well if held under the best conditions as long as three or four weeks.

5. When incubating turkey eggs in a forced-draft incubator, the machine should be run at the temperature which has given the best hatches with chicken eggs, or that which is recommended by the manufacturer.

6. In a sectional incubator the best hatches may be secured when the machine is run at 100½°, 101½°, 102½°, 103° F., the first, second, third and fourth weeks, respectively. The bottom of the bulb of the thermometer should be 1⅞″ inches above the floor of the egg tray, that is, even with the top of the turkey eggs.

7. Turkey eggs require as much if not more moisture during incubation than chicken eggs, especially during the last four days of incubation. The relative humidity should be about 60% until just before the eggs pip, when it should be increased to 70%. If the forced-draft incubator is run at 99½° F. (dry bulb) this means a wet bulb temperature of about 87° F., until the 25th day and about 90° F., while the eggs are hatching. If the hatching machine dry-bulb temperature is 97°, the wet-bulb should read 88° to obtain 70% relative humidity.

8. The rate of evaporation may be determined by weighing a few eggs when set, marking them so they may be easily removed from the tray later and weighing them on the 6th, 12th, 18th and 24th days. If the loss in weight is too great the incubator does not contain sufficient moisture, that is, the relative humidity is too low.

9. Turkey eggs should be turned at least three times daily, from the second through the 22nd to 24th day of incubation. The turnings should be evenly spaced from early morning until the last turning at night. If the eggs are turned an odd number of times (preferably five) daily, they will not remain in the same position overnight on any two successive nights. Turning every 3 hours day and night is best. Check automatic turning devices daily.

10. All dry and wet-bulb thermometers should be checked for accuracy before each hatching season.

11. Turkey eggs should never be hatched in the same incubator or incubator room with chicken eggs. This practice is believed to be the chief means of the spread of pullorum, typhoid, and paratyphoid groups of diseases from chickens to turkeys. However, if the chicken and turkey eggs both are from flocks showing no reaction to

Table 14.3—Incubation Trouble-Shooting Chart

Prepared by W. M. Insko, Jr.

Symptoms of Trouble	Probable Causes	Suggestions
Many clear eggs no development	1. Not enough or too many toms. 2. Toms too large. 3. Interference of toms during mating 4. Holding eggs too long	1. Use one tom to 10-12 hens, fewer hens for old toms. 2. Toms should not be more than twice as heavy as hens. 3. Use fewer toms, rotate toms by group, keeping each group together. 4. Set eggs within 1-10 days after laying. Hold eggs at a temperature of 50 to 60 degrees.
Clear eggs—blood rings or slight development when broken	1. Chilled or heated before setting 2. Irregular heat	1. Warn flock owners and shipping agencies to guard against heating or chilling of eggs. See (4) above. 2. Check temperature controls. Avoid extremes of temperature.
Many dead germs	1. Old eggs or those held at extreme temperatures 2. Improper turning of eggs 3. Improper ventilation 4. Close inbreeding 5. Incomplete ration for breeding stock	1. See (4) above. 2. Turn at least 3 and preferably 5 or more times in 24 hours. 3. Increase ventilation in incubator room. Avoid drafts. 4. Low hatchability is inherited. Discard families showing low hatchability. 5. Feed balanced ration to breeding stock.
Pipped eggs not hatching	1. Insufficient moisture	1. Increase moisture by sprays or increased moisture surface. First 24 days—wet bulb 87-88 degrees, 4 day hatching period —90-92 degrees, wet bulb. Increase moisture on 24th day; do not wait for pipping to start.
Hatching too early	1. Too high temperature throughout incubation	1. Use correct temperature. Read dry bulb when pilot light goes off. Do not let temperature go above manufacturer's recommendation.
Slow hatching	1. Insufficient heat 2. Bottom eggs too cold (natural draft incubator)	1. Use correct temperature. 2. Warm incubator room or lessen ventilation of incubator.
Dead in shell 1. Much unabsorbed yolk 2. Embryo normal—pipping slow	1. Too high temperature last week of incubation 2. Too low temperature or current interruption	1. Check temperature of hatcher after pilot light goes off. 2. Check operating thermometer —allow extra time for hatch if interruption.
Cripples and malpositions (spraddles)	1. Temperature too high 2. Improper turning 3. Improper setting 4. Hatching trays too smooth	1. See (1) above. 2. Turn at least 3 and preferably 5 or more times in 24 hours. 3. Set eggs large end up. 4. Use wire bottom trays or crinoline.
Rough Navels	High Temperature	Check temperature.

From "Turkey World"—January, 1955 (by permission).

the pullorum test, the risk from incubating the eggs of the two species together is considerably lessened.

12. Clean the incubator after each hatch. First use vacuum cleaner, then fumigate with wet bulb at 90° F. for 1 hours, with vents closed the first 15 minutes (with or without eggs in machine). If there are any eggs in the machine between the 1st and the 6th day of incubation, formalin should not be used at heavier concentrations than when poults are hatching.

NATURAL INCUBATION AND BROODING

There are certain disadvantages of natural incubation which cannot be overcome except by the adoption of artificial methods. By the natural method the flock owner is dependent upon the whims of the turkey or chicken hens for he cannot set them until they go broody. Consequently, the number and age of poults to be raised is limited by the number of available broody hens. If the turkey hen is used for natural incubation there is the loss of eggs that she might have layed during the time she is setting and brooding her poults. Usually the flock owner who is to use natural methods wishes to set the first turkey hens that go broody and consequently loses hatching eggs from among the best layers, those hens which are the first to lay during the spring. When a turkey hen hatches eggs and broods poults there is a loss of 8 to 10 weeks from production. If the eggs had been set under chicken hens or in incubators this turkey hen should have laid at least 15 to 20 eggs, which would have a value of $2 to $5. If the turkey hen was set on 15 eggs it would mean a cost of 10 cents or more per egg (the value of eggs that would have been layed had the hen been broken up) for the use of the turkey hen in incubating the eggs. Where custom-hatching of the turkey eggs in incubators is available it can be done for less than the cost of natural incubation with turkey hens. Chicken hens may be used more cheaply than turkeys, but even then there is the loss of about 3 dozen eggs that the chicken would have laid during the hatching and brooding period. This loss under natural conditions would be the least if the turkey eggs laid up to June 1st are set under chicken hens and the turkey hens allowed to set in June or early July.

Hens Carry Disease: Another distinct disadvantage of the natural method of incubation and brooding is that the turkey or chicken hen may be a carrier of diseases and parasites particularly blackhead, fowl pox, coccidiosis, typhoid, or lice and give these to the poults soon after they hatch. The danger from lice can be lessened by dusting the hens with sodium fluoride at the time they are set and again a week before the eggs hatch.

Setting the Hen: If chicken hens are to be used, only the heavier breeds such as the Plymouth Rock, Orpington, Red, New Hampshire or Wyandotte should be set. Turkey eggs should not be set under broody Leghorn hens. The chicken hen should be set with the number of eggs that she can cover easily. This is usually about 10, whereas a turkey hen can cover from 15 to 18. The hen should be broody for at least two days before the eggs are placed under her. If she is moved to another coop (a setting hen should be moved after dark) some nest eggs should be put in the new nest and care taken to be sure she sets the first night on them before placing the hatching eggs under her. Broody hens should be set in some cool secluded place in shallow weatherproof nesting boxes (about 2 feet square) or barrels that are protected from rats and other predatory animals. Setting hens should be away from the rest of the flock so that other birds cannot lay in the nest or break the eggs by disturbing them. Plenty of feed and water should be available at any time during the day that the broody hen may choose to leave the nest. A mixture of corn and wheat or barley should be left in an open hopper and if green feed is not available a little clover or alfalfa hay left nearby. Turkey eggs should not be washed if it can possibly be avoided. However, if the hen breaks an egg or otherwise soils the nest, the lesser of two evils is to wash those eggs that are left sticky and dirty. First the dirt should be soaked loose in lukewarm water (about 90° to 100° F.) and then gently sponged off, allowing the eggs to dry, not wiping them. The straw, hay or excelsior used as nesting material should be changed when soiled and again always about 2 days before the eggs are due to hatch. If mites are found in the cracks of the nest box, a diluted solution of 40% nicotine sulphate (such as Black Leaf 40) or chlordane (see Chap. 15) should be painted or sprayed on the sides and bottom of the coop and the painted portions covered with plenty of nest material, making sure the solution does not touch the eggs. Shallow nests with side entrances are preferable as the hen is apt to break eggs when returning to a deep nest. An excellent nest may be made with a level, flat piece of sod for a bottom and a little straw in the corners.

Set Several Hens: A desirable practice is to set the hens two at a time and after hatching give all of the poults to one of the hens. The other hen can then be broken up and should soon go back to laying. Early in the season it is not wise to trust more than 12 to 15 poults to a big chicken hen and 20 poults to a turkey hen. The poults should be confined with the mother hen in the coop at first. After a day or two the poults may be let out in a small wire enclosure in front of the brood coop, but the hen should be confined to the brood coop for several weeks. As soon as the poults have learned to come in and out of the coop to the mother hen, the yard can be made larger or done away with as the poults will not roam too far away from the coop containing the mother hen. The coop containing the hen and poults should be moved 4 or 5 yards away at least once a week. The feeding and drinking utensils should also be moved at the same time. This keeps the poults constantly ranging on fresh ground and takes them away from the contaminated area before the danger from diseases and parasites carried through the droppings becomes great.

Combination of Natural and Artificial Methods: Some few turkey raisers, who have a great deal of confidence in the turkey hen as a mother and none too much in themselves in caring for a brooder, combine the natural and artifiicial methods. They do this by keeping one turkey hen in a coop that is placed inside the brooder house. The poults can get in and out of the coop containing the mother hen but she cannot get loose in the brooder house. The coop in which she stays has a wire bottom of ½-inch mesh hardware cloth which prevents the poults picking at the mother hen's droppings. However, it should be covered with paper on which is placed sand for litter for the first week or 10 days.

A few of the poults will nestle around the mother hen at night and the others will spread themselves around the brooder stove. The mother hen tends to keep the poults contented and thus avoids piling. She also teaches the poults to eat when they are first placed in the brooder house, this being a difficult task in some cases where there is no hen around; although scattering rolled oats or brightly colored marbles on top of the mash serves to tempt even the most stubborn poults.

Irrespective of whether a chicken or turkey hen is used for natural brooding it is wise after the first 2 or 3 days to feed the turkey hen corn and wheat (or barley) inside the coop and feed the poults the starting mash in open troughs or hoppers outside the brood coop or at least out of reach of the mother hen. At the start the mash is placed in reach of both hen and poults until the poults eat mash readily. Starting mash is too expensive to use for the mother hen beyond the first two or three days and she will waste it no matter what precautions are taken.

The equipment that is necessary for the natural method of brooding consists of a good brood coop (See U.S.D.A. and other Bulletins), for each mother hen and brood; some one-inch mesh wire hurdles or frame to make small yards in front of the brood coop; feed hoppers and water fountains. If it is not desired to use frames for the wire, one-inch mesh wire one to three feet high can be arranged in front of the brood coop horse-shoe shape and fastened by 2" by 2" posts which are driven into the ground. Such a fence may easily be shifted when the coop is moved.

Stolen Nests: If the hen has stolen a nest and started setting before the nest is discovered, she should not be moved. Water and grain (shelled corn or wheat) should be placed nearby daily so the hen need not roam far in search of food. Provision should be made to place the hen and poults in a coop after the hatch is over to avoid losses from predators and inclement weather.

Natural Hatching and Brooding Program: Farmers who plan to use the natural method can learn many valuable lessons from game breeders who use bantam hens to hatch and brood quail and pheasants. The following suggestions are drawn from the authors' experience and contacts with successful raisers of game birds.

Practical Rules

1. Wire-bottom coops similar to quail coops* but large enough to accommodate a large chicken hen and poults may be used.

2. As the chicken hens become broody they should be removed (after dark) to these coops. Two hens may be set in one coop if a temporary partition is placed in the center to prevent quarrelling; but care must be taken to check up after feeding time to see that both hens are not on the same nest.

3. The hens should be set in the hatching coops on discarded chicken or turkey eggs until several hens may be set at one time.

---

*Plans for quail or pheasant rearing coops can be secured from the State Conservation Dept. of many states or the Wildlife Service, U. S. Dept. of Interior, Chicago, Ill.

The coop should be on the ground during the hatching period, and on legs or stilts thereafter.

4. The hatching coop should be sprayed with carbolineum, chlordane, 40% nicotine sulphate, or any other good mite killer, before the hens are moved into it.

5. The hens if lousy should be dusted with sodium fluoride or any other equally good lice power when transferred and again 3 weeks later.

6. Clean straw, hay, or excelsior should be used for nesting material and changed 2 days before the poults hatch.

7. Water and shelled corn and wheat (or barley) should be available to the setting hen so she may eat and drink whenever she chooses to leave the nest.

8. When the eggs first pip, the slatted front door of the nest box should be closed to prevent the hen leaving the nest before the hatch is over. If the hen is gentle the shells may be removed from the nest occasionally as the hatch progresses. Twenty-nine days of incubation should be allowed to elapse before discarding the unhatched eggs.

9. If the hen shows a desire to leave the nest before the hatch is over, 8 or more of the newly hatched poults may be removed and placed in a chick or poult shipping box or a shoe box with 6 to 10 holes the size of a penny in the lid and excelsior, shavings, or cotton in the bottom. This box should be kept in a warm place (not hot) out of drafts, until the hatch is over and the poults are returned to the mother hen.

10. When the hatch is over, if there were 2 hens in the coop, remove one hen and give all the poults to the other; all the unhatched eggs, nesting material and litter should be removed from the next box and a little dry sand scattered on the floor or on papers on top of a hardware cloth bottom. At least twice a week (preferably every other day) the sand and droppings should be brushed out of the nest box with an old whiskbroom or small broom with the handle cut to about 6 inches in length. If brushed forward in the coop, the soiled litter should fall through the hardware cloth floor of the coop yard or run.

11. Water for the mother hen and poults should be supplied in fountains or crocks similar to those recommended for artificial brooding that permit the baby poults to drink without getting their feet or droppings in the water.

12. Grain should be available for the mother hen at all times in a small hopper, self-feeder or a can nailed in one corner. A good starting mash should be fed in self-feeders similar to those illustrated, and placed just out of reach of the mother hen, under the board that is close to the nest box on top of the run. If rain dampens the mash, the poults should clean up the wet mash before adding more dry mash to the self-feeders.

13. The poults may be kept in this coop until the mother hen no longer hovers them. The hen may then be removed and if the coop is not needed for another broody hen the poults may remain until 6 to 8 weeks of age, when they should be transferred to clean range that has not been used by any poultry for 3 years.

14. When transferring to range the necessary precautions should be taken to protect the poults from predatory enemies and rainstorms.

CHAPTER 15

DISEASE PREVENTION AND SANITATION

OUTLINE

MORTALITY

The havoc wrought by diseases and parasites is a limiting factor in turkey raising and a threat to every turkey enterprise. There are three losses from a dead bird. First, there is the loss of the turkey itself, and second there is the loss of feed that it has consumed up to the time of its death and third, there is the loss of an opportunity to have made a profit on that bird. Dr. Theobald Smith, the discoverer of the organism causing blackhead, once said that "nature not only abhors a vacuum, she abhors a crowd." Nowhere is the truthfulness of this statement borne out better than in turkey raising. Nature, if she may be personified, is equally as interested in parasites and disease germs as she is in those animals which contribute food for man. When man defies her laws he must evade the consequences or take them. He may evade the dangers involved in raising turkeys in a crowd by practicing sanitation to the "nth degree" and thus keep the disease germs and parasites under control. As the turkey industry grows, diseases and parasites seem to grow with it. New ones appear and old ones lose none of their virulence. In spite of this situation, mortality according to Bureau of Agricultural Economics estimates, has been reduced by about one-half in breeding stock and by over two-thirds in young stock since its high point in 1943 as shown in the following table.

Table 15.1—Estimated Mortality in Turkeys in the United States[1]

Year	Death Loss	
	Breeding stock	Young turkeys
	Per cent	Per cent
1937	11.8	27.0
1938	10.9	26.3
1939	10.4	26.0
1940[2]	10.3	25.9
1941	10.5	27.8
1942	11.1	28.8
1943	11.4	29.7
1944	10.4	25.4
1945	9.3	22.3
1946	8.2	21.0
1947	7.5	19.6
1948	2.7	13.5
1949	7.8	16.1
1950	7.9	15.6
1951	7.5	12.9
1952	5.7	12.9
1953	6.1	9.4

[1]From "Farm Production, Disposition, Cash Receipts, and Gross Income—Turkeys." U. S. Bur. Agrl. Econ. April 1954.
[2]Unpublished revision.

A study of the records from 1935 through 1944, of the University of Massachusetts diagnostic laboratory showed the greatest losses from blackhead (11.14%), then coccidiosis (11.14%), paratyphoid (10.24%), ulcerative enteritis (5.87%), pullorum disease (5.72%), enteritis (4.21%), rickets (3.91%), internal parasites (2.71%), perosis (2.26%), moniliasis (2.26%), and the following diseases less than 2% each; aspergillosis, avitaminosis, cholera, erysipelas, hexamitiasis, omphalitis, paralysis, pox, sinusitis, staphylococcosis, trichomoniasis. Approximately 10% non-specific and miscellaneous diseases and 10% which could not be diagnosed were among the specimens received.

DISEASE CONTROL

The secret of disease control lies not in the use of medicines or the so-called intestinal antiseptics but in the practice of sanitation. According to Webster, sanitation is "the practical application of sanitary science." This means putting to work in a practical way those rules which relate to the preservation of health. Sanitation

Fig. 15.1. Large flocks such as this one can be reared with little or no expenditure for medicines if the pastures are rotated. (Photo from Walter E. Grant, Hamilton, Va.)

must be practiced, it cannot be purchased. If turkey diseases cannot be prevented, there will be no profit in turkey raising, for the costs of medication and handling, and the interference with efficient feed utilization invariably will raise production costs to prohibitive levels, even though control or cure should be brought about. Heading off disease outbreaks through prompt diagnosis and treatment, where effective treatment is available, however, is highly desirable to prevent catastrophic losses. These treatments will be discussed later.

It has been found by research workers that the tiny cecal worm which lives within the appendix of the turkey and chicken may have within its eggs the organism which causes blackhead in turkeys. These cecal worm eggs are frequently expelled with the droppings of the chicken and turkey. Scientists found that the simplest way to produce blackhead in the young poults with which to experiment was to mix with the feed of the young turkeys some chicken droppings. In practically every case the chicken dropping carried worm eggs within which was the blackhead germ. In order to avoid blackhead this cycle of infection (shown on page 854) must be broken. The simplest way to break the cycle is to raise young turkey poults away from chickens and old turkeys and on ground where neither chickens, turkeys nor any other poultry has run during the preceding three years. Many farmers say this is virtually impossible on their farm. The answer that they will probably secure is that the raising of turkeys is nearly impossible if they do not practice range rotation, or else use strict confinement. Fortunately the practices which tend to avoid trouble from blackhead also result in controlling coccidiosis and minimize the possibility of trouble from tapeworms and other diseases and parasites. Since pigeons, sparrows, and possibly other birds may be carriers of some of the diseases of turkeys, they must be given consideration as possible sources of infection.

Land cannot be disinfected by any known practical means. Treatment by disinfectants including lime is useless. The only practical method of freeing soil of disease germs and parasites is to "rest" it by keeping all poultry and poultry droppings away from it for 3 years, allowing maximum exposure to direct sunshine. Cultivation is of doubtful value in destroying harmful organisms but it does cover them up so that turkeys cannot easily reach them. Poultry manure however can be freed of disease organisms and parasites and their eggs by storing it for 2 weeks in insulated, covered boxes. The heat thus generated is sufficient to kill all known disease organisms and parasites in all stages of development.

Millions of dollars have been spent in past years by turkey raisers for so-called blackhead, typhoid, cholera and other supposed "cures," and yet the losses from these and similar diseases were not materially reduced except where the fundamental principles of sanitation, segregation, and prevention were put into practice (see Fig. 15.2).

Accurate diagnosis of turkey diseases in the field is almost impossible except in such obvious troubles as fowl pox, blackhead, and pendulous crop. Blackhead often may be diagnosed by the layman on postmortem examination. The combination of the spotted liver and ulcerated ceca is definite proof of blackhead. A laboratory diagnosis by a competent bacteriologist or veterinarian is necessary to differentiate between most of the bacterial and protozoan diseases. The diagnostic laboratories of the Veterinary Department of the Agricultural Experiment Station in each state (see list in Appendix) are usually available to assist the turkey raiser with differential diagnosis. In many cases, the local veterinarian is equipped to make an accurate diagnosis.

With turkeys, as with other animals (including humans) diagnosis in the early stages of the disease is of vital importance in treatment. Correct diagnosis long after the disease has taken a foothold is in many cases of little value. All sick turkeys look more or less alike, even to the expert, and a post-mortem view of the internal organs is absolutely necessary to differentiate between even the most obvious diseases and bacteriological examination is necessary in the majority of cases. Such symptoms as ruffled feathers, listlessness, loss of appetite, and the tendency to stand with head drawn into the shoulders are observed with many widely varied diseases. A yellowish diarrhea is common with ills like cholera, staphylococcosis, typhoid, blackhead, coccidiosis, and trichomoniasis. The normal, light-brown, soft, fine-textured cecal droppings, which are voided at intervals, sometimes are mistaken for symptoms of disease. Hence, it is virtually impossible to accurately diagnose turkey diseases by external symptoms.

The general recommendations given in this chapter to eradicate those diseases for which no cure is known, should be put into practice to avoid the disease hazards

Fig. 15.2. A picturesque but hazardous watering spot that may serve as a place favoring the spread of disease and parasites. (Photo from U. S. Dept. of Agr.)

rather than depend on treatment. Even though research has already provided us with effective treatment for several of the more troublesome turkey diseases, still the adage will hold that "an ounce of prevention is worth a pound of cure." Whenever disease makes its appearance the sick-appearing turkeys should be removed immediately and isolated or killed for diagnostic purposes; unless they are sent to the laboratory. The litter, if litter is being used, should be changed immediately, as many diseases are filth borne and the cycle of infection may be broken by frequent changing of the litter. If infection appears when the flock is on range and roosting in the open or under a simple shelter, it is wise to move the flock and necessary equipment to new pasture.

LAXATIVES

Routine dosing with laxatives is not a desirable practice and is not recommended. While it may be beneficial in some cases, it sometimes causes harm to sick birds. Proper feeding will promote normal function of the digestive tract. When a laxative treatment or "flush" is indicated, which is rarely, a good method is to feed a moist crumbly mash of bran two parts and clabber milk or condensed buttermilk one part by weight; or bran 4 parts and molasses 1 part. This mash may be the only feed for one day and the treatment repeated five days to a week later. Another effective means of administering a laxative without disturbing growth is the well known "milk flush." The flushing mash is prepared by mixing 60 pounds of the mash the poults have been receiving with 40 pounds of dried skimmilk or dried buttermilk or 68 pounds with 32 pounds of dried whey.

Epsom salts may be given as a laxative but since salts tend to shrink the tender lining of the intestinal tract (while the bran-milk mash does not) they should not be given often or in too heavy doses. A good method with epsom salts is to dissolve 1 pound in each 4 gallons of drinking water, putting it in at night so birds will have it in the morning, then removing it at 11 a.m., replacing it then with water and giving the usual feed.

Castor oil is not generally recommended as it must be administered individually. The usual dosage consists of 4 or 5 drops for a week-old poult, 1 teaspoonful for a 5 or 6 pound poult, 1 tablespoon for a 10 to 12 pound (half grown) bird and 2 tablespoons for a mature tom.

ANTISEPTICS AND DISINFECTANTS

Experimental results fail to show that antiseptics placed in the drinking water serve to kill disease germs in the intestines. The only value of such antiseptics is that they serve to reduce the spread of these germs through the drinking water.

The **sodium hypochlorite** solutions are quite effective antiseptics if they contain at least 2.6% of available chlorine by weight. The chief disadvantages of these solutions are the cost and the instability of the chlorine which soon loses its germ-killing power in the presence of organic matter. These solutions may be used as antiseptics in the drinking water at the appropriate strength indicated on the container. They are not good general disinfectants but may be used to disinfect water pans, feed containers, egg trays, and other equipment which first has been thoroughly cleaned.

Potassium permanganate is an antiseptic fairly effective at the rate of a level teaspoon per gallon of drinking water but has no other uses on the turkey farm. It corrodes metal, so should be used only in crocks or wooden drinking vessels. It soon loses its germicidal value (oxidizing property) and when it has turned from a clear wine color to brown it is no longer of value. Copper sulphate, commonly called bluestone is not a good general antiseptic or disinfectant but will kill algae and many fungi.

Cresol. The most trustworthy disinfectant for general disinfection around the turkey pens, houses, coops, and porches is cresol. Cresol with a soap base (Liquor cresolis compositus U.S.P.) is sold under various trade brands. It is at least three times more powerful as a germ killer than carbolic acid, is much cheaper and less dangerous to handle. An official list of approved disinfectants, many of which are based on cresol, may be secured from the Agriculture Research Service, USDA, Washington 25, D. C., asking for the list of "disinfectants permitted for official use." These products must contain between 50 and 53% total phenol, not less than 21% soap, and must mix easily with water so that the killing power is retained in 3.1% solution. For 4 to 3.1% solutions a pint of disinfectant to 14 to 16 quarts of water should be used.

Stock or sheep dips (coal-tar cresol products) usually are good disinfectants if used as directed. The brand used should be selected from the "Permitted Disinfectants" list.

Lye, (sodium hydroxide) one pound to 5½ gallons of water (about 2%) often is recommended as a fairly effective and relatively cheap disinfectant. It is too corrosive to use as a spray or on plaster or concrete but can be used for disinfecting wooden or metal (except aluminum) equipment. Users should protect the hands and face. Vinegar is an antidote. As a cleansing and antiseptic agent use 1 ounce of lye per gallon of hot water. An hour or so after application, rinse off the lye with water. Its use is not especially recommended in turkey sanitation.

Sodium orthophenylphenate,* a nearly white solid, is effective in 1% water solution (1⅓ avdp. oz. (38 gm.) per gal. or 1 pint per 6 gals.). It has little odor and is very effective if used at a temperature of 60° F. or higher. It should be kept in an air-tight container until used. It is not very toxic to animals or man and is excellent as a fungicide, disinfectant, and deodorant for metal, wood, concrete, fabrics, leather, and human skin. It is reported effective against lice and mites.

Quaternary ammonium disinfectants (odorless and non-irritating to the skin) are valuable for disinfecting clothing, instruments, glassware, utensils, eggs and metals. They tend to be specific for certain diseases and should be used according to directions of the manufacturer in non-soapy solutions. They themselves have a definite cleansing action.

Heat applied with a fire gun is not an efficient practical germicide and is a fire hazard. Heat *via* boiling water or steam, is effective if applied direct for 10 minutes or longer. However, the cheapest disinfectant in summer is direct sunshine. Exposure of porches, equipment, water pans, and the like to the sun's rays constitutes an effec-

*Sold by one manufacturer under the trade name "Dowicide A".

tive method of disinfection if all particles of dirt are first removed.

Formalin, a 40% solution of formaldehyde, is a very good general disinfectant in a 10% water solution and as a gas in the presence of high humidity for fumigating incubators and other enclosed places. It is efficient in the presence of organic matter, not very toxic (although very irritating and disagreeable) to animals and birds, is reasonable in cost, and does not injure clothing or metallic equipment. Its chief uses are to disinfect brooders, incubators, feeders and waterers. Its irritating qualities make it difficult to apply hence its chief use is to fumigate enclosed places in which equipment to be fumigated may be placed.

Detergents. The common detergents, the "soapless soaps," sold for household use (Dreft, Swirl, Tide, Cheer, All, Fab, etc.) are excellent for cleaning and have considerable disinfecting action. Applied hot, 167° F. or more, they will kill most disease germs excepting spores. They work well in the pressure steam jenny or can be applied hot by a sprayer or a broom, 1 lb. to 40 gal. water.

Mercuric chloride, or corrosive sublimate, at 1:1000 in water is a powerful disinfectant, except in the presence of organic matter, but is not recommended for use on poultry plants. It is extremely poisonous and very corrosive to metals.

Quicklime (unslaked lime or calcium oxide) reacts violently when water is added liberating heat and oxygen. If handled with care, it is useful for disinfecting small areas of soil, drains, and droppings, and for making whitewash. **A whitewash** with disinfecting action may be made by dissolving 5 lbs. of alum and ½ lb. of salt in 3 gal. of hot water; then separately dissolve 25 lbs. of quicklime in 4 gallons of hot water, and combine the two mixtures. To the resulting gallons add 2½ lbs. of cement and ½ can of lye (Jull, 1947). The disinfecting value may be increased by adding 1 lb. of chlorinated lime to 40 gallons of whitewash.

Hydrated lime (slaked) has mild deodorizing and disinfecting action. Its chief use is on manure, 100 to 200 lbs. per ton, to prevent maggots, cut down odors, and help preserve its fertilizing value.

Germicidal lamps and glycol vaporizers have little practical value around a poultry plant.

Colloidal iodine (iodine suspensoid) appears to be a superior germ killer being effective against coccidia which are especially difficult to kill. It will kill disease organisms embedded in fecal matter when used in a 0.5% solution (Mallman, 1947), something few other disinfectants will do. Its cost and strong, lasting odor limit its use.

CLEANING AND DISINFECTING

Disinfectants seldom have any direct effect on a disease outbreak and have a limited use in routine sanitation practices. However, where an infectious disease is encountered, a thorough cleanup followed by disinfection is desirable. To clean up a building in which an infectious disease has appeared, the following steps are suggested. Settle the dust by spraying lightly with the selected disinfectant the roof, walls, equipment and litter. Remove all litter and droppings and burn them or remove beyond any possible contact with poultry. If the floor has no drain, remove all portable equipment to a cleaning floor with a drain and with a hose scrub and rinse it clean and allow to dry then spray with disinfectant. If no floor drain is available, dry cleaning or cleaning with a steam jenny (a portable steam generator) should be done. For dry cleaning use the broom, scraper, and shovel to gather all loose dust and dirt for removal and safe disposal. Board or concrete floors can then be thoroughly sprayed along with roof, walls, and equipment. Enough of the disinfecting solution can be used to cover the floor lightly allowing it to dry up. Dirt floors are almost impossible to disinfect thoroughly; a good soaking with lye water or a common detergent solution may help. A vacuum cleaner is very useful in dry cleaning, especially of incubators. The tank-type cleaner probably is the most satisfactory.

The pressure steam "jenny" properly used probably provides the best method of disinfecting a building with or without a drain. Suggestions for the use of this equipment are made in "Cleaning and Disinfecting on the Farm" by C. R. Schroeder, in Proceedings of the U. S. Livestock Sanitary Association, 1948*. Thorough mechanical cleaning must be done first. The common household detergents are cheap, harmless to the operator, and efficient in the steam jenny, 1 lb. to 40 gal. of water. No other disinfectant need be used. Lye is efficient but damaging to paint, concrete, aluminum, and to the operator. Many disinfectants are efficient in the jenny but may be too expensive, smelly, and/or injurious to the operator. If the building to be cleaned has a drain, leave the equipment in it and use the hose combined with brush, broom and scraper, and wash down the building from peak of roof to floor, and the equipment from all angles. Broom all surplus water into the drain and allow the building to dry, then, if disease has been present or there is any other reason for using a disinfectant, apply with a pressure sprayer the selected disinfectant at the proper dilution—or better, use the steam jenny with a detergent or suitable disinfecting solution. Permanent buildings should be built with water connections and drain, otherwise efficient cleaning will be very difficult to accomplish.

PREVENTION PAYS

A rearing program embodying the essentials of sanitation presents the only practical solution to the disease problem. To grow turkeys that are healthy, thrifty, and free from disease and increase profits through reduced mortality, the following practices are recommended:

---

*Obtainable from Dr. R. A. Hendershott, D.V.M., Secy., Trenton, N. J.

Clean Poults

Obtain eggs or poults only from healthy pullorum-tested stock. Eggs should be incubated in pullorum-free machines.

Clean Brooder Houses

The brooder house should be cleaned, scraped, swept out, and then scrubbed; using water (preferably hot, 167° or more) containing a 13-ounce can of lye to 13 gallons of water for wooden floors, or a detergent for concrete or any other type of floor. Reliable disinfectants such as the coal-tar products, i.e., cresol solution, may be sprayed on the floor and side walls, but this would not be necessary unless disease has been present. Emphasis should be placed on the cleaning more than the disinfecting. All equipment used for brooding should first be cleaned thoroughly. All cleaning of brooder houses should be done before moving to clean range. Wire sun porches with ¾ to 1-inch mesh wire or slats for floors, should be used unless the brooder house can be moved easily or clean ground provided in some other way.

Clean Ground

To avoid soil-borne diseases the turkey grower should provide:

Complete separation of turkeys from chickens.

Land not used by poultry for three years.

Land not contaminated by drainage water or poultry manure.

Land well separated from previous range.

Land naturally well drained.

A range with grass, ladino clover, alfalfa, or other good pasture, or else confinement equipment.

Clean Litter

The litter should be changed often enough to keep the floor clean, dry, and warm (weekly in most cases). It must be dry at all times. A dry, clean, inexpensive litter should be used. Built-up litter results in considerable saving of labor and if a highly absorbent litter is used there is no increase in disease hazard. However, for turkeys the same litter should be used only during one brooding period, and in no case should chicken litter be used. Wire floors aid in the prevention of soil-borne diseases.

Clean Feed and Water

Make feeders and waterers contamination-proof.

All feed and water pans should be on wire platforms or on wire floors if practicable.

The area around water pans always should be dry.

Move outdoor feeders to a clean spot each week (unless the whole flock is being moved weekly).

Clean Management

Old and young stock should be kept separated at all times, and turkeys raised away from all other poultry.

Cross-drainage and transfer of contamination by visitors, animals, vehicles, etc., should be prevented.

Fig. 15.3. Beltsville Whites raised well away from chickens, yet not too far from farm house.

In most situations the moving of the flock and its equipment every 7 to 21 days within a clean area of ground is necessary (see Fig. 15.3).

Used Feed Sacks

The actual role of re-used feed bags in spreading poultry diseases is unknown but direct evidence that they do spread disease is lacking. However, experiments showed that pullorum, Newcastle, infectious bronchitis, and other disease germs survived well on burlap, feathers, and shavings for 28 to 97 days.* When sterile burlap samples were inoculated with disease germs, tightly baled, and the bales fumigated with cyanide for 48 hours, pullorum germs and coccidia still survived in the center of the bales. Some other organisms were killed. Ultraviolet light at 12 inches for up to 10 minutes exposure for each side of burlap strips failed to kill pullorum germs. Burlap strips impregnated with various organic mercurial compounds with or without a water-shedding substance showed good bacteria-inhibiting power but did not kill the bacteria and viruses tested. Dielectric heat (as employed in radar cookery and diathermic physiotherapy) showed the most promise of all treatments tried. At 200-215° F. applied for at least 5 minutes, it killed pullorum and micrococcus organisms but not a spore forming aerobic bacillus.

It can be assumed that dielectric heat properly used will kill on burlap bags all viruses and other microbiological agents of potential disease-making power except spore forming organisms.

Uses of DDT: The new spray material DDT (dichlorodiphenyltrichloroethane) developed during the period from 1941 to 1945 will prove quite useful to the turkey grower in the control of insects. However, tests with this chemical in the control of chicken lice indicate that it is

*From "Studies on sanitizing used feed bags," by E. Jungherr, Univ. of Conn., Storrs. Jour. Am. Vet. Med. Assn., 117 (883): 324-326 (October 1950).

no more effective than sodium fluoride or nicotine sulfate (Black Leaf 40). It is of less value in the control of poultry mites than the anthracene oils and some other new insecticides. Undoubtedly its principle place with turkeys will be in the control of flies, beetles, and other insects that frequent the brooder houses and dropping pits and serve as intermediate hosts and mechanical carriers of poultry diseases and parasites.

Precautions must be taken in the use of DDT around chicks and turkeys as the chemical is toxic if taken internally. Work with growing turkeys at the Beltsville Station (1947) indicates that 0.075% of technical DDT dissolved in peanut oil in the feed is toxic, while 0.038 or 0.019% is not. The Utah Station (1950) reported that growing turkeys were susceptible to DDT dusting powder as used in dusting alfalfa and were killed by a DDT level of 200 parts per million or 0.02% in the feed. Whether or not flies and other insects killed by DDT would be toxic to poults was studied in California. McNeil and Hinshaw (1945) conducted careful studies, spraying the brooder house walls and ceilings as well as the outside front and sides of the building and the lane between two brooder houses. One pen was sprayed three hours before the poults were put under the hover. The solutions used (both with good results) contained .024 and .048% DDT in water and were applied at the rate of 1 gal. per 100 sq. ft. of surface. No injurious effects on the poults were noted even though they ate some of the dead flies. One, four and six weeks after spraying there were no flies in the brooder house. At 10 weeks control was no longer effective. Spraying of the ground was effective for 3 weeks. Liberal spraying of droppings also was effective in fly control. A much more common dilution recommended where the maximum residual effect is desired is 2.5% of DDT.

DDT is available as a wettable powder, an oil solution, or as a gas (in gas bombs); although the gas has no residual effect. Only the best product prepared by reliable firms should be used. Complete instructions for the dilution and use of DDT can be secured from the State Entomologist or Entomology Department of any State College of Agriculture or Agricultural Experiment Station (address in Appendix).

Other **new insecticides** are useful in controlling insect life around the poultry plant. Among these are chlordane and benzene hexachloride (B.H.C.). The former has the advantage of being odorless while B.H.C. has a disagreeable musty odor which may remain as long as two weeks and which may become incorporated in poultry flesh and eggs. Both of these products are believed to be less toxic than DDT to animals, birds, and man. Aldrin (compound 118), an insecticide widely used for grasshopper control, was found to be highly toxic to growing turkeys. Levels of 25 p.p.m. or higher of technical aldrin were highly fatal. At a level of 3 p.p.m., growth of males was depressed significantly. (Dom. Exp. Sta., Swift Current, Sask.)

CHAPTER 16

DEFORMITIES, INJURIES AND VICES

OUTLINE

DEPRAVED APPETITE

A depraved appetite or even cannibalism is likely to result from subjecting the birds to highly artificial conditions or from failure to provide an adequate diet. The precautions suggested for the prevention of feather picking will usually forestall the development of a desire to eat abnormally large quantities of litter, earth, or droppings, or to indulge in feather eating, toe-picking, or cannibalism. Poults normally eat quantities of little stones, earth, and litter (where litter is used) so that when this occurs in moderation no concern need be felt. It is partly on account of this characteristic that it is so necessary to keep the inside and outside environment sanitary. Precautions should be taken to see that feed is available when the poults are placed under the brooder, otherwise they are likely to eat the litter to excess. Sand or fine gravel, or wire, minimizes this trouble.

Turkeys sometimes eat such oddities as shiny nails, screws, pieces of wire, broken glass, broken dishes and buttons. In one instance a sick turkey upon autopsy had a spark plug in its gizzard. The pointed end had penetrated the gizzard wall. Sharp metallic objects when swallowed often cause the crop, esophagus, proventriculus or gizzard to be penetrated usually with fatal results. Glass and chinaware usually cause no harm—they function as grit and are ground to smoothness. Presumably well-fed turkeys are less likely to have abnormal appetites. However the eating of indigestible materials is more likely to be undertaken by the turkey to supply grit and may happen even when plenty of good grit and feed are available. When a new plant is built or repairs are made, it might save a few turkey lives to clean up and remove all shiny metallic materials small enough to be eaten. It also would be advisable to prevent turkeys from ranging on dump-ground.

BLUEBACK

The deposition of bluish black pigment in the skin of dark-plumaged turkeys results in a condition known as blueback. The skin is not clear white or yellowish white but suffused with black or bluish black pigment which becomes part of the skin and can seldom be removed. The back is most often affected but in some cases this pigment is found on wings, breast, and sides. The most common causes are treading by the males, feather picking, and other damage to the feathers. The best explanation so far advanced to determine the cause of blueback is that of R. D. Taylor of Escondido, California. His remarks are quoted herewith:

"If a feather anywhere on the body is broken or damaged, the surrounding skin will absorb the bluish black coloring matter which ordinarily would feed the feather. The feather does not have to be pulled out completely to cause this discoloration. The discoloration develops in 10 to 25 days after damage to the feather.

"Among young toms, we sometimes find an epidemic of tail picking and in consequence have birds five to seven months old in beautiful condition otherwise, with a circle of dark blue around the oil bag two to three inches in diameter. In breeder hens, we find that the toms will wear away and damage the feathers on the back, and that contact with the ground and the sides of the nest will damage the feathers along the front and the rear of the keel and the sides of the turkey. Wherever this damage occurs, we get the dark blue stain. The small amount of pigment which is left in the skin of immature birds when they are dressed will not spread in the dead flesh, although they make one little pin point of dark color; but a stain from one broken feather will blacken an area around three-quarters of an inch in diameter."

In young birds, blueback usually is found only in flocks that have been subject to serious feather picking or flocks in which the females have become sexually mature so that treading occurs.

The Purina Research Farm (1946) reported that bluebacks were "the result of exposure to sunlight after feathers were picked or broken, which allowed the skin to become exposed to the sun. Spreading the feathers apart to expose the skin produced the same result. When feathers were artificially broken or pulled out of birds housed in darkened sheds, or if the exposed area was covered with canvas, the condition could not be produced. However, as soon as these same birds were removed from the house and the skin exposed to the sun, the discoloration appeared. When birds, which had been ranged in the presence of sunlight and had developed blue-backs, were placed in a darkened house, the color disappeared in from three to four weeks. Thus, this demonstrated that the color was not 'fast' and could be removed by the very simple management practice of removing the birds from sunlight.

"It was also found that bluebacks seldom appear in birds under 14 or 15 weeks of age and that, in the presence of sunlight, it required approximately 18 days after feather injury for the discoloration to appear. Usually bluebacks are not common during the winter months when the sunlight is at a minimum."

Sometimes the dark pigment contained in the pinfeathers of colored varieties becomes spread over the skin during picking. This can be prevented by careful picking and by avoiding the killing of birds with very short pinfeathers.

White varieties of course are not susceptible to blueback since their plumage contains no pigment.

CROOKED BACK

This type of deformity ranges in severity from a slightly crooked condition to a hunchback in which the backbone is curved both laterally and dorso-ventrally. Birds possessing it are of comparatively infrequent occurrence but a small percentage, usually less than 1 per cent may be considered normal. There is a possibility that injury or lack of vitamin D may cause it. That heredity may play a part is doubtful since a pen of crooked-backed turkeys at the University of California produced no deformed offspring.

However, it is good practice to eliminate from the breeding flock all individuals with crooked backs. Such birds will be graded down when marketed due to their deformed appearance, therefore they may well be consumed on the home table or marketed locally.

ROACHED BACK

A roached back is a deformity which is described as a non-lateral curvature of the spine giving the bird's back an arched appearance when viewed from the side. The curvature varies from slight to severe and is difficult to detect in the live bird. It is of fairly frequent occurrence and when severe may reduce the carcass possessing it to B grade. The cause is not known. Due to possible genetic involvement it is advisable to examine the backs of all prospective breeders very carefully and eliminate from the breeding flock all deformed birds and where possible, families producing them.

TORN BACKS

These injuries occur in the hens as a result of mating attempts. They seldom are fatal but often result in loss of egg production in breeders and loss of grade in young market stock. In the latter case, separation of the sexes prior to about 20 weeks of age will prevent trouble in early hatched market birds which, due to long hours of daylight, warm weather, and sunshine, come into sexual maturity earlier than normal. Such birds should be marketed before they lay to avoid loss of grade due to poor finish and old-bird appearance. Saddling of breeder hens largely prevents torn backs and is very widely practiced. In small varieties back injuries are much less common than in large varieties. Small-type breeder hens if mated to males of the same strain do not have to be saddled but many breeders use saddles in preference to treating the toenails of the toms. "Debeaking" all the toenails of small-type breeder males to a point even with the flesh will, for all practical purposes, suffice to prevent back injuries to the females. Cutting off the toenails up to the "quick" but not into it, and filing the ends smooth has the same effect. In large-type turkeys, saddling the hens is the best practice.

BEAK DEFORMITIES

Crossbill. Turkeys with the upper and lower beaks permanently crossed occur occasionally in most turkey flocks. The deformity varies greatly in severity and seems to be acquired during the poult's development in the egg, and appears to have no genetic basis. Crossbilled poults usually can eat well enough to grow fairly well under modern methods of self-feeding of complete diets. Crossbill is one of the deformities for which the breeder should be on the lookout to eliminate from breeding flocks.

Ostrich beak. In the deformity known as ostrich beak, the upper half of the beak fails to develop the proper thickness and curvature, or else in some way the horny material becomes severely worn, giving the appearance of a beak shaped like that of the ostrich. Inadequate diet, the shape of the feeders, or reaching through wire flooring or fencing especially when the poults are small, may be contributing factors. Birds possessing this deformity never should be used for breeding even if they recover from it, as the tendency may possibly be inherited.

Pressure necrosis is a term applied to a deformed necrotic condition of the beak resulting from impaction of feed. It may occur in turkeys of all ages but is more common in young poults. Its cause is believed to be the eating of mash that possesses a high percentage of finely ground floury material. The presence of hygroscopic (water-absorbing) feeds aggravated the condition. The Swift Current (Sask.) Station (1948) found that significant amounts of pressure necrosis were found in poults started on experimental feeds that contained 50% of ground wheat of which 49% or more would pass through a 3/64th inch screen. When oat groats were used pressure necrosis occurred when 30% or more of the groats (which comprised 50% of the mash) would pass through the screen.

BREAST BLISTERS, CALLUSES, AND ABSCESSES

These deformities seem to originate from injuries to the skin caused by turkeys bumping themselves when slipping as they walk along the roosts; as a result of rough handling; by flying against objects; or by sitting on peaked roosts or protruding wires or nails. The injuries apparently permit entrance of pus-forming disease organisms such as staphylococci which produces the deformities which may vary in severity from a small blister or callus to a large abscess. Prevention is the only remedy; methods are suggested by the causes listed. Poorly designed roosts probably cause many calluses and blisters; stampedes and rough handling no doubt are contributors. Because of the possibility that hereditary factors may be involved, birds with well-developed blisters, calluses or abscesses should not be selected as breeders.

BREASTBONE DEFORMITIES

Turkeys with well-developed breast musculature seldom show serious deformities of the keel, or breastbone probably due to better support of the keel by the breast muscles. Broad-breasted turkeys do frequently develop roost dents of slight to moderate depth if they are retained much beyond the usual market age. This however is of little economic significance.

A crooked breastbone is a deformity consisting of a lateral curvature of the keel. A roost dent is a nonlateral depression or dent in the keel. Deformed breastbones fall into three general classes. Class 1 consists of slightly crooked or "curved" keels and slightly dented keels, both often occurring in the same bird. These deformities do not detract seriously from the appearance of the carcass or interfere noticeably with slicing of the breast meat. Breastbones of class 1 do not prevent the carcass possessing them from grading A or ordinary No. 1 but do prevent them grading the equivalent of AA unless the curve or dent is "very slight" (1/8" or less in depth).

Class 2 consists of moderately to deeply dented and moderately but not severely crooked keels, often both occurring in the same bird. This type of deformity detracts from the appearance of the carcass and interferes slightly with the slicing of the breast meat. Class 3 keels cause reduction in grade to No. 2, or U. S. Commercial, Grade C.

In many strains of turkeys, the broad-breasted types excepted, a small percentage of crooked breastbones is to be expected. With the more susceptible strains from

1 to 2% of Class 3 breastbones (over two-thirds of them on males) and 1 to 2% of Class 2 breastbones (half or more on males) may be considered in the light of present day knowledge as non-preventable except possibly by selective breeding. Class 1 breastbones are of frequent occurrence in birds of both sexes. Since they do not under most conditions of marketing reduce the price received, they are not of great economic significance. However, the nearer straight the breastbones are, the better the turkeys sell, or other factors being equal. As far as weight loss during dry picking and full-drawing is concerned, crooked-keeled birds do not differ from straight-breasted birds.

Deformities of the breastbone may start to develop at any time during the poult's life. It has been observed in standardbred Bronze that a small percentage of poults may be hatched with breastbones very slightly dented or curved. Crooked breastbones have their beginnings in slight curvatures or dents which may or may not develop into moderate to severe deformities. Development from slight to moderate or severe deformities usually proceeds rather rapidly, just how long not being known, except that it takes less than two weeks.

According to the Wyoming and Nevada Station workers and others, the principal cause of crooked breastbones in non-broad-breasted turkeys appears to be faulty nutrition with heredity, disease, management, and narrow or otherwise unsuitable roosts playing a less significant, though sometimes important, part. Any serious interference with growth (due to outbreaks of disease, nutritional deficiency, or any other cause) can and frequently does give rise to deformed breastbones. Lack of vitamins especially vitamin D and mineral imbalance (particularly calcium, phosphorus, and manganese) are probably specific causes. The Nevada Station observed that more than 10% ash in the diet caused crooked keels. Roosts narrower than 2 inches and deficiency of calcium and other essential nutritional factors may cause crooked breastbones to develop late in the growing period, especially in large toms. Roosts made of sawed lumber laid in a horizontal position have been observed to cause Class 1 (curved and slightly dented) breastbones. Early roosting will not cause crooked or dented keels if the roosts are of a good type and the diet is well balanced and complete.

The following precautions will aid in decreasing the proportion of deformed keels; (1) feeding of a well-balanced diet; (2) prevention of disease; (3) using of roosts made (preferably) of poles 2 to 4 inches thick or of sawed lumber 2½ to 4 inches wide, upper edges beveled and tilted (except for large broad breasted types) at an angle of 15 to 20° (12 to 16″ rise to 4′ of run); 2x3's flat side up are good; (4) saving for breeding only birds possessing normal keels, and eliminating whenever possible, families producing them. For Broad Breasted Bronze it is better to have no roosts at all or to use low roosts of 2x2's, 2x3's, or 2x4's, narrow or wide side up, edges beveled or not, and laid flat, not tilted. The heavy breast musculature seems to support the keel and the level surface enables the heavy birds to keep their balance. Roost wire (to form a droppings pit) should be nailed to the undersides of the roosts; which should be 12″ to 18″ above the ground.

"Very slight" roost dents, ⅛ inch deep or less, do not influence the dressed grade in any way. Dents over ⅛ and not exceeding ¼ inch deep (called "slight") influence the grading only to the extent of keeping the specimen possessing them out of the AA grade or its equivalent. Dents over ¼ and not exceeding 1 inch in depth (called "moderate" to "deep") necessitate a reduction to grade B and dents over 1 inch deep (called "very deep") reduce the specimen to the C grade.

The suggestions dealing with the cause and prevention of crooked breastbones may be applied equally well to dented breastbones with special emphasis on the use of proper pole roosts or tilted roosts. Roosts of sawed lumber placed horizontally will cause dents with varieties other than the heavy, broad-breasted types of turkeys, hence the recommendations for tilting.

CANNIBALISM

Cannibalism in the strictest sense, refers to the habit of picking when it results in the drawing of blood. Turkeys are not very susceptible but occasionally young poults start picking each other especially around the vent resulting in "pickouts." Electrical debeaking at an early age has been suggested as a preventive. As with feather picking, this type of cannibalism seems to be the result of overcrowding, too-large brooding units, deficient diet (rarely), or diarrhea producing disease causing reddened inflamed vents which encourage vent picking.

Tailhead picking of breeding toms by the hens sometimes occurs. Severely picked toms are useless as breeders. Fighting among well-grown or adult toms may result in a few severely picked heads and necks. As with other vices, overcrowding and too-close confinement are predisposing factors. Injured turkeys if discovered in time may be saved by applying pine tar (preferably with bitter aloes added), chick-pick remedy, or "scarlet balsam" to the raw and bleeding areas. These remedies have a protective effect and are distasteful to the birds. Scarlet Balsam is a product used by veterinarians in the treatment of sores on animals. Electrical debeaking or installation of hog-ring type bits usually results in controlling picking of all types.

PENDULOUS CROPS

This abnormality is variously called crop bound, sour crop, baggy crop, water crop, drop crop, and sagging crop. The crop is enlarged, pendulous, and lacking in muscular activity; its contents pass out too slowly or not at all and keep accumulating. The bird often carries a tremendous load, most of which is below the outlet of the crop (see Fig. 16.1). In the early stages, the crop contents usually are liquid and little or no tissue change occurs.

Fig. 16.1. An extreme case of pendulous crop. (Photo from U. S. Dept. of Agr.)

In the later stages the crop contents are liquid or semi-liquid, often including dirt and other extraneous matter. The lining of the crop becomes thickened, ulcerated, and often necrotic. The contents invariably have an extremely disagreeable odor. Sometimes the crop and its contents compose as much as 30% of the total weight of the bird. The death of afflicted birds is, according to the University of California, principally due, (1) to lacerations in the crop area which results in the rupture of the crop, (2) to injury by pen mates, and (3) to mechanical pneumonia. A fair percentage of all afflicted birds recover naturally.

Pendulous crops usually are made possible by an inherited weakness plus heavy consumption of liquids. Both are prerequisites. More males than females are affected but birds of both sexes are susceptible. Turkeys inheriting the weakness will not show it unless they over-consume liquids. The California Station recently has found that the tendency to develop pendulous crops is inherited in a complex manner. Some strains carry this deformity producing gene or genes while others do not. Stock from the former will develop pendulous crops if grown in a hot environment or one that in some other way causes liquids to be drunk in large quantities. If on the other hand this same stock is given a cool environment and not permitted to drink liquid milk in great amounts, it will not develop the deformity. Stock that does not carry the genes or determiners for pendulous crops will not develop them in significant numbers even if grown in an unfavorable environment. These facts have been well established so that control of this malady is clearly indicated, i.e., breeding out the weakness and controlling the environment as much as possible. Preventing exposure of the turkeys to excessive heat, providing ample shade, feeding no liquid milk or whey, and providing cool drinking water easily accessible are management practices that will help to prevent pendulous crops developing even in susceptible stock. In hot climates, stock that is free from the hereditary weakness should be used whenever possible, water supply should be regular, kept cool, and easily available at all times, and every effort should be made to protect the birds from the heat.

Occasional cases of pendulous crop, apart from any hereditary or climatic basis, may be caused by mechanical obstructions such as coarse, tough roughage. These are impactions rather than true pendulous crops and are discussed in connection with brooding and rearing. Crop worms, diseases such as mycosis and trichomoniasis, and injuries, also may cause crop disorders superficially similar to the simple pendulous condition but differing usually in that the crop walls are thickened and the crop contests are not the usual watery mixture but may feel mushy or lumpy. Control in these cases is dictated by the results of diagnosis.

To prevent impactions, avoid unsuitable litter and green feed such as wilted lawn clippings and coarse, stringy range plants, not forcing or allowing birds to eat much of either, and supplying ample feed and water.

The California Station tried various methods of treatment of birds with pendulous crops but none was found to be efficacious and at the same time practical. Applying a tight vest made of a woman's stocking after the crop has been emptied may afford some relief. At the Beltsville Research Center some success was obtained by sewing the skin fairly tightly over the crop in a "brisket stitch," sewing through the two thicknesses of the skin but not through the crop tissue, which reduced the pendulous area and left the surplus skin hanging in brisket-like fold. Stitches can be removed just prior to slaughter.

Liquid-milk feeding frequently causes pendulous crops to develop. It also is insanitary and is not recommended. However when its use is desired (for economic reasons) it is suggested that water be fed along with the liquid milk. Possibly one container of water to several of milk is the best compromise when the milk is depended upon to supply most of the supplementary protein and riboflavin in starting, laying and growing diets or nearly all of them in growing, fattening, or wintering diets. If the milk is depended upon to supply only a half or less of the supplementary protein and riboflavin, then milk and water should both be available equally. The composition of the mash fed and the nature of the green feed supplied with milk appear to exert some influence on the incidence of sour crop. There is, however, practically no information available that is useful in helping to plan the best possible feeding program to include liquid milk and it is partly for this reason that its use is not recommended except under conditions where its availability at very low cost or lack of other better sources of protein and riboflavin justifies its inclusion in the diet.

From the genetic standpoint it is well to consider susceptibility to this deformity when selecting breeding stock. Pedigree breeders will do well to eliminate from their breeding flocks families and individuals that produce offspring developing pendulous crops. However, for some reason, possibly one involving structural differences, broad-breasted turkeys appear not to be very susceptible to the pendulous crop deformity. It is therefore a rather minor problem in broad-breasted strains though breeders always should be on their guard against the occurrence of susceptible individuals and strains.

CROWDING AND PILING

Turkey poults are especially susceptible to crowding and piling which may be caused by; (1) dampness of the brooder house floor, (2) insufficient heat in brooder house or under the hover, (3) malnutrition, (4) disease, especially coccidiosis, (5) fright, (6) overcrowded conditions, (7) warm spots at points other than around the brooder itself; such as radiators, heating stoves or patches of sunshine, (8) corners not rounded, (9) objects in the brooder room around which the poults crowd (coal buckets, feed-cans, etc.), (10) insufficient shaded areas.

Preventive measures will suggest themselves when the causes of crowding are studied. Crowding and piling are ways in which turkey poults can be and are frequently lost without warning and in large numbers especially during the early brooding period. Therefore, good brooding conditions are extremely important in turkey raising. Warmth, dryness, sanitation, protection, from winds and drafts, adequate diet, and continual care are important. Rats, crows, hawks, dogs, and cats may cause serious losses, sometimes indirectly through frightening. Protection from them must be provided. Low-flying airplanes, unless the birds are used to them, may cause losses from piling.

CURLED TONGUE

This deformity while of rather sporadic occurrence may cause severe losses in young poults at one to seven weeks old. The tip of the tongue is folded back upon itself causing drooling and a lump in the throat just under the beak which are characteristic symptoms. This condition results in lowered feed consumption, runty condition, and sometimes death. The Virginia Station (1952) in a preliminary investigation using Beltsville White poults, observed that: (1) A small incidence of the curled tongue condition existed at hatching time; (2) a greater incidence occurred among poults fed a finely ground mash than among those fed a coarser type diet; (3) it is not caused by a deficiency of the amino acids leucine, isoleucine, or phenylalanine in the poult diet; and (4) curled tongue was not prevented by terramycin or bacitracin. The presence of hygroscopic (water absorbing) feed ingredients may possibly aggravate the condition. The fact that some curled tongue was noted in day-old poults suggests the possibility of (1) deficiency in the diet of breeding stock, (2) adverse incubation conditions, or (3) genetic involvement. Hamilton (Profitable Turkey Management, 1951) suggests that the feeding of an insoluble grit on top of the mash 3 times weekly or daily feeding of grain along with plenty of drinking water may help. The Ohio (1953) and Maryland (1954) Stations suggest the use of crumbled starting feed to prevent curled tongue (submandibular displacement of the tongue). Overfeeding of minerals, especially rock phosphate, is another possible cause of curled tongue, according to the Ohio Station (1952).

Injury to the tongue may follow any attempt to trim or cut off any part, even the tip, of the lower beak.

EYE DEFORMITIES

Blindness and deformity may be due to heredity or to injury. Gray, blue, or mottled eyes are due to a hereditary factor or factors not to disease, so far as is known. Selectors of breeding stock should be on the lookout for these abnormalities of the eyes as well as for indications of sinusitis.

FAULTY FEATHER GROWTH AND ACHROMA

Rough and broken feathers appear to be caused either by a serious nutritional disturbance which itself may be caused by disease or by unbalance of nutritional factors. Abnormal feathering, therefore, is a symptom, not a disease, and when it occurs, turkey raisers may conclude that something is or has been wrong with the diet or that disease is or has been present. Sometimes the shafts of the feathers are weakened, causing the feathers to break off. The cause sometimes can be determined by checking the diet and examining the birds for signs of disease.

A bar of white, accompanied usually by narrowing and weakening of the feather structure, is often seen in the wings of young poults of the Bronze, Narragansett, and Black varieties. In the white varieties only the effect on the feather structure will be noted. As the feathers grow, this bar will move out toward the ends and unless it is very wide and conspicuous will not reappear when the juvenile feathers are moulted. However, the white bar or achroma, often is very wide and its effects will be seen as mottling in the wings of the fully grown bird. Much speculation has been made as to cause of this white barring, but nutritional experiments have shown recently that the true cause is deficiency of the amino acid lysine. Poults need about 1.3% lysine in their diet which may be provided by 20% soybean meal or 12% fishmeal, along with that supplied by the other normal diet ingredients. However, the use of a B_{12} supplement (not believed necessary for poults) may increase the need for lysine as does the use of an antibiotic, to 1.8 or 1.9%.

Sunshine promotes smooth feathering. Birds not exposed to sunshine usually have rather rough feathering. The Nebraska Station has reported that wheat bran contains a factor that promotes normal feathering.

FEATHER PICKING

Feather picking, which is a mild form of cannibalism, is a vice to which turkeys are very susceptible. Usually it is not important in commercial flocks raised with access to some range but occasionally it leads to loss of grade because of pinfeathers or the development of blueback.

It often is important in flocks from which exhibition birds and breeding stock are to be sold as it ruins turkeys for exhibiting and decreases their salability as breeders. In well-fed flocks this vice seems to start in most cases; (1) by birds attempting to clean their beaks on the feathers, particularly the tail feathers, of their companions; (2) as a result of enforced inactivity as in birds raised in confinement or being shut up in range houses or small enclosures during the early morning hours; and (3) in confined flocks by the birds attempting to satisfy their desire for roughage by eating feathers. In poorly fed flocks it may start through a craving of the birds for something lacking in the diet, presumably minerals or protein. Overcrowding in any situation is likely to be a contributing cause. In some cases the cause is not apparent. It may start very early in life, when the birds are only a few weeks old. More often, however, it becomes noticeable at the age of 12 to 16 weeks. Feather picking ranges in severity from a slight injury to the tips of the large tail feathers and coverts to the almost total destruction of the larger tail feathers and severe injury to the wing and back feathers. Occasionally feather picking starts at the front end of the bird, resulting in partial to almost complete removal of the neck feathers. Wires inside the feeders placed so that the birds continually rub their necks on them while eating have been known to cause this type of feather injury.

In general, feather picking is much more likely to be serious in flocks that are kept under artificial conditions than it is in flocks that are kept under natural conditions where green range is provided. The precautions to be observed in preventing it in commercial or farm flocks are as follows: (1) tightly stretched or spring-attached, strong, 11- to 12½-gauge wires for beak cleaning should be placed on top or inside of all the mash-feeders except those used by poults under 2 or 3 weeks old; (2) overcrowding at all ages should be avoided; (3) adequate feeder space, 2-2½ lineal inches per bird, should be provided; (4) an adequate diet should be fed; (5) the mash should be of medium grind and not gummy when wet; (6) scratch grain may be fed beginning at the age of one to six weeks and if poults are closely confined, one-third to two-thirds of the grain fed should consist of oats or all of it barley; (7) the practice of shutting the birds up in the house or on the roosts in the early morning hours or for long periods at any other time of day should be avoided as this practice may cause extensive feather picking in a flock otherwise well managed; (8) alfalfa hay or corn fodder may be kept available for poults, if confined, from the age of about 3 weeks to maturity; (9) plenty of roosts of suitable types be provided; (10) access to a large range containing ample green feed and shade may be furnished; (11) pellets or wet mash instead of dry mash may be fed; (12) drinking water should be of sufficient depth so turkeys can drink easily and wash their beaks at the same time; (13) the sexes may be separated at 8 to 16 weeks of age; (14) the beak may be "tipped" (see next page); (15) a netting clamp or hog ring (1¼″) or turkey bit may be inserted between the jaws of turkeys and into the nostrils but not the nasal separation; (16) limit confinement rearing groups to 200 turkeys; (17) the outer ⅓ of the upper beak may be removed with an electric debeaker; (18) common salt, 1 teaspoon per gallon of drinking water, or 2% in the mash, may be given temporarily; (19) prevent over-consumption of grit; (20) breeder-hen saddles of appropriate size will prevent back picking.

The feeding of liquid milk may or may not encourage feather picking. No data can be offered. The authors have seen liquid milk and whole grain fed to flocks that were completely free of feather picking. However, the plumage is likely to be spotted by the feeding of liquid milk, hence this practice probably is not a good one to use where it is necessary to avoid feather damage.

Breaking turkeys of feather picking is difficult unless they are "bitted" or debeaked. The application of pine tar or a chick-pick remedy* to the affected areas affords some relief.

Removal of the outer portion of the upper beak may be performed by cutting a nick in both sides of the upper beak about a quarter-inch or so back from the end. The cut should be about 1/16 inch to 1/8 inch deep (even with the roof of the mouth). By pressing on the tip of the beak the cut end can be broken off. This leaves a little of the "quick" exposed which allows eating but prevents feather picking. Electric debeaking, where it can be done, would be preferred to tip removal. The installation of blunt-tipped hog-ring-type bits is about as effective as debeaking. Special turkey bits of the hog-ring type can be purchased along with a special tool for applying them. If applied at about 14 weeks of age, the "adult turkey" size can be used for both sexes and all types of turkeys. The rings should be closed to the point where the blunt tips lightly touch but do not penetrate the nasal septum. If closed too tightly the rings may cause infection to develop and if closed too loosely the rings will be pulled. They can be removed by reverse action of long-nosed pliers and may be re-used if necessary.

INSECT ATTACK

Direct attack of a serious nature on turkeys by normally non-parasitic insects is rare. However the South Dakota Station (1948) reported serious injury to turkeys by gnats which go under the feathers to bite the birds. Colored turkeys were more seriously affected than white ones. Mosquitoes sometimes bite the heads of turkeys in sufficient numbers to cause trouble. Housing, even an open-sided range shelter, seems to protect turkeys against the attacks of both gnats and mosquitoes.

*Such a remedy may be made up by mixing 4 oz. of vaseline, ¼ oz. of carmine, and ¼ oz. of bitter aloes.

LEG AND FOOT DISORDERS

Bumblefoot

Bumblefoot, according to authors' observations is of two types. Type I or hard bumblefoot consists of a hard swelling in the large center foot pad. This firm swelling may have a horny center core, it may be solid without a hard core, or it may be roughened with one or more deep cracks which may bleed. There is no sign of pus formation. This type seems to be caused by hard and/or rough-surfaced slat or wire floors aided by too-high roosts. It occurs mostly in confinement-reared toms especially heavy-type toms and develops after the turkeys are about four months old. This type of bumblefoot, if severe, retards growth and reduces the ultimate weight attained. Infection does not seem to play a large part in the development of hard bumblefoot—it seems to be the body's reactions to continual bruising of the footpad by a heavy bird on hard or rough floor surfaces. To reduce or prevent hard bumblefoot the floor of the confinement rearing quarters should be smooth and free from rough edges and projecting points. Wide-slatted floors, slats 2½ to 3 inches wide and one inch apart have possibilities in the prevention of hard bumblefoot as well as some other leg troubles.

Along with the wide slats, the roosts either should be eliminated or placed with their top surfaces within 6 or 8 inches of the floor. In this way the turkeys are prevented from injuring their feet by jumping from the roosts onto the hard floors.

Wire floors seem to cause more bumblefoot than do slat floors. Floor wire coated with a plastic or rubber material may be an advantage particularly in a 1½ to 2″ square mesh.

Turkeys severely affected should not be retained as breeders. Moderately to slightly affected birds may be retained as breeders if they are badly needed. If placed on well-littered floors with or without access to dirt yards, the hard bumblefoot will not get worse and often gets better without any treatment, which is useless for this type of disorder.

Type II, or soft bumblefoot, is that described by California workers (1951) as abscesses of the footpads. These lesions are found on the toe-pads with or without involvement of the large center foot-pad and seem to be the result of an infection following an injury. It affects both hens and toms of all varieties. The California Station suggests that it occurs in turkeys kept continuously on dirty, muddy yards. The abscesses tend to be soft-centered with yellowish-white, pussy appearing edges and cause the toes to become enlarged laterally. Treatment suggested is to clean the affected parts, removing any pus accumulations, swabbing with tincture of iodine, or applying an ammoniated mercury or sulfa ointment, and then, putting the birds in clean dry quarters.

The authors have observed, in confinement-reared turkeys, a low incidence of Type II bumblefoot which apparently seldom becomes as serious as the other type. Affected birds usually are marketable without treatment but should not be retained as breeders as the infection tends to get worse unless treated, which is not worthwhile unless the affected birds are very valuable.

Enlarged Hock Disease

This type of leg trouble occurs as a slight, often imperceptible, hock enlargement at the age of about 2 weeks which returns to normal at about 4 weeks, then recurs in a severe form at 14 to 16 weeks, resulting in severe leg weakness, breast blisters and abscesses, sore feet and legs, and almost a total loss to the grower at market time. Cornell investigators believe the disorder to be distinct from true perosis and similar to encephalomalacia and exudative diathesis in chicks, and of muscular dystrophy in guinea pigs and rabbits. However, vitamin E deficiency in turkeys does not cause encephalomalacia or muscular dystrophy (muscle degeneration). The enlarged

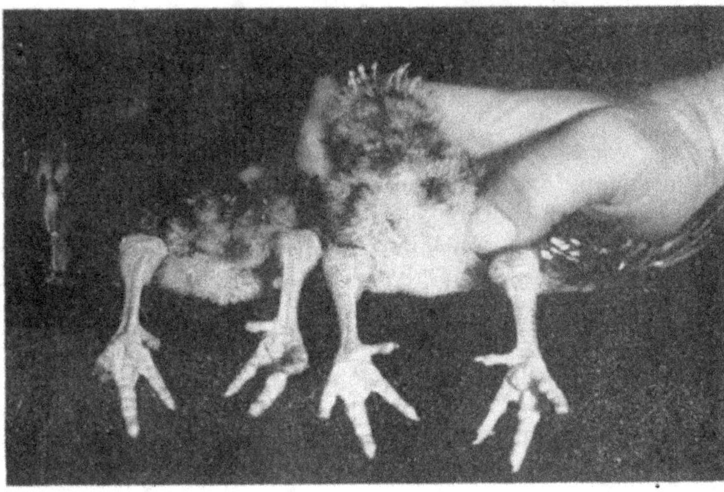

Fig. 16.2. Typical perosis produced by feeding a ration deficient in choline. (Cornell University Photo.)

hock disorder of turkeys, which may be a form of nutritional muscular dystrophy, appears to be due to improper metabolism of creatine, an essential muscle compound which is excreted in abnormally large amounts. This process with the resulting crippling effect appears to be accelerated by the feeding of fish-liver oil. Restricting the quantity of fish-liver oil to 0.25% or less and adding 5% dried brewers' yeast or, better, 20 mg. of niacin and 2.5 mg. of alphatocopheryl acetate per pound of feed in turkey starting diets, are recommended by the Cornell Station. Commercial supplements containing these products are available. In growing mashes the disorder has been prevented by the addition of 4% yeast or 20 mg. of niacin and 2.5 mg. of alphatocopheryl acetate per pound.

Heavy turkeys (mostly males) are affected to a much greater degree and extent than are the medium and small-sized strains. An antibiotic in the diet appears to have a beneficial effect but cannot be considered a satisfactory preventive or treatment. The North Carolina Station (1953) fed 7% topsoil in a 28% protein growing mash fed with grain (net 3½% soil of total feed intake) and reduced swelled hocks (at 28 weeks of age) from 85% (the control) to 29%. Removal of roosts increased the trouble to 96%; 11.9 gms. terramycin hydrochloride per ton reduced it to 79%.

In the later stages of the enlarged hock disorder, invasion by certain staphylococci may occur. No proof has been presented showing the staphylococcus organism to be a causative agent in this disease.

Perosis or Slipped Tendon

Leg deformities may affect turkeys from hatching to maturity. The term perosis is used to describe leg deformities, symptoms of which are slipped tendon, twisted shank, bowlegs, misshapen hock, and lameness. Perosis of day-old poults is often of a distinct type termed "spraddle-leg." Perosis involving softened leg bones and enlarged hocks is a symptom of rickets. Perosis may result from biotin or riboflavin deficiency and from choline or mineral deficiency or imbalance. The type of leg deformity called enlarged hock disorder by Cornell University (see preceding discussion) may possibly be a type of perosis. Nutritionists disagree as to its status.

Perosis in poults over 2 weeks old usually is caused by malnutrition, particularly manganese deficiency, which may be aggravated by too much phosphorus in the diet and by lack of direct sunshine. Breeders on deficient diets may produce poults that are perotic at hatching time. Hereditary tendencies also play a part, some families producing many more perotic offspring than others. Treatment of perosis is of no avail, except possibly spraddle-legs which may be tied together. Prevention consists of supplying well balanced diets including proper quantities of vitamins, manganese, calcium, and phosphorus. No minerals, particularly bonemeal, should ever be fed ad libitum. "Alkali water" may induce perosis (consult mineral and choline requirements). Turkeys on wire or slats or other hard floors tend to be more susceptible than those on soil or litter. Elimination of individuals and families producing individuals susceptible to perosis is necessary for its complete control. In typical perosis one leg (sometimes both) is bowed or twisted, the hock joint is slightly enlarged and misshapen, and the larger tendon in rear of the hock is slipped to one side.

Swelled hocks and/or feet, (but not perosis) may be caused by the disease organisms Salmonella enteritidis (Higgins, 1944) and Staphylococcus aureus and S. citreus (Jungherr, 1941) in well-grown turkeys.

Spraddle-Leg

This term is given to day-old perotic poults that have lost control of one or both legs so that the limb is held out sidewise at a right angle to the body. When both legs of a bird are spraddled (the true spraddle-legged condition) the poult is helpless and cannot move about at all. When one leg is spraddled the bird is able to hop about to a limited extent. Poults crippled in this manner are of frequent occurrence in poor hatches, but under favorable incubation conditions seldom are numerous enough to be of economic significance. The fact that inbreeding increases the number of spraddle-legged poults in some families indicates that the tendency to develop the deformity may be inherited. However, it is believed that faulty incubation is the principal cause and that dietary deficiency (lack of biotin, choline, manganese, or riboflavin) in the feed of the parent stock may play a part. Cornell workers found that the number of poults crippled at hatching (including spraddles) increased progressively with decrease of incubation temperature below normal. At the Kentucky Station, it was found that a lower-than-normal incubation temperature increased the percentage of crippled and weak poults. Whether or not abnormally high incubation temperatures, lack of moisture during incubation, or lack of proper turning causes spraddled legs is not clear. Early removal of the shells and unhatched eggs may increase the incidence of spraddle-legged poults; unless the tray bottom is of wire or contains sanitized mosquito netting (crinoline) or other material which the poults can grip with their claws.

When spraddle legs occur in good hatches, it seems probable that the cause is hereditary or nutritional. Preventive measures consist of (1) supplying an adequate diet to the breeding stock (2) providing proper care of the hatching eggs and favorable conditions of incubation, (3) eliminating from the breeding flocks strains or families of birds that consistently produce spraddles even when the eggs are produced and handled under favorable conditions, and (4) making sure that newly hatched poults are not placed on newspapers, cardboard, or other slick surfaces where they cannot get a toehold, as it will aggravate the tendency toward spraddled legs.

Treatment sometimes is effective. Usually it is possible to save about one-third of the spraddle-legged poults by tying their legs together with soft twine so they are held in the normal position and allow the poults freedom enough to walk.

STAMPEDING

Turkeys are subject to fright at night, especially on moonlight nights. If housed, they will fly violently against the walls and roof of the shelter and often pile up in the corners; if not housed, they will scatter over the surrounding country. Severe losses sometimes result from straying, smothering, bruising, broken wings, and broken legs. Prowling animals, strange noises, or sudden bright lights may cause a stampede. Night lights supplied by electricity or cannon-ball torches or flares at the roosting quarters afford the best protection; especially if a watchman is on guard or sleeping nearby. In large, enclosed roosting sheds, partitions should be placed at sufficient intervals so that not over 150 birds are contained in one section. Wire directly under the roosts helps to prevent smothering during stampedes; and the wire serves to discourage predators. If the roosting quarters are surrounded by wire fence close by, the wire should be placed on the outside of the posts (posts between roosts and fence) and not fastened tightly at the bottom. This will allow stampeding birds, pushing against the wire, to push on out of the enclosure, under the fence and thus avoid a pile-up. The same arrangement often is used with the temporary fences used to confine turkeys that are moved to clean ground at frequent intervals.

WEATHER HAZARDS

Sleet freezing on beaks and nostrils can seal the bird's breathing apparatus and cause smothering. Fine wind-driven snow has much the same effect and also will cause the birds to pile and smother. Prolonged rains often cause heavy mortality in unhoused young turkeys and may cause loss of egg production with lowered fertility in breeding stock. Even with housing, turkeys often must be driven to shelter in the face of strong winds and snow or cold, continuous rain. Adequate housing and never-ending watchfulness are essential to prevent losses from weather hazards. Weather and accident insurance can be obtained for turkeys at reasonable rates to offset some of the loss.

CHAPTER 17

DIAGNOSIS AND CONTROL OF DISEASES AND PARASITES

OUTLINE

DISEASES AND PARASITES OF TURKEYS

Actinomycosis

A preliminary report* describes a blackhead-like disease of turkeys believed to be caused by a species of *Actinomyces,* a mold organism. Symptoms in 6-week-old poults were thin, whitish droppings which changed to yellowish as death neared. Upon post-mortem, the livers were found to be speckled with pinpoint spots in early cases, in more advanced cases by hard, well-defined, rounded yellowish-white nodules resembling lesions of tuberculosis. These lesions became larger, irregularly rounded, and protruded from the surface of the liver. They were uniformly caseated and surrounded by a well defined capsule. Most cases had cecal lesions also, which resembled those of blackhead. When these cores disappeared in birds surviving the early infection, protruding nodules were formed in the cecal wall, and in a few cases also in the large intestine, in the leg muscle, on the ribs, and in the esophagus. The presence of histomonads or trichomonads could not be demonstrated in affected birds when sent to several laboratories. However, when liver nodules were ground and placed in potassium hydroxide overnight, the club-shaped organisms and filaments could be identified as a species of Actinomyces. Further investigation is needed to establish the facts concerning this disease. It may possibly be atypical blackhead or trichomoniasis.

Air Sac Infection

This term often is used loosely to indicate any infection in the air sacs which may be due to a number of causes. Certain molds, especially *Aspergillus fumigatus* may cause spots or thickened areas on the normally thin and transparent air-sac membranes. Chronic fowl cholera or Newcastle disease may cause air-sac involvement. More recently Boney, et al (Jour. Am. Vet. Med. Assn. 1952)

*"Actinomycosis—a new disease of turkeys," by J. F. Olney, Veterinary Medicine 45 (10): 392-4 (Oct. 1950.)

reported the isolation of an infective nonbacterial agent differing from that causing the air-sac form of turkey sinusitis. The turkeys affected were of market and breeding age, showed typical "air-sac" symptoms, and had cloudy air sacs, a fibrinous pericarditis, and livers covered with a thin fibrinous exudate. Mortality was not heavy, but an affected breeding flock went out of production. The agent causing the trouble has been identified as the virus of ornithosis; this agent is also infective for other domestic birds, many wild ones, some small laboratory animals, and man. (Pate, D. D., et al., 1954, J. A. V. M. A., 125: 476-8.) However, in turkeys, air-sac trouble is most likely to be a form of sinusitis which may occur at any age from day-old on through breeder age. Apparently the organism responsible for typical infectious sinusitis is identical with that which causes the so-called "air-sac colds" or "chronic respiratory disease" in chickens. This adds another good reason for segregating turkeys from chickens. For further discussion of air-sac trouble see "sinusitis."

Blepharitis

Blepharitis or inflammation of the eyelids sometimes occurs in the breeding stock, particularly the hens. Its cause is not known. Sodium propionate, an anti-bread-mold chemical, has been suggested as a treatment for this and other eye infections. Antibiotic ointments also may be effective in relieving the condition. Debeaking may aid in preventing its spread. Chronic erysipelas and fowl pox sometimes causes sore eyes.

Blackhead

(Histomoniasis or Enterohepatitis)

The sluggish circulation which usually follows infection with blackhead sometimes results in a dark appearance around the head. This is the reason the disease was first called blackhead. However, this symptom also accompanies other diseases and is not a dependable diag-

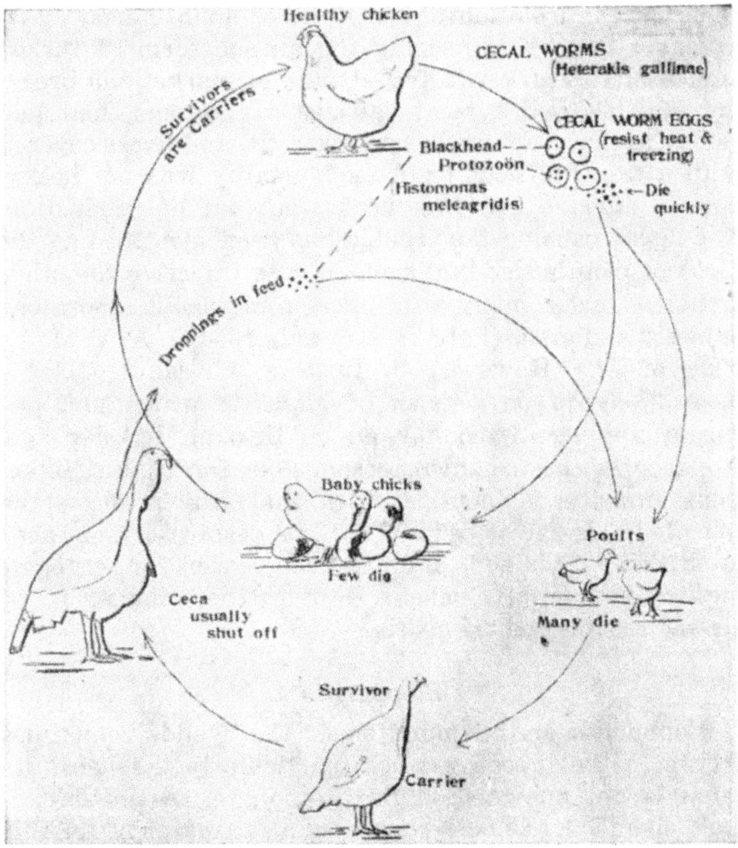

Fig. 17.1. The diagram illustrates the part the chicken plays in the transmission of blackhead. Starting at the top of the chart, with the healthy chicken, and following the arrows, it will be noted that cecal worms (Heterakis gallinae) which are in the chicken's ceca produce eggs that are given off in the droppings. Some of these eggs contain the germs (Histomonas meleagridis) which cause blackhead. Cecal worm eggs resist the heat of summer and freezing of winter and consequently the blackhead germs within them are protected and kept alive. Those blackhead germs which are emitted in the poultry droppings and are not within cecal worm eggs die quickly. The cecal worm eggs may be picked up either by young poults or baby chicks. If they are picked up by poults most of the poults die of blackhead as soon as the cecal worm eggs hatch and permit the blackhead germ to infect the turkey. However, an occasional turkey survives and is a carrier of blackhead, although the ceca are so badly diseased that they are occluded. Those infected turkeys which grow to maturity pass the disease on again by poults. If baby chicks pick up the cecal worm eggs which have within them the blackhead germ, few of these chicks die. They grow to maturity carrying the blackhead germs as well as the cecal worms in their ceca. These chickens then serve as a source of spreading both the cecal worms and the blackhead germs. To control blackhead the cycle must be broken, and the simplest way to break it is to keep chicken and turkey droppings out of the reach of young poults. This, of course, means raising turkeys on clean ground, away from all poultry or in confinement away from all other poultry.

nostic aid. Mortality from blackhead in young turkeys, if not controlled, is almost always high, above 50%. In the "blackhead country" which is almost all of the United States and Canada east of the Great Plains States, entire flocks have been wiped out completely. Turkeys older than about 28 weeks have considerable resistance to blackhead, few losses being sustained as a rule between about November 1 and June 1, after which losses up to about 10 to 15% may occur if birds are not marketed, treated, or rotated on clean range. Blackhead affects the liver and ceca (or blind guts) and is caused by a protozoan parasite *Histomonas meleagridis.*

The early symptoms of the disease are drooped wings, drowsiness, ruffled feathers, weakness; a thin watery liquid may run from the bird's mouth when held head down; and orange or sulphur-colored diarrhea may occur which often soils feathers. Turkeys of all ages are susceptible, starting early as 2 weeks in rare cases. The most susceptible period probably is 6 to 12 weeks. Another mortality peak sometimes occurs at 18 to 28 weeks but this may possibly be due to trichomoniasis of the lower digestive tract (pg. 916). Fevers of 2 to 3° may occur followed by subnormal temperatures of about 2° in the advanced stages. Younger birds lose flesh rapidly.

Post-mortem examination is essential for positive diagnosis. The liver is almost invariably spotted with yellow, yellowish-green, or grayish, irregularly round areas which usually are slightly sunken and resemble rotten spots in an apple. These spots extend into the liver tissue. The lesions also may be lumpy, granular, and slightly raised above or on the level with, the liver surface and are yellowish or cream-colored. This type of lesion is, according to the Beltsville theory, due to *Trichomonas gallinarum* infection which also causes cecal ulcers as does *Histomonas.* However most investigators do not recognize the existence of a disease of the lower digestive tract by a trichomonad—the different type of lesion being considered merely as a variation in the lesions of Histomoniasis. The ceca usually have ulcers in the walls and are more or less filled with yellowish or greenish yellow cheesy material. The spotted liver and ulcerated ceca are definitely diagnostic of blackhead, according to majority opinion even where liver lesions are elevated and lumpy. Other post-mortem findings sometimes observed are: inflamed upper small intestine (duodenum), enlarged heart and liver, thin heart walls, and flesh around the abdomen brownish and flabby.

Swales and Frank (1948) found that, in poults exposed to blackhead by means of chicken-house litter (which proved to be infective) 90% died from the disease and no immunity (by acquiring a lowgrade infection) was established. The poults became susceptible starting at 3 weeks of age and the average expected period between exposure and death was 22 days.

Breeding hens (rarely breeding toms) are susceptible to blackhead but losses usually are relatively light and as a rule seem to occur after about June 1 in most sections of the United States. Therefore most breeding turkeys have been sent to market before blackhead becomes a problem. Rotation of ranges, sanitation around feeders, waterers, and roosting areas, as with young stock, are effective in preventing the disease. Occasionally blackhead causes heavy losses of breeding hens early in the season. Treatment with ANT* appears to be effective as a preventive (see under "Treatment").

Prevention and Control. The management recommendations set forth in Chapters 2 and 14 usually are effective in the practical solution of this perplexing problem and among these, complete separation from chickens and older turkeys is of fundamental and paramount importance (see Fig. 17.1). Even when all the "rules of the game" are followed there may occasionally be some upset. This was clearly illustrated to the authors when a turkey raiser who was following the sanitation recommendations

---

*Abbreviation for 2-amino 5-introthiazole.

"to the letter" reported trouble which later proved to be blackhead. These turkeys were being raised on clean ground a quarter of a mile away from the chickens. They were on a slope away from the barnyard and all of the feed was carried to them in an old car. A careful study of all possible avenues of infection revealed that the car that was used for the hauling was parked in the garage near some chickens and that occasionally chickens roosted above it. Undoubtedly, the chicken droppings were carried to the turkey range on the car and as the car turned around they were dropped off and left there as a source of blackhead. When this source was stopped infection ceased. This illustrates the point that one cannot guard too carefully the many avenues of infection. Another source is by means of the shoes or on the feet of not only the farmer himself, but also of any farm animals, particularly the dog that might run in the chicken yard and then later come to the turkey range. Any person entering the range should clean and wipe his shoes off carefully or use rubbers or galoshes. Such a practice is highly desirable for the turkey grower himself, especially if he also cares for chickens or mature turkeys.

Prompt removal and isolation of droopy birds is an important control measure. Fly prevention and range rotation are other important means of producing a sanitary environment. Raising poults on well-designed slat or wire platforms virtually eliminates blackhead and other soil-borne diseases and parasites as do stone yards, a system of weekly moving of the turkeys and their gear on 4-year rotated ranges, and rearing on built-up litter.

Treatment: Swales and Frank (1948) tested a number of drugs including sulfamerazene, sodium sulfamethazine, atabrine dihydrochloride, Formula 144 (a quaternary ammonium compound) and "Stibophen" (a stable antimony compound) and found all of them ineffective as a preventive or therapeutic agent against blackhead in poults infected through contaminated soil. The drug Vioform (iodochloroxy quinaline) was reported by DeVolt and Holst (1948) as showing promise in the prevention and treatment of blackhead. Likewise favorable results were reported by Morehouse and McGuire (1950) on the use of four arsonic acids for the prevention of blackhead. Two of them (2-nitrobenzenearsonic and 4-nitrobenzenearsonic) appeared to have lasting effects.

The drug 2-amino 5-nitrothiazole (abbreviated ANT) was first shown to be effective both as a preventive and cure for blackhead by McGregor (1949)*. It is marketed under several trade names among such as "Enheptin," "Antihep," "Histosep-S," and "Heptrol" and can be administered in the feed, in the water, or by mouth to individual turkeys. The latter seems to be indicated for birds already noticeably sick. Directions of the manufacturer should be followed in each case. The preventive level is 0.05% of total feed; the therapeutic or curative level is 0.10%. The Ontario Station (1951) reported that naturally infected poults 13 weeks old when fed the drug at 0.05% sustained a 2 to 4% mortality while the treatment was in progress but died at a high rate after treatment given for 14, 28, and 42 days to 3 different groups, was discontinued. These results indicate that ANT is an effective preventive while being fed but that there is little or no carry-over effect.

While treatment with ANT has been shown to be very effective, it also is expensive. Prevention of blackhead through proper management is no less important since the discovery of this reliable remedy. In 1955 the American Cyanamid Company announced a cheaper and more effective drug "Enheptin A" (2 acetylamino-5-nitrothiazole).

The Delaware Station (1950), raising Broad Breasted Bronze turkeys on infected ground from 8 to 24 weeks without moving the range equipment reported no loss from blackhead in 220 poults fed ANT continuously at 0.05% of the diet; 5.21% loss when no ANT was fed. Average weight of survivors was 16.0 lbs. for the 0.05%

*McGregor, J. K., 1949. Canadian Jour. Comp. Med. 13:25.

group, 16.1 lbs. for the 0.025% group, and 16.8 lbs. for the untreated group. Feed utilization appeared to be improved slightly by both levels of ANT but average weight was reduced.

Swales at MacDonald College (1952) further confirmed the efficacy of ANT in controlling outbreaks of blackhead and also its effect in reducing the numbers of the larvae of the cecal worm which appears to be important in controlling natural infections of the disease. Medication of drinking water with 0.03% of soluble ANT and individual dosage of such poults at 48-hour intervals were found to be at least as effective as was treatment via the feed.

On the other hand McGregor at the Ontario Station (1952) reported that, although ANT was shown to have value as a preventive and cure for blackhead in growing turkeys, losses caused by extremely virulent strains of Histomonas, still could be crippling in spite of treatment with ANT.

The Texas Station (1952)* reported that ANT in the feed of 7-week-old turkeys on contaminated range was effective in preventing losses from blackhead at concentrations of 0.05, 0.04 and 0.03% for 13 weeks provided medication was started at the time of exposure to the disease. As a cure, treatment started upon outbreak of the disease was not effective at 0.1% for 7 days, 0.1% for 14 days, or at 0.1% intermittently. When ANT was fed to laying turkeys on contaminated ground at 0.1% for 18 weeks blackhead was prevented but egg production, fertility and hatchability were reduced. Blackhead was prevented and performance was not reduced, at levels of 0.05 and 0.03% (0.1% apparently could be used without ill effects for 2 weeks). The drug 2-acetylaminonitrothiazole gave similar results at the same concentrations except that no ill-effects were observed at 0.1%.

McGuire and Morehouse (1952) reported that 4-

---

*L. C. Grumbles et al in Am. Jour. Vet. Res. 13(49):572-4.

nitrobenzenearsonic acid when given in the feed of turkeys at 0.0125 to 0.03% stimulated growth and was highly effective in preventing histomoniasis. In the drinking water at 0.006 to 0.04% it was highly effective in preventing the disease provided that (as with the feed medication) treatment was started prior to time of infection. It also was found that relapses occurred if the medication has stopped while the birds were exposed to the disease and were still at a susceptible age.

Experiments proved ANT (enheptin) highly effective in the suppression of experimentally-induced blackhead and natural infection as well. Its effectiveness was greatest when given prior to exposure, but death loss was greatly reduced when it was administered immediately following exposure. It was still effective, although less so,

Table 17.1—Results of Enheptin Medication in Enterohepatitis (blackhead) Produced by Oral Administration of *Heterakis Gallinae* Ova in 4- to 6-Week-Old Poults

Days after inoculation before treatment was started	No. days treated	% Enheptin in diet	% survival 6 or 7 weeks after inoculation	
			Treated	Controls
0	14	0.15	100	12.5
1	14 14	0.1 0.05	100 100	30 30
3	14	0.15	100	12.5
4	14 14	0.1 0.05	70 40	30 30
5	14	0.15	62.5	12.5
7	14 14	0.1 0.05	44.4 20	30 30
13*	8	0.10	20**	20
13*	8 7	0.15 0.05	60	30

*Treatment begun when symptoms appeared.
**60% of deaths occurred after treatment was withdrawn.

when treatment was given extremely sick birds. This is clearly illustrated by the report of the Stamford Research Laboratories (Waletsky*) given in Table 17.1.

Furazolidone (NF180) recently (1954) has been shown to be effective as a preventive of and treatment for blackhead by mixing in the feed.

Cecal Worms

(Heterakis gallinae)

Cecal worms are small, whitish, pin-size worms 0.3 to 0.75 inch long that live in the ceca (blind pouches or blind guts) of turkeys, chickens, and other fowl. Upon post-mortem the worms are likely to be found in the closed ends of the ceca. They do not sap much nourishment from the turkey as is the case with the tapeworm. However, the cecal worms frequently are carriers of the blackhead protozoan and thus are dangerous to turkeys. They appear to aid in the carryover of the organism. Prevention is possible by power management (See Ch. 2) and is much preferred to control by other methods. The dye phenothiazine is an efficient vermicide for these parasites and as such may be an aid in blackhead control where cecal worms are numerous although it has no effect on the disease itself. The dose is one per cent in the mash for about seven days, repeated at 2- to 4-week intervals if necessary. This treatment may be given as a preventive of cecal worm infestations and an aid in blackhead control. It would be of value only where cecal worms are known to be present. Tobacco dust, 2% nicotine content, although not so efficient as phenothiazine, may be fed in the mash at a 4% level as long as needed, after poults are 4 or 5 weeks old, to keep down the cecal worm population. However, cecal worms themselves seldom become serious enough to warrant continual worming. Their presence in large numbers should be reliably demonstrated before treatment is attempted.

---

*Waletsky, E., 1950. Turkey World 25(5):16-17, 60-65.

Cholera

Fowl cholera, caused by the bacterium *Pasteurella multocida* (formerly avicida), sometimes causes heavy, sudden losses, usually in well-grown market turkeys or breeding stock. Transmission appears to be mainly by means of carrier turkeys-survivors of a cholera outbreak. The infection may be introduced on equipment, feed sacks, crates, etc. Chickens and some other birds also are susceptible and sometimes are sources of cholera outbreaks in turkeys. Houseflies may be mechanical carriers. Cholera usually is a cold-weather disease, appearing in the fall during changeable weather.

Symptoms: In the extremely acute form few symptoms are noted; birds that appear healthy the night before, may be found dead under the roosts. In the acute form the bird appears ill a day or two before it dies, showing a yellow to green diarrhea sometimes tinged with blood. There may be a 2 to 3° fever, darkened head, nasal and mouth discharge, and pinkish skin. Other symptoms are weakness, nervousness, droopiness, loss of appetite, thirst, and labored breathing. Occasionally the snood of males is swollen. In the chronic form the joints may become inflamed; there may be paralysis; and a loss of weight occurs. Although cholera outbreaks are uncommon among turkeys, severe losses sometimes occur. Birds 6 to 8 months old are most susceptible, although birds of all ages are susceptible. Breeding stock may be carriers but transmission of the disease within the egg has not been proven, though transmission on the shell surface appears possible. The germs are carried in the nasal passages and shed in nasal discharges. New grains, particularly wheat, has been suspected of causing cholera. The North Dakota Station says they do not actually cause it but may predispose birds to the disease if fed alone, due to the inadequate diet thus provided.

Post-mortem findings: The only certain test is an examination of the blood or tissues for the bipolar organisms of Pasteurella multocida (or avicida). If spots appear on the liver they have small white centers (pin-point size). Breast muscles may be reddened; heart enlarged, with whitish exudate and flaky yellowish fluid in heart-sac; putrid odor of viscera; spleen may be enlarged; yellowish deposits in air-sacs.

Treatment and Control: Quick action is necessary. All dead and ailing birds should be killed by dislocating the neck to avoid spilling blood which contains the living organism. Disinfection and general sanitation practices should be followed. Sulfaquinoxaline, sulfamerazine, sulfamethazine or streptomycin given under proper veterinary supervision may help. The Illinois Station (Alberts and Graham, 1950) reported that a flock of 6-month-old turkeys naturally infected with cholera was placed on a 0.5% sulfamerazine-mash mixture for 4 days then marketed at once with a mortality of 4.1% and a slight loss in quality. In another flock of 5-month-old turkeys naturally infected with cholera, 0.2% of sodium sulfamerazine was added to the drinking water on the first, second, third, fifth, eighth, and twelfth days. The disease recurred on the 29th day after treatment when another 3-day course of treatment was given and the birds marketed. Total loss was 7.4%. In a flock of 2971 cholera-infected tom turkeys given a sulfamerazine-mash mixture (0.5%) for 5 consecutive days, mortality on the first day of treatment was 22, second day 19, third day 6, fourth day 0, and fifth to eighth days, 4. In a control untreated lot, 50% died. Three days after discontinuing the treatment, the disease recurred, and a second treatment with 0.5% sulfamerazine mash was given for 7 days. Three per cent died in an untreated control group. For 12 days following the second treatment, no losses occurred, then the cholera reappeared necessitating a third treatment. Total losses during this 37-day period were 11.9% among the treated birds which appeared to eat the medicated mash readily. No permanent immunity apparently was acquired, so the sulfamerazine treatment is recommended only as a stopgap to save turkeys so they can be marketed with safety to the consumer and avoidance of crippling losses to the growers. Daily moving to clean ground is recommended along with the treatment.

The California Station (McNeil and Hinshaw, 1948) found that streptomycin in the calcium chloride complex inoculated intravenously at the rate of 25,000 units per kilogram (2.2 lbs.) of body weight (150,000 per well-grown turkey hen) prevented mortality and largely prevented the treated birds from becoming carriers if treatment was given before or at the time of the experimental inoculation with the cholera germs. When the streptomycin treatment was delayed 6 to 24 hours, mortality was still prevented but there were more carriers among the survivors and more hock-joint involvement.

Survivors from a cholera epidemic should be sold for market when the birds have regained their health, and not retained as breeding stock. If extremely valuable breeding stock has been affected, the survivors that are retained should be completely isolated so as not to transmit the disease to young stock.

Coccidiosis

Symptoms: Drooping wings, ruffled feathers, listless appearance, diarrhea light brown and mucoid rather than bloody (the common symptom in chicks) although blood occasionally appears as the disease continues to run its course. Coccidiosis makes its appearance in poults from 2 to 16 weeks of age, most commonly at 3 to 6 weeks, although occasionally as early as 2 weeks. It is more common in poults naturally brooded as the mother hen serves as the source of infestation. This disease does not cause as severe losses in poults as in chicks and if observed in time can be checked with less of a setback to the flock. The seven species which infect turkeys are different from those commonly troublesome to chickens so cross-infection does not occur. Only two species, *Eimeria melegrimitis* and *E. adenoeides* cause serious losses in turkeys. The five other are *E. melegridis*, *E. dispersa*

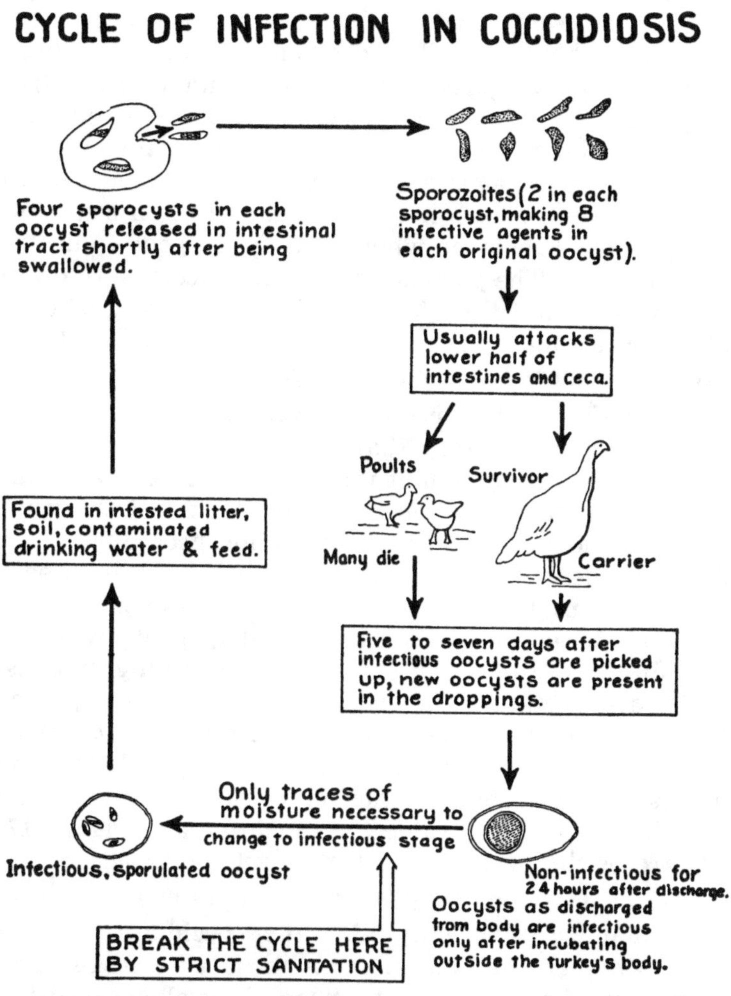

Fig. 17.2. Cycle of coccidiosis (*Eimeria*) of turkeys; adapted from Barger and Card. The occysts are the eggs of the coccidia which incubate the moist surroundings such as damp litter.

(also in quail and pheasant), *E. gallopavonis, E. innocua* and *E. subrotunda.* (Moore et al, 1954, Poul. Sci., 33: 925-29).

Post-mortem findings: Positive diagnosis depends upon microscopic examination of intestinal contents or contents of the ceca (blind pouches) for the one-celled protozoon. This should be followed by bacteriological examination. Coccidia are passed in the oocyst stage in the droppings of affected birds. The ceca may contain a yellowish-brown fecal material and frequently the cecal walls show signs of hemorrhages and ulcers. The lower part of the intestinal tract shows inflammation and often contains a milky white substance which is a fairly good indication of coccidiosis. The presence of whitish ulcers in the walls of the small intestines is indicative of one type of coccidiosis affecting turkeys. In these cases the walls are greatly weakened and may easily rupture.

Prevention and Control: Prevention is cheaper and much more satisfactory than treatment. To prevent disease, avoid introduction of infection by visitors and eliminate all contact between breeding stock and the young poults. Follow recommended sanitation procedures. Wire floors properly constructed are a fairly certain preventive. Radiant heat and dry litter are preventive aids. Only one brood per season is a great help in preventing coccidiosis. In the oocyst stage coccidia are highly resistant to disinfectants. The principle of control is to prevent reinfection through the droppings. Daily removal of the litter and droppings for a week or 10 days is efficacious and is often the only practicable method of control. "Dry cleaning" is of more value than disinfection. Figure 17.2 illustrates the place to break the cycle of infection.

The heat of fermentation (120° to 140° F.) of barnyard manure apparently kills the coccidia in a short time, whereas they may live in the oocyst stage in the dampness of a crack in the corner of a well-sprayed floor many months. Coccidia are not transmitted through the egg nor need one fear infection from small bits of dirt adhering to the eggshell of eggs in the incubator. The immediate removal of heavily infected birds is recommended. Such birds should be killed and burned.

Attempts to treat coccidiosis with special feeds and flushes have failed. However, the drug sulfaguanidine recently has been found to possess preventive and curative properties. It can be administered as a preventive at the rate of 0.5% in the dry mash for 4 or 5 consecutive days, or at 1% fed one day out of 4 for a series of 3 treatments; water and only the medicated feed being given. Curative dose of sulfaguanidine is 1% in the mash for 3 consecutive days (no repeat).

Sulfaquinoxaline also was found to be effective as a preventive of coccidiosis by the Cornell Station (1949). At 0.03% of the diet, it, and also sulfamerazine at 0.5%, effectively prevented coccidiosis when fed one day out of 4 for a series of 3 treatments to 6-week-old poults put on infected litter. Acetylsulfaquinoxaline at 0.0175 or 0.05% in the feed is suggested as a preventive or early treatment (AVMA Proceedings Book, 1953).

Sulfaquinoxaline in soluble form can be used in the drinking water as a treatment at 1:4000 (1 oz. per 30 gal.). Give treatment 2 or 3 days, skip 3 days, then give for 1 day only, skip 3 days and give one more day (California Station, 1951).

When the sulfas are given in coccidiosis outbreaks, frequent cleaning out of litter is not indicated, in fact it may be a disadvantage. A preventive treatment during the 3rd to 5th weeks of brooding sometimes is indicated in situations where coccidiosis can be counted on to occur.

Colibacillosis

A disease of young poults caused by organisms of the coli-aerogenes group was reported by Scrivner (1945) as being common in the turkey area around Wellman, Iowa. Poults 4 to 26 days old were susceptible. Diagnosis was based on cultural studies of coliform organisms in organs of recently killed poults along with absence of other recognizable causes. The poults showed diarrhea and enteritis. A successful treatment with sulfathalidine was developed,—4 pounds per ton of mash fed for a week or 10 days as soon as the first symptoms were noted.

Crop Worms

(Capillaria or hair worms)

Crop worms are slender, hairlike, whitish roundworms ½ to about 3″ long. *Capillaria contorta* is transmitted direct, and *C. annulata* requires the earthworm as an intermediate host. Both infest the crop and esophagus.

Gongylonema ingluvicola (the gullet worm, life history unknown) is found rarely in the crop and esophagus of chickens and turkeys where it causes only local damage. *Capillaria columbae* and *C. caudinflata* are hairworms found in the intestine. These worms are not common parasites but sometimes they cause serious losses in poults 3½ to 4½ months old. No satisfactory treatment is known for cropworms but for intestinal hairworms carbon tetrachloride, 1/cc. per bird in gelatin capsules a week apart is effective; 5% flowers of sulphur in the mash for 3 weeks may be helpful in cropworm control but may harm the birds. Prevention may be accomplished by sanitation, range rotation, and the use of well-drained ground. Infested land must be abandoned for an indefinite period, probably 3 years or longer.

Deficiency Diseases

(See Chapter 5)

Dehydration

Dehydration due to lack of sufficient intake of water may sometimes be mistaken for a disease. It may be caused by failure of day-old poults to learn to drink, sudden changes in watering equipment, or the use of medicines in the water. Pomeroy (1952) warns against the indiscriminate use of sulfa drugs, bluestone, and disinfectants. Symptoms of dehydration are listlessness, sunken appearance in the crop region, whitish crystals around the vent, crop empty or containing dry feed. Breast muscles are dry and reddened and the skin sticks to them. The lining of the intestine is very reddened and the organ empty. Water consumption may drop when adding anything to the drinking water, especially if there is a decided change in color or taste. Individual dosage of poults with water or fresh milk using a pipette or medicine dropper may save dehydrated birds.

Enteritis
(Hemorrhagic and Non-Specific)

Turkeys, particularly poults under 4 months old, may become droopy, show nervousness, fail to eat, or get diarrhea. Upon autopsy they may show inflamed intestines. Advanced cases may be very thin with dark congested liver, flabby heart, and dark, non-clotting blood. Possible causes are: (1) unknown infection, (2) unsuitable feed, (3) sudden changes, (4) mismanagement, (5) lack of water, (6) contaminated water. Control: good husbandry; find and remedy the causes; provide protection from inclement weather.

Erysipelas

This disease though uncommon seems to be on the increase. It may cause heavy losses in turkeys, mostly males, near market age, birds under 4 months seldom being affected. It is caused by the swine erysipelas organism, *Erysipelothrix rhusiopathiae,* and may be transmitted by sheep and swine.

Symptoms: Listlessness, drooping, aloofness, sulfur-colored droppings, cyanotic (dark blue) head, nasal catarrh, swelling of the snood or nasothelium, which may fall off, and fevers of 2 or 3° above the normal of about 106.5° F. Lameness may occur.

Post-mortem findings: Bloody areas in breast muscles; pink skin over the breast; mucus in nasal passages; enlarged, friable congested liver; reddened large intestine; mottled friable spleen.

Treatment: Isolate affected birds promptly. They may be saved by prompt intramuscular injections of penicillin in proper dosage; 100,000 to 200,000 units in a single dose, followed by a second dose when needed (Mass. Station). Streptomycin (a single dose of 140,000 mcg.) injected intramuscularly proved 100% effective while 80% of the untreated controls died in a test by C. G. Grey (U.S.D.A., 1947). An effective bacterin is now available.

Prevention: Keep turkeys from contact with sheep and swine especially where erysipelas has occurred. Early debeaking and de-snooding along with separation of the sexes have been suggested as preventives. Use rubber gloves with sick birds as erysipelas is transmissable to humans and may cause erysipeloid, a type of inflammation of the skin, or even more extensive and severe infection.

Chronic erysipelas was reported by the Washington Station (1950) in what apparently is the first report of this disease (called diamond skin disease in swine) in turkeys. The turkey examined was a Broad Breasted Bronze breeding tom from a flock suffering from a "sore-eye" infection which seems to be on the increase in the Northwest. At the time of examination one eye was destroyed, the other intact; there were numerous scabs on head and neck resembling pox lesions; the throat, wattle and snood were swollen and necrotic appearing; and there was a large scabby brown wheal covering the breast. In addition there were smaller scabby areas on other parts of the skin and soles of feet and toes were necrotic, eroded, and swollen. Upon autopsy the internal organs appeared normal. Culturing and tests with mice showed the lesions not to be pox but erysipelas. The authors do not state positively that the sore-eye condition in breeding turkeys is all due to chronic erysipelas but they suggest that it appears plausible. Penicillin in large doses has been observed to be of some value in treating sore-eye condition. Debeaking and de-snooding may help to retard its spread. The new bacterin injected subcutaneously at 8 to 12 weeks will prevent serious outbreaks until market age. Breeders should be revaccinated when selected (Bottorff, Turkey World, May 1955).

Flukes

The fluke *Collyriclum faba,* a small leaf-life organism is reported as the cause of cysts formed on the skin of young turkeys in the abdominal area (Calif. Manual No. 3). The cysts are fairly large and shiny. The few cases found have been in flocks reared near streams, lakes, or ponds where the flukes passed through a portion of their life cycle. Marshy meadows were also found to harbor the intermediate hosts (snails and dragon-fly nymphs) of the flukes.

Other flukes (trematodes), usually affecting the intestinal tract, such as *Echinoparyhium recurvatum,* have been found in turkeys and in a few instances have caused serious losses. Aquatic snails, tadpoles, and frogs act as intermediate hosts. The usual sanitary practices, including avoidance of natural sources of water and of marshy areas, will prevent loss from flukes. There is no effective treatment for any of the fluke infestations.

Fowl Pox

Fowl pox is caused by a filtrable virus affecting chickens and turkeys. Another variety of pox virus affects pigeons and another affects canaries. There is believed also to be a fourth strain of pox virus affecting turkeys only. For immunization purposes, fowl pox vaccine of chicken origin is best for turkeys.

Mortality from pox usually is light but the growing turkeys receive more or less of a setback from a pox attack while breeding flocks suffer greatly lowered egg production, fertility, and hatchability. Pox attacks vary greatly, from mild to very severe. Growing turkeys may not suffer much from mild attacks but an attack of any severity in a breeding flock is sure to result in great financial loss. Since there is no practicable and effective treatment, preventive vaccination should be practiced.

Usually if pox becomes established on a plant it becomes a permanent fixture requiring vaccination each year. In the Maryland area vaccination, usually in July, is sufficient. This protects the growing turkeys and apparently also those kept over for breeding. In some other areas, a second vaccination of breeders done about 6 months or a little longer after the first one, appears to be necessary.

Pox is spread by mosquitoes, animals, equipment, purchase of infected stock, and by vaccination (since a live vaccine is used).

Symptoms: In some cases the first symptom to cause one to suspect fowl pox in young turkeys is an abrupt drop in feed consumption. Lack of appetite is apparent almost as soon as the pox scabs form. In other cases there is no apparent drop in feed consumption and the appetite is not noticeably affected. Grayish-yellow, wart-like eruptions occur on the head and neck. At first they are soft, but later they harden and become brownish black. They sometimes appear on the legs and feet and in severe cases may occur on the skin of the body. In breeders, pox causes yellowish cheesy cankers in the mouth and under the eyelids. The eyes are watery and may have pus accumulations in the corners. These symptoms affecting the eyes and mouth should not be confused with sinus infection. Yellowish cheesy patches which form in the mouth are referred to as canker or diphtheria. These should not be confused with the curd-like cankers caused by fungi (see moniliasis).

Control Measures: Sanitation practices and separation from chickens minimize the probability of infection with fowl pox. However, turkeys that are kept under the best of conditions have in numerous cases been victims of a fowl pox epidemic when chickens and other turkeys in the neighborhood have been infected, or sparrows have transmitted the infection. Mosquitoes also can spread pox.

Vaccination with the same type of live fowl pox vaccine used to immunize chickens has been found effective in protecting turkeys. If there has been no fowl pox in

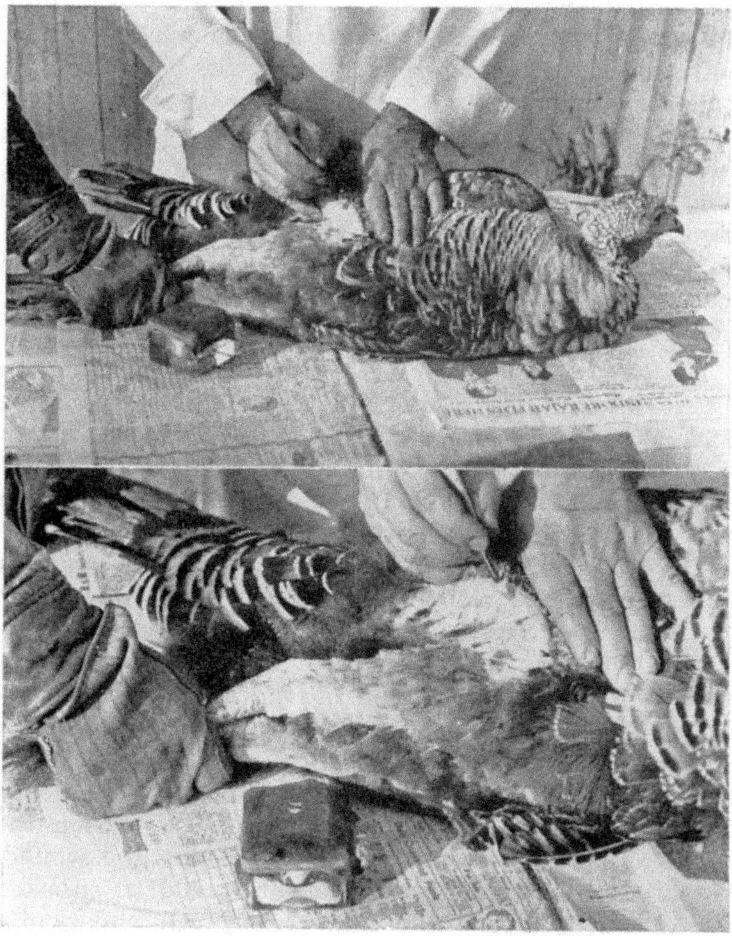

Fig. 17.3. Restraint of turkey for vaccination on upper thigh. Note that the table is covered with newspapers. This aids in preventing undue spread of vaccine. B. A close-up taken to show the suggested location. The long tuft of feathers that normally cover this naked area are being held back by the vaccinator's left hand. The heavy glass ink-well is a convenient holder for the vaccine. It is covered with a piece of rubber in which is cut a small opening to prevent excessive dust contamination. (Courtesy University of California.)

the immediate neighborhood and if it has not been present on the home farm either in chickens or turkeys, it is doubtful if the flock should be vaccinated. Vaccination has proved successful with turkeys from one day old on, but is not recommended for day-old poults that are to be shipped. Immunity from vaccination usually lasts 6 to 7 months so in some cases turkeys kept for breeders should be vaccinated again. All of the young turkeys (and young chickens on the same farm) should be vaccinated at the same time. After vaccination about a month is required for the establishment of immunity and the scab-like scars (takes) sometimes remain visible four to eight weeks.

The second vaccination for breeders, according to recent observations (California Station, 1951) should be made not earlier than 6 months after the first vaccination or until the immunity conferred by it has disappeared in the majority of birds. This may be determined by a test vaccination to see if a large percentage (75% or more) of "takes" is obtained. If the second vaccination is done too soon it is ineffective, and as a result, due to later loss of the first immunity, the turkeys may develop, during the breeding season, pox in the form of yellowish cankers in the mouth and in the mucous membranes of the eyes. Males are affected more than females and the infection is noticed as pus secretions in the eyes. Treatment by "sulfa" ointment may be effective. The mouth cankers should be removed by a blunt instrument and the ointment liberally applied and rubbed into the wound. For the eye infection the ointment should be inserted between the inner eyelid and the eyeball and gently spread with the fingers two or three times daily. The labor involved in such a proceeding obviously is prohibitive in large flocks hence prevention by proper vaccination procedures, allowing 6 months or more to elapse between vaccinations, is the only practicable solution to the problem. Climate and other factors may influence the need for vaccination because in some situations one vaccination at 1 to 3 months, preferably the latter, is sufficient to carry the birds through the breeding season. Experience will determine the need for a second vaccination but in general the warmer the climate the more likely it is to be needed.

Post-mortem: There are no characteristic intestinal lesions.

Treatment: Treatment of the scabs is useless.

Vaccination: Vaccination is cheap, easy, and effective. The stick method is used almost exclusively. This consists of penetrating the turkey's skin with a double-needle instrument usually furnished with the vaccine. If not furnished, one can be made from two sewing machine needles stuck through a cork so tips are even and about one-fourth inch apart. The instrument is dipped into the vaccine then forced into and through the turkey's skin. Troubles following wing-web vaccination make it appear desirable that it be abandoned in favor of thigh vaccination; the inner surface of the thigh might even be most desirable or safest. With the bird on its side, lay back the heavy fluff feathers exposing a ridge of skin. Insert the vaccinator into this ridge as shown in figure 17.3. To vaccinate by the follicle method, pull over 4 or 5 feathers from the upper thigh site and brush the vaccine mixture into the empty follicles with a very small stiff-bristled brush. Either large or small feathers will do.

Great care must be taken in the purchase and handling of the virus. It should be purchased from a reliable source and placed in the refrigerator immediately upon arrival. Only as much virus as will be used in a half day should be mixed at one time. It must be protected from the sun, heat and dust, to avoid a low percentage of "takes." The expiration date of the manufacturer should be noted upon receipt of the shipment.

All unused vaccine, together with the containers, should be burned or placed in disinfectant for several hours to prevent the spread of virus. All cotton, tape, threads, wraps, and papers should be removed and burned. Instruments should be put in some good disinfectant for several hours and then thoroughly rinsed with boiling water, or they may be boiled for 10 minutes.

Post-Vaccination Care of the Flock. If the flock is in good healthy condition, it need receive no special care after vaccination. Some time between the eighth and the twelfth day, about 10 per cent of the flock should be examined to determine whether the vaccination has been successful. If nearly all of the birds in such a sample have one or more of the small scab-like vaccination "takes," the flock can be considered properly immunized. If a sizeable percentage have not reacted, the entire flock should be handled, and the non-immunized birds should be revaccinated with a fresh lot of vaccine. As the birds cannot be considered immune until about one month after vaccination, precautions should be taken to prevent introduction of the disease until after that period. A few vaccinated birds may develop lesions on the head, but these usually clear up within a short time without any severe symptoms. It is well, however, to isolate such birds from the remainder of the flock to prevent spread by fighting before immunity is established.

Gapeworms

Symptoms: Heavy infestations cause turkeys to gasp or gape, as is the case with badly infested guineas or chickens. The gapeworm usually fastens itself in the lower half of the windpipe. The light red male is short (about ¼ in.) and relatively inactive all his life. The dark red female is longer (up to 1½ in.) and fastens her head to the windpipe beside the male, with the other end joined to him so that the two form a Y. While gapeworms (*Syngamus trachea*) usually cause little trouble in well-grown turkeys, in some cases mortality occurs in young birds. Wehr (1937) found these parasites in 15% of the windpipes of 936 turkeys raised and marketed in the vicinity of Washington, D. C., in 1936. He also observed that these nematodes persist as infective larvae in the earthworm, which is a carrier but not a true intermediate host. Earthworms are the means of transmission between the turkeys and chickens or between the old turkeys and the poults.

Treatment and Control: Mechanical removal is impractical. Treatment by exposing affected turkeys 15 to 20 minutes to 1 oz. of barium antimonyl tartrate agitated in a fairly tight chamber of 8 cu. ft. capacity is recommended in U.S.D.A. Leaflet 207 (1941).

Separation from chickens, range rotation on well-drained land, or confinement rearing are effective preventive measures.

Gizzard Worms

The nematode *Cheilospirura hamulosa,* a slender white gizzard worm is found in chickens and turkeys. The worms are round and ⅓ to ¾ inch in length. They inhabit the lining of the gizzard where nodules and tunnels are formed. Flour beetles, weevils and sandhoppers are the intermediate hosts of this parasite. The extent of damage caused by gizzard worms in turkeys is not known.

Prevention: If grasshoppers or other intermediate hosts are not eaten, or if the insects which are eaten do not feed on the droppings to any great extent, no serious trouble should be encountered. Separation of poults from chickens and their droppings is important. No satisfactory treatment is known.

Hexamitiasis

This disease, originally called *Infectious Catarrhal Enteritis* by California invesitgators and *trichomoniasis* by Missouri pathologists* is active chiefly in the western States where it is a serious menace, but it has caused losses in all sections of the country. In the East where it occurs sporadically it sometimes is referred to as "Cali-

*Mo. Exp. Sta. Bull. 456, Knight, McDougle and Durant (1942.)

fornia disease." The California Station found that the cause is a flagellated protozoon *Hexamita meleagridis* which causes a physiological starvation by inhibiting the birds' ability to utilize starches in the diet. Transmission is by the young birds eating infected droppings from breeders or from each other. Hinshaw (formerly of California) says Hexamita is a common inhabitant of the lower intestine where it does no harm. However, when it migrates to the upper intestine the disease called hexamitiasis results. The organism localizes in the bursa of young turkeys recovered from the disease. These birds remain carriers until the lymphoid bursa disappears at sexual maturity. (Hinshaw, Vet. Med. 48: 164, 1953.)

Symptoms: Nervousness followed by listlessness; body temperature subnormal; foamy, fetid, greenish to whitish thin diarrhea and frequently convulsions. Usually poults continue to eat, but do not assimilate food, so become emaciated. Duration of disease 1 to 6 days following period of incubation. May affect poults 1 to 16 weeks of age, seldom over 10, with heaviest mortality (20 to 90%) at 3-5 weeks.

A subacute type of the disease has been observed in older poults. In these cases listlessness and loss of weight are the most obvious symptoms. The mortality may be low but the financial loss from stunted individuals is great. Birds which recover may serve as carriers. Pheasants, quail, chukar partridge, and ducks are susceptible and may be carriers.

Post-mortem findings: Emaciation; dry skin; breast flesh dry and reddened; liver sometimes congested; intestinal inflammation with characteristic thin-walled bulbous areas filled with watery foamy material or thick mucus; lesions confined to the small intestine. Hexamita also has been found consistently in the bursa of Fabricius, the sac-like organ opening into the cloaca near the vent and elsewhere in the intestine from gizzard to vent. Microscopic examination of droppings alone is not sufficient for diagnosis. The intestines of fresh-killed sick birds should be examined.

Prevention and Control: A fairly satisfactory treatment is the feeding of 3% dried whey containing 50 to 70% lactose in a 1 to 2000 solution of bluestone ($CuSO_4$) instead of water for 5 to 7 days then 3 to 5 days on plain water, followed by another 5 to 7 days treatment gives promise of success. The mixture must be stirred very frequently since the whey settles out (McNeil, 1948). Plenty of heat and the best of care are helpful. Flushes are ineffective and dangerous. Finding a satisfactory method of preventing the transfer of droppings from adult carriers to young birds is the most effective means of preventing the disease. One brood a year helps. The aspirin, the sulfas, and mercuric chloride were found ineffective by the California Station (1948). The Texas Station (1955) suggests NF 180 (nitrofurazolidone) at 100 grams per ton of feed as a preventive and as treatment during an outbreak.

Aureomycin 25 to 50 grams per ton of feed has been suggested as a preventive treatment. Dibutyl tin maleate in the feed at 0.05% was reported successful as a practical preventive of hexamitiasis by McGuire and Morehouse (Dr. Salsbury's Labs.), 1952. These investigators also reported that a combination of terramycin and penicillin were effective but economically unprofitable. Almquist and Johnson (1951) reported success in treating artificially infected 9 weeks old BBB poults with ANT (2 amino-5-nitrothiazole) at 0.1% in the mash for 9 days and with penicillin G or terramycin, 25-mg. capsules, one per bird per day for 3 days.

A disease of young turkeys clinically and pathologically indistinguishable from hexamitiasis was reported from Scotland in 1945 by Campbell*. It was caused by a flagellated protozoan parasite of the genus *Cochlosoma* (Kotlin), family Trichomonadidae, in turkeys 2-10 weeks old.

*The British Vet. Jour. 101(12), 1945.

Symptoms were intense thirst, watery frothy diarrhea, disinclination to move, stilted gait, coma, and death within 2 or 3 days of appearance of symptoms. Bullae (distensions) in the intestines were consistent post-mortem findings. *Cochlosoma* also has been reported in turkeys in California and may be the same (*C. anatis*) as the one occuring in ducks. Individual dosage of a "000" capsule filled with an aureomycin-B_{12} supplement for 10-week-old poults—less for smaller ones—may be helpful.

Leucosis Complex

This term is used to include the condition variously called lymphomatosis, big-liver disease, range paralysis, leukemia, and fowl paralysis which causes great losses in flocks of chickens between four and eight months of age. It has been reported in turkeys a few times, in Europe and America according to Harnden and Smith (1943) but is not now considered a menace to turkeys. The leucosis-like lesions found in turkeys may be a by-product of blackhead infection.

Symptoms: Extremely variable; lameness, paralysis, emaciation, "glass eye," tumors, and diarrhea.

Post-mortem: Enlarged liver; tumors in skin, liver, intestines, esophagus, proventriculas, spleen, ovary, and kidneys. Nerve lesions were not observed by Huffman (Calif. Dept. Agric. Bull. 33, 1944).

Treatment: Losses may be held at a minimum by rearing young stock over 10 rods from mature stock.

Prevention: Turkey raisers should be ever on the alert against the possibility of an outbreak of this disease. Sanitation, range rotation, and separation of turkeys from chickens should be practiced faithfully, for no greater calamity could befall the turkey industry than for leucosis to become established in turkeys.

Leucocytozoon Infection

Leucocytozoon infection is a malaria-like disease sometimes called turkey malaria, gnat disease, or simply blood protozoan disease. *Leucocytozoon smithi,* the microscopic protozoan responsible for the disease appears in the peripheral blood in spindle shaped cells of unknown origin. It probably is more prevalent than realized and can be easily mistaken for fowl typhoid, fowl cholera, non-specific enteritis, or even respiratory infection, if symptoms and lesions are alone considered when making diagnosis.

Symptoms: Difficult breathing, thirst, yellow to greenish yellow feces, droopiness, decreased appetite; drunken posture—birds sometimes lie down and if disturbed move with difficulty; coarse coughing; running on tiptoe; usually affects young stock from a few weeks up to three months, sometimes up to four and five months. Infection usually occurs in midsummer. Mortality is rapid and heavy. In breeding stock: weakness sometimes accompanied by apoplexy and death; males seldom strut or mate.

Post-mortem findings: Liver enlarged and sometimes degenerated, with a catarrhal enteritis and a loose fibrous exudate yellow to ivory in color, heart sometimes enlarged, heart walls thin, spleen inflamed and enlarged, slight inflammation of duodenum. Flesh usually flabby and brownish. The only certain method of diagnosis consists of blood smears, blood being taken from any channel. Upon appropriate treatment of the blood smear (Johnson et al, 1938) the protozoan (*Leucocytozoon smithi*) known to cause this disease may be identified under the microscope.

Treatment and Control: Workers at the Nebraska and Virginia Stations found that the disease is transmitted from an infected turkey to others by an intermediate host, the blood sucking turkey gnat or black fly (of the genus *Simulium*). It cannot spread from bird to bird. Since these flies breed in running water the disease is more common around streams simply because the black fly is more likely to be present. Turkeys raised by the Virginia scientists in fly-proof shelters did not become infected although their pen mates in a nearby shelter not screened to exclude gnats suffered a heavy mortality. Birds which survive may serve as carriers the following year. Consequently it is unwise to carry over as breeding stock any survivors. If infection persists or one is in doubt, it might prove a wise investment to cover the shelters and sunporches with fine-mesh fly screen or cheese-cloth that will exclude gnats. The Manitoba Station found that flies other than the simuliidae were carriers.

Johnson suggests that control can be effectively accomplished by a microscopic examination of a stained blood smear from all breeder turkeys, and prompt disposal of any that are found to be carriers of this blood parasite. Another alternative which has been practiced where breeder turkeys are not of sufficient value to warrant their keeping beyond the first breeding season, is that they be disposed of before the young turkeys are large enough to be turned outside where they would be exposed to the turkey gnats. Thus, if the carrier turkey is eliminated, the source of infection for the flies is done away with, and this consequently becomes a very practicable way of controlling the disease.

Treatment: Sulfaquinoxaline used as in the control of other infectious turkey diseases may reduce losses in an infected flock and may have value as a preventive according to the South Carolina Station (1950).

This disease has been reported, so far, from California, Georgia, Minnesota, Nebraska, North Dakota, Virginia, South Carolina and Manitoba (Canada).

Lice

Lice of poultry are true insects with 3 pairs of legs and very much flattened bodies. They live on their hosts at all times either on the feathers or on the skin. They come to the turkey's vent for moisture and live on the dead portions of skin and feathers. Young turkeys hatched in incubators and brooded artificially seldom have lice; adult breeding turkeys, usually the toms, frequently acquire lice.

Symptoms: Young poults fail to grow normally, mature birds may be in poor flesh. Lice may be seen easily by examining the upper thigh feathers and the bases of the fluff feathers around the vent. The unhatched whitish louse eggs (nits) also may be seen in clumps about the base of the feathers. The life cycle of all turkey lice requires about 3 weeks. If turkeys are kept in close proximity to chickens, or if chicken hens are used in brooding, the common chicken body louse, *Eomenacanthus stramineus*; the chicken shaft louse, *Menopon gallinae*; and the head louse *Lipeurus heterographus* are likely to infest the turkeys. There are also at least two species of lice that are found chiefly on turkeys, *Goniodes meleagridis,* the large turkey louse, and *Lipeurus gallipavonis,* the slender turkey louse. Some species are 3/16″ in length. They are yellowish to grayish in color, sometimes possessing dark markings. Heavily infested toms may be poor breeders.

Control: Fortunately, lice are one of the pests that can be controlled, so that trouble from this source should be negligible. Artificial incubation and brooding break the cycle of infestation and if the poults are reared in sanitary quarters no trouble from lice should be experienced. If natural incubation or brooding is used the hens should be dusted thoroughly with sodium fluoride, commercial or pure, twice before they come in contact with young poults. Hens should be lightly dusted when set and two weeks later. Dusting should be done by the "pinch method." All the powder that is easily held between the thumb and forefinger constitutes a "pinch." The feathers should be parted and two pinches placed on the skin at different parts of the fluff below the vent. A pinch should be placed under each wing, holding the wing open. Two pinches should be placed on the back, one on each thigh, one on each side of the breast, one on the head and one on the neck. The bird should rest on a newspaper on top of a barrel or on a table while being dusted. The dust falling on the paper can be used on other birds. It is important to place the powder on the skin at the base of the feathers. One treatment, if thorough, should destroy all lice, including those that hatch from nits, but to insure complete elimination, repeat it in two weeks.

Other effective materials, which may be applied by the pinch method are; phenothiazine, fine sulphur, DDT, pyrethrum, derris, and orthophenylphenol (1 part to 4 parts of talc or fuller's earth). The first two are effective if applied liberally with a shaker. Malathion is excellent.

Baby poults should not be dusted with sodium fluoride or any strong louse powder until more than a week old, and even then the powder should be used quite sparingly. It is best to avoid louse trouble with the poults by treating of old birds. The Texas Station found one drop of nicotine sulphate solution on the skin one inch back from the rear of keel to be excellent for delousing turkeys.

Breeding stock should be examined during the fall and again in late winter. If lice are found on the turkeys, the top of the roosts may be painted lightly in the late afternoon with a thin line of 40% nicotine sulfate* solution, making sure that the turkeys roost on the treated surface. This treatment avoids handling the birds individually, but is too expensive if the birds spread out much at roosting time, so that a large amount of roosting surface must be treated in relation to the number of birds. Benzene hexachloride also is effective as a roost paint louse-killer (or dust ½%) but should not be used within 10 weeks of marketing due to possible danger of causing off-flavor in the flesh. Lindane is a new and effective roost-paint louse killer that has caused no objectionable flavor in poultry meat; 25% of the wettable powder in water is used to spray or paint on the roosts. Lindane dust (½%) used by the pinch method is effective in one treatment.

Another good method of treatment of well-grown turkeys is to dip the birds on a warm, sunny day in a solu-

*Such as Black Leaf 40. This treatment should not be used if the roosting quarters have been recently whitewashed or sprinkled with lime.

tion of 1 oz. of sodium fluoride or fluosilicate per gallon of warm water; 8 ounces of either drug dissolved in eight gallons of water furnishes enough solution to dip 100 turkeys if the birds are held over the tub a few moments after immersing to conserve the solution as it drips freely. The turkey must be held without struggling while immersed, another person parting the feathers so as to insure the solution reaching the base of the feathers. Young turkeys should not be dipped until well feathered and at least 12 weeks of age.

Manson's Eye Worm

This parasite (*Oxyspirura mansoni Cobbald*) is found in the South, where it affects chickens, turkeys, and wild birds. It is about ½ inch long, threadlike, and white and inhabits the eye under the nictitating membrane (the winking membrane of the fowl). Cockroaches are the intermediate host. Although reports of infestations of turkeys by this parasite are not common, considerable damage may result when an infestation does occur.

Symptoms: Winking, scratching, and eye discharge.

Prevention: The removal of rubbish under which cockroaches may hide, as well as other methods of controlling the roaches; treatment or destruction of affected birds; general sanitation.

Treatment: Two or three drops of a 5% solution of butyn should be dropped in the eye as an anesthetic. The nictitating membrane should be lifted to place one or two drops of a 5% solution of liquor cresolis on the worms. The eye should then be washed well with warm water.

Minor Infections

Micrococcus caseoliticus, a rather common bacterium was identified as the cause of heavy mortality in young poults in 1943 in Virginia. The organism in this case did not affect the breeding stock.

Pseudotuberculosis is a rare disease of turkeys and even man, caused by *Pasteurella pseudotuberculosis.* The trouble is more common in cold, wet weather. Diarrhea and droppiness are symptoms. The liver is enlarged with light-colored areas of varying sizes; enteritis, sometimes inflammation of kidneys and spleen. Diagnosis depends on laboratory identification of the causative organism. Sanitation, clean water supply, protection from weather, use of well-drained range are preventive measures for this and other diseases that are encouraged by unfavorable conditions.

Milkweed Poisoning

The whorled milkweed (*Asclepias galioides*) is common in western and southern Colorado, western Kansas, northern New Mexico, western and eastern Arizona, and southern Utah and may possibly be found in southeastern Nevada, northwestern Oklahoma, and the Texas panhandle. Field observations and laboratory tests by the U.S.D.A. (Stiles, 1942) indicate that this weed, especially the young tender plants, is toxic to turkeys and chickens. The dried weed also is poisonous. Symptoms are drowsiness, lack of coordination, convulsions, exhaustion, coma, and death. During convulsions the bird lies on its side or back, flaps its wings and makes jerky, sometimes running, movements with its legs. Head and neck are twisted, pupils of eyes are moderately dilated. After death, the crops are distended with a small amount of gas, heart congested, and lungs slightly edematous. Control consists of preventing turkeys from eating the weed. The availability of palatable and non-poisonous green feed would probably result in the turkeys failing to eat the milkweed. Eradication of the weed and fencing or herding the turkeys out of infested areas would also solve the problem.

Mites

Symptoms: Loss of vitality and color indicate a loss of blood due to the common red mite, *Dermanyssus gallinae*, which is a blood-sucking parasite frequently infesting chicken houses and turkeys that may be placed in them.

Since they prefer chickens, turkeys often remain free from these attacks. Because of this the common red mite seldom infests turkey premises. Mites crawl on the birds at night and suck blood. Before the birds leave the roost in the morning the mites crawl back into the cracks and crevices around the roost supports. A careful search of the roosting quarters will disclose the mites. The appearance of areas resembling a sprinkling of gray pepper indicates the presence of the eggs of the mite. Mites are not true insects but are members of the spider and scorpion order with 8 legs (*arachnida*).

Control: The sides and bottom, not the tops, of the roosts should be painted or sprayed with an anthracene oil (such as carbolineum), any coal-tar disinfectant, creosote, or crank-case drainings from the automobile. The drainings may be diluted with kerosene to make painting or spraying easier. The reason for not painting the top of the roost is to protect the breast and feathers of the turkey. The roost supports should also be painted or sprayed thoroughly. All painting or spraying should be done in the morning so that it may soak in and dry before the birds go to roost. Remove and burn all trash and litter. Malathion spray 0.5 to 1% is safe and effective.

Other mites: A depluming mite (*Cnemidocoptes gallinae*) occasionally infests turkeys. It burrows beneath the skin at the base of the feathers. Sulphur ointment rubbed on the affected areas will usually cure the trouble. One part powdered sulphur may be added to 4 parts vaseline. Dipping is a good method of control when infestation is general. A solution of 2 ounces of flowers of sulphur and ½ ounce of laundry soap per gallon of water is used. It may be added to the sodium fluoride dip. This solution will also kill the feather mite *Liponyssus sylviarum,* which lives in the feathers of the neck and rear portions of the bird. Nicotine sulphate on the roosts will kill feather mites. Apply it 3 times 3 days apart.

Infestation by *Cnemidocoptes mutans,* the scaly-leg mite, results in deposits under the scales of the shank causing the entire shank to appear rough, the bird sometimes becoming crippled. The shanks should be dipped in a solution of equal parts of crude petroleum and raw linseed oil or crude oil alone, care being taken to avoid wetting the feathers and skin.

Cytoleichus nudus (the air-sac mite) is a tiny whitish mite that lives in the bronchi, lungs, air-sacs, and the connecting bone cavities of chickens, turkeys, pheasants and pigeons. Its status as a harmful parasite is not clear and its life history is unsolved. Complete destruction of affected birds, cleanup and treatment of premises as suggested for red mites would be indicated for control.

Laminosioptes cysticola, the flesh mite, has been found in turkeys and other domesticated birds. These parasites burrow in the skin and cause yellowish nodules to form. They may also inhabit the muscles, viscera, lungs, and peritoneum. The tiny mites can be found in the nodules. Complete destruction of affected birds and treatment of premises as for red mites would be in order.

Syringophilus bipectinatus, the quill mite, affects chickens, turkeys, and pheasants causing loss of feathers, the stumps of which contain the relatively good-sized mites.

Eutrombicula alfreddugesi, the common chigger, occasionally infests turkeys. It is the larval stage of this mite that affects turkeys as well as man. In turkeys, skin blisters or abscesses may result from heavy attacks. The "spots" may be several dozen in number and from ⅛ to ½ inch in diameter, mainly on the breast and inner surface of the thighs. Frequent dusting of infested birds will aid in control, using powdered sulfur. However, elimination of the weedy or bushy areas that serve as breeding grounds of the chiggers is recommended. If this is impractical then treat the nearby ground with DDT (1% in pyrophyllite) at the rate of 96 pounds per acre, chlordane 2 pounds per acre or benzene hexachloride 25 to 50 pounds per acre.

Monocytosis—Mud Fever (Blue Comb)

The disease called "mud fever," "mud disease," enterotoxemia, toxemia, non-specific enteritis or Peterson's disease was first recognized in the state of Washington in 1944; usually occuring on muddy ranges. There is a striking similarity between it and blue comb or "pullet disease" in chickens. The original investigators (Peterson and Hymas) believe the two diseases to be identical and propose the term "monocytosis" (Turkey World, May 1951). Working apparently with the same disease in turkeys (as that encountered in the Pacific Northwest), Pomeroy has presented evidence of the infectiousness, or transmissibility, of the disease. The agent obtained from turkeys in areas in Minnesota where extensive outbreaks have occurred since 1951 appears to be a large virus. The agent appears sensitive to various antibiotics. (Pomeroy and Sieburth, 1953, Proceedings Book, A.V.M.A., 321-327). The cause still is un-named, though a virus is suspected. The disease first was observed to affect well-grown turkeys in the fall, but in 1950 it was recognized for the first time in young poults in the brooder house and on range. In the older birds the disease appears suddenly and seems to affect all birds at once. Mortality varies from negligible to about 15% averaging perhaps 10%. The flock goes off feed and loses much weight. Heads tend to be darkened, the crops are sour, and the droppings are brownish, fetid and loose. On post-mortem the flesh appears dark, dried out, and has a cooked appearance. There is a catarrhal mucoid-like enteritis of the small intestine and kidney inflammation sometimes with deposition of urates on the abdominal organs and fecal necrosis (small spots) on the liver. The pancreas may be chalky, there is evidence of dehydration, and the crop usually is empty. The course is about 2 weeks. In the young birds the symptoms and post-mortem findings are much like those in the older birds but with a higher mortality, up to 50%, and a great tendency to pile up, due to subnormal body temperature.

Prevention: Rather difficult, but treatment with the antibiotics terramycin or aureomycin appears to be effective and practical. In one flock of 600 poults suddenly affected, terramycin at the rate of 100 parts per million (p.p.m.) in the mash and 5 p.p.m. in the water for 48 hours kept losses to less than 5% compared to 50% losses in other non-medicated field outbreaks. In a laboratory experiment sick poults were treated individually by inserting the antibiotics into the crop twice daily for 4 days. The response was marked and rapid, the treated birds suffering no mortality compared to a 25% loss in the controls. Daily doses used were 500,000 units of penicillin or 40 mg. of aureomycin, or 20 mg. of terramycin, divided into the 2 daily (morning and evening) treatments. All the antibiotics were effective with terramycin the best. Penicillin treatment was considered to be impractical.

Flock treatment: One teaspoonful of potassium dichromate per 4 gallons of drinking water plus 3% molasses in the feed may help if treatment with an antibiotic such as terramycin is impracticable.

Another suggested treatment (Almquist 1952) for well grown turkeys 15-18 weeks old is 5% of aureomycin-B_{12} feed supplement in the feed with an epsom salt flush, (1 pound per 4 gallons of water) kept before them in many containers until all birds flush. If return to normal is delayed, give the bluestone and whey treatment as follows: Dissolve one pound of high lactose (50 to 70%) dried whey in 4 gal. of water. Add copper sulfate ($CuSO_4$) to the whey and water to make a 1-2000 dilution (use 1 tablespoonful of a stock solution—made with 1 lb. copper sulfate to 1 gallon of soft water—per gallon of water-whey mixture). Give this mixture—no other liquids—for 5 to 7 days, repeat if necessary. Force feeding of the mixture may sometimes be necessary to save the birds that will not eat or drink.

Terramycin 40 grams per ton in the feed during the brooding period and 5 grams per ton afterward has been suggested as a preventive of monocytosis. Severe outbreaks of the disease in young poults have been brought under control by administering terramycin at 50 grams per ton of feed, although 100 to 200 grams per ton is usually suggested for 5 days or as long as the flock is off feed.

Information recently reported from Minnesota (Pomeroy and Sieburth, 1955) suggests the following procedure for treating an outbreak. First, make sure of the diagnosis then flush for half a day with one pint or pound of molasses or epsom salts per 4 to 5 gallons of water. Follow flush with a flock treatment of antibiotic-treated feed or water or both. Use terramycin, penicillin, aureomycin, or streptomycin at 100 to 200 grams per ton of feed (up to 500 if necessary), or the equivalent in drinking water, for 4 or 5 days or as long as the flock is off feed. Provide adequate shelter, restrict the feeding of grain, and move the flock to clean, well-drained range. Provide chopped alfalfa and a milk product. Force-feed birds too sick to eat or drink with a gruel containing an antibiotic.

This disease may be acute, running its course in about 2 weeks or it may be chronic, persisting for several weeks. The chronic form may result from partial response to treatment.

Mycosis

Mycosis is a general term given to disease caused by parasitic fungi. When the fungi attack the skin of the unfeathered parts of the head the disease is called favus. Thrush or moniliasis is the result of a yeast-like fungus growth in the lining of the crop or mouth. Occasionally fungous growths occur in the air sacs or lungs and the disease is called aspergillosis. Insanitary conditions, mouldy litter and feed, dampness, over-crowding, and malnutrition are contributing causes. Man is susceptible. Jungherr (1944) reported two instances of encephalomalacia caused by molds affecting the brain.

Favus: Whitish, scaly lesions form on the head and neck and in extreme cases some of the feathered areas. This growth caused by the fungus *Achorion gallinae* is transmissible to man through cuts or scratches on the arms or hands, so care should be taken in handling affected birds. To treat, make an ointment of melted vaseline and 5% commercial formalin by mixing and shaking in a closed jar. Allow to cool and rub thoroughly into the lesions. Favus tends to be rare and sporadic. Infected birds should be destroyed or treated, cured, and marketed; and the quarters thoroughly cleaned, moving the flock to clean ground where practicable.

Thrush (Moniliasis): This disease, caused by fungi of the genus *Monilia* and usually affecting the crop, is not common but may cause losses in poults two to five months old. Listlessness, loss of appetite, and a sunken appearance of the chest are generally the visible symptoms. A post-mortem examination is essential for diagnosis. The disease affects primarily the crop, in which appear soft raised whitish-yellow ulcers of rose-like form; a solid mass having a curd-like appearance. Similar lesions may occur in the mouth cavity and the hard lining of the gizzard may become ulcerated. In older birds the liver occasionally shows a few grayish areas.

Control: Move the birds to clean ground, and clean up the equipment. Give copper sulfate (bluestone) in the drinking water at a dilution of 1 to 2,000 for a week at a time alternated with a week off for several weeks if necessary to control thrush. This solution will corrode metal, so should be given in crocks or wooden buckets. Turkeys do not like it hence should not have access to water when the solution is given. To make a 1:2000 solution, dissolve one pound in 1 gallon of soft water (soften hard water by adding 1 cup of vinegar per gallon) to make a stock solution in a glass bottle. Add 1 tablespoon of the stock solution to a gallon of water. In lieu of the treatment with copper sulfate solution, the following is suggested (Mullen); feed a mash medicated with 4 lbs. of bluestone and 2 lbs. of copperas (both powdered) per ton of feed.

While moniliasis (thrush) does not often assume serious proportions, it sometimes tends to become established in an area and when this happens, abandonment of the land for several years probably is necessary. Housing can be cleaned. Ordinary disinfectants are useless against these fungi but iodine preparations and copper sulfate are considered effective. The latter may be used in 0.5% solution as a disinfectant for equipment, floor and limited areas of ground.

Aspergillosis: In poults 5 days to 8 weeks old, heavy breathing and lesions in the mouth or windpipe indicate the presence of this mold growth usually caused by *Aspergillus fumigatus*. Because of the heavy breathing it is sometimes called brooder pneumonia. On post-mortem the lungs and air-sacs will have yellowish-white cheesy deposits, sometimes assuming typical cup-like form in the air sacs. In the later stages a greenish color may develop. Eyes frequently are affected, usually one eye, causing blindness. The Ohio Station (1952) reported that aspergillosis sometimes affects young poults, mostly males, and that the disease may be serious in male breeders (10-15 per cent) with only a few (2 per cent) of the breeding hens affected. Microscopic examination is necessary to identify fungus.

Control: Prevent the disease by avoiding sources of molds such as moldy feed and litter, and keep the litter dry particularly if beet pulp or bagasse (sugar cane) litter is used. Condensed buttermilk and skimmilk readily develop molds and if carelessly handled may be a source of infection. Cover the top of stored milk products with water to prevent mold growth and be careful with milk cans, milk barrels, and fountains. Damp spots around feeders and under waterers and permanent damp spots in the soil are danger spots. Treatment of soil and equipment and the turkeys is the same as that outlined above for moniliasis. The copper sulfate treatment is of little avail as a curative but has some value in preventing the spread of the fungus. Noticeably affected birds should be removed from the flock and treated separately. Affected birds should be killed and burned or buried deeply, as there is a possibility of transmitting the disease to humans. Sawdust litter may harbor the mold.

Newcastle Disease

The cause of Newcastle (*avian pneumoencephalitis*) is a filterable virus well adapted to growth in chickens but occurring occasionally in turkeys both young and adult. In turkey breeding stock the symptoms may be about the same as in chickens or they may be considerably milder. In chickens, eggs laid by affected breeders probably are infective for 30 days or so during and after an outbreak. Turkey eggs produced during the early phases of an outbreak have been shown to contain the virus, but the over all pattern of the virus becoming incorporated in the eggs is unknown (Moses, 1954).

It appears probable that the infective period in severe cases is shorter in turkeys and that turkey breeders with a mild infection in all probability do not transmit it through the eggs at all. Typically, breeding chickens may cough and gasp and show a quick and severe drop in feed consumption and in egg production. Muscular incoordination may occur and the few eggs laid may be off-shaped and thin-shelled with wobbly air-sacs. Recovery is rapid with light death losses. Turkey breeding birds may cough, sneeze, and gasp; may show reduced egg production, and may show nervous symptoms (paralysis), but often show no symptoms at all and their egg production rate not be slowed appreciably. Affected toms do not mate so fertility suffers. Sometimes the only sign of the disease is weak eggshells and lowered hatchability. The weakened shells result in a large percentage of cracks but the defective shell structure often cannot be seen except before the candle where it appears as light spots in the normal dark appearing shell. These weak spots may form a network, they may be quite large, or they may be small but anyone used to candling normal turkey eggs will spot the difference at once. As in chickens, recovery often is strikingly rapid with little or no death loss. Early in the course of the disease the virus may be found in chicken eggs and sometimes also in turkey eggs. Whether or not the virus may be transmitted to poults hatching from infected eggs and cause them to show symptoms later is not clear. There is a good chance that this may not occur, or if it does, only a very few eggs laid at the peak of the infection would be infective. DeLay (1947) isolated Newcastle virus from the yolk sac of 4-day-old chicks, so it is probable that the virus may also be found in poults; although there are no cases of actual isolation. Hofstad (1949) observed that in most cases of egg infection the embryo died during the incubation period. Doll, et al (1950) inoculated chicken eggs, but found no evidence of Newcastle in the subsequent chicks. Newcastle tends to be sporadic in both young and old turkeys. It is likely to appear suddenly and disappear in the same manner. It may affect poults from a few days old on up to market age but is most serious in poults under 8 weeks old. Poults show respiratory symptoms such as sniffling and gasping, and may be partly paralyzed, or may show nervous symptoms. Autopsy may reveal no special conditions though sometimes the air sacs are cloudy. Laboratory examination is necessary for correct diagnosis. Newcastle disease spreads by contact, by dust, and by eggs. Transmission by English sparrows also appears possible (Gustafson and Moses, 1952).

Treatment: As yet no drug has been found effective either as a preventive or a cure and some drugs are likely to cause harm. Vaccination is preventive only; there is no point in vaccinating a flock already sick with Newcastle.

Prevention: Due to the scarcity of reliable information on Newcastle disease in turkeys, the grower, breeder, or hatcheryman is handicapped in planning a prevention program. Whether or not to vaccinate turkey breeding stock is a problem for the individual. The present relative rarity of the disease in turkeys would militate against a general recommendation to vaccinate turkeys, either breeders or young stock. However, vaccination is effective and under proper veterinary advice and supervision the live virus vaccine could be given to turkey breeding birds mixed with pox vaccine at the time of selection or at least 6 weeks before egg production starts. Proper dilution would be important in mixing the two vaccines. Live-virus, web-stick vaccination of breeders or a natural outbreak of the disease in breeders is thought to protect the resulting young poults, as it does chickens, from the disease for the first 2 or 3 weeks of life. Killed virus vaccines might be helpful especially if given in two applications separated by several weeks and in flocks in which recent correction or elimination of disease might make the stress from live virus vaccination inadvisable.

Poults can be protected by live virus vaccination at one week of age, one drop in one eye, or by killed-virus vaccination prior to 10 days of age. As with other types of Newcastle vaccination the best available veterinary advice should be sought before undertaking it. The North Dakota Station recommends vaccination of day-old poults with killed-virus vaccine in cases where the disease has been diagnosed among turkeys. If chickens also are kept nearby, they too should be vaccinated. Two weeks are required for immunization to develop hence special care should be taken the first 2 weeks to prevent introduction of the disease.

Omphalitis—Navel Infection

In this disease of baby poults, the abdomen is enlarged, the navel unhealed, the poults feel flabby, and there is a putrid odor. Death losses start within 24 hours of hatching, reach a peak at 7 days and subside within 14 days or less. Cause is an infection with coliform bacteria or possibly a virus sometimes traceable to faulty incubation and/or poor hatchery sanitation. Mortality is often very high. The North Dakota Station reported that it sometimes is transmitted by breeding stock *via* the eggs.

A new, destructive virus disease resembling omphalitis was reported from North Dakota by Bolin and others

(1949). The virus was isolated from the ovaries of breeder hens whose offspring showed excessive mortality, from dead-in-shell poults, from fresh eggs and from poults dying soon after hatching. It was transmitted by means of brooders and by direct contact from poult to poult. There were two peaks of mortality, one on 5th and 6th days after hatching, the other on the 16th to 18th days. As with omphalitis, control probably is through hatchery sanitation.

Ornithosis (Psittacosis)

Ornithosis and psittacosis have been thought to be one and the same disease. However, recent work indicates that the virus which causes this infection in turkeys is different from the psittacosis virus commonly found in parrots and other related aviary birds. Ornithosis is sometimes found in flocks as a low-grade latent infection which causes no acute symptoms or troubles. However, such infection serves as a reservoir for infection of nearby flocks of turkeys and psittacine birds. Ultimate diagnosis of either ornithosis or psittacosis depends upon a serological test. Favorable results have been secured with the treatment of both ornithosis and psittacosis with antibiotics, particularly penicillin, aureomycin, and chloramphenicol. The virus frequently localizes in the lungs.

Turkeys have definitely been found to be reservoirs of the ornithosis virus. Sometimes the flock is not even suspected of carrying the disease until there is an outbreak among the workers involved in dressing the birds. Careful studies reveal that the workers most exposed are the ones involved in the killing and evisceration of the turkeys. The importance of locating carrier flocks is quite apparent to both the turkey raisers and the poultry processors. Accurate, differential diagnosis is possible only after careful laboratory diagnosis procedures. If the turkey owner has reasons to suspect the presence of this disease, he should, for the safety of his own family and fellow workers, as well as those who process the birds, make immediate contact either with his local veterinarian or the State Veterinary Diagnostic Laboratory, to determine the true cause of the trouble. There are frequently no visible symptoms in the turkeys, and sometimes the chief visible sign on post mortem is an enlarged spleen. A complete discussion of the distribution of this virus in the various species of birds and laboratory methods of determining the presence of the virus by Meyer (1952) will be found in "Diseases of Poultry," pages 569-618.* Dr. Meyer calls attention to the importance of exhausting the possibilities of probable ornithosis (or psittacosis) infection in humans, if persons with influenza-like illness or atypical pneumonia fail to yield to the usual medical treatment; particularly where such persons have been in touch with pigeons, turkeys, ducks, or psittacine birds.

Paratyphoid

The term paratyphoid designates the disease caused by bacteria of the genus Salmonella (except those causing fowl typhoid and pullorum disease); about 63 species of Salmonella have been found in turkeys. Of these, *S. typhimurium* has so far caused 50 to 70% of outbreaks in turkeys. Other troublesome species are *S. derby, S. montevideo, S. meleagrides.* Transmission is through eggs, on eggs *via* fecal material, from poult to poult, and by humans, reptiles, rodents, cats, wild birds, flies, hogs, dogs, rabbits, chickens, and waterfowl as infected or mechanical carriers. The fact that humans can contract and carry paratyphoid makes it doubly dangerous.

Paracolon organisms particularly Type 10 (Hinshaw and McNeil, 1945) sometimes cause disease symptoms resembling those of paratyphoid. The organisms may be transmitted through hatcheries and by turkeys and snakes. Control measures are similar to those suggested for paratyphoid. Gauger (1946) isolated a type-10 paracolon bacillus of the "Arizona" group from a pullorum-reactor turkey hen out of a Virginia flock, the poults from which had sustained heavy losses for 3 years previ-

*Published by Iowa State Col. Press, Ames, Ia.

ous. Circumstantial evidence indicated that the poult losses may have been caused by the paracolon organism. *S. newport,* and *S. orianenburg* along with *S. typhimurium* are especially pathogenic to turkeys. *S. enteritidis,* according to the Larro Research Farm (1944), caused a hock infection in market-age turkeys characterized by lameness with hocks swelled to a characteristic round shape. There was very little mortality. Watery, swollen blackish yolk sac, foul smelling and more liquid than normal, and peritonitis are found on autopsy. Control is by thorough cleaning and disinfection of hatchery buildings, equipment, and incubators, particularly an extra strong formaldehyde treatment in the incubators between hatches. Sulfa drugs may help reduce mortality.

The North Carolina Station (1948) showed that three types of Salmonella (paratyphoid organisms) were able to penetrate the shells of incubating turkey eggs after a period of time and suggested, as controls, the destruction, possibly by fumigation, of the organisms on shells of freshly laid eggs and the prevention of dirtying and rejection of dirty eggs in cases where paratyphoid infection exists. Turkey-egg albumen was shown to have little killing power over Salmonella organisms. The paratyphoid organisms apparently do not often get into eggs through the ovary as do the pullorum germs—they get on the egg from the digestive tract and sometimes penetrate the shell as noted above.

Turkeys of all ages are susceptible to paratyphoid but are affected most commonly from 3 or 4 days to one month. If poults are affected beginning the third or fourth days after hatching it is probable that the infection was derived from the eggs or from the incubator at hatching time. The disease runs its course in one to two weeks. If a paratyphoid infection starts two or three weeks after hatching it is probable that the infection was picked up in the brooding quarters or at least not in the hatchery or from the egg. Mortality varies from 5 to 80%.

Some types of *Salmonella* do not leave carriers among the survivors, while others including the important *S. typhimurium* do.

Symptoms: Unthriftiness, weakness, sitting posture or lying on side, legs dry and wrinkled, wings sag and feel cold, and high mortality in poults 3 to 30 days old. Diarrhea may be present. Poults may die overnight or be sick for several days; adults several days.

Post-mortem findings: Duodenum (upper small intestine) inflamed; liver may be congested and have pinpoint yellow spots; kidneys, gall bladder, and heart muscle congested; air sacs with very small cheesy abscesses; heart sac filled with straw-colored fluid, cheesy plugs in ceca. In adults, enteritis with occasional necrotic ulcers; liver and spleen congested and swollen.

Treatment and Control: No remedy of much practical value has been found, but the Washington Station found a fairly high preventive value for sulfamerazine at 0.5% of the mash fed for 24 hours prior to injection of experimental week-old poults with *S. typhimurium* germs. The poults were continued on the medicated mash for 72 hours after the last death occurred. Other recent reports (1954-5) have suggested treatment with furazolidone or water-soluble terramycin 100 p.p.m. for 3 days plus 250 gm. terramycin per ton of feed for 5 days.

The California Station (1951) suggests that, to reduce losses, start at the very first signs of the disease, to give sulfa drugs as follows: Sulfamethazine or sulfamerazine 0.5% in the mash for 4 days, then 0.25% for 3 days; or sulfaquinoxaline 0.03% in the mash for one week only. Since the sulfas are toxic to poults, their use should be cautious and undertaken only *after* a laboratory diagnosis and recommendation. Vaccination of day-old poults is considered futile; bacterins may be dangerous and their value is questionable. Good care will reduce losses. Control involves obtaining stock from disease-free flocks and incubators and eliminating carriers of all types including man, birds, animals and all turkey survivors of attacks. If valuable enough, infected breeding flocks may be tube-tested to spot reactors but this takes special technic and antigen. Recommendations for the control of paratyphoid are (1) to eliminate as breeders all survivors of outbreaks; or (2) to follow a good blood testing program; (3) to keep poults from exposure to animal reservoirs of the disease such as rats, mice, other rodents, reptiles, wild birds, and other animals listed on page 898.

Poisoning (Limberneck, Botulism, Wry-neck)

Occasionally turkeys will have access to spoiled food or the carcases of dead animals in which the deadly poison produced by certain bacteria (*Clostridium botulinum*) is present. The maggots of flies often contain dangerous amounts of this poison. The loss of the use of the neck muscles is usually the first symptom to be observed. This is followed by a sleepy appearance.

Treatment: Remove the cause, give birds protection from weather. Crops can be drained and washed out if time permits, and epsom salts and castor oil given. If birds are very valuable, botulinus antiserum (A. or C.) may be injected intraperitoneally.

Pseudomonas Infection

Pseudomonas aeruginosa, a bacterium, was identified as the cause of a virulent infection causing 80% mortality in White Holland poults 8 to 13 days old (Ontario Station, 1949). Sulfamerazine at 0.75% in the feed may be a satisfactory treatment if it is given early in the course of the disease. It appears impracticable, however, in most cases to obtain a diagnosis in time for treatment to be effective. The poults showed staggering gait and drooping wings with slimy brownish diarrhea. Upon post-mortem there were fluid in the abdominal cavity, gelatinous exudate under the skin, pale liver, swollen kidneys, pale watery bile, and enteritis with pin-point hemorrhages on the intestines. Other investgators have reported dark sirupy uncoagulated blood, small gray to green liver spots, and slightly enlarged spleen. The disease apparently spreads by the birds ingesting the organisms.

Pullorum Disease

Pullorum disease, caused by the bacillus *Salmonella pullorum* although primarily a disease of chickens now is well established in turkeys where it shows the same symptoms and runs the same course as it does in chickens. Likewise the variant or Canadian form of pullorum has appeared in turkeys (Snoeyenbos et al, 1952). Pullorum disease is widely distributed over the world and is likely to be found wherever there is poultry. It has been found in most types of domesticated birds and many wild species. Losses occur through mortality of the young birds, reduced reproductive performance of breeders and layers and retarded growth. In well grown and adult turkeys, pullorum tends to be of a chronic nature commonly centered in the ovary in females from which the infectious agent gains entrance to the developing egg. The North Carolina Station found pullorum in 4.68% of 256 eggs laid by pullorum-infected turkeys. In males the infection frequently is centered in the reproductive organs resulting in impaired sex function. The infection apparently does not spread to an appreciable extent by contact via the feces or otherwise in breeding stock, whereas it does in young poults. It spreads in incubators *via* the down, shells, and excreta, and such spread cannot be prevented by incubator disinfection no matter how efficient. However, formaldehyde disinfection of empty, cleaned incubators between hatches should be practiced.

Symptoms: Droopiness, huddling in warm places, skin on legs dry and wrinkled. The disease usually is acute, with losses in young poults from 10 to 100%. The most susceptible age is from the first week through the third week, although heavy losses have been experienced as late as 6 months of age, and as early as 2 days after hatching. Relapses or subsequent mortality peaks are not uncommon in affected broods or flocks. In less acute cases there often is a pasting up around the vent (white diarrhea) and labored breathing which indicates infection in the lungs.

Post-mortem: A bacteriological examination is necessary for accurate diagnosis. In poults, tiny cheesy abscesses or nodules in the lungs and heart and gizzard muscles are characteristic and help to distinguish pullorum from paratyphoid. The liver may be congested, enlarged, mottled, yellow to bronze in color, and have very small necrotic spots; intestines may have a voluminous mucus content and cecal cores may be found. In adults the ovary may be small and possess hard, discolored, undeveloped yolks; there may be an abdominal dropsy (water belly), the liquid containing cheesy masses.

Control: Pullorum control involves obtaining eggs from disease-free flocks (Fig. 17.4), hatching them in disease-free incubators separately from chicken eggs or infected turkeys, and brooding and rearing the poults away from chickens and contaminated land or equipment. Unless valuable for breeding purposes, infected flocks should be liquidated entirely, no survivors being retained for breeding. Valuable breeding flocks can be freed of pullorum by repeated tests according to rules set up by the National Turkey Improvement Plan and/or the State veterinary authorities. Incubator disinfection is of only minor value in pullorum control.

Blood testing of all breeding stock to remove reactors, males and females, is essential to an effective control program. For this the tube agglutination test is the only test recognized in most states but the rapid-serum and whole-blood plate tests with special turkey antigen and technic have been perfected and have been accepted in a few states for certain types of testing. The pullorum test also detects birds carrying typhoid germs and certain other disease organisms but the test itself does not tell which organism is involved. This is done by laboratory culturing of the infected tissues of the reactors. Turkeys do not need to be mature or in breeding condition to obtain an accurate test, in fact, it is better when they are not in breeding condition. (See National Turkey Improvement Plan, Federal Register of Dec. 9, 1954, Title 9 for information on pullorum control programs. Also see pg. 941.)

Treatment: Sulfa drugs may reduce losses if given at the very first signs of the disease. Use the treatment ment suggested for paratyphoid infections. The sulfas do not eliminate carriers, therefore survivors of an infected treated flock are as dangerous as those from an infected untreated flock. Bacterins are useless and may confuse blood-test results; flushing is useless and dangerous. The use of the blood test to remove reactors affords the most practical control measure (see Fig. 17.4).

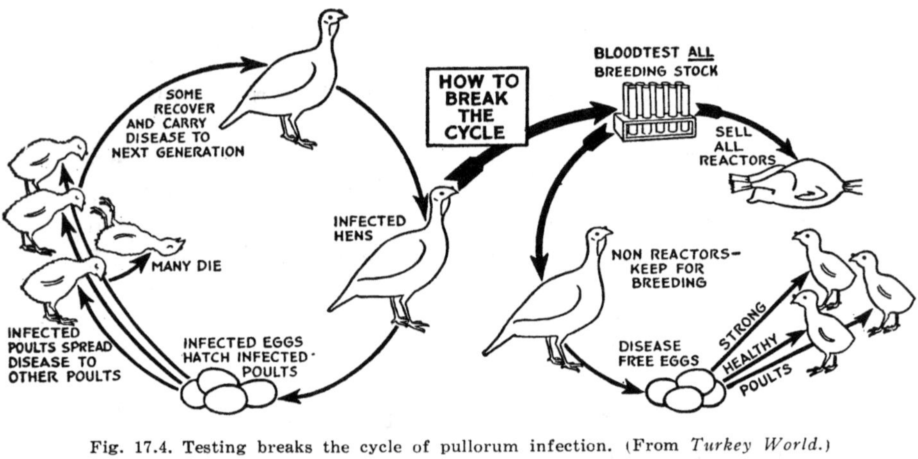

Fig. 17.4. Testing breaks the cycle of pullorum infection. (From *Turkey World.*)

Roundworms

Ascaridia galli, the large roundworm of poultry is commonly found in chickens, less commonly in turkeys, ducks, and geese. Turkeys of all ages may be affected but infestations are more likely to be serious in young birds. *Ascaridia dissimilis*, similar in appearance to *A. galli* but somewhat smaller, has been found in the wild turkey in U. S. and in the domesticated turkey in Cuba.

Symptoms: Unthriftiness, loss of weight, sagging wings, paleness and occasionally diarrhea. However, birds may be fairly heavily infested and yet appear healthy. In most sections of the United States they do not infest turkeys to any serious extent. In some areas they are troublesome. The eggs pass out through the droppings and soon infest other turkeys or chickens, after incubating in soil and litter, without an intermediate host.

Diagnosis and Treatment: Only by a post-mortem examination of the thin, emaciated birds can the presence of roundworm infestation be told. The mature worms are yellowish white, the diameter of a piece of string, and usually from 1 to 5 inches long. The intestinal tract should be slit open from the gizzard to the ceca and the contents emptied in a shallow pan of water. If roundworms (*Ascaridia galli*) are present they will be found free in the water, whereas tapeworms cling to the lining of the intestinal tract. Many roundworms in each turkey indicate a serious infestation, but a few worms in each bird should be considered no serious threat.

The Kansas Station has shown that turkeys are more resistant than chickens to roundworms. A sanitation program which avoids blackhead usually eliminates roundworm infestation. However, if roundworms are still troublesome, 2 to 4% of tobacco dust (1½% nicotine) in the mash, fed continuously for 3 or 4 weeks is considered to be an effective control. The use of worm remedies without a competent post-mortem examination to prove the presence of a serious worm infestation is to be discouraged. Routine "worming" is useless and expensive.

Sinusitis*

Infectious sinusitis is believed to be caused by a pleuropneumonia-like germ (the smallest forms of bacteria, having some features closely related to larger

*More recently referred to as "PPLO Infection" meaning pleuro pneumonia-like organisms.

viruses and rickettsia). There appear to be two forms of the disease, one involving the sinuses around the nasal passages (swellhead) and the other the lungs and air sacs. The disease appears to be egg transmitted and to be introduced also by infected individuals and contaminated equipment.

Control: Its spread may occur by contact and by the drinking water, feed sacks, crates, and clothing of attendants. Complete disposal of infected flocks followed by a thorough cleanup and resting of equipment should eliminate it from the premises. Prevention of contact between infected breeding stock and young poults is essential. Turkeys of all ages from one week to breeding age are susceptible. The Minnesota, Utah, and Connecticut Stations believe that circumstantial evidence accumulated over a period of years indicates strongly that the disease must be carried on the surface or in the egg. The Washington Station (1950) states that transmission by hatching eggs could occur, but that lice and droppings did not transmit it. The time elapsing between exposure and appearance has varied from 2 to 17 weeks, with three-fourths of the birds becoming infected during the 4th, 5th, and 6th weeks after first exposure. The place of inoculation seemed to determine the type of infection since sinus exudate injected into the trachea developed the lower respiratory or air-sac form of sinusitis. Although the sinus form frequently is noticed first, the air-sac form sometimes occurs first and sometimes there is no sinus involvement at all. Likewise there often is no lower respiratory trouble in connection with the sinus infection. Usually, however, there is some of each type.

The Texas Station (1952) reported the natural transmission of turkey sinusitis to chickens but not vice versa as "air sac" indicating that sinusitis of turkeys and "air-sac" or chronic respiratory disease of chickens are caused by the same organism. A blood-test may be feasible.

Symptoms: The swelled sinus (one or both sides of the face) is easily recognized. It is soft at first but may become hardened later. It is not certain however that the hardening, or caseation, is due to the sinusitis organism or to a secondary invader. The eye may be closed or may even be destroyed. Coughing, sneezing, and a thin nasal discharge usually occur preceding the appearance of the swelled sinus. In the lower respiratory form of the disease which is commonly called "air-sac," the early symptoms are coughing, sneezing, and gasping, especially when birds are excited. A bird held head down or forced to exercise violently may choke to death. Nervous symptoms rarely are seen but they do occur; likewise multiple abscesses and edema of face and neck.

Post-mortem: In the early stages of sinus infection the exudate is colorless or nearly so, and of the consistency of a thin jelly. Later it may or may not become caseated, hard, and yellowish white. The lower respiratory form of the disease may be indicated by spots on the lungs or more commonly by cloudiness or thickening of the air sacs with more or less caseation.

Diagnosis: While the sinus type of infection is readily recognized, the same cannot be said of the air-sac type when it occurs without the sinus symptoms. Only laboratory studies or transmission trials will show whether the trouble is due to sinusitis, aspergillus, or other causes. However, most air-sac trouble probably is due to the sinusitis organism according to the Minnesota Station (1952).

Treatment: For the sinus form, the needle of a hypodermic syringe should be inserted through the skin into the swollen sinus. The plunger of the syringe is then withdrawn, thus drawing the liquid from the swollen part through the needle (18 or 19 gauge 1½"). With another clean syringe (5 or 10 cc. size) about one-fourth teaspoonful (1cc.) of freshly prepared 4% silver nitrate solution (15% argyrol solution is almost as good) is inserted changing syringe barrels while leaving the needle inserted. This and other treatments are effective only in the early stages, before the material in the swollen sinus becomes hardened, according to the California Station (1951). It is effective and cheap except for labor costs.

Streptomycin is effective against the sinus form but is rather expensive. The Ontario Station (1950) reported 100% success from injecting turkeys 12 to 24 weeks of age with 150 mg. of streptomycin per swollen sinus of 300 market age turkeys and reported 90% recoveries in 5 to 7 days. The remaining 10% were re-treated and nearly all recovered. The cost was 10 to 25c per bird for the treatment. It was recommended that the exudate be first withdrawn from the sinus to be sure that the treatment went into the sinus and not into the surrounding tissues.

The Virginia Station (1953) found that 3.2 to 25 mg. terramycin in mineral oil in one dose injected into each swollen sinus without first withdrawing the exudate was highly effective. Some indication also was given that 150 to 250 mg. of streptomycin injected into the throat-wattle (dewlap) was effective against the air-sac and lung forms of sinusitis.

The California Station (1955) recommends injecting each sinus, even if one is not swollen, with about 100 mg. of streptomycin dissolved in ½ to 1 cc. of water or normal saline solution. Treat again birds having a swollen sinus after 7 to 10 days. For the air-sac form of sinusitis, inject 100 mg. of streptomycin per 5 pounds of body weight into the turkey's dewlap.

The North Dakota Station (1950) recommended a combination treatment of 1 to 2 milliliters (cc.) of 2% silver nitrate solution or 1% silver picrate injected into each drained sinus at the same time implanting subcutaneously (under the skin) at the rear of the head, a pellet containing 10,000 units of procaine penicillin G and 17 mg. of dihydrostreptomycin (equal to 125 mg. of strepto. base). The Utah Veterinary Laboratory reported that the implantation of 1, 2, or 4 pellets (each containing 5,000 units of procaine penicillin and 6.25 mg. of dihydrostreptomycin base) in 20-week-old turkeys was highly effective in reducing mortality from sinusitis and increasing the market quality of the turkeys. The dosage at each level was effective.

Recently the Texas Station (1952) reported that the antibiotic chloromycetin is highly effective against both sinus and air-sac forms in turkeys 6 to 24 weeks old when given at a level of 0.25% in the all-mash feed for 8 to 14 days. Recovered birds appear to act as carriers.

Air-sac infection in poults a few days old is, according to grower reports, amenable to treatment with soluble streptomycin in the drinking water plus extra A and D oil in the mash. One tablespoonful per gallon of drinking water of a stock solution of 50 gms. of soluble streptomycin per gallon of pure water has been used successfully. Other recommendations for "air sac" are 200 to 300 grams of terramycin or aureomycin per ton of feed for 1 to 3 weeks.

Mechanical sinusitis may be caused by foreign bodies such as barley awns (beards) and hulls of coarsely ground oats and barley. For this reason oats and barley should be fed whole or else finely ground, never cracked, rolled or coarsely ground, unless soaked before feeding; and bearded barley stubble avoided.

Sinusitis caused by vitamin-A deficiency may occur in poorly fed flocks and often cannot be told from the other types. Good feeding eliminates this possibility.

Spirochetosis

This blood disease was first reported in the United States by Hoffman et al (1946) in a flock of turkey breeder hens in California. Symptoms were sleepiness, thirst, high fever, loss of appetite and weight, yellow-green diarrhea, darkened head, hopping or stilted gait and occasionally paralysis. The spirochete *Borrelia anserina* is the cause.

On autopsy a mottled spleen due to hemorrhages was fairly characteristic; there also may be an enlarged heart, enlarged congested liver with necrotic spots very small to 0.4 inch diameter, enlarged pale kidneys, anemic intestines with catarrhal enteritis and bile-stained contents. The finding of spirochetes in stained blood smears is necessary for diagnosis.

Spread is by bluebugs, red mites, mosquitoes, lice; controlling these pests helps prevent the disease. A Brazilian vaccine for prevention has been reported. Penicillin 10,000 units for hens, 15,000 units intramuscularly per mature turkey is reported as an effective treatment. If given when symptoms first appear, it seems to run its course in about 8 weeks. Survivors may be carriers, hence complete liquidation of an infected flock would be indicated. Spread also may be direct via the droppings.

Synovitis

This malady is sometimes referred to as bursitis, staphylococcosis, staphylococcic arthritis, osteoarthritis or suppurative ostitis. A type of arthritis caused by the pus forming bacteria *Micrococcus pyogenes* var. aureus (formerly *Staphylococcus aureus*) and *M. pyogenes* var. citreus (formerly *Staphylococcus citreus*) was first described by Jungherr in 1933 and more recently by others. It is more prevalent in large tom turkeys but sometimes it is found in hens and in small-type turkeys, 10 weeks old to market age since it may be either acute or chronic. It seems to be more prevalent in the West. The swollen hocks, toes, and feet in chronic cases may be mistaken for slipped tendon, bumblefoot, or nutritional hock disease.

The birds may rest on the swollen hock joints and move with difficulty. The swellings are soft and hot and appear to be painful, and often are accompanied by breast blister from which the causative organism may be recovered. Acute cases may show little swelling in the joints but are likely to have watery sulphur-colored droppings. Liver may be enlarged, darkened, or have greenish streaks, less often, whitish necrotic areas; intestines may be congested with contents yellowish and foul smelling. The tendons may be destroyed and the bone injured.

Staphylococcus germs are widely distributed but it is not known how they gain entrance to the turkeys' bloodstream which they must do to cause disease. It has been suspected but not proved that entrance is through injuries to feet, legs, or keel bone, or possibly via the intestines if conditions are favorable. Mosquitoes and thistles have been suspected of being involved in the spread of this disease. Turkeys raised on wire often are affected, particularly if the wire is rough.

Prevention, Control, and Treatment: Prevention is indicated; treatment suggested by Fahey (Univ. of Toronto) inject terramycin into sick birds and add it to the feed, 200 grams per ton. Careful handling to avoid injuries, along with good nutrition and sanitation are preventive measures. Confinement pens should have smooth floors; it may be necessary to use part littered floors for the toms or grow them on range. Slats 2½ to 3″ wide and 1½ inches apart have been suggested in place of the narrower kind. No roosts, or very low ones of a suitable type, help to avoid injuries and the infections that sometimes follow.

Affected birds should be isolated and placed on littered floors or clean range. The worst ones might as well be killed. Never open the swelled joints which contain a yellowish semi-gelatinous or cheeselike flaky exudate.

Tapeworms

Tapeworms, or cestodes, are segmented parasitic flatworms that are attached to the wall of the small intestine of the host by the head from which growth emanates so that the segments farthest from the head are the oldest and usually contain ripe eggs. These ripe segments break off and are passed in the droppings for the continuation of the life process of the worm. So far as is known all tapeworms have an indirect life history and live part of their lives in an intermediate host, usually a small low form of life such as an insect, mollusc (snail or slug), or annelid worm such as the earthworm. To be infective to another turkey, the tapeworm egg in the ripe segments in the droppings must first be ingested by the intermediate host in which the eggs hatch and in a few weeks develop into small cysts which are infective to turkeys when they eat the intermediate host; thus completing the life cycle.

Turkeys are known to be hosts to 9 or 10 different species of tapeworms varying in size from those so small they can hardly be seen to those around 10 inches long. Most of these species are common to turkeys and chickens. Young turkeys 3 to 12 weeks old are most likely to be affected; as the birds grow older they become resistant to new infestations.

As with other worm parasites, sanitation is the key to the control of tapeworms. Routine "worming" for tapeworms especially is not desirable, because no really good tapeworm expeller is available at present.

Symptoms: Unthriftiness, paleness, loss of weight, appetite irregular, occasionally diarrhea. Tapeworms are a serious enemy of turkeys, and a threat practically everywhere turkeys are raised. However they seldom cause serious trouble in well-managed flocks.

Diagnosis and Treatment: A post-mortem examination is necessary to determine the presence of tapeworms. When the intestines are slit their entire length and placed in a shallow pan of water the flat, whitish, segmented worms may be found clinging to the lining of the intestinal tract. One species *Amoebataenia sphenoides* is so small (.08 to .14 inch long) that it scarcely can be seen with the naked eye. Common species in turkeys are: *Raillietina cesticillus,* .05 to .10″ by 5″; *Metroliasthes lucida,* narrow, up to 8″ long; *Choanotaenia infundibulum,* .125 by 2 to 9″; *Hymenolepis carioca,* .02 to .08″ by 1 to 3″; *Raillietina echinobothrida* and *R. tetragona,* up to 10″ long. *Raillietina williamsi,* up to 15 inches long and *R. ransomi* measuring up to 0.56 in. appear to be common parasites of the wild turkey and probably could also affect the domesticated variety. The tapeworm clings to the intestinal wall with its scolex, or head. No vermifuge has been found that will cause this scolex to release its hold. At present, the tapeworm remedies on the market merely expel segments. Other segments are quickly formed and the infestation continues the same as before treatment. Hexachlorophene however (p. 914) is believed to be fairly effective in removing the heads as well as the bodies of *R. cesticillus* from chickens.

The eggs of the tapeworms which infest turkeys are picked up by flies, slugs, earthworms, grasshoppers, ants, snails, and beetles, the tapeworm passing through a necessary part of its life cycle within an intermediate host.

Control: Control is based on preventing the turkeys from eating the intermediate hosts. The sanitation programs suggested in various places in this work will aid materially in this respect. Other precautions are to eliminate breeding places for flies and beetles in the vicinity of the turkey range. Dung beetles usually breed in manure and are not in themselves harmful to turkeys if they do not happen to be carriers of tapeworms. DDT will aid in insect control (See Chap. 15). Droppings that are inaccessible under the shelters may be sprayed with crankcase drainings or sprinkled with slaked lime or borax weekly to kill the fly larvae (maggots). Population of houseflies, beetles, ants, roaches, and other pests should be controlled or the turkeys kept a goodly distance from them.

Remedies: Kamala is no longer regarded as efficacious and in addition is a dangerous drug to feed to turkeys. Potassium dichromate has been used with fair results by the Manitoba Station. Prepare a saturated solution of the chemical then use a pint of the orange-colored stock solution in 2 gallons of drinking water in wooden or granite-wear containers. Pen and starve the turkeys for 12 to 14 hours (no feed or water) then allow them to drink the medicated water for 2 or 3 hours, restoring the usual feed and water thereafter. After a day or two the pen cleanings should be disposed of so as to destroy the ejected worms. Hexachlorophene has been reported effective in removing tapeworms from chickens at doses of 25-50 mg. per kilogram of body weight.

Ticks (Bluebugs)

The fowl tick *Argas persicus* is found mostly in the southern tier of states. It feeds on its host intermittently, mostly at night, and lays its eggs in sheltered places in buildings and the bark of trees. The tiny baby ticks at once look for a host upon which they live and feed for several days. At any stage, ticks can go without food for months and the adults for up to two years. Ticks often transmit parasites of other diseases notably spirochetosis and tularemia.

Symptoms: Loss of vitality and blood in the host bird. Ticks are arachnids (not insects) and have habits similar to those of their cousins the mites. The fowl tick is dark brown (bluish-red when full of blood) flat, and egg-shaped. Females are 1/4 by 7/16 inches in size, the males much smaller.

Control: Spray turkey buildings, wooden fence posts and nearby trees thoroughly with diesel oil, an anthracine oil such as "carbolineum"; or lindane house spray 0.3 to 0.5% emulsion. Remove all loose boards that may serve as hiding places. After the general spraying it may be necessary to paint the under sides of the roosts and the roost supports with carbolineum several times at 3- or 4-week intervals to destroy ticks attempting to reach the birds at night. Ticks are tough—the effectiveness of treatment for them never should be taken for granted.

Trichomoniasis

The only disease caused by a trichomonad that is generally recognized by pathologists is **trichomoniasis of the upper digestive tract** caused by a protozoan parasite identified as *Trichomonas gallinae*. The crop and sometimes the esophagus, and proventriculus of turkeys 16 to 30 weeks of age are ulcerated. The crops of affected birds are sunken as are their faces, and the birds assume an upright posture. There is a foul odor from the crop which does not occur in capillariasis. The crop ulcers are described by Hinshaw and Rosenwald (1951) as 1/8 to 3/5 inch in diameter at the base and may be as much as 1/5 inch high and extend about 1/8 inch into the tissue. They have a rough irregular surface and there is a ring of hemorrhagic tissue around them. Ulcers in the esophagus are like those in the crop but are usually much smaller. In the proventriculus the ulcers often run together until they form an almost solid ring of thickened, degenerated tissue. These lesions interfere with nutrition. Infection may spread through stagnant pools, old straw stacks, and generally moist, insanitary conditions. Houseflies probably aid in the spread. Well-drained land, sanitation, and proper range rotation should eliminate this disease as they do other soil-borne infections. Chickens and pigeons, doves, and other wild birds may be affected by trichomonads, and this suggests the importance of segregating these species.

Treatment: Copper sulfate 1:2000 solution given in place of regular drinking water for 2 or 3 days to a week, then repeated after a few days. Isolate obviously sick birds and care for them separately, preferably on wire or slat floors.

Another trichomonad, *Trichomonas gallinarium* often is found in the intestines, ceca, and sometimes the liver. Missouri investigators however believe that *Tr. gallinarum* causes a serious disease in turkeys 3 weeks to 18 weeks old, with greatest mortality at 10 to 12 weeks, lesions consisting of slightly inflamed lower intestines and ceca (with foamy brown droppings) and, rarely, spotted livers. There also is a marked vitamin A depletion. U.S.D.A. investigators' findings (F.B. 1652, 1949) tend to support the Missouri work and indicate that the Missouri disease and that investigated at Beltsville are one and the same and are caused by *Tr. gallinarium*. The disease is called *trichomoniasis of the lower digestive tract*. At Beltsville, turkeys 3 months to 18 months old were found to be susceptible thus extending the age range of this disease from 3 weeks to 18 months. In birds older than about 18 weeks, the trichomonads apparently cause more severe cecal lesions and invade the liver, causing spots superficially resembling those of blackhead (histomoniasis). Turkeys thus affected have liquid droppings, light yellow to orange or greenish-yellow in color. A fever up to 109° F. is frequently observed. On autopsy the liver spots are cream-colored, irregular-shaped, firm and granular, level with the liver surface or elevated slightly above it, and vary from pin-point to about an inch in diameter. This is in contrast to the typical description of blackhead liver spots as more or less circular and somewhat depressed below the liver surface. However, lesions of both types frequently occur in the same livers indicating a mixed infection. Cecal lesions frequently are similar in blackhead and trichomoniasis of older poults. Investigators opposed to the "trichomonas theory" consider that all of these cecal ulcers and liver spots are caused by the blackhead organism and the intestinal disturbances in young poults ascribed to *Trichomonas gallinarium* probably are caused by Hexamita. This situation of course is highly confusing. Fortunately, preventive methods are the same for all three diseases involved (hexamitiasis, trichomoniasis, and histomoniasis). Treatments however differ and the turkey grower should accept his veterinarian's or laboratory's diagnosis and suggestions for treatment.

Prevention and Control: This disease, like blackhead, can be prevented by sanitary practices involving separation from chickens and adult turkeys, range rotation, well drained land, contamination-proof feeders, clean water supply, fly-control, wire floors, slatted floors, or use of the clean built-up litter plan. As with blackhead, well-conducted confinement rearing usually eliminates the disease in turkeys over 3 months old. In range rearing, the use of well-drained land and separation from chickens and adult turkeys are the most important. In outbreaks, prompt isolation of sick birds is mandatory; removal of well birds to a clean environment is very desirable. Complete marketing of all birds exposed to the infection is a most important control measure, but where this is not possible, complete segregation of young and old stock will do much to prevent re-infection.

Treatment: Artificially induced fever cured 80% of infected birds at the Beltsville Station by 3 to 6 treatments; one every other day, of 1 to 2 hours each until internal temperature (thermometer in the vent) reached 110° to 112° F. The relative humidity of the chamber should be about 50%. This causes a rise in body temperature from a normal of 106½ to 111 or 112° F. A chamber similar to a forced-draft incubator would supply proper condition for treatment. Birds having histomoniasis (true blackhead) may be killed by the heat treatment. To be certain which type of infection the sick birds had, an incision was made in the side of the bird just back of the last rib and a strong light thrown onto the liver. Afterwards the skin is sewed up with no harm to the bird.

No treatment for this disease in turkeys over about 3 months old other than heat is known. Isolation of sick birds to comfortable, wire floored quarters with access to clean water and good feed will result in a fair percentage of natural recoveries. The effect of ANT has not been reported.

As a treatment for young turkeys (under 3 months) the Missouri Station reports fairly satisfactory flock treatment with 2% of baker's yeast or other live yeast added to 100 pounds of ground oats. This mixture is moistened to form a paste and allowed to ferment in a warm place for 12 to 15 hours after which time a pint of fish oil (3000A) is added and mixed thoroughly. The affected poults are fed all of this mash they will consume in 20 to 30 minutes, three or four times a day for a week or 10 days; as a supplement of the growing mash. For advanced cases, gentian violet in enteric coated tablets in doses of 1/16 grain for birds of one pound to 1¼ grains to birds of 12 lbs. is suggested in addition to the fermented mash, three doses on alternate days. Details of the treatment can be obtained from the University of Missouri, School of Veterinary Science.

Tuberculosis

Symptoms: In some cases the birds become lame and emaciated, the head becomes pale and diarrhea frequently is present. Such birds seem quite hungry and the disease is chronic. However, there are cases of tuberculosis in turkeys in which the birds remain in good flesh, and the disease is not even suspected in the flock until a bird is posted for some reason or lesions are observed in a bird slaughtered for the family table. Occasionally turkeys as young as 9 months are affected.

Post-mortem: Encapsulated lesions are usually found on the liver, spleen, ovaries, lungs, kidneys, intestines, or in the bone marrow. The lesions are raised growths in the form of nodules and are caused by *Mycobacterium tuberculosis avium.* These nodules are usually yellowish white, hard, and gritty and can be lifted with the point of a knife.

Treatment and Control: As with tuberculosis in other farm animals there is no cure. Diseased birds should be killed and burned or buried deeply. Suspicious birds in good flesh may be sold subject to condemnation, the carcass being safe for human consumption if no lesions are found. The tuberculin test may be applied to a suspected flock. Diseased birds may be located, with a high degree of accuracy, by the reaction to "tuberculin." Hinshaw recommends the edge of the web of the wing as the site of inoculation. The inoculation should be made and the subsequent swelling read by a veterinarian or other person trained in reading such reactions.

Spread of avian tuberculosis from diseased chickens to turkeys accounts for a large share of the outbreaks in turkeys. The separation of turkeys from all other poultry, mature stock as well as poults, will aid in the prevention of tuberculosis. Separation combined with sanitation practices should reduce loss from tuberculosis to a negligible quantity. It is seldom necessary to sell the entire flock and make a fresh start as is usually the case with chickens, although this procedure is desirable if the adult flock represents no special breeding or selection.

Typhoid

Fowl typhoid, caused by the bacterium *Salmonella* (formerly *Shigella*) *gallinarum*, regularly causes positive reactions to the pullorum test and is believed, like pullorum disease, to be transmissable through the egg although egg transmission apparently is not common. Chickens are the usual source of infection, but the disease may be spread by dogs and cats, from adult turkeys to young ones, and from poult to poult. Survivors of outbreaks should not be retained as breeders.

Symptoms: Listlessness, thirst, greenish-yellow diarrhea, loss of appetite, ruffled feathers, sagging wings; high body temperature, up to 112° F., until just before death when it may fall to 103°. Typhoid may be acute like cholera but is more likely to run its course in 4 to 10 days; birds huddle; head pale, held low or over back. Turkeys of all ages are susceptible; usual age is 20 weeks or over but birds 6 to 14 weeks often are affected. Typhoid is likely to recur one or more times, especially if birds are left on the same premises and/or have continued contact with chickens or older-turkey carriers. Losses are most severe in the first attack and in younger turkeys. Losses in four California outbreaks averaged 26% in well-grown turkeys. They are likely to run higher in younger birds.

Post-mortem: Mahogany-streaked, enlarged liver, sometimes greenish in color with tiny grayish spots; enlarged mottled spleen, parboiled-appearing firm lungs; heart swollen with small grayish areas, pin-point hemorrhages, or both, hemorrhagic enteritis of the duodenum, breast muscles congested with blood. The isolation of the causative germ, *Salmonella gallinarum* from blood or viscera is the only certain method of diagnosis. In young poults the liver is crumbly and mottled whitish and red with empty digestive tract.

Treatment and Control: The agglutination test for pullorum disease (both the rapid and tube methods) locates carriers of typhoid with a high degree of accuracy. The breeding stock can thus be freed of reactors. However, if a selected and tested breeding flock should suffer mortality from typhoid the Texas Station (1951) reported that the disease could be controlled successfully with sulfaquinoxaline. It was suggested that the drug be given in the drinking water at 0.04% (1:2500) for 2 or 3 days, then reduced to a level of 1:4000 and kept on it continuously until marketed. It was recommended that actively infected flocks be treated only long enough to find a market for the healthy birds. Egg production and hatchability were not affected by the treatment which however did not prevent the spread of the disease in the breeding flock or its transmission via the eggs laid.

In young turkeys the Texas Station was able to overcome mortality from typhoid in one week by feeding sulfaquinoxaline in the mash at 0.0125% (1:8000) and in the drinking water also, at 0.125%. The Illinois Station (1950) reported good results in treating typhoid-infected 12-week-old turkeys by feeding for 7 days a mash containing 0.5% sulfamerazine.

Sodium sulfathiazole given intermittently 3 days on, 2 days off, 1 day on, and 3 days off in the drinking water also has been recommended. If on range, they should be moved when each change in treatment is made; removing the sick ones to isolation quarters. Sick birds remain on the roosts later in the mornings than do the well birds and can be spotted then and removed. The available evidence regarding the use of vaccine is contradictory, but apparently a majority of pathologists feel that it is of little avail. A new treatment devised by the Texas Station (1954) is with furazolidone (N.F. 180) at 0.01% in the feed for 7-10 days, then move to clean ground and reduce to 0.005% for another 7-10 days. Repeat if birds get sick again. In large flocks not movable, the low level may be fed continuously. Birds too sick to eat may be given a 100-mg. capsule; if improved, another one the next day. Since turkeys usually become infected from chickens, the two species should not be mixed nor allowed to run on the same ground until it has been rested for 3 full years.

Vent Gleet

The literature apparently contains no reports of vent gleet in turkeys but a field report from Virginia indicates that it does occur in breeder flocks and that it can be treated successfully by swabbing with 3% chromic acid solution, every third day until recovery. However if the trouble is found in a breeder flock it probably would be preferable to inspect every bird carefully and remove all affected ones which should be treated until cured, then marketed.

Water Belly

This disorder, due to mineral imbalance, is discussed in Chap. 5.

REFERENCES

(This list is by no means complete, but contains salient references that the reader may wish to pursue further.)

Biester, H. E., and L. H. Schwarte, 1952. Diseases of Poultry (3rd Ed.). The Iowa State College Press, Ames, Iowa, (A 1,514-page book consisting of contributions by 32 scientists; edited by Biester and Schwarte).

Boney, W. A., Jr., 1951. A study of sulfaquinoxaline and sulfamethazine in fowl typhoid among breeder turkeys. Jour. Vet. Res. 12(43): 134-139.

Clark, C. H., 1946. Sulfamerazine in paratyphoid disease of poults and chicks. J. Am. Vet. Med. Assn. 109(835):279.

Dykstra, R. R., 1949. Animal Sanitation and Disease Control. Interstate, Danville, Ill. (808 pages illus.)

Eveleth, D. F., et al, 1948. Fowl cholera in North Dakota, N. D. Bi-monthly Bull. 11(2):36-38.

Francis, D. W., 1952. Effects of aureomycin on turkey sinusitis. 1952 Delaware Poultry Handbook 7:45-47.

Grey, C. G., 1947. Streptomycin in the treatment of *Erysipelothrix rhusiopathiae*-infected turkeys. Vet. Med. 42(6):216.

Grumbles, L. C. and W. A. Boney, Jr., 1950. Infectious sinusitis of turkeys. Proc. U. S. Livestock San. Assn. 54th Mtg. (Nov.): 166-170.

Grumbles, L. C. and W. A. Boney, Jr., 1950. Observations on the etiology and treatment of infectious sinusitis of turkeys. No. Amer. Vet. 31:523-529.

Grumbles, L. C. and W. A. Boney, Jr., 1951. Chloromycetin and terramycin for infectious sinusitis of turkeys. Jour. Amer. Vet. Med. Assn. 119(896):384-385.

Hall, Walter J., and E. E. Wehr, 1949. Diseases and parasites of poultry. USDA Farmers' Bull. 1652 (96 pages illus.).

Hinshaw, W. R. and A. S. Rosenwald, 1951. Turkey diseases. Calif. Exp. Sta. Manual No. 3, Berkeley, Calif.

Jaquette, D. S., and S. J. Marsden, 1947. Tests with mapharsen and metachloridine in the treatment of blackhead in turkeys. Poul. Sci. 26(2):218-219.

Jungherr, E., and R. Gifford, 1944. Three hitherto unreported turkey diseases in Connecticut. Cornell Veterinarian 34:214-26. (Conn. Sta.)

Marsden, S. J., and H. R. Bird, 1947. Effects of DDT on growing turkeys. Poul. Sci. 26(1):3-6.

McNeil, E., and W. R. Hinshaw, 1948. The effect of streptomycin on fowl cholera in turkeys. Cornell Veterinarian 38(3):239-246.

Moore, E. N., 1951. Disease now has Masters. Turkey World 26(3): 16-17, 101-104.

Peterson, E. H., and T. A. Hymas, 1951. Antibiotics in the treatment of an unfamiliar turkey disease (mud-fever). Poul. Sci. 30(3):466-468.

Prier, James E., 1953. Turkey diseases (1st ed.) Interstate, Danville, Ill. (150 pages illus.)

Stiles, Geo. W., 1946. Swine erysipelas in turkeys. Jour. Amer. Vet. Med. Assn. 109:65-69.

APPENDIX

SECTION 1

Table A1—U. S. Census Reports of Turkeys, 1890 to 1950[5]

State	1890[1] Turkeys on hand June 1		1900[2] Turkeys on hand June 1		1910[2] Turkeys on hand April 15		1920[3] Turkeys on hand January 1		1930[4] Turkeys raised in 1929		1940[4] Turkeys raised in 1939		1950[4] Turkeys raised in 1949	
	Rank	Number	Rank	Number	Rank	Number	Rank	Number	Rank	Number	Rank	Number	Rank	Number
Alabama	21	177,681	20	129,326	17	84,487	14	92,733	28	142,894	37	74,204	34	143,255
Arizona	46	2,744	42	6,043	42	5,498	37	13,937	36	83,818	42	54,256	43	72,631
Arkansas	26	118,816	19	140,661	18	83,550	22	73,368	41	55,635	40	68,376	29	264,445
California	13	287,799	17	158,356	13	116,602	4	174,708	4	1,246,993	2	2,593,406	1	7,005,334
Colorado	34	20,872	33	30,781	31	26,430	25	57,687	7	547,789	11	901,502	13	719,616
Connecticut	33	30,176	40	7,717	43	3,911	45	3,756	46	24,433	38	72,598	33	204,616
Delaware	29	70,578	36	19,045	38	8,181	40	7,977	40	60,714	35	92,654	38	103,903
Dist. Columbia	49	215	49	46	49	29	49	57	49	18	49	23	49	6
Florida	31	34,426	31	32,869	32	22,871	35	18,158	38	68,689	36	79,800	39	102,418
Georgia	25	148,797	26	103,416	20	79,634	20	76,892	34	85,731	39	69,124	30	240,052
Idaho	40	6,433	39	10,211	35	14,274	33	25,982	9	516,976	28	196,615	32	207,207
Illinois	1	1,043,947	3	446,020	4	189,418	10	107,876	32	91,487	18	407,592	19	612,974
Indiana	9	505,111	6	345,379	3	202,977	11	105,713	29	129,742	23	296,016	12	779,109
Iowa	2	940,849	4	424,306	10	124,164	19	78,126	31	111,981	5	1,436,256	3	2,303,432
Kansas	7	530,397	8	275,330	15	104,421	9	119,228	13	319,460	9	1,190,633	21	445,455
Kentucky	4	672,106	7	279,749	5	188,292	5	168,326	12	383,138	25	217,342	31	236,967
Louisiana	27	74,680	24	115,921	27	53,251	31	36,386	44	27,430	45	41,735	45	59,976
Maine	36	15,259	41	6,437	44	2,948	46	3,531	45	25,841	47	35,816	44	61,112
Maryland	14	278,522	27	101,782	26	60,260	27	55,323	14	278,579	21	340,442	25	341,774
Massachusetts	42	5,805	47	3,018	46	2,645	43	4,808	39	61,396	27	197,916	23	386,769
Michigan	20	185,847	13	191,863	19	81,263	21	74,780	19	229,640	19	395,641	16	674,370
Minnesota	23	151,459	12	193,143	8	147,335	7	142,896	3	1,306,058	3	2,508,830	2	3,435,444
Mississippi	19	194,398	15	189,698	16	92,448	18	80,236	35	85,010	41	67,538	42	78,950
Missouri	3	928,751	2	466,665	2	311,575	2	248,057	17	245,147	8	1,277,433	9	1,238,821
Montana	43	5,077	37	12,637	33	16,475	32	29,432	11	442,259	24	224,042	37	104,216
Nebraska	15	218,636	23	118,892	29	32,364	28	48,345	16	250,000	12	853,425	15	698,427
Nevada	45	4,193	46	3,618	45	2,839	44	4,569	33	89,573	43	47,039	48	23,749

Table A1—(Continued)

State	1890[1] Turkeys on hand June 1		1900[2] Turkeys on hand June 1		1910[2] Turkeys on hand April 15		1920[3] Turkeys on hand January 1		1930[4] Turkeys raised in 1929		1940[4] Turkeys raised in 1939		1950[4] Turkeys raised in 1949	
	Rank	Number	Rank	Number	Rank	Number	Rank	Number	Rank	Number	Rank	Number	Rank	Number
New Hampshire	38	10,207	48	2,386	47	1,949	47	2,363	47	19,460	46	38,515	40	97,134
New Jersey	22	162,270	32	32,378	37	10,647	41	7,925	42	31,609	32	129,045	27	289,461
New Mexico	48	928	44	3,805	40	7,308	38	13,771	30	124,361	44	46,622	46	46,768
New York	12	402,642	14	190,879	14	104,957	26	57,644	26	166,721	22	310,312	17	671,240
North Carolina	18	197,420	21	120,737	22	70,009	12	96,770	21	205,300	29	169,663	22	387,415
North Dakota	32	33,928	29	39,073	25	61,621	8	139,179	2	1,457,930	7	1,408,652	26	329,568
Ohio	8	521,171	5	362,924	6	162,664	13	93,928	23	177,322	17	601,659	10	1,142,102
Oklahoma	41	5,931	16	178,959	11	120,324	3	192,721	5	804,262	6	1,417,586	24	374,666
Oregon	44	43,555	30	36,031	30	26,689	30	38,373	6	600,359	4	1,677,851	6	1,790,413
Pennsylvania	6	535,828	9	259,824	9	136,942	16	87,404	25	175,089	15	606,830	8	1,478,717
Rhode Island	37	11,656	43	4,604	48	1,109	48	1,310	48	12,784	48	15,563	47	39,497
South Carolina	24	149,126	22	120,140	24	64,396	23	73,340	37	80,235	34	107,371	18	639,608
South Dakota	30	60,163	28	53,740	28	51,992	17	83,866	10	460,106	10	1,120,206	28	275,229
Tennessee	11	430,333	11	193,397	12	118,514	15	87,448	27	156,470	31	135,969	35	137,970
Texas	5	535,916	1	648,671	1	363,666	1	478,201	1	3,782,912	1	3,443,951	4	2,141,419
Utah	39	9,220	38	10,649	39	7,996	36	14,896	20	228,483	13	812,078	5	2,046,778
Vermont	28	72,164	35	22,689	36	11,693	42	5,514	43	28,549	33	121,361	36	105,226
Virginia	10	477,414	10	207,675	7	148,386	6	159,687	8	527,715	14	705,970	7	1,608,150
Washington	35	17,187	34	29,155	34	16,049	34	20,621	15	251,713	16	604,857	11	885,385
West Virginia	16	214,756	25	105,265	21	72,752	24	61,800	22	182,522	30	169,137	20	610,598
Wisconsin	17	206,230	18	155,121	23	65,020	29	44,820	24	177,116	20	345,322	14	704,305
Wyoming	47	2,441	45	3,664	41	5,883	39	12,560	18	233,016	26	204,982	41	86,847
Total U.S.	....	10,754,060	....	6,594,695	....	3,688,708	....	3,627,028	....	16,794,485	....	27,933,756	....	36,434,218

[1]Turkeys of all ages. Probably 65 to 70% of these reached the fall market.
[2]"Adult" turkeys; over 4 mo. of age, nearly all breeders.
[3]All turkeys; about half estimated to be breeding birds.
[4]When compared with AMS estimates of turkeys raised, census figures are smaller; the AMS figures are considered to be more accurate.
[5]From 1950 Census of Agriculture. General Report Livestock and Livestock Products, Vol. II Chap. VI. Superintendents of Documents, G. P. O., Washington 25, D. C. Price $1.00.

Table A2—Turkeys: Numbers Raised Annually on Farms by States and Sections at Three-Year Intervals According to Estimates by The U. S. Agricultural Marketing Service, Washington 25, D. C.

State and area	1929[1]		1932[1]		1935[1]		1938[1]		1941[2]		1944[2]		1947[3]		1950[4]		1953[5]	
	Thousands	Rank	Thousands	Rank	Thousands	Rank	Thousands	Rank	Thousands	Rank	Thousands	Rank	Thousands	Rank	Thousands	Rank	Thousands	Rank
Maine	29	45	32	46	37	46	43	47	45	46	49	45	64	45	74	45	394	30
New Hampshire	21	47	23	47	29	47	49	46	65	42	71	43	91	41	120	40	145	38
Vermont	32	43	34	45	57	45	107	37	149	34	142	37	137	35	120	39	127	40
Massachusetts	67	40	113	38	140	32	204	29	264	26	295	26	387	24	447	25	633	24
Rhode Island	14	48	18	48	21	48	23	48	22	48	30	48	39	47	47	47	53	47
Connecticut	27	46	56	44	60	43	90	41	105	41	78	31	184	32	269	32	340	34
New York	184	26	230	27	280	23	320	23	440	20	552	21	680	17	808	17	924	18
New Jersey	35	42	80	41	96	39	120	34	120	37	250	29	298	27	332	28	353	33
Pennsylvania	192	25	375	16	600	10	620	14	846	15	1,123	9	1,430	6	1,682	7	1,831	8
Total North Atlantic	601	VI	961	VI	1,320	VI	1,576	VI	2,056	VI	2,690	V	3,310	V	3,899	VI	4,800	VI
Ohio	195	23	470	11	576	11	640	13	815	16	979	11	1,213	9	1,361	10	1,972	6
Indiana	143	29	214	28	320	20	360	21	358	23	612	19	835	14	1,249	11	1,705	10
Illinois	100	32	160	32	270	24	420	19	630	18	742	16	818	12	865	16	899	20
Michigan	253	19	350	19	370	18	442	17	472	19	638	18	800	15	916	13	1,097	15
Wisconsin	195	24	350	20	300	21	355	22	426	21	582	20	559	20	977	12	1,592	11
Total East-North Central	886	V	1,544	V	1,836	V	2,217	IV	2,701	IV	3,553	IV	4,325	III	5,368	IV	7,265	IV
Minnesota	1,437	3	1,800	2	1,800	2	2,145	3	3,180	3	3,183	3	3,537	2	4,146	2	5,808	2
Iowa	123	31	310	22	840	7	1,386	6	1,782	4	2,312	4	2,438	4	2,956	3	3,673	4
Missouri	270	17	420	12	645	8	1,100	8	1,544	6	1,532	7	1,310	7	1,681	8	1,493	13
North Dakota	1,604	2	1,780	3	1,120	4	1,380	7	1,303	7	802	14	598	19	495	24	510	26
South Dakota	506	10	775	6	500	13	940	10	1,250	9	413	23	295	28	320	29	437	28
Nebraska	275	16	375	17	470	14	600	15	1,024	12	850	13	838	13	784	18	819	21
Kansas	351	13	420	13	410	16	960	9	1,157	10	749	15	599	18	713	21	698	22
West-North Central	4,566	III	5,880	II	5,785	I	8,511	I	11,240	I	9,841	II	9,615	II	11,095	II	13,438	II

[1]From mimeographed release dated April 1942.
[2]From mimeographed release dated April 1947.
[3]From printed release dated May 1952.
[4]From mimeographed release dated April 1952.
[5]From mimeographed release dated March 1955. Of the turkeys raised in 1953, about 27 percent were light-breed turkeys, almost all Beltsville Small Whites. In 1952 this figure was 29 percent; in 1951 it was 21 percent. The March 1955 estimate for 1954 was 29 percent light-breed turkeys.

Table A2—(Continued)

State and area	1929[1]		1932[1]		1935[1]		1938[1]		1941[2]		1944[2]		1947[3]		1950[4]		1953[5]	
	Thousands	Rank	Thousands	Rank	Thousands	Rank	Thousands	Rank	Thousands	Rank	Thousands	Rank	Thousands	Rank	Thousands	Rank	Thousands	Rank
Delaware	67	39	94	40	120	36	100	40	108	39	88	42	103	40	124	38	824	29
Maryland	307	14	375	15	400	17	370	20	405	22	426	22	396	23	438	26	492	27
Virginia	581	8	640	8	630	9	745	11	805	17	986	10	1,185	10	2,294	5	4,610	3
West Virginia	201	22	240	26	267	26	205	28	240	27	320	24	398	22	887	15	1,548	12
North Carolina	226	21	254	25	255	27	207	27	239	28	289	27	379	25	559	22	967	17
South Carolina	88	37	145	34	140	33	130	32	150	33	280	28	357	26	771	19	1,173	14
Georgia	95	34	136	35	138	34	105	38	115	38	144	35	190	30	316	30	372	31
Florida	76	38	120	37	130	35	104	39	106	40	96	40	109	38	131	36	186	36
Total South Atlantic	1,641	IV	2,004	IV	2,080	IV	1,966	V	2,168	V	2,629	VI	3,117	VI	5,520	III	10,172	III
Kentucky	421	12	520	10	460	15	300	24	278	24	231	30	198	29	314	31	371	32
Tennessee	172	27	205	29	225	28	174	31	180	30	151	33	139	34	186	34	164	37
Alabama	157	28	200	30	190	31	116	35	155	32	144	36	149	33	152	35	231	35
Mississippi	94	35	150	33	100	38	122	33	139	35	98	39	91	42	113	42	115	41
Arkansas	62	41	120	36	70	42	109	36	131	36	132	38	104	39	427	27	902	19
Louisiana	30	44	65	43	60	44	58	44	59	44	46	46	56	46	85	44	112	42
Oklahoma	884	5	1,000	5	1,000	5	1,540	4	1,276	8	653	17	522	21	545	23	621	25
Texas	4,161	1	4,820	1	2,900	1	3,750	1	3,651	1	3,761	2	2,722	3	2,927	4	3,222	5
Total South Central	5,981	I	7,080	I	5,005	II	6,169	III	5,869	III	5,216	III	3,981	IV	4,749	V	5,738	V
Montana	486	11	380	14	342	19	248	25	219	29	155	32	119	37	130	37	140	39
Idaho	569	9	317	21	300	22	196	30	276	25	318	25	189	31	247	33	95	45
Wyoming	256	18	255	24	200	30	210	26	173	31	145	34	129	36	117	41	100	44
Colorado	603	7	630	9	500	12	710	12	846	14	861	12	720	16	723	20	636	23
New Mexico	136	30	170	31	90	40	64	43	58	45	70	44	64	44	64	46	61	46
Arizona	92	36	112	39	105	37	65	42	61	43	89	41	90	43	85	43	103	43
Utah	251	20	375	18	215	29	520	16	1,042	11	1,642	6	1,281	8	1,673	9	1,734	9
Nevada	99	33	80	42	73	41	50	45	38	47	46	47	36	48	26	48	19	48
Washington	277	15	295	23	270	25	425	18	902	13	1,299	8	1,134	11	909	14	978	16
Oregon	660	6	750	7	900	6	1,460	5	1,726	5	2,300	5	1,639	5	1,985	6	1,963	7
California	1,372	4	1,500	4	1,800	3	2,500	2	3,527	2	4,762	1	4,226	1	7,202	1	9,899	1
Total Western	4,801	II	4,864	III	4,795	II	6,448	II	8,868	II	11,687	I	9,627	I	13,161	I	15,728	I
Total United States	18,476		22,333		20,821		26,887		32,902		35,616		33,975		43,792		57,141	

Table A3—U. S. Summaries by Years from U. S. Agricultural Marketing Service Estimates[1]

Year	Breeder hens on hand Jan. 1	Turkeys raised	Average live weight per bird sold			Live weights produced[2]	Price per pound received	Gross income	Death Loss	
			Hens	Toms	All				Breeding stock	Young turkeys
	Thousands	Thousands	Pounds	Pounds	Pounds	Thousand Pounds	Cents	Thousand dollars	Per cent	Per cent
1929	*	18,476	*	*	13.2	*	*	60,027	*	*
1930	*	17,419	*	*	13.4	*	*	51,224	*	*
1931	*	18,249	*	*	13.6	*	*	45,615	*	*
1932	*	22,333	*	*	13.8	*	*	41,197	*	*
1933	*	23,241	*	*	14.0	*	*	38,412	*	*
1934	*	21,702	*	*	14.1	*	*	45,082	*	*
1935	*	20,821	*	*	14.5	*	*	56,224	*	*
1936	*	27,981	*	*	14.7	*	*	64,970	*	*
1937	3,481	25,755	*	*	14.8	*	*	66,933	11.8	27.0
1938	3,222	26,887	*	*	14.9	*	*	69,453	10.9	26.3
1939	3,914	33,587	11.9	17.9	14.9	*	*	73,624	10.4	26.0
1940	4,607	34,047	12.1	18.1	15.1	505,527	15.2	80,843	10.3	25.9
1941	3,864	32,902	12.7	19.1	15.9	516,527	19.9	101,408	10.5	27.8
1942	4,003	32,805	13.0	19.6	16.3	525,899	27.5	148,257	11.1	28.8
1943	3,984	32,309	12.9	19.5	16.2	516,673	32.6	163,855	11.4	29.7
1944	4,294	35,616	13.2	20.4	16.8	591,342	34.0	202,295	10.4	25.4

[1]Formerly Bureau of Agricultural Economics. All are final revised estimates up to and including 1952.
[2]From turkeys sold plus those consumed in farm households, and the plus or minus change in inventory.

Table A3—(Continued)

Year	Breeder hens on hand Jan. 1	Turkeys raised	Average live weight per bird sold			Live weights produced[2]	Price per pounds received	Gross income	Death Loss	
			Hens	Toms	All				Breeding stock	Young turkeys
	Thousands	Thousands	Pounds	Pounds	Pounds	Thousand pounds	Cents	Thousand dollars	Per cent	Per cent
1945	4,505	42,900	13.5	21.4	17.4	740,462	33.7	245,347	9.3	22.3
1946	4,841	40,142	13.9	22.2	18.0	714,450	36.3	272,931	8.2	21.0
1947	3,779	33,975	13.8	22.5	18.1	610,737	36.5	235,823	7.5	19.6
1948	2,537	31,541	14.0	22.6	18.3	573,918	46.8	262,693	7.7	13.5
1949	3,148	41,266	14.2	23.3	18.8	769,351	35.2	267,397	7.8	16.1
1950	3,270	43,792	14.1	23.0	18.6	828,806	32.8	272,120	7.9	15.6
1951	3,301	52,476	13.6	22.2	17.9	935,412	37.4	345,276	7.5	12.9
1952	3,828	60,868	14.3	23.7	16.9[7]	1,021,941	33.6	346,212	5.7	12.6
1953[3]	3,364[6]	57,141	14.5	24.3	17.0[7]	966,718	33.6	325,181	6.1	9.4
1954[3]	3,421[5]	63,636	14.5	24.2	17.2	1,090,886	28.8	313,854	5.9	10.0
1955	3,434[5]	58,896[5]	*	*	*	*	*	*	*	*

[3]All figures for 1953 and 1954 except breeder hen estimates for 1953 are those reported in the release on March 1955 entitled "Farm Production, disposition, cash receipts, and gross income—Turkeys."

[5]From January 1955 intentions and breeder hen reports.

[6]From turkey production report dated March 1954.

[7]For 1952 and 1953, average turkey fryer live weights marketed were 7.6 and 8.0 lbs., respectively which reduced the average weight figures from previous years during which few fryers were produced. In 1954, average live weight of turkey fryers marketed was 8.2 lbs.

Table A4—Density of Turkey Population in the U. S. by States

State and geographical division	Total land area[1]		Turkeys raised in 1951[2]		Turkeys per square mile raised in 1951	
	Square miles	Rank	Number, Thousands	Rank	Number	Rank
Maine	31,040	38	133	40	4.28	38
New Hampshire	9,024	43	134	39	14.85	25
Vermont	9,278	42	133	41	14.33	26
Massachusetts	7,907	44	514	27	65.01	3
Rhode Island	1,058	48	51	47	48.20	8
Connecticut	4,899	46	285	32	58.18	5
New York	47,929	29	881	18	18.38	20
New Jersey	7,522	45	359	30	47.73	9
Pennsylvania	45,045	32	1,817	9	40.34	12
North Atlantic	163,702	VI	4,307	VI	26.31	II
Ohio	41,122	34	1,565	10	38.06	14
Indiana	36,205	37	1,436	11	39.66	13
Illinois	55,947	23	934	17	16.69	23
Michigan	57,022	21	962	16	16.87	22
Wisconsin	54,715	24	1,153	13	21.07	19
E. North Central	245,011	V	6,050	IV	24.69	III
Minnesota	80,009	13	4,644	2	58.04	6
Iowa	55,986	22	3,104	5	55.44	7
Missouri	69,270	18	1,849	8	26.69	16
North Dakota	70,054	16	619	24	8.84	34
South Dakota	76,536	15	352	31	4.60	37
Nebraska	76,653	14	862	19	11.25	28
Kansas	82,113	12	742	21	9.04	32
W. North Central	510,621	III	12,172	II	23.84	IV
Delaware	1,978	47	143	37	72.30	2
Maryland	9,887	41	460	28	46.53	10
Virginia	39,899	36	3,670	3	91.98	1
West Virginia	24,090	40	1,064	14	44.17	11
North Carolina	49,142	28	783	20	15.93	24
South Carolina	30,594	39	1,002	15	32.75	15
Georgia	58,518	20	518	26	8.85	33
Florida	54,262	25	151	36	2.78	41
South Atlantic	268,370	IV	7,791	III	29.03	I
Kentucky	40,109	35	392	29	9.77	30
Tennessee	41,961	33	205	34	4.89	36
Alabama	51,078	27	175	35	3.43	39
Mississippi	47,420	30	136	38	2.87	40
Arkansas	52,725	26	585	25	11.10	29
Louisiana	45,177	31	106	44	2.35	43
Oklahoma	69,283	17	627	23	9.05	31
Texas	263,644	1	3,220	4	12.21	27
South Central	611,397	II	5,446	V	8.91	VI
Montana	146,316	3	130	42	0.89	45
Idaho	82,808	11	222	33	2.68	42
Wyoming	97,506	8	129	43	1.32	44
Colorado	103,967	7	723	22	6.95	35
New Mexico	121,511	4	64	46	0.53	47
Arizona	113,580	5	85	45	0.75	46
Utah	82,346	10	2,075	7	25.20	17
Nevada	109,802	6	24	48	0.22	48
Washington	66,977	19	1,154	12	17.23	21
Oregon	96,349	9	2,382	6	24.72	18
California	156,803	2	9,507	1	60.63	4
Western	1,177,965	I	16,495	I	14.00	V
U. S. Totals	2,977,066		52,261		17.55	

[1]From Encyclopedia Brittanica World Atlas (1945).

[2]From Mimeo. release dated April 1952 by Poultry Section, Livestock and Poultry Statistics Branch, Agricultural Estimates Division, Agricultural Marketing Service, USDA, Washington 25, D. C.

Table A5—Turkeys Raised by States, AMS Data, U.S.D.A.)

State and Division	Average 1945–49	1950	1951	1954[1] Heavy breeds	1954[1] Light breeds	1954[1] Total
	Thousands	Thousands	Thousands	Thousands	Thousands	Thousands
Maine	60	74	133	148	344	492
New Hampshire	95	120	134	139	9	148
Vermont	143	120	133	118	10	128
Massachusetts	395	447	514	622	62	684
Rhode Island	39	47	51	48	5	53
Connecticut	206	269	285	309	31	340
New York	710	808	881	928	92	1,020
New Jersey	317	332	359	275	92	367
Pennsylvania	1,496	1,682	1,817	880[2]	916[2]	1,796
North Atlantic	3,463	3,899	4,307	3,467	1,561	5,028
Ohio	1,158	1,361	1,565	1,361	907	2,268
Indiana	931	1,249	1,436	1,482	325	1,807
Illinois	887	865	934	737	162	899
Michigan	834	916	962	873	246	1,119
Wisconsin	682	977	1,153	1,118	745	1,863
E. North Central	4,493	5,368	6,050	5,571	2,385	7,956
Minnesota	3,593	4,146	4,644	3,867	3,716	7,583
Iowa	2,615	2,956	3,222	3,652	498	4,150
Missouri	1,581	1,681	1,849	914	609	1,523
North Dakota	597	495	619	440	96	536
South Dakota	331	320	352	417	51	468
Nebraska	840	784	862	833	35	868
Kansas	668	713	742	645	123	768
W. North Central	10,226	11,095	12,290	10,768	5,128	15,896
Delaware	106	124	143	177	144[3]	321
Maryland	418	438	460	304	203	507
Virginia	1,368	2,294	3,670	1,637	3,480	5,117
West Virginia	486	887	1,064	718	952	1,670
North Carolina	404	559	783	851	213	1,064
South Carolina	471	771	1,002	967	358	1,325
Georgia	209	316	569	356	35	391
Florida	115	131	151	146	49	195
South Atlantic	3,578	5,520	7,842	5,156	5,434	10,590
Kentucky	238	314	392	312	78	390
Tennessee	153	186	205	173	43	216
Alabama	164	152	175	146	62	208
Mississippi	95	113	136	100	15	115
Arkansas	158	427	585	412	222	634
Louisiana	56	85	106	81	15	96
Oklahoma	540	545	627	386	291	677
Texas	2,947	2,927	3,220	2,682	589	3,271
South Central	4,352	4,749	5,446	4,292	1,315	5,607
Montana	137	130	130	123	17	140
Idaho	242	247	203	74	36	110
Wyoming	138	117	129	72	8	80
Colorado	784	723	723	715	80	795
New Mexico	65	64	64	49	9	58
Arizona	93	85	85	89	9	98
Utah	1,660	1,673	2,075	1,635	359	1,994
Nevada	36	26	24	10	1	11
Washington	1,223	909	1,154	699	299	998
Oregon	2,031	1,985	2,223	1,507	377	1,884
California	5,444	7,202	9,507	8,722	1,078	9,800
Western	11,853	13,161	16,317	13,695	2,273	15,968
U. S.	37,965	43,792	52,252	42,949	18,096	61,045

[1]Preliminary from release of Aug. 20, 1954.
[2]Revision will show about 1350 heavy and 560 light breeds.
[3]Revision will show about 1000 light-breed turkeys.

Table A6—Changes in Turkey Marketing 1940-1954 (Data Arranged by Kohls; Source—U.S.D.A.-AMS)

Area	Before Aug. 1	Aug.	Sept.	Oct.	Nov.	Dec.	Jan., later
	Percent of total						
United States, All							
1940	{——8.8——}				42.1	36.9	12.2
1954	9.9	6.8	11.1	17.3	30.2	19.6	5.1
United States, by breeds 1954:							
Heavy	4.6	6.4	11.5	17.5	32.9	21.1	5.6
Light	23.8	7.8	10.4	15.7	22.8	15.8	3.7
All, by Regions 1954:							
East North Central	4.1	6.1	9.2	16.8	37.0	23.2	3.6
West North Central	8.9	8.6	13.1	23.8	27.4	14.9	3.3
Western	9.4	7.5	11.7	18.7	28.1	18.4	6.2

Table A7—No. Turkeys Marketed in Canada, 1951. Data from Canadian Dept. of Agri.

Province	No. turkeys produced
British Columbia	305,000
Alberta	457,000
Saskatchewan	447,000
Manitoba	300,000
Ontario	700,000
Quebec	703,000
New Brunswick	37,000
Nova Scotia	48,000
Prince Edward Island	14,000
Total for Canada	3,011,000

SECTION 2

NATIONAL TURKEY IMPROVEMENT PLAN*

TITLE 9—ANIMALS AND ANIMAL PRODUCTS

Chapter 1—Agricultural Research Service, Department of Agriculture

Subchapter F—Poultry Improvement

PART 145—NATIONAL POULTRY IMPROVEMENT PLAN (CHICKENS AND CERTAIN OTHER POULTRY).

PART 146—NATIONAL TURKEY IMPROVEMENT PLAN (TURKEYS AND CERTAIN OTHER POULTRY).

PART 147—AUXILIARY PROVISIONS ON NATIONAL POULTRY AND TURKEY IMPROVEMENT PLANS.

Subchapter F-1—Animal Breeds

REVISION OF REGULATIONS AND REDESIGNATION OF SUBCHAPTER

On October 9, 1954, there appeared in the FEDERAL REGISTER (19 F. R. 6519) a notice that this Department was considering promulgating, under paragraph 101 (b) of the Department of Agriculture Organic Act of 1944, as amended (7 U. S. C. 429), a proposed revision of the National Poultry Improvement Plan and the National Turkey Improvement Plan and Auxiliary Provisions, recommended by the 1954 Conference of representatives of the State agencies cooperating in the administration of such plans.

In order to permit promulgation of such revised plans and provisions in a new Subchapter F—Poultry Improvement, of Chapter I, in Title 9, Code of Federal Regulations, present Subchapter F—Animal Breeds, of said chapter, is hereby redesignated as Subchapter F-1—Animal Breeds.

After due consideration of all relevant matters presented pursuant to the aforesaid notice, and under paragraph 101 (b) of the Department of Agriculture Organic Act of 1944, as amended, the provisions set forth in said notice as proposed Part 145—National Poultry Improvement Plan (Chickens and Certain Other Poultry), Part 146—National Turkey Improvement Plan (Turkeys and Certain Other Poultry), and Part 147—Auxiliary Provisions on National Poultry and Turkey Improvement Plans, are, with the modifications shown below, hereby adopted, to appear in 9 CFR Parts 145, 146 and 147, respectively, under new Subchapter F—Poultry Improvement as follows:

§ 146.1 *Definitions.* Except where the context otherwise requires, for the purposes of this Part the following terms shall be construed, respectively, to mean:

(a) *Plan.* The provisions of the National Turkey Improvement Plan contained in this Part.

(b) *Person.* A natural person, firm, or corporation.

(c) *Department.* The United States Department of Agriculture.

*From U. S. Federal Register, Dec. 9, 1954. Reprinted from "Title 9" available on request to U.S.D.A., Washington 25, D. C.

(d) *APH Branch.* The Animal and Poultry Husbandry Research Branch of the Agricultural Research Service of the Department.

(e) *State.* Any State, the District of Columbia, Alaska, Hawaii, or Puerto Rico.

(f) *Official State Agency.* The State authority recognized by the Department to cooperate in the administration of the Plan.

(g) *State Inspector.* Any person employed or authorized under § 146.11 (b) to perform functions under this part.

(h) *Authorized Agent.* Any person designated under § 146.11 (a) to perform functions under this part.

(i) *ROP Supervisor.* The person employed or authorized to perform functions under § 146.27.

(j) *ROP Inspector.* The person employed or authorized to perform functions under § 146.28.

(k) *Affiliated flockowner.* A flockowner who is participating in the Plan through an agreement with a participating hatchery.

(l) *Flock*—(1) *As applied to breeding.* All turkeys of one kind of mating (variety or combination of stocks) and of one classification on one farm;

(2) *As applied to disease control.* All of the turkeys on one farm, except that, at the discretion of the Official State Agency, any group of turkeys which is segregated from another group and has been so segregated for a period of at least 21 days may be considered as a separate flock.

(m) *Hatchery.* Hatchery equipment on one premises operated or controlled by any person for the production of poults.

(n) *Products.* Turkey breeding stock and hatching eggs, and poults.

(o) *Baby poults.* Poults that have not been fed or watered.

(p) *USROP or ROP.* U. S. Record of Performance.

(q) *Strain.* Turkey breeding stock bearing a given name produced by a breeder through at least five generations of closed flock breeding. Only (1) first generation progeny (F_1) of stock originating from eggs produced under the direct supervision of the breeder, or (2) stock multiplied by persons designated by the breeder, may be designated as of the breeder's strain.

(r) *Trade name or number.* A name or number compatible with State or Federal laws and regulations applied to a specified stock or product thereof, and which has been filed with the Official State Agency together with a description of such stock.

(s) *Broad-breasted.* A term used to describe a type of turkey which at the time of selection, and no later than 30 weeks of age, has a breast width at a point 1¾ inches above the keel of at least 3½ inches, for both toms and hens.

§ 146.2 *Administration.* The Department cooperates through a Memorandum of Understanding with Official State Agencies in the administration of the Plan. The Official State Agency shall carry out the administration of the Plan within the State according to the applicable provisions of the Plan and Memorandum of Understanding. The Official State Agency of any State may adopt regulations applicable to the administration of the Plan in such State further defining the provisions of the Plan or establishing higher standards compatible with the Plan.

§ 146.3 *Participation.* (a) Any person producing or dealing in poultry products may participate in the Plan when he has (1) demonstrated, to the satisfaction of the Official State Agency, that his facilities, personnel, and practices are adequate for carrying out the applicable provisions of the Plan; and (2) signed an agreement with the Official State Agency to comply with the general and the applicable specific provisions of the plan and any regulations of the Official State Agency under § 146.2. Affiliated flockowners may participate without signing an agreement with the Official State Agency.

(b) Each participant shall comply with the Plan throughout the operating year of the Official State Agency, or until released by such Agency.

(c) A participant in any State shall participate with all of his turkey hatching egg supply flocks and hatchery operations within such State.

(d) No person shall be compelled by the Official State Agency to qualify products for any of the classifications described in § 146.10 (a), (b), (c) and (d) as a condition of qualification for either of the classifications described in § 146.10 (e) and (f).

(e) Participation in the Plan shall entitle the participant to use the Plan emblem reproduced below:

Figure A1

§ 146.4 *General provisions for all participants.* (a) Records of purchases and sales and the identity of products handled shall be maintained in a manner satisfactory to the Official State Agency.

(b) Products, records of sales and purchases of products, and material used to advertise products shall be subject to inspection by the Official State Agency at any time.

(c) Advertising must be in accordance with the Plan and applicable rules and regulations of the Official State Agency and the Federal Trade Commission.

(d) Participants may not buy or receive for any purpose products from non-participants, or sell products of non-participants, except with the permission of the Official State Agency for use in breeding flocks or for experimental purposes.

(e) The minimum weight of turkey hatching eggs shipped interstate shall be 2⅓ ounces each for small varieties and 2½ ounces each for other varieties, unless otherwise specified by the purchaser of the eggs.

(f) Standard and broad breasted turkeys of the same variety may not be kept on the same premises or hatched in the same hatchery.

(g) When artificial insemination is used, its use and the extent of its use shall be stated in all performance reports.

§ 146.5 *Specific provisions for participating flocks.* (a) Poultry houses and the land in the immediate vicinity thereof shall be kept in sanitary condition acceptable to the Official State Agency.

(b) All flocks shall consist of birds that have been selected for health, vigor and freedom from physical deformities of economic importance by an Authorized Agent or State Inspector.

(c) A flock shall be deemed to be a participating flock at any time only if it has been officially blood tested within the past 12 months with no pullorum or typhoid reactors on the last test. (See § 146.14 relating to the official blood test.)

(d) Each bird shall be identified with a sealed and numbered band obtained through or approved by the official State Agency.

§ 146.6 *Specific provisions for participating hatcheries.* (a) Hatcheries shall be kept in sanitary condition acceptable to the Official State Agency.

(b) A hatchery which keeps started poultry (poultry that has been fed or watered) must keep such poultry separated from the incubator room in a manner satisfactory to the Official State Agency.

(c) All poults offered for sale under Plan terminology shall be normal and typical of the variety, cross, or other combination represented.

(d) Eggs incubated shall be sound in shell and reasonably uniform in shape. Eggs shall be trayed and poults boxed with a view to uniformity of size.

(e) All hatcheries operated under the same name, ownership, or management within a State shall participate in the Plan if any of them are to participate.

§ 146.7 *Specific provisions for participating dealers.* Dealers in turkey breeding stock or hatching eggs or poults shall comply with all provisions in this Part which apply to their operations.

§ 146.8 *Terminology and classification; general.* (a) The official classification terms defined in §§ 146.9 and 146.10 and the various designs illustrative of the official classifications as reproduced in § 146.10 may be used only by participants and to describe products that have met all the specific requirements of such classifications. (See also § 146.16 (e).)

(b) Products produced under the Plan shall lose their identity under Plan terminology when they are purchased for resale by, or consigned to, non-participants.

(c) Participating flocks and the eggs and poults produced from them may be designated by their own strain or trade name.

(d) Domesticated waterfowl, game birds and guineas meeting the requirements for participation for any of the classifications specified in § 146.10 may be designated as coming within such classification.

§ 146.9 *Terminology and classification; hatcheries.* Participating hatcheries and dealers shall be designated as "NTIP Participating."

§ 146.10 *Terminology and classification; flocks and products.* Participating flocks, and the eggs and poults produced from them, which have met the respective requirements specified in the § 146.10 may be designated by the following terms or illustrative designs:

(a) *U. S. Record of Performance.* Males and females meeting prescribed standards as provided in §§ 146.18 and 146.19. (Hatching eggs and poults designated as ROP must also meet the requirements specified in §§ 146.23 and 146.24, respectively.)

(b) *U. S. Performance Tested Parent Stock.* Flocks represented in both a turkey reproduction test as provided for in § 146.29 and in a central turkey meat production test as provided for in § 146.30 by entries that have met the following performance standards:

(1) Average poult production of at least 20 salable poults per hen in eight weeks;

(2) Poult livability of at least 90 percent to eight weeks of age. As an alternative, in case of excessive mortality in the entire test, the entry shall qualify for livability if the livability is above the average of the test.

(c) *U. S. Certified.* Flocks meeting one of the three following specifications:

(1) All males ROP or from ROP mass matings.

(2) Males and females from flocks composed of the following: ROP males or females from ROP mass matings mated to females from ROP qualified matings; or ROP mass matings; or ROP candidate matings in which 50 percent or more of the dam's family qualified for ROP. (Poults produced under this subparagraph for use in U. S. Certified flocks shall be properly identified at hatching time.)

(3) Males and females from Performance Tested Parent Stock. In case the tested stock is a cross of strains, the U. S. Certified flock shall be a combination of the same pure strains as used in the tested flock. (Poults produced under this subparagraph for use in U. S. Certified flocks shall be properly identified at hatching time.)

(d) *U. S. Approved.* All males and females selected by Authorized Agents according to standards prescribed by the Official State Agency or the State College of Agriculture.

(e) *U. S. Pullorum-Typhoid Passed.* Flocks in which no pullorum or typhoid reactors were found on the last official blood test provided for in § 146.5 (c).

(f) *U. S. Pullorum-Typhoid Clean.* Flocks in which no reactors were found on the first official blood test provided for in § 146.5 (c): *Provided,* That if reactors are found on the first test the flock may qualify with two consecutive official negative tests. In order to sell hatching eggs or poults of this classification, all hatching eggs and poults handled must be of this classification.

§ 146.11 *Supervision.* (a) The Official State Agency may designate qualified persons as Authorized Agents to do the selecting and blood collecting provided for in §§ 146.5, 146.10, and 146.14.

(b) The Official States Agency shall employ or authorize qualified persons as State Inspectors to perform, or supervise the performance of the selecting and testing of participating flocks and to perform the official inspections necessary to verify compliance with the requirements of the Plan.

§ 146.12 *Inspections.* (a) Each participating hatchery shall be inspected a sufficient number of times each year to satisfy the Official State Agency that the operations of the hatchery are in compliance with the provisions of the Plan.

(b) Each year at least 15 percent of the flocks selected and tested by Authorized Agents shall be inspected by a State Inspector. This must include the inspection of some flocks of each hatchery. Each flock inspection shall include the examination of a sufficient number of males and females to determine whether the work of the Authorized Agent was satisfactory and that the flock is qualified for participation.

§ 146.13 *Debarment from participation.* Non-compliance with the provisions of the Plan, or regulations of the Official State Agency under § 146.2, not corrected within the time specified by the Official State Agency, shall be grounds for the Official State Agency to bar a participant from further participation for a period to be determined in each case by the Official State Agency. Such action shall not be taken until a thorough investigation has been made by the Official State Agency and the participant has been given an opportunity for a hearing.

§ 146.14 *Blood testing.* (a) In the official blood test, the blood shall be drawn by an Authorized Agent or State Inspector and tested by an authorized laboratory, using either the standard tube agglutination or rapid serum test. The recommended procedures for conducting such tests are described in Subpart A, Part 147, of this chapter.

(b) There shall be an interval of at least 21 days between any official blood test and any previous test with pullorum-typhoid antigen.

(c) All turkeys to be used as breeders must be tested when more than four months of age.

(d) All domesticated fowl on the farm of the participant shall either be properly tested to meet the same standards as the participating flock or these birds and their eggs shall be separated from the participating flock and its eggs.

(e) All tests with Salmonella antigens of flocks participating in or candidates for participation in the Plan shall be reported to the Official State Agency within 10 days following the completion of such tests. All reactors shall be considered in determining the classification of the flock.

(f) Reactors may be submitted to a laboratory for autopsy and bacteriological examination. The laboratory and the number of reactors to be submitted shall be designated by the Official State Agency. The recommended minimum procedure for bacteriological examination is described in Subpart B, Part 147 of this chapter. In case such bacteriological examination fails to demonstrate pullorum or typhoid infection, the flock shall be deemed to have had no pullorum or typhoid reactors. If other members of the Salmonella group are isolated, the Official State Agency may disqualify the flock for participation, or require such other action as is deemed necessary with respect to the infection.

(g) After a flock has been classified by the Official State Agency, the results of any retesting of such flocks during the current breeding and hatching season shall not adversely affect its classification for the season, except that if in the opinion of the Official State Agency the amount of reaction found in any flock is such as to be dangerous or detrimental to the operation of the Plan, the Official State Agency shall require that such flock or flocks be retested after an interval of at least 21 days, or that use of such flocks as sources of eggs for participants be immediately discontinued. Furthermore, the Official State Agency may require that the hatching eggs from such flocks be removed from the incubator and destroyed prior to hatching.

(Paragraphs 146.15 to 146.28 deal with details of ROP.)

§ 146.29 *Turkey reproduction test.* A turkey reproduction test is a test of the reproductive (combination of egg production and hatchability) qualities of a representative sample of the entrant's stock conducted under the supervision of the Official State Agency, in accordance with the following provisions:

(a) The test shall be available to any participant whose flock is qualified for the U. S. Pullorum-Typhoid Clean classification.

(b) The entry shall consist of at least 200 hens, penned separately, which are representative of the entrant's supply flocks or his breeder replacement flocks.

(c) All eggs produced by the entry during a period of at least eight consecutive weeks designated in advance by the entrant, shall be incubated and complete records kept on the number of eggs produced, eggs set and salable poults hatched.

(d) The Official State Agency shall furnish the APH Branch, for publication, the following information for each entry:

(1) Kind of stock (variety, strain or cross; supply flock or breeder replacement);

(2) Number of breeder hens entered and total number in flocks which they represent;

(3) Average number of salable poults produced per hen entered at beginning of test.

§ 146.30 *Central turkey meat production test.* A central turkey meat production test is a test of a representative sample of the stock of two or more entrants, to fryer-roaster or mature marketing age, conducted under official supervision at a neutral location, in accordance with the following provisions:

(a) The test shall be available to any participant whose flock is qualified for the U. S. Pullorum-Typhoid Clean classification.

(b) The entry shall consist of at least 100 poults, 50 percent of each sex.

(c) A sufficient number of eggs, determined by the entrant, to produce the poults required for the entry shall be selected by a representative of the Official State Agency.

(d) The sample of eggs shall be taken at random from the entrant's supply flocks or from his breeder replacement flocks.

(e) The eggs from all entrants shall be incubated in one hatchery at the same time.

(f) Poults shall be banded in each wing with a sealed and numbered band for identification.

(g) Poults shall be brooded and reared either intermingled or separated by entries, at the option of the test management, provided that varieties of different size or color shall be separated.

(h) The duration of the test shall be determined by the test management.

(i) The following data shall be obtained and reported by the Official State Agency for each entry:

(1) Kind of stock (variety, strain or cross; supply flock or breeder replacement);

(2) Number of breeder hens in the flock or flocks from which egg sample was drawn;

(3) Mortality to two weeks of age; to eight weeks of age; and to the end of the test;

(4) Average live weight (i) of fryer-roaster entry at four weeks of age; at four week intervals thereafter; and at the close of the test; and (ii) of mature marketing entry at eight weeks of age; at 16 weeks of age; at 20 weeks of age; at two week intervals thereafter; and at the close of the test. (Such average live weights shall be based on individual weights and reported by sexes);

(5) Average eviscerated weight of all birds completing the test, by sexes;

(6) Average breast width, body depth and keel length of the New York dressed birds by sexes (The breast width shall be measured at the widest point 1¾ inches above the keel. The body depth shall be measured at the deepest point);

(7) Coefficient of variability of final live and eviscerated weights and body measurements;

(8) The number of birds in each U. S. Grade based on (i) fleshing; (ii) finish; and (iii) freedom from pin feathers, by sexes;[1]

(j) The results of the tests reported to the APH Branch by December 1 shall be published in a national summary alphabetically by test and entries, with no indication of rank.

§ 146.31 *On-the-farm turkey meat production test.* An on-the-farm turkey meat production test is a test to fryer-roaster or mature marketing age of a representative sample of the entrant's stock conducted under supervision of the Official State Agency to which the following provisions apply:

(a) The test shall be available to any participant whose flock is qualified for the U. S. Pullorum-Typhoid Clean classification.

(b) An entry shall consist of at least 100 poults, either straight-run or 50 percent of each sex.

(c) The sample of poults shall be taken at random from the entrant's supply flocks or from his breeder replacement flocks.

(d) Poults shall be banded in each wing with a sealed and numbered band for identification.

(e) Poults shall be brooded and reared separately from other birds.

(f) The duration of the test shall be determined by the Official State Agency.

(g) The following data shall be obtained and reported by the Official State Agency for each entry:

(1) Kind of stock (variety, strain or cross; supply flock or breeder replacement);

---

[1]The grading will be based on United States Classes, Standards, and Grades for Poultry as contained in 7CFR Part 70, Subpart B.

(2) Number of hens in the flock or flocks from which poult sample was drawn;

(3) Percentage of poults started that finished the test;

(4) Average live weight by sexes based on individual weights of all birds at the close of the test;

(5) Average eviscerated weight by sexes of all birds completing the test;

(6) Average breast width, body depth and keel length of the New York dressed birds, by sexes (The breast width shall be measured at the widest point 1¾ inches above the keel. The body depth shall be measured at the deepest point);

(7) Coefficient of variability of all weights and body measurements.

(h) The results of the test reported to the APH Branch by December 1, shall be published in a national summary alphabetically by entries, with no indication of rank.

PART 147—AUXILIARY PROVISIONS ON NATIONAL POULTRY AND TURKEY IMPROVEMENT PLANS.

SUBPART A—BLOOD TESTING PROCEDURES

§ 147.1 *The standard tube agglutination test.* (a) The blood samples should be collected and delivered as follows:

(1) The blood samples should be taken by properly qualified and authorized persons only, and in containers provided by the laboratory. The containers should be stout-walled test tubes preferable ⅜ by 3 inches, without lip, or small well-selected medicine vials, which have been thoroughly cleaned and dried in a hot-air drying oven. If stoppers are used they should be thoroughly cleaned and dried.

(2) Sufficient blood should be procured by making a small incision in the large median wing vein with a small sharp lancet and allowing the blood to run into the tube, or by the use of a small syringe (with 20 or 21 gage needle) which is properly cleansed between bleedings with physiological saline solution. To facilitate the separation of the serum the tubes should be placed in a slanted position until the blood has solidified. After the blood has completely clotted, they should be packed and shipped by mail (special delivery), rapid express, or by messenger, to the laboratory. All labeling must be clear and permanent, and may be done with a suitable pencil on etched portions of the tube, or by means of fast-gum labels.

(Paragraph 3 (a to g) describes details of the tube test, including antigen requirements.)

§ 147.2 *The rapid serum test.* (a) The procedure for the collection and delivery of blood samples in the rapid serum test is the same as that described in § 147.1 (a).

(b) The antigen should be prepared as follows: (1) The antigen should consist of representative strains of S. pullorum which are of known antigenic composition, high agglutinability, but are not sensitive to negative and nonspecific sera.

(Paragraphs 2 to 6 describe details of the serum test; paragraphs 147.3 the stained-antigen rapid whole-blood test and Subpart B the laboratory procedure for bacteriological examination of reactors. Subpart C sets forth the procedure for changing the National Plan and paragraph 147.28 therein concludes as follows:

The reporting and record keeping requirements contained herein have been approved by the Bureau of the Budget in accordance with the Federal Reports Act of 1942.

The foregoing provisions became effective December 9, 1954.

Table A8—Suggested minimum body weights for large- and medium-sized varieties and minimum and maximum body weights for small type turkeys as an aid in the selecting of breeding stock at various ages.

Sex and age in weeks	Minimum weights					Maximum weights	
	Broad Breasted Bronze or White, and Nebraskan	Standard bred Bronze	Standard varieties other than Bronze and Beltsville White	Jersey Buff	Beltsville Small White	Beltsville Small White	Jersey Buff
Toms:	pounds	pounds	pounds	pounds	pounds	pounds	pounds
12	7¼	6	5½	5	4½	7	7¼
14	9½	7½	7	6	5½	8½	9
16	11	9	8½	7½	7	10	10½
18	13	11	10	9	8½	11½	12½
20	15	12½	11½	10½	9½	13	14
22	17	14	13	12	11	14½	15½
24	19	15½	14½	13½	12½	16½	17½
26	21	17	16	15	14	18	19
28	23	18½	17½	16½	15½	19½	21
30	24½	19½	18½	17½	16	21	22½
32	26	20½	19½	18	16½	22	23½
34	27	21½	20½	19	17½	23	24½
36	28	22½	21½	19½	17½	23½	25
38	28	22½	21½	19½	17½	24	25½
Hens:							
12	5½	4½	4	3½	3½	5½	5½
14	7	5½	5	4½	4½	6½	6½
16	8½	6½	6	5	5	7½	7½
18	9½	7½	7	6	5½	8½	9
20	10½	8½	8	6½	6	9	9½
22	11½	9	8½	7	6½	9½	10
24	12½	10	9½	7½	7	10½	11
26	13	10½	10	8	7½	11	11½
28	13½	11	10½	8½	8	11½	12
30	14½	11½	11	9	8½	12	13
32	15	12	11½	9½	9	12½	13½
34	15½	12½	12	10	9½	13	14
36	16	13	12½	10½	9½	13½	14½
38	16	13	12½	10½	9½	13½	14½

Table A9—Summary of Hatchery Participation in the National Turkey Improvement Plan, 1946-47 to 1953-54, inclusive

		1946-47	1947-48	1948-49	1949-50	1950-51	1951-52	1952-53	1953-54
Number Hatcheries Hatching Egg Capacity		735 30,426,292	674 26,648,066	768 32,071,895	815 31,710,192	895 34,567,108	819 31,231,271	890 36,003,752	790 37,464,146
Percent Increase or Decrease	No. Cap.	+35.1 +57.3	− 8.30 −12.42	+11.39 +20.35	+6.12 −1.13	+9.82 +9.01	-8.49 -9.65	+8.67 +15.28	-11.24 +4.06
U. S. Approved	No. Cap. %	449 16,442,294 54.0	416 16,648,188 62.47	547 24,586,964 76.66	615 25,632,158 80.83	669 28,650,869 82.88	644 27,008,114 86.48	667 29,702,484 82.50	628 32,407,667 86.50
U. S. Certified	No. Cap. %	20 246,691 .8	21 310,528 1.17	22 408,345 1.27	22 529,551 1.67	31 872,015 2.52	20 683,050 2.19	21 879,800 2.44	16 550,370 1.47
Pullorum Classes only No Breeding Class	No. Cap. %	266 13,737,307 45.2	237 9,689,350 36.36	199 7,076,586 22.07	178 5,548,483 17.50	195 5,044,224 14.60	155 3,540,107 11.33	202 5,421,468 15.06	146 4,506,109 12.03
U. S. Pullorum-Tested	No. Cap. %	164 6,267,026 20.6	55 2,734,288 10.26	(eliminated)					
U. S. Pullorum-Controlled	No. Cap. %	241 8,596,613 28.2	206 7,565,080 28.39	203 10,758,402 33.54	164 7,531,997 23.75	161 6,544,767 18.93	85 4,162,425 13.33	34 1,452,620 4.04	16 579,636 1.55
U. S. Pullorum-Passed	No. Cap. %	50 2,090,515 6.9	87 3,401,316 12.76	159 6,090,042 18.99	193 6,595,006 20.80	148 4,918,411 14.23	143 4,405,761 14.11	132 5,652,949 15.70	98 3,591,232 9.58
U. S. Pullorum-Clean	No. Cap. %	280 13,472,138 44.3	326 12,947,382 48.59	406 15,223,451 47.47	458 17,583,189 55.45	586 23,103,930 66.84	591 22,663,085 72.56	724 28,898,183 80.26	676 33,293,278 88.87

Table A10—Summary of Flock Participation in the National Turkey Improvement Plan, 1946-47 to 1953-54, inclusive

		1946-47	1947-48	1948-49	1949-50	1950-51	1951-52	1952-53	1953-54
Number flocks		3,957	2,117	3,224	3,922	3,920	4,273	3,811	3,843
Number birds		1,890,879	1,095,938	1,849,907	2,132,644	2,264,369	3,063,677	2,595,701	2,766,354
Percent Increase or Decrease	Flocks	−14.12	−46.50	+52.29	+21.65	− .05	+9.01	-10.82	+0.84
	Birds	−19.27	−42.04	+68.80	+15.28	+6.13	+35.30	-15.24	+6.53
Av. No. Birds per flock		478	518	574	544	578	717	681	720
U. S. Approved	Flocks	2,926	1,561	2,441	3,213	3,204	3,367	3,244	3,321
	Birds	944,292	841,524	1,403,269	1,650,256	1,793,230	2,451,755	2,147,248	2,364,661
	%	49.94	76.79	75.86	77.38	79.20	80.03	82.69	85.48
U. S. Certified	Flocks	62	66	165	205	177	315	127	96
	Birds	20,427	18,592	91,441	97,637	115,570	148,902	64,491	54,173
	%	1.08	1.70	4.94	4.58	5.10	4.86	2.48	1.96
Pullorum Classes only No Breeding Class	Flocks	969	490	618	504	539	591	440	426
	Birds	926,160	235,822	355,197	384,751	355,569	463,020	384,962	347,520
	%	48.98	21.51	19.20	18.04	15.70	15.11	14.83	12.56
U. S. Pullorum-Tested	Flocks	273	59	—	—	—	—	—	—
	Birds	81,496	21,579	—	—	—	—	—	—
	%	4.31	1.97	—	—	—	—	—	—
U. S. Pullorum-Controlled	Flocks	1,338	479	362	291	272	145	35	50
	Birds	530,626	211,370	223,982	148,955	122,170	75,194	37,897	51,052
	%	28.06	19.29	12.11	6.99	5.40	2.46	1.46	1.84
U. S. Pullorum-Passed	Flocks	386	273	677	797	665	404	381	160
	Birds	69,543	80,922	199,124	305,680	282,172	244.809	211,179	85,416
	%	3.68	7.38	10.76	14.33	12.46	7.99	8.13	3.09
U. S. Pullorum-Clean	Flocks	1,960	1,306	2,185	2,834	2,983	3,724	3,395	3,633
	Birds	1,209,214	782,067	1,426,801	1,678,009	1,860,027	2,743,674	2,347,625	2,629,886
	%	63.95	71.36	77.13	78.68	82.14	89.55	90.41	95.07
FIRST TEST	No. Flocks	4,310	2,486	3,782	4,717	4,163	4,728	4,164	4,225
	No. Birds	2,061,045	1,265,037	1,990,073	2,340,576	2,392,630	3,277,780	2,753,966	2,946,586
	No. Reactors	25,019	9,472	10,210	9,172	8,502	6,725	4,882	2,641
	Percent Reactors	1.21	.75	.51	.39	.36	0.21	0.18	0.09

SECTION 3. CONVERSION TABLES

Table A11—Miscellaneous Equivalents

Ounces in Decimals

Oz.	*Lbs.*
1	.0625
2	.1250
3	.1875
4	.2500
5	.3125
6	.3750
7	.4375
8	.5000
9	.5625
10	.6250
11	.6875
12	.7500
13	.8125
14	.8750
15	.9375
16	1.0000

Conversion of Temperature

C°	F°
0	32
20	68
37	98.6
37.8	100
38	100.4
38.3	101
38.9	102
39	102.2
39.4	103
40	104
41.4	106.5
52.8	127
100	212

To Convert

F =(9/5xC) +32

C =5/9(F-32)

Weights and Measurements

1 inch	=2.54 centimeters
1 meter	=39.37 inches
1 board foot	=1/12 cubic foot
1 bushel of oats	=32 lbs.
1 bushel of barley	=48 lbs.
1 bushel of shelled corn	=56 lbs.
1 bushel wheat	=60 lbs.
1 pint water (at 39.2 F)	=1.042 lbs.
1 gal. water (at 39.2 F)	=8.339 lbs.
1 gal. whole cow's milk	=8.606 lbs.
1 gal. skimmilk	=8.639 lbs.
1 gal. buttermilk	=8.639 lbs.
1 cu. ft. water	=62.4 lbs.
1 cu. ft. ice	=57.2 lbs.
1 cu. ft. ice	=57.2 lbs.
1 pint cod-liver oil	=.96 lb.
1 Imperial (British) gal.	=1 1/5 U.S. gal.

1 c.c. water =1 gram (approx.)

Metric Conversion Scale

1 grain	=.064799 gram
15.4324 grains	=1 gram
1 ounce avdp.	=28.35 grams
1 pound avdp.	=453.59 grams
2.2 pounds avdp.	=1,000 grams (1 Kilogram)
1 avdp. dram	=1.7718 grams
1 apoth. dram	=3.8879 grams
.5644 avdp. dram	=1 gram
.2572 apoth. (troy) dram	=1 gram

1 teaspoonful (t) =⅛ liquid oz. =3.69661 c.c.

4 teaspoons (liquid) =1 tablespoon (T) = ½ fluid oz.

3 teaspoons, dry (cookery) =1 tablespoon = ½ fluid oz. =14.79 c.c.

Measurements and Tables

Linear Measure

12 inches (in.)	=1 foot (ft.)
3 feet	=1 yard (yd.)
5½ yds. or 16½ ft.	=1 rod (rd.)
320 rds. or 5,280 ft.	=1 mile (mi.)

Square Measure

144 square inches (sq. in.)	=1 sq. ft.
9 square feet	=1 sq. yd.
30¼ square yards	=1 sq. rod
160 square rods	=1 acre (A)
640 acres	=1 sq. mi.
1 acre =4,840 sq. yds.	=43,560 sq. ft.

Cubic Measure

1,728 cubic inches (cu. in.)	=1 cu. ft.
27 cubic feet	=1 cu. yd.
128 cubic feet	=1 cord (cd.)
1 quart liquid	=57.75 cu. in.
1 quart dry	=67.20 cu. in.
105 dry quarts	=1 dry barrel
1 standard bushel	=2,150.4 cu. in.
1 cu. foot	=4/5 bushel (approx.)

Dry Measure

2 pints (pt.) =1 quart (qt.); 8 qts. =1 peck (pk.); 4 pecks =1 bushel (bu).

Weight

24 grains	=1 pennyweight
20 grains	=1 apoth. (troy) scruple
3 apoth. scruples	=1 apoth. dram
8 apoth. drams	=1 apoth. ounce
12 apoth. ounces	=1 apoth. (troy) pound
20 pennyweights	=1 apoth. (troy) ounce

1 avdp. dram	=27.3438 grains
16 avdp. drams	=1 avdp. ounce
16 avdp. ounces	=1 avdp. pound
2000 avdp. pounds	=1 short ton
2240 avdp. pounds	=1 long ton
2204.62 avdp. pounds	=1 metric ton

Table A11—(Continued)

Liquid Measure for Disinfectants

Dilution	Add per Qt.
1:10 (10% sol.)	6.4 T.
1:20 (5% sol.)	3.2 T.
1:25 (4% sol.)	2½ T.
1:50 (2% sol.)	1¼ T.
1:100 (1% sol.)	2½ t.
1:200 (½% sol.)	1¼ t.
1:400 (¼% sol.)	⅝ t.
1:500 (1/5% sol.)	½ t.
1:1000 (0.1% sol.)	¼ t.

T = Tablespoonful

64 T (liquid)= 1 qt.

t = teaspoonful (4 t. = 1 T.)

1 fluid oz. = 29.57 c.c.

Liquid

1 minim = .06161 c.c. = .00375977 cu. in. = 1 drop
60 minims = 1 fluid dram = 1 fluid teaspoonful
8 fluid drams = 1 fluid oz. = 1.80469 cu. in.
4 fluid (or liquid) ounces = 1 gill = 8 tablespoonfuls liquid = 7.21875 cu. in.
4 gills = 1 liquid pint = 473.167 milliliters or 473.1793 c.c. = 32 tablespoons
2 liquid pints = 1 liquid quart = .9463 liter
4 liquid quarts = 1 gallon = 231.00 cu. in.
31½ gallons = 1 barrel, liquid
2 barrels = 1 hogshead = 238.5 liters
42 gallons = 1 petroleum barrel
1 liter = 1000 milliliters = 1000.027 c.c.
1 c.c. = 0.06102 cu. in. = .99997196 ml.

Useful Data

To find diameter of circle multiply circumference by .31831.

To find the circumference of a circle multiply diameter by 3.1416.

To find area of a circle multiply square of diameter by .7854.

Doubling the diameter of a pipe increases its capacity four times.

Double riveting is from 16 to 20 per cent stronger than single.

One cubic foot of anthracite coal weighs about 53 pounds.

One cubic foot of bituminous coal weighs from 47 to 50 pounds.

One ton of coal is equivalent to two cords of wood for steam purposes.

A gallon of water (U.S. Standard) weighs 8⅓ pounds and contains 231 cubic inches.

A cubic foot of water contains 7½ gallons, 1,728 cubic inches, and weighs 62.4 pounds.

Steam rising from water at its boiling point (212 degrees) has a pressure equal to the atmosphere (14.7 pounds to the square inch).

To evaporate one cubic foot of water requires the consumption of 7½ pounds of ordinary coal, or about 1 pound of coal to 1 gallon of water.

Suggestions for Using Table A12

(1) To find oz. and oz.-per doz. equivalents for eggs weighing 82 gms., use direct reading; for 90.6 gm. eggs, add the equivalents for 90 gms. and for .6 gm. which total 3.19616 oz. or 38.354 oz. per doz.

(2) To find gram and oz.-per-doz. equivalents for 3¼ oz. eggs, use direct reading, which is 92.14 gms. or 39 oz. per doz.; to find equivalents for eggs weighing 3.273 oz., find equivalents for 3.280 oz. and subtract from them the equivalents for .007 oz.—the result is 92.802 gm. or 39.276 oz. per doz.

(3) To find equivalents for eggs weighing 32 oz. per doz., use direct reading, which is 75.61 gms. or 2.667 oz.; for eggs of 42⅓ oz. per doz., add the equivalents for 42 oz. and ⅓ oz. (per doz.), the result being 100.0075 gms. or 3.5278 oz.

Table A12—Three-Way Conversions for Egg Weights*

Conversion factors: { 1 gram=.035273957 oz.; 1 oz.=28.34953 grams
1 gram=.002204622 lb.; 1 lb.=453.5924277 grams }

Grams	Ounces	Ounces per doz.	Grams	Ounces	Ounces per doz.	Grams	Ounces	Ounces per doz.
.02835	.001	.012	19	.67021	8.043	77.96	2.750	33.00
.05670	.002	.024	20	.70548	8.466	78	2.751	33.01
.08505	.003	.036	21	.7408	8.889	79	2.787	33.44
.11340	.004	.048	22	.7760	9.312	80	2.822	33.86
.14175	.005	.060	23	.8113	9.736	80.31	2.833	34.00
.17010	.006	.072	24	.8466	10.16	81	2.857	34.28
.19845	.007	.084	25	.8819	10.58	82	2.892	34.70
.22680	.008	.096	26	.9171	11.01	82.70	2.917	35.00
.25515	.009	.108	27	.9524	11.43	83	2.928	35.14
.28350	.010	.120	28	.9877	11.85	84	2.963	35.56
.56699	.020	.240	29	1.023	12.27	85	2.998	35.98
.85049	.030	.360	30	1.058	12.70	85.05	3.000	36.00
1.1340	.040	.480	31	1.094	13.12	86	3.034	36.41
1.4175	.050	.600	32	1.129	13.55	87	3.069	36.83
1.7010	.060	.720	33	1.164	13.97	87.40	3.083	37.00
1.9845	.070	.840	34	1.199	14.39	88	3.104	37.25
2.2680	.080	.960	35	1.235	14.82	89	3.139	37.67
2.5515	.090	1.080	36	1.270	15.24	89.78	3.167	38.00
2.8350	.100	1.200	37	1.305	15.66	90	3.175	38.10
3.5437	(1/8).1250	1.500	38	1.340	16.08	91	3.210	38.52
4.0511	(1/7).1429	1.714	39	1.376	16.51	92	3.245	38.94
4.7249	(1/6).1667	2.000	40	1.411	16.93	92.14	3.250	39.00
5.6699	(1/5).2000	2.400	40.17	1.417	17.00	93	3.280	39.36
7.0874	(1/4).2500	3.000	41	1.446	17.35	94	3.316	39.79
8.0999	(2/7).2857	3.429	42	1.482	17.78	94.49	3.333	40.00
8.5049	(3/10).3000	3.600	42.52	1.500	18.00	95	3.351	40.21
9.4498	(1/3).3333	4.000	43	1.517	18.20	96	3.386	40.63
10.6311	(3/8).3750	4.500	44	1.552	18.62	96.87	3.417	41.00
11.340	(2/5).4000	4.800	44.88	1.583	19.00	97	3.422	41.06
12.150	(3/7).4286	5.143	45	1.587	19.04	98	3.457	41.48
14.175	(1/2).5000	6.000	46	1.623	19.48	99	3.492	41.90
16.200	(4/7).5714	6.857	47	1.658	19.90	99.22	3.500	42.00
17.010	(3/5).6000	7.200	47.26	1.667	20.00	100	3.527	42.32
17.718	(5/8).6250	7.500	48	1.693	20.32	101	3.563	42.76
19.845	(7/10).7000	8.400	49	1.728	20.74	101.58	3.583	43.00
20.250	(5/7).7143	8.571	49.61	1.750	21.00	102	3.598	43.18
21.262	(3/4).7500	9.000	50	1.764	21.17	103	3.633	43.60
22.680	(4/5).8000	9.600	51	1.800	21.60	103.96	3.667	44.00
23.625	(5/6).8333	10.000	51.97	1.833	22.00	104	3.668	44.02
24.298	(6/7).8571	10.286	52	1.834	22.01	105	3.704	44.45
24.806	(7/8).875	10.500	53	1.870	22.44	106	3.739	44.87
25.515	(9/10).9000	10.800	54	1.905	22.86	106.31	3.750	45.00
28.350	1.000	12.000	54.35	1.917	23.00	107	3.774	45.29
.2362	.00833	.10	55	1.940	23.28	108	3.810	45.72
.4726	.01667	.20	56	1.975	23.70	108.66	3.833	46.00
.5905	.02083	(¼) .25	56.70	2.000	24.00	109	3.845	46.14
.7087	.02500	.30	57	2.011	24.13	110	3.880	46.56
.7875	.02778	(⅓) .33	58	2.046	24.55	111	3.915	46.98
.9450	.03333	.40	59	2.081	24.97	111.05	3.917	47.00
1.1813	.04167	(½) .50	59.06	2.083	25.00	112	3.951	47.41
1.4175	.05000	.60	60	2.116	25.39	113	3.986	47.83
1.5751	.05556	(⅔) .67	61	2.152	25.82	113.40	4.000	48.00
1.6536	.05833	.70	61.43	2.167	26.00	114	4.021	48.25
1.7718	.06250	(¾) .75	62	2.187	26.24	115	4.057	48.68
1.8901	.06667	.80	63	2.222	26.66	115.75	4.083	49.00
2.1262	.07500	.90	63.79	2.250	27.00	116	4.092	49.10
.1	.00353	.0423	64	2.258	27.10	117	4.127	49.52
.2	.00705	.0847	65	2.293	27.52	118	4.162	49.94
.3	.01058	.1270	66	2.328	27.94	118.13	4.167	50.00
.4	.01411	.1693	66.15	2.333	28.00	119	4.198	50.38
.5	.01764	.2116	67	2.363	28.36	120	4.233	50.80
.6	.02116	.2540	68	2.399	28.79	120.49	4.250	51.00
.7	.02469	.2963	68.52	2.417	29.00	121	4.268	51.22
.8	.02822	.3386	69	2.434	29.21	122	4.303	51.64
.9	.03175	.3810	70	2.469	29.63	122.84	4.333	52.00
1.0	.03527	.4233	70.87	2.500	30.00	123	4.339	52.07
10	.35274	4.233	71	2.504	30.05	124	4.374	52.49
11	.38801	4.656	72	2.540	30.48	125	4.409	52.91
12	.42329	5.080	73	2.575	30.90	125.22	4.417	53.00
13	.45856	5.503	73.23	2.583	31.00	126	4.445	53.34
14	.49384	5.926	74	2.610	31.32	127	4.480	53.76
15	.52911	6.349	75	2.646	31.75	127.57	4.500	54.00
16	.56438	6.773	75.61	2.667	32.00	128	4.515	54.18
17	.59966	7.196	76	2.681	32.17	129	4.550	54.60
18	.63493	7.619	77	2.716	32.59	129.93	4.583	55.00

*See suggestions on preceding page.

Table A13—Weight Per Bushel (32 qts.) of Feedstuffs*

Feedstuff	Pounds	Feedstuff	Pounds
Alfalfa meal	19	Oatmeal	28
Alfalfa leaf meal	27	Oats, pulverized	26
Barley, whole	48	Oats, whole	32
Barley, ground	35	Peanut meal	42
		Peas, field or cow	60
Beans	60	Rice bran	25
Brewers grains, dried	19	Rice, rough or paddy	45
Buckwheat, whole	48	Rye, whole	56
Buttermilk, dried	38	Sardine meal	42
		Skim milk, dried	38
Corn, whole shelled	56	Soybean meal	54
Corn, ground	48	Spelt or emmer	40
Corn, gluten feed	41	Tankage	50
Corn gluten meal	54		
		Wheat bran	16
Cottonseed meal	48	Wheat flour middlings	38
Distillers grains, corn, dried	19	Wheat, ground	54
Grain sorghum; milo, kafir, etc.	56	Wheat standard middlings	26
Linseed meal	45	Wheat std. midds. with screenings	29
Meat scrap, 55% protein	51	Wheat whole	60
Millet, proso	48 to 50		
Navy beans	60	Whey dried	62
	Weight per dry quart		
Bonemeal, steamed	2 lbs.	Limestone, finely ground	2.8 lbs.
Granite grit, chick size	3 lbs.	Oyster shell, finely ground	2.3 lbs.
Mash	1 lb.	Salt, fine table	2.5 lbs.

*Such weights are inaccurate and should only be used where no scale is available. The feeds should be placed in the basket or measure and not packed tightly or heaped. The weights of the various grains are the standard for a bushel.

Table A14—Sixteenths to Tenths; Ounces to Tenths

Sixteenth inches or ounces	Inches or pounds
.5	.03125 (1/32)
.8	.050 (1/20)
1.0	.063 (1/16)
1.6	.100 (1/10)
2.0	.125 (1/8)
3.0	.1875
3.2	.200 (2/10)
4.0	.250 (1/4)
4.8	.300 (3/10)
5.0	.3125
6.0	.375 (3/8)
6.4	.400 (4/10)
7.0	.4375
8.0	.500 (1/2 or 5/10)
9.0	.5625
9.6	.600 (6/10)
10.0	.625 (5/8)
11.0	.6875
11.2	.700 (7/10)
12.0	.750 (3/4)
12.8	.800 (8/10)
13.0	.8125
14.0	.875 (7/8)
14.4	.900 (9/10)
15.0	.9375
16.0	1.000

Table A15—Data on Some of the Chemical Elements

Name	Symbol	Atomic Weight	Nature	Needed by Poultry?
Aluminum	Al	26.97	Metal	Prob. not
Arsenic	As	74.91	Metal	Prob. not
Barium	Ba	137.36	Metal	No
Bismuth	Bi	209.00	Metal	No
Boron	B	10.82	Non-metal	Prob. not
Bromine	Br	79.916	Gas	Prob. not
Calcium	Ca	40.08	Metal	Yes
Carbon	C	12.01	Non-Metal	Yes
Chlorine	Cl	35.457	Gas	Yes
Cobalt	Co	58.94	Metal	Yes
Copper	Cu	63.57	Metal	Yes
Fluorine	F	19.00	Gas	Prob. Yes
Gold	Au	197.2	Metal	No
Helium	He	4.003	Gas	No
Hydrogen	H	1.0080	Gas	Yes
Iodine	I	126.92	Non-metal	Yes
Iron	Fe	55.85	Metal	Yes
Lead	Pb	207.21	Metal	No
Lithium	Li	6.940	Metal	No
Magnesium	Mg	24.32	Metal	Yes
Manganese	Mn	54.93	Metal	Yes
Mercury	Hg	200.61	Metal	No
Nickel	Ni	58.69	Metal	Prob. not
Nitrogen	N	14.00	Gas	Yes
Oxygen	O	16.000	Gas	Yes
Phosphorus	P	30.98	Non-metal	Yes
Platinum	Pt	195.23	Metal	No
Potassium	K	39.096	Metal	Yes
Radium	Ra	226.05	Metal	No
Selenium	Se	78.96	Non-metal	No
Silicon	Si	28.06	Non-metal	Prob. Yes
Silver	Ag	107.88	Metal	No
Sodium	Na	22.997	Metal	Yes
Sulphur	S	32.06	Non-metal	Yes
Tin	Sn	118.70	Metal	No
Uranium	U	238.07	Metal	No
Zinc	Zn	65.38	Metal	Yes

Table A16—Percentage Relative Humidity at 30 inches (Normal sea-level and barometric pressure.)

	Dry-bulb temperature in degrees Fahrenheit													
	—10	0	10	30	35	45	55	60	65	96	98	100	102	106
.2	90	93	96	....	....	....	....	....	....	....	....	....	....	....
.5	73	83	89	94	95	96	97	97	97	....	....	....	....	....
1	46	67	78	89	91	93	94	94	95	96	96	96	96	97
2	....	33	56	78	81	86	88	89	90	93	93	93	93	93
3	....	1	34	67	72	78	82	83	85	89	89		90	90
4	....	....	13	56	63	71	76	78	80	86	86	86	86	87
5	....	....	....	46	54	64	70	73	75	82	83	83	83	84
6	....	....	....	36	45	57	65	68	70	79	79	80	80	81
7	....	....	....	26	36	51	59	63	66	76	76	77	77	78
8	....	....	....	16	27	44	54	58	61	73	73	73	74	75
9	....	....	....	6	19	38	49	53	56	69	70	70	71	72
10	....	....	....	....	10	31	43	48	52	66	67	68	68	69
11	....	....	....	....	2	25	38	43	48	63	64	65	65	66
12	....	....	....	....	....	18	33	39	44	61	61	62	62	64
13	....	....	....	....	....	12	28	34	39	58	58	59	60	61
14	....	....	....	....	....	6	23	30	35	55	56	56	57	58
15	....	....	....	....	....	....	19	26	31	52	53	54	55	56
16	....	....	....	....	....	....	14	21	27	50	50	51	52	53
17	....	....	....	....	....	....	9	17	24	47	48	49	49	51
18	....	....	....	....	....	....	5	13	20	44	45	46	47	49
19	....	....	....	....	....	....	0	9	16	42	43	44	45	46
20	....	....	....	....	....	....	....	5	12	39	40	41	42	44
21	....	....	....	....	....	....	....	1	9	37	38	39	40	42
22	....	....	....	....	....	....	....	....	5	35	36	37	38	40
25	....	....	....	....	....	....	....	....	....	28	29	30	32	34
27	....	....	....	....	....	....	....	....	....	24	25	26	28	30
30	....	....	....	....	....	....	....	....	....	18	19	21	22	24
33	....	....	....	....	....	....	....	....	....	12	14	15	16	19
36	....	....	....	....	....	....	....	....	....	7	8	10	11	14
39	....	....	....	....	....	....	....	....	....	2	3	5	7	10
42	....	....	....	....	....	....	....	....	....	....	....	1	2	5

STANDARD CONCRETE MIX

(Parts by measure)

Gravel or crushed rock up to ¾″ diam.	3
Sand, clean, fine enough to pass through ¼″ sieve	2
Cement (1 cu. ft. = 1 sack)	1
Total (parts)	6

Water requirements vary; usually one gallon per cu. ft. of the dry ingredients.

SECTION 4

PERIODICALS OF INTEREST TO TURKEY GROWERS

Popular Poultry Periodicals:*

American Egg and Poultry Review. Published monthly by the Urner-Barry Co. at 92 Warren St., New York 7, N. Y.

American Poultry Journal. Published monthly at 180 N. Wabash Ave., Chicago 1, Ill.

*No attempt has been made to make this list complete, neither has any periodical been purposely omitted. Periodicals of greatest interest and widely available to turkey growers are italicized.

Canadian Poultry Review. Published at 204 W. Richmond St., Toronto, Ont., Canada.

California Turkey News. Published annually by the California Turkey Growers' Assn., at 24 California St., San Francisco.

Everybody's Poultry Magazine. Published monthly at Hanover, Pa.

The Feathered World. (British.) Published weekly at 9 Arundel St., London, W.C. 2.

Feedstuffs. Pub. every Saturday at 118 So. 6th St., Minneapolis 2, Minnesota.

Hatchery and Feed. Published monthly at Mount Morris, Ill.

Minnesota Turkey Gobbles. Pub. monthly at 2200 Univ. Ave., St. Paul 14, Minn. by Minn. Turkey Growers Assn., Inc.

The National Poultry Journal. (British.) Published 13 times a year at Fyfield, Essex, England.

Northeastern Poultryman. Published the 15th of each month at 4 Park Street, Boston, Mass.

Norbest Turkey News. Published monthly at Salt Lake City 11, Utah, by Norbest Turkey Growers' Assn.

Northwestern Poultry Journal and Pacific Homestead. Published monthly at Salem, Oregon.

Nulaid News. Published monthly at 840 Battery St., San Francisco, Calif.

Pacific Poultryman. Published monthly at P. O. Box 521, Palo Alto, Calif. (Pacific edition of Poultry Tribune.)

Pacific Rural Press. Published weekly at 83 Stevenson St., San Francisco, Calif.

Poultry Craftsman and the Pacific Poultryman. Published monthly at 337 Central Ave., Los Angeles, Calif.

Poultry Herald. Published monthly at 310 Anchor Bldg., St. Paul, Minnesota.

Poultry Processing and Marketing. Published monthly at Mt. Morris, Ill., (formerly the U. S. Egg and Poultry Magazine).

Poultry and Egg National Board Reporter. Pub. monthly by P.E.N.B., 185 N. Wabash Avenue, Chicago 1, Ill.

Poultry Supply Dealer. Published monthly at 1230 Washington Blvd., Chicago, Ill.

Poultry Tribune. Published monthly at Mt. Morris, Ill.

The Poultryman. Published weekly at Vineland, N. J.

Southeastern Poultryman. Published monthly by Southeastern Poultry and Egg Association, 615 E. Franklin, Richmond 19, Va.

Turke-Gram, Official house organ of Natl. Turkey Federation. M. C. Small, Exec. Sec.-Treas., Mt. Morris, Ill.

Turkey World. Published monthly at Mount Morris, Ill.

Scientific Journals:*

American Journal of Hygiene. Pub. bi-monthly at Lancaster, Pa.

American Journal of Physiology. Published monthly at 19 W. Chase St., Baltimore 1, Md.

*No attempt has been made to make this list complete, neither has any periodical been purposely omitted. Periodicals most likely to contain turkey material are italicized.

American Journal of Veterinary Research. Published monthly at 600 So. Michigan Ave., Chicago 5, Ill.

American Naturalist. Published bi-monthly by Science Press, Lancaster, Pa.

Anatomical Record. Published monthly by Wistar Institute, 36th and Woodland Ave., Philadelphia 4, Pa.

Bibliography of Agriculture. Pub. monthly by the Library of the U.S.D.A., Washington 25, D. C.

Biological Abstracts. Published monthly Oct. to May and bi-monthly June to Sept., at 31 So. Frederick St., Baltimore 2, Md.

Canadian Journal of Agricultural Science. Published by Agr. Inst. of Canada, Ottawa, Ontario.

Canadian Journal of Research. Published monthly in 4 sections by the National Research Council at Ottawa, Ontario, Canada.

Cornell Veterinarian. Published quarterly at Ithaca, N. Y.

Genetics. Published bi-monthly at Brooklyn Botanic Garden, 1000 Washington Ave., Brooklyn, N. Y.

Journal of Agricultural Science. (British.) Published quarterly by Cambridge Univ. Press, E.C. 4, London, England. For U. S.; University, of Chicago Press.

Journal of American Veterinary Medical Association. Published monthly at 600 So. Michigan Ave., Chicago 5, Ill.

Journal of Biological Chemistry. Published monthly at Mt. Royal and Guilford Avenues, Baltimore 2, Md.

Journal of Experimental Biology. (British.) Published quarterly by Cambridge University Press, E.C. 4, London, England, and Academic Press, 123 E. 23rd St., New York, N. Y.

Journal of Experimental Zoology. Published nine time a year by Wistar Institute, 36th and Woodland Ave., Philadelphia, Pa.

Journal of Genetics. (British.) Published bi-monthly by Cambridge University Press, E.C. 4, London, England. For U. S.; University of Chicago (Ill.) Press.

Journal of Infectious Diseases. Pub. bi-monthly by the Univ. of Chicago Press, Chicago, Ill.

Journal of Heredity. Published monthly at 1507 M St., N.W., Washington, D. C.

Journal of Morphology. Published quarterly by Wistar Institute, 36th and Woodland Avenues, Philadelphia 4, Pa.

Journal of Nutrition. Published monthly by Wistar Institute, 36th and Woodland Avenues, Philadelphia, Pa.

Nature. (British.) Published weekly at St. Martin's Street, W.C. 2, London, England, and St. Martin's Press, 103 Park Ave., New York 17, N. Y.

North American Veterinarian. Published monthly at Box 872, Evanston, Ill.

Nutrition Abstracts and Reviews. (British.) Published quarterly by Rowett Institute, Bucksburn, Aberdeen, Scotland.

Poultry Science. Published bi-monthly at Menasha, Wis.

Proceedings of the Society for Experimental Biology and Medicine. Published monthly except Sept. at 104 Liberty St., Utica, N. Y.

Veterinary Medicine. Published monthly at Room 803, Livestock Exchange Bldg., Kansas City 15, Mo.

World's Poultry Science Journal. Pub. quarterly at Cornell University, Ithaca, N. Y.

SECTION 5

STATE COLLEGES OF AGRICULTURE OF THE UNITED STATES AND CANADA

(Except where listed separately, the Experimental Station is located at the same post office and is an integral part of the State College or University.)

(1) Correspondence dealing with diseases and parasites should be addressed to the Veterinary Department; where also diseased birds for diagnosis should be shipped express prepaid. (See list of Diagnostic Laboratories in Section 6.)

(2) Correspondence dealing with the general feeding and care of turkeys should be addressed to the Poultry Department. Every progressive turkey grower should have his name on the poultry mailing list of his own College of Agriculture.

(3) Matters dealing with the market situation should be addressed to the Department of Markets or Agr. Economics of either the Experiment Station or the State Department of Markets.

Alabama, Alabama Polytechnic Institute, *Auburn, Alabama.*

Arizona, University of Arizona, *Tucson, Arizona.*

Alaska, University of Alaska, *Palmer, Alaska.*

Arkansas, University of Arkansas, *Fayetteville, Arkansas.*

California, University of California, College of Agriculture or Vet. Med, *Davis, California.*

Colorado, The Colorado State College of Agriculture and Mechanic Arts, *Fort Collins, Colorado.*

Connecticut, University of Connecticut, *Storrs, Connecticut.*

Delaware, University of Delaware, *Newark, Delaware.*

Florida, University of Florida, *Gainesville, Florida.*

Georgia, University of Georgia, *Athens, Georgia.*

Hawaii, University of Hawaii, *Honolulu, Hawaii.*

Idaho, University of Idaho, *Moscow, Idaho.*

Illinois, University of Illinois, *Urbana, Illinois.*

Indiana, Purdue University, *Lafayette, Indiana.*

Iowa, Iowa State College, *Ames, Iowa.*

Kansas, Kansas State College, *Manhattan, Kansas.*

Kentucky, University of Kentucky, *Lexington, Kentucky.*

Louisiana, Louisiana State University, *Baton Rouge, Louisiana.*

Maine, University of Maine, *Orono, Maine.*

Maryland, University of Maryland, *College Park, Maryland.*

Massachusetts, University of Massachusetts, *Amherst, Mass.*

Michigan, Michigan State University, *East Lansing, Michigan.*

Minnesota, University of Minnesota, *University Farm, S. Paul, Minnesota.*

Mississippi, Mississippi State College, *State College, Mississippi.*

Missouri, University of Missouri, *Columbia, Missouri.*

Montana, Montana State College, *Bozeman, Montana.*

Nebraska, University of Nebraska, *Lincoln, Nebraska.*

Nevada, University of Nevada, *Reno, Nevada.*

New Hampshire, University of New Hampshire, *Durham, N. H.*

New Jersey, Rutgers University, *New Brunswick, New Jersey.*

New Jersey Turkey Experiment Station, *Millville, N. J.*

New Mexico, New Mexico College of Agriculture, *State College, New Mexico.*

New York, Cornell University, *Ithaca, New York.*

North Carolina, N. C. State College, *Raleigh, N. C.*

North Dakota, North Dakota Agricultural College, *Fargo, N. D.*

Ohio, Ohio State University, *Columbus, Ohio.*
Ohio Agricultural Experiment Station, *Wooster, Ohio.*
Ohio Animal Disease Diagnostic Laboratory, *Columbus, Ohio.*

Oklahoma, Oklahoma Agricultural and Mechanical College, *Stillwater, Oklahoma.*

Oregon, Oregon State Agricultural College, *Corvallis, Oregon.*

Pennsylvania, Pennsylvania State University, *University Park, Pa.*

Puerto Rico, University of Puerto Rico, *Rio Piedras, Puerto Rico.*
Federal Experiment Station, *Mayaguez, Puerto Rico.*

Rhode Island, Rhode Island State College, *Kingston, Rhode Island.*

South Carolina, Clemson Agricultural College, *Clemson, S. C.*

South Dakota, South Dakota College, *Brookings, S. D.*

Tennessee, University of Tennessee, *Knoxville, Tennessee.*

Texas, Agricultural and Mechanical College of Texas, *College Station, Texas.*

Utah, Utah State Agricultural College, *Logan, Utah.*

Vermont, University of Vermont, *Burlington, Vermont.*

Virginia, Virginia Polytechnic Institute, *Blacksburg, Virginia.*

Virgin Islands, Agricultural Experiment Station, *Christiansted, St. Croix, Virgin Islands.*

Washington, State College of Washington, *Pullman, Washington.*

West Virginia, West Virginia University, *Morgantown, W. Va.*

Wisconsin, University of Wisconsin, *Madison, Wisconsin.*

Wyoming, University of Wyoming, *Laramie, Wyoming.*

U. S. Department of Agriculture

(Laboratories with Poultry Research Underway)

Agricultural Research Center, USDA Station, *Beltsville, Md.*

North Central Poultry Testing Station, *Lafayette, Indiana.*

Regional Poultry Research Laboratory, *East Lansing, Michigan.*

Southwest Poultry Experiment Station, *Glendale, Arizona.*

Western Utilization Research Branch, *Albany, California.*

Canada

Dominion Department of Agriculture, Exp. Farms Service, Central Experimental Farm, *Ottawa, Ontario.* Dominion Experimental Farm, *Swift Current, Sask.*

Alberta, University of Alberta, *Edmonton, Alberta.*

British Columbia, University of British Columbia, *Vancouver, British Columbia.*

Manitoba, University of Manitoba, *Winnipeg, Manitoba.*

New Brunswick, Department of Agriculture, *Fredericton, N. B.*

Nova Scotia, College of Agriculture, *Truro, Nova Scotia.*

Ontario, Ontario Agricultural College, *Guelph, Ontario.*

Quebec, Macdonald College, *Macdonald College, Quebec.*

Saskatchewan, University of Saskatchewan, *Saskatoon, Sask.*

SECTION 6

DISEASE DIAGNOSTIC LABORATORIES IN U. S.

Instructions for Shipping

1. Ship express prepaid or take to laboratory.
2. Send 2 or 3 live specimens showing typical symptoms.
3. Send a letter (envelope glued or tacked on the coop) telling the conditions of feeding, brooding, care, and the age of the birds. Indicate flock symptoms and mortality to date.
4. Do not bring or ship to arrive on Saturday or Sunday unless arrangements have been made in advance (by telephone).
5. Send birds to your own State laboratory or a commercial laboratory.

STATE LABORATORIES WHERE BIRDS CAN BE SENT FOR DISEASE DIAGNOSIS

(Send birds to your own state laboratory)

Alabama—Veterinary College, Alabama Polytechnic Institute, Auburn, Ala.

Arizona—Veterinary Division, University of Arizona, Agri. Exp. Sta., Tucson, Ariz.

Arkansas—Dept. of Bacteriology and Veter. Science, University of Arkansas, Fayetteville, Ark.

California—(seven laboratories), State Dept. of Agri., U. S. Highway 101, Petaluma, Calif.; State Dept. of Agri., 1451 Mirasol St., Los Angeles, Calif.; Co. Livestock Dept., 4005 Rosecrans

Ave., San Diego, Calif.; Animal Path. Lab., State Office Bldg. No. 1, Sacramento, Calif.; University of Calif., College of Veterinary Science, Davis, Calif.; Poultry Pathological Lab., District Fair Grounds, Turlock, Calif.

Colorado—Diagnostic Lab., School of Veterinary Medicine, Colorado A. & M. College, Fort Collins, Colo.

Connecticut—Department of Animal Diseases, Storrs Agricultural Experiment Station, Storrs, Conn. (via Willimantic, Conn.)

Delaware—Poultry pathologist, State Board of Agriculture, Dover; Del. Substation, Georgetown, Poul. Dept., U. of Del., Newark, Del.

Florida—Poultry Disease Diagnostic Laboratory, P. O. Box 248, Dade City, Florida. (Laboratory located at Pasco County Fair Grounds); Florida Agricultural Experiment Station, Department of Veterinary Science, Gainesville, Florida.

Georgia—College of Vet. Med., University of Georgia, College of Agriculture, Athens, Ga.

Idaho—Veterinary Department, Experiment Station, Moscow, Ida.

Illinois—Diagnostic Laboratories, College of Vet. Med., University of Illinois, Urbana, Ill.

Indiana—Purdue Diagnostic Laboratory, Purdue University, Agri. Exp. Sta., Lafayette, Ind.

Iowa—Iowa Vet. Diagnostic Lab., Iowa State College, Ames, Iowa.

Kansas—Department of Pathology, School of Vet. Med., Kansas State College, Manhattan, Kans.

Kentucky—Veterinary Department, Agri. Exp. Sta., Lexington, Ky.

Louisiana—Louisiana Livestock Sanitary Board Diagnostic Lab., P. O. Box 1951, Baton Rouge, Louisiana; Branch Poultry Diagnostic Laboratory, Louisiana Polytechnic Institute, Ruston, Louisiana.

Maine—Department of Animal Pathology, University of Maine, Orono, Me.

Maryland—(six laboratories), Livestock Sanitary Service Laboratory, College Park, Md.: Salisbury, Md.; Bel Air, Md.; Centreville, Md.; Frederick, Md.; and Hagerstown, Md.

Massachusetts—Department of Veterinary Science, Massachusetts Agricultural Experiment Station, Amherst, Mass.

Michigan—Poultry Diagnostic Laboratory, Michigan State Univ., East Lansing, Mich.

Minnesota—Animal Diagnosis Lab., University Farm, St. Paul 1, Minn.

Mississippi—State Diagnostic Lab., 2531 No. West St., Jackson, Miss.

Missouri—Poultry Diagnostic Lab., Connaway Hall, U. of Missouri, Columbia, Mo.

Montana—(two laboratories), Veterinary Department, State College, Bozeman, Mont.; State Veterinarian, Livestock Sanitary Board, Helena, Mont.

Nebraska—Department of Animal Pathology, University of Nebraska, Lincoln, Nebr.

Nevada—Veterinary Control Service, University of Nevada, Reno, Nevada.

New Hampshire—Poultry Pathology Lab., University of New Hampshire, Durham, N. H.

New Jersey—(two laboratories), College Farm, New Brunswick, N. J.; Poultry Pathology Lab., Landis and Brewster Sts., Vineland, N. J.

New Mexico—N. Mex. State College Poultry Department, Mesilla Park, N. Mex.

New York—(two laboratories), New York State Veterinary College, Cornell University, Ithaca, N. Y.; State School of Applied Agri., Farmingdale, L. I., N. Y.

North Carolina—Poultry Disease Laboratory, State College, Raleigh, N. C.

North Dakota—Department of Animal Pathology, N. Dak. Agricultural College, State College Station, Fargo, N. Dak.

Ohio—Reynoldsburg Diagnostic Service Lab., Div. of An. Ind., P. O. Box 2159, Columbus, Ohio.

Oklahoma—College of Veterinary Science, Okla. A. & M., Stillwater, Okla.

Oregon—Department of Veterinary Medicine, Ore. Sta. Agri. Col., Corvallis, Ore.

Pennsylvania—Pa. Bureau of Animal Industry Laboratory, P. O. Box 403, Harrisburg, Pa. (Located at Summerdale in Cumberland County; mailing address is Harrisburg); Regional Poultry Diagnostic Laboratory, University of Pennsylvania, New Bolton Center, R.F.D. 1, Kennett Square, Pa.; Regional Poultry Diagnostic Laboratory, Pennsylvania State University, College of Agr., State College, Pa.; Regional Poultry Diagnostic Laboratory, Mounted Route 11, Butler, Pa.; Regional Poultry Diagnostic Laboratory, Tunkhannock, Pa.; Regional Poultry Diagnostic Laboratory, National Agricultural College, Farm School, Bucks County, Pa.

Rhode Island—Poultry Department, Rhode Island State College, East Farm, Kingston, R. I.

South Carolina—Livestock Sanitary Department, P. O. Box 1174, Columbia, S. C. (Lab. at Pontiac.)

South Dakota—Animal Health Laboratory, South Dakota College of Agriculture, Brookings, S. Dak.

Tennessee—Department of Veterinary Science, University of Tennessee, Knoxville, Tenn.

Texas—(two laboratories), Livestock Sanitary Commission, 304 Waggoner Building, Fort Worth, Tex.; Veterinary Hospital, A. & M. College of Texas, College Station, Tex.

Utah—Animal Pathology Laboratory, Utah Agri. Exp. Sta., Logan, Utah.

Vermont—Dept. of Animal Pathology, Hills Science Bldg., U. of Vermont, Burlington, Vt.

Virginia—(six laboratories), State Diagnostic Laboratory, Accomac, Harrisonburg, Ivor, Richmond (1102 State Bldg.), Warrenton and Wytheville, Va.

Washington—(two laboratories), Veterinary Diagnostic Lab., State College of Washington, Pullman, Wash.; Western Washington Experiment Station, Puyallup, Wash.

West Virginia—(three laboratories), Animal Path. Lab., West Virginia University, Morgantown, W. Va.; State Veterinarian, Charleston, W. Va.; State Dept. of Agr., Moorefield, W. Va.

Wisconsin—State Disease Control Lab., 2115 Linden Drive, Madison, Wisconsin.

Wyoming—Veterinary Department, University of Wyoming, Laramie, Wyoming.

CANADIAN DIAGNOSTIC LABORATORIES

Alberta—Prairie Area Branch Laboratory, Dominion Department of Agriculture, Lethbridge, Alta.; Veterinary Laboratory, Provincial Department of Agriculture, Edmonton, Alta.

British Columbia—Pacific Area Branch Laboratory, Dominion Department of Agriculture, Hut O-18, University of British Columbia, Vancouver, B. C.; Veterinary Laboratory, Provincial Department of Agriculture, Vancouver, B. C.

Manitoba—Veterinary Laboratory, Provincial Department of Agriculture, University of Manitoba, Winnipeg, Man.

Brunswick—Maritime Area Branch Laboratory, Dominion Department of Agriculture, Mount Allison University, Sackville, N. B.; Veterinary Laboratory, Provincial Department of Agriculture, Fredericton, N. B.

Nova Scotia—Veterinary Laboratory, Provincial Department of Agriculture, Truro, N. S.

Ontario and Quebec—Animal Diseases Research Institute, Dominion Department of Agriculture, Hull, Que.; Montreal Area Branch Laboratory, Dominion Department of Agriculture, Macdonald College, Que.; Ontario Veterinary College, Guelph, Ont.; School of Veterinary Medicine, St. Hyacinthe, Que.

Prince Edward Island—Veterinary Laboratory, Provincial Department of Agriculture, Charlottetown, P. E. I.

Quebec—Animal Diseases Research Institute, 100 Gamelin Blvd., Hull, Que.

Saskatchewan—Department of Veterinary Science, University of Saskatchewan, Saskatoon, Sask.

SECTION 7

DRAWING AND BONING

(Prepared by *Turkey World*)

Many turkey growers have increased their turkey profits by selling a large percentage of their turkeys dressed and drawn—ready for the oven. Correctly speaking, "dressed" means killed and picked. The bird is said to be "drawn" when the feet, head and neck and "insides" are removed.

When catering to this type of trade, particular care should be taken to turn out an attractive carcass, free from ugly cuts and tears. When properly dressed and drawn, a turkey is really a thing of beauty which excites the appetite, but otherwise, it may actually be repulsive and destroy the consumer's appetite for turkey.

The hair should be singed, by holding the bird over a hot flame. Care should be taken not to burn the skin.

If the bird has been cooled, the body will be stiff, and it will be necessary to flex the legs. Grasp the legs firmly, and flex the joints by bending the legs against the body.

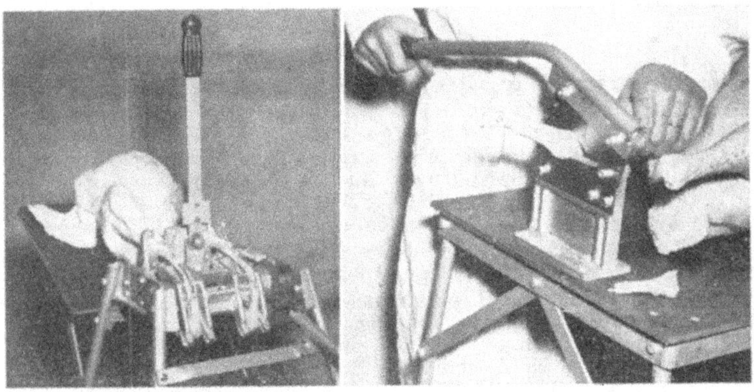

Fig. A2. Alligator jaw grip tendon puller (left) and dressing guillotine (right). (Photo from Stephen's Turkey Farm, Hamburg, N. Y.)

Removing the tough tendons from the legs will add to consumer satisfaction. This may be done by breaking the leg about two inches below the hock joint, placing the foot in a bracket that holds it firmly, grasping the thigh, and pulling downward with the hands until the tendons pull out. Commercial tendon pullers do the job more quickly and easily.*

Pulling Tendons

Another method of pulling tendons is as follows: Cut skin along side of shank to expose the main tendons, then cut skin in front of hock to expose tendons on front of leg. Tendons then can be removed one by one with a steel skewer or tendon puller.

Drawing

Slit the skin on the *back* of the neck from the shoulders to the head. The crop can then be reached. Work it loose with the hands and pull it out, bringing with it the windpipe and esophagus, all of which remain attached to the head. Now disjoint the head from the neck and remove it with the attached crop, windpipe, and esophagus. Cut through the neck muscles at the shoulders, and the neck may be twisted off easily.

There are two general methods of removing the entrails. The method in most common use is to make an incision crosswise of the abdomen about midway between the vent and the point of the

*The International Edge Tool Co. (Newark 3, N. J.) makes a single puller.

breastbone. Some prefer to make the incision about one and one-half inches to two inches from the vent. The hand is then thrust through this incision and the entrails loosened and pulled out. Cut the skin entirely around and slightly away from the vent and remove it with the entrails.

The second method of removing the entrails is to cut around the vent first and removing the entrails through this opening without making a second incision.

Carefully remove the gall bladder from the liver. Open the gizzard and remove the lining and contents. Separate the heart from the other organs. Wipe the carcass with a clean cloth. If it is to be prepared ready for the oven it should be washed. Place the neck, liver, gizzard, and heart inside of the carcass. They may be wrapped in wax paper. The skin of the neck should be folded over the back.

If the first method of drawing is used, there will be a strip of skin and flesh between the point where the first incision was made and where the vent was removed. The legs can then be thrust through this fold of skin, which helps to hold the desired shape. Clean butcher's twine tied around the legs and tail will serve the same purpose.

Boning a Turkey

Turkeys may be completely boned and a turkey roll made of the meat. This is a very fine method of preparing exceptionally large birds for cooking.

The process of boning a turkey is comparatively simple, but rather difficult to describe. First, remove a part of the wings by separating at the second joint. Slit the skin from neck to vent along a center line over the breast bone. Remove the skin in one piece by pulling from the flesh. Loosen with the knife where necessary. After the bird has been skinned, remove the legs by disjointing at the "hip" joint. Also remove the wings.

Remove the flesh from the carcass. Start at the end of the breast and pull and work the flesh from the bone. Remove the remainder of the flesh from the carcass. Stretch the skin out flat and distribute the flesh over it. The light and dark meat may be arranged in different ways as desired. Start at one side and roll as tightly as possible. Tuck the ends of the skin under and tie the roll with butcher's cord.

Several variations of the roll may be used. The roll may be made entirely of dark or of light meat. Dressing may be placed on the meat and skin before making the roll. This is the usual method of making the roll for immediate cooking. If it is to be frozen and stored, dressing should not be added.

The roll is then roasted. It can be sliced and served in many attractive way.

SECTION 8

SMOKING AND BARBECUING

Smoking*: Turkeys to be used for curing and smoking should be subjected to the usual overnight fasting period (with access to

*U. S. Dept. Agr. recommendations from A.H.D. 28.

water), then bled and brained through the mouth, and carefully dry-picked, or the feathers removed by the slack-scald method in water heated to 126° F. Care should be taken not to break the skin in the dressing and handling. Birds with badly torn skin should be rejected. The full or sub-scald is undesirable since the skin is likely to be injured when this method is used. Immediately after being picked, the birds should be drawn and then chilled to an internal temperature of 30° to 40° F. To prepare for curing, remove the head, viscera, neck, and feet, leaving the body cavity open at both the fore and rear ends. Then pull the tendons (as suggested in Chapter 6) so the drumstick meat will be improved in curing.

A suitable curing mixture consists of 6 pounds of salt, 3 pounds of sugar, and 2 ounces of saltpeter dissolved in 4½ gallons of water. This pickle contains approximately 13 per cent of salt and has a salinometer reading of about 70° at a temperature of 38° F. Experience has shown that about four times this indicated quantity of pickle is required to cover 100 pounds of moderately large dressed turkeys when packed carefully in a 50-gallon barrel.

The dressed turkeys** should be packed carefully and close together in a suitable container, such as a crock or a clean, well-soaked, odorless, hardwood barrel, and weighted down with a clean board and brick or stone so they will not float when the curing solution is added. Then pour the solution over the turkeys until they are covered with a slight excess of liquid. It is important that the temperature of both the meat and the pickle be approximately 38° F. when the curing process is begun and be kept at that point throughout the curing period. At weekly intervals the turkeys should be removed from the container and repacked in order to remix the pickle and to insure that it will come in contact with all parts of the bird.

Depending on the weight of the turkeys, the meat should be sufficiently cured in from 2 to 4 weeks to be removed from the pickle and prepared for smoking. It is suggested that the individual turkey remain in cure not less than 1⅓ days per pound of drawn weight; probably 1½ days per pound would be a good standard to follow.

The cured turkey should be washed in warm water, hung up until dry, and then smoked, using hardwood, at a suggested smokehouse temperature of about 110° F. During the smoking process it is believed the turkey should be hung by the wings in order to provide for maximum exposure of skin as well as an opportunity for further drainage of curing fluid, especially from the body cavity. About 20 hours in hardwood smoke gives a mild smoked flavor, although some persons may prefer to smoke the meat longer, even to the extent of having a fire under it 8 to 10 hours each day for several days. However, in addition to other effects, smoking also causes dehydration, and therefore long-continued smoking may result in the concentration of salt in the meat to a point where is becomes objectionable. Turkeys prepared in this manner must be cooked, much in the manner of smoked hams, before eating.

To produce a smoke-cooked turkey that is ready for eating without further cooking, the following procedure is recommended by the Extension Service, Texas A. & M. College, College Station,

**See page 453 for data on weight loss of smoked turkeys.

Texas, in Bulletin No. B-163. After selection and drawing as described above, proceed as follows:

"Step 1—Chill bird to remove animal heat before pumping for curing.

Step 2—Pump the bird with brine mixture by injecting 10 per cent of its weight. Stitch with a pressure pump using small or medium sized needle. Pump bird to give uniform distribution of brine in all muscles. This will require on each side of the bird three stitches in breast, one in thigh, one in drumstick, one in wing and one in back.

Brine Mixture for Curing: This mixture should give a reading of 45 to 50 per cent saturation when measured with a sodium chloride salinometer: 10 gallons of water; 9 pounds of salt; 1 pound prague powder; 1½ pounds sugar; 4 ounces ham spice emulsion.

Step 3—After bird has been pumped, cover it with the brine mixture. Be sure bird is covered with the solution. Keep it in this solution for three days.

Step 4—Remove bird after three days and drain thoroughly. Be sure none of the brine is left in the pockets of body cavities. Put it in stockinette and hang breast down.

Step 5—After it is about dry, put it in the smokehouse with the heat control set at 170° F. As soon as it is completely dry, smoke can be applied.

Step 6—Smoke to a light lustrous pecan nut brown. This usually takes 8 to 12 hours.

Step 7—When the desired color is obtained, increase the temperature in the smokehouse to 185° F. Cook the bird until the inside temperature at the thickest breast muscle area is 160° F. To determine this heat accurately it is necessary to insert an inside meat thermometer in the bird at the thigh joints and breast muscles. Approximately 20 per cent shrinkage may be expected from this process.

Step 8—This process will cook the bird sufficiently to be eaten without additional cooking. This cooked product will not keep in ordinary refrigerator temperatures any longer than other meats such as pre-cooked cured ham. If the birds are to be held longer than two weeks they should be frozen, packaged, and held at 0° F. temperature."

Barbecuing*: "To prepare the dish, one must first of all have a turkey trussed and dressed for roasting. Any sound stuffing material will do. The favorite dressing is a rich blend of wild rice, whole oysters, and half pecans, seasoned to taste. Another variant consists of tame rice—steamed—to which quartered apples, raisins, and pecans are later added.

In addition to the turkey, the requisite materials include five or six large cabbages, a dishpan, a burlap bag, some twenty bricks, a long-handled ditching shovel, a baker's dozen of large boulders, a lemon, four or five red-pepper pods of unimpaired vigor, an adequate supply of firewood, a strong back, some baling wire, and approximately ten hours.

*Quoted by permission of the editor and author, from the **Saturday Evening Post**, Jan. 19, 1935. Recipe of H. B. Deutsch.

Begin by digging a pit about two and a half feet deep, and wide enough to leave a clearance of eight inches or so between the dishpan and the clay walls. Floor this pit with bricks, on top of which you light your fire. Into the flames, when the blaze is well advanced, tumble all of the boulders. Add fuel liberally from time to time, tossing the chunks atop the boulders until these are at the heart of a roaring bonfire. Keep the blaze going for at least two hours, at the end of which period the stones will have acquired a cherry-red color. Dismember the cabbages, leaf by leaf, and line the dishpan—sides as well as bottom—to the depth of about three inches with the crisp leaves that remain.

Lay the turkey upon this bedding of cabbage. The turkey is covered over with more cabbage leaves to the approximate depth of 3 inches; while the lemon is sliced and laid atop the mound, together with the pepper pods. The burlap bag is then dipped in water, wrung out, and tied over the dishpan with its freight of cabbage-wrapped turkey, being wired into position for greater security. Four or five feet of wire should be left projecting beyond each handle of the dishpan when the job is finished.

Then, holding the pan by the free ends of wire attached to the handles, lower it into the pit until it rests upon the hot bricks. Toss over the pan the coarse cabbage leaves previously laid aside. On top of them place two or three armfuls of freshly cut grass. We use Spanish moss for this purpose, but it is not available in the North. With the shovel, tumble the hot boulders down on top of the grass and, without a moment's delay, fill the pit solidly with the earth that was dug to create it. Dig some more earth and pile it over all, tamping it down with the back of the shovel, until the hot spot has been well sealed and resembles a newly made grave. After the pit has remained sealed for five hours, shovel out dirt, grass, boulders and all until you bare the pan, which—using the two wires—you must lift out carefully, for it will be at least half full of a very rich gravy. Untwist the wires and lay back the slightly charred burlap bag. The aroma is, not to put too fine a point on it, something swell. Even the most skeptical of your guests will begin to look distinctly hopeful when that savory steam permeates the vicinage. The rest will soon be history."

SECTION 9

HINTS FOR PLANNING AND MANAGING A LIVE TURKEY SHOW

The secretary generally handles the following items:

(1) Solicits exhibitors

(2) Solicits donations for show expenses and prizes

(3) Puts out the catalog (a booklet or merely a handbill) listing the prizes, judges, classes to be shown, place and date of show, deadline for entries, rules of the show, fees, explanations, and advertising (which usually pays for the catalog).

(4) Advertises the show through newspapers, handbills, and signs

(5) Secures the judges
(6) Aids amateur exhibitors to select and prepare their birds
(7) Secures showroom, coopage, etc.
(8) Classifies and coops the birds
(9) Makes out list of awards as soon as the judges place the birds
(10) Pays the awards and judges' fees
(11) Crates and returns the birds
(12) Keeps records and accounts

The superintendent is the secretary's right hand man. His responsibilities may be:

(1) To assist the secretary in his work
(2) To inspect the birds as they are uncrated. Those showing evidences of disease should not be accepted but replaced in their crates and their owners notified at once. Anyone handling sick specimens should disinfect his hands and any equipment with which the sick birds have come in contact.
(3) To set up the coops just before the show opens
(4) To supervise cooping the birds as they arrive
(5) To assist the judges
(6) To place the award stickers
(7) To see that the birds are fed and watered regularly
(8) To keep the coops and showroom clean
(9) To clean up the showroom and properly dispose of the equipment after the show.

Some of the equipment needed in setting up the show is as follows:

(1) A showroom of sufficient size and well lighted, both with daylight and artificial light
(2) Tables, benches, or frames to hold the coops at a convenient height, 2 to 3 feet
(3) Uniform steel-wire exhibition coops of standard size for turkeys, the number based on the estimated number of entries plus 20 per cent
(4) Coop tags, preferably with exhibitor's name at bottom; to be out of sight until judging is completed.
(5) Numbered legbands for emergency use (to be furnished to the exhibitors in case they have not banded their birds). Even then a special series of bands is advisable
(6) Litter for coops (shavings, straw, hay, excelsior, or some commercial litter), unless wire mesh bottoms are used
(7) Water and feed cups to attach to side of coops (2 cups for each coop)
(8) Cleaning equipment
(9) Spray pump and disinfectant (3% or 4% compound cresol solution or 1% sodium orthophenylphenate.)
(10) Carpenter's tools for uncooping, recooping, and arrangeing the equipment

(11) Feed: A well-balanced mash and a scratch grain mixture. (Sometimes a local feed dealer will supply feed free for the advertising and good will.)
(12) Water supply
(13) Tincture of iodine or other standard antiseptic for treatment of wounds
(14) Colored and lettered award stickers made of gummed paper
(15) Colored silk award ribbons (one for each prize listed) to be given to the owner of the winning birds. They correspond in color to the award stickers and usually are lettered, telling the name, date, and place of the show and the letters "First Prize," "Champion," or whatever the award may be
(16) An accurate scale for use in weighing birds
(17) Sleeping quarters in the showroom so that a caretaker may be on duty day and night to prevent thievery, fire, or loss from other causes

A separate steel-wire exhibition coop should be provided for each bird, or, if exhibited in pairs, trios, or pens, one coop for each group. For hens, a coop 30″ x 30″ by 30″ high or one 36″ x 36″ by 33″ high is of adequate size. For toms, coops 36″ x 36″ by 33″ high are a good size. For very large toms, larger coops are preferred—of a size about 36″ x 42″ to 48″ by 36″ high. If one size only is used, the kind 36″ x 36″ by 33″ high is most practical. Well-constructed coops will be so built that the birds cannot fight through the ends. It is highly desirable to have coops in which the toms are kept separated to prevent fighting and feather damage. If special turkey coops cannot be obtained, the regular 5-bird (pen) size chicken exhibition coop (24″ x 48″ by 27″ high) can be pressed into service, each one to hold a tom or a hen. Single-bird chicken coops (24″ x 24″ by 28″ high will hold a turkey hen but her plumage is likely to be damaged. However, chicken-size cooping is at best only a makeshift and its use may cause experienced exhibitors to refuse to exhibit their best birds. The coops are set on wooden platforms which should have a strip around the edge to prevent loss of litter. Alfalfa hay or clean shavings make good litter. Straw, peat moss, dehydrated sugar cane, and other types may be used. Wire floors in place of litter are excellent. A feed can and water cup are hooked to the side of each coop.

The coops should not be double-decked if it possibly can be avoided. The usual height of the platforms from the floor is about 2½ feet. Usually two rows of coops are placed back to back with a five-foot alleyway between each double row. To prevent the birds fighting through the backs of the coops when placed in double row, unless "fight-proof" coops are used, the rows are placed a foot or more apart, or, if placed close together, a strip of cloth or heavy paper is placed between them. In cold weather, the temperature of the live-bird showroom usually is maintained at about 60° F. At this temperature, both birds and spectators will be comfortable. Too high a temperature may increase the possibility of the birds catching cold when they are taken from the showroom, and dressed exhibits will not keep (see remarks on handling of dressed exhibits). If the showroom is lighted at night, the lights should be dimmed gradually to accustom the birds to the darkness in a strange place.

Each bird, pair, trio, or pen entered is provided with an entry tag upon which is written the legband (or possibly wingband and tattoo) number, the entry (or exhibitor's) number, and the variety, age, sex, and weight. Each bird should be weighed when it is first cooped and the weight recorded on the entry tag. Judges, exhibitors, and spectators all are interested in the weights. Sometimes the exhibitor's name is not written on the tag or otherwise disclosed until after the judging has been done. After that, it is essential to the best interests of the show to do so. Exhibitors should be required to *place a legband or wingband on each bird* or better, the show management should, for its own protection to prevent a possible mixup or dispute over ownership. This is a very important point for the show management to observe. Legbands cost only a very small sum and can be applied very quickly. Each entry when it arrives is written up in the entry book with all necessary details (such as band numbers, class, sex, variety, name and address of exhibitor, and entry number.) The entry number is the key to all the birds entered by each exhibitor. The shipping or carrying coops of each exhibitor are stored during the show, often

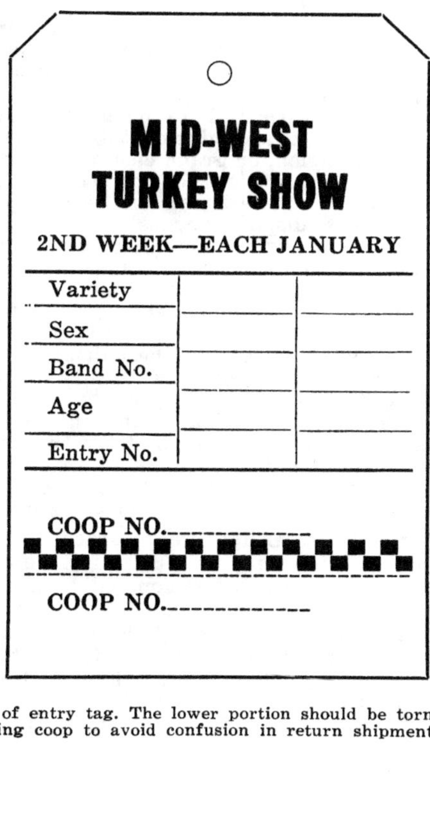

Fig. A3. Sample of entry tag. The lower portion should be torn off and fastened on the shipping coop to avoid confusion in return shipments.

by placing them under the exhibition coops, just below the birds that were shipped in them. The exhibitor's name and address should be marked on the carrying coops before the birds are removed. (See Fig. A3.)

Usually the birds of each variety are cooped together, preferably each class or subclass separately. For example, all the Bronze turkeys exhibited are grouped together and all the Bronze young toms placed in adjacent coops, all the Bronze young hens in adjacent coops, and so on with each variety and sex. Live market classes and special classes (such as 4-H Club entries) should be cooped separately, but in accord with the same general scheme. Attention to these details makes for better judging and greater interest in the show by spectators and exhibitors alike.

Colored labels should be provided for the prize-winning birds, and pasted on the corresponding entry tag. Champion or Grand Champion labels usually are in the form of deep purple rosettes and Reserve Champion a light purple. First-prize labels are deep blue; second prize, red; third, yellow; fourth, green; fifth, white; sixth, light blue; seventh, pink; eighth, orange; ninth, brown, and tenth, gray. Lower placings than these should be identified by appropriate figures written on the card by the judge or clerk. Proper labeling of prize winners is most essential, and should be done as quickly as possible after judging.

Awards. The number of placings to be made in each class is determined by the show management. Where classes are large, many placings, enough to include three-fourths or more of the birds shown in each class, frequently are made, although prizes and ribbons seldom are furnished for placings below tenth. Small shows usually award first, second, and third prizes in each class of each variety; a champion, or grand champion and reserve champion of the show; champion prizes for the best bird in each variety or class; and a prize for the largest and best display or in some cases merely the largest display. Largest and best displays may be determined by the point system, counting the number of placings won by each exhibitor and giving each a value approximately the reverse of the placing. This value is then multiplied by the number of entries exhibited in the particular class to obtain the number of points which are then added. The exhibitor obtaining the highest sum total of points wins largest and best display. For example, if five prizes are awarded, a first prize is valued at 5 points, a second prize at 4 points, a third prize at 3 points, a fourth prize at 2 points, and a fifth prize at 1 point. If an exhibitor wins say third prize in a class where 10 birds are exihibited the winning is worth 30 points while a first prize in a class of three would be worth only 15 points. Special prizes such as the best-shaped bird, or the best-colored bird in the show are sometimes awarded.

Champion prizes generally are awarded in each class, all varieties competing. There may be a prize for the champion young tom, champion yearling tom, champion old tom, champion young hen, champion yearling hen, and champion old hen. The champion for each class is selected from among the first-prize birds of that class in all varieties and is expected to be the highest-scoring bird of all the first-prize birds of the class. The grand champion is the highest-scoring bird in the show, all varieties and classes competing. This award may go to a male or female of any age or variety.

The reserve grand champion is the second highest-scoring bird. In competing for champion (sweepstake) awards the solid-colored varieties are handicapped slightly as follows: White Holland, 2 points; Black, 1½ points; Slate, ½ point. Handicapping is the practice of subtracting the number of points indicated from the scores of all solid-colored individuals when comparisons are made for awards.

Selecting and Preparing Birds for Showing

Selection. Robinson long ago noted that "an expert exhibitor is selecting, preparing his exhibition birds from the time they are hatched. . . . In large measure, success in exhibiting poultry depends upon having specimens just right for the shows at which they are to be exhibited. . . . The seasoned exhibitor plans for it."

Amateur exhibitors have neither the time nor the facilities to expertly select and condition their birds for showing. Nor would this be justified. But a reasonable amount of planning and thought on the subject will result in keener competition and more interest in shows.

Some exhibitors, through lack of foresight, fail to select good quality birds or exhibit good birds in poor condition. Several weeks before the show, the exhibitor should be on the lookout for the best specimens in his flock and band them. This also identifies the best prospects for breeding and sales. In fact, it is well to handle the whole flock at some convenient time well in advance of the show. The most promising birds may then be marked and retained. This applies to the selection of breeding and sales stock and to showing in the meat classes as well as in the exhibition classes.

Young turkeys, if well-raised, are in prime condition for exhibition or meat showing at about 7 or 8 months of age; old turkeys are in good condition from late October to March under most conditions, unless lighter for winter production.

Feeding. Sunflower seed is said to enhance the natural gloss of the plumage. Soybean meal and corn gluten meal added to the mash appear to favor feather development. These feeds in many sections are economical for general use as well as desirable in promoting feather development. Linseed meal, though favored as a feather conditioner by some poultrymen, has not been shown to be of any special value in turkey feeding and is not recommended. Special feeding methods seldom are justified; a good standard feeding program should be sufficient.

Putting old turkeys on a maintenance ration of grain supplemented by abundant green feed from the range during the summer will encourage early moulting, so that most of them will have their new plumage by September. Following the maintenance period with the ad libitum feeding of a mixed grain diet supplemented by abundant green feed or a mash and grain diet fed in separate self-feeders in the fall will cause weight to be put on rapidly, and prime condition is essential whether the birds are exhibited alive or dressed. As with young birds, a natural environment with some range is best, especial care being taken to avoid overcrowding and insanitary conditions.

Feather picking. Feather picking ruins birds for showing. It occurs as a rule only in young growing birds. It can be prevented to a large extent by the precautions listed in Chapter 16.

Conditioning: Conditioning young birds for showing in live classes consists mainly in providing proper care, good feeding, and favorable environment throughout the growing season. Turkeys grown in an uncrowded, clean, natural environment, and handled gently, develop best and are most likely to attain maturity with fine market finish and undamaged plumage.

Artificial conditioning consists of (1) treating the plumage, head, and legs and (2) removing feathers defective in color or form. The removal of a few defective contour feathers (which are defined as all the feathers except the filoplumes,* main tail, and main wing feathers) is generally considered as legitimate practice; in fact, it is good judgment to go over each bird carefully. If many defective feathers are found, the bird should be discarded and marketed. If only a few defective contour feathers are found they may be plucked. Large feathers that are defective should not be plucked; the bird possessing them should be marketed unless the defects are few and of a minor nature. It requires time to go over birds carefully but no good exhibitor will show birds that have not had a thorough examination. This examination should be conducted well in advance of the show so that no ruinous last-minute changes have to be made in the show lineup. If the birds can be given a preliminary examination at least two months in advance of the show, and deformed or defective feathers pulled out, there will still be time (6 weeks the minimum) for new ones to grow before the show.

For summer and early fall shows the plumage of turkeys is usually not washed. For late fall and winter shows white turkeys are commonly washed; colored turkeys usually are not washed except in special cases. If the weather is cold, a warm room is desirable for the washing. For the drying a warm room is absolutely necessary. The temperature of the drying room should be 80° F. and then gradually lowered, so that by the end of 48 hours the room is not artificially heated. Slow drying of the feathers is preferred, about 36 to 48 hours being required to finish off a washed turkey with a presentable appearance.

For all shows, the head and legs if dirty should be washed with mild soap, rinsed, and dried. A small brush may be used to remove the dirt. Dirt showing conspicuously under the scales of the legs may be removed with a wooden toothpick. The legs and feet after drying thoroughly may be treated with olive oil and polished.

Stubs, feather quills, or feathers are seldom found on the shanks, or feet of turkeys. When present below the hock joint they are a disqualification, consequently the birds should be examined very carefully before showing. A conscientious breeder and showman will discard birds possessing them because the tendency for stubs to grow on the shanks undoubtedly is inherited. Furthermore, it is unfair and unethical to show or sell as purebreds, birds that have possessed any inherited disqualification.

Training: If the growing stock is encouraged to be gentle, little trouble will be experienced in exhibiting. It is good practice, however, to try out the birds to be shown a week or two in advance of the show. Handling and confining to cages for a while each day

---

*See glossary of terms, page 971.

during this time will help to relieve the nervousness of birds at the time of judging. Well-trained birds have the advantage over untrained birds in the show room.

Crating and shipping. Live turkeys when transported for show or sales purposes must be properly crated to prevent serious damage. Valuable toms should be crated separately or placed in individual compartments in a large crate. For small and medium-sized toms, a crate or a compartment about 42 inches long, 12 inches wide, and 36 inches high is needed. Less valuable males may be shipped in two-tom crates. These are built about 42 inches long, 24 inches wide, and 36 inches high. The toms are put in with their heads at opposite ends of the crate. Hens may be crated in groups of two to five but high-class exhibition birds should be crated separately. If coops are too wide, turkeys will try to turn about in them. Individual hen coops may be made 36 inches long, 10 inches wide and 24 inches high. The same style of coop made 20 inches wide will hold two hens; a coop about 48 inches long, 30 inches wide, and 24 inches high willl hold 5 hens. The coops usually are constructed of lightweight wood, with the ventilation at the top but the bottoms, sides, and ends of solid construction so feathers or other parts cannot protrude and thus be liable to injury.

If shipped, a drinking can must be wired to one corner of each compartment for use in case of delay or of long-distance shipments. If the time en route is to exceed 24 hours, it is well to include also a can for feed which can be filled with grain at the start.

SECTION 10

DEFINITIONS AND DESCRIPTIONS

Apteria (-ium, singular): The bare or comparatively sparsely feathered areas of a bird's body.

Axial or key feather: The short feather in the center of each wing. It separates the primaries from the secondaries and is considered as a secondary, but does not count toward disqualifying young turkeys for off-colored wing feathers.

Aftershaft: An accessory plume arising from the shaft of the feather on the side nearer the body at the point of junction of quill and rachis. It is well developed in the contour feathers and poorly developed in the quill feathers. The stem of the aftershaft is called the hyporachis.

Beak: In poultry terminology, the horny mouth parts of land birds, consisting of the upper and lower mandibles. (The upper beak of a 25-lb. tom measures about 2.5 in. from corner of mouth to tip; of a 14-lb. hen, about 2.0 in.)

Bill: In poultry terminology, the horny mouth parts of waterfowl. Ornithologists use this term to denote the horny mouth parts of any bird.

Beard: In turkeys, a cluster of coarse, hairlike, feather structures attached to the upper part of the breast of all well grown, normal males, and some females, notably white ones. The beard of adult males is commonly 5 to 7 inches long; in hens it occasionally reaches 3 inches in length. (Absence of beard does not disqualify turkey males in the show.) Several beards lineally arranged sometimes occur.

Breastbone: (See keel.)

Breeding pen: (1) A mating; (2) a shelter, usually including a fence, designed to segregate a group of mated birds; (3) a tom and 3 or 4 hens shown together.

Capon: A male chicken or turkey from which the testes have been removed.

Caruncles: Fleshy prominences on the head and upper neck of turkeys. These along with the skin are normally bright red or pale flesh color but those on the head and to a lesser extent those on the neck are changeable to bright blue and white.

Condition: In the showroom, the term applied to the state of health and cleanliness of a bird.

Contour feathers: All of the soft feathers of the bird except the filoplumes. The large quill feathers of the tail and wing are not included.[1]

Coverts: The feathers covering the base of the main tail feathers are the tail coverts; those covering the base of the large wing feathers are the primary and secondary wing coverts, respectively.

Creaminess: The color of cream, a defect sometimes found in white plumage.

Crop, or craw, or ingluvies: The sac-like organ in which the food is accumulated before it passes into the glandular stomach (or proventriculus) and gizzard. The crop contains a few glands and muscles. Its function is to store and soften feed.

Defect: An imperfection not serious enough to disqualify birds in the showroom.

Dewbill or snood: The tubular, fleshy appendage on top of the head near the front. It can be extended or contracted at will and is much larger and more elastic in males than in females. (In 25-lb. males the length varies from 4½ to 6 inches when fully extended. When contracted the length is only 1 to 1½ inches. In 12-lb. to 16-lb. hens it rarely measures more than 2 inches when fully extended and the usual length is 1 to 1½ inches.)

Dewlap: The fold of the skin on the neck below the beak (also called the wattle or throat wattle).

Disqualification: A deformity or fault so serious as to bar from competition the specimen possessing it.

Down: A term applied (1) to the hairlike feather-covering of newly hatched birds; (2) to those of the birds' contour feath-

[1]The U. S. Biological Survey counted approximately 4,600 contour feathers on a 16-lb. yearling Black hen, 5,600 on a 27-lb. adult Black tom, 5,350 on a 26-lb. adult wild tom (domesticated), and an average of 5,523 on two 11-lb. yearling, small-type, white hens. In these counts the contour and quill feathers were first picked, and sacked, and later counted. It is probable that during this procedure a fair number of feathers were uncounted. A ten per cent addition to each of the above figures would probably bring the count closer to the actual number and would include the filoplumes, which are real feathers and are an integral part of the plumage. Apparently the area of the skin influences the number of feathers. It also appears that white turkeys have more feathers per square inch of skin than do colored turkeys.

ers, the plumules, that are soft and apparently quill-less or nearly so; and (3) to the downy feathers sometimes found on feet and shanks of clean-legged varieties of poultry. Down on legs or feet is a standard disqualification. Turkeys have relatively few plumules or downy feathers whereas waterfowl have many.

Drawn: A market term describing turkey carcasses from which blood, feathers, crop, esophagus, trachea, entrails, head, feet, and preen gland have been removed and giblets cleaned and returned to the body cavity. The neck usually is severed about 2 inches below the skull and at or close to its base (or junction with the back) and placed in the body cavity with the giblets.

Dressed or New York-dressed: A market term descriptive of turkey carcasses from which blood and feathers have been removed.

Faking: In the showroom, any unfair altering of a bird that is intended to cover up or remove a serious defect or disqualification. Evidence of faking is a standard disqualification.

Fan feathers: The primary coverts which sometimes are left in the wings of dressed turkeys.

Filoplumes: The hairlike feathers of the bird. They usually are removed by singeing or wax picking.

Flight feathers or primaries: The large outer feathers of the wing located between the pinion feathers and the axial feathers. Turkeys usually have 10 primaries on each wing, although occasionally there are 9 and infrequently one wing contains one more primary than the other. More than 10 primaries have not been observed. (In adult wild hens, the longer primaries are about 14 inches long, in adult males about 17 inches. Of the total length given for females, about 1⅜ inches is buried in the flesh; in the males, about 1⅝ inches is in the flesh.)

Fluff: A term applied (1) to the soft feathers found in the region of the vent; and (2) to the fluffy portion of the feathers.

Heterozygous: Not true-breeding; a condition where a gene and its allel are determiners for opposite characters.

Homozygous: Of a genetic make-up to breed true; a condition where both genes at the same loci (on homologous chromosomes are determiners for the same character.

Hormone: A complex substance produced by one of the organs of internal secretion (such as the thyroid, testes, or adrenals). Hormones are poured into the bloodstream and transported by way of the blood and possibly in some cases by the lymph, to other organs or cells which are in turn excited to activity or restrained.

Horn color: A term, now obsolete, used by ornithologists to denote a rather pale bluish-gray, somewhat translucent color found in the bills, or beaks, and feet of some birds.

Keel: The median, longitudinal process of the sternum or breastbone. The keel and that part of the sternum to which it is directly attached are called the metasternum. The keel is commonly called the breastbone in fowls but in reality is only part of the breastbone. The length varies considerably. In 25-lb. males the average is about 8 inches; in 15-lb. hens, about 6 inches.

Knob: The forward tip of the keel raised above the surrounding breast fleshing in an unsightly manner.

Livability: See viability.

Longevity: Ability to live to old age; in poultry it usually refers to long-continued productive ability.

Main tail feathers: The large, stiff feathers of the tail. The turkey normally has 18 main tail feathers, although occasionally there are 17, 19, or 20. When the count is uneven, the larger number is usually but not always on the bird's right. The longer main tail feathers of adult wild males and 25-lb. domesticated yearling males are about 15 inches in length, of which about 1⅜ inches is buried in the flesh. In adult wild hens and in 15-lb. domesticated yearling Bronze hens, the longer main tail feathers measure about 13 inches over all, about 1⅛ inch of which is buried in the flesh.

Mating (noun): In poultry terminology, a breeding pen or group of birds placed together for the purpose of reproduction. A mating may vary in size from two birds (one male and one female) up to a thousand or more hens with an appropriate number of toms. The term also is used to refer to the act of copulation.

To mate (transitive verb): In poultry terminology, to select and segregate the matings, or breeding pens.

To mate (intransitive verb): To associate, couple, or pair for breeding.

Overflow: A mottled, greenish-gray color caused by the running together of the black and white bars of the back and tail feathers of Bronze turkeys.

Oyster: The term applied to those portions of the thigh muscles that fill the two concavities that exist in each half of the hip bone of the turkey. The flesh of the oyster is tender, dark meat of fine flavor.

Penciling: Narrow, parallel, linear markings or bars that run across the feather or tend to follow the outline of the feather.

Pinion feathers: The small group of rather short, stiff feathers attached (very firmly) to the very last joint, or thumb, of the wing. They are the same color as the primaries in most color patterns.

Poult: The term applied to immature turkeys. It is generally used only until sex can be distinguished, after which the young birds are properly called young toms (or cockerels) and young hens (or pullets). The age of 16 weeks is suggested as the dividing point because sex can be distinguished accurately at that age by secondary sexual characters in all varieties of turkeys.

Preen, coccygeal, or uropygial gland: The oil gland at the base of the tail. It is normally soft and small in size, but occasionally becomes enlarged and caseated.

Primaries: See flight feathers.

Pterylae (-a, singular): The heavily feathered areas of a bird's body, or feather tracts.

Quill (calamus, or barrel): The hollow, proximal (inner) portion of the stem of a feather.

Quill feathers: The large, stiff-stemmed feathers of wings and tail.

Rachis: The solid, distal (outer) portion of the stem of a feather. To the rachis is attached the web or vane of the feather.

Remiges: The quill feathers of the wing of a bird, i.e., the primaries, secondaries, and (where present) the tertiaries.

Rectrices: The main tail feathers of a bird.

Secondaries: The large feathers of the wing located between the axial feather and the body. Each wing of the turkey has 18 to 20 (usually 19 or 20) secondaries, including the axial feather and all of the small secondaries next to the back. There are 13 to 15 (usually 14) full-sized secondaries counting the axial feather, and 5 to 7 shorter secondaries graduated in size down to feathers no longer than the adjoining back feathers. The full-sized secondaries of adult wild hens measure about 14 inches in length over all, and those of toms about 17 inches.

Shaft or stem: The bole or trunk of a feather the inner (hollow) portion of which is the quill and the outer (solid) portion the rachis.

Shank: The metatarsus of the turkey, which is the unfeathered portion of the leg between the foot and the hock. The length varies considerably. In turkeys of normal proportions, the shank of well-matured, 25-lb. toms, when measured from back of hock to ball of foot, and held at a right angle to the leg (drumstick), averages about 8 inches in length. In well-matured, 15-lb. hens the average length is about 6½ inches.

Slipped wing: A wing permanently deformed so that when it is closed the primaries slip down from their normal position under the secondaries. This deformity is a standard disqualification.

Snood: See dewbill.

Split wing: A wing permanently deformed so that there is a noticeable gap between primaries and secondaries. This deformity is a standard disqualification.

Tertiaries: The feathers carried on the first or humeral joint of the wing. In turkeys the tertiaries form an inconspicuous tuft of extra-long contour feathers that lies in the space between the feathers of the back and those of the wings and tends to join the two.

Throat wattle: See dewlap.

Trachea: The windpipe.

Undercolor: The color of the feathers (or down in the case of day-old poults) below the surface. The undercolor cannot be seen when feathers (or down) are in normal position.

Viability: Capacity to remain alive. Frequently called livability by poultrymen.

Vigor: Active force or strength. Vigor in turkeys may be measured by the following: Fertility, hatchability, resistance to disease and parasites, growth rate, rate of feathering, egg laying, viability, and longevity.

Wattle: See dewlap.

Wingspread: The distance between the tips of the longest primaries when both wings are fully extended as in flight. A

36-lb. Broad Breasted Bronze tom measures 60", an 18-lb. hen 48"; a 21-lb. Beltsville Small White tom 56", a 12-lb. hen 37".

SECTION 11

COOKING TURKEYS*

Roasting the Heavy Turkey

First of all, inspect for clean crop and inside. This takes but a moment and establishes with certainty that all visceral material such as intestines, lungs, etc., have been completely removed, since any such material remaining can be removed at this time. Although normal kidneys are edible most people prefer to remove and discard them.

Stuffing—For institutional use, it is recommended that turkey be roasted without stuffing, because of the extra labor involved in stuffing, extra time required to roast the bird, extra time required to remove dressing, and the added cost caused by this work. It might also be added that many more cases of food poisoning are reported from stuffing which cooks slowly in the bird than have been reported from contaminated turkey meat itself.

The quantity of dressing needed is usually greater than the capacity of the turkey and the quality of the dressing is just as good if it is cooked separately. Dressing baked in pans is much easier served than when stuffed into the bird. A satisfactory alternative to stuffing the turkey is to rub the inside of the bird with a mixture of salt, pepper, poultry seasoning, and a little onion salt which permeates the meat as it cooks.

If, however, the roasting of the stuffed turkey is desired, arrangements should be made to place the stuffing in the bird at a time as close as possible to the roasting period. Stuffing made in advance and particularly that made with hot water should be promptly chilled and held in the refrigerator until it is used.

Rub inside cavity of bird with salt before stuffing.

While the quantity of stuffing will vary in accordance with the service ideas of the restaurant operator, a general rule which can be used as a guide is that three pounds of stuffing will be required for a ten pound eviscerated bird; four pounds for a fifteen pound bird, and four and one-half to five pounds for heavier birds.

Roasting—Correct roasting is slow cooking, with temperature from 250° to 300° F. by dry heat, on a rack in an open pan. It requires no cover, no initial searing.

Low temperatures assure better flavor and appearance, less shrinkage and less loss of juice.

The turkey is done when meat thermometer placed in the center of inside thigh muscle registers 190° F. If no thermometer is used, test by pressing fleshy part of drumstick with fingers, protecting them with cloth or paper. If the meat feels soft, the

*From the "Turkey Handbook" of the National Turkey Federation (Mt. Morris, Ill.), a beautifully illustrated booklet selling for 50 cents.

Time Table for Roasting

(Unstuffed Birds)

Weight of Turkey	Oven Temperature	Minutes Per Pound	Cooking Time Hours Per Bird
8 to 10 pounds	250°F.	25 to 30	3 to 3½
10 to 14 pounds	250°F.	18 to 20	3½ to 4
14 to 18 pounds	250°F.	15 to 18	4 to 4½
20 pounds	250°F.	12 to 15	5 to 6

turkey is done. Or move the drumstick up and down. If the leg joint gives readily, or breaks, the turkey is done. Check tenderness of meat under wing in shoulder muscle by inserting fork.

Remove turkey from oven and keep on warm platter until served. If, for special occasions, the bird is to be served whole and carved at the table, limit the garnish to a little parsley or watercress in order not to inconvenience the carver.

Generally, however, the turkey may be carved in the kitchen and attractively arranged on a platter for serving.

For small institutions, and where individual portions are served, the entire bird is carved and kept hot, servings being removed from platter as called for.

For institutions of any size, birds may be roasted, cooked, carved cold, portioned out by weight, reheated, and served with equally good results.

(Temperatures should never be higher than 300° F. Above table is for 250° F. unstuffed birds. If higher temperatures are used, roasting time will be shortened).

How to Roast a Small Turkey

Young turkeys of any size may be roasted, with or without stuffing. If stuffed, stuff turkey just before roasting. Stuffing prepared in advance must be refrigerated. Allow ¾ to 1 cup stuffing per pound ready-to-cook weight.

Stuffing and Trussing—Rub cavity of bird with ½ to 1 teaspoon salt.

Stuff body and wishbone cavities lightly.

Close by placing skewers across the body opening and lacing shut with cord.

Tie drumsticks securely with tail.

Fasten neck skin to back with skewer. Shape wings "akimbo" style, bringing tips onto back.

Roasting Technique—Brush skin with melted fat. Poultry seasoning or other herbs may be sprinkled over the surface.

Place trussed bird, breast down, on a rack at least ½ inch high in a shallow open pan.

Lay fat-moistened thin cloth (large enough to drape down the sides) over top of turkey.

If desired the string holding the legs may be cut when the bird has been roasted for one and one-half hours. Cutting the string makes it easy to test for doneness.

Slow Roasting—Roast at low temperature. See time table below. Do not sear. Do not add water. Do not cover.

Turn bird breast up when about ¾ done. Roast until tender.

If cloth dries during cooking, moisten cloth with fat from bottom of pan.

Time Table for Roasting

Stuffed Weight	Oven Temperature	Total Cooking Time Approximate Hours
4 to 6 lb.	325°F.	3 to 3½
6 to 8 lb.	325°F.	3½ to 4

These times are approximate. Roasting time will vary slightly with individual birds.

Test for Doneness—Move leg by grasping end of the bone. The drumstick-thigh joint breaks or moves easily. Or, press drumstick meat between fingers. Meat should be very soft when done. Do not pierce meat with fork.

BROILED TURKEY

Broiled turkey—either whole or parts—is deliciously appetizing and may well be featured. Where a restaurant is equipped for preparing whole broiled turkeys, the eye and aroma appeal will soon make this a favorite dish.

Whole Broiled Turkey—To broil is to cook by direct radiant heat from hot coals, a gas flame, or an electric element. Prepare the turkey as for roasting, omitting stuffing. Hang on broiling rack over grill. Turn frequently until done.

How to Broil a Small Turkey

Choose a young turkey about 4 lbs. Have turkey split in half lengthwise for broiling. Snap drumstick, hip and wing joints to keep bird flat during broiling.

Flatten Halves—skewer leg to body. Fold wing tip back under wing. Skewer wing flat against cut edge of backbone.

Season each half turkey with about ¼ teaspoon salt and a sprinkling of pepper.

Place turkey in broiler pan (not on rack). Brush with melted fat and then place skin side down.

Broil Slowly—Place in broiler 7 to 10 inches under heat source. Regulate heat or pan position so that turkey just begins to brown lightly in 15 minutes. Cook slowly.

Turn and brush with fat. Baste frequently during broiling to brown and cook evenly. Total cooking time, about 60 minutes for 4 pound ready-to-cook turkey.

To Test For Doneness—The turkey is done when meat on the thickest part of the drumstick cuts easily and there is no pink color visible.

Serve on warm platter with pan drippings poured over turkey.

If giblets are served with broiled turkey, coat the liver and precook heart and gizzard with fat, season and broil just long enough to brown.

To Add a Gourmet Touch—Turkey may be sprinkled with one of the following during broiling: chopped chives or parsley, powdered dried rosemary, thyme, ginger, mace or poultry seasoning.

BARBECUED TURKEY

To barbecue is to roast meat slowly on a gridiron, spit, or over coals in a specially prepared trench. The meat may be left whole or cut into pieces. While cooking it is usually basted with a highly seasoned sauce.

Barbecuing is a method of cooking meat that has been used for many ages. The entire carcass of the ox or hog slowly turning on a spit over the glowing coals to become deliciously "barbecued" was a part of festive occasions in times past.

In recent years barbecuing has been used by many clever restaurant operators with a flair for showmanship; excellent results in boosting sales volume are reported. The popularity of the barbecuing method has resulted in the commercial development of a number of mechanical devices or machines. Basically the purpose of all is to rotate the meat in the glowing heat. A choice of these machines is available to the restaurant operator.

One highly successful operator who favors a smoked flavor on his barbecue item states: "Barbecuing today is an accepted method of cooking which cannot be duplicated in any other way. The smoke imparts to the meat that smoke-like flavor of hickory and oak, pecan, orange, and any of the hardwoods of the great southwest. All of these woods transmit a delectable flavor to barbecued foods."

A top-quality barbecue sauce is regarded as essential. The operator may have his own personal recipe for barbecue sauce. This may be used to add a touch of individuality to the product and to boost repeat sales. Sauce poured over a portion of roast meat does not truly constitute a "barbecued" dish, though this is often erroneously called "barbecued."

No longer a holiday or seasonal bird, plump, tasty turkeys are ideal barbecue fare. And the new, small-sized birds are perfect for broiling or grilling with tender results a sure thing. A spicy, savory sauce brushed on the whole or quartered turkey from time to time as it broils penetrates all parts of the meat. The end result is a delectable meal that pleases your customer and makes him "want to come back for more."

Barbecued Turkey: To serve twenty-four, obtain six broiler turkeys weighing 3 to 4 pounds each or four turkeys, 6 to 8 pounds each. The turkeys may be barbecued whole, or split into quarters, removing neck and any extra skin. Wash turkey or pieces and dry thoroughly. Sprinkle with a dash of monosodium glutamate, salt and pepper.

To Barbecue in Oven: Place oven rack approximately five inches from the source of broiling heat.

Slow Grill Cooking: This brings best results. The grill rack should be eight to twelve inches above the coals.

Grease the grill rack thoroughly before placing the turkey quarters skin side down on the rack. Broil slowly 15 minutes, turn, brush with barbecue sauce. Repeat until turkey is done (about one hour for a 2½ pound ready-to-cook turkey, proportionately longer for larger birds).

Test by cutting into the thickest part of meat on drumstick. It should cut easily and there should be no pink color visible. Serve extra sauce with turkey.

Commercial Barbecue Machine: See individual manufacturer's specifications. A variety of models are obtainable—some are made of stainless steel and aluminum, some have heat resistant glass employed, some are open in front. There are three things that they all have in common: (1) motion; (2) eye appeal, which shows the customers the cooking; and (3) tantalizing odors which carry for some distance. Capacities vary, of course; one large machine which has a small hardwood fire in the bottom (to give meat a smokey flavor) will take care of 550 pounds of meat or 28 turkeys.

Turkey Barbecue Sauce: Use four ounce servings per portion. Many excellent commercial preparations are available, or the enterprising restauranteur may formulate his own distinctive sauce.

One formula for barbecue sauce is as follows:

Quantity	Ingredients
6	8 oz. can tomato sauce
3	cups turkey stock or broth
¾	cup minced onion
3	tbsp. Worcestershire sauce
1½	tsp. dry mustard
3	tsp. salt
¾	tsp. black pepper
1½	tsp. monosodium glutamate
3	tsp. chili powder
¾	tbsp. oregano or marjoram
6	tbsp. butter or margarine

Combine all ingredients; simmer until reduced by about one-quarter.

A Word of Caution Concerning Barbecuing: Sanitary conditions in a barbecuing operation are often difficult. It is not true that the operations per se are difficult. In outdoor barbecuing operations satisfactory sanitary conditions are difficult to maintain, and this has been so in some localities with the "pit" barbecuing method. Years ago, numerous cases of sickness were traced to improperly handled outdoor barbecued foods. This caused health authorities to look with disapproval upon that form of barbecuing. We now know that all foods to be barbecued should be kept under 50° F. (refrigeration) or above 135° F. to prevent spoilage and arrest bacteriological growth. When too much is cooked ahead of time, cool promptly and keep under refrigeration until needed. Then reheat and serve promptly.

POACHING-SIMMERING OF CUT-UP TURKEY

After the whole turkey has been cut up, there are several good methods of cooking the parts. Perhaps the best of these for general use is the poaching-simmering method. By this process one or several turkeys can be cooked in the oven, or on top of the range in 1½ to 3 hours, regardless of the size of the bird or birds.

The parts to be sliced—breast and/or fillets, thighs, and wing sticks (2nd joint)—are poached, this taking about 1½ hours. The remaining parts—drumsticks, back strip (lower back), flat wings, wing tips, neck, gizzard, heart, and all trimmings of skin and carcass—are simmered; this takes about 3 hours.

The stock is concentrated, and after cooling, the cooked parts are placed in it for several hours, thus, re-absorbing their natural moisture plus fine turkey flavor. Stock that is not used immediately may be kept in the refrigerator for later use in soaking, soups, or in many dishes. Unused meat portions should also be kept under refrigeration until used. (Stock should be promptly chilled, held under refrigeration and held not more than 24 to 48 hours. Use cooked turkey meat within three days.)

The slicing parts are sliced by machine or by hand, care being taken to slice with the grain and to obtain the largest slices practical. Wing sticks and flat wings are used in appropriate dishes, such as simmered flat wing, braised wing, fricasses, etc. The simmered meat is used for salads, sandwiches, loaves, soups, a la king dishes.

POACHING-SIMMERING TEST

Date: May 8, 1951, Chicago, Illinois.

Conducted by: Colonel Paul P. Logan, Mrs. Kathryn Bele Niles, Miss Jeannette MacLuckie, and C. J. Clement.

Product: 24-pound 7-ounce (391 ounce) Grade A broad-breast tom turkey.

Purpose: To determine best cooking times and method of cut-up turkey by poaching-simmering method for institutional use, number of portion yields, and shrinkage percentage.

Procedure: The turkey was cut up in 16 parts. The parts were divided into two groups (1) those for poaching and (2) those for simmering.

Poaching—Ingredients: Turkey thigh (2nd joint), breast, and serving.

NOTE: Breast, thighs and wingsticks can be hand or machine sliced.

Simmering: Ingredients: Portions not used for poaching—these include drumsticks cutlets, flat wings, wing tips, giblets, neck and carcass. Cover with water. For each gallon of water add 1 tablespoon salt, 2 ounces each of carrots, celery, and onions, chopped coarsely, 2 cloves, and 4 peppercorns.

Procedure: (1) Place turkey parts to be simmered (drumsticks, back, neck, gizzard, heart and all trimmings of skin) in heavy kettle or steam kettle. (2) Add enough warm water to cover. (Keep track of quantity used so seasoning may be measured). Add salt to water. (3) Add weighed and measured ingredients (may be tied in cheesecloth bag if desired). (4) Bring water to boiling point and remove surface froth. (5) Reduce heat to simmering. (Bubbles barely break and temperature is 180° F.) (6) Cover tightly and **simmer** until meat is fork-tender. Cooking time: 2 to 3 hours counting time after heat is reduced to simmering. (7) Remove meat. Set meat aside to cool. (8) Continue simmering broth to ⅓ original volume. Cool rapidly. (9) Add meat to this broth and hold in refrigerator several hours or overnight.

Comments: The poaching-simmering method saves:

1. Utensils and stove space—more turkeys can be poached in a single turkey pan at the same time.
2. Cooking time is greatly reduced. A turkey which normally takes 4 to 6 hours to roast as a whole bird can be cooked after cutting up in 1 hour 30 minutes.
3. This method gives the greatest amount of slicable meat when the slicing is done either by hand or machine.

Subjective: Turkey cooked by the poaching-simmering method was tested by a panel of eight in comparison with turkey cooked in the conventional whole oven-roast method. In the opinion of all of the testers, the turkey was tasty, tender and juicy. Appearance was excellent. Every part of the turkey was usable when cooked this way. Even the skin, which should never be thrown away but should be ground up and added to makeup dishes, is the most flavorable part of the turkey. This official test was subsequently confirmed by two additional different test panels.

Results: Uncooked weight of turkey, 360 ounces (30 ounces bone); cooked weight, 216 ounces (30 ounces bone). loss, 30%. Breast meat: Uncooked weight, 112 ounces; cooked weight, 86 ounces. Loss, 23%. Thighs: Uncooked weight, 68 ounces; cooked weight, 48 ounces. Loss, 30%.

After poaching, the cooked portions were soaked in a light turkey stock overnight to recover a great amount of the weight lost in the poaching process. This poached turkey is then ready to serve. It has been found, however, that it not only increased in weight, but took on an exquisite flavor if soaked in a gelatinized turkey broth overnight.

TREATMENT OF DRESSED BIRDS

1. **Dressed Poultry:** Dressed poultry is customarily handled as ice-packed stock, box-packed unfrozen stock, or box packed frozen stock. Ice-packed poultry should be handled rapidly, since the ice will melt, permitting the temperature of the product to rise, with consequent development of deteriorative changes. In addition, bacteria flourish more abundantly on wet surfaces than on dry surfaces.

2. **Storage of Fresh-Dressed Birds:** Box-packed unfrozen poultry should be held in a cooler maintaining a temperature of about 34° F., particularly if the product is to be held for several days. At a temperature of 34° F., fresh-chilled, dressed (undrawn) poultry keeps one week, but is ordinarily used within two or three days. Temperatures above 40° F. are considered unsafe for fresh-chilled poultry. Dressed birds should be drawn as soon as possible and, when drawn, should be kept loosely wrapped in the refrigerator.

3. **Storage of Frozen-Dressed Birds:** Box-packed frozen stock can be kept in the freezer for a very long period if the temperature is sufficiently low and the product is protected from "freezer burn."

4. **Defrosting Dressed Birds:** Unwrap birds and place in refrigerator for 24 hours or leave at room temperature overnight. Or, unwrap the bird and place under a stream of slow running water for 3 or 4 hours. An electric fan will hasten thawing.

A defrosted turkey, like an unfrozen one, may be held safely 24 hours at 38° F. or less. Complete defrosting before cooking is preferable, but, if the cooking process begins before the bird is completely defrosted, more time must be allowed for cooking.

Once defrosted, frozen poultry should be cooked exactly like fresh poultry. It is advisable not to refreeze poultry.

DR. BILLINGS' FAMOUS TURKEY DRESSING

It is now several years since I took it upon myself to do something about this turkey dressing business. I love turkey, but I can't go for the filler that is usually served with it. I used to wonder why the turkey dressing was always slyly hid under a thin slab or two of turkey, but there was a reason . . . they were ashamed of it, I guess. So I finally set about to develop a dressing recipe to end all other dressing recipes. After literally thousands of trials, I think I have succeeded, but the verdict rests with you.

So many people have asked for this recipe lately, we will repeat it for the benefit of those who came in late. I warn you that since this recipe has become so popular, there are many, many imitators. Some jealous souls even go so far as to claim it as their own.

Every year, usually at holiday time, most magazines and newspapers print scads of dressing recipes. Most of them use bread as a base and then start throwing in everything they can find around the kitchen, including oysters, pork, beans, celery, raisins, giblets, chestnuts, prunes, sausage, apples, etc., etc. Several years ago I recall finding a piece of someone's bridgework. This sort of mixture, they call a "dressing" . . . it comes out of the bird a dark, sticky, soggy mess that lays like a ton of brick in your stomach. This may be an excellent way to get rid of kitchen leftovers, but as a dressing . . .ugh. We've had New Deals in nearly everything, so why can't we have a new deal in dressing. It's way past due now. After all . . . the dressing is only a "chaser" for the turkey, not a complete dinner in itself.

Here's the recipe . . . even a child could make it. (1) Take enough of yesterday's bread (unsliced, if possible) and cut off the outside crust. Now, cut the trimmed loaf into hunks just big enough to make a good handful. With a fairly coarse grater, the bread is now finely crumbled into a big dishpan. This will give you a panful of nice fluffy grated bread. It will take about three small loaves for a 15 to 18 pound turkey. (2) Have someone who does not cry easily, chop a medium-sized onion into very fine particles. One onion is usually plenty . . . it won't hardly taste, but gives the dressing a certain amount of what we men call "oomph." (3) Stir the chopped onion evenly into the grated bread and (4) add just enough salt to taste right. (5) Then add the powdered sage . . .

shaking it on lightly as you stir it into the bread. And, please, do not use too much sage . . . just enough to taste tangy, but not to gag you. (6) This next step will test your generosity. Melt up one-half to one pound, depending on size of turkey, of good butter (don't be Scotch here) and have your helper pour the melted butter slowly over the mass while you stir it in evenly throughout. That's all there is to it . . . the dressing is now made. Don't add another single, solitary thing. Do not moisten with water. The finished product will appear a golden color, still light and fluffy although it is slightly moistened by the butter. It's so tasty, you can almost eat it as is.

How To Put This Dressing into The Bird

This is where some fall down. The procedure is very simple. Have someone tip the helpless turkey up on the back of his neck with the opening upward. Now, you will spoon the dressing into the bird, shaking the carcass slightly to settle the dressing. Do not pack the dressing down . . . don't even push it down with your spoon. Merely spoon in all the bird will hold and stop there and sew up the incision. That's all there is. People who eat pie for breakfast probably won't like this dressing . . . others love it. Good luck to you.

P.S. This dressing won't go as far as that other soggy stuff because everybody will be hollering for more. Should you have some "makings" left after you have filled the bird, you could place this in a casserole and on top, lay the neck, gizzard, liver and heart. Slide this into the oven toward the end of the turkey roasting period and you will have a bit extra for the second helpers. It won't be as good as the dressing inside the bird, but it will come in handy.

INDEX

L

M

N

O

P

T

W

X

Y

Z

www.ingramcontent.com/pod-product-compliance
Lightning Source LLC
LaVergne TN
LVHW052201200926
848481LV00001B/7
* 9 7 8 1 9 3 8 0 9 9 0 5 2 *